C0-ALP-130

Advances in DNA Damage and Repair

Oxygen Radical Effects, Cellular Protection, and Biological Consequences

NATO ASI Series

Advanced Science Institutes Series

A series presenting the results of activities sponsored by the NATO Science Committee, which aims at the dissemination of advanced scientific and technological knowledge, with a view to strengthening links between scientific communities.

The series is published by an international board of publishers in conjunction with the NATO Scientific Affairs Division

A	**Life Sciences**	Plenum Publishing Corporation
B	**Physics**	New York and London
C	**Mathematical and Physical Sciences**	Kluwer Academic Publishers Dordrecht, Boston, and London
D	**Behavioral and Social Sciences**	
E	**Applied Sciences**	
F	**Computer and Systems Sciences**	Springer-Verlag
G	**Ecological Sciences**	Berlin, Heidelberg, New York, London,
H	**Cell Biology**	Paris, Tokyo, Hong Kong, and Barcelona
I	**Global Environmental Change**	

PARTNERSHIP SUB-SERIES

1. Disarmament Technologies	Kluwer Academic Publishers
2. Environment	Springer-Verlag
3. High Technology	Kluwer Academic Publishers
4. Science and Technology Policy	Kluwer Academic Publishers
5. Computer Networking	Kluwer Academic Publishers

The Partnership Sub-Series incorporates activities undertaken in collaboration with NATO's Cooperation Partners, the countries of the CIS and Central and Eastern Europe, in Priority Areas of concern to those countries.

Recent Volumes in this Series:

Volume 300 — Targeting of Drugs 6: Strategies for Stealth Therapeutic Systems
edited by Gregory Gregoriadis and Brenda McCormack

Volume 301 — Protein Dynamics, Function, and Design
edited by Oleg Jardetzky and Jean-François Lefèvre

Volume 302 — Advances in DNA Damage and Repair: Oxygen Radical Effects, Cellular Protection, and Biological Consequences
edited by Miral Dizdaroglu and Ali Esat Karakaya

Volume 303 — Molecular and Applied Aspects of Oxidative Drug Metabolizing Enzymes
edited by Emel Arinç, John B. Schenkman, and Ernest Hodgson

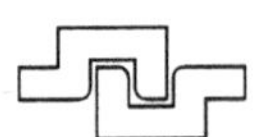

Series A: Life Sciences

Advances in DNA Damage and Repair

Oxygen Radical Effects, Cellular Protection, and Biological Consequences

Edited by

Miral Dizdaroglu

National Institute of Standards and Technology
Gaithersburg, Maryland

and

Ali Esat Karakaya

Gazi University
Ankara, Turkey

Kluwer Academic / Plenum Publishers
New York, Boston, Dordrecht, London, Moscow
Published in cooperation with NATO Scientific Affairs Division

Proceedings of a NATO Advanced Study Institute on
DNA Damage and Repair: Oxygen Radical Effects, Cellular Protection, and Biological Consequences,
held October 14 – 24, 1997,
in Tekirova, Antalya, Turkey

NATO-PCO-DATA BASE

The electronic index to the NATO ASI Series provides full bibliographical references (with keywords and/or abstracts) to about 50,000 contributions from international scientists published in all sections of the NATO ASI Series. Access to the NATO-PCO-DATA BASE is possible via a CD-ROM "NATO Science and Technology Disk" with user-friendly retrieval software in English, French, and German (©WTV GmbH and DATAWARE Technologies, Inc. 1989). The CD-ROM contains the AGARD Aerospace Database.

The CD-ROM can be ordered through any member of the Board of Publishers or through NATO-PCO, Overijse, Belgium.

Library of Congress Cataloging-in-Publication Data

Advances in DNA damage and repair : oxygen radical effects, cellular protection, and biological consequences / edited by Miral Dizdaroglu and Ali Esat Karakaya.
p. cm. -- (NATO ASI series. Series A, Life sciences ; v. 302)
"Published in cooperation with NATO Scientific Affairs Division."
"Proceedings of a NATO Advanced Study Institute on DNA Damage and Repair : Oxygen Radical Effects, Cellular Protection, and Biological COnsequences, held October 14-24, 1997 in Antalya, Turkey"--CIP t.p. verso.
Includes bibliographical references and index.
ISBN 0-306-46042-4
1. DNA damage--Congresses. 2. DNA repair--Congresses. 3. Active oxygen--Pathophysiology--Congresses. I. Dizdaroglu, Miral. II. Karakaya, Ali Esat. III. NATO Advanced Study Institute on DNA Damage and Repair : Oxygen Radical Effects, Cellular Protection, and Biological Consequences (1997 : Antalya, Turkey) IV. Series.
QH465.A1A36 1999
572.8'6--dc21 98-47311
CIP

ISBN 0-306-46042-4

233 Spring Street, New York, N.Y. 10013

10 9 8 7 6 5 4 3 2 1

A C.I.P. record for this book is available from the Library of Congress.

Printed in the United States of America

PREFACE

Damage to DNA by both exogenous and endogenous sources is increasingly regarded as highly important in the initiation and progression of cancer and in the occurrence of other pathological events. DNA damage caused by reactive oxygen-derived species, also called oxidative DNA damage, is the most frequent type encountered by aerobic cells. Mechanistic studies of carcinogenesis indicate an important role of this type of damage to DNA. There is also strong evidence to support the role of oxidative DNA damage in the aging process.

DNA damage is opposed *in vivo* by repair systems. If not repaired, DNA damage may lead to detrimental biological consequences. Therefore, the repair of DNA damage is regarded as one of the essential events in all life forms. In recent years, the field of DNA repair flourished due to new findings on DNA repair mechanisms and the molecular basis of cancer. In 1994, DNA repair enzymes have been named *Science* magazine's *Molecule of the Year*.

There is an increasing awareness of the relevance of DNA damage and repair to human health. A detailed knowledge of mechanisms of DNA damage and repair, and how individual repair enzymes function may lead to manipulation of DNA repair in cells and ultimately to an increase in the resistance of human cells to DNA-damaging agents. Our knowledge in this field has increased vastly in recent years. The time was ripe to convene a NATO Advanced Study Institute of scientists of international standing from the fields of biochemistry, molecular biology, enzymology, biomedical science, and radiation biology to analyze these questions in detail and to teach the student participants the basics and new developments of the field of DNA damage and repair.

The NATO Advanced Study Institute on "DNA Damage and Repair: Oxygen Radical Effects, Cellular Protection, and Biological Consequences" was held October 14–24, 1997, in Tekirova, Antalya/Turkey. During the meeting, the invited lecturers presented and discussed the state-of-the-art knowledge and recent developments in this research field, and its pertinence to human health. The interactions between the lecturers and participants were synergistic and challenging, and contributed greatly to dissemination of scientific knowledge and the formation of international scientific collaborations. This book contains the papers presented by invited lecturers and the abstracts of the posters presented by the student participants.

This meeting was sponsored by the NATO Scientific Affairs Division, Belgium. Additional support was obtained from Biotechnology Division, National Institute of Standards and Technology (USA), Danish Center for Gerontology, Scientific and Technical Research Council of Turkey, and Turkish Society of Toxicology.

CONTENTS

Advances in DNA Damage and Repair

Oxygen Radical Effects, Cellular Protection, and Biological Consequences

1

MAPPING REACTIVE OXYGEN-INDUCED DNA DAMAGE AT NUCLEOTIDE RESOLUTION

Steven Akman,[1] Regen Drouin,[2] Gerald Holmquist,[3] and Henry Rodriguez[4]

[1]Bowman Gray School of Medicine of Wake Forest University
Winston-Salem, North Carolina, 27157
[2]Hôpital Saint-Francois d'Assise
Québec, Canada G1L 3L5
[3]Beckman Research Institute of the City of Hope Medical Center
Duarte, California, 91010
[4]Biotechnology Division, National Institute of Standards and Technology
Gaithersburg, Maryland 20899

1. ABSTRACT

The frequency of reactive oxygen (ROS)-induced DNA base damage along the human *p53* and *PGK1* genes was determined at nucleotide resolution by cleaving DNA at ROS-modified bases with the Nth and Fpg proteins from *Escherichia coli* and then using the ligation-mediated PCR (LMPCR) technique to map induced break frequency. Damage was induced either *in vivo* by exposing cultured human fibroblasts to H_2O_2 or *in vitro* by exposing purified genomic DNA to H_2O_2 /ascorbate in the presence of Cu(II), Fe (III), or Cr(VI) metal ions. With the exception of a few footprints observed in the promoter region of *PGK1* at transcription factor binding sites, all four base damage patterns from either *in vivo* or *in vitro* treatments were nearly identical in both regions of the genome. Guanines in the triplet d(pCGC) were the most commonly damaged base. Isolated nuclei suffered little ROS-induced DNA base damage in the presence of ascorbate and H_2O_2; damage was restored by addition of transition metal ions, suggesting that during *in vivo* exposure of cells to H_2O_2, metal ions (or metal-like ligands) are freed from extranuclear sites to supply redox-cycling ligands to the nucleus. These data simplify the complexity of H_2O_2-induced DNA damage and mutagenesis studies by demonstrating the commonality of damage catalyzed by different transition metal ions and by showing that the pattern of H_2O_2-induced DNA base damage is determined almost entirely by the primary DNA sequence, with chromatin structure having a limited effect.

Advances in DNA Damage and Repair, edited by Dizdaroglu and Karakaya. Kluwer Academic / Plenum Publishers, New York, 1999.

2. INTRODUCTION

DNA damage induced by reactive oxygen species (ROS) is an important intermediate in the pathogenesis of human conditions such as cancer and aging (Ames, 1987; Guyton and Kensler, 1993). ROS-induced DNA damage products are both mutagenic and cytotoxic [reviewed in Wallace, 1994].

One commonly studied ROS is that produced by H_2O_2 in the presence of transition metal ions. The mutational spectra of H_2O_2 (Moraes et al., 1989; Akman et al., 1991)and the transition metal ions Fe, Cu (Loeb et al., 1988; Tkeshelasnvili et al., 1991), and Cr (Kawanishi et al., 1994) have been studied in model systems, but the relationship of induced DNA damage to these spectra remains unknown. Knowledge of this relationship is crucial in order to extrapolate from model mutational analysis systems to endogenous mutational spectra generated *in vivo*. This is of significant scientific and medical interest since elucidation of the role oxidatively-induced DNA damage has in carcinogenesis could ultimately enable one to develop rational therapeutic interventions that might be beneficial in the prevention of human malignancies.

Until recently, progress in this area has been hampered by the lack of damage measurement techniques with nucleotide resolution. The ligation-mediated polymerase chain reaction (LMPCR) is a genomic sequencing method for mapping of rare DNA single-stranded breaks. In a previous report, we produced a map of DNA base damage caused by Cu(II)/ascorbate/H_2O_2 in purified DNA *in vitro* at nucleotide resolution (Rodriguez et al., 1995). We showed that the DNA base damage pattern was distributed non-uniformly in the promoter region of the human *PGK1* gene, with certain sequence motifs being highly susceptible to damage. Certain questions remain. How does the *in vitro* Cu(II)/ascorbate/H_2O_2 base damage pattern compare to the base damage pattern induced *in vivo* by exposure of intact target cells to H_2O_2? Can metal ions other than copper produce a similar mutational pattern? Furthermore, if other metal ions or metal-like ligands contribute to the *in vivo* damage pattern, do they normally reside in proximity to DNA or are they recruited from extranuclear sites? These questions are addressed in this report.

3. MATERIALS AND METHODS

3.1. *In Vivo* H_2O_2 Treatment of Human Skin Fibroblasts

Human male foreskin fibroblasts were grown in 150 mm dishes to confluent monolayers in Dulbecco's Modified Eagle Medium containing 10% (v:v) fetal bovine serum. Fibroblasts were treated with serum-free Minimum Essential Medium with 1 mM sodium phosphate containing 50 mM H_2O_2 at 37 °C for 30 minutes. After washing, cells were harvested and DNA was isolated as previously described (Drouin et al., 1996).

Human male skin fibroblast DNA was prepared for *in vitro* assay as previously described (Rodriguez et al., 1995). After phenol/chloroform extraction, the DNA was precipitated in ethanol, redissolved in 10 mM HEPES, 1 mM EDTA, pH 7.4 at 70 μg/ml, then dialyzed against distilled water overnight at 4 °C.

3.2. Metal Ion-Ascorbate-H_2O_2 Treatment

Ten μg of dialyzed DNA dissolved in 161 μl of H_2O was incubated at room temperature for 30 min with 50 μM $CuCl_2$, 100 μM $FeCl_3$, or 100 μM $K_2Cr_2O_7$. Chelex®-treated potassium phosphate, pH 7.5 (± 0.3 M sucrose), ascorbate, and H_2O_2 were added to final concentrations of 1 mM, 100 μM, and 5 mM, respectively. After 30 min at room temperature with gentle rocking, the reaction was quenched by the addition of EDTA to 2 mM, followed by precipitation of DNA in 0.3 M sodium acetate, pH 7.0 and 2 volumes of cold ethanol.

3.3. Isolation and Treatment of Fibroblast Nuclei

Human male foreskin fibroblasts were grown in 150 mm dishes to confluent monolayers in Dulbecco's Modified Eagle Medium containing 10% (v:v) fetal bovine serum. After removal of medium and washing with 25 ml of a 154 mM NaCl solution, cells were lysed by the addition of 10 ml buffer A (0.3 M sucrose, 60 mM KCl, 15 mM NaCl, 60 mM Tris-HCl, pH 7.4, 2 mM EDTA) containing 0.5% Nonidet-P40. Nuclei were collected by centrifugation at 1000 x *g* for 10 min at room temperature, then gently resuspended in sucrose/phosphate buffer (1 mM potassium phosphate, pH 7.5, 60 mM KCl, 15 mM NaCl, 0.3 M sucrose). After 2 additional washes in sucrose/phosphate buffer, nuclei were exposed to 50 mM H_2O_2 ± 50 μM Cu(II) or 50 μM Fe(III) ± 100 μM ascorbate in sucrose/phosphate buffer for 30 min at 37°C. Reactions were quenched by the addition of EDTA or desferrioximine (for iron-containing samples) to 2 mM. Treated nuclei were then collected by centrifugation, resuspended in 4 ml buffer A and DNA was isolated as described previously (Drouin et al., 1996).

3.4. Ligation-Mediated Polymerase Chain Reaction (LMPCR)

Digestion of treated DNA with Nth and Fpg proteins (Rodriguez et al., 1995) and fragment size analysis by glyoxal gel electrophoresis (Drouin et al., 1996) have been described in detail elsewhere. LMPCR is a six step process. The steps are as follows: (1) primer extension of an annealed gene-specific oligonucleotide (upstream primer 1) to generate blunt ends; (2) ligation of a universal asymmetric double-strand linker onto the blunt ends; (3) PCR amplification using a second gene-specific oligonucleotide (upstream primer 2) along with a linker primer (downstream linker primer); (4) separation of the DNA fragments on a sequencing polyacrylamide gel; (5) transfer of the DNA to a nylon membrane by electroblotting; (6) hybridization of a radiolabeled probe prepared by repeated primer extension using a third gene-specific oligonucleotide (upstream primer 3). Air-dried membranes were exposed to Kodak XAR-5 X-ray films for 0.5 to 8 h with intensifying screens at -70 °C. On the final autoradiogram, each band represents a nucleotide position where a break was induced, and the signal intensity of the band reflects the number of DNA molecules with ligatable ends terminating at that position. The intensity of the bands was confirmed by PhosphorImager scans. The intensity of each band was plotted on a scale from "0" (no band), to "5" (the most intense band) after normalization to the corresponding Maxam-Gilbert sequence lane, in order to allow for nonlinearity of the primer extension and linker ligation steps.

3.5. Enhancement of LMPCR Sensitivity by Genomic Gene Enrichment

This technique is described in detail elsewhere (Rodriguez and Akman, 1998). Briefly, 300 μg of purified, metal ion-ascorbate-H_2O_2-treated human genomic DNA, digested to completion with *Bam H1*, were loaded onto a Model 491 Prep Cell (Bio Rad Laboratories) continuous elution electrophoresis apparatus (CEE) containing a 0.65% preparative agarose gel and a 0.25% agarose stacking gel. Electrophoresis was performed

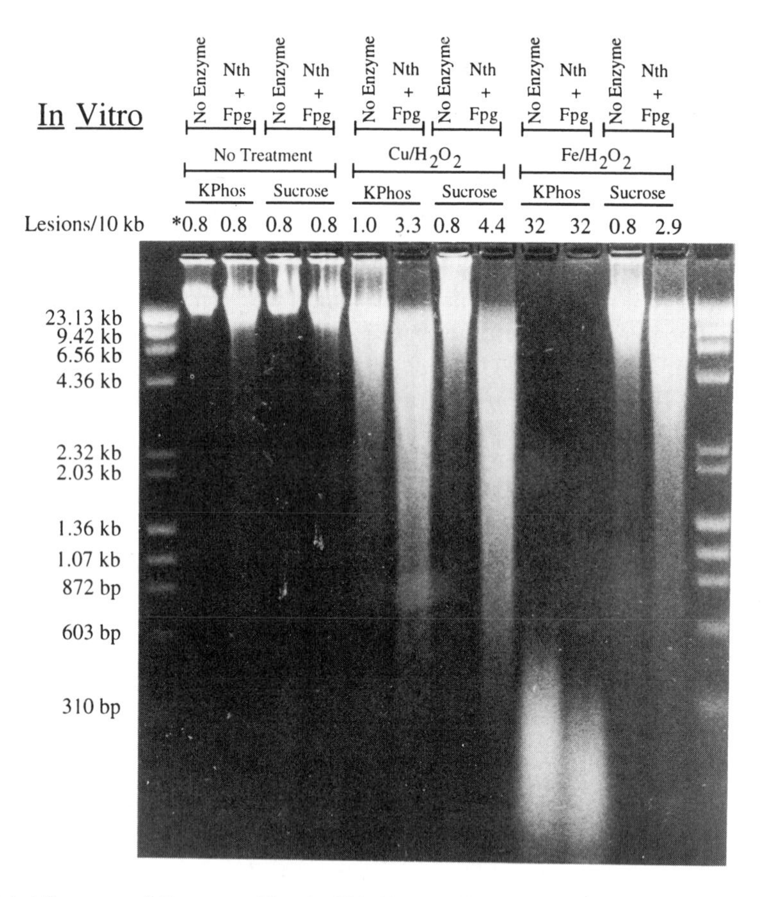

Figure 1. Global frequency of direct strand breaks ("No Enzyme" lanes) and direct strand breaks plus modified bases ("Nth + Fpg" lanes) observed in dialyzed human genomic DNA (lanes 2–5), or after treatment with 50 μM Cu(II) (lanes 6–9) or 50 μM Fe(III) (lanes 10–13) + 100 μM ascorbate + 5 mM H_2O_2 in potassium phosphate buffer ±0.3 M sucrose. Treated DNA was denatured with dimethylsulfoxide/glyoxal then electrophoresed through a 1.5% neutral agarose gel. The first and last lanes of this and all subsequent gels contain *Hind III* digested lambda phage + *Hae III* digested X174 DNA molecular weight standards. $1/\langle M_n \rangle$ values derived from this gel are shown above the appropriate lanes. (*) 0.8 is an upper limit of $1/\langle M_n \rangle$ in these lanes; a more precise $1/\langle M_n \rangle$ value could not be determined. Reproduced by permission.

at 50 volts, 4 °C, in 1X TBE buffer. The elution flow rate was maintained at 50 μl/min; 1 ml fractions were collected. Fractions containing the gene of interest, determined by dot blot analysis, were pooled and used as substrate for LMPCR analysis.

4. RESULTS

4.1. Estimate of Global DNA Damage Induced by *in Vitro* Cu(II), Fe(III), or Cr(VI) Plus H_2O_2/Ascorbate

Analysis of DNA fragment mobility distribution on denaturing agarose gels is a useful method for determining the frequency of strand breaks. In our analyses, strand breaks

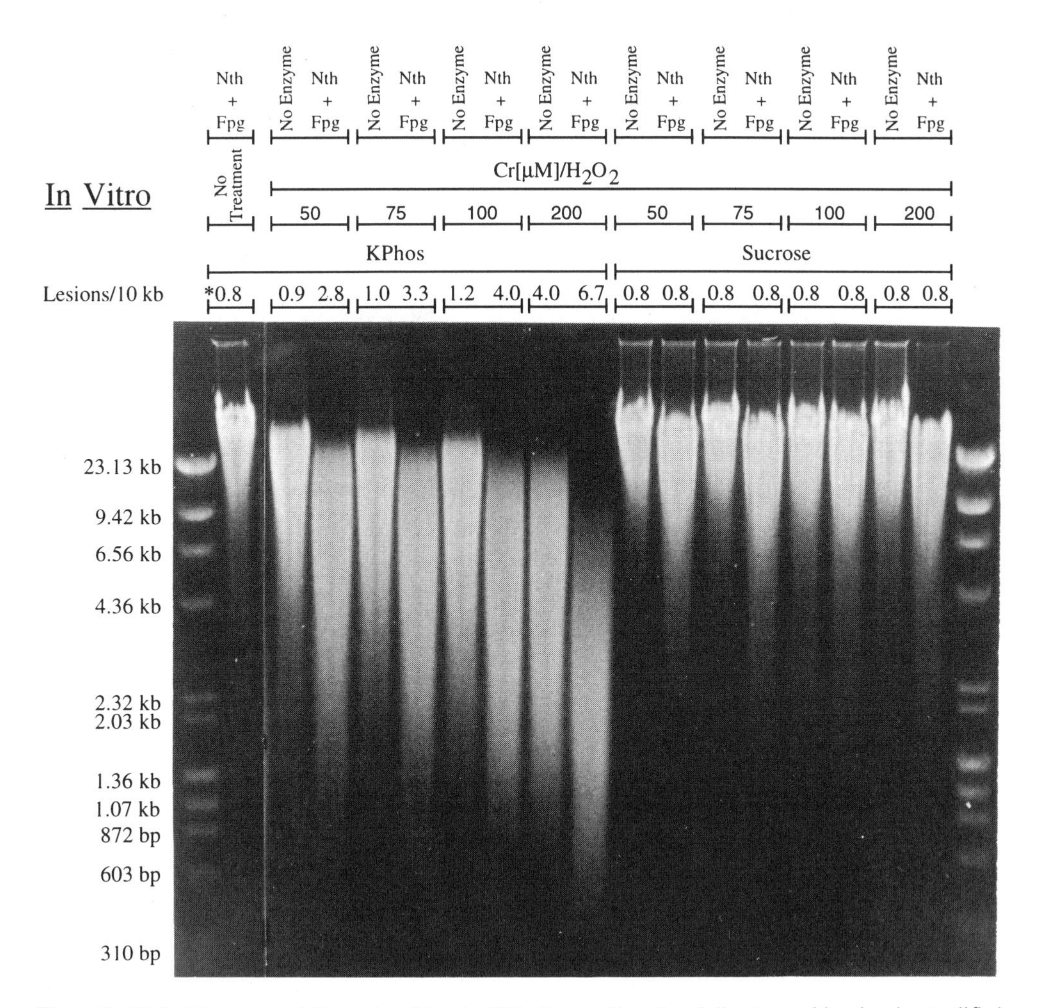

Figure 2. Global frequency of direct strand breaks ("No Enzyme"lanes) and direct strand breaks plus modified bases ("Nth + Fpg" lanes) observed in dialyzed human genomic DNA after treatment with the indicated concentrations of Cr(VI)/100 μM ascorbate/5 mM H_2O_2 in potassium phosphate buffer ± 0.3 M sucrose. Electrophoresis was carried out as described in the Legend to Figure 1. $1/<M_n>$ values derived from this gel are shown above the appropriate lanes. (*) 0.8 is a lower limit of $1/<M_n>$ in these lanes; a more precise $1/<M_n>$ value could not be determined. Reproduced by permission.

may be created directly by the action of transition metal ion/H_2O_2/ascorbate, or may result from the AP lyase activity of Nth or Fpg proteins which accompanies their DNA glycosylase activity. In such a gel mobility analysis, the weight average molecular weight $<M_w>$ coincides with the mobility of maximum peak of ethidium fluorescence (Willis et al., 1988). $<M_n>$, the number average molecular weight, is equal to $<M_w>/2$, provided cleavage sites are randomly distributed (Tanford, 1961). The frequency of DNA strand breaks is $1/<M_n>$.

Exposure of dialyzed human genomic DNA to 50 μM Cu(II)/100 μM ascorbate/5 mM H_2O_2 in phosphate buffer caused principally Nth- and Fpg protein-sensitive base lesions at a frequency of 2.3 per 10 kb (Figure 1, lanes 6–7) (Rodriguez et al., 1997b). Previous work had indicated that abasic sites were only a minor component of this Cu(II) induced damage (Rodriguez et al., 1995). Inclusion of 0.3 M sucrose did not affect the induced damage frequency (Figure 1, lanes 8–9). In contrast, 50 μM Fe(III)/100 μM ascorbate/5 mM H_2O_2 in potassium phosphate buffer caused a preponderance of direct strand breaks ($<M_n> <$ 100 bp, Figure 1, lanes 10–11) (Rodriguez et al., 1997b). Inclusion of 0.3 M sucrose inhibited most of the direct strand breaks, permitting detection of Nth/Fpg protein-sensitive lesions at a frequency of 2.1 per 10 kb (Figure 1, lanes 12–13).

In contrast to both Cu(II) and Fe(III), both direct strand breakage and base damage mediated by Cr(VI) was inhibited by 0.3 M sucrose (Figure 2, lanes 11–18) (Rodriguez et al., 1997b). In potassium phosphate buffer without sucrose, 50–200 μM Cr(VI)/100 μM ascorbate/5 mM H_2O_2 caused a concentration-dependent increase in direct strand breaks from <1 per 10 kb to 4 per 10 kb and Nth or Fpg protein-sensitive base lesions from 1.9 per 10 kb to 2.7 per 10 kb (Figure 2, lanes 3–10).

4.2. Distribution of DNA Damage Induced *in Vitro* by Cu(II), Fe(III), or Cr(VI) Plus H_2O_2/Ascorbate and *in Vivo* by H_2O_2

The distribution of oxidative damage induced in exons 5 and 9 of human *p53*, as well as the promoter region of human *PGK1*, was assessed by LMPCR (Rodriguez et al., 1997b). The autoradiograms indicating the damage distributions induced in the region of the human *PGK1* gene covered by primer set A (see reference 23 for primer set designations) and in exon 9 of the human *p53* gene are shown in Figures 3 and 4, respectively. In these regions of the genome, the base damage frequency distributions induced in dialyzed DNA *in vitro* by Cu(II) and Fe(III) plus ascorbate/H_2O_2 are nearly identical. Sucrose was included in the Fe(III) reaction to suppress the direct strand break signal. Sucrose has no effect on the LMPCR-derived damage distribution signals (Figure 3). The base damage distribution associated with these two transition metal ions is non-uniform, confirming our previous observations (Rodriguez et al., 1995) with Cu(II). Prominent base damage hotspots are observed in *PGK1* and in *p53* exon 9 (Figure 3–4).

Tabulations of the damage frequencies induced in exons 5 and 9 of *p53* and the region of the *PGK1* promoter covered by primer sets A, F, and G are shown in Figure 5 (Rodriguez et al., 1997b). The distribution of DNA base damage occurring in both of these regions by *in vitro* Cu(II)- and Fe(III)-mediated damage, were nearly identical. The distribution of damage caused by Cr(VI)/ascorbate/H_2O_2 was similar, but not identical, to that mediated by copper or iron ions in these regions. The unique chromate sensitive positions were often thymines, whereas thymines were uncommonly modified in the presence of Fe(III) or Cu(II), (Figures 3–4).

The distribution of DNA base damage occurring *in vivo* induced by exposure of cultured human male fibroblasts to 50 mM H_2O_2 was also determined in *PGK 1* and *p53* (Rodriguez et al. 1997b). Here, 50 mM H_2O_2 induces a global damage frequency in human male fibroblast DNA *in vivo* equivalent to the damage frequency induced *in vitro* by 50 μM Cu(II)/100 μM ascorbate/5 mM H_2O_2 (data not shown). Figures 3–5 demonstrate that the base damage distribution induced *in vivo* in the assessed regions of *PGK1* and *p53* were identical to the damage distributions induced *in vitro* by Cu(II) or Fe(III) plus H_2O_2/ascorbate and similar to the damage distribution induced *in vitro* by Cr(VI)/H_2O_2/ascorbate. Control experiments, such as extending the number of cell rinses with Chelex®-treated phosphate-buffered saline after H_2O_2 exposure to as many as 9 times, resulted in no alteration of the damage frequency. This demonstrates that the putative *in vivo* damage distribution was not a post-DNA extraction artifact (data not shown). Furthermore, the

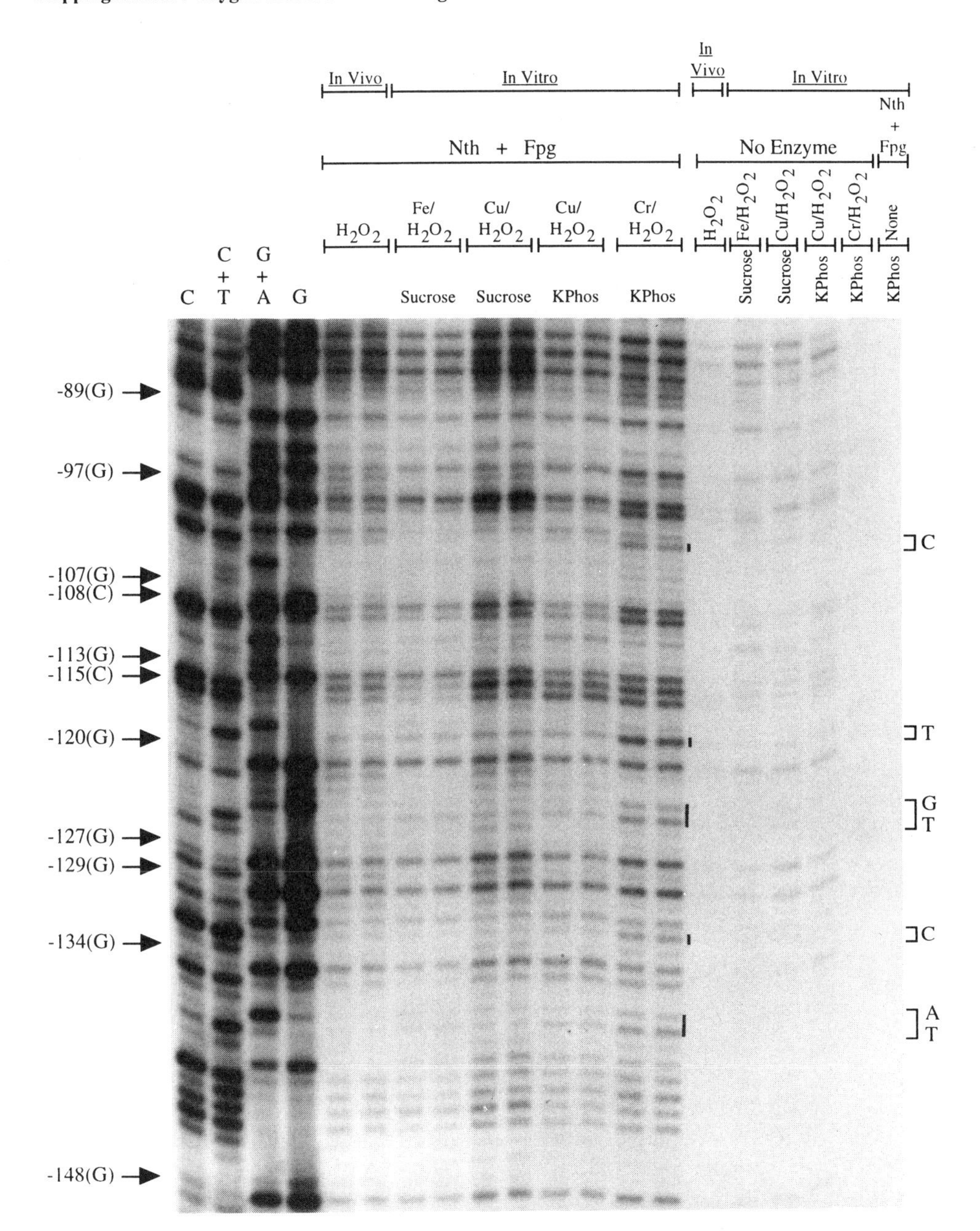

Figure 3. LMPCR analysis of damage induced in the promoter region of human *PGK1* using primer set A (transcribed strand). *Lanes 1–4*, DNA treated with standard Maxam-Gilbert cleavage reactions. *Lanes 5–6, 15*, DNA recovered from intact human foreskin fibroblasts exposed to 50 mM H_2O_2. *Lanes 7–8, 16*, dialyzed genomic DNA treated with 100 μM Fe(III)/100 μM ascorbate/5 mM H_2O_2 in the presence of 0.3 M sucrose. *Lanes 9–10, 17*, DNA treated with 50 μM Cu(II)/100 μM ascorbate/5 mM H_2O_2 in the presence of 0.3 M sucrose. *Lanes 11–12, 18*, DNA treated with 50 μM Cu(II)/100 μM ascorbate/5 mM H_2O_2. *Lanes 13–14, 19*, DNA treated with 100 μM Cr(VI)/100 μM ascorbate/5 mM H_2O_2. *Lane 20*, DNA incubated in potassium phosphate buffer. The DNA in lanes 5–14 was digested with Nth and Fpg proteins after treatment; the DNA in lanes 15–20 were incubated in digestion buffer alone after treatment. Positions of high damage frequency bases are marked with arrows to the left of lane 1. The sequence of positions heavily damaged in the presence of chromium, but not copper or iron, is denoted by the open rectangles to the right of lane 20. Reproduced with permission.

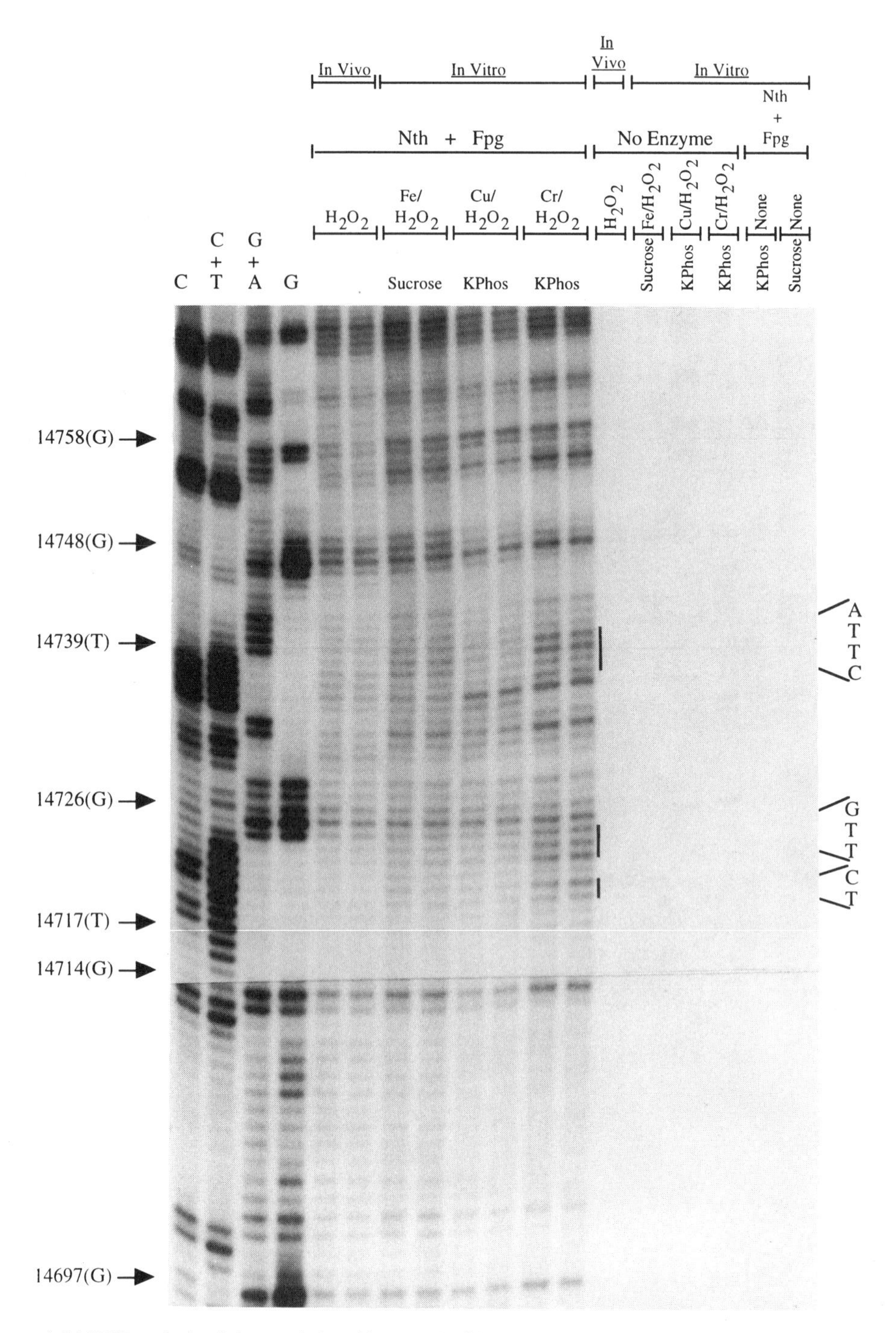

Figure 4. LMPCR analysis of damage induced in exon 9 of human *p53*. *Lanes 1–4*, DNA treated with standard Maxam-Gilbert cleavage reactions. *Lanes 5–6, 13*, DNA recovered from intact human foreskin fibroblasts exposed to 50 mM H_2O_2. *Lanes 7–8, 14*, dialyzed genomic DNA treated with 100 μM Fe(III)/100 μM ascorbate/5 mM H_2O_2 in the presence of 0.3 M sucrose. *Lanes 9–10, 15*, DNA treated with 50 μM Cu(II)/100 μM ascorbate/5 mM H_2O_2. *Lanes 11–12, 16*, DNA treated with 100 μM Cr(VI)/100 μM ascorbate/5 mM H_2O_2. *Lane 17*, DNA incubated in potassium phosphate buffer. *Lane 18*, DNA incubated in potassium phosphate buffer containing 0.3 M sucrose. The DNA in lanes 5–12 was digested with Nth and Fpg proteins after treatment; the DNA in lanes 13–18 was incubated in digestion buffer alone after treatment. Positions of high damage frequency are marked with arrows to the left of lane 1. Vertical bars to the right of lane 12 indicate positions heavily damaged in the presence of Cr(VI), but not Fe(II) or Cu(II). The sequences of these positions are denoted to the right of lane 18. Reproduced with permission.

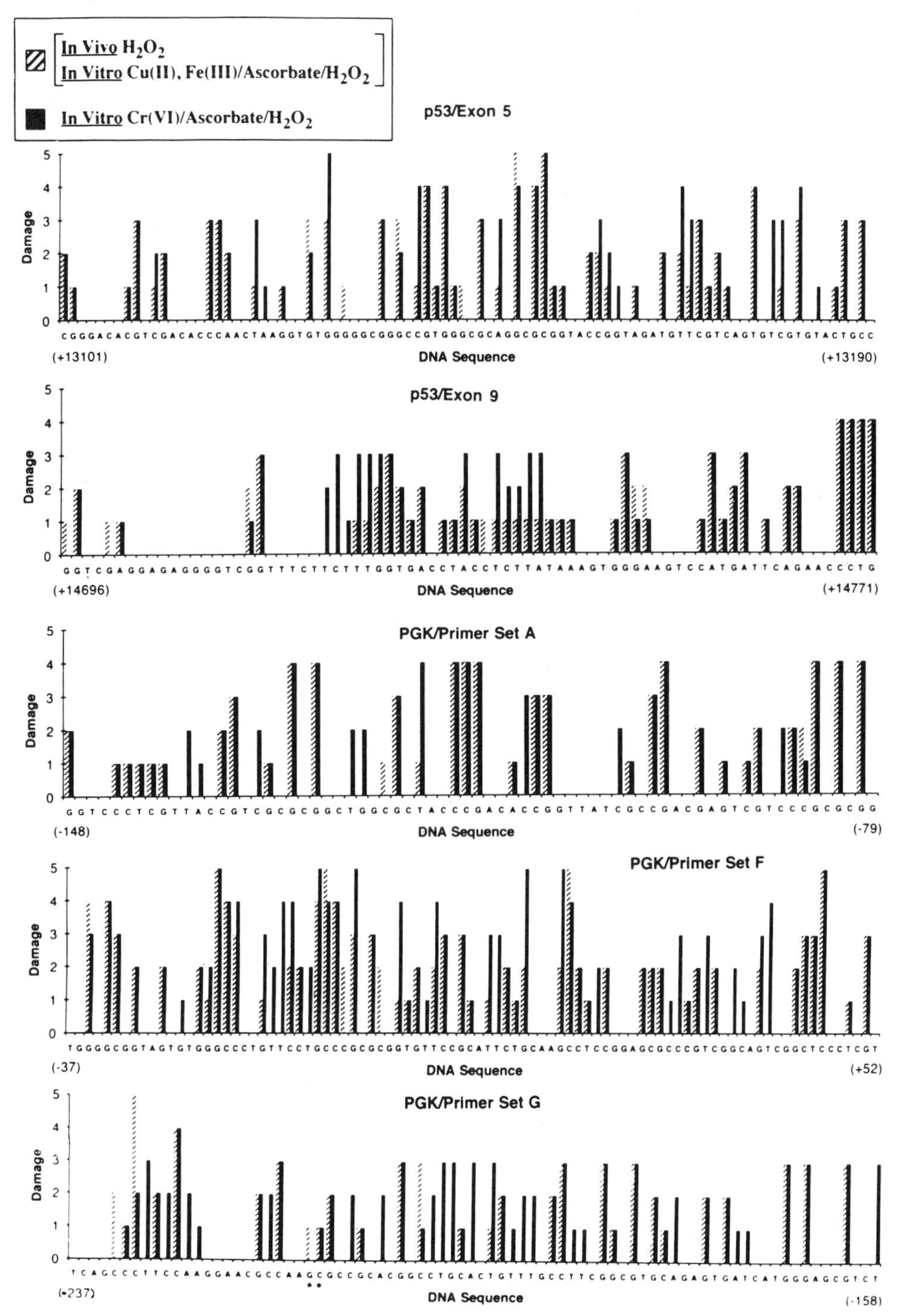

Figure 5. Histogram representing damage frequency at positions in exons 5 and 9 of *p53,* and in the *PGK1* promoter covered by primer sets A, F, and G. LMPCR autoradiograms were quantified by PhosphorImager analysis; signal intensities were divided into quintiles from the most (5) to the least (1) intense. The frequency of Cu(II) or Fe(III)/ascorbate/ H_2O_2-induced damage *in vitro* and H_2O_2-induced damage *in vivo* are represented by a single bar at each position because they were identical. Reproduced with permission.

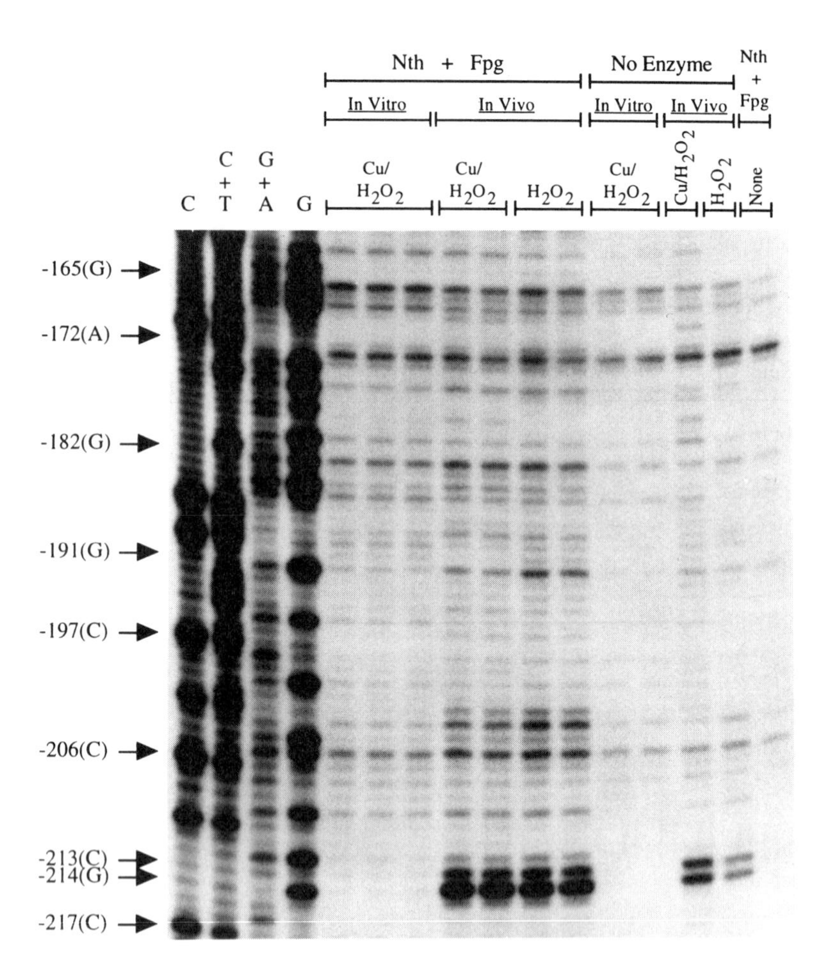

Figure 6. H_2O_2-induced footprints in human foreskin fibroblasts along the transcribed strand of the *PGK1* promoter from 5' G_{-165} to C_{-217} (GGGTA CTAGTGAGAC **GTGCGGCTTC CGTT***TGTCAC GTCCGGCACG CC***GC*G*AAC**$_{-217}$) using primer set G . Positions G_{-214} (214 bases upstream of transcription initiation) and C_{-213} (* bases) show strong positive footprints. These are included in the DNase I footprint which extends from -187 to -223 (boldtype bases) (Pfeifer and Riggs, 1991). The minimal HIF-1 enhancer domain (underlined bases) (Pfeifer et al., 1991) extends from -194 to -211. The arrows point to the location on the autoradiogram of the base positions denoted. Reproduced with permission.

concentration of H_2O_2 in the medium at the end of the 30 min incubation period was well below that required to observe damage signals by LMPCR.

Comparison of damage intensity among sequence contexts were made by applying sequential Wilcoxson rank-sum tests (Rodriguez et al., 1997b). Guanine is the most easily modified base associated with H_2O_2-mediated DNA damaging reactions both *in vivo* and *in vitro*. The triplet d(C*G*C) is the principal hotspot sequence.

4.3. Footprints of ROS-Induced DNA Damage *in Vivo*

The *in vivo* damage pattern induced in human fibroblasts by H_2O_2 did show a few deviations from the *in vitro* pattern (*e.g.* bases G_{-214}, C_{-213}, C_{-212}, C_{-202}, Figure 6) representing mainly positive footprints (Rodriguez et al., 1997a). These are coincident with transcription factor binding sites as previously determined by other footprinting techniques (Pfeifer et al., 1991; Pfeifer and Riggs, 1991).

The prominent positive footprint shown in Figure 6 occurs at positions G_{-214} and C_{-213} in the hypoxia-inducible factor 1 (HIF-1) binding site (Firth et al., 1994) of the transcribed strand, but not the non-transcribed strand (data not shown). Damage at the two positions has two discernable components, base damage (57%) and strand breaks (43%). The percentage of breaks is somewhat greater than the global *in vivo* or *in vitro* break frequency of 19% as determined from analysis of DNA fragment migration on neutral denaturing agarose gels. The *in vivo* G_{-214} signal represents 1/113 molecules with a strand break or oxidized Gua and 1/344 molecules with a strand break. *In vivo* G_{-214} shows a 47 fold increase in base damage and a 10 fold increase in strand breaks relative to the damage at that position *in vitro*, as quantified with a Molecular Dynamics PhosphorImager. Both the DNaseI footprint (Pfeifer and Riggs, 1991) and the H_2O_2-footprint (data not shown) are absent from the *PGK1* allele on the inactive X-chromosome of females.

Exposure of human male fibroblasts to a concentration of H_2O_2 several orders of magnitude higher than those generated *in vivo* under basal metabolic conditions was required in order to generate sufficient DNA base damage for the purpose of damage frequency mapping by LMPCR. Therefore, we assessed to what extent artifacts caused by exposure to high concentration H_2O_2 contributed non-physiologic distortion of the observed damage frequency patterns. Disruption of chromatin structure was assessed by the following experiment: After exposure to 50 mM H_2O_2 for 30 min at 37 C in serum-free medium, human fibroblasts were immediately exposed to 10,000 joules/m^2 of ultraviolet B light. The ultraviolet B light-induced cyclopyrimidine dimer frequency pattern (measured by LMPCR as described in Pfeifer et al., 1991; Pfeifer and Riggs, 1991) induced *in vivo* in fibroblasts exposed to H_2O_2 was compared to the pattern induced *in vivo* in control (no H_2O_2 exposure) fibroblasts and to the pattern induced *in vitro* in purified human genomic DNA. Figure 7 shows the cyclopyrimidine dimer frequency pattern determined using primer set A, which maps a portion of the transcribed strand; these data are representative of all the primer sets tested. The position of photofootprints induced *in vivo* was identical in H_2O_2-exposed *vs.* control fibroblasts indicating that chromatin structure was not disrupted by exposure to 50 mM H_2O_2.

4.4. Damage Induced by H_2O_2 in Isolated Fibroblast Nuclei

Figure 8 shows that isolated nuclei behave similar to naked DNA in that neither H_2O_2 alone nor Cu(II)/Fe(III) + ascorbate cause detectable DNA damage in isolated human male fibroblast nuclei (Figure 8) (Rodriguez et al., 1997b). However, nuclei behave differently from naked DNA in that significant DNA damage was observed if H_2O_2 and Cu(II)/Fe(III) were added [$1/<M_n> = 33$ per 10 kb for global copper ion-associated DNA damage (lane 11); $1/<M_n> = 3.6$ per 10 kb for iron ion-associated DNA damage (lane 13)]. In fact, base damage induced by H_2O_2/copper ion in the isolated nuclei was greater than that induced *in vivo* by equimolar H_2O_2 with or without supplemental 50 μM Cu(II) in the medium (data not shown). Thus, isolated nuclei have the following two notable properties: (1) They contain endogenous reducing agents capable of reducing transition metals such that the metals redox cycle in the presence of H_2O_2, and (2) They do not contain sufficient bound metals (or metal-like ligands) to cause significant base damage in the presence of H_2O_2.

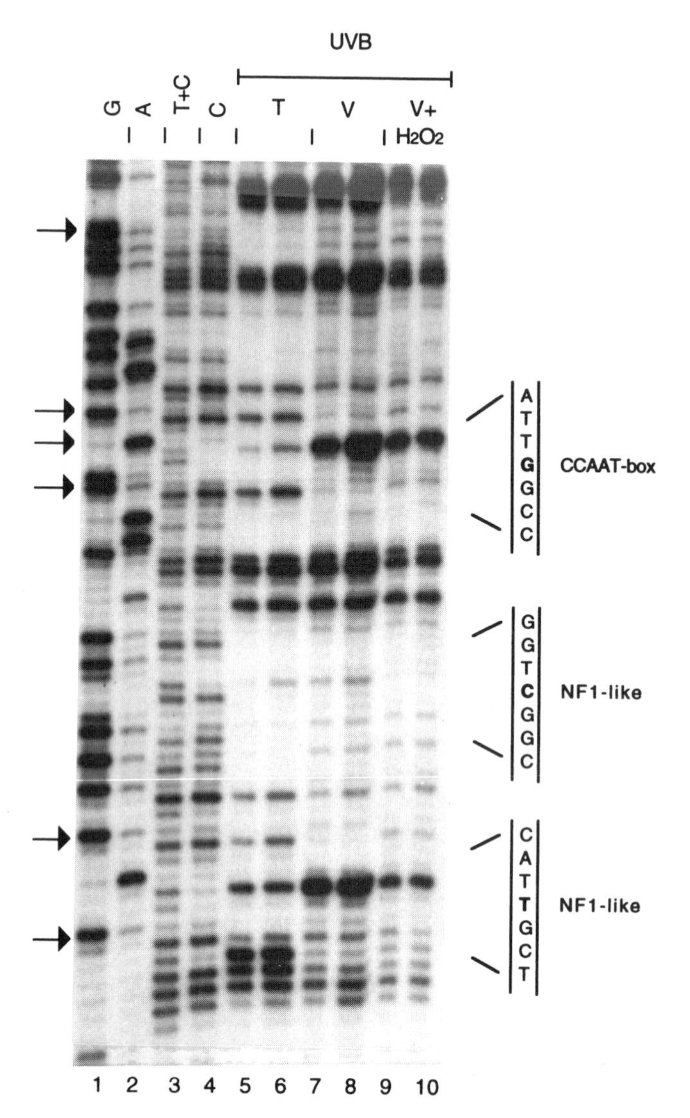

Figure 7. Footprint pattern of ultraviolet B light-induced cyclopyrimidine dimers in human foreskin fibroblasts along the transcribed strand of the *PGK1* promoter using primer set A . Lanes 5 and 6 (T) represent the cyclopyrimidine dimer frequency pattern induced in purified human genomic DNA *in vitro*; lanes 7 and 8 (V) represent the cyclopyrimidine dimer frequency pattern induced in human fibroblasts *in vivo*; lanes 9 and 10 (V + H_2O_2) represent the cyclopyrimidine dimer frequency pattern induced in human fibroblasts *in vivo* after exposure to ultraviolet B light plus 50 mM H_2O_2. The position of footprinted bands (both positive and negative) are pointed out by the arrows to the left of the autoradiogram. The minimal enhancer domains and the DNA-binding proteins assigned to them are denoted to the right of the autoradiogram. Reproduced by permission.

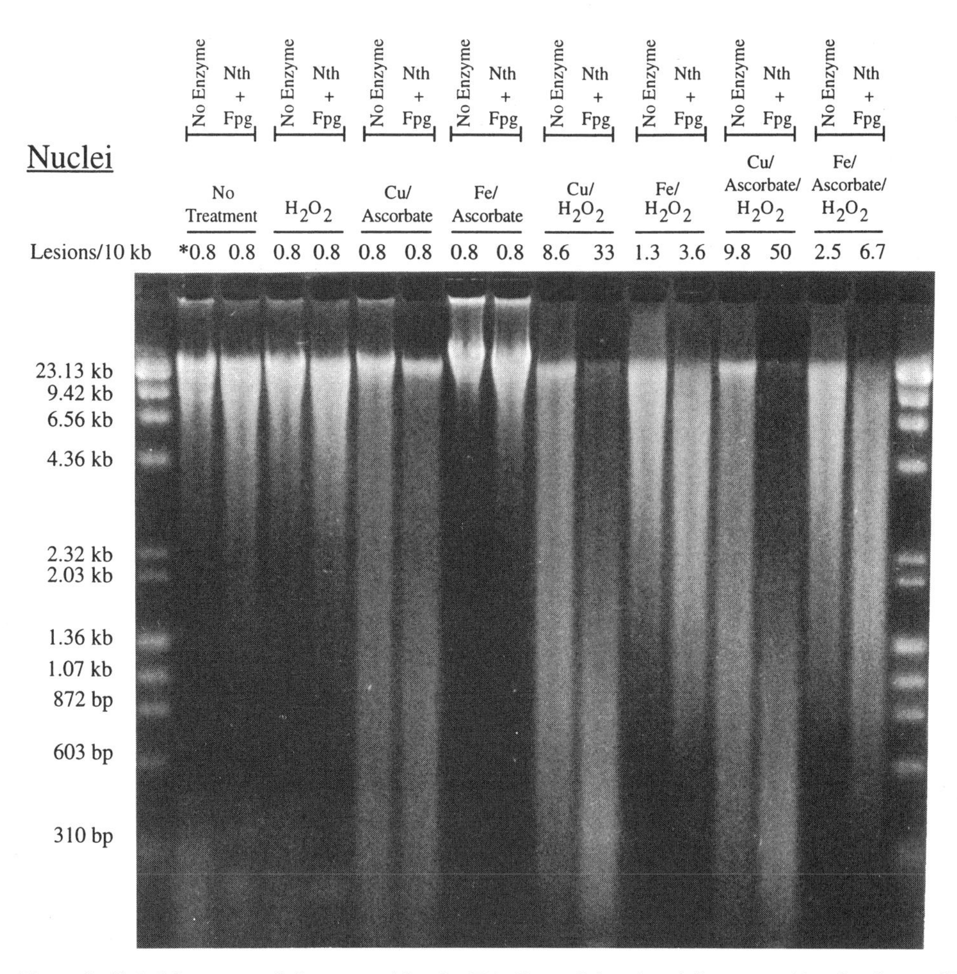

Figure 8. Global frequency of direct strand breaks ("No Enzyne" lanes) and direct strand breaks plus modified bases ("Nth + Fpg" lanes) observed after exposure of isolated human male fibroblast nuclei to 50 mM H_2O_2 (lanes 4–5), 50 μM Cu(II) + 100 μM ascorbate (lanes 6–7), 50 μM Fe(III) + 100 μM ascorbate (lanes 8–9), or combinations of these reagents (lanes 10–17). Nuclear isolation was performed as described in "Materials and Methods". Exposures were for 30 min at 37°C, after which DNA was isolated as described in "Materials and Methods". Electrophoresis was carried out as described in the Legend to Figure 1. (*) 0.8 is a lower limit of $1/\langle M_n \rangle$ in these lanes; a more precise value for $1/\langle M_n \rangle$ could not be determined. Reproduced with permission.

4.5. Enhancement of LMPCR Damage Detection Sensitivity by Genomic Gene Enrichment

The requirement for exposure of fibroblasts to cytotoxic concentrations of H_2O_2 in order to map ROS-induced DNA base damage *in vivo* by our current LMPCR protocol spurred us to develop methods to enhance the sensitivity of LMPCR. One approach to enhancing LMPCR-generated base damage signal intensity is to increase the relative copy number of the target gene in the substrate genomic DNA. This was accomplished by size-fractionating restriction endonuclease-digested, ROS-exposed genomic DNA by CEE through a preparative agarose gel (Rodriguez and Akman, 1998). Pooled fractions contain-

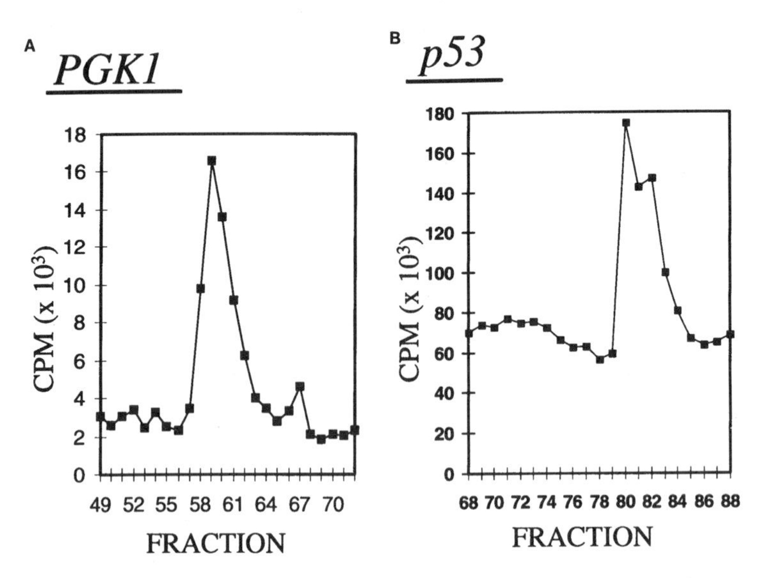

Figure 9. Determination of CEE fractions containing target genes. Aliquots of every CEE fraction were dot-blotted to nylon membranes, then hybridized with [^{32}P]-labeled probes ([A]: *PGK 1* probe; [B]: *p53* probe). Hybridization signal intensities were quantified by PhosphorImager analysis.

ing the *p53* and *PGK1* genes, determined by dot blot analysis (Figure 9), were used as LMPCR substrates. Use of the target gene-enriched substrate DNA as LMPCR substrate resulted in an average 24-fold enhancement of LMPCR-derived base damage signal intensity as compared to non-enriched total genomic DNA (Figure 10). The enhancement of LMPCR-derived damage signal intensity by target gene enrichment is sufficient to permit mapping of base damage induced by non-cytotoxic exposures of fibroblasts to H_2O_2 *in vivo*; such studies are currently in progress.

5. DISCUSSION

We have observed the frequency patterns of DNA base damage in two genes on separate chromosomes induced *in vivo* by H_2O_2 and *in vitro* by H_2O_2 in the presence of copper, iron, or chromate ions plus ascorbate. The data indicate two generalizations: (1) DNA base damage caused by any one of the three metal ions, Cu(II), Fe(III), or Cr(VI), *in vitro* is remarkably similar to each other, as well to the *in vivo* base damage induced by H_2O_2, and (2) DNA base damage frequency *in vivo* is enhanced by metals or metal-like ligands which reside at extranuclear sites, but are mobilized by H_2O_2 exposure, and migrate to the nucleus.

DNA-reactive species generated from H_2O_2 in association with soluble (Drouin et al., 1996)or loosely-bound redox cycling ligands (Luo et al., 1994) predominately cause frank strand breaks, whereas cycling ligands bound intrahelically to DNA bases predominately cause modified bases. *In vivo* H_2O_2 treatment predominantly causes base damage, suggesting that damage induced in cultured human fibroblasts by H_2O_2 involves redox cycling ligands bound to DNA bases in the helix. The near identity of the *in vivo* and *in vitro* transition metal ion-catalyzed damage patterns suggest that these DNA-bound redox cy-

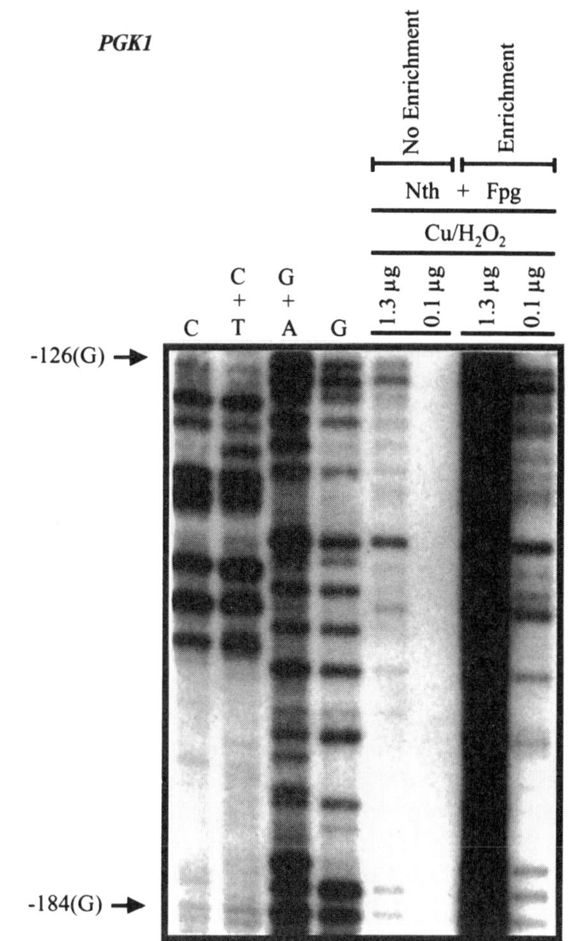

Figure 10. LMPCR analysis (primer set G, transcribed strand) of damage induced in the promoter region of human *PGK 1* by treating purified genomic DNA with 50 µM Cu(II)/ 100 µM ascorbate/ 0.5 mM H_2O_2 in phosphate buffer for 30 min at 37 C. Lanes 5–6, non-enriched treated DNA, digested with Nth and Fpg proteins after treatment; lanes 7–8, treated DNA, digested with Nth and Fpg proteins after treatment, then enriched by CEE fractionation; lanes 1–4, non-enriched, non-treated DNA subjected to standard Maxam-Gilbert cleavage reactions.

cling ligands are transition metal ions. Additionally, the damage frequency patterns indicates that the principal determinant of the probability of a H_2O_2-associated damaging event occurring at any position is the primary DNA sequence. Chromatin structure, with the exception of the transcription factor footprints, is only a minor determinant of base damage probability.

With regard to the DNA damage induced *in vitro*, iron and chromate ions caused more sucrose-inhibitable frank strand breakage than did copper ions. Sucrose probably inhibits strand break formation by competitively binding loosely bound metal ions, as proposed by Gutteridge (Gutteridge, 1984). Physical chemical data suggest that, relative to copper ions, ferrous ions prefer to interact electrostatically with the DNA phosphate backbone (Sissoeff et al., 1976). Radical scavenging by sucrose may also be contributing to strand break inhibition.

The nucleotide-resolution maps of DNA base damage induced *in vitro* in the presence of Cu(II), Fe(III), or Cr(VI) transition metal ions in two genes, were remarkably similar. This similarity suggests a model in which the local binding site occupancy rate and the local geometry of the metal ion-DNA-peroxo coordination complex determine the probability of a damage event at each position.

Experiments with isolated fibroblast nuclei indicated that nuclei have insufficient bound transition metal to induce base damage detectable by neutral denaturing gel electrophoresis; exogenous transition metal ions must be added to the nuclei to produce measurable base damage. Sufficient DNA-bound redox cycling ligands are likely present *in vivo*, but are normally sequestered at extranuclear sites. The extreme oxidative stress caused by exposure of cells to 50 mM of H_2O_2 may have liberated normally sequestered extranuclear transition metal ions into the nucleus, allowing chromatin to show a transition metal ion-saturated pattern. Oxidant-mediated liberation of cytoplasmic sequestered transition metal ions has been observed in other model systems (Calderaro et al., 1993). Larger non-sequestered transition metal ion pools caused by H_2O_2 exposure would most likely facilitate DNA binding at low affinity binding sites to a greater extent than high affinity binding sites, which may become saturated with metal ions; therefore, this effect may tend to "smooth out" the differences in DNA damage frequency between certain nucleotide positions.

6. CONCLUSION

Analysis of H_2O_2-mediated *in vivo* DNA damage frequency patterns in human fibroblasts at nucleotide resolution indicate that the principle determinant of the local DNA damage probability is the primary DNA sequence. Chromatin structure has a very minor effect. Furthermore, damage caused by H_2O_2 *in vivo* is mediated by DNA-bound transition metal ions (or other redox cycling ligands with similar characteristics), that in addition to residing within a cell's nucleus, also reside at extranuclear sites. The damage frequency patterns do not permit definitive determination of which transition metal ions are involved in *in vivo* DNA damage production because the damage frequency pattern is independent of the type of transition metal ion, at least with respect to copper, iron, and chromium ions. Our data simplifies the field of oxidative base damage by showing that the pattern of base damage is independent of the redox cycling metal (or metal-like ligand).

ACKNOWLEDGMENT

The authors wish to thank Mr. Steven Bates for supplying cultured human fibroblasts.

REFERENCES

S.A. Akman, G.P. Forrest, J.H. Doroshow and M. Dizdaroglu (1991) Mutation of potassium permanganate- and hydrogen peroxide-treated plasmid pZ189 replicating in CV-1 monkey kidney cells. *Mutat. Res.* **261**, 123–130.

B.N. Ames Oxidative DNA damage, cancer, and aging. (1987) Ann. Intern. Med. **107**, 526–545

M. Calderaro, E.A. Martins and R. Meneghini (1993) Oxidative stress by menadione affects cellular copper and iron homeostasis. *Mol. Cell Biochem.* **126**, 17–23.

R. Drouin, H. Rodriguez, S. Gao, Z. Gebreyes, T.R. O'Connor, G.P. Holmquist and S.A. Akman (1996) Cupric ion/ascorbate/hydrogen peroxide-induced DNA damage: DNA-bound copper ion primarily induces base modifications. *Free Radic. Biol. Med.* **21**, 261–273.

J.D. Firth, B.L. Ebert, C.W. Pugh, and P.J. Ratcliffe (1994) Oxygen-regulated control elements in the phosphoglycerate kinase I and lactate dehydrogenase A genes: Similarities with erythropoetin 3' enhancer. *Proc. Natl. Acad. Sci. USA.* **91**, 6496–6500.

J.M.C. Gutteridge (1984) Reactivity of hydroxyl and hydroxyl-like radicals discriminated by release of thiobarbituric acid-reactive material from deoyx sugars, nucleosides, and benzoate. *Biochem. J.* **224**, 761–767.

K.Z. Guyton and T.W. Kensler Oxidative mechanisms in carcinogenesis. (1993) *British Med. Bull.* **49**, 523–544.

S. Kawanishi, S. Inoue and K. Yamamoto (1994) Active oxygen species in DNA damage induced by carcinogenic metal compounds. *Environ. Health Perspect.* **102 Suppl 3**, 17–20.

L.A. Loeb, E.A. James, A.M. Waltersdorph and S.J. Klebanoff (1988) Mutagenesis by the autoxidation of iron with isolated DNA. *Proc. Natl. Acad. Sci. USA.* **85**, 3918–3922.

Y. Luo, Z. Han, S.M. Chin and S. Linn (1994) Three chemically distinct types of oxidants formed by iron-mediated Fenton reactions in the presence of DNA. *Proc. Natl. Acad. Sci. USA.* **91**, 12438–12442.

E.C. Moraes, S.M. Keyse, M. Pidoux and R.M. Tyrrell, R.M. (1989) The spectrum of mutations generated by passage of a hydrogen peroxide damaged shuttle vector plasmid through a mammalian host. *Nucleic Acids Res.* **17**, 8301–8312.

G.P. Pfeifer, R. Drouin, A.D. Riggs and G.P. Holmquist (1991) Binding of transcription factors creates hot spots for UV photoproducts in vivo. *Mol. Cell. Biol.* **12**, 1798–1804.

G.P. Pfeifer and A.D. Riggs (1991) Chromatin differences between active and inactive X chromosomes revealed by genomic footprinting of permeabilized cells using Dnase I and ligation-mediated PCR. *Genes Dev.* **5**, 1102–1113.

H. Rodriguez and S.A. Akman (1998) Large scale isolation of genes as DNA fragment lengths by continuous elution electrophoresis through an agarose matrix. *Electrophoresis* **19**, 646–652.

H. Rodriguez, R. Drouin, G.P. Holmquist and S.A. Akman (1997a) A hot spot for hydrogen peroxide-induced damage in the human hypoxia-inducible factor 1 binding site of the PGK 1 gene. *Arch. Biochem. Biophys.* **338**, 207–212.

H. Rodriguez, R. Drouin, G.P. Holmquist, T.R. O'Connor, S. Boiteux, J. Laval, J.H. Doroshow, and S.A. Akman (1995) Mapping of copper/hydrogen peroxide-induced DNA damage at nucleotide resolution in human genomic DNA by ligation-mediated polumerase chain reaction. J. *Biol. Chem.* **270**, 17633–17640.

H. Rodriguez, G.P. Holmquist, R. D'Agostino, J. Keller and S.A. Akman (1997b) Metal ion-dependent hydrogen peroxide-induced DNA damage is more sequence specific than metal specific. *Cancer Res.* **57**, 2394–2401.

I. Sissoeff, J. Grisvard, and E. Guille (1976) Studies on metal ions-DNA interactions: specific behaiour of reiterative DNA sequences. *Prog. Biophys. Molec. Biol.* **31**, 165–199.

L.K., McBride, T., Spence, K., and Loeb, L.A. (1991) Mutation spectrum of copper-induced DNA damage. *J. Biol. Chem.* **266**, 6401–6406.

S.S. Wallace DNA damage processed by base excision repair: biological consequences. (1994) *Int. J. Radiat. Biol.* **66**, 579–584.

C.K. Willis, D.G. Willis and G.P. Holmquist (1988) An equation for DNA electrophoretic mobility. *Appl. and Theor. Electr.* **1**, 11–18.

2

APPLICATION OF OXIDATIVE DNA DAMAGE MEASUREMENTS TO STUDY ANTIOXIDANT ACTIONS OF DIETARY COMPONENTS

Okezie I. Aruoma,[1,2] Andrew Jenner,[2] and Barry Halliwell[2,3]

[1]Universidade de São Paulo-Ribeirão Preto
Faculdade de Ciências Farmacêuticas de Ribeirão Preto
Departamento de Análises Clínicas
Toxocologicas e Bromatólogicas
Avenida do Café s/nº-Monte Alegre
CEP 14040-903, Ribeirão Preto-São Paulo, Brasil
[2]The Pharmacology Group, King's College London
Manresa Road, London, SW3 6LX, Great Britain
[3]Department of Biochemistry, National University of Singapore
Kent Ridge Crescent, Singapore 119260

1. ABSTRACT

Plant-derived antioxidants such as flavonoids, vitamin E, vitamin C, β-carotene or rosemary and sage extracts are increasingly proposed as important dietary antioxidant factors. In this article, assays for characterizing the potential prooxidant and antioxidant actions of food additives, antioxidant supplements, antioxidant drug molecules and nutrient components based on the assessment of products of DNA oxidation are reviewed.

2. INTRODUCTION

Excessive production of reactive oxygen species (ROS), beyond the antioxidant defence capacity of the body can cause oxidative stress (Halliwell and Gutteridge 1998, Sies 1985). Oxidative stress may also be mediated by activation of phospholipases, increased activity of radical generating enzymes (e.g. xanthine oxidase) and/or increased levels of their substrates (e.g. hypoxanthine), disruption of electron transport chains and increased electron leakage of superoxide radical ($O_2^{\cdot -}$), release of "free" metal ions from sequestered

Advances in DNA Damage and Repair, edited by Dizdaroglu and Karakaya. Kluwer Academic / Plenum Publishers, New York, 1999.

sites, activation of cyclooxygenase and lipoxygenase, and, release of heme proteins (hemoglobin, myoglobin).

The free radical oxidation of the lipid components in foods by the chain reaction of lipid peroxidation is a major problem for food manufacturers (Hudson 1990, Loliger 1991). The extent to which oxidation of fatty acids and their esters occurs in foods depends on the chemical structure of the fatty acids, presence of trace metals, levels of antioxidants and, food processing and handling conditions. An antioxidant may be defined as a substance which when present at low concentrations compared with those of an oxidizable substrate such as fats, proteins, carbohydrates or DNA, significantly delays or prevents oxidation of the substrate (Halliwell and Gutteridge, 1998). An alternative definition is "any acidic compound (including phenols) used in foods which can readily donate an electron or a hydrogen atom to a peroxyl or alkoxy radical to terminate a lipid peroxidation chain reaction or to regenerate a phenolic compound, or which can effectively chelate a pro-oxidant transition metal". Antioxidants indigenous to foods and oil-soluble antioxidants in foods such as butylated hydroxyanisole (BHA), butylated hydroxytoluene (BHT), tertiary butylhydroquinone (TBHQ), esters of 3,4,5-tri-hydroxybenzoic acid (propyl, octyl and dodecyl esters), ethoxyquin (6-ethoxy-1,2-dihydro 2,2,4-trimethyl quinoline) (used mostly in animal feeds), dl-α-tocopherols and flavonoids are well known (Anton 1988, Loliger 1991, Hudson 1990, Pratt 1993, Pappas 1993). However, some of the flavonoids e.g. quercetin and phenolic antioxidants such as carnosic acid or carnosol from rosemary have been shown to stimulate free radical damage to non-lipid components such as DNA, proteins and carbohydrates using *in vitro* assays (Aruoma et al 1990, 1992, Halliwell 1990, Laughton et al 1989, Aruoma 1991, 1994a,b). This has led to the development of experimental tools for characterizing the potential pro-oxidant and/or antioxidant activities. The emerging *in vitro* data would help in delineating the *in vivo* contribution of antioxidants to the modulation of the pathological consequences of free radicals in the human body, as well as helping in the evaluation of potential applications of natural antioxidants in food processing.

3. PROOXIDANT ACTION AND ITS MEASUREMENT *IN VITRO*

Hydroxyl radical (OH) is often generated in the test tube using a reaction mixture containing ascorbate, hydrogen peroxide (H_2O_2) and Fe^{3+}-EDTA at pH 7.4 (equations 1 and 2). The addition of ascorbic acid greatly increases the rate of OH generation by reducing iron and maintaining a supply of Fe^{2+} (Halliwell, 1990).

$$Fe^{3+}\text{-EDTA} + \text{Ascorbate} \rightarrow Fe^{2+}\text{-EDTA} + \text{Oxidized Ascorbate} \quad (1)$$

$$Fe^{2+}\text{-EDTA} + H_2O_2 \rightarrow Fe^{3+}\text{-EDTA} + OH^{\cdot} + OH^{-} \quad (2)$$

The hydroxyl radicals that escape being scavenged by the chelator molecule (EDTA in this example), will become available in free solution. This is similar to the generation of free radicals during ionizing radiation. The critical point here is that compounds that are able to mimic the actions of ascorbate will increase free radical generation in this test system. The extent of inhibition by antioxidants however, will be dependent on the effective concentrations of the molecule compared with the target molecule and on its rate constant for reaction with OH . Inhibitors or scavengers of superoxide radicals would be ineffective in this system since superoxide is not involved in the mechanism. Catalase and other

Table 1. Antioxidant activity of Herbor 025 and Spice Cocktail Provençal: Antioxidant index (Rancimat/110°C)

	Antioxidant index			
Extract	Chicken fat	Lard	Soya oil	Sunflower oil
Herbor 025				
0.05%, w/w	1.6	3.6	1.3	1.2
0.1%, w/w	2.3	5.9		
Spice Cocktail Provençal	8.3	9.1	1.8	2.0
Rosemary	12.6	11.4	2.1	2.3
Sage	8.4	8.5	1.8	1.8
Thyme	5.7	4.8	1.2	1.3
Oregano	3.4	2.9	1.3	1.3
Ginger	2.4	2.9	1.1	1.1
Turmeric	1.8	1.6	1.1	1.1
Cayenne pepper	1.2	1.1	1.0	1.1

The principle of this test is to bubble air through heated oil and to monitor continuously the conductivity of water in which the effluent gas is trapped. The test was performed with 5 g fat and 0.5-1% (w/w) concentration of antioxidant mixture at an air flow rate of 20 litres/hr and a temperature of 110°C. The antioxidant index was calculated as: Antioxidant index = Induction period of fat with extract/Induction period of fat alone. Spice cocktail, rosemary, sage, thyme, oregano, ginger, turmeric and cayenne pepper were added at a concentration of 1% (w/w).

scavengers of hydrogen peroxide will inhibit the reactions. By contrast, when $O_2^{\cdot-}$ hypoxanthine and xanthine oxidase are used to generate $OH^{\cdot}$, superoxide dismutase, catalase and scavengers of $O_2^{\cdot-}$ and H_2O_2, will be effective in inhibiting damage to biomolecules by Fenton-type reactions (if the $O_2^{\cdot-}$ is able to reduce the metal ion). When ascorbate is omitted from the reaction mixture, the ability of added compounds to reduce the Fe^{3+}-EDTA complex (reaction 2) can be tested. This idea led to the proposal to use assays involving DNA damage to specifically test for the abilities of dietary antioxidants to exert prooxidant actions, different from their intended abilities to minimize oxidation of lipids. Table 1 shows typical examples of natural antioxidants that are potent inhibitors of lipid oxidation in the Rancimat test.

4. OXIDATIVE DNA DAMAGE AND MEASUREMENT OF PROOXIDANT ACTIONS

DNA is prone to oxidative attack. DNA damage is often measured as single strand-breaks, double strand breaks or chromosomal aberrations (Breimer 1990, Aruoma and Halliwell, 1998a). Mechanisms involving the Fenton system, ionizing radiation and nuclease activation have been suggested to account for much of the DNA damage that occurs in biological systems (Stoewe and Putz 1987, Halliwell and Aruoma 1991, Breen and Murphy 1995, Marnet and Burchman 1993, von Sonntag 1987, Aruoma and Halliwell 1998b, Byrnes 1996). In the Fenton mechanism, oxidative stress could cause release of catalytic copper or iron within cells which could then bind to DNA. Metal ions are naturally occurring metal constituents of the cell nucleus (Pezzano 1980, Bryan et al 1981). Generation of $OH^{\cdot}$ by reaction of H_2O_2 with the transition metal ions already bound onto the DNA would lead to strand breakage, base modification and deoxysugar fragmentation. In the nuclease activation mechanism, oxidative stress leads to the inactivation of Ca^{2+}-binding

Table 2. Damage to DNA bases by hydroxyl radicals generated from Fe(III)-EDTA, H_2O_2 and ascorbate or vanillin at pH 7.4. Values for the products formed are the means of 3 determinations that agreed to within 10%. Control experiments showed that Trolox C, mannitol, hypotaurine, vanillin or dimethylsulphoxide (DMSO) did not themselves cause any base modification. RM - reaction mixture; RMA -reaction mixture plus ascorbate

Systems tested		Products monitored (nmol/mg DNA)					
		A	B	C	D	E	Total
DNA, Fe^{3+}-EDTA, H_2O_2 (RM)		0.08	0.02	0.06	0.10	0.43	0.69
RM + ascorbate (RMA)	100 μM	1.13	0.98	1.76	8.20	2.94	15.01
RMA + mannitol	100 mM	0.23	0.06	0.12	0.33	0.31	1.05
RMA + Trolox C	20 mM	0.31	0.11	0.17	0.88	0.21	1.68
RMA + DMSO	100 mM	0.14	0.01	0.08	0.19	0.15	0.57
RMA + hypotaurine	20 mM	0.19	0.04	0.18	0.41	0.42	1.24
RM + vanillin	20 mM	0.40	0.18	0.26	1.20	0.50	2.30

Key: A = Thymine glycol; B = 5,6-Dihydroxycytosine; C = 4,6-Diamino-5-formamidopyrimidine; D = 2,6-Diamino-4-hydroxy-5-formamidopyrimidine; E = 8-Hydroxyguanine.

by endoplasmic reticulum, inhibition of plasma membrane Ca^{2+}-extrusion systems, and the release of Ca^{2+} from mitochondria resulting in increases in the levels of intracellular free calcium ions. The resulting endonuclease activation leads to DNA fragmentation without the base modification observed in the Fenton mechanism.

Hydroxyl radicals ($OH^{\cdot}$) induce extensive damage to all the four bases in DNA to yield a variety of products. The ability of antioxidants to induce $OH^{\cdot}$-dependent base modification may therefore be used as a tool for assessing pro-oxidant potentials. It also follows that antioxidants (protecting lipids against oxidation, Table 1) may be assessed for their ability to affect DNA base modifications *in vitro* as illustrated in Table 2. Incubation of DNA with Fe(III)-EDTA, ascorbate and H_2O_2 led to significant rises in the amounts of several oxidized bases: this is characteristic of attack by $OH^{\cdot}$ (Halliwell and Arouma, 1991). The $OH^{\cdot}$ scavengers mannitol, dimethylsulfoxide and hypotaurine decreased the levels of the oxidized bases (Table 2). Trolox C, the water soluble analogue of vitamin E and a good scavenger of $OH^{\cdot}$ also inhibited the base modification (Table 2). Omission of ascorbate from the reaction mixture greatly decreased the DNA base modification (Table 2, first line) but high concentrations of vanillin restored some of it (Table 2, last line). It is important to note the high concentrations required. The exact concentration of vanillin *in vivo* is not known but could reach high micromolar levels. Nevertheless, the high concentrations of vanillin required for the prooxidant effects suggests that this action may not present adverse physiological consequences. The results of GC-MS analysis of modified bases in DNA can be expressed as nanomoles (nmols) of modified bases per milligram of DNA (equivalent to pmol/μg DNA). However, it is easy to convert these data into the actual number of bases modified. Dividing the amount of nmol bases/mg DNA by 3.14 (or multiplying by 0.318) gives the number of modified bases per 10^3 bases in DNA, i.e. 1 nmol/mg DNA corresponds to about 318 modified bases per 10^6 DNA bases.

5. THE BLEOMYCIN-IRON DEPENDENT OXIDATION OF DNA DAMAGE

Bleomycin, an anti-tumor antibiotic, binds to DNA using its bithiazole and terminal amine residues, and complexes with metals (such as iron) using the β-aminoalaninepyrimidine-β-hydroxy-histidine portion of the molecule. Bleomycin binds iron ions and the bleomycin-iron complex will degrade DNA in the presence of O_2 and a reducing agent such as ascorbic acid. The reaction occurs by attack of a ferric bleomycin peroxide (BLM-Fe(III)-O_2H^-) on the DNA. The ferric peroxide can be formed by direct reaction of ferric-bleomycin with hydrogen peroxide, or from a BLM-Fe(III)-O_2^- complex. It is possible that under certain conditions the BLM-Fe(III)-O_2^- might decompose to yield $O_2^{\cdot-}$, and BLM-Fe(III)-O_2H^- to release $OH^\cdot$ (Sugiura et al. 1982, Petering et al 1990). Hydroxyl radical is not necessarily the major DNA damaging species in the bleomycin system. The bleomycin-iron(III) complex by itself is inactive in inducing damage in DNA. Oxygen and a reducing agent or hydrogen peroxide are required for the damage to DNA to occur.

DNA cleavage by bleomycin releases some free bases and base propenals in amounts that are stoichiometric with strand cleavage (Burger et al 1981, Giloni et al 1981). When heated with thiobarbituric acid (TBA) at low pH, base propenals rapidly decompose to give malondialdehyde (MDA), which combines with TBA to form a pink $(TBA)_2$MDA adduct. A positive test is obtained when compound is able to reduce bleomycin-Fe^{3+}-DNA complex to the more active bleomycin-Fe^{2+}-DNA complex (in the presence of oxygen) in the absence of added ascorbate in the reaction mixture, resulting in DNA damage.

6. THE COPPER-PHENANTHROLINE DEPENDENT OXIDATION OF DNA

Another method for assessing pro-oxidant activity is the copper-phenanthroline mediated DNA damage. The copper -1,10-phenanthroline complex has nuclease activity and has been used for structural studies upon DNA (Sigman 1986, Thederahn et al 1989) as it can induce strand breakage. Hydrogen peroxide is implicated in the mechanism of the DNA damage by the copper-phenanthroline system. For this reason, it is important to ascertain that the compounds under test do not themselves scavenge H_2O_2 to the extent that would affect the outcome of the assay. Hydroxyl radicals are involved in the damage to DNA caused by the copper-phenanthroline system (Que et al 1980, Dizdaroglu et al 1990). Unlike the bleomycin-iron mediated damage to DNA, damage in the copper-phenanthroline system is confined mainly to the DNA bases (Dizdaroglu et al 1990). The small amount of DNA sugar damage is what the copper-phenanthroline assay measures (Table 3). When a reducing agent is omitted from the reaction mixture, no damage to deoxyribose in DNA occurs in this system. Increasing the concentrations of reducing agents such as ascorbate and/or mercaptoethanol, lead to increased deoxyribose damage. This is in agreement with earlier observations that the nuclease activity of copper-phenanthroline complex is enhanced by thiols, a superoxide radical generating system, xanthine-xanthine oxidase and NADH in the presence of hydrogen peroxide (Reich et al 1981).

The assays involving DNA damage to assess prooxidant actions have unique features. The positive prooxidant actions in the deoxyribose system rely on the ability of the compounds to promote reduction of Fe^{3+} to Fe^{2+} chelates and hence $OH^\cdot$ formation in the presence of H_2O_2. Assays involving DNA rely on the ability reductants to reduce either the iron-bleomycin-DNA or copper-1,10,phenanthroline-DNA complex. If the compound under test promotes the two reactions described, it possesses a prooxidant property . A compound might be prooxidant in the deoxyribose system and/or the DNA systems but sometimes not in both. There are often solubility problems. Fortunately organic solvents do not affect the outcome of DNA-dependent assays. Thus where the deoxyribose assay

Table 3. Effect of Herbor 025 and Spice Cocktail Provençal on copper-1,10-phenanthroline-dependent DNA damage

Addition to reaction mixture (%, v/v)	Extent of DNA damage (A532 nm)
None	0.000
Ascorbate (240 μM)	0.201
Herbor 025	
0.05	0.090
0.10	0.159
0.20	0.244
Spice Cocktail Provençal	
0.05	0.035
0.10	0.075
0.20	0.122

Values are the means from duplicate experiments that varied no more than 5%. Ascorbate at a final concentration of 0.24 mM was used as a positive control. Herbor 025 and the spice cocktail at the concentrations tested, weakly stimulated DNA damage. Although this may be a feature of the constituent compounds, the plant extract are potent inhibitors of lipid peroxidation. Products of DNA oxidation may make a minor (if any) contribution to food stability. *In vivo*, the low levels of the absorbed components would suggest that the prooxidant effects are unlikely to present physiological problems.

cannot be performed due to solubility restriction, the copper-phenanthroline assay might still be applicable. Circumventing potential prooxidant action could contribute to increased protective ability of dietary antioxidants towards susceptible substrates. For example, proteins protect DNA against the prooxidant actions of some flavonoids and polyphenolic compounds *in vitro* (reviewed in Aruoma 1994).

7. *IN VIVO* IMPLICATIONS OF PROOXIDANT ACTIONS

The interaction between food additives and nutrient components within the food matrix on the one hand, coupled with other interactions when the food and plant materials are consumed, is an area of current interest. Antioxidant and prooxidant actions would have different implications for the food matrix and the biological systems (*in vivo*). For the food matrix, the emphasis is on minimizing oxidation to the lipid component of the food. Different factors have to be considered *in vivo*. Iron ions catalytic for free radical reactions are safely sequestered in the human body. However, they can become available at sites of tissue injury e.g. in advanced atherosclerotic lesions (Smith et al 1992). Ehrenwald et al (1994) have claimed that caeruloplasmin containing one redox active copper per protein molecule can oxidize LDL in the presence of cells. Indeed, increased serum copper levels are associated with accelerated progression of carotid atherosclerosis in humans. The body iron store ferritin has been reported to be associated with carotid atherosclerosis. That transition metal ions within human atherosclerotic lesions can stimulate LDL oxidation by macrophages (Lamb et al 1995) suggests that use of natural antioxidants needs to be approached with caution.

Dietary antioxidants can promote increased synthesis of endogenous antioxidant defenses by up-regulation of their biosynthesis and/or increased gene expression (Ushakova et al 1996). Dietary supplements can modify gene expression induced by heat shock *in vivo* as well as protecting animal tissues against oxidative stress by enhancing the level of endogenous antioxidants and inducing hsp (heat shock protein)-70 gene expression (Ushakova et al 1996).

The dietary antioxidants may act directly by scavenging reactive oxygen species *in vivo*. For the proposed antioxidant to have a physiologically meaningful effect *in vivo* it must become absorbed and presented to the site of intended action at a concentration that actually exerts an antioxidant effect. However, the feasibility of a compound exerting a direct antioxidant effect can be evaluated by *in vitro* tests that investigate how the putative antioxidant react with biologically-relevant oxygen-derived species. This may then be extrapolated to *in vivo* situations (Aruoma, 1996, Aruoma and Cuppett 1997, Halliwell 1996). As far as the ability of nutrients and drugs to act as antioxidants *in vivo* is concerned, specific assays to measure rates of oxidative damage to DNA would enable scientists to assess steady-state and total body oxidative damage to these molecular targets, providing the technology to examine the effects of antioxidants *in vivo*.

REFERENCES

R. Anton (1988). Flavonoids and traditional medicine, in *Plant Flavonoids in Biology and Medicine II: Biochemical and Medicinal Properties*, (eds. V. Cody, E. Middleton, J.B. Harborne, and A. Bevertz), Alan R Liss, New York, pp 423–438.

O.I. Aruoma (1993). Use of DNA damage as a measure of pro-oxidant actions of antioxidant food additives and nutrient components, in *DNA and Free Radicals*, (eds. B. Halliwell and O.I. Aruoma), Ellis Horwood, London, pp. 315–327.

O.I. Aruoma (1994a). Nutrition and health aspects of free radicals and antioxidants, *Fd. Chem. Toxicol.* **32**: 671–683.

O.I. Aruoma (1994b). Deoxyribose assay for detecting hydroxyl radicals, *Meth. Enzymol.* **233**: 57–66.

O. I. Aruoma (1996). Characterization of drugs as antioxidant prophylactics. *Free Rad. Biol. Med.* **20**: 675–705.

O. I. Aruoma and S. Cuppett (1997). *Antioxidant Methodology: In Vivo and In Vitro Concepts*, AOCS Press, Champaign.

O.I. Aruoma, and B. Halliwell (1998b). *DNA and Free Radicals; Techniques, Mechanisms, and Applications*, OICA International: Saint Lucia.

O.I. Aruoma, and B. Halliwell (1998b). *Molecular Biology of Free Radicals in Human Diseases*. OICA International: Saint Lucia.

O. I. Aruoma, P.J. Evans, H. Kaur, L. Sutcliffe and B. Halliwell (1990). An evaluation of the antioxidant and potential pro-oxidant properties of food additives and trolox c, vitamin E and probucol. *Free Rad. Res. Commun.* **10**: 143–157.

O.I. Aruoma, B. Halliwell, R. Aeschbach, and J. Löliger (1992). Antioxidant and pro-oxidant properties of active rosemary constituents: carnosol and carnosic acid, *Xenobiotica* **22**: 257–268.

R. M. Burger, J. Peisach, and S.B. Horwitz (1981). Mechanism of bleomycin action: in vitro studies, *Life Sci.* **28**: 715–727.

A. P. Breen, and J.A. Murphy (1995). Reactions of oxyl radicals with DNA, *Free Rad. Biol. Med.* **18**: 1033–1077.

L. H. Breimer (1990). Molecular mechanisms of oxygen radical carcinogenesis and mutagenesis. The role of DNA base damage, *Mol. Carcinogen.* **3**: 188–197.

S. E. Bryan, D.L. Vizard, D.A. Beary, R.A. LaBiche, and K.J. Hardy (1981). Partitioning of zinc and copper within subnuclear nucleoprotein particles, *Nuc. Acids. Res.* **9**: 5811–5823.

R. W. Byrnes (1996). Evidence for involvement of multiple iron species in DNA single-strand scission by H_2O_2 in HL-60 cells, *Free Rad. Biol. Med.* **20**: 399–406.

M. Dizdaroglu (1991). Chemical determination of free radical induced damage to DNA, *Free Rad. Biol. Med.* **10**: 225–242.

M. Dizdaroglu, O.I. Aruoma, and B. Halliwell (1990). Modification of bases in DNA by copper-ion-1,10-phenanthroline complexes, *Biochemistry* 29: 8447–8451.

E. Ehrenwald, G.M. Chisolm, and P.L. Fox (1994). Intact human ceruloplasmin oxidatively modifies low density lipoprotein, *J. Clin. Invest.* **93**: 1493–1501.

L. M. Giloni, Takeshita, F. Johnson, C. Iden, and A.P. Grollman (1981). Bleomycin-induced strand scission of DNA: mechanism of deoxyribose cleavage, *J. Biol. Chem.* **256**: 8608–8615.

B. Halliwell (1990). How to characterize a biological antioxidant, *Free Rad. Res. Commun.* **9**: 1–32.

B. Halliwell (1996). Oxidative stress, nutrition and health. Experimental strategies for optimization of nutritional antioxidant intake in humans, *Free Rad. Res.* **25** : 57–74.

B. Halliwell and O.I.Aruoma (1991). DNA damage by oxygen-derived species. Its mechanism and measurement in mammalian systems, *FEBS Letters* **281**: 9–19.

B. Halliwell, and O.I. Aruoma (1993), *DNA and Free Radicals*, Ellis Horwood, London.

B. Halliwell, and J. M. C. Gutteridge (1998). *Free Radicals in Biology and Medicine*, Oxford, Clarendon Press, 3rd Edition.

B. J. F. Hudson (1990) *Food Antioxidants.* Elsevier Applied Science, London, New York.

D. J. Lamb, M.J. Mitchinson, and D.S. Leake (1995). Transition metal ions within human atherosclerotic lesions can catalyse the oxidation of low density lipoprotein by macrophages. *FEBS Lett.* **374**: 12–16.

M. J. Laughton, B. Halliwell, P.J. Evans, and J.R.S. Hoult (1989). Antioxidant and pro-oxidant actions of the plant phenolics quercetin, gossypol and myricetin, *Biochem. Pharmacol.* **38**: 2859–2865.

J. Löliger (1991). The use of antioxidants in food, in Free Radicals and Food Additives, (eds. O.I. Aruoma, and B. Halliwell), Taylor & Francis, London, pp. 121–150.

L. J. Marnet and P. P. Burcham, (1993). Endogenous DNA adducts: potential and paradox, *Chem. Res Toxic.* **6** 771–785.

A. M. Papas (1993). Oil-soluble antioxidants in foods. *Toxicol. Ind. Health* **9**: 123–149.

H. Pezzano and F. Podo (1980), Structure of binary complexes of mono- and polynucleotides with metal ions of the first transition group, *Chem. Rev.* **80**: 365–401.

D. H. Petering, R.W. Byrnes, and W.E. Antholine (1990). The role of redox-active metals in the mechanism of action of bleomycin, *Chem. Biol. Interact.* **73**: 133–182.

D. E. Pratt (1993). Antioxidants indigenous to foods, *Toxicol. Ind. Health* **9**: 63–75.

B. G. Que, K.M. Downey, and A.G. So (1980). Degradation of deoxyribonucleic acid by a 1,10-phenanthroline-copper complex: the role of hydroxyl radicals, *Biochemistry* **19**: 5987–5991.

K. A. Reich, L.E. Marshall, D.R. Graham, and D.S. Sigman (1981). Cleavage of DNA by the phenanthroline-copper ion complex. Superoxide mediates the reaction dependent on NADH and hydrogen peroxide, *J. Am. Chem. Soc.* **103**: 3582–3584.

D. M. Schaefer, Q. Liu, C. Faustman, and M.-C. Yin (1995). Supranutritional administration of vitamins E and C improves oxidative stability of beef, *J. Nutr.* **125**: 1792S-1798S.

Y. T. Sugiura, Suzuki, J. Kuwahara, and H. Tanaka (1982). On the mechanism of hydrogen peroxide-, superoxide-, and ultraviolet light-induced DNA cleavages of inactive bleomycin iron-(III) complex, *Biochem. Biophys. Res. Commun.* **105**: 1511–1518.

D. S. Sigman (1986). Nuclease activity of 1,10-phenanthroline-copper ion, *Acc. Chem. Res.* **19**: 180–186.

C. Smith, M.J. Mitchinson, O.I. Aruoma, and B. Halliwell (1992). Stimulation of lipid peroxidation and hydroxyl radical generation by the contents of human atherosclerotic lesions, *Biochem. J.* **286**: 901–905.

A. J. St. Angelo (1992). *Lipid Oxidation in Food.* American Chemical Society, Series 500, Washington .

R. Stoewe, and W.A. Prütz (1987). Copper-catalyzed DNA damage by ascorbate and hydrogen peroxide: kinetics and yield. *Free Rad. Biol. Med.* **3**: 97–105.

T. B. Thederahn, T.B., D.M. Kuwabra, T.A. Larson, and D.S. Sigman (1989). Nuclease activity of 1,10-phenanthroline-copper: kinetic mechanism, *J. Am. Chem. Soc.* **11**: 4941–4946.

T. H. Ushakova, H. Melkonyan, L. Nikonova, N. Mudrik, V. Grogvadze, A. Zhukova, A.I. Gaziev, and R. Bradbury (1996). The effect of dietary supplements on gene expression in mice tissues, *Free Rad. Biol. Med.* **20**: 279–284.

C. von Sonntag (1987). The Chemical Basis of Radiation Biology. Taylor & Francis, London.

3

DNA REPAIR AND TRANSCRIPTION IN PREMATURE AGING SYNDROMES

Vilhelm A. Bohr, Adabalayam Balajee, Robert Brosh, Jan Nehlin, Amrita Machwe, Michele Evans, Grigory Dianov, and David Orren

Laboratory of Molecular Genetics
National Institute on Aging, NIH
4940 Eastern Ave.
Baltimore, Maryland 21224

1. ABSTRACT

The human progeroid disorders Cockayne syndrome (CS) and Werner syndrome (WS) exhibit several clinical features that are associated with normal aging. With the recent cloning of the Werner syndrome (*WRN*) gene, and with the information that this gene, the CS complementation group B (*CS-B*) gene, and some *XP* gene products are putative helicases and involved in nucleic acid metabolism, further understanding the molecular deficiency in these disorders is a high priority. Helicases are involved in a number of DNA metabolic activities including transcription, replication and DNA repair. These human disorders provide excellent model systems for studies on aging. The patients have many signs and symptoms of normal aging, but they are segmental progeroid diseases, indicating that some features of normal aging are not seen. The function of the CS-B and WRN proteins appear to be at the crossroads of aging, DNA repair, DNA replication, and transcription, and hence these studies nicely combine our mechanistic interest in basic processes with our interest in aging. CS (Group B), WRN, and other age-related disorders (Bloom, Xeroderma pigmentosum Groups B and D) carry mutations in related genes characterized by conserved motifs of sequence homology. Some proteins of this family have been demonstrated to be DNA-dependent ATPases, a subset of which have also been shown to be helicases. Proteins of this family are involved in various aspects of chromosome metabolism. The molecular defects responsible for the clinical phenotypes of these diseases remain to be determined, but presumably relate to the functional activities of these conserved proteins. In addition, specific protein-DNA and protein-protein interactions are likely to play critical roles in cellular function. Cells derived from CS patients are deficient in a special type of DNA repair, transcription coupled DNA repair, but they also ap-

Advances in DNA Damage and Repair, edited by Dizdaroglu and Karakaya. Kluwer Academic / Plenum Publishers, New York, 1999.

pear to be defective in basal transcription. The diverse functions of the CSB protein are under intense study. Werner syndrome cells may have subtle defects in DNA repair, and possibly also in transcription. The biochemical clarification of the precise role(s) of these gene products is likely to provide very significant clues into the mechanism of aging.

2. INTRODUCTION

2.1. DNA Repair Processes

Living cells are constantly exposed to environmental agents and endogenous processes that inflict damage in DNA. Several complex enzymatic mechanisms have evolved to repair DNA lesions, and lately there has been a tremendous progress towards more understanding of the mechanisms involved. There are several major pathways of DNA repair, and the particular pathway used depends in part upon the type of DNA damage that is being removed. Most types of cellular stress or damage induce a large spectrum of DNA lesions. Endogenous metabolic processes generate oxygen radicals that are removed from the DNA mainly by the base excision repair (BER) pathway. UV exposure generates two major lesions, or adducts, in the DNA: the pyrimidine dimer (PD) and the 6–4 photoproduct (6–4 PP). Both of these adducts and other bulky lesions are removed by the nucleotide excision repair (NER) pathway.

NER is deficient in the human genetic disorder xeroderma pigmentosum (XP). This condition involves hyperpigmentation of the skin and the continuous development of skin cancers. In XP there are 7 different complementation groups representing different genes involved in the disorder. These genes have now been cloned and characterized, leading to a clearer understanding of the NER process in humans. There are several recent reviews in which the pathway of NER is discussed in detail (Friedberg (96b), Sancar (96), Wood (96)).

2.2. DNA Repair Heterogeneity and the Link to Transcription

Another area of very active research in the field of DNA repair concerns the heterogeneity of the DNA repair process. In the last 10 years, evidence has accumulated that the repair process differs significantly in efficiency between different regions of the genome. DNA repair occurs preferentially in genes, and particularly those that are actively transcribed. The active component of the genome is only about 1% of the total DNA, but it appears to be much more efficiently repaired than the rest of the genome which is largely inactive. A component of this repair is directly linked to the basal transcription process. This pathway is also termed transcription-coupled DNA repair (TCR) or "strand specific" DNA repair since the transcribed strand of the active genes is preferentially repaired. The molecular link between DNA repair and transcription involves the basal transcription factor TFIIH which, among its 9 components, contains at least two DNA repair genes, *XPD* and *XPB*.

It is of widespread interest to resolve the mechanism of TCR. The signaling for entering this pathway remains elusive. It is generally thought that the process involves a transcription blocking lesion, and that RNA polymerase II arrests at the site, followed by the recruitment of various repair factors including TFIIH. This hypothesis can not yet be tested in *in vitro* assays since such experiments can only be done with inactive rather than actively transcribed DNA.

2.3. Changes with Aging, Premature Aging Syndromes

In aging research it is important to choose a good model system for the more mechanistic studies. While there has been much work on cell lines that grow to senescence, there are now human natural mutant syndromes that have signs and symptoms of premature aging. Xeroderma pigmentosum (XP) has been characterized as such a condition, and others include Cockayne syndrome (CS) and Werner syndrome (WS). With the recent cloning of the Werner syndrome (WRN) gene, and with the information that this gene, the CS complementation group B (CS-B) gene, and some XP genes have helicase motifs, further understanding the molecular deficiency in these disorders is a high priority. These human disorders provide excellent model systems for studies on aging. The patients have many signs and symptoms of normal aging, but they are segmental progeroid diseases, indicating that some features of normal aging are not seen. The function of the CSB and WRN proteins appear to be at the crossroads of aging, DNA repair, DNA replication, and transcription, so these studies nicely combine a mechanistic interest in basic processes with an interest in aging. The genes that are defective in CS (Group B), WRN, and other age-related disorders (Bloom, Xeroderma pigmentosum (Groups B and D)) are characterized by conserved motifs of sequence homology. A number of proteins of this family have been demonstrated to be DNA-dependent ATPases, a subset of which have also been shown to be helicases, indicating that these proteins are involved in various aspects of chromosome metabolism. The molecular defects responsible for the clinical phenotypes of these diseases remain to be determined, but presumably relate to the functional activities of these conserved proteins. In addition, specific protein-DNA and protein-protein interactions are likely to play critical roles in diverse cellular functions.

2.4. Cockayne Syndrome

Cells from CS patients are sensitive to UV light, exhibit a delay in the recovery of DNA and RNA synthesis following irradiation, and are defective in preferential repair and strand-specific repair of active genes (Friedberg (96a)). Complementation studies demonstrated at least two genes involved in CS, designated *CSA* and *CSB*. CSA protein contains amino-acid repeats (WD), characteristic for a group of proteins that are required for cell metabolism, and interacts with CSB and the transcription factor TFIIH (Iyer et al (96)). CSB protein, by sequence comparison, belongs to the SNF2 family of proteins, which have roles in transcriptional regulation, chromosome stability and DNA repair. The mechanism of TCR in eukaryotes remains to be elucidated, and the CSB protein appears to be an important component.

The cellular and molecular phenotype of CS include a significantly increased sensitivity to a number of DNA-damaging agents including ultraviolet (UV) radiation (Schmickel et al (79); Friedberg (96a); Naumovski and Friedberg (86)). Studies in CS cells were initially confined to DNA repair in the general, overall genome, where no defect was found (reviewed in (Friedberg et al,95)). The observation that CS cells are deficient in the resumption of RNA synthesis after UV irradiation (Mayne and Lehmann (82)) led to the idea that these cells might be deficient in the repair of transcriptionally active genes. With the development of gene specific repair assays (Bohr (91)) CS cells were indeed shown to be defective in the preferential repair of active genes and in the preferential repair of the transcribed strand of such genes (Venema et al (90); Evans and Bohr (94)) . This defect in transcription coupled repair (TCR) in CS is not only found after UV exposure but also after gamma radiation (Leadon and Cooper (93)).

Transfection of the *CSB* gene into hamster cells with the CS-B phenotype completely restores TCR and UV resistance to normal levels (Orren et al, 96), demonstrating that the TCR defect in CS-B is due to mutation in that gene.

The complex clinical phenotype of CS, however, suggests that DNA repair may not be the primary defect. Moreover, recent evidence from this laboratory demonstrated that CSB cells are defective in basal transcription, supporting the notion that reduced gene-specific repair of CS is a consequence of a transcription deficiency. We are pursuing whether the defect in CS is a primary DNA repair defect or a primary transcription defect.

We find a defect in basal transcription in CS both *in vivo* and *in vitro* (Balajee et al (97)). This transcription defect is seen in CS-B lymphoblastoid cells and fibroblasts without any exposure to stress such as UV light. In collaboration with Dr. Friedberg at UT, Dallas, we find that the transcription defect may be associated with a special sensitivity of the transcription apparatus in CS towards DNA damage or special DNA modification (Dianov *et al.* 1997). This is currently under study, and we are also analyzing DNA from CS patients for DNA base modifications. In collaboration with Dr. Dizdaroglu, we are analyzing DNA from CS patients together with DNA from young and old individuals. The goal is to investigate whether any form of oxidative DNA base modifications accumulate with aging and/or in CSB. The DNA samples are analyzed by GC/MS, and preliminary results (Stierum *et al.*, in prep) indicate that certain minor lesions rather than 8-OH guanosine are at high levels in the CS DNA. In these assays, we have found that the DNA extraction procedure is of critical importance, and this variable makes it necessary to sample many individuals over many days to assure that statistical significance is reached. This has not yet been accomplished. There are many types of oxidative base modifications in DNA that are seen in human cells, and the accumulation of any one single lesion could possibly be a very important factor in the development of the molecular phenotype.

In support of our results, a previous study found that expression of a metalloproteinase was reduced to 50% in CS cells (Millis et al (92)), and very recently it was reported that the CSB protein affects transcription as a purified component *in vitro* (Selby et al (97a)).

We also determined the effects of lysis with different concentrations of Triton X-100 on transcription rates in chromatin from normal and CS-B lymphoblasts (Balajee et al (97)). Increasing the Triton concentration appeared to disrupt chromatin structure more in CS-B cells than in normal cells. In normal cells, transcription rates expressed as pmol UMP incorporation /10^6 cells remained essentially the same after lysis with 0.05% and 0.5% Triton X-100. In contrast, the transcription rate in CS-B cells was significantly reduced (by 75%) when the Triton concentration was increased to 0.5%. In normal cells, the factors involved in the organization of the transcription complex are apparently tightly associated with chromatin and hence cannot be easily solubilized with Triton X-100. This confirms earlier observations that transcription complexes are firmly attached to the nucleoskeleton (Pardoll et al (80); Robinson et al (82)). The same may not be true for CS-B cells where the increased removal of proteins resulted in a decrease in transcription rate. The association between the transcription complex and chromatin may be disrupted in CS-B cells, and their chromatin organization appears to be looser than that in normal cells. It is possible that CS-B cells have a defective chromatin organization, which could account for the reduced transcription.

The predicted amino acid sequence of the CSB protein shows homology with the ATPase subunit of the yeast SWI/SNF transcription activation complex (Richmond and Peterson (96)) which is composed of at least 12 different subunits (Cairns et al (94)). In fact, the yeast homologue of CSB, Rad26, has been shown to exhibit DNA dependent AT-

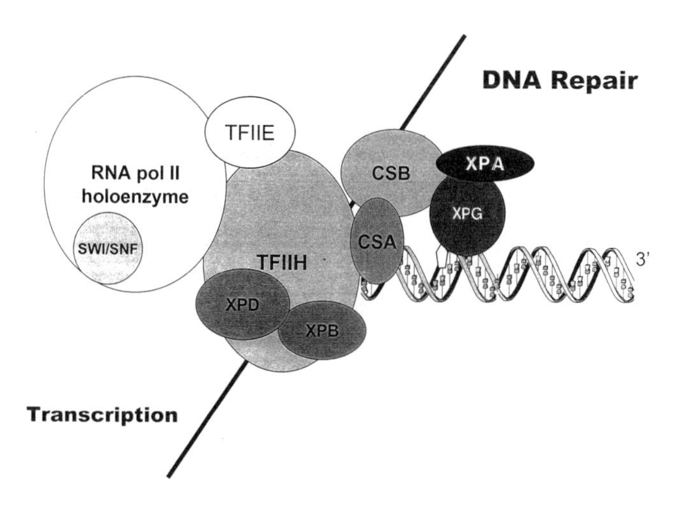

Figure 1. Interactions of the CSB protein, at the crossroads of DNA repair and transcription.

Pase activity *in vitro* (Guzder et al (96)), and recently it was shown that the human CSB protein is a DNA-stimulated ATPase (Selby and Sancar (97b)). The SWI/SNF complex activates transcription by facilitating access of the transcriptional machinery to the promoter regions of active genes. Recent work showed that the SWI/SNF complex may be an integral component of the RNA pol II holoenzyme (Wilson et al (96)). The homology of CSB to one of the subunits of the SWI/SNF complex suggests a role for CSB protein in transcription. Hence, mutations in the *CSB* gene could affect the overall efficiency of transcription. If the CSB protein is involved in chromatin remodeling, mutational inactivation might impair this event not only during transcription, but also during NER.

The CSB protein is at the crossroads of DNA repair, transcription and the molecular aging process. This concept, together with some of its molecular interactions are schematizised in Fig. 1.

2.5. Werner Syndrome (WS)

Werner Syndrome (WS) is a homozygous recessive disease characterized by early onset of many characteristics of normal aging, such as wrinkling of the skin, graying of the hair, cataracts, diabetes, and osteoporosis. Cancers, particularly sarcomas have been seen in these patients with increased frequency. The symptoms of WS begin to appear around the age of puberty, and most patients die before age 50. Because of the acceleration of aging in WS, the study of this disease will hopefully shed light on the degenerative processes that occur in normal aging.

Cells from WS patients grow more slowly and senesce at an earlier population doubling than age-matched normal cells, possibly because these cells appear to lose the telomeric ends of their chromosomes at an accelerated rate. Telomeric shortening is an established marker for cellular senescence. A hallmark molecular defect of WS is genomic instability arising from karyotypic abnormalities including inversions, translocations and chromosome losses. These effects could potentially be the result of defects in DNA repair, replication, and/or recombination, although the actual biochemical defect remains unknown. In addition, the mutation rate of transformed WRN fibroblasts is elevated 50-fold, largely due to gross deletions. It has been hypothesized that the clinical features and in-

- *Slow growing, extended S-phase*
- *Hypersensitivity to 4NQO (?)*
- *Elevated genomic instability*
  - Deletions, insertions, rearrangements
  - Compromised ligation fidelity
  - Accelerated telomere loss
- *Defective DNA repair*
  - Reduced 5-hydroxymethyl fluoracil glycosylase (Duker)
  - Transcription coupled repair (Webb et al.)
  - Mismatch repair (Kunkel)
  - Recombination abnormalities ?
- *Defective transcription*
  - Global (Balajee et al., Machwe et al.)
  - Gene specific ?

Figure 2. Werner syndrome cellular phenotype.

creased probability of tumors in non-epithelial tissue of WS patients is a consequence of this increased genomic instability. The gene that is defective in WS, the *WRN* gene, has recently been identified (Yu et al (96)). The amino acid sequence suggests that the *WRN* gene is a member of a large family (RecQ) of helicases with the putative ability to unwind DNA or RNA duplexes. Thus, the genetic evidence also points to a role for the WRN protein in some aspect of DNA metabolism.

The molecular cause(s) of genomic instability in WS remains to be defined. The cellular phenotype involves the elements shown in Figure 2. Earlier studies from our laboratory showed subtle defects in telomeric repair and TCR of UV-induced cyclobutane pyrimidine dimers in WS cells (Kruk et al (97); Webb et al (96)). A role for WRN protein in chromosomal metabolism is consistent with its sequence homology with known DNA helicases, a class of enzymes with essential roles in various processes of nucleic acid metabolism. The WRN protein has been demonstrated to have intrinsic helicase activity (Gray *et al.*, 1997). It may also have other additional activities. Importantly, the determination of the biologically relevant DNA substrates upon which the protein acts will help to elucidate the molecular deficiencies of WS.

Our research is directed towards elucidating the DNA metabolic defect, and how this defect causes an accelerated aging phenotype. We have previously reported a defect in TCR in some (but not all) WS cell lines (Webb et al (96)). Our observations of a TCR defect is in transformed lymphoblastoid cell lines. A recent observation from Kunkel's laboratory reported a defect in mismatch repair in WS cell lines (Bennett et al (97)) but this is seen only in fibroblasts. We have recently observed a defect in transcription in WS cells. This is seen *in vivo* in intact cells and also in chromatin prepared from WS cell lines. This defect in basal transcription is also observed in experiments using an *in vitro* transcription assay, and we are currently examining the mechanisms involved and the interactions between the WRN protein and other proteins involved in DNA repair and transcription. Use of stably transfected cells involves the establishment of a functional assay to detect that the gene is active after the transfection. No functional assay for the *WRN* gene has yet been reported. We have purified the WRN protein and are currently investigating its molecular biochemistry and functions.

3. PERSPECTIVES

There is an exciting development in the area of understanding the molecular dysfunction in the premature aging syndromes. This has been greatly facilitated by the cloning and identification of a number of the genes responsible for these phenotypes. A number of these genes have ATPase/helicase sequences, and helicases are involved in various important DNA metabolic processes such as transcription, DNA repair and replication. It is not yet clear how important the deficient helicase enzymatic activity is in the complex molecular phenotypes, and other functional activities of the genes may turn out to be more important than the helicase activities.

ACKNOWLEDGMENTS

We appreciate the interaction with the Danish Center for Molecular Gerontology.

REFERENCES

A.S. Balajee, A. May, G.L. Dianov, E.C. Friedberg, V.A. Bohr: Reduced RNA polymerase II transcription in intact and permeabilized Cockayne syndrome group B cells. (1997). *Proc Natl Acad Sci USA* **94:** 4306.

S.E. Bennett, A. Umar, J. Oshima, R.J.J. Monnat, T.A. Kunkel: Mismatch repair in extracts of Werner syndrome cell lines.(1997). *Cancer Res.* **57:** 2956.

V.A. Bohr: Gene specific DNA repair (1991). *Carcinogenesis* **12**: 1983.

B.R. Cairns, Y. Kim, M.H. Sayre, B.C. Laurent, R.D. Kornberg: A multinsubunit complex containing the SWI1/ADR6, SWI2/SNF2, SWI3, SNF5, and SNF6 gene products isolated from yeast (1994). *Proc.Nat.Acad.Sci.USA* **91** :1950.

G.L. Dianov, J.F. Houle, J.F. Iyer, V.A. Bohr, E.C. Friedberg: Reduced RNA polymerase II transcription in extracts of Cockayne syndrome and xeroderma pigmentosium/Cockayne syndrome cells (1997) *Nucl. Acids Res.* **18**: 3636–3642.

M.K. Evans, V.A. Bohr: Gene-specific DNA repair of UV-induced cyclobutane pyrimidine dimers in some cancer-prone and premature-aging human syndromes.(1994). *Mutat. Res.* **314**: 221.

E.C. Friedberg: Cockayne syndrome--a primary defect in DNA repair, transcription, both or neither? (1996a). *Bioessays* **18**:731.

E.C. Friedberg: Relationships between DNA repair and transcription (1996b). *Ann.Rev.Biochem.* **65**:15.

E.C. Friedberg , G.C. Walker and W. Siede : *DNA repair and mutagenesis*, Washington, D.C., ASM Press, 1995.

M.D. Gray, J.C. Shoe, A.S. Kamath-Loeb, A. Blank, G.M. Martin, J. Oshima, L.A. Loeb: The Werner Syndrome protein is a DNA helicose (1997). *Nat. Genetics* **17**:100.

S.N. Guzder, P. Sung, L. Prakash, S. Prakash: Nucleotide excision repair in yeast is mediated by sequential assembly of repair factors and not by a preassembled repairosome (1996). *J.Biol.Chem.* **271**:8903.

N. Iyer, M.S. Reagan, K. J. Wu, B. Canagarajah, E.C. Friedberg: Interactions involving the human RNA polymerase II transcription/nucleotide excision repair complex TFIIH, the nucleotide excision repair protein XPG, and Cockayne syndrome group B (CSB) protein. (1996). *Biochemistry* **35**:2157.

P.A. Kruk, D.K. Orren, V.A. Bohr: Telomerase activity is elevated in early S phase in hamster cells. (1997). *Biochem. Biophys. Res. Comm.* **233**:717.

S.A. Leadon, P.K. Cooper: Preferential repair of ionizing radiation-induced damage in the transcribed strand of an active human gene is defective in Cockayne's syndrome(1993). *Proc Natl Acad Sci* USA **90**:10499.

L.V. Mayne, A.R. Lehmann: Failure of RNA synthesis to recover after UV irradiation: an early defect in cells from individuals with Cockayne's syndrome and xeroderma pigmentosum (1982). *Cancer Res.* **42** (4):1473–1478 .

A.J. Millis, M. Hoyle, H.M. McCue, H. Martini: Differential expression of metalloproteinase and tissue inhibitor of metalloproteinase genes in aged human fibroblasts (1992). *Expt.Cell.Res.* **201**:373.

L. Naumovski, E.C. Friedberg: Analysis of the essential and excision repair functions of the *RAD3* gene of Saccharomyces cerevisciae by mutagenesis(1986). *Mol. Cell. Biol.* **6**:1218.

D.K. Orren, G.L. Dianov, V.A. Bohr: The human CSB (ERCC6) gene corrects the transcription-coupled repair defect in the CHO cell mutant UV61.(1996). *Nucl. Acids Res.* **24**:3317

D. M. Pardoll, B. Vogelstein, D.S. Coffey: A fixed site of DNA replication in eukaryotic cells.(1980). *Cell* **19**:527.

E. Richmond, C. L. Peterson: Functional analysis of the DNA stimulated ATP-ase domain of yeast SWI2/SNF2(1996). *Nucl. Acids Res.* **24**:3685.

S. H. Robinson, B. D. Nelkin, B. Vogelstein: The ovalbumin gene is associated with the nuclear matrix of chicken oviduct cells.(1982). *Cell* **28**:99.

Sancar: DNA excision repair (1996). *Ann.Rev.Biochem.* **65**:43.

R. Schmickel, E. Chu, J. Trosko, C. Chang: Cockayne's syndrome: a cellular sensitivity to ultraviolet light (1979). *Pediatrics* **60**:135.

C. P. Selby, R. Drapkin, D. Reinberg, A. Sancar: RNA polymerase II stalled at a thymine dimer: footprint and effect of excision repair (1997a). *Nucl. Acid.Res.* **25**:787.

C. P. Selby, A. Sancar: Human transcription-repair coupling factor CSB/ERCC6 is a DNA stimulated ATPase but is not a helicase and does not disrupt the ternary transcription complex of stalled RNA polymerase II.(1997b). *J.Biol.Chem.* **272**:1885.

J. Venema, L. H. Mullenders, A. T. Natarajan, A. A. van Zeeland, L. V. Mayne: The genetic defect in Cockayne syndrome is associated with a defect in repair of UV-induced DNA damage in transcriptionally active DNA(1990). *Proc.Natl.Acad.Sci.U.S.A.* **87**:4707.

D.K. Webb, M. K. Evans, V. A. Bohr: DNA repair fine structure in Werner's syndrome cell lines.(1996). *Exp.Cell Res* **224**:272.

C. J. Wilson, D.M. Chao, A.N. Imbalzano, G.R. Schnitzler, R.E. Kingston, R.A. Young: RNA polymerase II holoenzyme contains SWI/SNF regulators involved in chromatin remodelling (1996). *Cell* **84**:235.

R.D. Wood: DNA repair in eukaryotes (1996). *Ann.Rev.Biochem.* **65**:135.

C.E. Yu, J. Oshima, Y.H. Fu, E.M. Wijsman, F. Hisama, R. Alisch, S. Mathews, J. Nakura, T. Miki, S. Ouais, G.M. Martin, J. Mulligan, G.D. Schellenberg: Positional cloning of the Werner's syndrome gene (1996). *Science* **272**:258.

4

EXCISION REPAIR OF 8-OXOGUANINE IN EUKARYOTES

The Ogg1 Proteins

Serge Boiteux and J. Pablo Radicella

Laboratoire de Radiobiologie du DNA
UMR217 CNRS-CEA
60 Avenue du Général Leclerc
BP6, F-92265 Fontenay aux Roses, France

1. ABSTRACT

7,8-Dihydro-8-oxoguanine (8-OxoG) is a major mutagenic lesion produced on DNA by the endogenous oxidative stress. In eukaryotes, this modified base can be removed by specific DNA glycosylases/AP lyases, the Ogg1 proteins. The *OGG1* gene of *Saccharomyces cerevisiae* encodes a protein of 376 amino acids with a molecular mass of 43-kDa. The yOgg1 protein catalyses the excision of 8-OxoG and the nicking of DNA at AP sites using a β-elimination reaction. The *ogg1* mutant strains exhibit a mutator phenotype characterized by an exclusive increase in G/C to T/A transversions, attributed to unrepaired 8-OxoG in DNA. Using the yeast *OGG1* sequence, two human cDNAs coding for proteins showing >30% identity with yOgg1 were identified. The predicted proteins have 345 and 424 amino acids and a molecular mass of 39- and 47-kDa, respectively. These two proteins, α-hOgg1 and β-hOgg1, result from an alternative splicing of a single transcript. The α-hOgg1 protein has nuclear localization and is ubiquitously expressed in human tissues. Both, α-hOgg1 and β-hOgg1, release 8-OxoG and nick DNA at AP sites. Moreover, when they are expressed in a bacterial *fpg mutY* mutant strain, they complement the mutator phenotype. The human *OGG1* gene was localized to chromosome 3p25. The mouse *Ogg1* gene was also identified and localized to chrosome 6E. The mOgg1 protein contains 345 amino acids and is 84% identical to α-hOgg1. In addition to hOgg1 proteins, human cells also possess a MutY homolog and possibly use nucleotide excision repair to excise 8-OxoG from DNA. The results show that repair enzymes involved in the excision of 8-OxoG are highly conserved in eukaryotes and suggest an important impact of oxidative DNA damage on the process of carcinogenesis.

Advances in DNA Damage and Repair, edited by Dizdaroglu and Karakaya. Kluwer Academic / Plenum Publishers, New York, 1999.

2. INTRODUCTION

Considerable interest has arisen in recent years in the study of the biological consequences and the repair of oxidative damage to DNA. This interest derives in part from the possibility that oxidative DNA damage is implicated in the etiologies of human pathologies such as cancer and aging (Ames, 1983; Halliwell and Gutteridge, 1989; Breimer 1990; Feig et al., 1994). In the case of cancer, oxidative damage to DNA is thought to cause mutations that activate oncogenes or inactivate tumor suppressor genes. Oxidative DNA damage is the result of the reaction between DNA and reactive oxygen species (ROS) which are formed as byproducts of normal metabolism and during exposure to physical and chemical environmental agents. As many as 50 different DNA modifications have been identified after exposure to reactive species such as the hydroxyl radical ($OH^{\cdot}$) (Dizdaroglu, 1991; Cadet et al., 1997). Amongst these lesions, an oxidatively damaged form of guanine, 7,8-dihydro-8-oxoguanine (8-OxoG) is abundantly produced (Cadet et al., 1997). The presence of unrepaired 8-OxoG in DNA may be mutagenic since this modified base preferentially pairs to adenine during replication (Shibutani et al., 1991). In *Escherichia coli* two DNA glycosylases cooperate to prevent mutagenesis by 8-OxoG: the Fpg protein which excises 8-OxoG in DNA and the MutY protein which excises the adenine residues incorporated by DNA polymerases opposite 8-OxoG. Inactivation of any of those two genes leads to a mutator phenotype characterized by the exclusive increase in G/C to T/A transversions (Radicella et al., 1988; Nghiem et al., 1988; Cabrera et al., 1988). Moreover, inactivation of both Fpg and MutY activities results in a strong mutator phenotype showing that these two enzymes act sinergistically to prevent G/C to T/A events (Michaels et al., 1992; Duwat et al., 1995). This has lead to the proposal of a sophisticated cellular system for the avoidance of the mutations induced by 8-OxoG in bacteria (Michaels and Miller, 1992; Grollman and Moriya, 1993; Boiteux and Laval, 1997). Additionaly, the strong spontaneous mutator phenotype of bacterial strains deficient in the repair of 8-OxoG suggests that the endogenous oxidative stress is a major threat to genome stability.

In eukaryotes, it has been proposed that a mutator phenotype might be involved at some point in the multistage process of carcinogenesis (Loeb, 1991). This model has been actually confirmed by the finding that the hereditary nonpolyposis colorectal cancer (HNPCC) is associated to defects in the gene coding for a homolog to the bacterial mismatch repair protein MutS (Fishel et al., 1993; Leach et al., 1993). Indeed, cells from these tumors have a hypermutator phenotype and the biochemical defect in the mismatch repair process was established (Parsons et al., 1993). These findings have been extended to other tumors with mutations in a second MutS homolog as well as in the homologs of MutL, another essential component in the biochemical mismatch repair machinery of *E. coli* (Bronner et al., 1994; Jiricny, 1994). According to this model and the data from the bacterial system, inactivation of the homologs of *fpg* and *mutY* genes in eukaryotes should cause a mutator phenotype which could contribute to carcinogenesis. The aim of this review is to summarize recent findings dealing with the identification and characterization of eukaryotic DNA repair functions responsible of the elimination of 8-OxoG from oxidatively damaged DNA.

3. RESULTS

3.1. The Ogg1 Protein of *Saccharomyces cerevisiae*

The biological properties of *fpg* and *mut*Y mutant strains in *Escherichia coli* prompted us to study the DNA repair mechanisms involved in the excision of 8-OxoG

Table 1. Comparison of the properties of the yeast and human Ogg1 proteins

	yOgg1	α-hOgg1	β-hOgg1
Gene localization	Chromosome XIII	3p25	3p25
Protein (amino acids)	376	345	424
Molecular mass (kDa)	42.8	38.8	47.2
Isoelectric point	8.83	8.89	6.52
Putatif motifs			
HhH	Yes	Yes	Yes
NLS	Yes	Yes	No
Mitochondrial	No	Yes	Yes
Enzymatic activities			
DNA glycosylase	Yes	Yes	Yes
Me-Fapy-Gua/C	Yes	Yes	Yes
Fapy-Gua/C	Yes	nd	nd
8-OxoG/C	Yes	Yes	Yes
8-OxoG/A	No	No	No
AP lyase	Yes: β-elimination	Yes: β-elimination	Yes: β-elimination
AP/C	Yes	Yes	Yes
AP/A	No	No	nd
Mutant phenotype	mutator G/C to T/A	nd	nd
Complementation			
E.coli fpg mutY	Yes	Yes	Yes
S.cerevisiae ogg1	Yes	Yes	nd
DNA damage inducible	No	nd	nd

nd: not determined.
The data reported in this table are collected from, Auffret van der Kemp, et al. (1996); Nash et al. (1996); Thomas et al. (1997); Girard et al. (1997); Radicella et al. (1997); Roldan-Arjona et al. (1997); Arai et al. (1997); Aburatani et al. (1997); Rosenquist et al. (1997); Lu et al. (1997); Bjoras et al. (1997); Karahalil et al. (1998).

from DNA in *Saccharomyces cerevisiae*. Yeast is a simple organism which can be used as a paradigm for DNA repair in all eukaryotic cells and whose complete genome sequence is available. Moreover, amino acid sequence of DNA repair proteins is often well conserved from yeast to man. The complementation of the mutator phenotype of the *fpg mutY* strain of *E.coli* was used as a genetic assay to clone yeast homologs of these bacterial functions. Using this genetic assay, we have cloned the *OGG1* gene of *S.cerevisiae* (Auffret van der Kemp et al., 1996). The yeast *OGG1* gene was localized on chromosome XIII and codes for a protein of 376 amino acids with a molecular mass of 43-kDa (Fig.1 and Table 1). The yOgg1 protein is a DNA glycosylase that excises 8-OxoG and 2,6-diamino-4-hydroxy-5-N-methylformamidopyrimidine (Me-Fapy-Gua) from damaged DNA. The yOgg1 protein displays a marked preference for DNA duplexes containing 8-OxoG paired with a cytosine or a thymine (Girard et al., 1997). In contrast, the yOgg1 protein does not release 8-OxoG from DNA when this lesion is placed opposite a guanine or an adenine (Girard et al., 1997). The substrate specificity of the yOgg1 protein was analysed using DNA exposed to γ-radiation as substrate. The results show that yOgg1 efficiently excises 8-OxoG and 2,6-diamino-4-hydroxy-5-formamidopyrimidine (Fapy-Gua) from the irradiated DNA (Karahalil et al., 1998). Fourteen other modified bases, including a variety of damaged pyrimidines and purines, were not significantly excised. The fact that yOgg1 does not excise 4,6-diamino-5-formamidopyrimidine (Fapy-Ade) reveals significant differences in the substrate specificities between yOgg1 and Fpg (Boiteux, 1993). The yOgg1 protein is also endowed with a lyase activity that incises DNA at AP sites *via* a β-elimination reaction

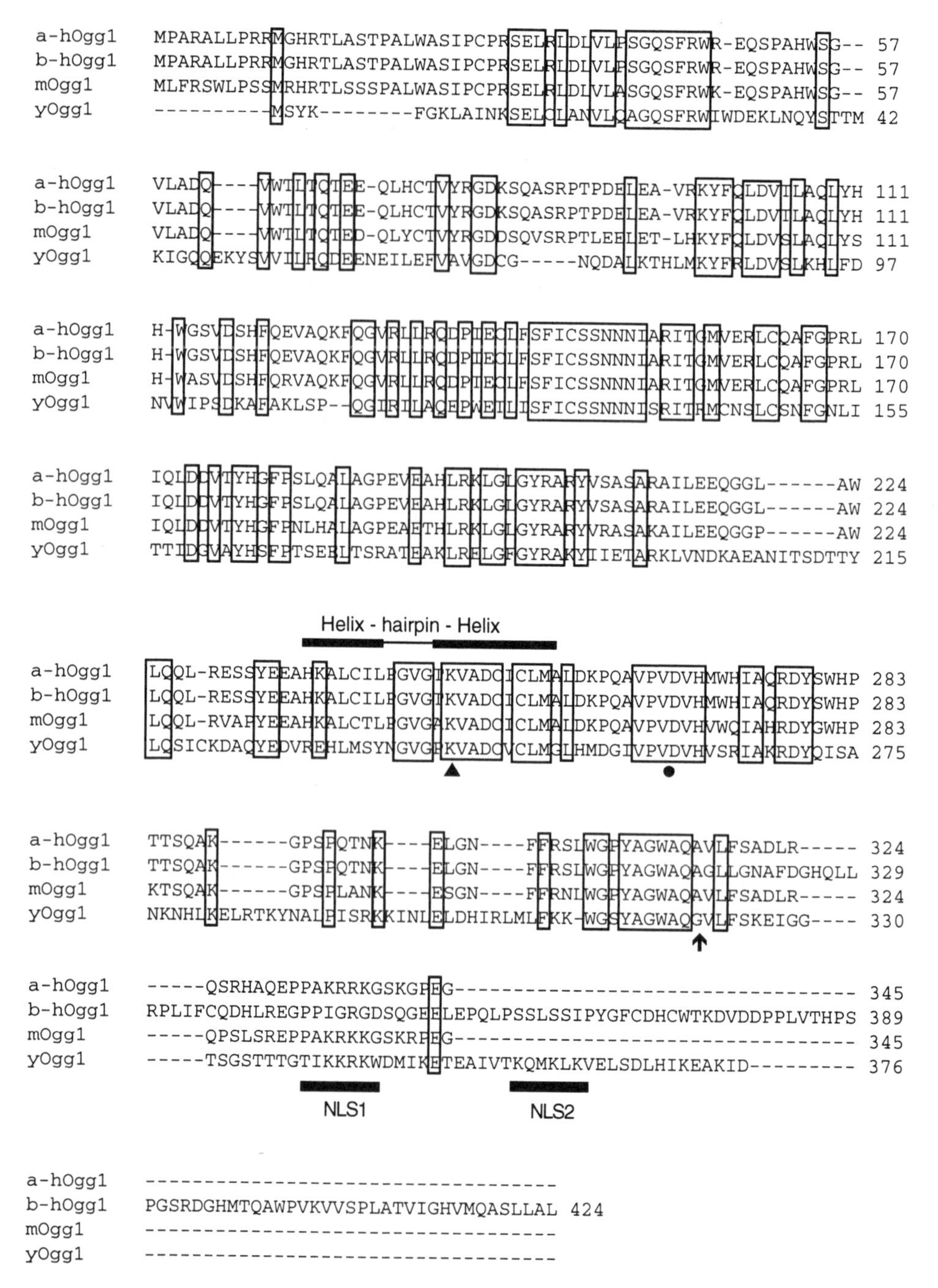

Figure 1. Alignment of amino acid sequences of mouse (m), α- and β-human (h) and yeast (y) Ogg1 proteins. The amino acids identical in the four sequences are boxed. The consensus sequence for the HhH and NLS motifs are indicated. The catalytically important amino acids, Lys241 (yOgg1) or Lys249 (hOgg1) and Asp260 (yOgg1) or Asp268 (hOgg1) are also shown using a black triangle or a black circle, respectively. The site of divergence between α- and β-hOgg1 generated by alternative splicing is shown by an arrow. The data are from: Radicella et al. (1997); Roldan-Arjona et al. (1997); Arai et al. (1997); Aburatani et al. (1997); Rosenquist et al. (1997); Lu et al. (1997); Bjoras et al. (1997); Boiteux et al. (1998).

(Auffret van der Kemp et al., 1996, Nash et al., 1996, Girard et al., 1997). As previously described for the removal of 8-OxoG, the yOgg1 protein preferentially incises DNA containing an AP site placed opposite a cytosine (Table 1). The rank order for cleavage at a preformed AP site is; AP/C>>AP/T>>AP/G and AP/A (Girard et al., 1997). Furthermore, the catalytic mechanism of yOgg1 involves the formation of a transient DNA-enzyme covalent intermediate which is converted into a stable adduct in the presence $NaBH_4$ (Nash et al., 1996, Girard et al., 1997). These results demonstrate that the yOgg1 protein is a member of the family of DNA repair enzymes possessing associated DNA glycosylase / AP lyase activities (Nash et al.,1996).

The ability of the *OGG1* gene to complement the mutator phenotype of the bacterial *fpg mutY* mutant and the enzymatic properties of the yOgg1 protein suggest that *OGG1* is a yeast homolog of the bacterial *fpg*. However, the sequence analysis does not reveal any obviously conserved region between yOgg1 and Fpg, Auffret van der Kemp et al. (1996). Both, the highly conserved [PELPEVET] sequence at the amino terminal-end and the zinc-finger motif at the carboxy terminal-end of bacterial Fpg proteins are absent in the yOgg1 protein (Boiteux and Laval, 1997). The amino acid sequence of the yOgg1 protein reveals the presence of a nuclear translocation signal (NLS) and of the helix-hairpin-helix (HhH) DNA binding motif which is the hallmark of a familly of DNA glycosylases / AP lyases sharing a common ancestor gene with the endonuclease III of *E.coli* (Fig.1). This consensus sequence shows two highly conserved residues, Lys241 and Asp260, corresponding to Lys120 and Asp138 of endonuclease III, respectively, Tayer et al. (1995). Mutation of Lys241 to Gln has been obtained after site directed mutagenesis of *OGG1*. This mutation, (K241Q), completely abolishes catalytic activities of the yOgg1 protein. However, the (K241Q) mutant of yOgg1 still binds DNA containing a 8-OxoG/C base pair (Girard et al., 1997). These results indicate that the free amino group of Lys241 is involved in the catalytic mechanism of yOgg1 as previously described for Lys120 of the endonuclease III (Tayer et al., 1995).

To investigate the biological role of the *OGG1* gene in *S.cerevisiae*, mutant strains were constructed by partial deletion of the open reading frame and insertion of marker genes. The results show that *OGG1* is not essential and that *ogg1* extracts have no detectable activity that can repair 8-OxoG (Thomas et al., 1997). The *ogg1* mutant is not unusually sensitive to DNA damaging agents such as H_2O_2 , UVC-light or methylmethane sulfonate. However, the *ogg1* mutant exhibits a spontaneous mutator phenotype (Thomas et al., 1997). When compared to those of a wild-type strain, the frequencies of mutation to canavanine resistance and reversion of an ochre mutation to Lys+ are 7- and 10-fold higher for the *ogg1* mutant, respectively. Moreover, using a specific tester system, it was shown that the yOgg1-deficient strain displays a 50-fold increase in spontaneously occurring G/C to T/A transversions compared to the wild-type strain. In contrast, the five other base substitution events are not affected by the disruption of the *OGG1* gene (Thomas et al., 1997). Therefore, the properties of the yeast *ogg1* mutant are very similar to that of the *fpg-* mutant of *E.coli*. These results in *S.cerevisiae* demonstrate that the endogenous oxidative stress causes genetic instability in cells where the base excision repair of 8-OxoG has been compromised.

3.2. The Mammalian Ogg1 Proteins

Using the yeast *OGG1* sequence to screen expressed sequence tags (EST) databases, several groups have recently isolated human cDNAs coding for homologs of the yOgg1 protein of *S.cerevisiae* (Radicella et al., 1997; Roldan-Arjona et al., 1997; Arai et al.,

1997; Aburatani et al., 1997; Rosenquist et al., 1997; Lu et al., 1997; Bjoras et al., 1997). Although the homology spans the whole sequence (38% identity) several regions with identities higher than 60% can be found (Fig.1). The analysis of these sequences revealed two kinds of cDNAs with open reading frames coding for peptides of 345 and 424 amino acids, α-hOgg1 and β-hOgg1, respectively (Fig.1). Comparison of α-hOgg1 and β-hOgg1 sequences shows that these two proteins have identical 316-first amino acids at the amino terminal-end. In contrast, the carboxy terminal-ends are completely different, the human α-hOgg1 one showing strong similarity with yOgg1 (Fig.1). It should be noticed that the cDNA coding for the β-hOgg1 protein has been obtained from a library using messenger RNA isolated from brain tissue (Roldan-Arjona et al., 1997). The two forms of hOgg1 seem to correspond to the messengers RNA of 1.7 and 2.1 kb found in most tissues analyzed by Northern blot analysis (Arai et al., 1997; Radicella et al., 1997a). RT-PCR experiments confirmed the presence of both transcripts in most tissues, being form α the most abundant (Aburatani et al., 1997; Arai et al., 1997). These results suggest an alternative splicing being at the origin of the α- and β-hOgg1 forms in human cells. Using *in situ* hybridization the *OGG1* gene was localized to chromosome 3p25 suggesting the presence of a single gene. Furthermore, the analysis of the genomic sequence of *OGG1* confirmed the hypothesis of an alternative splicing at the origin of the two forms of hOgg1, Aburatani et al. (1997). The human α-*OGG1* transcribed region spans less than 5 kb of genomic sequences and is composed of seven exons. On the other hand, β-*OGG1* shares the six first exons with the α-form, but as a consequence of an alternative splicing, it carries an eighth exon located 8 to 9 kb downstream of the last exon of the α-form, Aburatani et al. (1997). Interestingly, the 3' last-exons (exons 7 and 8) from the human *OGG1* gene display homology to the antisense strand of the calmodulin kinase I (*CaMK1*) gene. Human hOgg1 protein sequences reveal several highly conserved regions (Fig.1). Among those, the Helix-hairpin-Helix (HhH) motif which is characteristic of the active site of a family of DNA glycosylases / AP lyases, Nash et al. (1996). A putative NLS present at the carboxy terminal-end of the protein is also conserved in yOgg1 and α-hOgg1 (Fig.1). The nuclear localization of the α-hOgg1 protein has been demonstrated, Bjoras et al. (1997). In contrast, the β-hOgg1 protein has no NLS which may suggest another cellular localization, possibly mitochondrial.

A cDNA coding for a mouse homolog of the yeast yOgg1 protein has also been identified. The predicted open reading frame of the mouse cDNAs code for a peptide of 345 amino acids (Fig.1). Comparison of the sequences shows that mOgg1 is 84% identical to the human α-hOgg1 (Rosenquist et al., 1997; Lu et al., 1997;Boiteux et al., 1998). To the best of our knowledge, the β-form of the mOgg1 protein has not yet been identified. Finally, the mouse *Ogg1* gene was localized on chromosome 6E and its 3' region overlaps with a previously unidentified gene with strong homology to both the rat and the human *CaMK1* (Boiteux et al., 1998).

As suggested by their sequences, α-hOgg1, β-hOgg1 and mOgg1 are DNA glycosylases / AP lyases that excise 8-OxoG and Me-Fapy-Gua and nick DNA at AP sites *via* a β-elimination reaction (Table 1) (Radicella et al., 1997; Aburatani et al., 1997; Roldan-Arjona et al., 1997; Rosenquist et al., 1997; Lu et al., 1997; Bjoras et al., 1997; Boiteux et al., 1998) (Table 1). Moreover, all these studies confirm that 8-OxoG is efficiently repaired when paired with a cytosine or a thymine. The results also show that 8-OxoG is not released when placed opposite an adenine (Table 1). Further analysis of the catalytic properties of the human α-hOgg1 shows the formation of a covalent enzyme-DNA complex in the presence of $NaBH_4$, Lu et al. (1997). The mammalian Ogg1 proteins have a Lys249 residue corresponding to the Lys241 of the yOgg1 protein (Fig.1). Directed

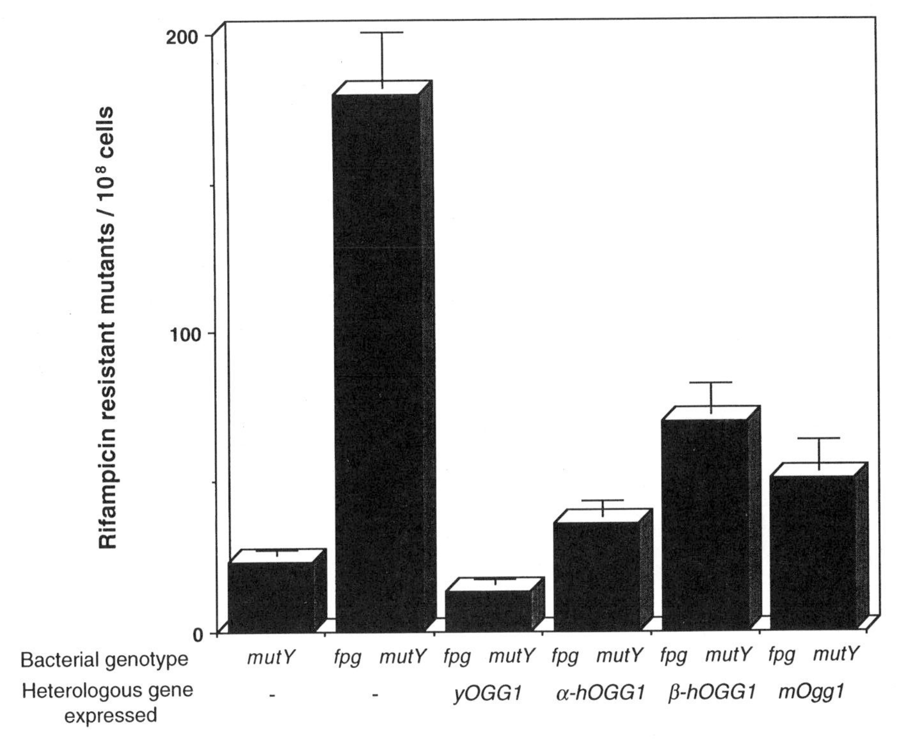

Figure 2. Complementation of the mutator phenotype of *fpg mutY* double mutant strain of *E.coli* by plasmids harboring sequences coding for yOgg1, α-hOgg1, β-hOgg1 and mOgg1, respectively. The data are from our laboratory except the β-hOgg1 which is from Roldan-Arjona et al. (1997).

mutagenesis of Lys249 yielding the (K249Q) mutant of the α-hOgg1 protein demonstrates that this residue is essential for the catalytic activities but not for DNA binding (Lu et al., 1997; Nash et al., 1997). Together, these results show that α-hOgg1 and β-hOgg1 have very similar catalytic properties (Fig.1). One can speculate that the biological function of α-hOgg1 and β-hOgg1 could be to protect nuclear and mitochondrial DNA from oxidative damage. Indeed, an 8-OxoG repair enzyme activity has been purified from rat liver mitochondria (Croteau et al., 1997).

The biological function of the Ogg1 proteins in mammalian cells has not yet been investigated directly. However, the efficient repair of 8-OxoG/C and the capacity to suppress the mutator phenotype of bacterial *fpg mutY* (Fig.2) and yeast *ogg1* strongly suggest that one of the biological role of the mammalian Ogg1 proteins is to prevent mutations by endogenous oxidative DNA damage (Radicella et al., 1997; Roldan-Arjona et al., 1997; Arai et al., 1997; Aburatani et al., 1997; Rosenquist et al., 1997; Lu et al., 1997; Bjoras et al., 1997).

3.3. The Ribosomal S3 Protein of *Drosophila melanogaster*

A recent study suggests that Ogg1 is not the only 8-OxoG DNA glycosylase in eukaryotes. In *Drosophila melanogaster*, the ribosomal protein S3 possesses an enzymatic activity that catalyses cleavage of DNA containing an 8-OxoG/C base pair (Yacoub et al., 1997). Furthermore, the expression of the S3 protein in a *fpg*⁻ mutant of *E.coli* suppresses the spontaneous mutator phenotype. Although not demonstrated, *Drosophila melanogaster* may also possess an Ogg1 homolog.

3.4. Repair of 8-OxoG by Alkylpurine DNA Glycosylases

Human and rodent 3-methyladenine DNA glycosylases display an unexpectedly broad substrate range. *In vitro* experiments have shown that the human and mouse recombinant alkylpurine DNA glycosylases (Aag) are able to excise 8-OxoG from DNA (Bessho et al., 1993). However it has recently been reported that tissue extracts from *Aag* -/- mice, although completely lacking 3-methyl-adenine glycosylase activity, maintain the wild-type level of 8-OxoG DNA glycosylase activity (Engelward et al., 1997, Hang et al., 1997). This implies that Aag is not a major DNA glycosylase for 8-OxoG removal in mammalian cells, consistent with the results from the yeast system in which *ogg1* strains have no detectable 8-OxoG excision repair activity.

3.5. The MutY Proteins in Eukaryotes

The bacterial model shows that Fpg and MutY cooperate to protect DNA from the mutagenic action of 8-OxoG in DNA. In *E.coli*, MutY is an adenine DNA glycosylase that acts on DNA containing either G/A or 8-OxoG/A base pairs (Au et al., 1988; Michaels et al., 1992). In *S.cerevisiae*, neither sequence homology searches, nor complementation assay of the mutator phenotype of *fpg mutY* bacteria have permitted the identification of a MutY homolog. Moreover, we have not been able to detect an enzymatic activity that mimics the action of MutY in yeast cell free extracts, (unpublished data). In contrast, a mammalian homolog of the *E.coli* MutY protein has been identified (Slupska et al., 1995). The human *hMYH* gene encodes a protein of 535 amino acids that displays 41% identity to the *E.coli* MutY protein. Furthermore, nuclear extracts from human cells contain an enzymatic activity that cleaves DNA at adenine residues when paired with 8-OxoG or guanine (Mc Goldrick et al., 1995).

3.6. Excision of 8-oxoG by Nucleotide Excision Repair (NER)

Several studies show that oxidized DNA bases can also be eliminated *via* the nucleotide excision repair (NER) pathway mediated by the UvrABC complex in *E.coli* (Kow et al., 1990, Czeczot et al., 1991). However, genetic experiments strongly suggest that NER has a marginal role in the repair of 8-OxoG in DNA. In *E.coli*, the introduction of a NER mutation (*uvrA*) in a *fpg mutY* mutator strain does not significantly enhance the spontaneous mutation frequencies (Wagner et al., 1997). In *S.cerevisiae*, the double mutant *rad14 ogg1* does not exhibit a significant increase in spontaneous mutation frequencies when compared to the *ogg1* mutant, (unpublished data). However, a recent *in vitro* study shows that the reconstituted human excision nuclease efficiently excises 8-OxoG from DNA templates containing a single 8-OxoG/C lesion (Reardon et al., 1997).

4. DISCUSSION

In yeast, the inactivation of the *OGG1* gene results in a spontaneous mutator phenotype with the exclusive accumulation of G/C to T/A transversions probably due to unrepaired 8-OxoG in DNA. These results demonstrate that endogenously formed oxidative

DNA damage causes genetic instability in eukaryotes. The fact that the Oggl proteins have highly conserved sequences and enzymatic properties from yeast to man, suggests a conserved biological role which could be the elimination of mutagenic oxidative DNA damage resulting from the endogenous oxidative stress. Several observations indicate that Oggl proteins protect the genome from a stress generated by the cellular metabolism. First, *OGG1* is ubiquitously expressed in human tissues. Second, *OGG1* gene is expressed in both embryonic and adult tissues. Third, Oggl may be present in nucleus and mitochondria. However, the elucidation of the biological function of the mammalian Oggl proteins will await the isolation of mutant cell lines.

If the inactivation of the *OGG1* gene in mammalian cells also causes a mutator phenotype, one can expect biological consequences in term of carcinogenesis as previously reported for other mutator genes such as *MSH2*. The working hypothesis developped in our laboratory suggests that the inactivation of *OGG1* is one of the steps in the course of the multistage process of carcinogenesis. The validation of this hypothesis requires the identification of human tumors where both alleles of the *OGG1* has been inactivated. We have used two criteria to select the kind of human tumors possibly affected in *OGG1*: i) tumors harboring deletions of the chromosome 3p where *OGG1* is localized, ii) tumors where changes in p53 are G/C to T/A transversions. A combination of cytogenetic and molecular studies have implicated loss of heterozygosity or deletions in chromosome 3p in all major types of lung cancers (Naylor et al., 1987; Yokota et al., 1987; Hibi et al., 1992). The analysis of the sequence changes in p53, a tumor supressor gene very commonly mutated in many cancers, showed that, unlike in most other tumors, in lung cancer there is a strong bias for the presence of G/C to T/A tranversions (Hollstein et al., 1996). This type of mutations would be expected in mammalian cells incapable of eliminating 8-OxoG from their DNA and therefore likely to be deficient in the *OGG1* gene. Consequently, we are currently analysing the expression of the *OGG1* gene and the nucleotide sequence of the *OGG1* cDNAs in lung human tumors. In parallel, the development of an animal model where the *OGG1* gene has been inactivated could allow the assesment of the impact of oxidative DNA damage on the triggering of the carcinogenic process.

ACKNOWLEDGMENTS

The authors thank the Centre National de la Recherche Scientifique and the Commissariat à l'Energie Atomique, UMR217 CNRS-CEA and ACC-SV8.

REFERENCES

H. Aburatani, Y. Hippo, T. Ishida, T., et al. (1997) Cloning and characterization of mammalian 8-hyroxyguanine-specific DNA glycosylase/apurinic, apyrimidinic lyase, a functional MutM homologue. *Cancer Res.* **57**, 2151–2156.

B.N. Ames (1983) Dietary carcinogens and anticarcinogens. Oxygen radicals and degenerative diseases. *Science* **221**, 1256–1264.

K. Arai, K. Morishita,, K. Shinmura et al. (1997) Cloning of a human homolog of the yeast *OGG1*gene that is involved in the repair of oxidative DNA damage. *Oncogene.* **14**, 2857–2861.

K.G. Au, M. Cabrera, J.H. Miller. and P.Modrich (1988) The *Escherichia coli mut*Y gene encodes an adenine glycosylase active on G:A mispairs. *Proc. Natl. Acad. Sci.* USA. **85**, 8877–8881.

P. Auffret van der Kemp, D. Thomas, R. Barbey, R. De Oliveira, and S.Boiteux (1996) Cloning and expression in *E. coli* of the *OGG1* gene of *S. cerevisiae* which codes for a DNA glycosylase that excises 7,8-dihydro-8-

oxoguanine and 2,6-diamino-4-hydroxy-5-N-methylformamidopyrimidine. *Proc. Natl. Acad. Sci.* USA. **93**, 5197–5202.

T. Bessho, R. Roy, K. Yamamoto, H. Kasai, S. Nishimura, K. Tano and S. Mitra. (1993) Repair of 8-hydroxyguanine in DNA by mammalian N-methylpurine-DNA glycosylase. *Proc. Natl. Acad. Sci. USA*, **90**, 8901–8904.

M. Bjoras, L. Luna, B. Johnsen, E. Hoff, T. Haug, T. Rognes and E. Seeberg (1997) Opposite base-dependent reactions of a human base excision repair enzyme on DNA containing 7,8-dihydro-8-oxoguanine and abasic sites. *EMBO J.* **16,** 6314–6322.

S. Boiteux (1993) Properties and biological functions of the Nth and Fpg proteins of *Escherichia coli*: two DNA glycosylases that repair oxidative damage in DNA. *Photochem. Photobiol. B.* **19**, 87–96.

S. Boiteux and J. Laval (1997) Repair of oxidized purines in DNA. *Base Excision Repair of DNA Damage.* I.D. Hickson ed. Landes Bioscience-Springer. 31–44.

S. Boiteux, C. Dherin, F. Reille, F. Apiou, B. Dutrillaux, and J.P. Radicella (1998) Excision repair of 8-hyroxyguanine in mammalian cells: The mouse Ogg1 protein as a model. *Free Rad. Res.* (in press)

L.H. Breimer (1990) Molecular mechanisms of oxygen radical carcinogenesis and mutagenesis, the role of base damage. *Mol. Carcinogenesis* **3**, 188–197.

C.E. Bronner, S.M. Baker, P.T.Morrison et al. (1994), Mutation in the DNA mismatch repair gene homologue *hMLH1* is associated with hereditary non-polyposis colon cancer. *Nature* **368**, 258–261.

M. Cabrera, Y. Nghiem and J.H. Miller (1988) *mutM*, a second mutator locus in *E.coli* that generates G:C to T:A transversions. *J. Bacteriol.* **170**, 5405–5407.

J. Cadet, M. Berger, T. Douki, and J.L. Ravanat (1997) Damage to DNA: Formation, measurement, and biological significance. *Rev. Physiol. Biochem. Pharmacol.* **131**, 1–87.

D.L. Croteau, C.M.J. Rhys, E.K. Hudson, G.L. Dianov, R.G. Hansford and V.A. Bohr (1997) An oxidative damage specific endonuclease from rat liver mitochondria. *J. Biol. Chem.* 272, in press.

H. Czeczot, B. Tudek, B. Lambert, J. Laval and S. Boiteux (1991) *Escherichia coli* Fpg protein and UvrABC endonuclease repair DNA damages induced by methylene blue plus visible light *in vivo* and *in vitro*. *J. Bacteriol.* **173,** 3419–3424.

M. Dizdaroglu (1991) Chemical determination of free-radical induced damage to DNA. *Free Radic Biol Med.* **10**, 225–242.

P. Duwat, R. De Oliveira, S.D. Ehrlich, and S. Boiteux (1995). Repair of oxidative DNA damage in Gram-positive bacteria: the *Lactococcus lactis* Fpg protein. *Microbiology.* **141**: 411–417.

B. P. Engelward, G. Weeda, M. D. Wyatt, J. L. M. Broekhof , J. de Wit, I. Donker, J. M. Allan, B. Gold, J. H. J. Hoeijmakers, and L. D. Samson. (1997) Base excision repair deficient mice lacking the Aag alkyladenine DNA glycosylase. *Proc. Natl. Acad. Sci.* USA. **94**, 13087–13092.

D.I. Feig, T.M. Reid and L.A. Loeb (1994) Reactive oxygen species in tumorigenesis. *Cancer Res Suppl.* **54**, 1890–1894.

R. Fishel, M.K.Lescoe, M.R.S. Rao, N.G. Copeland, N.A. Jenkins, J. Garber, M. Kane, and R. Kolodner (1993) The human mutator gene homolog *MSH2* and its association with hereditary nonpolyposis colon cancer. *Cell* **75**, 1027–1038.

P.M. Girard, N. Guibourt and S. Boiteux, S. (1997) The Ogg1 protein of *Saccharomyces cerevisiae* : a 7,8-dihydro-8-oxoguanine DNA glycosylase/AP lyase whose lysine 241 is a critical residue for catalytic activity. *Nucleic Acids Res.* **25**, 3204–3211.

A.P. Grollman and M. Moriya (1993) Mutagenesis by 8-oxoguanine: an ennemy within. *Trends Genet.* **9**, 246–249.

B. Halliwell and J.M.C. Gutteridge (1989) *Free Radicals in Biology and Medicine.* Oxford University Press 2nd edn.

B. Hang, B. Singer, G. P. Margison and R. H. Elder (1997) Targeted deletion of alkylpurine-DNA-N-glycosylase in mice eliminates repair of 1,N6-ethenoadenine and hypoxanthine but not of 3,N4-ethenocytosine or 8-oxoguanine. *Proc. Natl. Acad. Sci.* USA **94**, 12869–12874.

K. Hibi, T. Takahashi, K.Yamakawa et al. (1992) Three distinct regions involved in 3p deletion in human lung cancer. *Oncogene* **7**, 445–449.

M. Hollstein, B. Shomer, M. Greenblatt, T. Soussi, E. Hovig, R. Montesano and C.C. Harris (1996) Somatic point mutations in the p53 gene of human tumors and cell lines: updated compilation. *Nucl. Acids Res.* **24**, 141–146.

J. Jiricny (1994) Colon cancer and DNA repair: have mismatches met their match? *Trends Genet.* **10**, 164–168.

B. Karahalil, P.M. Girard, S.Boiteux and M. Dizdaroglu (1998) Substrate secificity of the Ogg1 protein of *Saccharomyces cerevisiae*: Excision of guanine lesions produced in DNA by ionizing radiation- or hydrogen peroxide/metal ion-generated free radicals. *Nucl. Acids Res.* **26**, 1228–1232.

Y.W. Kow, S.S. Wallace and B. Van Houten (1990) UvrABC endonuclease complex repairs thymine-glycols and oxidative DNA base damage. *Mut. Res.* **235,** 147–156.

F.S. Leach, N. Nicolaides, N. Papadopoulos, et al. (1993) Mutations of a *mutS* homolog in hereditary nonpolyposis colorectal cancer. *Cell* **75**, 1215–1225.

L.A. Loeb (1991) Mutator phenotype may be required for multistage carcinogenesis. *Cancer Res.* **51**, 3075–3079.

R. Lu, H.M. Nash and G.L.Verdine (1997) A mammalian DNA repair enzyme that excises oxidatively damaged guanines maps to a locus frequently lost in lung cancer. *Curr. Biol.* **7**, 397–407.

J.P. Mc Goldrick, Y. Yeng-Chen, M. Solomon, J. Essigman and A.L. Lu (1995) Characterization of a human homolog of the *E.coli* MutY repair protein. *Mol. Cell. Biol.* **15**, 989–996.

M.L. Michaels, C. Cruz, A.P. Grollman and J.H.Miller (1992) Evidence that MutY and MutM combine to prevent mutations by an oxidative damaged form of guanine. *Proc. Natl. Acad. Sci.* USA. **89**, 7022–7025. 69.

M.L. Michaels and J.H. Miller (1992) The GO system protects organisms from the mutagenic effect of the spontaneous lesion 8-hydroxyguanine (7,8-dihydro-8-oxoguanine). *J. Bacteriol.* **174**, 6321–6325.

H.M. Nash, S.D. Bruner, O.D. Scharer, T. Kawate, T.A. Addona, E. Spooner, W.S. Lane and G.L. Verdine (1996) Cloning of a yeast 8-oxoguanine DNA glycosylase reveals the existence of a base excision repair protein superfamily. *Current Biol.* **6**, 968–980.

H.M. Nash, L. Rongzhen, W.S. Lane and G.L. Verdine (1997) The critical active-site amine of the human 8-Oxoguanine DNA glycosylase, hOgg1: direct identification, ablation and chemical reconstitution. *Chemistry & Biology* **4**, 693–702.

S.L. Naylor, B.E. Johnson, J.D. Minna and A.Y. Sakaguchi (1987) Loss of heterozygosity of chromosome 3p markers in small-cell lung cancer. *Nature* **329**, 451–454.

Y. Nghiem, M. Cabrera, C.G. Cupples and J.H. Miller, (1988) The *mutY* gene: a mutator locus in *Escherichia coli* that generates G:C to T:A transversions. *Proc. Natl. Acad. Sci.* USA **85**, 2709–2713.

R. Parsons, G.M. Li, M.J. Longley, N. Papadopoulos, J. Jen, A. de la Chapelle, K.W. Kinzler, B. Vogelstein and P. Modrich (1993) Hypermutability and mismatch repair deficiency in RER+ tumor cells. *Cell* **75**, 1227–1236.

J.P. Radicella, E.A. Clark, E.A. and M.S. Fox, (1988) Some mismatch repair activities in Escherichia coli. *Proc. Natl. Acad. Sci.* USA **85,** 9674–9678

J.P. Radicella, C. Dherin, C. Desmaze, M.S. Fox and S. Boiteux (1997) Cloning and characterization of *hOGG1*, a human homolog of the *OGG1* gene of *Saccharomyces cervisiae. Proc. Natl. Acad. Sci.* USA **94**, 8010–8015.

J.T. Reardon, T. Bessho, H.C. Kung, P.H. Bolton and A. Sancar (1997) In vitro repair of oxidative DNA damage by human nucleotide excision repair system: possible explanation for neurodegeneration in *Xeroderma pigmentosum* patients. *Proc. Natl. Acad. Sci.* USA **94**, 9463–9468.

T. Roldan-Arjona, Y-F. Wei, K.C. Carter, A. Klungland, C.Anselmino, R-P.Wang, M. Augustus and T. Lindahl (1997) Molecular cloning and functional expression of a human cDNA encoding the antimutator enzyme 8-hydroxyguanine-DNA glycosylase. *Proc. Natl. Acad. Sci.* USA **94**, 8016–8020.

T.A. Rosenquist, D.O. Zharkov and A.P. Grollman, A.P. (1997) Cloning and characterization of a mammalian 8-oxoguanine DNA glycosylase. *Proc. Natl. Acad. Sci.* USA **94**, 7429–7434.

S. Shibutani, M.Takeshita and A.P. Grollman (1991) Insertion of specific bases during DNA synthesis past the oxidation-damaged base 8-OxodG. *Nature* **349**, 431–434.

M.M. Slupska, C. Baikalov, W.M. Luther, J.H. Chiang, Y-F.Wei. and J.H. Miller, (1996) Cloning and sequencing a human homolog (*hMYH*) of *the Escherichia coli mutY* gene whose function is required for the repair of oxidative DNA damage. *J. Bacteriol.* **178**, 3885–3892.

M.M. Tayer, H. Ahern, D. Xing, R.P. Cunningham and J.A. Tainer (1995) Novel DNA binding motifs in the DNA repair enzyme endonuclease III crystal structure. *EMBO J.* **14,** 4108–4120.

D. Thomas, A.D. Scott, R. Barbey, M. Padula and S. Boiteux (1997). Inactivation of *OGG1* increases the incidence of GC-->TA transversions in *Saccharomyces cerevisiae*: evidence for endogenous oxidative damage to DNA in eukaryotic cells. *Mol Gen Genet.* **254**, 171–178.

J. Wagner, K. Hiroyuki and R.P.P. Fuchs (1997) Leading versus lagging strand mutagenesis induced by 7,8-dihydro-8-oxo-2'-deoxyguanosine in *E.coli. J. Mol. Biol.* **265**, 302–309.

A. Yacoub, L. Augeri,L., M.R. Kelley. et al. (1996) A Drosophila ribosomal protein contains 8-oxoguanine and abasic site DNA repair activities. *EMBO J.* **15**, 2306–2312.

J. Yokota, M. Wada, Y. Shimosato, M. Terada and T. Sugimura, T. (1987) Loss ofheterozygosity on chromosomes 3, 13, and 17 in small-cell carcinoma and on chromosome 3 in adenocarcinoma of the lung. *Proc. Nat. Acad. Sci.* USA **84**, 9252–9256.

5

OXIDATIVE BASE DAMAGE TO DNA

Recent Mechanistic Aspects, Measurement and Repair

J. Cadet*, M. Bardet, M. Berger, T. Berthod, T. Delatour, C. D'Ham, T. Douki, D. Gasparutto, A. Grand, A. Guy, F. Jolibois, D. Molko, M. Polverelli, J. -L. Ravanat, A. Romieu, N. Signorini, and S. Sauvaigo

Département de Recherche Fondamentale sur la Matière Condensée, SCIB/LAN
CEA/Grenoble, F-38054 Grenoble Cedex 9, France

1. ABSTRACT

Emphasis was placed in this short survey on recent chemical and biochemical aspects of oxidative base damage to DNA. Significant progress has been accomplished in a better understanding of the oxidation reactions mediated by ·OH radical, singlet oxygen and one-electron oxidation of the guanine moiety of DNA and model compounds. As a second major topic, relevant information was recently gained on the origin of major artifacts in the measurement (gas chromatography-mass spectrometry, high performane liquid chromatography-[^{32}P]-postlabeling and immunological assays) of oxidative base lesions in cellular DNA. Prepurification of the targeted base damage prior to derivatization and enzymatic phosphorylation is a requisite to prevent artifactual oxidation to occur in the two former assays. Finally insight is provided on the specificity of repair by DNA-glycosylases using site specific modified oligonucleotides

2. INTRODUCTION

Oxidation reactions of DNA, a critical cellular target, are ubiquitous and are involved in mutagenesis, carcinogenesis, aging and cellular lethality. These deleterious processes may arise under various conditions of oxidative stress associated for example with either the presence within the cells of endogenously generated oxidants or the expo-

* To whom correspondence should be addressed

Advances in DNA Damage and Repair, edited by Dizdaroglu and Karakaya. Kluwer Academic / Plenum Publishers, New York, 1999.

sure to physical exogenous agents such as ionizing and solar radiations [for recent reviews, see Ames and Gold, 1991; Sies, 1991; Lindahl, 1993]. Various reactive oxygen species including hydroxyl radicals, hydrogen peroxide, peroxynitrite, ozone and singlet oxygen together with one-electron processes may give rise to the formation of several classes of DNA damage [von Sonntag, 1987; Cadet, 1994; Cadet *et al*, 1997a]. These include oxidized bases, abasic sites, oligonucleotide strand breaks, DNA-protein cross-links and aldehyde adducts to nucleobases.

In the present survey, the recently become available information on the major oxidation products of guanine and their mechanism of formation is reviewed. In addition, the main chemical and biochemical methods aimed at singling out the formation of oxidized bases within cellular DNA and in biological fluids are critically evaluated. In this respect, emphasis was placed on the gas chromatography-mass spectrometric (GC-MS) assay whose application was until recently a major matter of debate due to the occurrence of significant drawbacks [Cadet *et al*, 1997b]. Another major aspect which is discussed deals with the repair of oxidized bases using specifically modified oligonucleotides as DNA substrates.

3. RESULTS

Guanine is a critical DNA target of the bulk of the oxidation reactions. As streaking features, the guanine base reacts specifically with singlet oxygen and presents the lowest ionization potential among the nucleic acid components [Cadet and Vigny, 1990; Cadet *et al*, 1997c]. An almost complete description of the radical reactions mediated by both ·OH radicals and one electron oxidation is now possible for guanine compounds in aerated aqueous solution. It should be mentioned that relevant information on the main radical oxidation reactions of adenine, cytosine, thymine and 5-methylcytosine, the other DNA bases, is now available [see for example, Wagner *et al*, 1990, 1994; Raoul *et al*, 1995; Bienvenu *et al*, 1996].

3.1. Oxidation Reactions of the Guanine Moiety

3.1.1. Hydroxyl Radical-Mediated and One-Electron Oxidation Reactions. The main decomposition pathways involved in the ·OH radical and one-electron oxidation of the guanine moiety of isolated DNA and model compounds are depicted in Fig. 1. This was inferred from both earliest investigations on the transient radicals involved in these reactions [for comprehensive reviews, see von Sonntag, 1987; Steenken, 1989] and the recent characterization of the related final decomposition products [for a review, see Cadet *et al,* 1997a]. The two overwhelming oxidation products of the purine moiety of 2'-deoxyguanosine (**1**) as resulting from both the reaction of ·OH radical and the conversion of the guanine radical cation (one-electron oxidation pathway) have been isolated and characterized as 2,2-diamino-4-[(2-deoxy-ß-D-*erythro*-pentofuranosyl)amino]-5(2*H*)-oxazolone (**6**) and its 2-amino-5-[(2-deoxy-ß-D-*erythro*-pentofuranosyl)amino]-4*H*-imidazol-4-one (**5**) precursor [Cadet *et al*, 1994a; Buchko *et al*, 1995a, Raoul *et al*, 1996]. The mechanism of formation of the two oxidation products may be rationalized in terms of transient production of the highly oxidizing oxyl radical **3**. The latter intermediate may arise from either dehydration of the C-4 ·OH adduct **2** or deprotonation of the guanine radical cation **4**. Then, the resulting neutral radical 3 or rather one of its several tautomeric forms is implicated in a rather complicate multi-step decomposition pathway. This involves initial

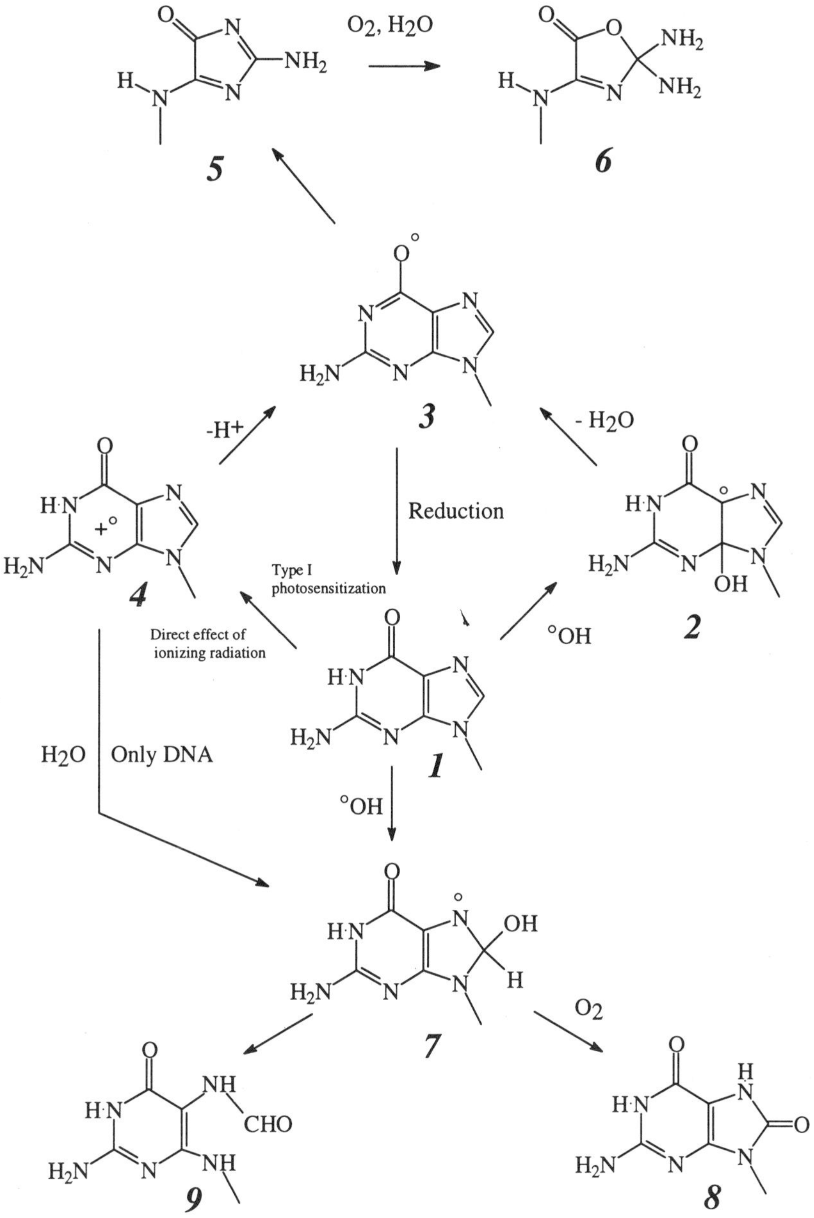

Figure 1. ·OH radical and one-electron oxidation reactions of tthe guanine moiety of DNA and model compounds.

opening of the pyrimidine ring at the C5-C6 bond and further decarboxylation subsequent to the transient formation of a peroxyl radical arising from the addition of molecular oxygen to the likely tautomeric C(5) carbon centered radical. In a following step, nucleophilic addition of a water molecule across the 7,8-ethylenic bond was suggested to occur. This leads after subsequent rearrangement to the formation of the unstable imidazolone **5** (half-life $\approx$10 h in aqueous solution at 20°C) which in turn is quantitatively hydrolyzed into the oxazolone **6** [Cadet *et al*, 1994a; Buchko *et al*, 1995a, Raoul *et al*, 1996].

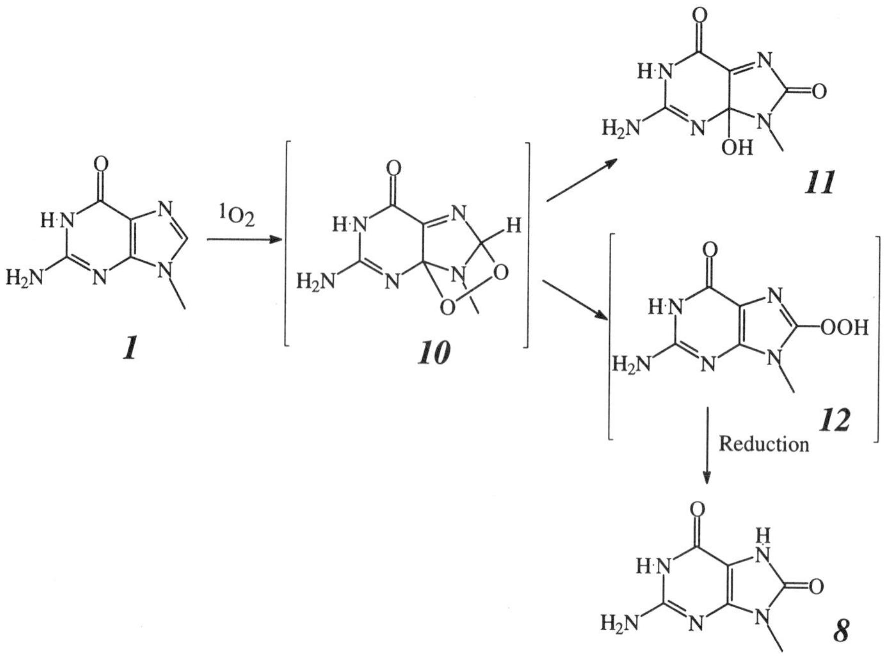

Figure 2. Main oxidation pathways of the guanine moiety of DNA and model compounds by singlet oxygen.

The formation of 8-oxo-7,8-dihydro-2'-deoxyguanosine (8-oxodGuo) (**8**) is a minor process when 2'-deoxyguanosine (**1**) is exposed to ·OH radicals in aqueous aerated solution. However, the level of 8-oxodGuo **8** significantly increases at the expense of the oxazolone derivative **6** in double-stranded DNA [Cadet *et al*, 1997a]. This is even more striking when considering the transformation reactions of the guanine radical cation **4** within native DNA. Under the latter conditions, a significant hydration reaction which does not occur within free dGuo **1** and short single-stranded oligonucleotides was found to give rise to 8-oxodGuo **8** through the transient formation of 7,8-dihydro-2'-deoxyguanos-7-yl radical (**7**) [Kasai, *et al* 1992; Angelov *et al*, 1997]. It should be added that the latter radical was found to be the precursor of the related formamidopyrimidine (Fapy) residue **9** as the result of the opening of the imidazole ring of the purine moiety. However, the latter competitive reduction reaction of **7** is a minor process (10%) in aerated aqueous solution. The overall decomposition pathway depicted in Fig. 1 illustrates the complexity of the radical oxidation reactions of the guanine moiety of DNA. Interestingly; the decomposition pathways associated with the hydroxyl radical and the one-electron process present close similarities.

3.1.2. Singlet Oxygen Cycloaddition Reaction. Relevant information is also available on the main singlet oxygen oxidation reactions of the guanine moiety of DNA and model compounds. The two main products of $^1O_2$ oxidation of 2'-deoxyguanosine (**1**) were identified as the 4R* and 4S* diastereoisomers of 4-hydroxy-8-oxo-4,8-dihydro-2'-deoxyguanosine (**11**) [Buchko *et al*, 1992; Ravanat *et al*, 1992; Ravanat and Cadet, 1995]. A reasonable mechanism for the formation of the latter specific oxidation products involves a [4 + 2] cycloaddition of $^1O_2$ to the guanine moiety according to a Diels-Alder mechanism (Fig. 2). This leads to the formation of unstable 4,8-endoperoxides **10** [Sheu

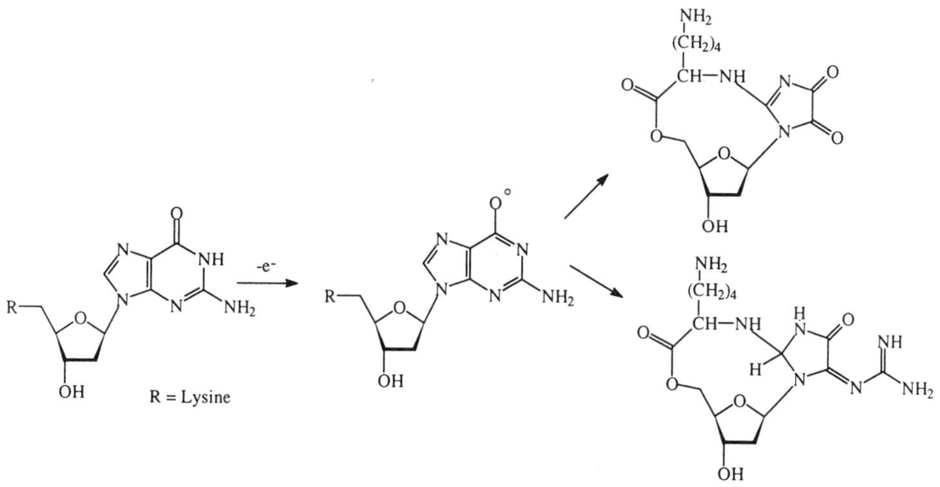

Figure 3. Nucleophilic adducts including a lysine residue to the guanine radical cation or subsequent neutral radical.

and Foote, 1993] which were characterized in the case of 8-methylguanosine derivative by ^1H NMR at low temperature. It should be added that 8-oxodGuo **8** is also produced by the reaction of $^1O_2$ with dGuo **1**, but in an only 1:7 ratio of **11**. However, in double-stranded DNA, the formation of 8-oxodGuo **8** becomes predominant [Cadet *et al*, 1994b; Ravanat and Cadet, 1995] at the expense of 4-hydroxy-8-oxo-4,8-dihydro-2'-deoxyguanosine (**11**), through a likely isomerization of the 4,8-endoperoxide precursors **10** into 8-hydroperoxy-2'-deoxyguanosine (**12**). From these model studies, it may be concluded that 8-oxodGuo **8** is not a specific marker of oxidative stress since ·OH radical, $^1O_2$, one-electron oxidation process and also peroxynitrite [Douki and Cadet, 1996] may be involved in the formation of this ubiquitous oxidative purine lesion. It should be added that **8** may be efficiently oxidized by $^1O_2$, yielding 4-hydroxy-8-oxo-4,8-dihydro-2'-deoxyguanosine (**11**) together with the oxazolone **6** and *N*-(2-deoxy-β-D-*erythro*-pentofuranosyl)cyanuric acid [Buchko *et al*, 1995b; Raoul and Cadet, 1996].

3.1.3. Guanine-Lysine Adduct Formation. The discovery of the occurrence of a nucleophilic addition of a water molecule in the formation of **6** (*vide supra*) has provided strong impetus in exploring the mechanism of the generation of DNA-protein cross-links under radical oxidative conditions. In this respect, various model systems involving either inter- or intra-molecular addition of hydroxyl and amino groups to the type I photosensitized have been investigated [Morin and Cadet, 1994; 1995a; 1995b]. Interestingly, it was shown as depicted in Fig. 3 that the amino group of a lysine residue tethered to the 5'-hydroxyl group of dGuo is able to covalently bind to the C-8 position of the purine base subsequent to one-electron oxidation [Morin and Cadet, 1995b].

3.2. Measurement of Oxidative Base Damage to Cellular DNA and in Biological Fluids

Major efforts were made in the last decade to measure within either cells or individuals the level of oxidative base damage to DNA [for comprehensive reviews, see Halli-

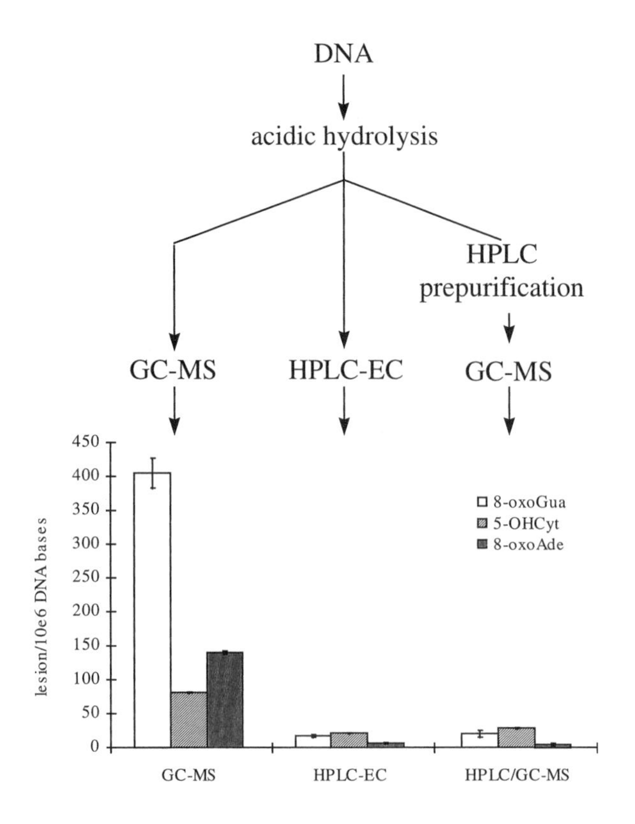

Figure 4. Comparative measurement of three oxidized bases (8-oxo-7,8-dihydroaguanine, 5-hydroxycytosine and 8-oxo-7,8-dihydroadenine) by GC/MS, HPLC/EC and HPLC/GC/MS assays.

well and Dizdaroglu, 1992; Cadet and Weinfeld, 1993]. The measurement of oxidative base damage in tissue and cellular DNA remains a challenging analytical problem [Cadet *et al*, 1997b, Collins *et al*, 1997]. This may be due, at least partly, by the high level of sensitivity that is required together with the multiplicity and, sometimes, the lability of the lesion. In addition, one of the main limiting factors deals with the occurrence of autoxidation reactions during the work-up of DNA which may induce a significant level of oxidized base damage. This is particularly critical for 8-oxodGuo **8**, whose formation is significantly enhanced under Fenton reaction conditions. Most of the assays aimed at singling out oxidized bases within DNA are based on either chemical hydrolysis or enzymatic digestion of the extracted DNA. In a subsequent step, the complex mixture of released DNA components is resolved by efficient separative methods such as high performance liquid chromatography (HPLC), gas chromatography (GC) or capillary electrophoresis (CE). A sensitive and specific method of detection is required at the output of the analytical column since the level of sensitivity should be, at least, of 1 lesion per 10^5 normal bases in a sample size of less than 50 μg of DNA. In this respect, the association of the electrochemical detection (ECD) with HPLC, proposed already more than 10 years ago [Floyd *et al*, 1986], constitutes the most suitable method for monitoring the formation of 8-oxoGua and a few other electroactive oxidized bases including 8-oxo-7,8-dihydroadenine, 5-hydroxycytosine and 5-hydroxuracil [Cadet *et al*, 1997a; 1997b].

3.2.1. Gas-Chromatography-Mass Spectrometry Assays: Artifacts and Improved Method. The GC-mass spectrometry assay is much more versatile since it can be applied to a wider range of damage [Dizdaroglu, 1991; 1993]. However, the earliest version of the method was recently found to induce significant levels of artifactual background of 8-oxodGuo **8** during the silylation step [Hamberg and Zhang, 1995; Ravanat *et al*, 1995] which is required to make the samples volatile. This early observation is likely to be applicable to any oxidized base since 8-oxo-7,8-dihydroadenine, 5-hydroxycytosine and 5-(hydroxymethyl)uracil were found to be generated during the derivatization of adenine, cytosine and thymine respectively [Douki *et al*, 1996b] Interestingly, an optimized version of the GC-MS assay which allows an accurate measurement of several oxidized purine and pyrimidine bases is now available. The improved method requires prepurification of the target oxidized base either by HPLC (Fig. 4) or immunoaffinity chromatography prior to the derivatization step in addition to the use of [M + 4] isotopically labeled internal standards [Ravanat *et al*, 1995; Douki *et al*, 1996a; 1996b]. Efforts are currently made to minimize the formation of the artifactual oxidation of the overwhelming normal purine and pyrimidine components during extraction and work-up of DNA. It should be added that attention should be paid to the conditions of hydrolysis of oxidized DNA. In this respect, it was recently shown that the formamidopyrimidine derivative of guanine (Fapy-Gua) is almost completely converted into guanine under the usual conditions of formic acid hydrolysis [Douki *et al*, 1997]. The use of hydrogen fluoride stabilized in pyridine as a milder hydrolyzing agent represents a suitable alternative. The availability of such improved assays should allow to reassess both the radiation-induced formation of DNA base damage and the specificity of DNA repair enzymes.

3.2.2. Artifacts Associated with the HPLC/ ^{32}P Postlabeling and Immunological Assays. The most sensitive method of detection currently available is based on the radioactive postlabeling of nucleoside 3'-monophosphates which are substrates for polynucleotide kinases. One example is illustrated by the HPLC-^{32}P-postlabeling assay that was devised for the measurement of adenine N^{-1} oxide in cellular DNA exposed to hydrogen peroxide [Mouret *et al*, 1990]. The sensitivity of the method is of one modification per 10^7 normal bases in a sample size of 1 μg of DNA. Attempts are currently made to make the method more quantitative by using an internal standard. Application of the assay has been extended to the measurement of 5-(hydroxymethyl)uracil and 8-oxo-7,8-dihydroguanine [for a review, see Cadet *et al*, 1992]. It should be noted that HPLC prepurification of the targeted nucleotide is a requisite non only for enhancement purpose prior to the enzymatic labeling, but also to avoid the formation of artifacts.Work-up of the bulk of the DNA samples and self radiolysis induced by [^{32}P]-2'-deoxyguanosine 5'-monophosphate led to a siginificant increase in the level of 8-oxodGuo. This is likely to explain the about 40-fold higher values of 8-oxodGuo **8** in control samples of cellular DNA when the prepurification was not applied [Devaboyina and Gupta, 1996].

Various attempts were recently made to prepare mono- and polyclonal antibodies directed against 8-oxodGuo. In most cases, the specificity of the antibodies was not high enough (1 oxodGuo **8** per 10^4 2'-deoxyguanosine residues) due to cross-reactivity with **1**, preventing their use for monitoring 8-oxodGuo **8** in whole DNA and isolated cell [Park *et al*, 1992; Girault *et al*, 1996]. It should be noted that application of such an immunoassay has led to an overstimated level of 1 8-oxodGuo **8** per 10 2'-deoxyguanosine residues (**1**) in 0.1 mM H_2O_2-treated samples of cellular DNA [Musarrat and Wani, 1994]. Further work is required to increase the specificity of antibodies against oxidative base lesions.

3.2.3. Measurement of Oxidative Base Damage in Single Cells. Interesting developments of the single cell electrophoretic method, the so-called "comet assay" [Collins *et al*, 1996] and the alkaline elution technique [Epe, 1995] lie with the use of DNA repair glyco-

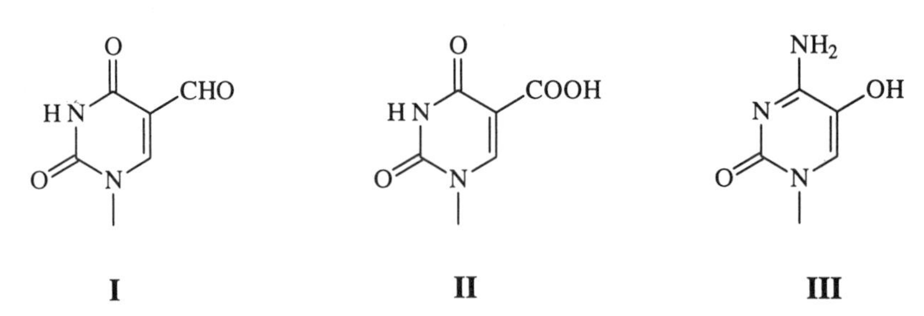

Figure 5. Structure of the base moieties (I = 5-formyluracil, II = 5-carboxyuracil and III = 5-hydroxycytosine) which have been inserted into defined sequence oligonucleotides.

sylases (Endonuclease III, Fpg protein) [Boiteux, 1993; Hatahet *et al*, 1994; Tchou *et al*, 1994]. This, which allows the conversion of oxidized base damage into additional strand breaks, makes the two above assays more specific. A major advantage of the two methods is the almost lack of DNA work-up which is expected to minimize artifactual oxidation of normal nucleosides. This is illustrated in a recent work where the radiation-induced formation of 8-oxodGuo in the DNA of tumor cells was assessed by HPLC-EC in one hand and the Fpg-Comet assay on the other hand (Pouget et al., in press). Interestingly a linear increase in the number of Fpg-sensitive sites (mostly 8-oxoGua) is observed within the 2 - 5 Gy range. In contrast a ten higher dose of gamma rays is required in order to observe a significant increase in the level of 8-oxodGuo **8** as measured using the reliable HPLC-EC assay. Further work is required to better delineate the quantitative aspects and the specificity of these promising enzymatic assays. Another possibility which is offered by the comet assay is the use of specific antibody against a defined base lesion. An interesting application of this new approach was provided by the measurement of UVB-induced cyclobutane pyrimidine dimers in isolated cells using highly specific monoclonal antibodies [Sauvaigo *et al*, 1997]. However, the method cannot be yet applied to the measurement to oxidative base damage since there is a lack of highly specific antibodies available.

3.2.4. Non-Invasive Assays. Evaluation of the effects of oxidative stress on DNA in humans requires the development of non-invasive assays. Attempts are being currently made to use the release of oxidized bases and nucleosides in urine as an index of DNA damage. For this purpose, accurate HPLC-EC and GC-MS assays have been applied to the measurement of several oxidized DNA compounds including mostly 8-oxo-7,8-dihydroguanine, 5-(hydroxymethyl)uracil and their corresponding nucleosides [Bianchini *et al*, 1996; Faure *et al*, 1996]. Interestingly, a significant increase in the release of 8-oxodGuo **8** was observed in human urine and leukocytes of cigarette smokers [for a review, see Loft and Poulsen, 1996]. These few examples illustrate the potentiality of using oxidized bases and related nucleosides as indicators for epidemiological studies aimed at correlating dietary and life habits with cancer risks for example. However, the biological validation of such approaches which should require the simultaneous measurement of several oxidized bases and nucleosides is still needed.

3.3. Evaluation of the Biological Role of Single and Complex Base Damage to DNA

Other classes of DNA damage such as clustered lesions as suggested by Ward [1994] are likely to play a major role in the biological effects of ionizing radiation. Attempts are currently made to characterize the latter complex damage (two base lesions, base damage associated with a strand break ...) which may result from multiionization processes. Specific synthesis of modified oligonucleotides that contain multiple damage sites (8-oxoGua with either adjacent formylamine or another vicinal 8-oxoGua residue, for example) constitutes an interesting approach to evaluate the biological role (repair, mutagenesis) of these lesions and single oxidized base damage. In this respect, a program of insertion of several oxidized bases at specific sites of short DNA fragments was set up in the laboratory. It should be added that the conformational changes associated to the presence of a specifically incorporated urea residue within a defined sequence oligonucleotide were recently inferred from a detailed ^{1}H NMR study [Gervais *et al*, 1998].

3.3.1. Preparation of Defined Sequence Oligonucleotides that Contain One or Two Oxidized Bases. Relatively stable oxidized bases under the conditions of oligonucleotide synthesis can be inserted into short oligonucleotides. In this respect, 5-formyl-2'-deoxyuridine [Berthod *et al*, 1996a), 5-carboxy-2'-deoxyuridine [Berthod *et al*, 1996b] and 5-hydroxy-2'-deoxycytidine [Romieu *et al*, 1997] were recently inserted (Fig. 5) into defined sequence oligonucleotides using the phosphoroamidite approach.

The preparation of oligonucleotides that contain labile oxidized nucleosides including the oxazolone derivative and 5,6-dihydroxy-5,6-dihydrothymidine is not possible using the above synthetic approach. In order to overcome the problem of instability of the latter oxidized nucleosides, particularly during the alkaline deprotection step, specific oxidation reactions of defined sequence oligonucleotide were applied. Permanganate oxidation was used for the preparation of the thymidine glycol whereas oxazolone containing oligonucleotides were prepared by type I photosensitization reaction [Gasparutto *et al*, in press].

3.3.2. Oxidized Base Excision Repair Studies. These probes are currently used for assessing the specificity of available DNA repair glycosylases (Endonuclease III, Fpg and AlkA) and the mutagenic potential of the related lesions (works in cooperation with Serge Boiteux and Alain Sarasin). It was recently found that 5,6-dihydroxy-5,6-dihydrothymine (thymine glycol) is a much better substrate than 5,6-dihydrothymine for the *N*-glycosylase activity of *E. coli* Endonuclease III as shown in Fig. 6.

4. CONCLUSION

Significant progress was accomplished in the last decade for a better understanding of the mechanisms of formation of oxidative damage to DNA. In addition there is an increasing attention devoted to the assessment of the biological significance of the latter lesions. However, as already outlined, there is still a paucity of data on the formation of the bulk of oxidized base lesions within isolated and cellular DNA. The availability of powerful analytical tools such as capillary electrophoresis and HPLC coupled with electrospray ionization mass spectrometry provides new possibilities which have to be further explored. It may be added that the growing implication of theoretical approaches [Jolibois *et al*, 1996, 1998a, 1998b] should give a strong impetus to the studies of oxidative damage to DNA (reactivity, conformational features involved in mutagenesis and DNA repair ....).

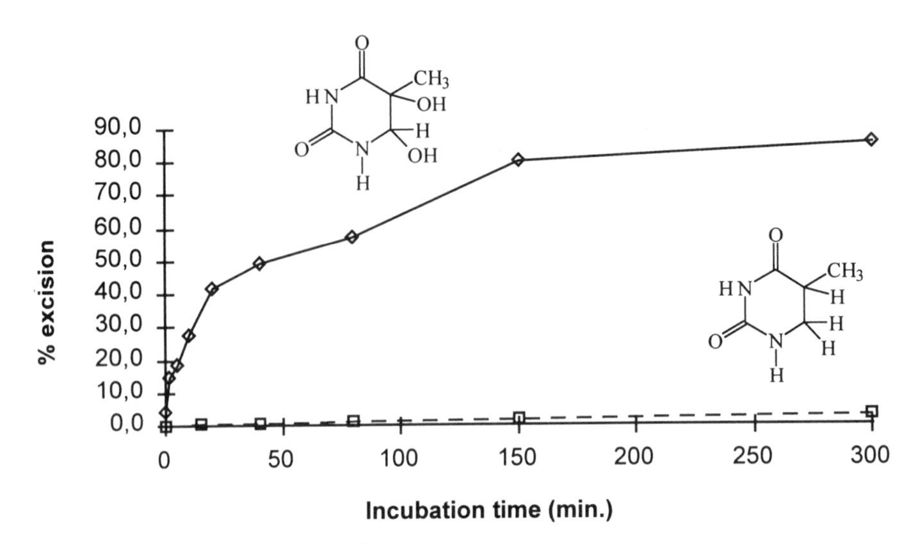

Figure 6. *E. coli*-mediated excision of 5,6-dihydroxy-5,6-dihydrothymine and 5,6-dihydrothymine from site-specific synthesized oligonucleotides with increasing periods of time at 37°C (4 μM 33-mer, 4 mg . L^{-1} enzyme).

REFERENCES

B.N. Ames and L.S. Gold (1991) Endogenous mutagens and the causes of aging and cancer. *Mutat. Res.*, **250**, 3–16.

D. Angelov, A. Spassky, M. Berger & J. Cadet (1997) High-intensity UV laser photolysis of DNA and purine 2'-deoxyribonucleosides: formation of 8-oxopurine damage and oligonucleotide strand cleavage as revealed by HPLC and gel electrophoresis studies. *J. Am. Chem. Soc.*, **119,** 11373–11380.

T. Berthod, Y. Petillot, A. Guy, J. Cadet, E. Forest and D. Molko (1996a) Synthesis and mass spectrometry analysis of oligonucleotides bearing 5-formyl-2'-deoxyuridine in their structure. *Nucleosides Nucleotides*, **15**, 1287–1305.

T. Berthod, Y. Petillot, A. Guy, J. Cadet and D. Molko (1996b) Synthesis of oligonucleotides containing 5-carboxy-2'-deoxyuridine at defined sites. *J. Org. Chem*, **61**, 6075–6078.

F. Bianchini, J. Hall, F. Donato and J. Cadet (1996) Monitoring urinary excretion of 5-hydroxymethyluracil for assessment of oxidative DNA damage and repair. *Biomarkers*, **1**, 178–184.

C. Bienvenu, J.R. Wagner and J. Cadet (1996) Photosensitized oxidation of 5-methyl-2'-deoxycytidine by 2-methyl-1,4-naphthoqinone: characterization of 5-hydroperoxymethyl-2'-deoxycytidine and stable methyl group oxidation products. *J. Am. Chem. Soc.*, **118**, 11406–11411.

S. Boiteux (1993) Properties and biological functions of the NTH and FPG proteins of *Escherichia coli*: two glycosylases that repair oxidative damage in DNA. *J. Photochem. Photobiol. B: Biol.*, **19**, 87–96.

G.W. Buchko, J. Cadet, M. Berger and J.-L. Ravanat (1992) Photooxidation of d(TpG) by phthalocyanines and riboflavin. Isolation and characterization of dinucleoside monophosphates containing the 4R* and the 4S* diastereoisomers of 4,8-dihydro-4-hydroxy-8-oxo-2'-deoxyguanosine. *Nucleic Acids Res.*, **20**, 4847–4851.

G.W. Buchko, J. Cadet, B. Morin and M. Weinfeld (1995a) Photooxidation of d(TpG) by riboflavin and methylene blue. Isolation and characterization of thymidylyl-(3',5')-2-amino-5-[(2-deoxy-β-D-*erythro*-pentofuranosyl)amino]-4*H*-imidazol-4-one, and its primary decomposition product, thymidylyl-(3',5')-2-diamino-4-[(2-deoxy-β-D-*erythro*-pentofuranosyl)amino]-(2*H*)-oxazolone. *Nucleic Acids Res.*, **19**, 3954–3961.

G.W. Buchko, J.R. Wagner, J. Cadet, S. Raoul and M. Weinfeld (1995b) Methylene blue-mediated photooxidation of 8-oxo-7,8-dihydro-2'-deoxyguanosine. *Biochim. Biophys. Acta*, **1263**, 17–24.

J. Cadet (1994) DNA damage caused by oxidation, deamination, ultraviolet radiation and photoexcited psoralens. In *DNA adducts: Identification and biological significance*.(eds. K. Hemminki, A. Dipple, D.G.E. Shuker, F.F. Kadlubar, D. Segerbäck & H. Bartsch) Lyon: International Agency for Research on Cancer; IARC Scientific Publications, No. 125, pp. 245–276.

J. Cadet and P. Vigny (1990) The photochemistry of nucleic acids. In *Bioorganic Photochemistry* (ed. H. Morrison) Vol. 1. New York: Wiley and Sons, pp. 1–272.

J. Cadet and M. Weinfeld (1993) Detecting DNA damage. *Anal. Chem.*, **65**, 675A-682A.

J. Cadet, F. Odin, J.-F. Mouret, M. Polverelli, A. Audic, P. Giacomoni, A. Favier and M.-J.Richard (1992) Chemical and biochemical postlabeling methods for singling out specific oxidative DNA lesions. *Mutat. Res.*, **275**, 343–354.

J. Cadet, M. Berger, G.W. Buchko, P.C. Joshi, S. Raoul and J.-L. Ravanat (1994a) 2,2-Diamino-4-[(3,5-di-O-acetyl-2-deoxy-β-D-*erythro*-pentofuranosyl)amino]-5-(2*H*)-oxazolone: A novel and predominant radical oxidation product of 3',5'-di-O-acetyl-2'-deoxyguanosine. *J. Am. Chem. Soc.*, **116**: 7403–7404.

J. Cadet, J.-L. Ravanat, G.W. Buchko, H.C. Yeo and B.N. Ames (1994b) Singlet oxygen DNA damage: Chromatographic and mass spectrometric analysis of damage products. *Methods Enzym.*, **234**, 79–88.

J. Cadet, M. Berger, T. Douki and J.-L. Ravanat (1997a) Oxidative damage to DNA: formation, measurement, and biological significance. *Rev. Physiol. Biochem. Pharmacol.*, **131**, 1–87.

J. Cadet, T. Douki & J.-L. Ravanat (1997b) Artifacts associated with the measurement of oxidized DNA bases. *Environ. Health Perspect.*, **105**, 1034–1039.

J. Cadet, M. Berger, T. Douki, B. Morin, S. Raoul, J.-L. Ravanat and S. Spinelli (1997c) Effects of UV and visible radiations on DNA - Final base damage. *Biological Chem.*, **378**, 1275–1286.

A. Collins, J. Cadet, B. Epe and C. Gedik (1997) Problems in the measurement of 8-oxoguanine in human DNA. *Carcinogenesis*, **18**, 1833–1836.

U-s. Devanaboyina and R.C. Gupta (1996) Sensitive detection of 8-hydroxy-2'-deoxyguanosine in DNA by ^{32}P-postlabeling assay and the basal levels in rat tissues. *Carcinogenesis*, **17**, 917–924.

M. Dizdaroglu (1991) Chemical determination of free radical-induced damge to DNA. *Free Radical Biol. Med.*, **10**, 225–242.

M. Dizdaroglu (1993) Quantitative determination of oxidative base damage in DNA by stable isotope-dilution mass spectrometry. *FEBS Lett.*, **315**, 1–6.

T. Douki and J. Cadet (1996) Peroxynitrite mediated oxidation of purine bases of nucleosides and isolated DNA. *Free Rad. Res.*, **24**, 369–380.

T. Douki, T. Delatour, F. Bianchini and J. Cadet (1996a) Observation and prevention of an artifactual formation of oxidised DNA bases and nucleosides in the GC-EIMS method. *Carcinogenesis*, **17**, 347–353.

T. Douki, T. Delatour, F. Paganon and J. Cadet (1996b) Measurement of oxidative damage at pyrimidine bases in γ-irradiated DNA. *Chem. Res. Toxicol.*, **9**, 1145–1151.

T. Douki, R. Martini, J.-L. Ravanat, R.J. Turesky and J. Cadet (1997) Measurement of 2,6-diamino-4-hydroxy-5-formamidopyrimidine and 8-oxo-7,8-dihydroguanine in gamma irradiated DNA. *Carcinogenesis* **18**, 2385–2391.

B. Epe (1995) DNA damage profiles induced by oxidizing agents. *Rev. Physiol. Biochem. Pharmacol.*, **127**, 223–249.

H. Faure, C. Coudray, M. Mousseau, V. Ducros, T. Douki, F. Bianchini, J. Cadet and A. Favier (1996) 5-Hydroxymethyluracil excretion, plasma TBARs and plasma antioxidant vitamins in adriamycin-treated patients. *Free Rad. Biol. Med.*, **20**, 979–983.

R.A. Floyd, J.J. watson, P.K. Wong, D.H. Altmiller and R.C. Rickard (1986) Hydroxyl free radical adduct of deoxyguanosine: sensitive detection and mechanism of formation. Free Radic. Res. Commun., 1, 163–172.

V. Gervais, J.A.H. Cognet, A. Guy, J. Cadet, R. Téoule and G.V. Fazakerley (1998) Solution structure of *N*(2-deoxy-D-*erythro*-pentofuranosyl)urea frameshifts, one intrahelical and the other extrahelical, by nuclear magnetic resonance and molecular dynamics. *Biochemistry* (in press).

I. Girault, D.E.G. Shuker, J. Cadet and D. Molko (1996) Use of morpholino-nucleosides to conjugate oxidized DNA bases to proteins. *Bioconjugate Chem.*, **7**, 445–450.

B. Halliwell and M. Dizdaroglu (1992) The measurement of oxidative damage to DNA by HPLC and GC/MS techniques. *Free Radic. Res. Commun.*, **16**, 15–28.

M. Hamberg and L.-Y. Zhang (1995). Quantitative determination of 8-hydroxyguanine and guanine by isotope dilution mass spectrometry. *Anal. Biochem.*, **229**, 336–344.

Z. Hatahet, Y.W. Kow, A.A. Purmal, R.P. Cunningham and S.S. Wallace (1994) New substrates for old enzymes. *J. Biol. Chem.*, **269**, 11814–18820.

F. Jolibois, L. Voituriez, A. Grand and J. Cadet (1996) Conformational and electronic properties of the two *cis* (5S,6R) and (5R,6S) diastereoisomers of 5,6-dihydroxy-5,6-dihydrothymidine: X-ray and theoretical studies. *Chem. Res. Toxicol.*, **9**, 298–305.

F. Jolibois, C. D'Ham, A. Grand, R. Subra and J. Cadet (1998a) *Cis*-5-Hydroperoxy-6-hydroxy-5,6-dihydrothymine: Crystal structure and theoretical investigations of the electronic properties by DFT. *Theochem.* **427**, 143–155.

F. Jolibois, J. Cadet, A. Grand, V. Barone, N. Rega and R. Subra (1998b) Structures and spectroscopic characteristics of 5,6-dihydro-6-thymyl radicals by an integrated quantum mechanical approach including electronic, vibrational and solvent effects. *J. Am. Chem. Soc.* **120**, 1864–1871.

H. Kasai, Z. Yamaizumi, M. Berger and J. Cadet (1992) Photosensitized formation of 7,8-dihydro-8-oxo-2'-deoxyguanosine (8-hydroxy-2'-deoxyguanosine) in DNA by riboflavin: a non singlet oxygen mediated reaction. *J. Am. Chem. Soc.*, **114**, 9692–9694.

T. Lindahl (1993) Instability and decay of the primary structure of DNA. *Nature*, **362**, 709–715.

S. Loft and H.E. Poulsen (1996) Cancer risk and oxidative DNA damage in man. *J. Mol. Med.*, **74**, 297–312.

B. Morin and J. Cadet (1994) Benzophenone photosensitisation of 2'-deoxyguanosine: Characterization of the *2R* and *2S* diastereoisomers of 1-(2-deoxy-β-D-*erythro*-pentofuranosyl)-2-methoxy-4,5-imidazolidinedione. A model system for the investigation of photosensitized formation of DNA-protein crosslinks. *Photochem. Photobiol.*, **60**, 102–109.

B. Morin and J. Cadet (1995a) Type I benzophenone mediated nucleophilic reaction of 5'-amino-2',5'-dideoxyguanosine. *Chem. Res. Toxicol.*, **8**, 792–799.

B. Morin and J. Cadet (1995b) Chemical aspects of the benzophenone photosensitized formation of two lysine-2'-deoxyguanosine crosslinks. *J. Am. Chem. Soc.*, 117, 12408–12415.

J.-F. Mouret, F. Odin, M. Polverelli and J. Cadet (1990) ^{32}P-Postlabeling measurement of adenine N-1-oxide in cellular DNA exposed to hydrogen peroxide. *Chem. Res. Toxicol.*, **3**, 102–110.

J. Musarrat and A.A. Wani (1994) Quantitative immunoanalysis of promutagenic 8-hydroxy-2'-deoxyguanosine in oxidized DNA. *Carcinogenesis*, **15**, 2037–2043.

E.U. Park, M.K. Shigenaga , P. Degan, T.S. Korn, J.W. Kitzler, C.M. Wehr, P. Kolachana and B.N. Ames (1992) Assay of excised oxidative DNA lesions: isolation of 8-oxoguanine and its nucleoside derivatives from biological fluids with a monoclonal antibody column. *Proc Natl Acad Sci USA*, **89**, 3375–3379.

S. Raoul and J. Cadet (1996) Photosensitized reaction of 8-oxo-7,8-dihydro-2'-deoxyguanosine: Identification of 1-(2-deoxy-β-D-*erythro*-pentofuranosyl)-cyanuric acid as the major singlet oxygen oxidation product. *J. Am. Chem. Soc.*, **118**, 1892–1898.

S. Raoul, M. Bardet and J. Cadet (1995) Gamma irradiation of 2'-deoxyadenosine in oxygen-free aqueous solutions: Identification and conformational features of formamidopyrimidine nucleoside derivatives. *Chem. Res. Toxicol.*, **8**, 924–933.

S. Raoul, M. Berger, G.W. Buchko, P.C. Joshi, B. Morin, M. Weinfeld and J. Cadet (1996) ^{1}H, ^{13}C and ^{15}N NMR analysis and chemical features of the two main radical oxidation products of 2'-deoxyguanosine: Oxazolone and imidazolone nucleosides. *J. Chem. Soc. Perkin Trans. 2*, 371–381.

J.-L. Ravanat, M. Berger, F. Benard, R. Langlois, R. Ouellet, J.E. van Lier and J. Cadet (1992) Phthalocyanine and naphthalocyanine photosensitized oxidation of 2'-deoxyguanosine: distinct type I and type II products. *Photochem. Photobiol.*, **55**, 809–814.

J.-L. Ravanat and J. Cadet (1995) Reaction of singlet oxygen with 2'-deoxyguanosine and DNA. Isolation and characterization of the main oxidation products. *Chem. Res. Toxicol.*, **8**, 379–388.

J.-L. Ravanat, R.J. Turesky, E. Gremaud, L.J. Trudel and R.H. Stadler (1995) Determination of 8-oxoguanine in DNA by gas chromatography - mass spectrometry and HPLC-electrochemical detection: overestimation of the background level of the oxidized base by the gas chromatography-mass spectrometry assay. *Chem Res Toxicol*, **8**, 1039–1045.

A. Romieu, D. Gasparutto, D. Molko and J. Cadet (1997) A convenient synthesis of 5-hydroxy-2'-deoxycytidine phosphoroamidite and its incoporation into oligonucleotides. *Tetrahedron Lett.*, 38, 7531–7534.

S. Sauvaigo, N. Signorini, N. Emonet, M.-J. Richard and J. Cadet (1997) Immunofluorescent detection of cyclobutane pyrimidine dimers by a modification of the comet assay. *Mutat. Res.*, **379s**, 134.

H. Sies H. (1991) *Oxidative Stress, Oxidants and Antioxidants*, New York, Academic Press Inc., 1991.

C. Sheu and C.S. Foote (1993) Endoperoxide formation in a guanosine derivative. J. *Am. Chem. Soc.*, **115**, 10446–10447.

S. Steenken (1989) Purine bases, nucleosides, and nucleotides: aqueous solutions redox chemistry and transformation reactions of their radical cations and e- and OH adducts. *Chem. Rev.*, **89**, 503–520.

C. von Sonntag (1987) *The Chemical Basis of Radiation Biology*, London: Taylor Francis.

J. Tchou, V. Bodepudi, S. Shibutani, I. Antoshechkin, J. Miller, A.P. Grollman and F. Johnson (1994) Substrate specificity of Fpg protein. *J. Biol. Chem.*, **269**, 15318–15324.

J.R. Wagner, J.E. van Lier, C. Decarroz, M. Berger and J. Cadet (1990) Photodynamic methods for oxy radical-induced DNA damage. *Methods Enzym.*,**186**, 502–511.

J.R. Wagner, J.E. van Lier, M. Berger and J. Cadet (1994) Thymidine hydroperoxides: Structural assignement, conformational features, and thermal decomposition in water *J. Am . Chem. Soc.*, **116**, 2235–2242.

J.-F. Ward (1994) The complexity of DNA damage: relevance to biological consequences. *Int. J. Radiat. Biol.*, **66**, 427–432.

6

ROLES OF AP ENDONUCLEASES IN REPAIR AND GENETIC STABILITY

Bruce Demple,* Elisabeth Bailey, Richard A. O. Bennett, Yuji Masuda, Donny Wong, and Yong-jie Xu

Department of Cancer Cell Biology, Harvard School of Public Health
Boston, Massachusetts 02115

1. ABSTRACT

Abasic sites arise in DNA by many routes, including the excision of modified bases and direct attack by free radicals. The key enzymes in repair of these sites are AP endonucleases, which can incise abasic residues or remove abasic deoxyribose fragments from oxidative strand breaks. We have examined the role of AP endonucleases in repair of oxidative damage caused by exogenous agents, and in limiting the spontaneous mutation rate that arises from endogenously produced AP sites. The AP endonuclease proteins comprise two structural families with overlapping but distinct substrate specificity. Endonuclease IV of *E. coli* has a critical role in repairing oxidative lesions produced by nitric oxide or bleomycin; for the latter, the enzyme is required for repair of 4'-oxidized abasic residues. The yeast Apn1 protein, an endonuclease IV homolog, is key for maintaining normal genetic stability in that organism. The main AP endonuclease of human cells, Ape1 protein, is a homolog of *E. coli* exonuclease III. Ape1 is a robust enzyme for cleaving abasic sites, and it can replace Apn1 to control mutagenesis in yeast. We have studied Ape1 protein in its interaction with DNA and cleavage of abasic sites. These studies show how the enzyme engages its substrate, and point to specific interactions that coordinate steps of the base excision repair pathway. We have shown that Ape1 protein acts effectively in the incision of oxidized abasic residues produced by bleomycin. Ape1 protein also stimulates human DNA polymerase β in the excision of 5'-terminal abasic residues produced by Ape1 cleavage in vitro.

2. INTRODUCTION

Oxidative damage to DNA arises from many sources, as documented in other chapters in this volume. The array of oxidative lesions is large, comprised of a hundred or so distinct damages (von Sonntag, 1987). The repair pathways for these damages are likely

* Tel. 617-432-3462, Fax 617-432-2590/-0377, E-mail: bdemple@hsph.harvard.edu

Advances in DNA Damage and Repair, edited by Dizdaroglu and Karakaya. Kluwer Academic / Plenum Publishers, New York, 1999.

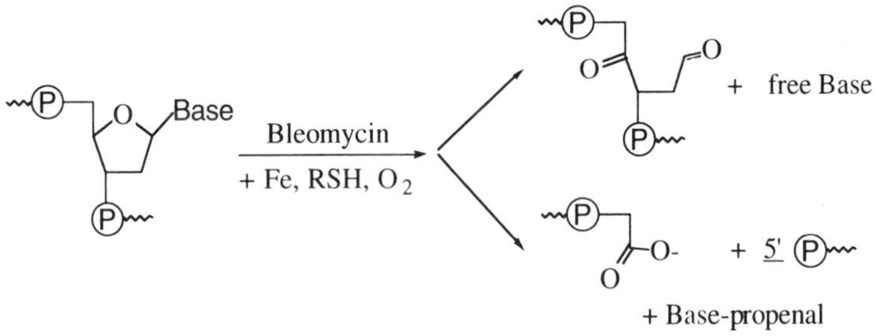

Figure 1. Oxidized abasic sites (OAS) generated by bleomycin. A mixture of two different lesions is made at each site, in a ratio dependent on the oxygen concentration and the precise sequence at the target. The upper branch of the pathway shows the formation of the KA lesion (2-deoxypentos-4-ulose, shown in the keto-aldehyde form) in an unbroken DNA strand. The lower branch shows the formation of a 3'-phosphoglycolate (3'-PG) ester accompanying a strand break, a process that also releases the DNA base attached to a propenal fragment derived from the deoxyribose.

also complex, as suggested by the recent observation that a nucleotide excision repair protein (XP-G) has a role in excision of oxidized bases (Cooper et al., 1997). A substantial fraction (up to one-third) of radiation-induced lesions is comprised of various types of oxidized abasic sites (OAS) with or without accompanying strand breaks (von Sonntag, 1987). Other agents introduce more restricted types of damage; for example, the antitumor drug bleomycin (Povirk, 1996) oxidizes the 4'-position of DNA sugars to produce 4-keto-1-aldehyde (KA) abasic sites (2-deoxypentos-4-ulose residues) and strand breaks terminated by 3'-phosphoglycolate (3'-PG) esters (Fig. 1).

Abasic sites represent a loss of the genetic information on the affected strand, and those generated by DNA glycosylases are indeed mutagenic (Loeb, 1985). Although the mutagenic consequences of oxidative strand breaks are not known, these lesions can activate poly(ADP-ribose) polymerase (Lindahl et al., 1995) and p53 (Nelson and Kastan, 1994). The potentially negative biological effects of OAS indicate that efficient repair pathways must operate on them.

Base excision repair proteins are likely involved in the correction of these OAS, and prominent among the likely enzymes are the apurinic/apyrimidinic AP endonucleases (Demple and Harrison, 1994). Here we will consider only the hydrolytic AP endonucleases, which form two broad protein families related either to endonuclease IV of *E. coli*, or to *E. coli* exonuclease III (Fig. 2). The hydrolytic AP endonucleases have varying amounts of 3'-diesterase activity for abasic residues and synthetic deoxyribose fragments, and for normal nucleotides. Cellular, molecular, and genetic studies have helped define the biological roles of some of these enzymes, particularly in microorganisms (Demple and Harrison, 1994).

3. BIOLOGY OF AP ENDONUCLEASES: SPECIFIC ROLES IN DNA REPAIR AND LIMITING SPONTANEOUS MUTAGENESIS IN *E. COLI* AND *S. CEREVISIAE*

The *APN1* gene encodes the primary AP endonuclease protein of *Saccharomyces cerevisiae*. Apn1 protein is a close homolog of *E. coli* endonuclease IV, but has a C-termi-

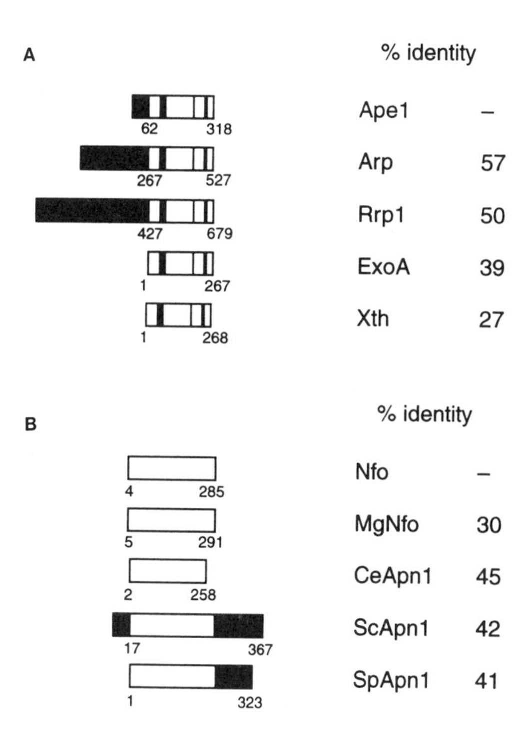

Figure 2. The two families of hydrolytic AP endonuclease proteins. The gray regions indicate polypeptide segments that are not conserved within the families. Panel A shows a schematic alignment of the Ape1/exonuclease III family, with the numbers below each box indicating the polypeptide segments used to calculate sequence identity compared to Ape1 (shown on the right). All of these proteins have demonstrated AP endonuclease activity in vitro (see Demple and Harrison, 1994). Arp, protein from *Arabidopsis thaliana*; Rrp1, protein from *Drosophila melanogaster*; ExoA, protein from *Streptococcus pneumoniae*; Xth, exonuclease III from *E. coli*. In the Ape1/exonuclease III family, key residues analyzed by mutagenesis are indicated by black vertical lines. Panel B shows the Apn1/endonuclease IV family, with sequence identities calculated against *E. coli* endonuclease IV (Nfo). MgNfo, endonuclease IV homolog from *Mycoplasma genitalium*; CeApn1, and Apn1 homolog from *Caenorhabditis elegans*; ScApn1, *S. cerevisiae* Apn1 protein; SpApn1, *Schizosaccharomyces pombe* Apn1 homolog. Among these, only *E. coli* Nfo and ScApn1 have been reported to have AP endonuclease activity in vitro.

nal extension of ~80 residues that includes functional nuclear localization signals (Ramotar et al., 1993). Apn1 has a clear function in DNA repair in yeast: Apn1-deficient strains are hypersensitive both to monofunctional alkylating agents (which generate numerous AP sites) and to oxidative agents such as H_2O_2 (Ramotar et al., 1991). Most strikingly, *apn1-Δ1* strains have a sharply increased spontaneous mutation rate (Ramotar et al., 1991; Kunz et al., 1994), consistent with the need for AP endonucleases to counteract the DNA damage by-products of cellular metabolites. *In vitro*, Apn1 protein has essentially the same range of activities, and at similar levels, as *E. coli* endonuclease IV (Johnson and Demple, 1988), and Apn1 can functionally substitute for endonuclease IV (Ramotar et al., 1991). Conversely, endonuclease IV equipped with the Apn1 nuclear localization signals can efficiently suppress the mutator phenotype of *apn1-Δ1* yeast strains (Ramotar and Demple, 1996).

E. coli endonuclease IV acts on specific types of oxidative DNA damage that are not efficiently handled by other enzymes. For example, enzyme-deficient *nfo* mutants are hypersensitive to bleomycin, even though the amount of endonuclease IV in *E. coli* cells is normally quite low (50–100 molecules). Chromosomal DNA isolated from bleomycin-treated *nfo* mutants contains unrepaired OAS that are cleaved preferentially in vitro by endonuclease IV (Levin and Demple, 1996), which suggests that the enzyme has a special role in OAS repair. Two other lines of evidence support this view. First, endonuclease IV is inducible as part of the *soxRS* oxidative stress regulon, and in that context is critical for bacterial survival of macrophages producing nitric oxide (Nunoshiba et al., 1993). It appears that other enzymes cannot replace endonuclease IV in repair of a key class of nitric oxide-dependent damage, most likely OAS. Second, endonuclease IV was far more effective than *E. coli* exonuclease III in cleaving randomly introduced bleomycin damage in vitro (Häring et al., 1994), again consistent with a specialized role for the enzyme.

4. FUNCTIONS OF Ape1, THE MAIN AP ENDONUCLEASE OF MAMMALIAN CELLS

The family of hydrolytic AP endonucleases related to *E. coli* exonuclease III includes the major enzymes of mammalian cells (Fig. 2). The human enzyme, encoded by the *APE1* gene* (also called *HAP1*, *APEX*, or *REF1*), is a potent AP endonuclease, with a turnover number for AP sites ~10-fold higher than that for the bacterial enzymes (Chen et al., 1991). In contrast to exonuclease III, Ape1 protein is a poor 3'-repair diesterase on synthetic OAS analogs (Chen et al., 1991) or site-specific 3'-PG residues (Winters et al., 1994). Expression of Ape1 in *S. cerevisiae* restores cellular resistance to alkylating agents (which generate AP sites) but not oxidative agents (generating OAS that include many oxidative strand breaks), consistent with the in vitro preference of the enzyme (Wilson III et al., 1995). Notably, even modest Ape1 expression in *apn1-Δ1* yeast restores a normal spontaneous mutation rate, in keeping with the high-level AP endonuclease activity of the enzyme.

4.1. Ape1 Interaction with DNA

As a complement to the above investigations, we have studied the in vitro enzymology of Ape1 in order to understand the enzyme's biochemical specificity. Analysis of a collection of synthetic AP site analogs showed that key substrate determinants include the DNA duplex just 5' of the abasic residue, and the space that corresponds to the abasic residue, but there seem to be few constraints on the structure of the abasic residue itself (excepting large substituents at C4) (Wilson III et al., 1995).

More insight into Ape1 recognition of its target sites came from in vitro binding experiments using the electrophoretic mobility shift assay (EMSA) and filter binding (Wilson III et al., 1997). Ape1 bound with high affinity to DNA containing either a regular AP site generated by a DNA glycosylase, or a tetrahydrofuran analog, with ~50% of the abasic DNA bound at an Ape1 concentration of 25 nM (Wilson III et al., 1997). The binding preference of Ape1 for DNA with an abasic site was much higher than that for unmodified DNA. Under our conditions, the Ape1•DNA complexes had limited stability and exhibited a half-life of ~50 sec at 0°C; notably, the half-life was reduced to <10 sec when Mg^{2+} was added to allow DNA cleavage (Wilson III et al., 1997). This destabilizing effect evidently corresponds to product release by the enzyme, while the limited stability of Ape1 complexes with DNA containing uncleaved AP sites is consistent with the need for enzymes such as Ape1 to locate rare sites of damage within large stretches of DNA.

In this work, footprinting studies of the Ape1•AP-DNA complex revealed a site of pronounced hypersensitivity to Cu-orthophenanthroline (Wilson III et al., 1997). Such hypersensitivity indicates a structural distortion of the DNA backbone precisely at the abasic site, indicative of a specific pre-cleavage structure present in the protein-DNA complex. Contact sites to the DNA within this complex were mapped by methylation and ethylation interference, and these sites lie in both strands across a DNA face that includes the abasic site nearly in the center. These studies have begun to evolve a more detailed picture of the recognition mechanism of the AP endonuclease for its target DNA sites, with binding determinants including the DNA duplex on the 5' side of the abasic residue, the abasic residue itself, and limited contacts on the 3' side (Wilson III et al., 1997).

---

* We have added a number to the name for this gene and its protein (Ape1) to accommodate naming of future enzymes of this type.

We generated several mutant forms of Ape1 based on comparisons of the polypeptide sequences of proteins of this family. The targeted sites focused on acidic residues and histidines that exhibited absolute conservation in the Ape1/exonuclease III family, with the mutagenesis designed to convert aspartate-283 and aspartate-308 to alanines (designated D283A and D308A), and histidine-309 to asparagine (H309N). Independently, Hickson and coworkers generated the same mutations, and measured very low or undetectable AP endonuclease activity for each (Barzilay et al., 1995). Unexpectedly, we have found that both the D283A protein and the D308A protein retained very high AP endonuclease activity (10–20% of the value measured for wild-type Ape1 protein). As the assays used by both groups are similar, based on DNA glycosylase-generated AP sites in a synthetic duplex oligonucleotide, the reasons for these differences in activity are unclear and merit further study.

We have employed the mutant Ape1 proteins to probe the interaction with DNA by using EMSA (Masuda et al., 1998a). These studies optimized the binding conditions further and employed the longer oligonucleotide fragment also used by S. Wilson's group (Prasad et al., 1996), with 50% DNA binding observed at 4 nM wild-type Ape1 (a concentration ~10-fold lower than found for the original binding conditions; see above). Most strikingly, the D283A, D308A, and D283A/D308A (double-mutant) proteins all complexed with DNA containing an intact abasic site with affinities as high or higher than that of wild-type Ape1. Only the H309N protein showed moderately reduced binding affinity. The H309N protein is more poorly recovered from cell extracts than are the other mutant proteins (Masuda et al., 1998a), which could indicate some instability of Ape1 containing this modification. As found under the original conditions, the binding of unmodified DNA by any of the proteins was very low or undetectable. Measurement by EMSA of the dissociation of the mutant Ape1 proteins showed greater differences, with the complexes containing the D283A, D308A and D283A/D308A proteins exhibiting half-lives of ~6 min, compared to the ~50 sec at 0°C (Masuda et al., 1998). The value for wild-type Ape1 dissociation was identical to that found using the previous conditions (Wilson III et al., 1997).

Experiments examining the Ape1 interaction with incised AP sites revealed some unusual features of the mutant proteins (Masuda et al., 1998). The incision did not significantly change the binding affinities of wild-type or the mutant Ape1 proteins compared to the binding of DNA with intact AP sites, except for H309N protein, which did not produce detectable complexes. As seen previously for wild-type Ape1, the stability of complexes of the mutant proteins with the incised DNA was diminished several fold by magnesium. The dissociation from incised DNA was complex and included a faster and a slower phase (Masuda et al., 1998b), as seen earlier for wild-type Ape1 (Wilson III et al., 1997). The greatest difference was seen for dissociation of the D308A protein, which had a slow phase corresponding to a half-life of 3–4 min (Masuda et al., 1998b). These observations suggest that it might be possible to generate altered forms of Ape1 that bind tightly enough to intact or incised AP sites to interfere with normal repair processes.

4.2. Ape1 and Repair of Oxidative DNA Damage

We showed recently that Ape1 protein interacts with human DNA polymerase ß (Polß) to recruit the polymerase to abasic sites, and that the interaction accelerates the excision of 5'-terminal deoxyribose-phosphate residues by Polß (Bennett et al., 1997). This observation suggested that these proteins might cooperate in a coordinated pathway of base excision repair. We investigated whether these interactions might extend to the repair of bleomycin-induced OAS in DNA (Fig. 1).

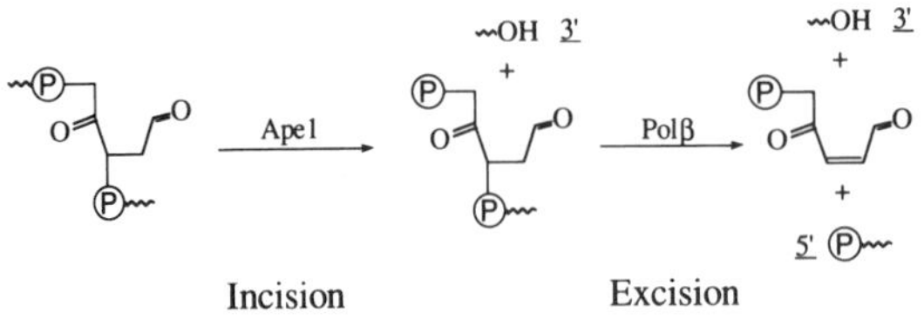

Figure 3. Ape1 and DNA polymerase ß in the repair of OAS. The two-step process of removal of a KA site is shown. Ape1 protein catalyzes effective incision on the 5' side of KA and generates a normal 3'-hydroxyl group. Polß excises the 5'-terminal KA residue formed by Ape1 incision; this activity is stimulated in the presence of Ape1. The proposed ß-elimination product of Polß is shown.

We generated appropriate substrates using a duplex oligonucleotide with a single major target site (a GpT sequence) for bleomycin attack (Xu et al., 1998). Treatment with bleomycin produced damage predominantly at the main target site in one strand, with direct strand breaks containing 3'-PG termini and KA sites (the latter revealed as strand breaks after treatment of the DNA with hydrazine) in a ratio of ~1:1.5. Incubation of this DNA with purified Ape1 protein showed efficient incision of the KA sites, with a cleavage rate at most a few fold lower than that for hydrolytic AP sites in the same sequence. In contrast, Ape1 was inefficient in removing 3'-PG residues; quantitation indicated that the 3'-PG diesterase activity of Ape1 was <4% of that for KA sites. The low 3'-repair diesterase activity of Ape1 in these experiments was in keeping with prior results using a synthetic analog of 3'PG (Chen et al., 1991), and was in contrast with the robust 3'-PG diesterase activity of bacterial endonuclease IV protein acting on the bleomycin-treated substrate (Xu et al., 1998). Endonuclease IV also cleaved KA sites at a rate similar to its activity on hydrolytic AP sites.

A key question was whether Polß, which excises hydrolytic AP sites after cleavage (Matsumoto and Kim, 1995), could also participate in the excision of oxidized residues. Again using the bleomycin-damaged substrate, we showed that human Polß excises KA residues following cleavage by Ape1 or endonuclease IV (Fig. 3). Product analysis of the material excised by Polß from double-labeled substrates (^{32}P in the phosphate, ^{3}H in the deoxyribose) showed that both labels were retained, consistent with KA excision. Excision of the 5'-KA residues was also observed for *E. coli* Fpg (MutM) protein, consistent with its activity on hydrolytic abasic residues (Graves et al., 1992). We were unable to detect such activity for *E. coli* RecJ protein (Dianov et al., 1994), either on the incised KA sites or on incised AP sites.

As noted earlier, Ape1 stimulates the 5'-excision activity of Polß acting on a standard hydrolytic AP site, and this stimulation by Ape1 was also observed for excision of KA: essentially complete removal of the KA was achieved by Polß at a ~3-fold lower concentration in the presence of Ape1 than in the presence of *E. coli* endonuclease IV. Thus, Ape1 and Polß have activities that would place them centrally in the repair pathways for an important class of oxidative DNA damage, the OAS (Fig. 3). Participation by these two enzymes in the repair of AP sites generated by DNA glycosylases that remove oxidized bases would further expand the roles of Ape1 and Polß in repairing oxidative DNA damage.

ACKNOWLEDGMENTS

Work in the authors' laboratory was supported by grants to B.D. from the U.S. National Institutes of Health (GM40000, CA71993, and ES03926). E.B. was supported by a National Research Service Award F32-ES05731).

REFERENCES

Barzilay, G., Walker, L. J., Robson, C. N. and Hickson, I. D. (1995) Site-directed mutagenesis of the human DNA repair enzyme HAP1: identification of residues important for AP endonuclease and RNase H activity. *Nucl. Acids Res.* **23**, 1544–50.

Bennett, R. A. O., Wilson, D. M. III, Wong, D. and Demple, B. (1997) Interaction of human apurinic endonuclease and DNA polymerase ß in the base excision repair pathway. *Proc. Natl. Acad. Sci. USA* **94**, 7166–7169.

Chen, D. S., Herman, V. and Demple, B. (1991) Two Distinct Human Diesterases that Act on 3'-Fragments of Deoxyribose in Radical-Damaged DNA. *Nucl. Acids Res.* **19**, 5907–5914.

Cooper, P. K., Nouspikel, T., Clarkson, S. G. and Leadon, S. A. (1997) Defective transcription-coupled repair of oxidative base damage in Cockayne syndrome patients from XP group G. *Science* **275**, 990–993.

Demple, B. and Harrison, L. (1994) Repair of oxidative damage to DNA: enzymology and biology. *Ann.Rev.Biochem.* **63**, 915–948.

Dianov, G., Sedgwick, B., Daly, G., Olsson, M., Lovett, S. and Lindahl, T. (1994) Release of 5'-terminal deoxyribose-phosphate residues from incised abasic sites in DNA by the *Escherichia coli* RecJ protein. *Nucl. Acids Res.* **22**, 993–8.

Graves, R. J., Felzenszwalb, I., Laval, J. and O'Connor, T. R. (1992) Excision of 5'-terminal deoxyribose phosphate from damaged DNA is catalyzed by the Fpg protein of *Escherichia coli. J. Biol. Chem.* **267**, 14429–14435.

Häring, M., Rudiger, H., Demple, B., Boiteux, S. and Epe, B. (1994) Recognition of oxidized abasic sites by repair endonucleases. *Nucl. Acids Res.* **22**, 2010–5.

Johnson, A. W. and Demple, B. (1988) Yeast DNA 3'-Repair Diesterase is the Major Cellular Apurinic/Apyrimidinic Endonuclease: Substrate specificity and kinetics. *J. Biol. Chem.* **263**, 18017–18022.

Kunz, B. A., Henson, E. S., Roche, H., Ramotar, D., Nunoshiba, T. and Demple, B. (1994) Specificity of the mutator caused by deletion of the yeast structural gene (*APN1*) for the major apurinic endonuclease. *Proc. Natl. Acad. Sci. USA* **91**, 8165–9.

Levin, J. D. and Demple, B. (1996) In vitro detection of endonuclease IV-specific DNA damage formed by bleomycin in vivo. *Nucl. Acids Res.* **24**, 885–9.

Lindahl, T., Satoh, M. S., Poirier, G. G. and Klungland, A. (1995) Post-translational modification of poly(ADP-ribose) polymerase induced by DNA strand breaks. *Trends Biochem. Sci.* **20**, 405–11.

Loeb, L. A. (1985) Apurinic sites as mutagenic intermediates. *Cell* **40**, 483–4.

Masuda, Y., Bennett, R. A. O. and Demple, B. (1998a) Dynamics of the interaction of human apurinic endonuclease (Ape1) with its substrate and product. *J. Biol. Chem.* **273**, in press.

Masuda, Y., Bennett, R. A. O. and Demple, B. (1998b) Rapid dissociation of human apurinic endonuclease (Ape1) from incised DNA induced by magnesium. *J. Biol. Chem.* **273**, in press.

Matsumoto, Y. and Kim, K. (1995) Excision of Deoxyribose Phosphate Residues by DNA Polymerase Beta During DNA Repair. *Science* **269**, 699–702.

Nelson, W. G. and Kastan, M. B. (1994) DNA strand breaks: the DNA template alterations that trigger p53-dependent DNA damage response pathways. *Mol. Cell. Biol.* **14**, 1815–1823.

Nunoshiba, T., deRojas-Walker, T., Wishnok, J. S., Tannenbaum, S. R. and Demple, B. (1993) Activation by nitric oxide of an oxidative-stress response that defends *Escherichia coli* against activated macrophages. *Proc. Natl. Acad. Sci. USA* **90**, 9993–7.

Povirk, L. F. (1996) DNA damage and mutagenesis by radiomimetic DNA-cleaving agents: bleomycin, neocarzinostatin and other enediynes. *Mutat. Res.* **355**, 71–89.

Prasad, R., Singhal, R. K., Srivastava, D. K., Molina, J. T., Tomkinson, A. E. and Wilson, S. H. (1996) Specific interaction of DNA polymerase ß and DNA ligase I in a multiprotein base excision repair complex from bovine testis. *J. Biol. Chem.* **271**, 16000–16007.

Ramotar, D. and Demple, B. (1996) Functional expression of *Escherichia coli* endonuclease IV in apurinic endonuclease-deficient yeast. *J. Biol. Chem.* **271**, 7368–7374.

Ramotar, D., Kim, C., Lillis, R. and Demple, B. (1993) Intracellular localization of the Apn1 DNA repair enzyme of *Saccharomyces cerevisiae*. Nuclear transport signals and biological role. *J. Biol. Chem.* **268**, 20533–9.

Ramotar, D., Popoff, S. C. and Demple, B. (1991) Complementation of *E. coli* Oxidant-Sensitive Mutants by a Yeast DNA Repair Enzyme. *Mol. Microbiol.***5**, 149–155.

Ramotar, D., Popoff, S. C., Gralla, E. B. and Demple, B. (1991) Cellular role of yeast Apn1 apurinic endonuclease/3'-diesterase: repair of oxidative and alkylation DNA damage and control of spontaneous mutation. *Mol. Cell. Biol.***11**, 4537–4544.

von Sonntag, C. (1987) The Chemical Basis of Radiation Biology. Taylor and Francis Ltd., London.

Wilson III, D. M., Bennett, R. A. O., Marquis, J. C., Ansari, P. and Demple, B. (1995) Trans-complementation by human apurinic endonuclease (Ape) of hypersensitivity to DNA damage and spontaneous mutator phenotype in *apn1*⁻ yeast. *Nucl. Acids Res.* **23**, 5027–33.

Wilson III, D. M., Takeshita, M. and Demple, B. (1997) Abasic site binding by the human apurinic endonuclease, Ape, and determination of the DNA contact sites. *Nucl. Acids Res.* **25**, 933–939.

Wilson III, D. M., Takeshita, M., Grollman, A. P. and Demple, B. (1995) Incision activity of human apurinic endonuclease (Ape) at abasic site analogs in DNA. *J. Biol. Chem.* **270**, 16002–7.

Winters, T. A., Henner, W. D., Russell, P. S., McCullough, A. and Jorgensen, T. J. (1994) Removal of 3'-phosphoglycolate from DNA strand-break damage in an oligonucleotide substrate by recombinant human apurinic/apyrimidinic endonuclease 1. *Nucl. Acids Res.* **22**, 1866–1873.

Xu, Y.-j., Kim, E. Y. and Demple, B. (1998) Excision of C4'-oxidized deoxyribose lesions from double-stranded DNA by human AP endonuclease (Ape1 protein) and DNA polymerase ß. *J. Biol. Chem.* **273**, in press.

7

MECHANISMS OF OXIDATIVE DNA DAMAGE; LESIONS AND THEIR MEASUREMENT

Miral Dizdaroglu

Chemical Science and Technology Laboratory
National Institute of Standards and Technology
Gaithersburg, Maryland 20899

1. ABSTRACT

Free radicals produce a number of lesions in DNA such as base lesions, sugar lesions, single-strand breaks, double-strand breaks, abasic sites and DNA-protein cross-links by a variety of mechanisms. Hydroxyl radical, hydrated electron and H atom react with the heterocyclic bases in DNA by addition. In oxygenated systems, adduct radicals of pyrimidines are converted into corresponding peroxyl radicals with oxygen. Hydroxyl radical reacts with purines at diffusion-controlled rates by addition to C4-, C5- and C8-positions. A fraction of hydroxyl radicals reacts with the sugar moiety in DNA by abstraction of H atoms from all five carbon atoms, producing sugar radicals. Further reactions of DNA radicals result in formation of numerous products. The types and yields of DNA modifications profoundly depend on the free radical-generating system, experimental conditions, and the presence or absence of oxygen. In chromatin, DNA-protein cross-links are formed by combination of two radicals, or by radical addition reactions. A number of analytical techniques have been used to identify and quantify a variety of products of DNA. Many pyrimidine and purine lesions have been identified and quantified in cells and tissues. Evidence indicates that most products arise as a result of reactions of hydroxyl radical with DNA constituents in cells. A number of DNA lesions possess premutagenic properties. The biological consequences of DNA modifications in cells are largely unknown and await further exploration.

2. INTRODUCTION

Oxidative DNA damage is produced when reactive oxygen species are generated in cells by normal aerobic metabolism or by exogenous sources such as ionizing radiation or carcinogenic compounds, and this type of damage in cells is implicated in mutagenesis, carcinogenesis and aging (Halliwell and Gutteridge, 1990). Among oxygen-derived spe-

Advances in DNA Damage and Repair, edited by Dizdaroglu and Karakaya. Kluwer Academic / Plenum Publishers, New York, 1999.

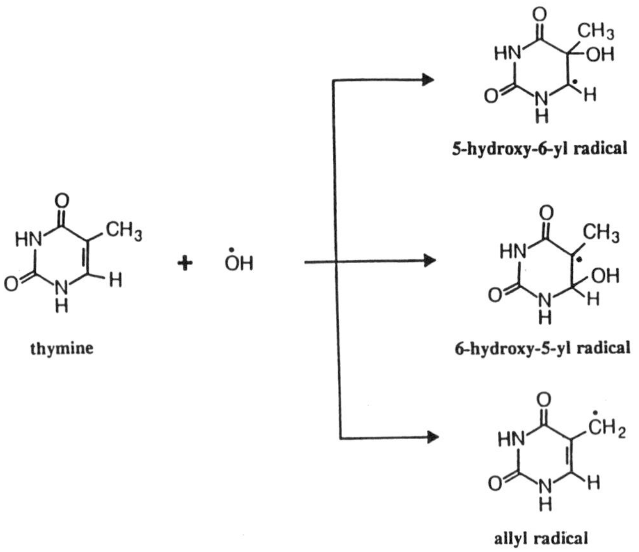

thymine

5-hydroxy-6-yl radical

6-hydroxy-5-yl radical

allyl radical

Figure 1. Reactions of •OH with thymine.

cies, hydroxyl radical (•OH) is highly reactive and can produce a number of modifications in DNA including base lesions, sugar lesions, DNA-protein cross-links and single-strand breaks, double-strand breaks and abasic sites by a variety of mechanisms (Téoule and Cadet, 1978; von Sonntag, 1987; Dizdaroglu, 1992; Breen and Murphy, 1995). Hydrated electron (e_{aq}^-) and H atom produced by ionizing radiation from water in cells also react with DNA bases (von Sonntag, 1987).

3. REACTIONS OF FREE RADICALS WITH DNA

3.1. Reactions with DNA Bases and Sugar Moiety

Hydroxyl radical adds to the C5-C6 double bond of pyrimidines at diffusion-controlled rates ($\approx 5 \times 10^9$ $M^{-1} s^{-1}$) (von Sonntag, 1987), and produces 5-hydroxy-6-yl and 6-hydroxy-5-yl radicals. The percentage of addition depends on the compound. Thus, •OH adds to the C5 and C6 of thymine to the extent of ≈60% and ≈30%, respectively, and to the C5 and C6 of cytosine to the extent of ≈90% and ≈10%, respectively (Fujita and Steenken, 1981; Hazra and Steenken, 1983). Ten percent of •OH abstract an H atom from the methyl group of thymine (Fujita and Steenken, 1981). Figure 1 illustrates the structures of thymine radicals.

Hydrated electron reacts with pyrimidines at diffusion-controlled rates ($1.3–1.7 \times 10^{10}$ $M^{-1} s^{-1}$) (von Sonntag, 1987). As a result, electron adducts (radical anions) are produced and these protonate in water to give 6-hydro-5-yl radicals (Hissung and von Sonntag, 1979) (Figure 2). Similar radicals are also generated by reactions of H atom with pyrimidines at rate constants $1–5 \times 10^8$ $M^{-1} s^{-1}$ (von Sonntag, 1987). Oxygen reacts with pyrimidine adduct radicals at diffusion-controlled rates to give peroxyl radicals.

Hydroxyl radical reacts with purines at diffusion-controlled rates ($5–9 \times 10^9$ $M^{-1} s^{-1}$) by addition to C4-, C5- and C8-positions (von Sonntag, 1987; Steenken, 1989). The structures of the OH-adduct radicals of guanine are shown in Figure 3. C4-OH- and C5-OH-ad-

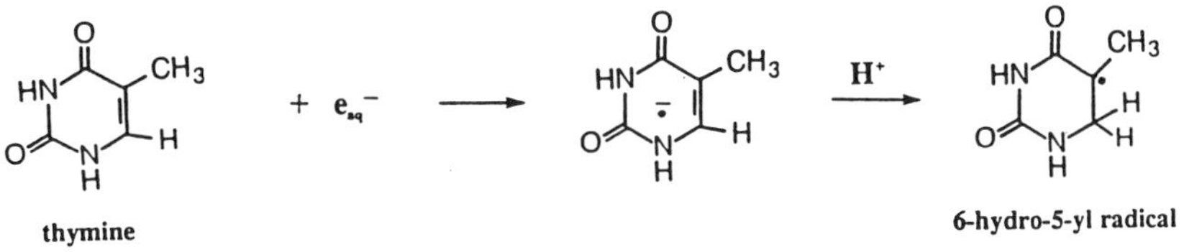

Figure 2. Reaction of e_{aq}^- with thymine.

duct radicals dehydrate and C8-OH-adduct radicals undergo unimolecular opening of the imidazole ring (Vieira and Steenken, 1990; Steenken,1989). Hydrated electron reacts with purines at diffusion-controlled rates (0.9–1.3 x 10^{10} M^{-1} s^{-1}) and the resulting electron-adducts undergo protonation reactions (von Sonntag,1987; Steenken, 1989). The majority of purine adduct radicals may not react with oxygen (von Sonntag, 1987; Steenken, 1989).

Further reactions of radicals of DNA base moieties result in formation of numerous products by a variety of mechanisms (Téoule and Cadet, 1978; von Sonntag, 1987; Téoule, 1987; Dizdaroglu, 1992; Breen and Murphy, 1995). Radicals can be oxidized or reduced depending on their redox properties and their reaction partners. For example, the oxidation of 5-hydroxy-6-yl radicals of pyrimidines followed by addition of OH^- (or addition of water followed by deprotonation) leads to formation of pyrimidine glycols (Téoule and Cadet, 1978; Téoule, 1987; von Sonntag, 1987) (Figure 4). The reduction, followed by protonation of the 6-hydro-5-yl radicals, gives rise to 5,6-dihydropyrimidines. 5-Hydroxy-6-hydro- and 6-hydroxy-5-hydropyrimidines may be formed by reduction of 5-hydroxy-6-yl and 6-hydroxy-5-yl radicals, respectively. Oxidation of the allyl radical of thymine followed by reaction with water (addition of OH^-) yields 5-hydroxymethyluracil. In deoxygenated systems, H-abstraction by 5-hydroxy-6-yl radicals from neighboring sugar moieties may occur, leading to DNA strand cleavage (Lemaire et al., 1984).

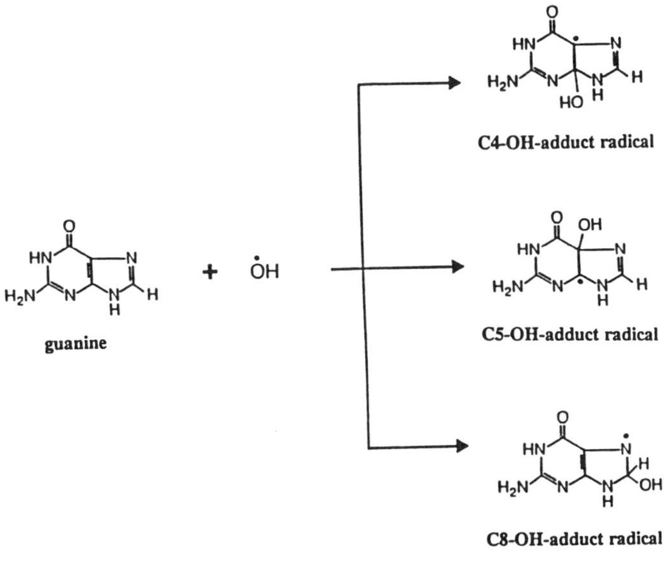

Figure 3. Reactions of •OH with guanine (Adapted from Steenken, 1989).

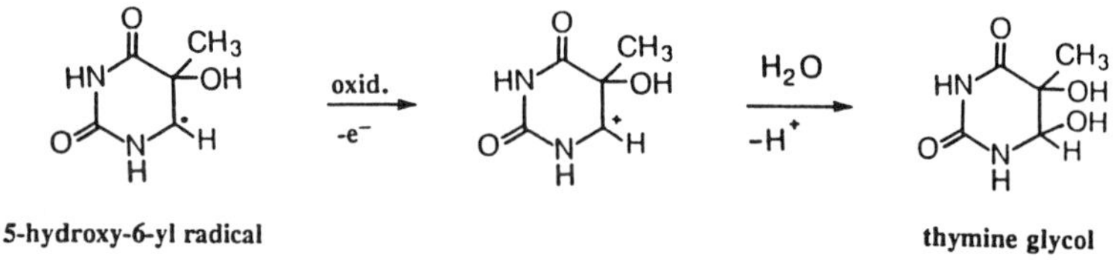

Figure 4. Formation of thymine glycol in the absence of oxygen.

5-Hydroxy-6-yl radicals of pyrimidines react with molecular oxygen to give 5-hydroxy-6-peroxyl radicals, which may eliminate $O_2^{\bullet-}$ followed by reaction with water (addition OH^-), leading to pyrimidine glycols (Téoule, 1987) (Figure 5). 5-hydroxymethyluracil from the allyl radical of thymine may be formed by a similar mechanism. 6-Hydroperoxy-5-hydroxy- and 5-hydroperoxy-6-hydroxy derivatives of pyrimidines and 5-hydroperoxymethyluracil may be formed by reduction followed by protonation of corresponding peroxyl radicals (Wagner et al., 1994). Thymine peroxides decompose to give thymine glycol, 5-formyluracil and 5-hydroxymethyluracil (Wagner et al., 1994). 5-Formyluracil has been identified in γ-irradiated DNA (Kasai et al., 1990). Intramolecular cyclization of 5-hydroxy-6-hydroperoxides of cytosine yields *trans*-1-carbamoyl-2-oxo-4,5-dihydroxyimidazolidine as a major product of cytosine in oxygenated aqueous solution (Téoule, 1987; Wagner, 1994), although this compound appears to be a minor product in DNA (Wagner, 1994; Wagner et al., 1996). Ring-reduction products such as 5-hydroxyhydantoins, urea, *N*-formyl urea and pyruvamide (released) are also formed in DNA in the presence of oxygen (Téoule and Cadet, 1978; Téoule, 1987).

Pyrimidine glycols and 5-hydroxymethyluracil are formed both in the presence and absence of oxygen, whereas 5,6-dihydropyrimidines are only produced in the absence of oxygen due to scavenging of e_{aq}^- or H atoms by oxygen at diffusion-controlled rates. Similarly, scavenging of 5-hydroxy-6-yl radicals by oxygen inhibits formation of 5-hydroxy-6-hydropyrimidines (Téoule, 1987; Dizdaroglu, 1992).

Products of cytosine may be converted into uracil derivatives. Thus, cytosine glycol yields 5-hydroxycytosine by dehydration, uracil glycol by deamination, and 5-hydroxyuracil by deamination and dehydration (Téoule, 1987; Dizdaroglu et al., 1986, 1993). Indeed, only 5-hydroxycytosine and 5-hydroxyuracil were found in derivatized hydrolysates of DNA exposed to ionizing radiation or other •OH-producing systems (Dizdaroglu, 1992; Dizdaroglu et al., 1993; Wagner, 1994). Uracil derivatives 5,6-dihydrouracil, 5,6-dihydroxyuracil (enol form of isodialuric acid) and 5-hydroxy-6-hydrouracil are formed by deamination of corresponding cytosine products (Téoule, 1987; Dizdaroglu et al., 1993). Alloxan is also a product of cytosine in DNA and decarboxylates by acidic treatment to give rise to 5-hydroxyhydantoin (Dizdaroglu, 1993; Dizdaroglu et al., 1993).

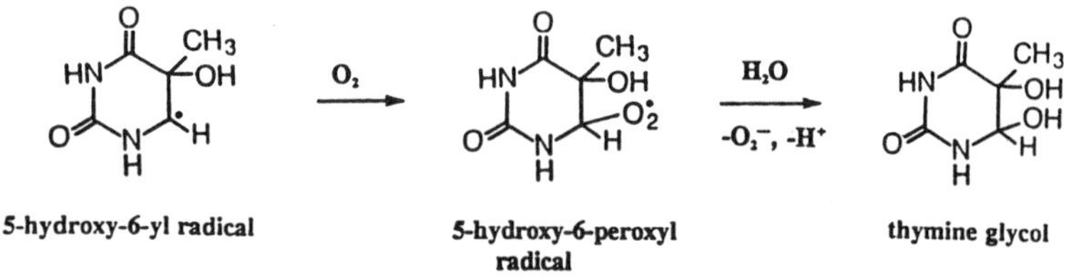

Figure 5. Formation of thymine glycol in the presence of oxygen.

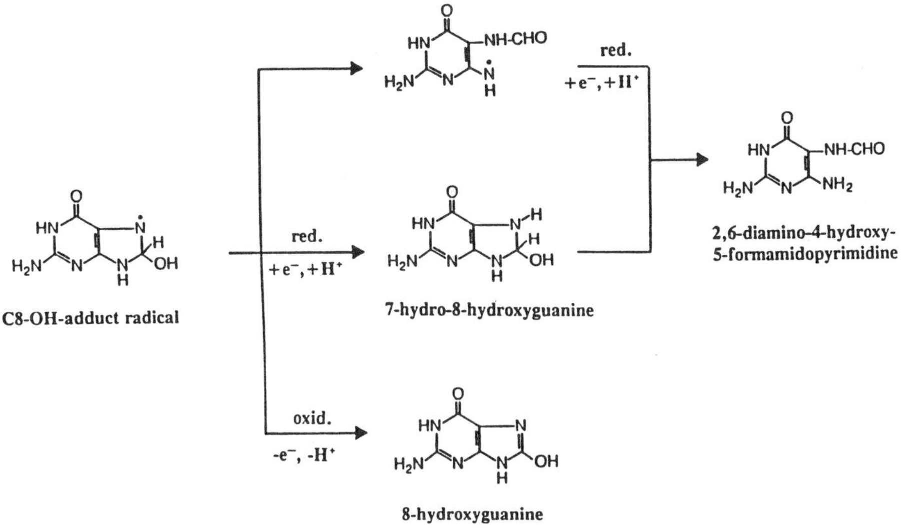

C8-OH-adduct radical

7-hydro-8-hydroxyguanine

2,6-diamino-4-hydroxy-5-formamidopyrimidine

8-hydroxyguanine

Figure 6. Formation of guanine products from C8-OH adduct radical (Adapted from Steenken (1989)).

Both C4-OH- and C5-OH-adduct radicals of guanine can lose water to give the oxidizing Gua(-H)• radical, which may be reduced and protonated to form guanine (O'Neill and Chapman, 1985; Steenken, 1989). The C4-OH-adduct radical of adenine is converted to adenine by a similar mechanism (Viera and Steenken, 1990). The Gua(-H)• radical may react with oxygen to give an oxazolone derivative (Kasai et al., 1992) or a cyclonucleoside in the case of 2′-deoxyguanosine (Buchko et al., 1993).

The C8-OH-adduct radicals of purines undergo one-electron oxidation to give 8-hydroxypurines (7,8-dihydro-8-oxopurines), and one-electron reduction to form formamidopyrimidines (Hems, 1960; Bonicel et al., 1980; Chetsanga and Grigorian, 1983; Kasai et al., 1984; Dizdaroglu, 1985a,b; Steenken, 1989). The one-electron reduction of the C8-OH-adduct radicals without the ring-opening also may take place resulting in formation of 7-hydro-8-hydroxypurines (Steenken, 1989). These compounds are hemiorthoamides and may be converted into formamidopyrimidines. Figure 6 illustrates these reactions in the case of guanine. 8-Hydroxypurines [8-hydroxyguanine (8-OH-Gua) and 8-hydroxyadenine (8-OH-Ade)] and formamidopyrimidines [2,6-diamino-4-hydroxy-5-formamidopyrimidine (FapyGua) and 4,6-diamino-5-formamidopyrimidine (FapyAde)] are formed both in the absence and presence of oxygen, although the formation of 8-hydroxypurines is favored in the presence of oxygen (Fuciarelli et al., 1990; Gajewski et al., 1990; Dizdaroglu, 1992). Nackerdien et al. (1991a) first described the formation of 2-hydroxyadenine in DNA. Later, Kamiya and Kasai (1995) also found this compound in DNA. A possible mechanism may include an •OH attack at the C2-position of adenine in DNA followed by oxidation of the thus-formed C2-OH-adduct radical. Figure 7 illustrates the structures of some of the base-derived products in DNA.

8,5′-Cyclopurine-2′-deoxynucleosides are formed in DNA by addition of the C5′-centered sugar radical to the C8-position of the purine ring of the same nucleoside followed by oxidation of the thus-formed adduct radical (Keck, 1968; Raleigh et al., 1976; Mariaggi et al., 1976; Fuciarelli et al., 1985; Dizdaroglu, 1986; Dirksen et al., 1988). Both 5′R- and 5′S-stereoisomers of 8,5′-cyclopurine-2′-deoxyguanosine and 8,5′-cyclopurine-

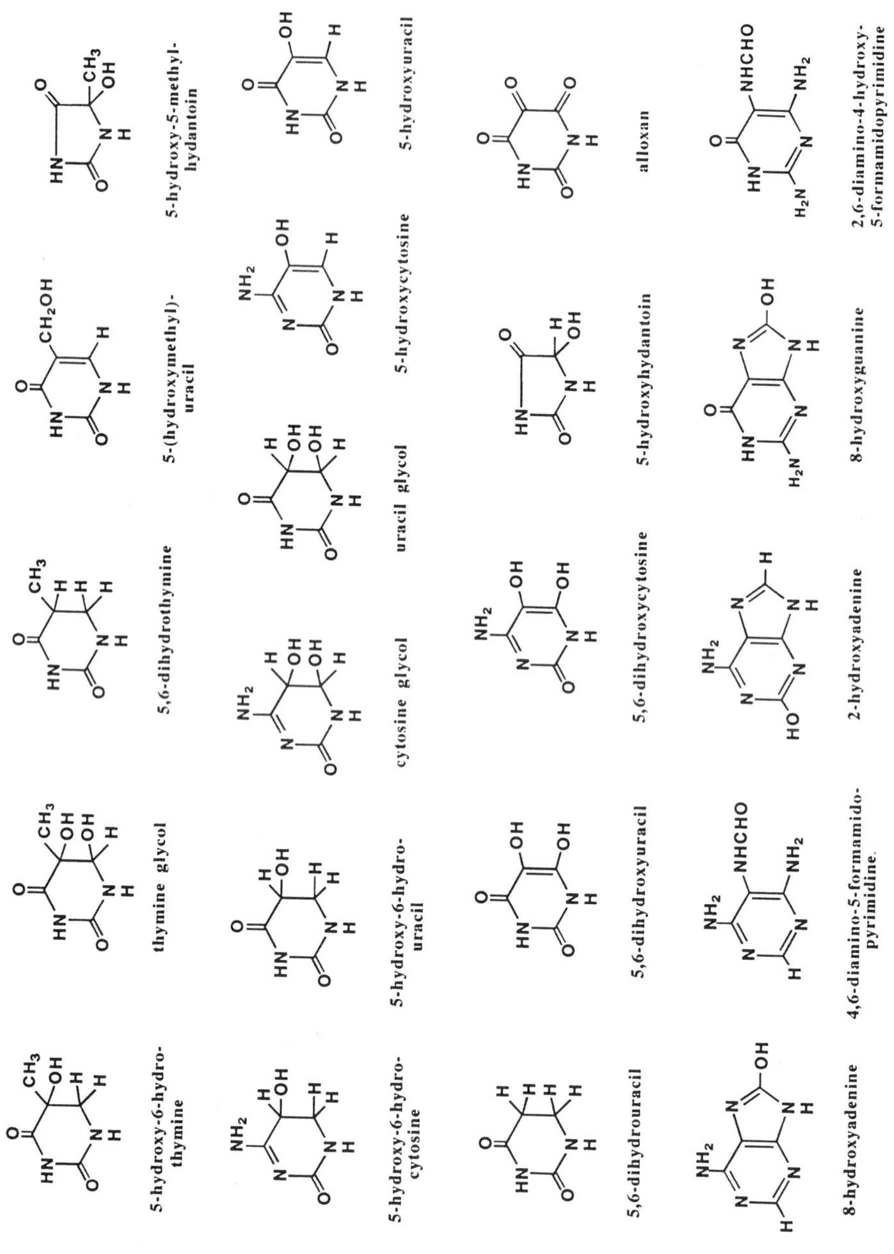

5-hydroxy-6-hydrothymine | thymine glycol | 5,6-dihydrothymine | 5-(hydroxymethyl)uracil | 5-hydroxy-5-methylhydantoin

5-hydroxy-6-hydrocytosine | 5-hydroxy-6-hydrouracil | cytosine glycol | uracil glycol | 5-hydroxycytosine | 5-hydroxyuracil

5,6-dihydrouracil | 5,6-dihydroxyuracil | 5,6-dihydroxycytosine | 5-hydroxyhydantoin | alloxan

8-hydroxyadenine | 4,6-diamino-5-formamidopyrimidine | 2-hydroxyadenine | 8-hydroxyguanine | 2,6-diamino-4-hydroxy-5-formamidopyrimidine

Figure 7. Structures of modified DNA bases.

2′-deoxyadenosine have been observed in DNA (Dizdaroglu, 1986; Dirksen et al., 1988). Oxygen inhibits the formation of these compounds by reacting with the C5′-centered sugar radical.

Hydroxyl radical reacts with 2′-deoxyribose of DNA by abstracting H-atoms from all five carbon atoms. Reactions of thus-formed sugar radicals give rise to altered sugars, which are either released from the DNA chain or bound to DNA with one or both phosphate linkages. An extensive review of mechanistic aspects of sugar radical reactions and resulting products can be found elsewhere (von Sonntag, 1987).

3.2. DNA-Protein Cross-Links

Hydroxyl radical produces DNA-protein cross-links in cells or in nucleoprotein *in vitro* exposed to free radical-producing systems (Mee and Adelstein, 1981; Oleinick et al., 1987; Lesko et al., 1982; Nackerdien et al., 1991b). DNA-protein cross-links may be formed by combination of a DNA radical and a protein radical, or by addition of a DNA radical to an aromatic amino acid in proteins (or addition of a protein radical to a DNA base). A thymine-tyrosine cross-link was identified in mammalian chromatin *in vitro* or in cells exposed to free radical-generating systems (Dizdaroglu et al., 1989; Nackerdien et al., 1991b; Olinski et al., 1992; Toyokuni et al., 1995; Altman et al., 1995). The chemical structure of the thymine-tyrosine cross-link was elucidated by Margolis et al. (1988). Recently, Charlton et al. (1997) confirmed the structure of the thymine-tyrosine cross-link and described its synthesis. Other DNA-protein cross-links involving thymine and aliphatic amino acids, and cytosine and tyrosine were also identified in nucleoprotein exposed to ionizing radiation in the absence of oxygen (Gajewski et al., 1988; Dizdaroglu and Gajewski, 1989; Gajewski and Dizdaroglu, 1990).

4. MEASUREMENT OF DNA PRODUCTS

Various analytical techniques including immunochemical techniques, postlabeling assays, NMR spectroscopy, and HPLC with radioactivity, absorbance and electrochemical (EC) measurements have been used extensively to chemically characterize and quantify free radical-induced products of DNA. It is beyond the scope of this article to review the vast literature in the field of measurements by the techniques mentioned above. When using these techniques, the detection of a product is carried out with no specific structural evidence. Generally, one product or a limited number of products are measured at a time. The measurement of a single product such 8-hydroxy-2'-deoxyguanosine (8-OH-dG) as done by HPLC-EC may be misleading because oxidative damage may generate many lesions in DNA at the same time as was outlined in the previous sections of this article. On the other hand, the technique of gas chromatography-mass spectrometry (GC/MS) is capable of positive identification and quantification of a large number of base products from all four DNA bases in a single DNA sample at the same time (Dizdaroglu, 1985a, 1991, 1994). Sugar products in DNA and DNA-protein cross-links can also be identified and quantified by this technique (for a review see Dizdaroglu, 1991). The application of GC/MS to the measurement of oxidative DNA base damage is briefly discussed below.

4.1. Measurement of DNA Base Products by GC/MS

More than a decade, GC/MS has been shown to be a well-suited technique for the measurement of free radical-induced base damage to DNA (Dizdaroglu, 1984,1985a).

GC/MS can be applied to DNA itself or directly to chromatin. Thus far, it is the only technique used for the measurement of multiple products from all four bases in DNA and chromatin damaged *in vitro* or *in vivo*. The yields of products and ratios of their yields to one another substantially depend on the DNA-damaging agent among other reaction conditions such as the presence or absence of oxygen (reviewed in Dizdaroglu, 1992). For this reason, the measurement of multiple products of all four DNA bases permits the precise comparison of DNA damage caused by a variety of DNA-damaging agents (Dizdaroglu, 1992), and also prevents possible misleading results obtained by measurement of a single product as done by other techniques such as HPLC-EC. An excellent example of the latter case is a recent work by Podmore et al. (1998), who clearly demonstrated the necessity of measurement of multiple products at the same time. They found both a pro-oxidant and antioxidant behavior of vitamin C *in vivo* by measurement of 8-OH-Gua and 8-OH-Ade using GC/MS. 8-OH-Gua, which is widely measured as its nucleoside 8-OH-dG by HPLC-EC, was found to decrease whereas 8-OH-Ade increased. The measurement of 8-OH-dG only would have led to misleading results on the activity of vitamin C *in vivo*.

For GC/MS analysis, DNA is hydrolyzed by acid to release modified bases from DNA. A mixture of enzymes is used to hydrolyze DNA to nucleosides. After hydrolysis, released bases and nucleosides are derivatized to obtain volatile derivatives. Derivatized components of hydrolysates are separated on a fused silica capillary column in the GC instrument. The column effluents are directed to the ion source of a mass spectrometer through an interface between the gas chromatograph and the mass spectrometer. Electron-ionization mass spectra of trimethylsilyl derivatives of modified bases and nucleosides of DNA are characteristic for this class of compounds and can be used for unequivocal identification (Dizdaroglu, 1984, 1985a). The mass spectra contain an intense molecular ion ($M^{+\cdot}$ ion), an intense $[M - 15]^+$ ion and other characteristic ions (reviewed in Dizdaroglu, 1991). The measurement at low concentrations (*e.g.,* low femtomole) of DNA products is performed using the technique of GC/MS with selected-ion monitoring (SIM) (Watson, 1990; Dizdaroglu, 1993). The quantification of organic compounds by GC/MS is best achieved by isotope-dilution mass spectrometry (IDMS) (Watson, 1990). The use of this technique for quantification of DNA base products has recently been introduced (Dizdaroglu, 1993). Stable isotope-labeled analogues of modified bases, which contain ^{13}C, ^{15}N and ^{2}H, are added as internal standards to DNA samples prior to hydrolysis. Reference compounds and their stable isotope-labeled analogues are available commercially (e.g., from Cambridge Isotope Laboratories) or can be synthesized (Dizdaroglu, 1993; Hamberg and Zhang, 1995; Ravanat et al., 1995; Wagner et al., 1996). The DNA amount in a given DNA sample can also be quantified by GC/MS in addition to the use of absorbance measurement by UV spectroscopy. For this purpose, 2′-deoxyguanosine-$^{15}N_5$ is routinely used in our laboratory as an internal standard for guanine in DNA. This compound yields guanine-$^{15}N_5$ upon hydrolysis, which serves as an internal standard for guanine in DNA to precisely quantify the DNA amount in samples. Additionally, Hamberg and Zhang (1995) have also reported the use of labeled guanine such as guanine-$^{13}C_3$ for quantification of DNA by GC/MS.

4.2. Facts about the Artifacts in the Measurement of DNA Base Damage by GC/MS

Recently, several papers reported an artifactual formation of a number of modified bases from intact DNA bases during derivatization of DNA hydrolysates to be analyzed by GC/MS (Hamberg and Zhang, 1995; Ravanat et al., 1995; Douki et al. 1996). These reports dealt with 8-hydroxyguanine, 5-hydroxycytosine, 8-hydroxyadenine, 5-hydroxymethyluracil and 5-formyluracil among 20 or so modified bases that can be measured by GC/MS (see Figure 7). The findings have been reviewed (Cadet et al. 1997; Collins et al., 1997; see also the chapter by Cadet in this book). The oxidation of guanine in DNA hydrolyzates during trimethylsilylation to give 8-OH-Gua has been reported (Hamberg and Zhang, 1995; Ravanat et al., 1995). The presence of oxygen in derivatization mixtures increased the amount of 8-OH-Gua by oxidation of guanine (Hamberg and Zhang, 1995; Ravanat et al., 1995). Purging the mixtures with nitrogen prevented oxidation of guanine, although not completely at longer derivatization times. However, no evidence was provided determining whether oxygen was completely removed from derivatization mixtures by purging with nitrogen (Ravanat et al., 1995). The procedures for hydrolysis of DNA and derivatization of DNA hydrolyzates used in these papers substantially differed from the procedures described previously for GC/MS analysis of modified DNA bases. The differences may have led to the high values observed for 8-OH-Gua following derivatization at high temperature. Furthermore, the derivatization at room temperature as was performed by Hamberg and Zhang (1995) may have been incomplete since 8-OH-Gua is not soluble under the conditions used, although trifluoroacetic acid was added to derivatization mixtures to solve the purines. Prepurification of 8-OH-Gua by HPLC or immunoaffinity chromatography prior to GC/MS led to levels of 8-OH-Gua, which were comparable to those measured by HPLC-EC (Ravanat et al., 1995). It was assumed that HPLC-EC provided correct values for 8-OH-Gua level in DNA although the drawbacks of this technique such as possible incomplete enzymatic DNA hydrolysis and other factors have been known for some time (Halliwell and Dizdaroglu, 1992; Collins et al., 1997). The levels of 8-OH-Gua were compared to only a few previous values in the literature obtained with calf thymus DNA, neglecting to cite the vast majority of measurements, which had much lower values for 8-OH-Gua amounts obtained with DNA isolated from cultured cells or tissues (e.g., Malins and Haimanot, 1990, 1991; Dizdaroglu et al., 1991a; Nackerdien et al., 1992; Olinski et al., 1992a; Kasprzak et al., 1992; Mori et al., 1993; Malins and Gunselman, 1994; Toyokuni et al., 1994; Mori and Dizdaroglu, 1994; Olinski et al. 1995). Some of the later work done using GC/MS (Jaruga and Dizdaroglu, 1996; Kasprzak et al., 1997) also reported levels of 8-OH-Gua, which were close to the values reported by Hamberg and Zhang (1995) and Ravanat et al. (1995). Other researchers measured the background level of 8-OH-Gua by other techniques and without derivatization and reported values similar to those reported originally using GC/MS (Finnegan et al., 1995; Herbert et al., 1996; Kaur and Halliwell, 1996). For example, Finnigan et al. (1995) and Herbert et al. (1996) used guanase digestion to remove guanine from formic acid-hydrolyzates of DNA and then applied GC/MS and HPLC-EC to measure 8-OH-Gua. The values obtained for untreated calf thymus DNA or γ-irradiated DNA by both techniques were very similar, but these values were about 4.5 times higher than the level of 8-OH-dGuo measured by HPLC following enzymatic hydrolysis. These facts strongly indicates the possibility that the enzymes do not completely release 8-OH-dGuo from DNA (Halliwell and Dizdaroglu, 1992; Collins et al., 1997).

An artifactual formation of 5-hydroxycytosine, 8-hydroxyadenine and 5-hydroxymethyluracil during derivatization for GC/MS analysis has been reported (Douki et al., 1996; reviewed in Cadet et al., 1997). Removal of intact DNA bases by prepurification of calf thymus DNA hydrolyzates using HPLC was shown to prevent artifactual formation of these modified bases during derivatization. The levels found before and after prepurification are given in Table 1 along with levels of these modified bases measured in our laboratory or by other groups in DNA isolated from cultured cells or tissues. Douki et al. (1996), however, did not follow the same procedures for hydrolysis and derivatization de-

Table 1. Levels of four DNA lesions (number of lesions/10^6 DNA bases) as reported in the literature

	5-OH-Cyt	8-OH-Ade	8-OH-Gua	5-OHMeUra
Douki et al. (1996), Cadet et al (1997)				
direct GC/MS	70	157	410	31
prepurification, and then GC/MS	30	12	25	-
Podmore et al. (1998)	-	16	67	-
Kasprzak et al. (1997)	14, 16	17, 20	18, 22	6, 12
Lui et al. (1996)	15	15	70	-
Herbert et al. (1996)	-	-	50	-
Olinski et al. (1996)	22	16	48	-
Jaruga and Dizdaroglu (1996)	4	10	26	2
Zastawny et al. (1995a)	22	22	90	10
Olinski et al. (1995)	32	38	32	22
Mori and Dizdaroglu (1994)	20	5	22	14
Kasprzak et al. (1994)	10, 17, 27	11, 16, 18	30, 22, 48	1, 15, 1
Toyokuni et al. (1994)	32	26	26	8
Altman et al. (1994)	19	38	96	-
Malins and Gunselman (1994)	-	25	45	-
Mori et al. (1993)	8	16	37	7
Dizdaroglu et al. (1993b)	10	6	60	7
Kasprzak et al. (1992)	5, 5, 12	10, 6, 20	25, 26, 29	4, 5, 6
Olinski et al. (1992a)	7, 10	4, 7	9, 13	4
Nackerdien et al. (1992)	24	29	77	7
Dizdaroglu et al. (1991)	10	27	35	3
Malins and Haimonot (1990, 1991)	-	6, 10	32, 30	-

scribed in the literature for analysis of these and other modified bases in DNA by GC/MS. In their discussion, these authors compared the levels of 8-hydroxyadenine and 5-hydroxycytosine measured by GC/MS prior to prepurification to only two values given in the literature. Unfortunately, they ignored and failed to cite a large number of relevant papers where the levels of these modified bases in DNA of various sources had been reported. The same holds true for the review by Cadet et al. (1997). Some of the values in the literature are given in Table 1 and compared to the levels reported by Douki et al. (1996) and reviewed by Cadet et al. (1997). Interestingly, the levels of 8-hydroxyadenine and 5-hydroxycytosine reported in the literature were not as high as those reported by Douki et al. (1996) prior to prepurification. Most values for the level of 5-hydroxycytosine were even lower than the level measured by Douki et al. (1996) after prepurification. Levels of 8-hydroxyadenine were never as high as 157 lesions/10^6 DNA bases, either, but they were quite close to, or even the same as, or smaller than the value given by Douki et al. (1996) after prepurification. The same holds true for 5-hydroxymethyluracil and 8-OH-Gua as the values in Table 1 clearly show. In fact, all the levels of 5-hydroxymethyluracil in the literature shown in Table 1 were smaller than that measured by Douki et al. (1996) prior to prepurification. The levels of 8-OH-Gua were not as high as 410 lesions/10^6 DNA bases. A number of these values were quite close to those obtained by Douki et al. (1996) after prepurification. 5-Formyluracil was not measured in our laboratory. Thus it is not included in Table 1. All these facts raise the question of the validity of the claims about the measurement of DNA base damage by GC/MS presented in the paper by Douki et al. (1996) and in the review by Cadet et al. (1997).

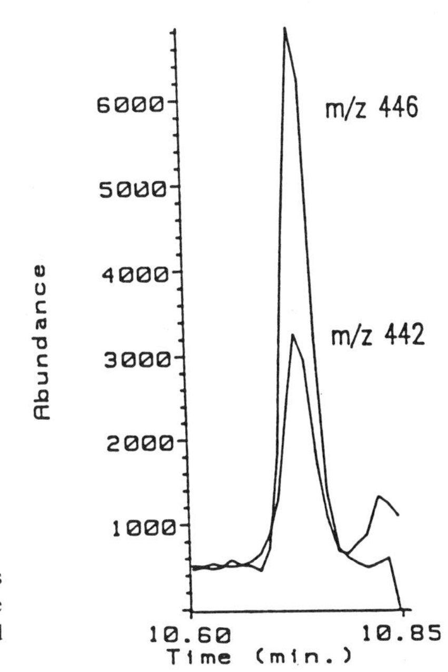

Figure 8. Ion-current profiles of the trimethylsilyl derivatives of FapyGua (m/z 442) and its stable isotope-labeled analogue (m/z 446) recorded during GC/IDMS analysis of γ-irradiated chromatin (from Dizdaroglu, 1993).

A recent paper reported a complete destruction of 2,6-diamino-4-hydroxy-5-formamidopyrimidine (FapyGua) and 4,6-diamino-5-formamidopyrimidine (FapyAde) by formic acid (88% or 60%) under the conditions of DNA hydrolysis (Douki et al., 1997; reviewed in Cadet et al., 1997). The complete destruction of FapyGua and FapyAde by formic acid is in complete disagreement with the data on these compounds in the literature. These two compounds were measured by GC/MS following formic acid hydrolysis for many years in our laboratory and by other researchers with no difficulties. Even the first publication on the application of GC/MS to the measurement of oxidative DNA base damage contained GC-profiles of these compounds and the mass spectra of their trimethylsilyl derivatives recorded during GC/MS analysis (Dizdaroglu, 1985a). Subsequent works reported the formation and measurement of these compounds in many systems including in isolated DNA, isolated chromatin, or in DNA isolated from cultured cells or tissues. It is the beyond the scope of this review article to cite all those works, but a few examples will be presented to clarify some recent misleading claims (Douki et al., 1996; Cadet et al., 1997). Figure 8 illustrates the ion-current profiles of the TMS derivatives of FapyGua and its stable isotope-labeled analogue recorded during GC/MS analysis of a formic acid-hydrolysate of γ-irradiated chromatin (Dizdaroglu, 1993). A dose-yield plot of FapyGua formed in chromatin is shown in Figure 9. These data are in contrast to the claim by Douki et al. (1997) that the only information on the formation of FapyGua in γ-irradiated DNA had been done without the use of the labeled internal standard by citing an earlier work from our laboratory (Fuciarelli et al., 1990). Also, there exists another work in the literature that showed the formation of FapyGua and other purine and pyrimidine lesions in isolated chromatin by exposure to γ-irradiation under various gaseous conditions (Gajewski et al., 1990). The formation of FapyGua in cultured human cells exposed to ionizing radiation has also been demonstrated (Nackerdien et al., 1992), as Figure 10 illustrates. FapyGua and FapyAde were shown to be formed in the DNA of wild fish exposed to toxic chemicals (Malins and Gunselman, 1994). Recently, we demonstrated the formation and

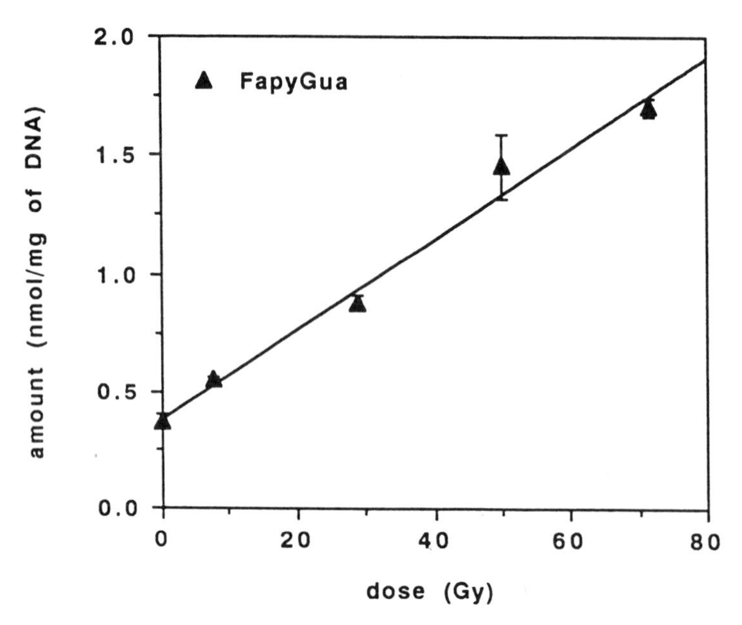

Figure 9. Radiation dose-yield plot of FapyGua formed in γ-irradiated chromatin (from Dizdaroglu, 1993).

cellular repair of FapyGua and FapyAde, and other lesions in cultured human cells and mouse forebrain using GC/MS with formic acid hydrolysis of cellular DNA (Jaruga and Dizdaroglu, 1996; Lui et al., 1996). Figure 11 illustrates formation by ischemia-reperfusion and repair of FapyGua in mouse forebrain. In similar studies, Spencer et al. (1996) and Abalea et al. (1998) showed formation and cellular repair of these compounds among other lesions in human respiratory tract epithelial cells and in primary rat hepatocyte cultures, respectively. There are many other examples of the application of GC/MS to the measurement of these compounds with no difficulty following formic acid hydrolysis of DNA. The hydrolysis procedure used by Douki et al. (1997) included the removal of formic acid under vacuum at room temperature, whereas other studies used lyophilization at low temperature for removal of formic acid. Unfortunately, Douki et al. (1997) neglected to mention the differences between the procedures, and also failed to cite many studies in the literature that successfully measured FapyGua and FapyAde by GC/MS using formic acid hydrolysis. These facts clearly raises the question of validity of the claims made in the paper by Douki et al. (1997) and in the review article by Cadet et al. (1997) about the previous measurements of FapyGua and FapyAde by GC/MS. At present, it is not clear as to why Douki et al. were not able to prevent the destruction of these compounds during formic acid hydrolysis.

It should be pointed out that statistically significant differences in the levels of any base product observed between control and treated samples should not be due to any artifactual formation even if an artifactual formation of the base product occurs during analysis. This fact was well recognized by Ravanat et al. (1995), but was ignored by Douki et al. (1996) and Cadet et al. (1997). The background levels of modified bases may also depend on the isolation procedure of DNA from cells.

On another note, Ward (see the chapter by J. Ward in this book) calculated the yield of damaged bases in mammalian cells using the yield of DNA single strand breaks and "making estimates of the relative amounts of sugar damage and strand break damage from

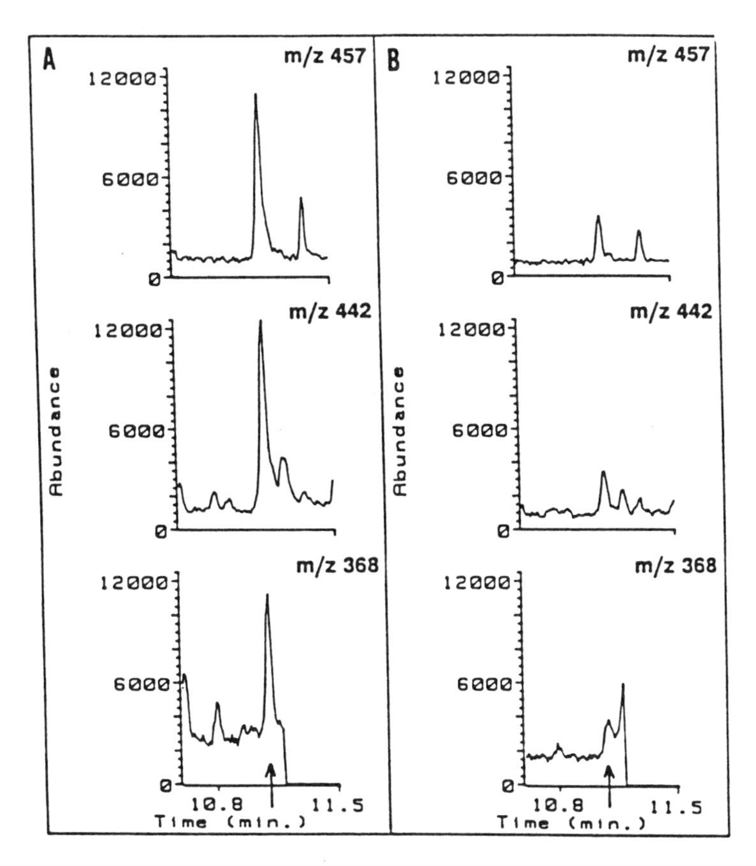

Figure 10. Ion-current profiles of characteristic ions of the trimethylsilyl derivatives of FapyGua recorded during GC/IDMS analysis of chromatin isolated from γ-irradiated human cells (A) and from unirradiated cells (B) (from Nackerdien et al., 1992).

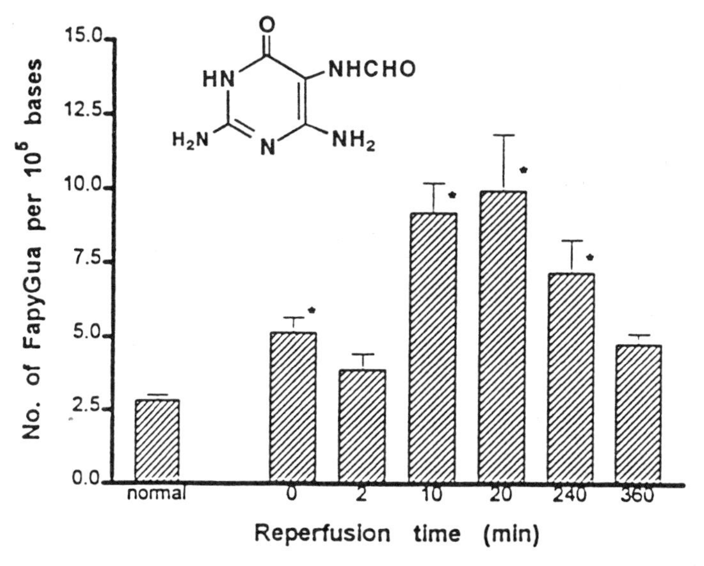

Figure 11. Formation by ischemia-reperfusion and repair of FapyGua in mouse forebrain (from Lui et al., 1996).

direct ionization of DNA and from •OH reactions." The calculated yield was comparable to the actual yields of DNA base damage in mammalian cells, which were measured by GC/MS in our laboratory (Nackerdien et al., 1992; Mori et al., 1993; Mori and Dizdaroglu, 1994).

Recently, DNA repair enzymes such as *E.coli* Fpg protein and Nth protein (endonuclease III) have been used for measurement of DNA base damage (Häring et al., 1994; Pflaum et al., 1997; Collins et al., 1997). These enzymes generate abasic sites in DNA by removing modified purines and pyrimidines. Subsequently, abasic sites are measured by alkaline elution, alkaline unwinding or comet assays. This method allows quantification of substrates of Fpg and Nth proteins without specifically determining the lesions involved. Furthermore, there is no clear evidence that a complete removal of DNA lesions is achieved. Interestingly, the levels of DNA lesions measured by this method appear be at least 10-fold lower than those measured by HPLC-EC (reviewed in Collins et al., 1997).

It is clear that different results may be obtained in different laboratories. Furthermore, different techniques may provide different results as was discussed here and elsewhere (Halliwell and Dizdaroglu, 1992; Collins et al., 1997). Data obtained in various laboratories should be compared to one another by discussing all relevant data in the literature and all scientific facts including experimental differences between laboratories without ignoring the vast majority of previous work.

4.3. Measurement of DNA Repair by GC/IDMS

The concept of measurement of DNA repair by GC/MS was introduced by us more a decade ago (Dizdaroglu, 1985a). Since then, GC/MS has extensively been used for the study of substrate specificities DNA glycosylases such as *E. coli* Fpg protein, Nth protein, T4 endonuclease V, *E. coli* and human uracil glycosylases, *S. pombe* and human Nth proteins, *S. cerevisiae* Ogg1 protein and *Drosophila* ribosomal S3 protein (Boiteux et al., 1992; Dizdaroglu et al., 1993; Zastawny et al., 1995b; Dizdaroglu et al., 1996a,b; Karakaya et al., 1997; Deutsch et al., 1997; Karahalil et al., 1998a,b). Since GC/MS with isotope-dilution technique can simultaneously identify and quantify a multitude of DNA base lesions in the same DNA sample, it facilitates the determination of lesions that are excised or not excised from DNA by a given DNA glycosylase under the same conditions. This technique also permits the simultaneous measurement of excision rates of substrate lesions from DNA (Karakaya et al., 1997; Deutsch et al., 1997; Karahalil et al.,1998a,b). Such repair experiments involving DNA with multiple lesions may be more representative of the circumstances in a cell than those with an oligonucleotide containing a single lesion, as was generally done in the past.

DNA containing a large number of lesions is incubated with a DNA glycosylase and then precipitated with ethanol. DNA pellets and supernatant fractions are separated. Pellets are analyzed by GC/IDMS after hydrolysis and supernatant fractions without hydrolysis because a DNA glycosylase releases free modified bases. Removal of substrate lesions from pellets and appearance of lesions in supernatant fractions unequivocally prove the activity of the enzyme on corresponding lesions. Typical Michaelis-Menten kinetics of excision and temperature dependence of excision can also be determined (Karakaya et al., 1997; Deutsch et al., 1997; Karahalil et al., 1998a,b). Figure 12 illustrates excision of 8-OH-Gua and FapyGua from γ-irradiated DNA by *Drosophila* ribosomal S3 protein. Here, the amounts removed from the pellet fractions by the enzyme are accounted for by the appearance of these lesions in the supernatant fractions of the same samples, unequivocally proving that 8-OH-Gua and FapyGua are substrates for this enzyme. As a further example,

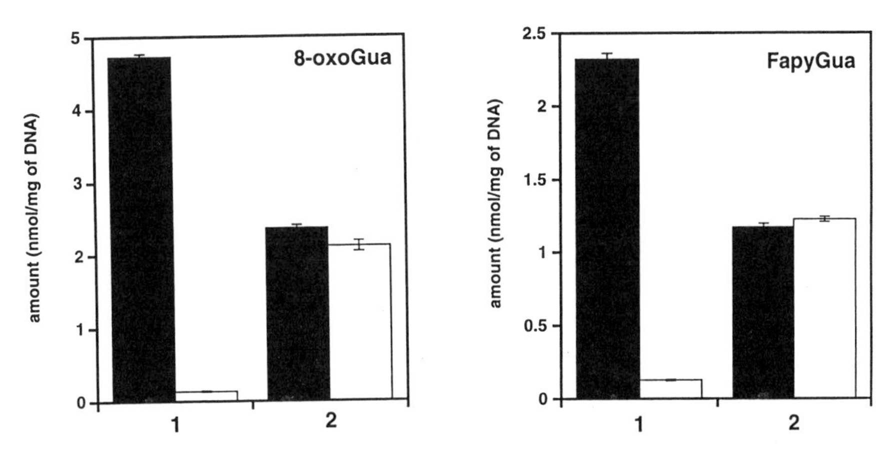

Figure 12. Excision of 8-OH-Guand and FapyGua by *Drosophila* ribosomal S3 protein from γ-irradiated DNA. Dark columns, pellets; light column, supernatant fractions (from Deutsch et al., 1997).

Lineweaver-Burk plots for excision of 5-hydroxy-6-hydrothymine (5-OH-6-HThy) and 5-hydroxycytosine (5-OH-Cyt) from DNA by *S. pombe* Nth protein are illustrated in Figure 13.

The GC/IDMS technique has also been applied to the determination of DNA repair in cells (Jaruga and Dizdaroglu, 1996; Liu et al., 1996; Spencer et al., 1996; Abelea et al., 1998). Figure 14 illustrates formation and repair of FapyGua and 8-OH-Gua in human cells exposed to H_2O_2. Half-lives of repair of these and other DNA lesions in human cells were determined (Jaruga and Dizdaroglu, 1996).

A recent review article (Cadet et al., 1997) questioned the validity of the determination of substrate specificities of DNA glycosylases by GC/MS, without giving any data to the contrary. As was explained above, the excision of a lesion by a given DNA glycosylase

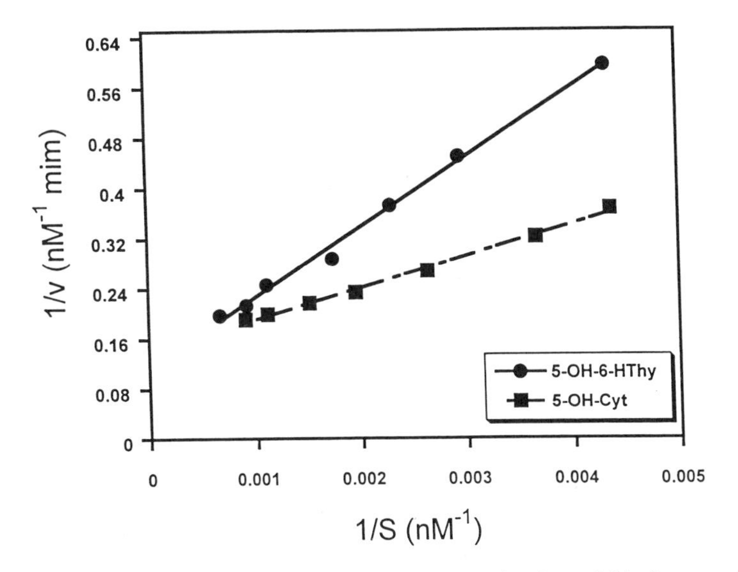

Figure 13. Lineweaver-Burk plot for excision of 5-hydroxy-6-hydrothymine and 5-hydroxycytosine from γ-irradiated DNA (from Karahalil et al., 1998a).

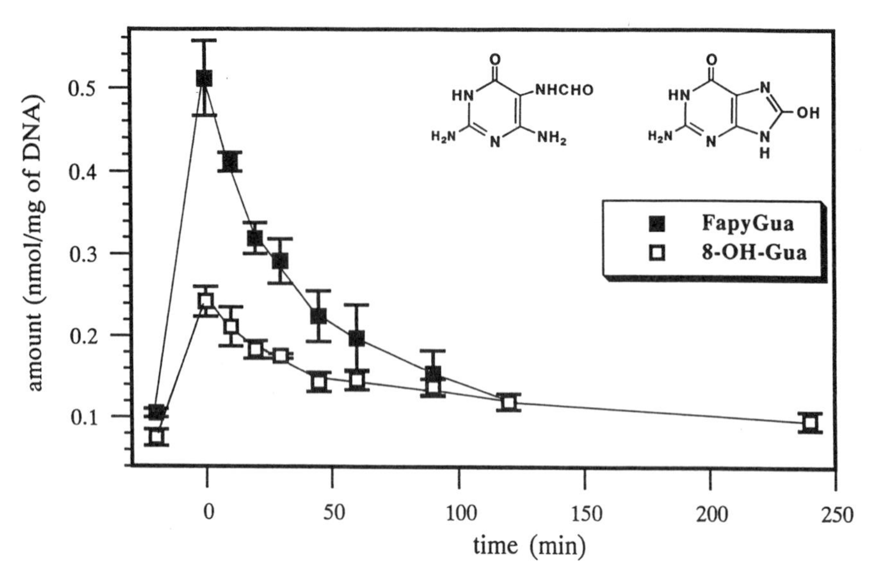

Figure 14. Formation and repair of FapyGua and 8-OH-Gua in human cells exposed to H_2O_2 as determined by GC/IDMS (from Jaruga and Dizdaroglu, 1996).

and kinetics of excision are determined by measurement of the amounts of the lesion in supernatant fractions. These fractions do not contain DNA, which is precipitated by ethanol. Furthermore, supernatant fractions are not hydrolyzed, but derivatized following lyophilization to be analyzed by GC/IDMS. Therefore, it is not possible to artifactually generate any of substrate lesions that are quantified. Only the lesions that are excised by the enzyme are detected and quantified by this procedure as was clearly explained in the relevant published papers.

5. CONCLUSIONS

Free radicals such as •OH, e_{aq}^- and H atoms react with components of DNA by addition or abstraction. Addition of oxygen reacts to base or sugar radicals of DNA results in peroxyl radicals. DNA radicals further react to give a large number of final products by a variety of mechanisms. Products and their yields depend on the type of radicals, their reaction partners and reaction conditions. DNA radicals react with proteins or protein radicals to yield DNA-protein cross-links in chromatin. Various analytical techniques exist to identify and quantify DNA base and sugar products and DNA-protein cross-links. Different techniques do not always provide similar results. Products identified in DNA *in vitro* are also formed in the DNA of cultured cells and mammalian tissues *in vivo* upon treatment by free radical-generating agents.

ACKNOWLEDGMENT

Certain commercial equipment or materials are identified in this paper in order to specify adequately the experimental procedures. Such identification does not imply recommendation or endorsement by the National Institute of Standards and Technology, nor does it imply that the materials or equipment identified are necessarily the best available for the purpose.

REFERENCES

V. Abalea, J. Cillard, M.-P. Dubos, J.-P. Anger, P. Cillard and I. Moral (1998) Iron-induced oxidative DNA damage and its repair in primary rat hepatocyte culture. *Carcinogenesis* **19**, 1053–1059.

S. A. Altman, T. H. Zastawny, L. Randers, J. Remacle, M. Dizdaroglu and R. Rao (1994) *tert*-Butyl hydroperoxide-mediated DNA base damage in cultured mammalian cells. *Mutat. Res.* **306,** 35–44.

S. A. Altman, T. H. Zastawny, L. Randers-Eichhorn, M. A. Cacciuttolo, S. A. Akman, M. Dizdaroglu and R. Rao (1995) Formation of DNA-protein cross-links in cultured mammalian cells upon treatment with iron ions. *Free Rad. Biol. Med.* **19,** 897–902.

A. P. Breen and J. A. Murphy (1995) Reactions of oxyl radicals with DNA. *Free Rad. Biol. Med.* **18**, 1033–1077.

S. Boiteux, E. Gajewski, J. Laval and M. Dizdaroglu (1992) Substrate specificity of the *Escherichia coli* Fpg protein (formamidopyrimidine-DNA glycosylase): excision of purine lesions in DNA produced by ionizing radiation or photosensitization, *Biochemistry* **31,** 106–110.

A. Bonicel, N. Mariaggi, E. Hughes and R. Téoule, R. (1980) *In vitro* γ-irradiation of DNA: Identification of radioinduced chemical modifications of the adenine moiety. *Rad. Res.* **83**, 19–26.

G. Buchko, J. Cadet, J.-L. Ravanat and P. Labataille (1993) Isolation and characterization of a new product produced by ionizing radiation and type I photosensitization of 2'-deoxyguanosine in oxygen-saturated aqueous solution: (2S)-2,5'-anhydro-1-(2'-deoxy-β-D-erythro-pentofuranosyl)-5-guanidinylidine-2-hydroxy-4-oxoimidazolidine. *Int. J. Rad. Biol.* **63,** 669–676.

J. Cadet, T. Douki and J.-L. Ravanat (1997) Artifacts associated with the measurement of oxidized DNA bases. *Envir. Health Pers.* **105**, 1034–1039.

T. S. Charlton, D. Black, D. C. Craid, K. E. Mason, P. G. Harris and M. W. Duncan (1998) Synthesis and quantification of a thymine-tyrosine adduct. *Free Rad. Biol. Med.* (in press).

C. J. Chetsanga and C. Grigorian (1983) A dose-response study on opening of imidazole ring of adenine in DNA by ionizing radiation. *Int. J. Rad. Biol.* **44,** 321–331.

A. Collins, J. Cadet, B. Epe and C. Gedik, C. (1997) Problems in the measurement of 8-oxoguanine in human DNA. Report of a workshop, DNA oxidation, held in Aberdeen, UK, 19–21 January, 1997. *Carcinogenesis* **18**, 1833–1836.

W. A. Deutsch, A. Yacoub, P. Jaruga, T. H. Zastawny and M. Dizdaroglu (1997) Characterization and mechanism of action of *Drosophila* ribosomal protein S3 DNA glycosylase activity for the removal of oxidatively damaged DNA bases. *J. Biol. Chem.* **272**, 32857–32860.

M.-L. Dirksen, W. F. Blakely, E. Holwitt and M. Dizdaroglu (1988) Effect of DNA conformation on hydroxyl radical-induced formation of 8,5′-cyclopurine 2′-deoxynucleoside residues in DNA. *Int. J. Rad. Biol.* **54**, 195–204.

M. Dizdaroglu (1984) The use of capillary gas chromatography-mass spectrometry for identification of radiation-induced DNA base damage and DNA base-amino acid cross-links. *J. Chromatogr.* **295**, 103–121.

M. Dizdaroglu (1985a) Application of capillary gas chromatography-mass spectrometry to chemical characterization of radiation-induced base damage in DNA implications for assessing DNA repair processes. *Anal. Biochem.* **144**, 593–603.

M. Dizdaroglu (1985b) Formation of an 8-hydroxyguanine moiety in deoxyribonucleic acid on γ-irradiation in aqueous solution. *Biochemistry* **24**, 4476–4481.

M. Dizdaroglu (1986) Free radical-induced formation of an 8,5′-cyclo-2′-deoxyguanosine moiety in deoxyribonucleic acid. *Biochem. J.* **238**, 247–254.

M. Dizdaroglu (1991) Chemical determination of free radical-induced damage to DNA. *Free Rad. Biol. Med.* **10**, 225–242.

M. Dizdaroglu (1992) Oxidative damage to DNA in mammalian chromatin. *Mut. Res.* **275**, 331–342.

M. Dizdaroglu (1993) Quantitative determination of oxidative base damage in DNA by stable isotope-dilution mass spectrometry. *FEBS Lett.* **315,** 1–6.

M. Dizdaroglu (1994) Chemical determination of oxidative DNA damage by gas chromatography-mass spectrometry. *Methods Enzymol.* **3,** 3–16.

M. Dizdaroglu and E. Gajewski (1989) Structure and mechanism of hydroxyl radical-induced formation of a DNA-protein cross-link involving thymine and lysine in nucleohistone. *Cancer Res.*, **49**, 3463–3467.

M. Dizdaroglu, E. Gajewski, P. Reddy and S. A. Margolis (1989) Structure of a hydroxyl radical induced DNA-protein cross-link involving thymine and tyrosine in nucleohistone. *Biochemistry* **28**, 3625–3628.

M. Dizdaroglu, E. Holwitt, M. P. Hagan and W. F. Blakely (1986) Formation of cytosine glycol and 5,6-dihydroxycytosine in deoxyribonucleic acid on treatment with osmium tetroxide. *Biochem. J.* **235,** 531–536.

M. Dizdaroglu, A. Karakaya, P. Jaruga, G. Slupphaug and H. E. Krokan (1996b) Novel Activities of Human Uracil DNA *N*-Glycosylase for Cytosine-Derived Products of Oxidative DNA Damage. *Nucl. Acids Res.* **24,** 418–422.

M. Dizdaroglu, J. Laval and S. Boiteux (1993a) Substrate specificity of *Escherichia coli* endonuclease III: Excision of thymine- and cytosine-derived lesions in DNA produced by radiation-generated free radicals. *Biochemistry* **32,** 12105–12111.

M. Dizdaroglu, Z. Nackerdien, B.-C. Chao, E. Gajewski and G. Rao (1991) Chemical nature of in vivo DNA base damage in hydrogen peroxide-treated mammalian cells. *Arch. Biochem. Biophys.* **285**, 388–390.

M. Dizdaroglu, R. Olinski, J. H. Doroshow and S. A. Akman (1993b) Modification of DNA bases in chromatin of intact target human cells by activated human polymorphonuclear leukocytes. *Cancer Res.* **53,** 1269–1272.

M. Dizdaroglu, T. H. Zastawny, J. R. Carmical and R. S. Lloyd (1996a) A Novel DNA *N*-Glycosylase Activity of *E. coli* T4 Endonuclease V that Excises 4,6-Diamino-5- formamidopyrimidine from DNA, a UV-Radiation- and Hydroxyl Radical-induced Product of Adenine. *Mut. Res.* **362,** 1–8.

T. Douki, T. Delatour, F. Bianchini, F. and J. Cadet, J. (1996) Observation and prevention of an artifactual formation of oxidized DNA bases and nucleosides in the GC-EIMS method. *Carcinogenesis* **17,** 347–353.

T. Douki, R. Martini, J.-L. Ravanat, R. J. Turesky and J. Cadet (1997) Measurement of 2,6-diamino-4-hydroxy-5-formamidopyrimidine and 8-oxo-7,8-dihydroguanine in isolated DNA exposed to gamma radiation in aqueous solution. *Carcinogenesis* **18,** 2385–2391.

M. T. Finnegan, K. E. Herbert, M. D. Evans, S. Farooq, P. Farmer, I. D. Podmore and Lunec, J. (1995) Development of an assay to measure 8-oxoguanine using HPLC with electrochemical detection. *Biochem. Soc. Trans.* **23,** 431S.

A. F. Fuciarelli, G. G. Miller and J. A. Raleigh (1985) An immunochemical probe for 8,5'-cycloadenosine-5'-monophosphate and its deoxy analog in irradiated nucleic acids. *Rad. Res.* **104**, 272–283.

A. F. Fuciarelli, B. J. Wegher, W. F. Blakely and M. Dizdaroglu (1990) Yields of radiation-induced base products in DNA: effects of DNA conformation and gassing conditions. *Int. J. Rad. Biol.* **58**, 397–415.

S. Fujita and S. Steenken (1981) Pattern of OH radical addition to uracil and methyl- and carboxyl-substituted uracils. Electron transfer of OH adducts with N,N,N′,N′-tetramethyl-p-phenylenediamine and tetranitromethane. *J. Amer. Chem. Soc.* **103**, 2540–2545.

E. Gajewski and M. Dizdaroglu (1990) Hydroxyl radical induced cross-linking of cytosine and tyrosine in nucleohistone. *Biochemistry* **29**, 977–980.

E. Gajewski, A. F. Fuciarelli and M. Dizdaroglu (1988) Structure of hydroxyl radical-Induced DNA-protein cross-links in calf thymus nucleohistone *in vitro*. *Int. J. Rad. Biol.* **54**, 445–459.

E. Gajewski, G. Rao, Z. Nackerdien and M. Dizdaroglu (1990) Modification of DNA bases in mammalian chromatin by radiation-generated free radicals. *Biochemistry* **29**, 7876–7882.

B. Halliwell and M. Dizdaroglu (1992) Commentary: The measurement of oxidative damage to DNA by HPLC and GC/MS techniques. *Free Rad. Res. Commun.* **16,** 75–87.

B. Halliwell and J. M. C. Gutteridge (1990) Role of free radicals and catalytic metal ions in human disease: an overview. *Methods Enzymol.* **186**, 1–85.

M. Hamberg and L.-Y. Zhang (1995) Quantitative determination of 8-hydroxyguanine and guanine by isotope dilution mass spectrometry, *Anal. Biochem.* **229,** 336–344.

M. Häring, H. Rüdiger, B. Demple, S. Boiteux and B. Epe (1994) Recognition of oxidized abasic sites by repair endonucleases, *Nucl. Acids Res.* **22,** 2010–2015.

D. K. Hazra and S. Steenken (1983) Pattern of OH radical addition to cytosine and 1-, 3-, 5-, and 6-substituted cytosines. Electron transfer and dehydration reactions of the OH adducts. *J. Amer. Chem. Soc.* **105,** 4380–4386.

G. Hems (1960) Chemical effects of ionizing radiation on deoxyribonucleic acid in dilute aqueous solution. *Nature* **186**, 710–712.

K. E. Herbert, M. D. Evans, M. T. Finnegan, S. Farooq, N. Mistry, I. D. Podmore, P. Farmer and J. Lunec (1996) A novel HPLC procedure for the analysis of 8-oxoguanine in DNA. *Free Rad. Biol. Med.* **20,** 467–473.

A. Hissung and C. von Sonntag (1979) The reaction of solvated electrons with cytosine, 5-methylcytosine and 2′-deoxycytidine in aqueous solutions. The reaction of the electron adduct intermediates with water, p-nitroacetophenone and oxygen. A pulse spectroscopic and pulse conductometric study. *Int. J. Rad. Biol.* **35,** 449–458.

P. Jaruga and M. Dizdaroglu (1996) Repair of products of oxidative DNA base damage in human cells. *Nucl. Acids Res.* **24,** 1389–1394.

P. Jaruga, T. H. Zastawny, J. Skokowski, M. Dizdaroglu and R. Olinski (1994) Oxidative DNA base damage and antioxidant enzyme activities in human lung cancer. *FEBS Lett.* **341,** 59–64.

H. Kamiya and H. Kasai (1995) Formation of 2'-hydroxydeoxyadenosine triphosphate, an oxidatively damaged nucleotide, and its incorporation by DNA polymerases. *J. Biol. Chem.* **270**, 19446–9450.

A. Karakaya, P. Jaruga, V. A. Bohr, A. P. Grollman and M. Dizdaroglu (1997) Kinetics of Excision of Purine Lesions from DNA by Fpg Protein. *Nucl. Acids Res.*, **25,** 474–479.

B. Karahalil, T. Roldán-Arjona and M. Dizdaroglu (1998a) Substrate Specificity of *Schizosaccharomyces pombe* Nth protein for products of oxidative DNA damage. *Biochemistry* **37**, 590–595.

B. Karahalil, P. -M. Girard, S. Boiteux and M. Dizdaroglu (1998b) Substrate Specificity of the Oggl protein of *Saccharomyces cerevisiae*: Excision of guanine lesions produced in DNA by ionizing radiation- or hydrogen peroxyde/metal ion-generated free radicals. *Nucl. Acids Res.* **26**, 1228–1232.

H. Kasai, A. Iida, Z. Yamaizmi, S. Nishimura and H. Tanooka, H. (1990) 5-Formyldeoxyuridine: A new type of DNA damage induced by ionizing radiation and its mutagenicity to Salmonella strain TA102. *Mutat. Res.* **243**, 249–253.

H. Kasai, H. Tanooka and S. Nishimura (1984) Formation of 8-hydroxyguanine residues in DNA by X-irradiation. *Gann* **75**, 1037–1039.

H. Kasai, Z.Yamaizumi, M. Berger and J. Cadet (1992) Photosensitized formation of 7,8- dihydro-8-oxo-2'-deoxyguanosine (8-hydroxy-2'-deoxyguanosine) in DNA by riboflavin. A non singlet oxygen mediated reaction. *J. Amer. Chem. Soc.* **114,** 9692–9694.

K. S. Kasprzak, B. A. Diwan, J. M. Rice, M. Misra, R. Olinski and M. Dizdaroglu (1992) Nickel(II)-mediated oxidative DNA base damage in rat renal chromatin *in vivo*. Possible relevance to nickel(II) carcinogenesis. *Chem. Res. Toxicol.* **5,** 809–815.

K. S. Kasprzak, T. H. Zastawny, S. L. North, C. W. Riggs, B. A. Diwan, J. M. Rice and M. Dizdaroglu (1994) Oxidative DNA base damage in renal, hepatic, and pulmonary chromatin of rats after intraperitoneal injection of cobalt(II) acetate. *Chem. Res. Toxicol.* **7**, 329–335.

K. S. Kasprzak, P. Jaruga, T. H. Zastawny, S. L. North, C. W. Riggs, R. Olinski, R. and M. Dizdaroglu (1997) Oxidative DNA base damage and its repair in kidneys and livers of nickel(II)-treated male F344 rats. *Carcinogenesis* **18**, 271–277.

H. Kaur and B. Halliwell (1996) Measurement of oxidized and methylated DNA bases by HPLC with electrochemical detection, *Biochem. J.* **318,** 21–23.

K. Keck (1968) Bildung von Cyclonucleotiden bei Bestrahlung wässriger Lösungen von Purinnucleotiden. *Z. Naturforsch.* **23b**, 1034–1043.

D. G. E. Lemaire, E. Bothe and D. Schulte-Frohlinde (1984) Yields of radiation- induced main chain scission of poly U in aqueous solution: Strand break formation via base radicals. *Int. J. Rad. Biol.* **45**, 351–358.

S. A. Lesko, J.-L. Drocourt and S.-U. Yang (1982) Deoxyribonucleic acid-protein and deoxyribonucleic acid interstrand cross-links induced in isolated chromatin by hydrogen peroxide and ferrous ethylenediaminetetraacetate chelates. *Biochemistry* **21,** 5010–5015.

P. K. Liu, C. Y. Hsu, M. Dizdaroglu, R. A. Floyd, Y. W. Kow, A. Karakaya, L. Rabow and J.-K. Cui (1996) Damage, repair and mutagenesis in nuclear genes of the brain after forebrain ischemia and reperfusion. *J. Neuroscience* **16,** 6795–6806.

D. C. Malins and S. J. Gunselman (1994) Fourier transform infrared spectroscopy and gas chromatography mass spectrometry reveal a remarkable degree of structural damage in the DNA of wild fish exposed to toxic chemicals. *Proc. Natl. Acad. Sci. USA* **91**, 13038–13041.

D. C. Malins and R. Haimanot (1990) 4,6-Diamino5-formamidopyrimidine, 8-hydroxyguanine and 8-hydroxyadenine in DNA from neoplastic liver of English sole exposed to carcinogens. *Biochem. Biophys. Res. Commun.* **173**, 614–619.

D. C. Malins and R. Haimanot (1991) The etiology of cancer: hydroxyl radical-induced DNA lesions in histologically normal livers of fish from a population with liver tumors. Aquatic Toxicol. **20**, 123–130.

S. A. Margolis, B. Coxon, E. Gajewski and M. Dizdaroglu (1988) Structure of a hydroxyl radical induced cross-link of thymine and tyrosine. *Biochemistry* **27,** 6353–6359.

L. K. Mee and S. J. Adelstein (1981) Predominance of core histones in formation of DNA-protein cross-links in γ-irradiated chromatin. *Proc. Natl. Acad. Sci. USA* **78,** 2194- 2198.

T. Mori and M. Dizdaroglu (1994) Ionizing radiation causes greater DNA base damage in radiation-sensitive mutant M10 cells than in parent mouse lymphoma L5178Y cells. *Rad. Res.* **140,** 85–90.

T. Mori, Y. Hori and M. Dizdaroglu (1993) DNA base damage generated *in vivo* in hepatic chromatin of mice upon whole body γ-irradiation. *Int. J. Rad. Biol.* **64,** 645–650.

Z. Nackerdien, R. Olinski and M. Dizdaroglu (1992) DNA base damage in chromatin of γ-irradiated cultured human cells. *Free Rad. Res. Commun.* **16,** 259–272.

Z. Nackerdien, K. S. Kasprzak, G. Rao, B. Halliwell and M. Dizdaroglu (1991a) Nickel(II)- and cobalt(II)-dependent damage by hydrogen peroxide to the DNA bases in isolated human chromatin. *Cancer Res.* **51**, 5837–5842.

Z. Nackerdien, G. Rao, M. A. Cacciuttolo, E. Gajewski and M. Dizdaroglu (1991b) Chemical Nature of DNA-protein cross-links produced in mammalian chromatin by hydrogen peroxide in the presence of iron or copper ions. *Biochemistry* **30,** 4873–4879.

H. M. Novais and S. Steenken (1986) ESR studies of electron and hydrogen adducts of thymine and uracil and their derivatives and of 4,6-dihydroxypyrimidines in aqueous solution. Comparison with data from solid state. The protonation at carbon of the electron adducts. *J. Amer. Chem. Soc.* **108,** 1–6.

N. L. Oleinick, S. Chiu, N. Ramakrishnan and L. Xue (1987) The formation, identification, and significance of DNA-protein cross-links in mammalian cells. *Brit. J. Cancer* **55 (Suppl. VIII),** 135–140.

R. Olinski, Z. Nackerdien and M. Dizdaroglu (1992b) DNA-protein cross-linking between thymine and tyrosine in chromatin of γ-irradiated of H_2O_2-treated cultured human cells. *Arch. Biochem. Biophys.* **297**, 139–143.

R. Olinski, T. H. Zastawny, J. Budzbon, J. Skokowski, W. Zegarski and M. Dizdaroglu (1992a) DNA base modifications in chromatin of Human Cancerous Tissues. *FEBS Lett.* **309,** 193–198.

R. Olinski, T. H. Zastawny, M. Foksinski, A. Barecki and M. Dizdaroglu (1995) DNA base modifications and antioxidant enzyme activities in human benign prostatic hyperplasia. *Free Rad. Biol. Med.* **18**, 807–813.

R. Olinski, T. H. Zastawny, M. Foksinski, W. Windorbska, P. Jaruga and M. Dizdaroglu (1996) DNA base damage in lymphocytes of cancer patients undergoing radiation therapy. *Cancer Lett.* **106**, 207–215.

P. O'Neill (1983) Pulse radiolytic study of the interaction of thiols and ascorbate with OH-adducts of dGMP and dG. Implications for DNA repair processes. *Rad. Res.* **96,** 198–210.

P. O'Neill and P. W. Chapman (1985) Potential repair of free radical adducts of dGMP and dG by a series of reductants: A pulse radiolysis study. *Int. J. Rad. Biol.* **47,** 71–80.

M. Pflaum, O. Will and B. Epe (1997) Determination of steady-state levels of oxidative DNA base modifications in mammalian cells by means of repair endonucleases. *Carcinogenesis* (in press).

I. D. Podmore, H. R. Griffiths, K. E. Herbert, N. Mistry, P. Mistry and J. Lunec (1998) Vitamin C exhibits both a pro-oxidant and antioxidant behavior in vivo. *Nature* **392**, *559.*

J. A. Raleigh, W. Kremers and R. Whitehouse (1976) Radiation chemistry of nucleotides: 8,5'-cyclonucleotide formation and phosphate release initiated by hydroxyl radical attack on adenosine monophosphates. *Rad. Res.* **65**, 414–422.

J.-L. Ravanat, R. J. Turesky, E. Gremaud, L. J. Trudel and R. H. Stadler (1995) Determination of 8-oxoguanine in DNA by gas chromatography-mass spectrometry and HPLC-electrochemical detection: overestimation of the background level of the oxidized base by the gas chromatography-mass spectrometry assay. *Chem. Res. Toxicol.* **8,** 1039–1045.

J. P. E. Spencer, A. Jenner, O. I. Aruoma, C. E. Cross, R. Wu and B. Halliwell (1996) Oxidative DNA damage in human respiratory tract epithelial cells. Time course in relation to DNA strand breakage. *Biochem. Biophys. Res. Commun.* **224**, 17–22.

S. Steenken (1989) Purine bases, nucleosides, and nucleotides: Aqueous solution redox chemistry and transformation reactions of their radical cations and e^- and OH adducts. *Chem. Rev.* **89,** 503–520.

R. Téoule (1987) Radiation-induced DNA damage and its repair. *Int. J. Rad. Biol.* **51,** 573–589.

R. Téoule and J. Cadet (1978) Radiation-induced degradation of the base component in DNA and related substances-final products, in *Effects of Ionizing Radiation on DNA*, edited by J. Hüttermann, W. Köhnlein, R. Téoule and A. J. Bertinchamps (New York: Springer-Verlag), pp. 171–203.

S. Toyokuni, T. Mori and M. Dizdaroglu (1994) DNA base modifications in renal chromatin of wistar rats treated with a renal carcinogen, ferric nitrilotriacetate. *Int. J. Cancer* **57,** 123–128.

S. Toyokuni, T. Mori, H. Hiai and M. Dizdaroglu (1995) Treatment of wistar rats with a renal carcinogen, ferric nitrilotriacetate causes DNA-protein cross-linking between thymine and tyrosine in renal chromatin. *Int. J. Cancer* **62,** 309–313.

A. J. S. C. Vieira and S. Steenken (1990) Pattern of OH radical reaction with adenine and its nucleosides and nucleotides. Characterization of two types of isomeric OH adduct and their unimolecular transformation reactions. *J. Amer. Chem. Soc.* **112,** 6986–6994.

C. Von Sonntag (1987) *The Chemical Basis of Radiation Biology* (New York: Taylor & Francis), pp. 117–166, 221–294.

J. R. Wagner (1994) Analysis of oxidative cytosine products in DNA exposed to ionizing radiation. *J. Chimie Phys.* **91**, 1280–1286.

J. R. Wagner, B. C. Blount and M. Weinfeld (1996) Excision of oxidative cytosine modifications from γ-irradiated DNA by *Escherichia coli* endonuclease III and human whole-cell extracts. *Anal. Biochem.* **233**, 76–86.

J. R. Wagner, J. E. van Lier, M. Berger, M. and J. Cadet (1994) Thymidine hydroperoxides: Structural assignment, conformational features, and thermal decomposition in water. *J. Amer. Chem. Soc.* **11**, 2235–2242.

J. T. Watson (1990) Selected-ion measurements. *Methods Enzymol.* **193**, 86–106.

T. H. Zastawny, S. A. Altman, L. Randers-Eichhorn, R. Madurawe, J. A. Lumpkin, M. Dizdaroglu and G. Rao (1995a) DNA base modifications and membrane damage in cultured mammalian cells treated with iron ions. *Free Rad. Biol. Med.* **18,** 1013–1022.

T. H. Zastawny, P. W. Doetsch and M. Dizdaroglu (1995b) A novel activity of uracil DNA *N*-glycosylase: excision of isodialuric acid (5,6-dihydroxyuracil) from DNA, a major product of oxidative DNA damage. *FEBS Lett.* **364,** 255–258.

8

DROSOPHILA RIBOSOMAL PROTEIN S3 CONTAINS N-GLYCOSYLASE, ABASIC SITE, AND DEOXYRIBOPHOSPHODIESTERASE DNA REPAIR ACTIVITIES

Walter A. Deutsch

Pennington Biomedical Research Center, Louisiana State University
Baton Rouge, Louisiana 70808

1. ABSTRACT

The DNA repair activities possessed by *Drosophila* ribosomal protein S3 (dS3) are summarized in this report. Originally, the dS3 protein was found to possess AP lyase activity similar to that observed for the human homologue of S3. Subsequent tests using a heavily UV-irradiated 5′ end-labeled oligonucleotide suggested that dS3 was acting on a guanine photoproduct that was determined to be 2,6-diamino-4-hydroxy-5-formamidopyrimidine. The dS3 protein was also found to act on 5′ end-labeled oligonucleotides containing a single 8-oxoguanine residue. That dS3 was acting as an N-glycosylase to process these lesions was confirmed using DNA substrates prepared by γ-irradiation under N_2O and analyzed by gas chromatography/isotope-dilution mass spectrometry. We went on to demonstrate the *in vivo* significance of this DNA repair activity by showing the ability of dS3 to abolish completely the mutator phenotype of *Escherichia coli mutM* (Fpg-) caused by 8-oxoguanine-mediated G to T transversions. The dS3 protein was also able to rescue the alkylation sensitivity of an *E. coli* mutant defective for the hydrolytic AP endonuclease activities associated with exonuclease III and endonuclease IV. That an AP lyase could be a significant source of DNA repair activity for the repair of an AP site came from studies that determined that dS3 also possessed deoxyribophosphodiesterase activity not only for the removal of 5′-incised AP sites, but notably, it was determined that dS3 could also excise *trans*-4-hydroxy-2-pentenal-5-phosphate from substrates containing 3′ incised AP sites. Taken together, our results suggest that dS3 is able to create a one nucleotide gap for efficient filling by β polymerase by utilizing its N-glycosylase/AP lyase activity to create a 3′ terminal AP site that can then be liberated by the dRpase activity possessed by dS3.

Advances in DNA Damage and Repair, edited by Dizdaroglu and Karakaya. Kluwer Academic / Plenum Publishers, New York, 1999.

2. INTRODUCTION

Reactive free radicals can arise from ionizing radiation, chemical carcinogens, and as a consequence of normal oxygen metabolism. These free radicals are capable of interacting with DNA, and thus are responsible for the formation of a number of different DNA lesions. One of the most abundant forms of oxidative DNA damage is 7, 8-dihydro-8-oxoguanine (8-oxoG), which is highly mutagenic if allowed to persist in DNA since adenines are commonly misincorporated opposite it during DNA replication (Wood et al, 1990; Shibutani et al, 1991; Moriya et al., 1991).

The Fpg protein was the first enzyme found to efficiently remove 8-oxoG from DNA, although it was originally purified and characterized in *Escherichia coli* as an N-glycosylase for the removal of 2, 6-diamino-4-hydroxy-5-(N-methyl)-formamidopyrimidines (Chetsanga and Lindahl, 1979). The enzyme also possesses an AP lyase activity, processing abasic sites via a concerted β, δ elimination reaction (Bailly et al., 1989; O'Connor and Laval, 1989).

Several other products of oxidative DNA damage have been found to be a substrate for Fpg, namely 2, 6-diamino-4-hydroxy-5-formamidopyrimidine (FapyGua) and 4, 6-diamino-5-formamidopyrimidine (FapyAde) (Karakaya et al., 1997). The Fpg protein also possesses an activity that excises 2-deoxyribose-5-phosphate at 5′-incised AP sites via a β elimination reaction (Graves et al., 1992).

The Fpg protein has been found to be encoded by the *E. coli mutM* locus (Michaels et al., 1992). The lack of Fpg in this bacterial strain results in a mutator phenotype brought on by 8-oxoG mediated GC to TA transversion events (Cabrera et al., 1988). This mutator phenotype is even more pronounced when combined in bacterial strains also deficient for MutY, which excises adenine residues incorporated opposite 8-oxoG. This doubly-deficient mutator strain was recently exploited for the cloning of the *OGG1* gene of *Saccharomyces cerevisiae* (van der Kemp et al., 1996), which encodes an N-glycosylase/AP lyase activity for the repair of 8-oxoG (van der Kemp et al., 1996; Nash et al., 1996). The enzyme also possesses activities for the removal of me-Fapy (van der Kemp et al., 1996) and FapyGua, but not for FapyAde. Notably, the yeast protein can remove 5′-incised AP sites similar to the Fpg protein but, unlike the *E. coli* enzyme, Ogg1 can also process *trans*-4-hydroxy-2-pentenal-5-phosphate at a 3′ incised AP site via a hydrolytic mechanism dependent on the presence of Mg^{++} (Sandigursky et al., 1997b).

While Fpg and yeast Ogg1 are functional homologues of one another, no significant sequence similarity appears to exist between the two. Another protein that was thought to fit this description was the *Drosophila* ribosomal protein S3 (dS3), which contains activities similar to the bacterial and yeast proteins, yet was considered to lack any significant homology to the two. In this report the biochemical characteristics of dS3 are summarized. Evidence that dS3 may indeed share with the yeast enzyme those amino acids thought to be critical for catalyzing the removal of modified or nonconvential bases in DNA is also presented.

3. RESULTS

3.1. *Drosophila* S3 Contains N-Glycosylase/AP Lyase Activities

That a ribosomal protein may in fact possess DNA repair activities came from studies performed in collaboration with Dr. Mark Kelley, Indiana University School of Medicine, where our groups were studying *Drosophila* ribosomal protein PO which appeared to have a role in DNA repair (Kelley et al., 1989; Grabowski et al., 1991; Grabowski et al., 1992; Yacoub et al., 1996b). That ribosomal protein S3 may also participate in both DNA repair and protein translation came from studies originally performed in the laboratory of Dr. Stuart Linn, University of California, Berkeley on the human enzyme (Kim et al., 1995). We therefore cloned the *Drosophila* S3 gene in anticipation that it might also be active in DNA repair in this organism. The *Drosophila* S3 cDNA was subsequently placed into a pGEX-3X vector so that it could be overexpressed in *E. coli* as a fusion with glutathione S-transferase (GST-S3). The GST-S3 fusions were in turn purified by glutathione agarose affinity chromatography and used for the production of antibodies and for biochemical characterization.

The human S3 protein described by the Linn laboratory contained AP lyase activity, so our original tests were to determine if indeed the *Drosophila* enzyme possessed a similar type of activity. The position of enzymatic cleavage relative to an AP site was first determined using a poly-(dA-dT) substrate containing sporadic ^{32}P-labeled AP sites. These experiments revealed that dS3 was acting as a β elimination catalyst, cleaving DNA 3′ and adjacent to an abasic site (Wilson et al., 1994). The reactions were inhibited by antibody specific to dS3, suggesting that the AP lyase activity we were detecting was due dS3 and not some bacterial contaminate.

Since many AP lyases also contain associated N-glycosylase activity, we initiated a test in collaboration with Dr. Paul Doetsch, Emory University School of Medicine, to see if dS3 was active on a 5′ end-labeled oligonucleotide that had been heavily UV-irradiated. The reactions separated on a DNA sequencing gel revealed that dS3 was acting on a guanine photoproduct that was suspected to be FapyGua.

In order to eliminate the possibility of bacterial Fpg contamination as a reason for our results, subsequent tests for biochemical activity used GST-S3 fusions that were overexpressed in *E. coli mutM*. The purified fusions were next examined for the repair of 8-oxoG that was positioned in a 5′ end-labeled DNA duplex oligonucleotide (8-oxoG-37mer). The results of these experiments revealed that GST-S3 was able to efficiently process the 8-oxoG-37mer in a reaction that was dependent on both the incubation time and the amount of GST-S3 protein introduced into the experiments (Yacoub et al., 1996a). It also appeared that the products of the reaction produced by dS3 were generated by both a β and δ elimination reaction. To more fully understand the mechanism by which dS3 was acting at an abasic site, we compared dS3 to *E. coli* endonuclease III, a known β elimination catalyst (Bailly and Verly, 1987), and *E. coli* Fpg, which has been shown to carry out a concerted β, δ elimination reaction (Bailly et al., 1989; O'Connor and Laval, 1989). We used in this case a 5′ end-labeled 37mer containing a single AP site, and the reaction products separated on a 16% polyacrylamide DNA sequencing gel. The results obtained for *E. coli* endonuclease III and *E. coli* Fpg were in complete agreement with previous findings. These experiments were also in agreement with previous conclusions that dS3 cleaved AP DNA via a β elimination reaction (Wilson et al., 1994). However, longer incubation times or increasing amounts of protein also revealed that a δ elimination product was being produced. This most likely is due to a second binding event following the β elimination reaction at an AP site, and therefore appears to be similar to that suggested by Dr. S. Lloyd for the T4 UV endonuclease enzyme acting on pyrimidine dimers (Latham and Lloyd, 1995).

Although our results up to that time were suggestive of dS3 containing N-glycosylase activity, we had yet to show the liberation of the 8-oxoG base from the DNA substrate. To test for authentic N-glycosylase activity, we examined the ability of dS3 to liberate 8-oxoG, as well as 15 other damaged bases formed by γ irradiation of calf thymus DNA under anoxic conditions. The reaction products combining dS3 and γ-irradiated calf thymus DNA were subsequently identified by gas-chromatography/isotope-dilution mass spectrometry. This technique revealed that 8-oxoG and FapyGua were substrates for dS3, with no real preference for the excision of one modified base over the other (Deutsch et al., 1997). No other modified base was liberated by dS3, including FapyAde, which is a substrate for *E. coli* Fpg (Karakaya et al., 1997).

Many DNA repair enzymes that have a combined N-glycosylase/AP lyase activity undergo a Schiff base intermediate that can be trapped by $NaBH_4$ to form a covalent enzyme-DNA complex (Dodson et al., 1994). Similar properties were found for dS3 (Deutsch et al., 1997), thus confirming the mechanism by which dS3 acts at sites of DNA damage.

3.2. *Drosophila* S3 Contains Deoxyribophosphodiesterase (dRpase) Activity

In view of dS3 showing similar, but not exact, specificities towards oxidatively-damaged DNA bases as *E. coli* Fpg and yeast Ogg1, it seemed reasonable to examine if the *Drosophila* protein also possessed dRpase activities known to be associated with the above bacterial and yeast N-glycosylases/AP lyases. Tests were originally performed on DNA substrates containing a 5′ - incised AP site. Removal of the 2-deoxyribose-5-phosphate as 4-hydroxy-2-pentenal-5-phosphate was indeed observed as an activity possessed by dS3 (Sandigursky et al., 1997a) and similar to that associated with Fpg, Ogg1, and rat DNA polymerase β (Matsumoto and Kim, 1995). Surprisingly, dS3 was also able to liberate *trans*-4-hydroxy-2-pentenal-5-phosphate from a substrate containing 3′ incised AP sites via a hydrolytic reaction requiring Mg^{++} (Sandigursky et al., 1997a). This activity is not retained by the *E. coli* Fpg protein (Sandigursky et al., 1997b), but is associated with the yeast Ogg1 protein (Sandigursky et al., 1997).

In summary, our results suggest that dS3 (and yeast Ogg1) acts on FapyGua or 8-oxoG to liberate the free base and form an AP site in its place. The abasic site in turn can be removed by a combined β elimination reaction to form an incision on the 3′ side of the lesion, followed by a Mg^{++}-dependent hydrolytic reaction to release the existing 3′ terminal *trans*-4-hydroxy-2-pentenal -5-phosphate. This would leave a one nucleotide gap in the DNA bordered by a 3′ OH and 5′ phosphate terminus. Thus, contrary to previous doubts concerning the significance of AP lyases, it now appears from the results gathered on dS3 and yeast Ogg1 that AP lyases can be extremely efficient in generating a one nucleotide gap that is a substrate for β polymerase to fill in as part of the Base Excision Repair pathway.

3.3. *In Vivo* Properties of dS3

We have determined using Western analysis that dS3 is in fact located in the cell in places that suggest it carries out more than one function. Specifically, we used antibody prepared and purified against S3 to follow its expression throughout *Drosophila* development and found that roughly 60% - 75% of S3 could be found in the cytoplasm, depending on the stage of development, whereas 25% - 40% could be found in the nucleus (Wilson et al., 1994). Moreover, Western blot analysis of subnuclear preparations of *Drosophila* embryos revealed that dS3 was tightly bound to chromatin and the nuclear matrix (Wilson et al., 1994). Thus, significant amounts of dS3 is indeed present in the nucleus where it could play a role in DNA repair. The presence of a nuclear localization signal for dS3 (Wilson et al., 1993) is consistent with these observations.

We also have presented data that shows that dS3 is able to act *in vivo*. For example, dS3 was able to reduce to wild-type levels the number of lac^+ revertants that are found in *mutM* (Yacoub et al., 1996a) due to specific 8-oxoG mediated GC-TA transversion events (Cabrera et al., 1988). We were also able to show, both by survival curves and gradient plate analysis, that dS3 was able to restore to wild-type levels the modest sensitivity of *mutM* strains to H_2O_2 (Yacoub et al., 1996a).

The ability of dS3 to rescue the methyl melthane sulfonate (MMS) sensitivity of *E. coli* RPC501 strains (Cunningham et al., 1986) that are deficient for the hydrolytic AP endonuclease activities associated with exonuclease III (*xth*) and endonuclease IV (*nfo*) has been tested. Notably, dS3 was able to reverse the sensitivity of RPC501 to MMS at low concentrations (Yacoub et al., 1996a). This finding was rather surprising at the time considering the uncertainty of the contribution of AP lyases to the repair of AP sites. However, considering that dS3 contains dRpase activity for the removal of a blocked 3′ terminus produced by β elimination (Sandigursky et al., 1997a) makes these results not so unexpected.

3.4. Functional and Sequence Similarities to Other DNA Repair Proteins

When functional comparisons are made between Fpg, Ogg1, and S3, it appears that the yeast and *Drosophila* proteins are more similar than Fpg is to either. For example, Fpg possesses N-glycosylase activity for FapyAde, whereas the yeast and *Drosophila* proteins do not. Secondly, Fpg carries out a concerted β, δ elimination reaction at an abasic site. The dS3 protein generates predominantly a β elimination product except at high protein concentrations or longer incubation times, where a δ elimination product is then revealed (Yacoub et al., 1996a). In our hands, the yeast enzyme appears to carry out only a β elimination reaction. Perhaps the most significant difference between these activities is that Fpg is unable to remove *trans*-4-hydroxy-2-pentenal-5-phosphate from a DNA substrate containing a 3′ - incised AP site, whereas the yeast (Sandigursky et al., 1997b) and *Drosophila* (Sandigursky et al., 1997a) enzymes clearly retain this function.

While the yeast Ogg1 and *E. coli* Fpg have been concluded to be functional homologues of one another, as noted above they share no significant sequence homology. For dS3 , our original sequence homology searches failed to show any obvious homology between Fpg and Ogg1. However, the recent solving of the crystal structures for *E. coli* endonuclease III (Thayer et al., 1995) and *E. coli* 3-methyladenine - DNA glycosylase II (Yamayata et al., 1996; Labahn et al., 1996) have provided some suprising insight into the possible relatedness of dS3 and yeast Ogg1.

Even though *E. coli* endonuclease III and *E. coli* 3-methyladenine-DNA glycosylase II do not share similarities in substrate specificity, they nevertheless were found to be similar in structure and active site location that was suggested to be fundamental for them in carrying out base excision repair (Labahn et al., 1996). For both enzymes, the active site pocket is bracketed by Lys120 and Asp138 (endonuclease III nomenclature). The Asp residue is thought to function as a nucleophile or as a general base that activates a nucleophilic water molecule, whereas Lys120 could act in a transimination reaction (Thayer et al., 1995), and would therefore be found in N-glycosylases with associated AP lyase activity. Gln41 is also conserved and is positioned at the mouth of the pocket and may be important for DNA binding and is perhaps involved in nucleotide-flipping.

Identification of these conserved amino acids using the BLAST network failed to show any correlation of the Fpg and Ogg1 protein to the Drosophila S3 protein. It should be noted, however, that these searches were done prior to the completion of the crystal structures and site directed mutagenesis that were performed on *E. coli* endonuclease III and 3-methyladenine-DNA glycosylase II. Importantly, these homologies to S3 are detected using a Hidden Markov statistical modelling method that recognizes similar, but divergent family members presnt in the databases. Indeed, the conserved amino acids noted above are in complete alignment when Ogg1 and S3 are compared with endonuclease III and 3-methyladenine-DNA glycosylase II. While this supports the notion that S3 is therefore both functionally and structurally related to Ogg1, it should be noted that the helix-hairpin-helix motif (Thayer et al., 1995) present in the endonuclease III/Ogg1 family of DNA repair proteins is absent in S3.

4. DISCUSSION

4.1. Multifunctional Proteins as DNA Repair Proteins

This report has reviewed those studies that show that S3 has DNA repair activities that were revealed through the use of *in vitro* biochemical studies, the use of bacterial mutants to demonstrate its *in vivo* activity, and cell biology techniques showing that significant amounts of dS3 can be found in the nucleus associated with the nuclear matrix. On the other hand, S3 is also involved in the initiation of protein translation. Consistent with this is that the S3 antigen can be found associated with ribosomes. Moreover, S3 maps to the *Minute* chromosomal location that most likely encodes the S3 ribosomal protein (Wilson et al., 1994). Unfortunately, *Minute* mutants are homozygous lethal, therefore preventing a detailed analysis of the possible sensitivity of these mutants to various DNA damaging agents.

A few proteins have now been identified that participate in the coupling of transcription to DNA repair, therefore providing an additional example of multifunctional proteins involved in DNA repair. The emerging story on the major AP endonuclease endonuclease present in humans (APE/Ref-1) is yet another example of a protein involved not only in the DNA repair of AP sites, but also functions in an important role in regulating the redox state of a number of proteins, including the recently reported ability to convert inert p53 to an active form (Jayaraman et al., 1997). Interestingly, the AP endonuclease activity of APE/Ref-1 can be inactivated by phosphorylation (Yacoub et al., 1997), thus providing a hint as to how one unrelated function possessed by the protein may be converted to another. No such clues have yet emerged in understanding how S3 may carry out one role vs another.

4.2. Concluding Remarks

We have yet to establish whether the biochemical properties possessed by S3 is in fact active in protecting *Drosophila* from the deleterious consequences of *in vivo* oxidative DNA damage. In view of dS3's tight association with the nuclear matrix, this observation makes it likely that dS3 is indeed involved in base excision repair in *Drosophila* cells. The lack of genetics, however, prevents a firm conclusion in dS3's participation *in vivo*.

In human cells, the AP lyase activity associated with S3 is absent in xeroderma pigmentosam (XP) group D fibroblasts (Kim et al., 1995). Notably, we have found that

Drosophila S3 protects Fanconi's anemia (FA) cells from the toxic consequences of mitomycin C (Kelley and Deutsch, unpublished observations), which produces not only DNA cross links, but also oxygen free radicals (Clarke et al., 1997). It does not appear, however, that the S3 structural gene is defective for either XP or FA, so it remains to be seen how S3 might play a role in the etiology of these diseases.

We have found DNA repair activities associated with another ribosomal protein, namely PO (Yacoub et al., 1996b). Notably, others have found that the mRNA's of PO and S3 are elevated in adenomatous polyps and colorectal carcinomas (Pogue-Geile et al., 1991). As noted previously (Yacoub et al., 1996a), it is conceivable that the overexpression of these ribosomal proteins may confer a selective advantage if these same proteins contribute to an elevated DNA repair capacity.

ACKNOWLEDGMENTS

Many different laboratories have made a significant contribution to the work presented here. Dr. Mark Kelley and members of his laboratory have conducted the majority of the molecular and cell biology studies reviewed here; Dr. Paul Doetsch and Laura Augeri performed some of the original experiments showing dS3 activity on 8-oxoG; Dr. Bill Franklin and Margarita Sandigursky did the dRPase work described here; and Dr. Miral Dizdaroglu and his collaboraters showed that dS3 possessed authentic DNA glycosylase activity. I would particularly like to thank Dr. S. Mian, University of California, Santa Cruz, for bringing to our attention the similarities between dS3 and Ogg1. The work described here was supported by grants from the National Institutes of Health.

REFERENCES

V. Bailly and W.G. Verly (1987) *Escherichia coli* endonuclease III is not an endonuclease but a β-elimination catalyst. *Biochem. J.* **242**, 565–572.

V. Bailly, W.G. Verly, T. O'Conner and J. Laval (1989) Mechanism of DNA strand nicking at apurinic/apyrimidinic sites by *Escherichia coli* [formamidopyrimidine] DNA glycosylase. *Biochem. J.* **262**, 581–589.

M. Cabrera, Y. Nghiem and J.H. Miller (1988) MutM, a second mutator locus in *Escherichia coli* that generates G•C→T•A transversions. *J. Bacteriol.* **170**, 5405–5407.

C.J. Chetsanga and T. Lindahl (1979) Release of 7-methylguanine residues whose imidazole rings have been opened from damaged DNA by a DNA glycosylase from *Escherichia coli. Nucleic Acids Res.* **6**, 3673–3684.

A.A. Clarke, N.J. Philpott, E.C. Gordon-Smith and T.R. Rutherford (1997) The sensitivity of Fanconi anaemia group C cells to apoptosis induced by mitomycin C is due to oxygen radical generation, not DNA crosslinking. *British J. Haematology* **96**, 240–247.

R.P. Cunningham, S.M. Saporito, S.G. Spitizer and B. Weiss (1986) Endonuclease IV (*nfo*) mutant of *Escherichia coli. J. Bacteriol.* **168**, 1120–1127.

W.A. Deutsch, A. Yacoub, P. Jaruga, T.H. Zastawny and M. Dizdaroglu (1997) Characterization and mechanism of action of *Drosophila* ribosomal protein S3 DNA glycosylase activity for the removal of oxidatively damaged DNA bases. *J. Biol. Chem.* **272**, 32857–32860.

M.L. Dodson, M. Michaels and R.S. Lloyd (1994) Unified catalytic mechanism for DNA glycosylases. *J. Biol. Chem.* **269**, 32709–32712.

D.T. Grabowski, W.A. Deutsch, D. Derda and M.R. Kelley (1991) *Drosophila* AP3, a presumptive DNA repair protein, is homologous to human ribosomal associated protein PO. *Nucleic Acids Res.* **19**, 4297.

D.T. Grabowski, R.O. Pieper, B.W. Futscher, W.A. Deutsch, L.C. Erickson and M.R. Kelley (1992) Expression of ribosomal phosphoprotein PO is induced by antitumor agents and increased in Mer⁻ human tumor cell lines. *Carcinogenesis* **13**, 259–263.

R.J. Graves, I. Felzenszwalb, J. Laval and T.R. O'Conner (1992) Excision of 5′ - terminal deoxyribose phosphate from damaged DNA is catalyzed by the Fpg protein of *Escherichia coli. J. Biol. Chem.* **267**, 14429–14435.

L. Jayaraman, K.G.K. Murthy, C. Zhu, T. Curran, S. Xanthoudakis and C. Prives (1997) Identification of redox/repair protein Ref-1 as a potent activator of p53. *Genes and Development* **11**, 558–570.

A. Karakaya, P. Jaruga, V.A. Bohr, A.P. Grollman and M. Dizdaroglu (1997) Kinetics of excision of purine lesions from DNA by *Escherichia coli* Fpg protein. *Nucleic Acids Res.* **25**, 474–479.

M.R. Kelley, S. Venugopal, J. Harless and W.A. Deutsch (1989) Antibody to a human DNA repair protein allows for cloning of a *Drosophila* cDNA that encodes an apurinic endonuclease. *Mol. Cell. Biol.* **9**, 965–973.

J. Kim, L.S. Chubatsu, A. Admon, J. Stahl, R. Fellows and S. Linn (1995) Implication of mammalian ribosomal protein S3 in the processing of DNA damage. *J. Biol. Chem.* **270**, 13620–13629.

J. Labahn, O.D. Scharer, A. Long, K. Ezaz-Nikpay, G.L. Verdine and T.E. Ellenberger (1996) Structural basis for the excision repair of alkylation-damaged DNA. *Cell* **86**, 321–329.

K.A. Latham and R.S. Lloyd (1995) δ-Elimination by T4 endonuclease V at a thymine dimer site requires a secondary binding event and amino acid glu-23. *Biochemistry* **34**, 8796–8803.

Y. Matsumoto and K. Kim (1995) Excision of deoxyribose phosphate residues by DNA polymerase β during DNA repair. *Science* **269**, 699–702.

M.L. Michaels, L. Pham, C. Cruz and J.H. Miller (1991) MutM, a protein that prevents G•C→T•A transversions, is formamidopyrimidine-DNA glycosylase. *Nucleic Acids Res.* **19**, 3629–3632.

M. Moriya, C. Ou, V. Bodepudi, F. Johnson, M. Takeshita and A.P. Grollman (1991) Site-specific mutagenesis using a gapped duplex vector:a study of translesion synthesis past 8-oxodeoxyguanosine in *E.coli. Mutat. Res.* **254**, 281–288.

H.M. Nash, S.D. Bruner, O.D. Scharer, J. Karvate, T.A. Addona, E. Spooner, W.S. Lane and G.L. Verdine (1996) Cloning of a yeast 8-oxoguanine DNA glycosylase reveals the existence of a base-excision DNA-repair protein superfamily. *Current Biology* **6**, 968–980.

T.R. O'Conner and F. Laval (1989) Physical association of the 2,6-diamino-4-hydroxy-5*N*-formamidopyrimidine-DNA glycosylase of *Escherichia coli* and an activity nicking DNA at apurinic/apyrimidinic sites. *Proc. Natl. Acad. Sci. USA* **86**, 5222–5226.

T.R. O'Connor and F. Laval (1990) Isolation and structure of a cDNA expressing a mammalian 3-methyladenine-DNA glycosylase. *EMBO J.* **9**, 3337–3342.

K. Pogue-Geile, J.R. Geiser, M. Shu, C. Miller, I.G. Wool, A.I. Meisler and J.M. Pipas (1991) Ribosomal protein genes are overexpressed in colorectal cancer:isolation of a cDNA clone encoding the human S3 ribosomal protein. *Mol. Cell. Biol.* **11**, 3842–3849.

M. Sandigursky, A. Yacoub, M.R. Kelley, W.A. Deutsch and W.A. Franklin (1997a) The Drosophila ribosomal protein S3 contains a DNA deoxyribophosphodiesterase (dRpase) activity. *J. Biol. Chem.* **272,** 17480–17484.

M. Sandigursky, A. Yacoub, M.R. Kelley, Y. Xu, W.A. Franklin and W.A. Deutsch (1997b) The yeast 8-oxoguanine DNA glycosylase (Ogg1) contains a DNA deoxyribophosphodiesterase (dRpase) activity. *Nucleic Acids Res.* **25,** 4557–4561.

S. Shibutani, M. Takeshita and A.P. Grollman. (1991) Insertion of specific bases during DNA synthesis past the oxidation-damaged base 8-oxodG. *Nature* **349**, 431–434.

M.H. Thayer, H. Ahern, D. Xing, R.P. Cunningham and J.T. Tainer (1995) Novel DNA binding motifs in the DNA repair enzyme endonuclease III crystal structure. *EMBO J.* **14**, 4108–4120.

P.A. van der Kemp, D. Thomas, R. Barbey, R. deOliveira and S. Boiteux (1996) Cloning and expression in *Escherichia coli* of the *Ogg*1 gene of *Saccharomyces cerevisiae,* which codes for a DNA glycosylase that excises 7,8-dihydro-8-oxoguanine and 2,6-diamino-4-hydroxy-5-N-methylformamidopyrimidine. *Proc. Natl. Acad. Sci. USA* **93**, 5197–5202.

D. Wilson, W.A. Deutsch and M.R. Kelley (1994) *Drosophila* ribosomal protein S3 contains an activity that cleaves DNA at apurinic/apyrimidinic sites. *J. Biol. Chem.* **269**, 25359–25364.

D. Wilson, W.A. Deutsch and M.R. Kelley (1993) Cloning of the *Drosophila* ribosomal protein S3: another multifunctional ribosomal protein with AP endonuclease DNA repair activity. *Nucleic Acids Res.* **21**, 2516-?.

M.L. Wood, M. Dizdaroglu, E. Gajewski and J.M. Essigman (1990) Mechanistic studies of ionizing radiation and oxidative mutagenesis genetic effects of a single 8-hydroxyguanine (7-hydro-8-oxoguanine) residue inserted at a unique site in a viral genome. *Biochemistry* **29,** 7024–7032.

A. Yacoub, L. Augeri, M.R. Kelley, P.W. Doetsch and W.A. Deutsch (1996a) A *Drosophila* ribosomal protein contains 8-oxoguanine and abasic site DNA repair activities. *EMBO J.* **15**, 2306–2312.

A. Yacoub, M.R. Kelley and W.A. Deutsch (1996b) *Drosophila* ribosomal protein PO contains apurinic/apyrimidinic endonuclease activity. *Nucleic Acids Res.* **24**, 4298–4303.

A. Yacoub, M.R. Kelley and W.A. Deutsch (1997) The DNA repair activity of human redox/repair protein APE/Ref-1 is inactivated by phosphorylation. *Cancer Res.* **57,** 5457–5459

Y. Yamagata, et al. (1996) Three-dimensional structure of a DNA repair enzyme, 3-methyladenine DNA glycosylase II, from *Escherichia coli*. *Cell* **86**, 311–319.

9

BYPASS OF DNA DAMAGE BY RNA POLYMERASES

Implications for DNA Repair and Transcriptional Mutagenesis

Paul W. Doetsch,[1] Anand Viswanathan,[2] Wei Zhou,[3] and Jiang Liu[1]

[1]Departments of Biochemistry and Radiation Oncology
[2]Graduate Program in Genetics and Molecular Biology
Emory University School of Medicine
4123 Rollins Research Center
Atlanta, Georgia, 30322–3050
[3]The Johns Hopkins Oncology Center
424 North Bond St.
Baltimore, Maryland 21231-1001

1. ABSTRACT

We have placed various DNA base damage products into single, defined locations on the template or nontemplate strands of model gene segments and carried out in vitro transcription experiments with phage RNA polymerases and E. coli RNA polymerase. The DNA damages investigated are representative of several major classes of base modifications and include uracil (cytosine deamination product), dihydrouracil (pyrimidine ring saturation product), abasic sites (depurination product), 8-oxoguanine (purine oxidative damage product), and O^6-methylguanine (alkylation damage product). In addition to transcription elongation, we have also investigated the effects of some of these damages on promoter clearance. All of these DNA base damages are bypassed efficiently under both single and multiple round transcription conditions by RNA polymerase (with or without brief pausing at the lesion site) and result in mutagenic insertions in the resulting transcripts. We term this process *transcriptional mutagenesis* and speculate that it may have several important biological consequences with respect to DNA repair as well as the mechanisms which lead to the generation of mutant proteins.

Advances in DNA Damage and Repair, edited by Dizdaroglu and Karakaya. Kluwer Academic / Plenum Publishers, New York, 1999.

2. INTRODUCTION

2.1. Transcription-Coupled Repair

More than a decade ago, Hanawalt and colleagues made the seminal observation that DNA repair in mammalian cells is heterogeneous and occurs at different rates depending on the genomic location of the damage (Bohr et al., 1985; Hanawalt, 1994). Subsequent studies by Hanawalt's laboratory and several other groups have shown that the repair of template strands of actively transcribed genes in both prokaryotes and eukaryotes occurs more rapidly compared to the nontranscribed strand as well as to the overall genome (Mellon et al., 1987; Hanawalt, 1994). Most of these investigations used cyclobutane pyrimidine dimers (CPDs), the major UV light-induced DNA photoproduct, as the model DNA lesion. These investigations suggested that the nucleotide excision repair machinery was coupled to the cellular transcription machinery for targeting repair to critical regions of the genome and that this pathway was present in all cells. In studies by Selby and Sancar, it was determined that in E. coli, the Mfd protein (also called transcription coupling repair factor or TRCF) was responsible for targeting components of the nucleotide excision repair pathway to the vicinity of a RNA polymerase arrested at the site of a bulky lesion present on the template strand of a transcribed gene (Selby and Sancar 1991a; 1991b; 1993a; 1993b). In human cells, an analogous function has been ascribed to the CS-A and CS-B proteins and was deduced from transcription-coupled repair (TCR) studies in cells from patients with Cockayne's syndrome (van Hoffen et al., 1993; Henning et al., 1995). In addition, a role for E. coli (MutS and MutL) and human (hMSH2 and hMLH1) mismatch repair proteins in TCR of CPDs in certain E. coli (lac operon) and human (DHFR) genes has been recently demonstrated (Mellon and Champe, 1996; Mellon et al., 1996). It appears that the linkage of transcription and nucleotide excision repair (NER) may involve the participation of several different factors which may mediate roles in the repair of specific genes or classes of genes (Mellon et al., 1996). Recent studies by Leadon, Cooper and colleagues have shown that in yeast and human cells, the repair of oxidative DNA damages handled primarily by the base excision repair (BER) system such as thymine glycol are also coupled to transcription (Leadon et al., 1995; Cooper et al., 1997). Thus, the repair of DNA damages by either the NER or BER systems may be coupled to transcription.

2.2. RNA Polymerase Arrest and TCR

A key event that determines whether a particular DNA damage will be subject to TCR is the ability of that lesion (located on the template strand) to arrest RNA polymerase during the elongation stage of transcription (Selby and Sancar, 1993a). The majority of studies (with various eukaryotic cell lines or bacterial strains) on TCR have focused on CPDs and other bulky lesions. In vitro prokaryotic and eukaryotic transcription systems utilizing template strands containing singly placed CPDs or other bulky lesions have shown that such damages permanently arrest elongating RNA polymerases at the damage site and allow a direct correlation with in vivo studies demonstrating preferential removal of such damages from the template strands of actively transcribed genes (Hanawalt, 1994; Donahue et al., 1996). A prediction from these studies is that a DNA lesion which fails to arrest an elongating RNA polymerase will not be subject to preferential, strand specific repair.

2.2.1. In Vitro Transcription Studies with Non-Bulky Lesions. Until recently, little information existed concerning the in vitro and in vivo effects of less bulky DNA damages such as uracil and abasic sites on the transcriptional elongation machinery. Our group has conducted a number of in vitro transcription studies with defined prokaryotic RNA polymerase templates containing specific damages placed downstream from the start of transcription. We have determined the effects of uracil, dihydrouracil, 7,8-dihydro-8-oxoguanine (8-oxoguanine), O^6-methylguanine, abasic sites, strand breaks, and strand gaps on the transcriptional elongation complex and have found that many of these damages do not arrest RNA polymerases and cause various types of misinsertion events when they are bypassed (Zhou and Doetsch, 1993; 1994a; 1994b; 1996; Zhou et al., 1995; Liu et al., 1995; Liu and Doetsch, 1996; Viswanathan and Doetsch, 1998). Using a similar type of in vitro transcription system, Chen and Bogenhagen have shown that 8-oxoguanine does not arrest T7 RNA polymerase and causes predominantly adenine (A) misinsertions in the resulting transcript (Chen and Bogenhagen, 1993). Another major oxidative and radiation-induced DNA base damage product, thymine glycol, efficiently blocks RNA polymerases in vitro (Htun and Johnson, 1992). Likewise, the relatively bulky guanine C-8 aminofluorene (AF) adduct only causes brief pausing by RNA polymerase II followed by efficient bypass (Donahue et al., 1996). These studies indicate that the size of a DNA lesion cannot be used a priori to predict its effect on the transcription elongation machinery. However, there appears to be a good correlation between the ability to arrest RNA polymerase in vitro and template strand specific repair in vivo. It should also be pointed out that in general, most damages on the non-transcribed strand do not block progression of RNA polymerase and are not preferentially repaired.

2.2.2. In Vivo Transcription-Coupled Repair Studies with Non-Bulky Lesions. A simple prediction from in vitro studies showing that a particular lesion does not arrest an elongating RNA polymerase would be that such damage does not show strand-specific preferential repair in living cells. Conversely, the lack of in vivo strand specific repair for a particular type of lesion predicts that such lesions will not be strong blocks to RNA polymerases in vitro. In the past several years a number of groups have investigated strand-specific repair of less bulky lesions such as those caused by alkylating agents (Scicchitano and Hanawalt, 1990; Pirsel et al., 1993; Wang et al., 1995). For example, dimethyl sulfate (DMS)- and methyl methanesulfonate (MMS)-induced N-methyl purines are not repaired preferentially on the transcribed strands of active genes and suggests that base damages such as 7-methylguanine and 3-methyladenine do not arrest RNA polymerase. In contrast, thymine glycol is preferentially removed from the transcribed strands of active genes in yeast and this observation correlates with its known blocking effects on RNA polymerases (Htun and Johnson, 1992; Leadon et al., 1995).

It is important to distinguish between the ability of lesion to cause *brief pausing* versus *permanent arrest* of RNA polymerase. It appears that lesions such as guanine C-8 aminofluorene adducts that only briefly pause RNA polymerase do not exhibit strand-specific repair in vivo (Donahue et al., 1996). Those damages such as CPDs and thymine glycols that will arrest progression of the transcription elongation complex are the class of DNA damages that are preferentially removed on the transcribed strands.

2.3. Transcriptional Mutagenesis and Its Potential Biological Consequences

With regard to the effects of DNA damage on the transcriptional machinery, most investigations on TCR have emphasized the biological relevancy for targeted removal of lesions that block transcriptional elongation complexes (Hanawalt, 1994; Friedberg et al.,

1995). Repair of such damages is of obvious importance and is carried out to maintain normal physiological expression of various housekeeping and other types of genes. However, the situation where (1) RNA polymerase is not arrested at the lesion site and (2) such damage causes continuous misinsertions or deletions in the resulting transcript has potentially profound implications for both DNA repair pathways and the routes leading to the generation of mutant proteins. In vitro studies showing that RNA polymerases are capable of bypassing a variety of DNA damages on the template strand together with in vivo investigations of template strand-specific repair of these damages have provided direct and indirect evidence for the following concepts.

2.3.1. Potential Inhibition of Repair by Damages which Do Not Block RNA Polymerase. A key feature in the TCR mechanism is the ability of the lesion to permanently arrest RNA polymerase at or near the damage site. It is likely that brief stalling by RNA polymerase followed by efficient bypass is not sufficient for triggering a TCR response; a notion that has recently been confirmed using guanine C-8 aminofluorene adducts as a model lesion (Donahue et al., 1996). Transcribed genes which contain DNA damages that do not permanently arrest RNA polymerase may actually reduce the ability of the cell to repair such lesions by preventing access to DNA repair proteins due to RNA polymerase occupancy of that gene. Such a situation might lead to an increase in mutations arising from unrepaired damage on the transcribed strand of a gene during a subsequent DNA replication cycle and may explain the recent finding that transcriptionally active DNA and enhanced spontaneous mutation rates are associated in yeast (Datta and Jinks-Robertson, 1995). In other situations, such as when RNA polymerase bypasses a base damage product such as 8-oxoguanine in vitro without pausing (Chen and Bogenhagen, 1993; Viswanathan and Doetsch, 1998), preferential repair of the template strand does not occur in vivo (Taffe et al., 1996).

2.3.2. Transcriptional Mutagenesis. Efficient RNA polymerase bypass of damages, if it occurs in vivo, could lead to the generation of mutant proteins via a route involving a transcriptional miscoding mechanism (*transcriptional mutagenesis*). For many of the damages bypassed, analysis of the base misinsertion events and taking codon usage into account leads to the prediction that the majority of the base substitutions will lead to amino acid sequence changes in the resulting protein (Zhou and Doetsch, 1993; Liu et al., 1995 . Such transcriptional behavior leading to the generation of mutant proteins could have a number of important biological outcomes, especially in a population of nondividing cells. As discussed below, such mutant proteins could exert toxic effects leading to impaired function or death or initiate events leading to the cell's entry into a replicative cycle and cell division.

3. RESULTS

The DNA damages investigated in these studies were uracil, dihydrouracil, 8-oxoguanine, O^6-methylguanine, abasic sites, and a variety of single strand breaks and gaps of various sizes. The potential for these damages to arrest transcription elongation or be bypassed was determined for a series of RNA polymerases ranging from simple, single polypeptide phage polymerases (T7 and SP6) to a mutiple subunit prokaryotic polymerase (E. coli RNA polymerase).

3.1. Generation of Transcription Templates

The DNA templates used for these studies contain a RNA polymerase promoter and a defined type of DNA damage placed at a single location on the transcribed (or nontranscribed) strand downstream from the start of transcription. The approach previously used by most investigators was to prepare the DNA template by using restriction enzyme digestion to generate a DNA fragment containing an RNA polymerase promoter from a plasmid. This DNA fragment is subjected to treatments with various chemical or physical agents, which introduce DNA damage randomly throughout the entire template (Cullinane and Philips, 1992; Sanchez and Mamet-Bratley, 1994). Such methods are relatively simple and straightforward and have been used successfully by several groups to determine the overall effects of several bulky DNA lesions on transcription. However, the heterogeneous nature of the DNA template generated by the above approach makes it difficult to attribute a specific type of transcription product to a defined type and location of DNA damage. An additional problem with such methods is that it is difficult to determine the extent of RNA polymerase blockage by, or bypass of, a particular DNA lesion. These problems can be circumvented by utilizing DNA templates containing a specific DNA damage at a defined location. However, even this approach can be difficult due to the template construction method which involves multiple ligation steps in order to place a synthetic oligonucleotide containing the DNA damage into a plasmid, a procedure which often results in extremely low DNA template yields. For our studies, we have employed methods which provide relatively large amounts of duplex template for direct determinations of the interaction of the transcription elongation complex with various DNA damages under either single or multiple round transcription conditions. One method is PCR-based (Zhou and Doetsch, 1993; 1994a) while the second method relies on direct annealing of single stranded oligonucleotides (Zhou and Doetsch, 1994b; Zhou et al., 1995; Liu et al., 1995; Liu and Doetsch, 1996). Both techniques result in the generation of duplex templates containing the appropriate RNA polymerase promoter and a singly placed damage on the template (or nontemplate) strand at a defined location downstream from the transcriptional start site. Prior to their utilization in transcription experiments templates were analyzed to verify the location and authenticity of the base modification. We have constructed such templates containing SP6, T7 promoters for phage RNA polymerases and the tac promoter for E. coli RNA polymerase. A general summary of these constructions is shown in Figure 1.

3.2. In Vitro Transcription Systems and Transcript Sequence Analysis

Studies have been carried out with defined, damage-containing templates under both multiple round (template reutilization/recycling) and single round (one promoter-dependent elongation cycle per template-RNA polymerase complex) transcription conditions (Zhou and Doetsch, 1993; 1994a; 1994b; Zhou et al., 1995; Liu et al., 1995; Liu and Doetsch, 1996; Zhou and Doetsch, 1996). Single round transcription conditions are particularly informative for revealing whether a DNA lesion causes temporary polymerase pausing before bypassing the damage and continuing elongation.

Sequence analysis of transcripts was carried out by utilizing either an RT-PCR method (Zhou and Doetsch, 1993) or direct sequence analysis of 5' end labeled transcripts by base-specific RNases (Liu et al, 1995).

3.3. Phage RNA Polymerase Transcription Studies

Using various templates (construction described in detail in Zhou and Doetsch, 1993; 1994a; 1994b; Zhou et al., 1995; Liu et al., 1995; Zhou and Doetsch, 1996), we

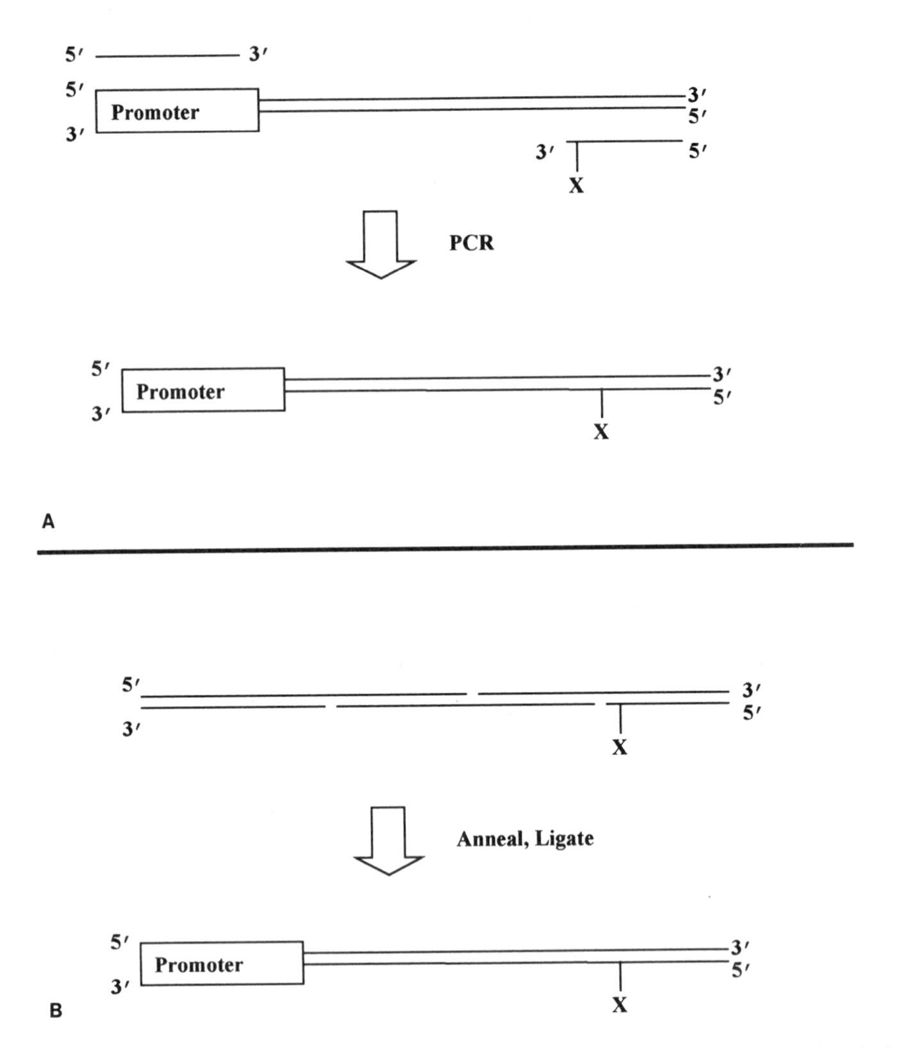

Figure 1. General scheme for template construction for in vitro transcription studies. (A) PCR-based method utilizing upstream and downstream primers and duplex DNA segment containing an appropriate RNA polymerase promoter (references cited in text). Downstream primer oligonucleotide contains specific DNA damage (X) at defined location and is present in excess amounts during PCR cycles to generate PCR product containing singly placed damage on the transcribed strand at a defined location downstream from the transcription start site. (B) Annealing/ligation-based method involving assembly of a set of oligonucleotides via annealing with one of the component oligos containing a specific damage (X). Ligation and purification of product results in generation of duplex template containing singly placed damage on the transcribed strand (references cited in text).

have determined the effects of the following DNA modifications on SP6 and T7 RNA polymerase elongation: uracil, abasic sites, dihydrouracil, strand breaks and strand gaps.

3.3.1. Uracil. Both SP6 and T7 RNA polymerases efficiently bypass uracil with no indication of blockage or pausing. Transcript sequence analysis showed that these phage RNA polymerases inserted exclusively adenine opposite to uracil and although not a particularly surprising finding, indicates that uracil is mutagenic at the level of transcription (Zhou and Doetsch, 1993; 1994b; Liu et al., 1995).

3.3.2. Abasic Sites. Single round transcription elongation experiments revealed that abasic sites cause phage RNA polymerases to briefly pause at the site of damage followed by efficient bypass and extension of transcripts into full length, runoff products (Zhou and Doetsch, 1993; Zhou and Doetsch, 1994a; 1994b). Transcript sequence analysis indicates that adenine is preferentially inserted opposite to abasic sites. The biological importance of such an insertional bias will be discussed below. We have also placed this lesion into several different DNA sequence contexts on the template strand and find that the bypass and insertion events are the same (Zhou and Doetsch, 1994a). Placement of an abasic site on the nontemplate (nontranscribed) strand is without effect on elongation.

3.3.3. Dihydrouracil. 5,6-Dihydrouracil (dihydrouracil) is a pyrimidine ring saturation product and serves as a model for base damage products of the type recognized by E. coli endonuclease III (and functional homologs in other species). Dihydrouracil is a major base damage product in DNA exposed to ionizing radiation under anoxic conditions (Dizdaroglu et al., 1993). Both SP6 and T7 RNA polymerases briefly pause at sites of dihydrouracil on the template strand but subsequently efficiently bypass this lesion with preferential insertion of adenine over guanine (Liu et al., 1995). Such an insertion preference indicates that dihydrouracil is also mutagenic at the level of transcription.

3.3.4. Strand Breaks and Gaps. DNA single strand breaks (ssbs) occur frequently in cells as a consequence of exposure to various physical and chemical agents or generated by nucleases involved in certain nucleic acid transactions. Ssbs are quite variable with regard to the nature of the 3' and 5' termini which flank the break site. Template strand discontinuities in the form of nicks or gaps of various sizes are strong blocks to the progression of DNA polymerases. We have carried out several detailed studies of the events that occur when phage RNA polymerases encounter single strand breaks and gaps on the DNA template strand. We have found that ssbs are bypassed by phage RNA polymerases to various extents depending on the nature of the termini flanking the break site (Zhou and Doetsch, 1994b). For example, bypass efficiency is low (~6%) for ssbs with flanking 3' and 5'-phosphoryl groups and high (~75%) for ssbs with flanking 3' and 5'-hydroxyl groups. We have examined many different types of termini flanking ssbs and the results are summarized in Table 1 of Zhou and Doetsch (1994b). A surprising result from these studies was the observation that single nucleotide gaps were bypassed and that the resulting full length, runoff transcript was exactly one nucleotide shorter (FL-1) than that produced from the corresponding control, template containing no gaps (FL). Sequence analysis of the FL-1 transcript revealed that it contained the correctly templated nucleotides minus the gap. Subsequent studies with gapped templates ranging from small to large sizes (up to 24 nt) indicate that (1) as the gap size of the template increases, the bypass efficiency and production of "full length" runoff transcript decreases and (2) the length of the runoff transcript resulting from gap bypass is equal to the length of the transcript generated from the control, unbroken template (containing no gaps) minus the length of the gap. These studies have revealed important information regarding the mechanism of elongation, particularly with regard to the role of the nontemplate strand in this process (Zhou and Doetsch, 1994b; Zhou and Doetsch, 1995).

3.4. E. coli RNA Polymerase Transcription Elongation Studies

The results of our studies with phage RNA polymerases have served as informative, simple models to predict the effects of various DNA damages on more complex, mul-

Table 1. Effects of nonbulky base damage on RNA polymerases[1]

	RNA Polymerase		
DNA Lesion	SP6	T7	E. coli
Abasic Sites	bp (A)	bp (A)	bp (A)
Uracil	bp (A)	bp (A)	bp (A)
Dihydrouracil	bp (A, G)	bp (A, G)	bp (A)
8-Oxoguanine	nd	bp (A, C)	bp (A, C)
O^6-Methylguanine	nd	nd	bp (U)
Ssb	bp or arr	bp or arr	bp or arr
Gap	variable bp	variable bp	variable bp

[1]DNA lesions which are bypassed (bp) or cause arrest (arr) of SP6, T7, or E. coli RNA polymerases. Bases inserted opposite to bypassed lesions indicated in parentheses; nd: not determined.

tisubunit RNA polymerases. We have conducted a series of studies with E. coli RNA polymerase with templates containing uracil, abasic sites, dihydrouracil, strand breaks and gaps, 8-oxoguanine, and O^6-methylguanine. The results of these studies are summarized together with the results from our phage RNA polymerase studies in Table 1.

3.4.1. Uracil and Abasic Sites. Both uracil and abasic sites are efficiently bypassed by E. coli RNA polymerase (Zhou and Doetsch, 1993). As in the case with phage RNA polymerases, abasic sites cause E. coli RNA polymerase to temporarily pause followed by efficient bypass and production of full length, runoff transcripts. Transcript sequence analysis indicates that E. coli RNA polymerase primarily inserts adenine opposite to uracil and abasic sites.

3.4.2. Single Strand Breaks and Single Nucleotide Gaps. Nicks generated at abasic sites by various DNA repair proteins or chemical treatment are bypassed poorly by E. coli RNA polymerase (Zhou and Doetsch, 1993). However, single nucleotide gaps flanked by 3'- and 5'-hydroxyl groups are bypassed by this enzyme albeit less efficiently (~50%) compared to phage polymerases (80–100%) (Liu and Doetsch, 1996). The full length, runoff transcripts generated by E. coli RNA polymerase bypass of one nucleotide gaps are exactly one nucleotide shorter compared to those generated from unbroken, control templates. Sequence analysis of the one nucleotide shortened transcript indicates a one nucleotide deletion in the transcript sequence at the position of the gap. We conclude from these results that prokaryotic (phage or E. coli) RNA polymerases are capable of bypassing certain types of template strand discontinuities and produce transcripts containing deletions equivalent to the gap size.

3.4.3. Dihydrouracil Effects on RNA and DNA Polymerase within Same Sequence Context. We have begun to examine the potential similarities and differences between RNA and DNA polymerases in vitro with respect to their behaviors when encountering the same DNA lesion placed within an identical sequence context. We have utilized the same template sequence in which dihydrouracil was embedded and have made a direct comparison between E. coli RNA and DNA polymerases. E. coli RNA polymerases briefly pauses at the site of, then efficiently bypasses dihydrouracil (Liu and Doetsch,1998). E. coli DNA polymerase (Klenow fragment) bypasses dihydrouracil with no apparent pausing or arrest and generates full length products in primer extension-type experiments. Both E. coli

RNA and DNA polymerase insert adenine opposite to dihydrouracil, suggesting that dihydrouracil is mutagenic with respect to both replication and transcription.

3.4.4. 8-Oxoguanine and O^6-Methylguanine. As discussed above, all of the DNA damages examined in our phage polymerase studies were also studied in the E. coli RNA polymerase-driven in vitro transcription system. In addition, we have also examined two other important DNA lesions in the E. coli system, 8-oxoguanine and O^6-methylguanine. Neither 8-oxoguanine nor O^6-methylguanine cause significant pausing or arrest of E. coli RNA polymerase and templates containing these lesions are efficiently transcribed in vitro to produce full length, runoff transcripts (Viswanathan and Doetsch, 1998). Transcript sequence analysis indicates that E. coli RNA polymerase inserts primarily either adenine or cytosine opposite to 8-oxoguanine and thymine opposite to O^6-methylguanine. These results provide yet two more examples of DNA base damage products that are likely to be mutagenic at the level of transcription. Interestingly, the base insertion preferences of E. coli RNA polymerase opposite to 8-oxoguanine and O^6-methylguanine are the same as those observed for DNA polymerases (Friedberg et al., 1995).

3.5. E. coli RNA Polymerase Promoter Clearance Studies

To date, the majority of our studies have focused on the effects of DNA damage on the elongation stage of transcription. Obviously, the other stages of transcription (i.e. initiation, termination) are also potential targets for perturbation by the presence of DNA damage in the template or nontemplate strand. We have begun studies that address whether the transition of the RNA polymerase from the initiation to elongation state (promoter clearance) is affected by DNA base damage. Previous reports have suggested that the ability of RNA polymerase to successfully make this transition depends on the exact nucleotide sequence of the initially transcribed region of DNA (Hernandez et al., 1996). We have built 8-oxoguanine into a template similar to those used for elongation studies except that the lesion is positioned close (4 nt downstream) from the start of transcription (Viswanathan and Doetsch, 1998). When 8-oxoguanine is present on the template strand, it reduces the level of runoff transcript production (under single round transcription conditions) to ~50% that of control template. The results of these initial experiments indicate that 8-oxoguanine has a negative effect on promoter clearance and may reduce the level of gene expression when present on the template strand near the transcription start site. Furthermore, these results demonstrate that the nature of the effect of base damage on transcription (reduction in levels of normal transcript versus production of normal levels of mutant transcript) is determined by the exact location of the damage on the template strand.

4. DISCUSSION

The results of our studies with various types of DNA damage showing lesion bypass and transcriptional miscoding by RNA polymerase have at least four important implications related to DNA repair and mutagenesis mechanisms: (1) lack of transcription-coupled repair of such lesions, (2) potential net decreased repair due to RNA polymerase occupancy of template, (3) transcriptional mutagenesis, and (4) "fixation" of mutation by DNA polymerase under certain circumstances when replication over the same, unrepaired damage occurs.

4.1. Implications for Transcription-Coupled Repair

The results of these studies indicate that various types of DNA base damage, when located on the template strand of a transcribed sequence, can be efficiently bypassed by prokaryotic RNA polymerases which insert incorrect bases opposite to these lesions. These findings have several important implications for transcription-coupled repair mechanisms. The current model for the preferential removal of DNA damage from the template strand of an actively transcribed gene involves arrest of an elongating RNA polymerase at the lesion site (Selby and Sancar, 1991a; 1991b). In vitro transcription studies with DNA templates containing cyclobutane pyrimidine dimers (Selby and Sancar, 1990) on the transcribed strands show that RNA polymerase is permanently arrested at the site of damage. Cyclobutane pyrimidine dimers are preferentially removed from the transcribed strands of genes (Mellon and Hanawalt, 1989; Selby and Sancar, 1991b) and an arrested elongation complex is thought to be part of the signalling system that directs the repair machinery to the damage. Our findings that prokaryotic RNA polymerases efficiently bypass several different types of DNA damages including uracil, abasic sites, dihydrouracil, 8-oxoguanine, and O^6-methylguanine suggest that these lesions may not be subject to strand specific, transcription-coupled repair.

4.2. Implications for Reduced Lesion Processing by "Global" Repair Machinery

If DNA damages that do not cause RNA polymerase arrest are not targeted by the transcription-coupled repair machinery, then they should ultimately be repaired by the "global" DNA repair machinery that is not associated with transcription. Processing of DNA damage by global repair will depend on several factors, including how accessible a particular damage might be to the global repair machinery and, for base damages whose repair is primarily initiated by the BER pathway, on the copy number and turnover number of various BER N-glycosylases. It can be envisioned that under some circumstances, multiple elongating RNA polymerase molecules occupying a template containing a non-arresting damage might decrease accessibility of that damage for the appropriate repair proteins and thus decrease the rate of repair for certain lesions compared to their rate of repair in other areas of the genome.

4.3. Implications for Transcriptional Mutagenesis

The ability of prokaryotic RNA polymerases to efficiently bypass and insert incorrect bases opposite to a variety of different DNA damages suggests that mutant transcripts and subsequently mutant proteins might be produced in substantial amounts via miscoding during transcription. What is the likely outcome of some of the base insertion preferences we have observed in our in vitro transcription studies? For example, uracil and abasic sites are frequently occurring spontaneous lesions in cells and are generated via deamination of cytosine and depurination most often at sites of guanine, respectively (Friedberg et al., 1995). For example, RNA polymerase bypass of depurinated guanine resulting in insertion of adenine opposite to abasic sites would be equivalent to a C to A transversion mutation on the resulting transcript. Taking into account E. coli codon usage (Wada et al, 1991), if such an event occurred with an equal probability at any guanine site on the template strand of a gene, it would lead to a transcript missense mutation in about 74% of the cases (Table 2). For cytosine conversion into uracil via deamination (or into dihydrouracil via ir-

Table 2. Predicted mutagenic effects of replacing guanine or cytosine with adenine in E. coli mRNA

DNA damage site precursor base on template strand	RNA base sequence change in codon	Mutation type[1]	Number of possible different mutations[2]	Occurence (%)[3]
G	C to A	silent	10	20.4
"	"	missense	34	74.0
"	"	nonsense	4	5.6
C	G to A	silent	15	36.0
"	"	missense	31	63.0
"	"	nonsense	2	1.0

[1]Type of base substitution mutation in mRNA from adenine insertion by E. coli RNA polymerase opposite to an abasic site (originating from G) or opposite to uracil or dihydrouracil (originating from C) on the DNA template strand.
[2]Different mutations resulting from the replacement of cytosine or guanine with adenine in the first, second or third nucleotide position of a codon.
[3]Frequency of occurence of a particular mutation type taking into account the number of possible different mutations and the frequency of codon usage in E. coli. Values are normalized to 100% for each collection of codons containing cytosine and guanine.

radiation), the resulting insertion of adenine opposite to these lesions would be equivalent to a G to A transition mutation and would lead to a transcript missense mutation in about 63% of the cases. Similar calculations can be done for transcription-level base substitutions occurring opposite to sites of 8-oxoguanine and O^6-methylguanine. These calculations lead to the prediction that the most likely result of RNA polymerase bypass of these DNA base damages will be a transcript containing a missense mutation. If such a situation (transcriptional mutagenesis) occurs in vivo, it may be a major source of mutant proteins, particularly in non-dividing cells.

4.4. Implications for DNA Level Mutation Fixation

The base insertion preferences by prokaryotic RNA polymerases for most of the DNA damages examined in our in vitro transcription studies are identical to the base insertion preferences by prokaryotic DNA polymerases for the same damages from in vitro and in vivo studies (Friedberg et al., 1995). Thus lesions which miscode for RNA polymerases also miscode for DNA polymerases. One potential biological consequence of transcriptional mutagenesis is a situation where cells in a nongrowth state produce mutant proteins (via mutant transcripts) as a result of transcriptional miscoding at sites of RNA polymerase non-arresting DNA damage. Depending on the mutant protein produced, it is conceivable that a switch from a nongrowth to a growth state could take place with the initial round of DNA replication involving a DNA polymerase miscoding event at the same unrepaired lesion site in one of the daughter strands. This would effectively fix the mutation at the level of DNA and lead to the emergence of a proliferating cell population containing a mutant gene sequence (Fig. 2). We have termed this process "retromutagenesis" (since it arises initially from a transcriptional miscoding event). This could be one of the plausible pathways for the phenomenon of directed or adaptive mutation in bacteria (Bridges, 1996). In mammalian cells it may provide an explanation for how cells might overcome a growth-inhibitory environment and allow for a switch from a nongrowth to a growth state (Richards et al, 1997). Whether or not retromutagenesis occurs in cells is currently under investigation in our laboratory.

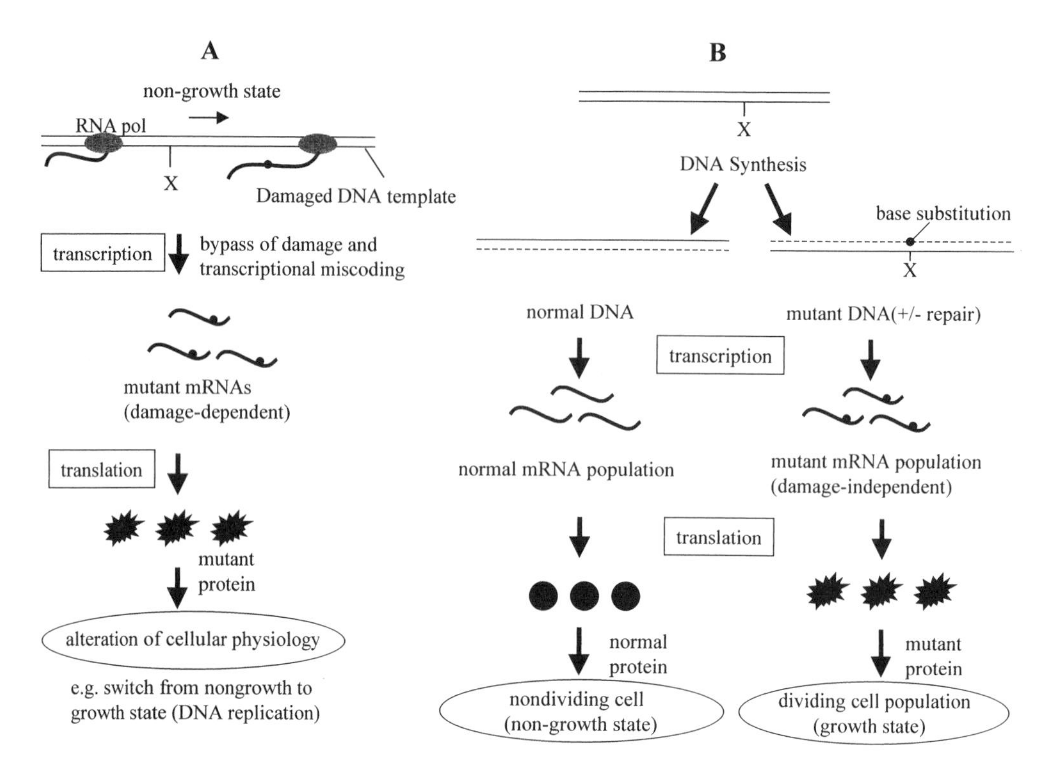

Figure 2. Retromutagenesis due to transcriptional miscoding at sites of DNA damage. One potential consequence of transcriptional mutagenesis is a switch from a cellular non-growth to a growth state. (A) Under non-growth conditions, transcription through a non-arresting DNA lesion which causes RNA polymerase miscoding results in generation of a population of mutant mRNAs which are translated into mutant proteins which, depending on the nature of the protein, could cause a switch to a growth state by allowing a round of DNA replication to occur. (B) During this first round of replication, the damage-containing template (still unrepaired) causes DNA polymerase miscoding and permanently fixes the mutation into one of the two daughter DNA strands (the other daughter strand/cell contains a normal, undamaged gene and remains in a non-growth state). A second round of replication will occur (whether or not the original damage is ultimately repaired) and will lead to the genesis of a dividing population of cells.

ACKNOWLEDGMENTS

We would like to thank the members of the Doetsch laboratory for their input and discussions related to this work. This work was supported by research grants CA42607, CA55896, and CA73041 from the National Institutes of Health.

REFERENCES

V.A. Bohr, C.A. Smith, D.S. Okumoto, and P.C. Hanawalt (1985) DNA repair in an active gene: Removal of pyrimidine dimers from the DHFR gene of CHO cells is much more efficient than in the genome overall. *Cell* **40**, 359–369.

B.A. Bridges (1995) *mutY* "directs" mutation? *Nature* **375**, 741–742.

Y-H. Chen and D.F. Bogenhagen (1993) Effects of DNA lesions on transcription elongation by T7 RNA polymerase. *J. Biol. Chem.* **268**, 5849–5855.

P.K. Cooper, T. Nouspikel, S.G. Clarkson and S.A. Leadon (1997) Defective transcription-coupled repair of oxidative base damage in Cockayne Syndrome patients from XP group G. *Science* **275**, 990–993.

C. Cullinane and D.R. Philips (1992) *In vitro* transcription analysis of DNA adducts induced by cyanomorpholino adriamycin. *Biochemistry* **31**, 9513–9519.

A. Datta and S. Jinks-Roberts (1995) Association of increased spontaneous mutation rates and high levels of transcription in yeast. *Science* **268**, 1616–1618.

M. Dizdaroglu, J. Laval and S. Bioteux (1993) Substrate specificity of the *E. coli* endonuclease III: Excision of thymine and cytosine-derived lesions in DNA produced by radiation-generated free radicals. *Biochemistry* **32**, 12105–111.

B.A. Donahue, R.P.P. Fuchs, D. Revier and P.C. Hanawalt (1996) Effects of aminofluorene and acetylaminofluorene DNA adducts on transcriptional elongation by RNA polymerase II. *J. Biol. Chem.* **271**, 10588–594.

E.C. Friedberg, G.C. Waller and W. Siede (1995) DNA Repair and Mutagenesis, American Society for Microbiology, Washington, D.C.

P.C. Hanawalt (1994) Evolution of concepts of DNA repair. *Environ. Molec. Mut.* **23(Suppl. 24)**, 78–85.

K.A. Henning, et al. (1995) The Cockayne syndrome group A gene encodes a WD repeat protein that interacts with CSB protein and a subunit of RNA pol II TFIIH. *Cell* **82**, 555–564.

V.J. Hernandez, L.M. Hsu and M. Cashel (1996) Conserved region 3 of Excherichia coli final sigma70 is implicated in the process of abortive transciption. *J. Biol. Chem.* **271**, 18775–18779.

H. Htun and B.H. Johnson (1992) Mapping adducts of DNA structural probes using transcription and primer extension approaches. *Methods Enzymol.* **212**: 272–294.

S.A. Leadon, S.L. Barbace and A.B. Dunn (1995) The yeast *RAD2* but not *RAD1* gene is involved in the transcription-coupled repair of the thymine glycols. *Mutat. Res.* **337**, 169–78.

J. Liu and P.W. Doetsch (1998) Escericha coli RNA and DNA polymerase bypass of dihydrouracil: Mutagenic potential via transcription and replication. *Nucleic Acids Res.* **26**, 1707–1712.

J. Liu and P.W. Doetsch (1996) Template strand gap bypass is a general property of prokaryotic RNA polymerases: Implications for elongation mechanisms. *Biochemistry* **35**, 14999–15008.

J. Liu, W. Zhou and P.W. Doetsch (1995) RNA polymerase bypass at sites of dihydrouracil: Implications for transcriptional mutagenesis. *Mol. Cell. Biol.* **15**: 6729–6735.

I. Mellon and G.N. Champe (1996) Products of DNA mismatch repair genes mutS and *mutL* are required for transcription-coupled nucleotide excision repair of the lactose operon in *E. coli*. *Proc. Natl. Acad. U.S.A.* ***93***:, 1292–97.

I. Mellon and P.C. Hanawalt (1989) Induction of the Escherichia coli lactose operon selectively increase repair of its transcribed DNA strand. *Nature* **342**, 95–98.

I. Mellon, D.K. Rajpal, M. Koi, C.R. Bolard and G.N. Champe (1996) Transcription-coupled repair deficiency and mutations in human mismatch repair genes. *Science* **272**: 557–560.

I. Mellon, G. Spivak and P.C. Hanawalt (1987) Selective removal of transcription-blocking DNA damage from the transcribed strand of the mammalian DHFR gene. *Cell* **51**, 241–249.

M. Pirsel and V.A. Bohr (1993) MMS adduct formation and repair in the DHFR gene and in mtDNA in hamster cells. *Carcinogenesis* **14**, 2105–08.

B. Richards, H. Zhang, G. Phear and M. Meuth (1997) Conditional mutator phenotypes in hMSH2-deficient tumor cell lines. *Science* **277**, 1523–1526.

A.Sancar (1994) Mechanisms of DNA excision repair. *Science* **266**, 1954–1956.

G. Sanchez and M.D. Mamet-Bratley (1994) Transcription by T7 RNA polymerase of DNA containing abasic sites. *Environ. Mol. Mutagen.* **23**, 32–36.

D.A. Scicchitano and P.C. Hanawalt (1990) Lack of sequence-specific removal of N-methylpurines from cellular DNA. *Mut. Res.* **233**, 31–37.

C.P. Selby and A. Sancar (1991a) Gene- and strand-specific repair *in vitro*: Partial purification of a transcription-repair coupling factor. *Proc. Natl. Acad. U.S.A.* **88**, 8232–8236.

C.P. Selby and A. Sancar (1991b) Escherichia coli mfd mutant deficient in "mutation frequency decline" lacks strand specific repair and *in vitro* complementation with purified coupling factor. *Proc. Natl. Acad. Sci., U.S.A.* **88**, 11574–11578.

C.P. Selby and A. Sancar (1993a) Molecular mechanism of transcription-repair coupling. *Science* **260**, 53–58.

C.P. Selby and A. Sancar (1993b) Transcription-repair coupling and mutation frequency decline. *J. Bacteriol.* **175**, 7509–7514.

C.P. Selby and A. Sancar (1990) Trancription preferentially inhibits nucleotide excision repair of template DNA strand *in vitro*. *J. Biol. Chem.* **265**, 21330–21336.

B.G. Taffe, F. Larminat, J. Laval, D.L. Croteau, R.M. Anson and V.A. Bohr (1996) Gene-specific nuclear and mitochondrial repair of formamidopyrimidine-DNA glycosylase-sensitive sites in Chinese hamster ovary cells. *Mutat. Res.* **364**, 183–192.

A. Van Hoffen, et al. (1993) Deficient repair of the transcribed strand of active genes in Cockayne's syndrome. *Nucleic Acids Res.* **21**, 5890–5895.

A. Viswanathan and P.W. Doetsch (1998) Effects of non bulky DNA base damages on Escherichia coli RNA polymerase-mediated elongation and promoter clearance. *J. Biol. Chem.* **273**, 21276–21281.

K. Wada, Y. Wada, H. Doi, F. Ishibashi, T. Gojobori and T. Ikemura (1991) Codon usage tabulated from the Gen Bank genetic sequence data. *Nucleic Acids Res.* **19**: 1981–1986.

W. Wang, A. Sitaram, & D.A. Scicchitano (1995) 3-Methyladenine and 7-methyladenine exhibit no preferential removal from the transcribed strand of the DHFR gene in CHO B11 cells. *Biochemistry* **34**, 1798–1804.

W. Zhou and P.W. Doetsch (1993) The effects of abasic sites and single strand DNA breaks in prokaryotic RNA polymerases. *Proc. Natl. Acad. Sci. U.S.A.* **90**, 6601–6605.

W. Zhou and P.W. Doetsch (1994a) Efficient bypass and base misinsertions at abasic sites by prokaryotic RNA polymerases. In: DNA Damage: Effects on DNA Structure and Protein Recognition. (eds. S.S. Wallace, B. Van Houten, and Y.W. Kow), The New York Academy of Sciences, New York. *Annals*, **726**, 351–354.

W. Zhou and P.W. Doetsch (1994b) Transcription bypass or blockage at single strand breaks on the DNA template strand: Effect of different 3' and 5' flanking termini on the T7 RNA polymerase elongation complex. *Biochemistry* **33**, 14926–934.

W. Zhou, Reines D. and P.W. Doetsch (1995) T7 RNA polymerase bypass of large gaps on the template strand reveals a critical role of the non-template strand in elongation. *Cell* **82**, 579–587.

W. Zhou and P.W. Doetsch (1996) Techniques for the introduction of specific DNA lesions downstream from prokaryoticRNA polymerase promoters and analysis of transcription products. In: Microbal Genome Methods (ed. Adolph, K.W.) CRC Press, Boca Raton, pp. 153–167.

10

THE CURRENT STATUS OF NUCLEOTIDE EXCISION REPAIR IN THE YEAST *SACCHAROMYCES CEREVISIAE*

Errol C. Friedberg[1], William J. Feaver,[1] Wenya Huang,[1] Michael S. Reagan,[1] Simon H. Reed,[1] Zhaoyang You,[1] Shuguang Wei,[1] Karl Rodriguez,[2] Jose Talamantez,[2] and Alan E. Tomkinson[2]

[1]Laboratory of Molecular Pathology, Department of Pathology
University of Texas Southwestern Medical Center
Dallas, Texas 75235
[2]University of Texas Health Science Center at San Antonio
San Antonio, Texas 78245

1. ABSTRACT

The removal of UV radiation-induced pyrimidine dimers and (6–4) photoproducts as well as other bulky base adducts from the DNA of higher eukaryotes relies on the concerted action of about 30 proteins which excise base damage and restore the DNA to its native state. This process is known as nucleotide excision repair (NER), and the proteins involved are highly conserved throughout the *eukaryotae*. There are indications that the initial steps of NER are effected by a large preformed multi-protein complex comprising the RNA polymerase II transcription factor IIH (TFIIH) and other NER proteins known to be required for damage recognition and DNA incision. Humans with inactivating mutations in the genes encoding NER proteins suffer from the cancer-prone syndrome xeroderma pigmentosum (XP). In yeast, a mode of NER which is apparently coupled to transcription requires the activity of Rad26 protein, while transcription-independent NER requires the activity of the Rad7 and Rad16 proteins. Defects in transcription-dependent NER are associated with the human hereditary disorder Cockayne syndrome (CS).

2. PROTEINS REQUIRED FOR THE EARLY STEPS IN NUCLEOTIDE EXCISION REPAIR IN YEAST

NER can be conveniently considered in six sequential steps: base damage recognition, localized unwinding of the DNA flanking such lesions, bimodal incision, excision of damage-

Advances in DNA Damage and Repair, edited by Dizdaroglu and Karakaya. Kluwer Academic / Plenum Publishers, New York, 1999.

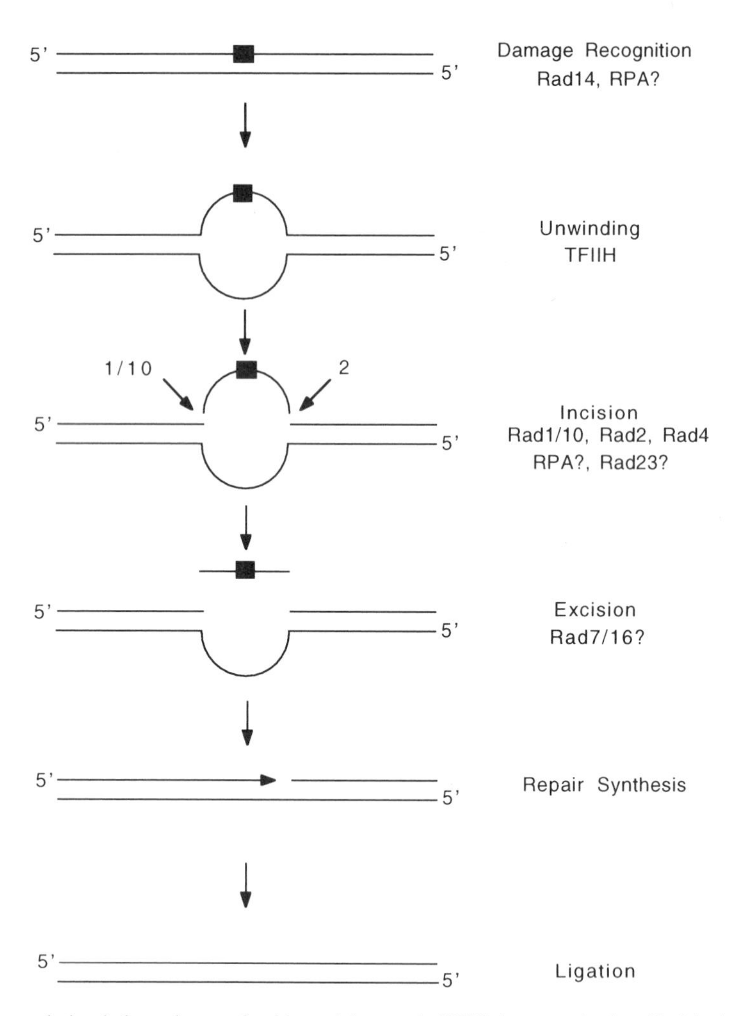

Figure 1. Transcription-independent nucleotide excision repair (NER) in yeast. As described in the text the NER reaction can be thought of as occurring in six steps. The yeast proteins believed to be required for each of the early steps are indicated. Since the exact roles of the Rad7 and Rad16 proteins are unknown their assignment to the excision step is tentative based largely on their requirement for repair synthesis, but dispensability for DNA incision *in vitro* (Wang et al., 1997; Reed et al., 1998).

containing oligonucleotides, repair synthesis, and finally ligation of the repaired regions (Friedberg et al., 1995; Figure 1). In the yeast *Saccharomyces cerevisiae* many of the genes involved in NER are known as *RAD* genes due to the sensitivity to ultraviolet (UV) *rad*iation they confer when mutated. Genes involved in the early steps of NER fall into three groups: 1) Those essential for NER but not cell viability in the absence of DNA damage (*RAD1*, *RAD2*, *RAD4*, *RAD10*, *RAD14*); 2) those essential for NER and cell viability in the absence of DNA damage (*RAD3*, *TFB1*, *TFB2*, *TFB3*, *TFB4*, *SSL1*, *SSL2*); and 3) those not essential for NER or cell viability in the absence of DNA damage (*RAD7*, *RAD16*, *RAD23*).

2.1. Damage Recognition and Incision

Mutants in *RAD* genes from group 1 have been shown to be defective for the incision step of NER, an early event in the process, indicating their involvement either in the incision reaction itself or in events leading to incision such as damage recognition. Early genetic studies in *S. cerevisiae* identified a number of genes that are essential for NER, namely *RAD1*, *RAD2*, *RAD3*, *RAD4*, *RAD10* and *RAD14* (reviewed in Hoeijmakers, 1993a,b). Mutations in these genes render cells extremely sensitive to UV irradiation and all have been shown to be totally defective in NER *in vitro*. Most of these proteins have been purified to homogeneity and their functions in NER defined biochemically. Rad14, a 35 kDa protein, is the yeast homolog of human xeroderma pigmentosum group A protein (XPA). Similar to the bacterial damage-recognition protein UvrA, Rad14 is a zinc metalloprotein which binds specifically to UV radiation-damaged DNA (Guzder et al., 1993). Independent studies have demonstrated the absolute requirement for Rad14 in the incision of damaged DNA (Guzder et al., 1995a). Rad14 is thought to be the primary damage recognition factor for NER.

Two other NER proteins, Rad1 (126 kDa) and Rad10 (28 kDa), have also been extensively studied. These proteins constitute an endonuclease required for both NER and mitotic recombination (Bardwell et al., 1992; Bardwell et al., 1993; Tomkinson et al., 1993; Bardwell et al., 1994a). Studies using model DNA substrates have demonstrated that the Rad1/Rad10 endonuclease specifically incises "bubbled DNA" at the 5' side of the unpaired region, indicating that during NER it functions as the endonuclease which cleaves 5' to damaged bases (Davies et al., 1995). The endonuclease that cleaves on the 3' side of the lesion is thought to be the Rad2 protein. Rad2 (130 kDa) is absolutely required for the incision of UV-damaged DNA and has been shown to possess endonuclease activity specific for single-stranded DNA (Habraken et al., 1993). Purified XPG protein, the human counterpart of Rad2, cleaves on the 3' side of "bubbled" DNA structures and together with Rad1/Rad10 can excise a fragment 24–27 nucleotides in length (Davies et al., 1995). The 5' and 3' incisions are therefore thought to be generated by Rad1/Rad10 and Rad2 respectively. Early genetic evidence indicated that the 5' and 3' incisions may be coordinated (Wilcox and Prakash, 1981). However, in humans uncoupled 5' and 3' incisions can be detected under some conditions (Matsunaga et al., 1995). It is conceivable that both endonucleases need to be positioned correctly with respect to the lesion and each other before either incision can occur.

Unlike the proteins discussed above, relatively little is known about the role of Rad4. *In vivo* and *in vitro* studies have revealed no detectable NER in the absence of Rad4 protein (Wilcox and Prakash, 1981; Svejstrup et al., 1995). Rad4 has been shown to co-purify with Rad23 protein (Guzder et al., 1995b) and interacts with the Rad7 and Rad23 proteins in the two-hybrid assay (Wang et al., 1997). Rad4, in addition to all of the other proteins in group 1, has been shown to be a component of the yeast nucleotide excision repairosome (Svejstrup et al., 1995), a large multi-protein complex believed to contain all proteins essential for NER (Section 3).

2.2. TFIIH

Mutants from the second group of NER genes are also incision-defective and have an additional role in cell viability in the absence of DNA damage. The latter role was identified following the discovery that these proteins exist as a complex known as TFIIH, which is required for transcription initiation by RNAP II as well as NER (Svejstrup et al., 1996a). TFIIH was identified and purified based on its ability to restore transcription activity to a heat-inactivated nuclear extract (Feaver et al., 1991a). The repair connection became apparent with the identification of Rad3 as a TFIIH subunit (Feaver et al., 1993). Subsequently a requirement for the entire TFIIH complex in NER was directly demonstrated (Wang et al., 1994). With the recent identification of genes encoding the Tfb2,

Table 1. Yeast and human TFIIH subunits*

Yeast TFIIH (Gene)	Human TFIIH (gene)	% Identity (similarity)
p105 / Ssl2 (SSL2)	p89 (XPB)	55.0 (72.1)
p85 / Rad3 (RAD3)	p80 (XPD)	53.0 (72.9)
p75 / Tfb1 (TFB1)	p62 (BTF2p62)	26.0 (49.3)
P55 / Tfb2 (TFB2)	p52 (BTF2p52)	39.7 (63.9)
p50 / Ssl1 (SSL1)	p44 (BTF2p44)	41.8 (60.1)
p47 / Ccl1 (CCL1)	p38 (Cyclin H)	31.5 (53.6)
p45 / " "	" "	—
p38 / Tfb3 (TFB3)	p32 (MAT1)	32.5 (58.8)
p37 / Tfb4 (TFB4)	p34 (BTF2p34)	32.6 (57.4)
p33 / Kin28 (KIN28)	p41 (MO15/CDK7)	47.5 (68.3)

*Adapted from Feaver et al. (1997). Correlation of yeast and human TFIIH subunits showing amino acid identity/similarity. Yeast subunits p47 and p45 are distinct isoforms of Ccl1 protein likely resulting from translation initiation at a downstream AUG codon (Svejstrup et al., 1996b). It is thought that each isoform individually associates with Kin28 protein to form a cyclin/cyclin-dependent kinase pair.

Tfb3 and Tfb4 subunits the molecular definition of yeast TFIIH has been completed (Feaver et al., 1997). Holo-TFIIH, the form of the factor active in transcription, is comprised of 9 subunits, each of which has a counterpart in human TFIIH (Table 1; Feaver et al., 1997; Marinoni et al., 1997). Under certain conditions holo-TFIIH can be dissociated into 2 sub-complexes known as core-TFIIH and TFIIK (Svejstrup et al., 1995). Core-TFIIH, comprised of the Ssl2, Rad3, Tfb1, Tfb2, Ssl1, Tfb4 and Tfb3 subunits, is the form of TFIIH thought to be active in NER (Svejstrup et al., 1995). TFIIK, the protein kinase which phosphorylates the C-terminal domain of the largest subunit of RNAP II during transcription initiation, is comprised of the cyclin/cyclin-dependent kinase pair Ccl1/Kin28 (Feaver et al., 1991b; Feaver et al., 1994; Svejstrup et al., 1996b). During purification it became apparent that Ssl2 and Tfb4 were weakly associated with the other subunits of core-TFIIH (Feaver et al., 1993; Svejstrup et al., 1994; Feaver et al., 1997). Similarly, a form of core-TFIIH lacking the Rad3 subunit has been observed (Sung et al., 1996; W.J.F. and E.C.F., unpublished). The functional significance of core-TFIIH sub-complexes is currently unclear and may simply represent purification artifacts.

Prior to the discovery of the NER repairosome (Section 3) the first evidence for higher order repair complexes in yeast cells in the absence of exogenous DNA damage was the observation that both Rad2 and Rad4 proteins co-immunoprecipitated with core-TFIIH (Bardwell et al., 1994c). Since this study much effort has gone into the identification of interactions between TFIIH subunits and between TFIIH subunits and other NER proteins. The results of these studies are summarized in Table 2. The elucidation of pair-wise interactions between various proteins has prompted us to propose models of TFIIH and repairosome structure (Figure 2). Additional interactions remain to be established. For example, it is not known which subunit of core-TFIIH interacts with Rad4 (Bardwell et al., 1994c). Similarly, co-purification has shown that Tfb2 is tightly associated with the other subunits of core-TFIIH in the absence of Ssl2, the only protein that has currently been shown to interact directly with Tfb2 (W.J.F. and E.C.F., unpublished).

Together with the elucidation of protein/protein interactions, genetic and biochemical approaches have been used to investigate the requirement of individual TFIIH subunits for NER. Previous studies had shown that mutations in *SSL2*, *RAD3*, *TFB1* and *SSL1* rendered cells sensitive to UV radiation and defective for NER *in vitro* (Wang et al., 1994; Wang et al., 1995a). More recently a similar analysis for a *TFB2* C-terminal deletion mu-

Table 2. Known interactions between yeast TFIIH and/or repairosome subunits and the methods used to identify them*

Interaction	Method	References
Tfb1/Ssl1	TH	Feaver et al., 1993; Bardwell et al, 1994b
Rad3/Ssl1	TH;CIP	Bardwell et al., 1994b
Rad3/Ssl2	CIP	Bardwell et al., 1994b
Rad2/Tfb1	CIP	Bardwell et al., 1994c
Rad2/Ssl2	CIP	Bardwell et al., 1994c
Tfb4/Ssl1	TH	W.J.F. and E.C.F., unpublished
Rad3/Tfb3	TH	W.J.F. and E.C.F., unpublished
Tfb2/Ssl2	TH	W.J.F. and E.C.F., unpublished
Tfb3/Kin28	TH	Feaver et al., 1997
Kin28/Ccl1	TH	Valay et al., 1993; Feaver et al., 1997
Rad4/Rad23	CP;TH	Guzder et al., 1995a; Wang et al., 1997
Rad7/Rad4	TH	Wang et al., 1997
Rad1/Rad10	CIP;TH	Bardwell et al., 1992; Bardwell et al., 1993
Rad7/Rad16	TH;GST;CP	Wang et al., 1997; Guzder et al., 1997

*TH, two-hybrid assay; CIP, co-immunoprecipitation; CP, co-purification; GST, GST pull down

tant (Feaver et. al., 1997) and temperature-sensitive alleles of *TFB3* and *TFB4* (W.J.F., W.H. and E.C.F., unpublished) have also revealed UV radiation sensitivity and defective NER *in vitro*. Thus, all of the subunits of core-TFIIH are required for NER *in vivo* and *in vitro*. In contrast, several temperature-sensitive alleles of both *KIN28* and *CCL1* failed to exhibit significantly enhanced killing by UV irradiation, suggesting that TFIIK does not play a role in NER (Z. Wang, G. Faye and E.C.F., unpublished). The apparent absence of TFIIK in the repairosome supports this conclusion (Svejstrup et al, 1995).

2.3. The Rad7, Rad16 and Rad23 Proteins

Genes from the third group are only moderately UV sensitive and are neither essential for NER nor cell viability (Friedberg et al., 1995). This has led to the suggestion that these proteins either affect the overall efficiency of NER or specifically operate in certain sub-pathways (Wang et al., 1997). Consistent with the latter conclusion it has been shown that *rad7* and *rad16* mutants are defective in the removal of cyclobutane pyrimidine dimers (CPD) in transcriptionally-silent regions of the genome, such as the mating type loci and several mating type genes (e.g., *MFA2* gene in α cells) (Bang et al., 1992; Verhage et al., 1994). Based on these results it was initially proposed that the role of Rad7 (64 kDa) and Rad16 (92 kDa) was to "open up" transcriptionally-silent regions of the genome. It is known that many of these silent regions exist in a nucleosome conformation similar to that of heterochromatin in higher eukaryotes (Nasmyth, 1982). It was suggested that in these regions the DNA conformation precludes access of the repair machinery to sites of DNA damage. The demonstration that Rad7 and Rad16 are required for CPD removal from the non-transcribed strand of transcriptionally-active genes (Verhage et al., 1994) provided further support for the existence of NER sub-pathways. The current consensus is that in the absence of transcription a global repair pathway operates which requires the Rad7 and Rad16 proteins, while in actively transcribing regions a subpathway known as transcription coupled repair (Section 5) takes place, which can occur in the absence of Rad7/16. In contrast to the NER proteins discussed above, Rad7 and Rad16 do not appear to be absolutely required for incision of damaged DNA (Section 4).

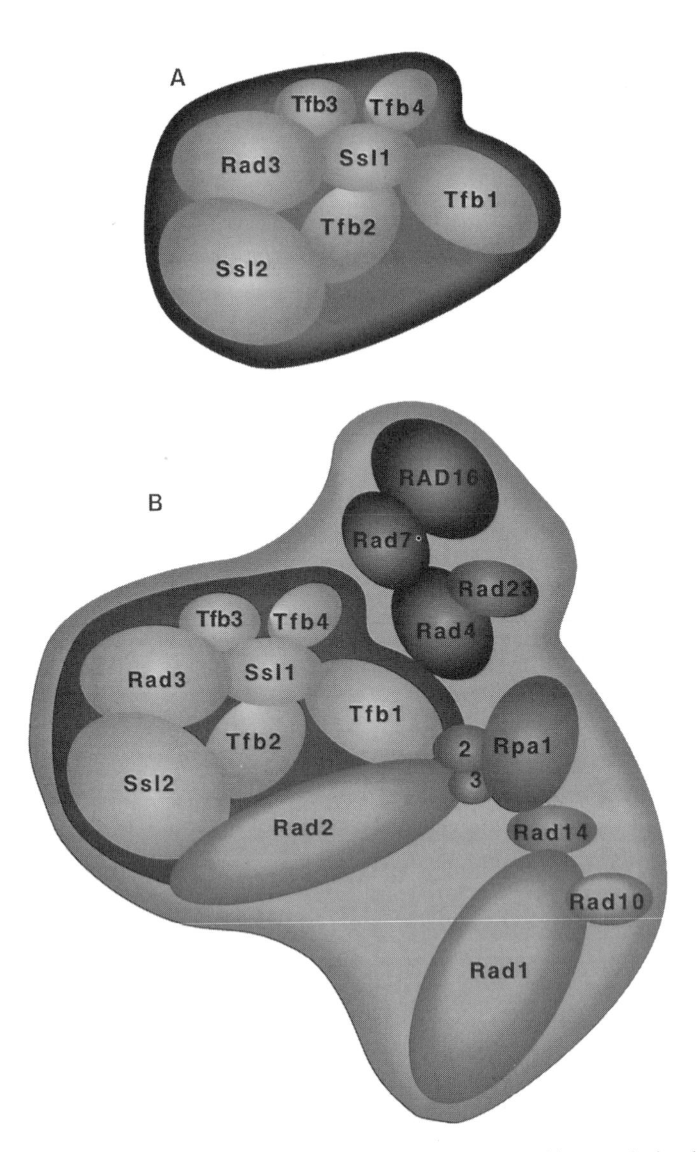

Figure 2. Structure of holo-TFIIH (A) and the NER repairosome (B). The architecture depicted is based on chromatographic properties and known pairwise interactions (Table 2). Additional interactions are inferred from the properties of the homologous human proteins. He et al. (1995) have shown by affinity chromatography and co-immunoprecipitation that RPA interacts with both XPA and XPG proteins.

Mutant strains defective in the *RAD23* gene also show an intermediate UV radiation sensitivity, similar to that shown by *rad7* and *rad16* mutants (Friedberg et al. 1995), and are defective in NER *in vitro* (Wang et al., 1997). However, in the case of *rad23* the repair defect is not confined to non-transcribed regions. The presence of a ubiquitin-like N-terminus suggests a possible role in ubiquitin-mediated protein degradation (Watkins et al., 1993). However, the precise role of Rad23 in NER remains elusive. As mentioned above, we have shown by two-hybrid analysis that Rad23 interacts with Rad4 protein and that Rad4 interacts with Rad7 protein (Wang et al., 1997). Additionally, functional complementation has shown that Rad23 is a component of the NER repairosome (Wang et al., 1997). We have constructed a yeast strain expressing histidine-tagged Rad23 protein in order to facilitate immobilized metal affinity chromatography (IMAC). We are currently using this strain to identify novel Rad23 interacting proteins.

2.4. DNA Replication Proteins Involved in the Early Steps of NER

In recent years three human DNA replication factors have been reported to be involved in NER. Yeast replication protein A (RPA), consisting of the three subunits RPA1 (70 kDa), RPA2 (34 kDa) and RPA3 (14 kDa), is essential for viability and contains a domain which preferentially binds single-stranded DNA (Brill and Stillman, 1989; Brill and Stillman, 1991). RPA has been shown to be important for maintaining the structure of replication foci (Wobbe et al., 1987; Wold and Kelly, 1988; Fairman and Stillman, 1988). He et al. (1995) have shown that RPA interacts with the NER proteins XPA and XPG in humans, suggesting a possible role in both damage recognition and incision. Additionally, purification and characterization of the NER repairosome using his-tagged Rad14 protein has identified RPA2 as a repairosome component (Rodriguez et al., in press), consistent with the association of RPA with other NER subunits. Independent lines of study using reconstituted NER systems have demonstrated that RPA is required for the incision of DNA damaged by either UV radiation or N-acetoxy-2-aminoacetylfluorene, indicating the involvement of RPA in an early step of NER (Guzder et al., 1995a). Recent studies in cell free extracts have confirmed that RPA is indispensable for NER (Huang et al., 1998).

Another replication factor recently identified as being involved in NER is proliferating cell nuclear antigen (PCNA). *In vivo* PCNA consists of a homotrimer of three 29 kDa subunits, and like RPA it is essential for cell viability. PCNA is recruited by replication factor C (RFC) and functions as a processivity factor for DNA polymerases δ and ε during DNA replication (reviewed in Kelman, 1997). A role for PCNA in NER has been demonstrated in human cells, presumably during the DNA resynthesis step (Shivji et al., 1992; Aboussekhra et al., 1995). In *S. cerevisiae* a number of PCNA mutant strains have been reported to be mildly sensitive to UV irradiation. However, extracts from these mutants fail to exhibit a deficiency in NER *in vitro* (Ayyagari et al., 1995). Recently the N-degron strategy (Dohmen et al., 1994) has been used to deplete yeast PCNA *in vivo*. Cell-free extracts from PCNA-depleted cells exhibit a marked deficiency in NER activity, suggesting a similar role *in vivo* (Huang et al., 1998). It remains to be determined whether PCNA functions solely during DNA resynthesis or also plays additional roles in earlier steps. Studies by Aboussekhra et al. (1995) have shown a requirement for RFC in a reconstituted human NER system, however, similar evidence has not yet been reported for yeast RFC.

3. THE NUCLEOTIDE EXCISION REPAIROSOME

An area of current controversy surrounds the existence or not of a preformed "repairosome" that can be isolated from eukaryotic cells in the absence of DNA damage. Svejstrup et al. (1995) demonstrated the presence of a large multi-subunit complex consisting of core-TFIIH and other repair proteins (Rad1, Rad2, Rad4, Rad10, Rad14 and Ssl2) that has been termed the repairosome (Figure 2B). Other investigators were unable to isolate this complex using a similar, but not identical, purification procedure. Instead they isolated sub-complexes which they suggested are sequentially assembled at sites of base damage in DNA (Guzder et al., 1996). The yeast repairosome was initially purified by modification of the existing TFIIH purification protocol, which took advantage of a histidine affinity tag on the Tfb1 subunit (Svejstrup et al., 1994). Recent studies from our laboratory have used a strain expressing histidine-tagged Rad14 protein to partially purify and characterize a large repairosome complex which includes the Rad7, Rad16 and RPA proteins in addition to those previously identified (Rodriguez et al., in press; Figure 2). Interestingly, the original histidine-tagged Tfb1 repairosome did not contain functional Rad7 or Rad16 (Wang et al., 1997). Perhaps Rad7 and Rad16 are loosely associated and were largely lost from the early repairosome preparations during purification. It remains to be determined if Tfb4 is a component of the repairosome. This seems likely based on the Tfb4 requirement for NER *in vitro* and the UV radiation sensitivity of a *tfb4* mutant strain *in vivo*. The recent isolation of a repairosome complex from human cells provides additional support for the existence of preassembled repair complexes in eukaryotic cells (He et al., 1997).

4. POSSIBLE ROLES OF *Rad7* AND *Rad16* IN TRANSCRIPTION-INDEPENDENT NER

The *in vitro* NER assay developed in our laboratory measures transcription-independent incorporation of radiolabeled dCTP into repair patches of a damaged plasmid (Wang et al., 1995b). The assay therefore measures a *post-incision* event in the repair reaction. Extracts prepared from *rad7* or *rad16* mutants are defective in this assay (Wang et al., 1997). Investigations to determine the cause of this defect initially focused on the possibility that addition of naked damaged template DNA to the yeast cell-free extracts results in nucleosome assembly, thus generating a heterochromatin-like DNA structure which requires the Rad7 and Rad16 proteins for DNA repair. Evidence for nucleosome assembly *in vitro* has not been detected using standard techniques (S.H.R and E.C.F). However, we cannot rule out the possibility that limited nucleosome structures exist which cannot be detected using current methods. While extracts of *rad7* or *rad16* strains are defective in repair synthesis, they are capable of incision of damaged DNA *in vitro* (Reed et al., 1998). We have also shown that incision is proficient *in vivo* in regions of the genome that are unrepaired in *rad7* or *rad16* mutants (such as the mating type loci).

The observation that the Rad7 and Rad16 proteins are components of the repairosome, and that neither is apparently required for DNA incision, has led us to consider the possibility that extracts from *rad7* or *rad16* mutants contain a defective repairosome complex which is capable of DNA incision *in vitro* but is blocked in later steps.

Evidence based on UV radiation survival and NER phenotypes are consistent with Rad7 and Rad16 proteins residing in a complex. We have demonstrated a direct interaction of these two proteins by the two-hybrid assay and by using a GST "pull-down" protocol (Wang et al., 1997). The observation that the Rad16 protein shares homology with the yeast DNA-dependent ATPase Snf2/Swi2 (Bang et al., 1992; Laurent et al., 1993) has stimulated interest in identifying a yeast Rad7/Rad16 complex containing novel components. Our observation that Rad7 and Rad16 interact has been independently confirmed in a study in which both proteins were co-overexpresed and purified from yeast (Guzder et al., 1997). These authors showed that the two proteins complex in a 1:1 stoichiometry and bind to UV radiation-damaged DNA in an ATP-dependent manner. This has led to the suggestion that the Rad7/Rad16 heterodimer acts as a sensor of DNA damage. Intriguingly, while binding of the complex to damaged DNA is dependent on the presence of ATP, ATP hydrolysis is not required.

We have constructed a histidine-tagged strain in which *RAD7* expression is under the control of its own promoter on a centromeric plasmid, and the chromosonal *RAD7* gene is deleted (Reed et al. 1998). Under these conditions Rad7 protein levels are comparable to that of the parental strain. We have purified tagged Rad7 protein and demonstrated copurification with Rad16. Our current model for the role of Rad7 and Rad16 is that during NER a repair intermediate is formed which requires the Rad7/16 complex for further processing. This intermediate may not be formed in actively transcribing regions, thereby precluding a requirement for Rad7 and Rad16 proteins in NER of transcriptionally-active DNA. Alternatively some other protein(s) related to transcription such as Rad26, may perform this function in transcriptionally-active DNA (see Section 5).

5. POSSIBLE ROLES OF *Rad26* AND *Rad28* IN TRANSCRIPTION-DEPENDENT NER

In humans defects in genes encoding NER proteins, including the XPB and XPD subunits of the transcription/repair factor TFIIH, have been directly implicated in the hereditary NER-defective disease XP. Patients with XP exhibit severe photosensitivity and are extremely prone to sunlight-induced skin cancer. Defects in the XPB and XPD proteins have also been implicated in a form of XP associated with the clinical features of Cockayne syndrome (CS). CS is a rare autosomal disease characterized primarily by growth and neurological defects. The difficulty of rationalizing the clinical features of CS in XP/CS patients exclusively in terms of defective NER has led to the "transcription hypothesis" for these diseases (Bootsma and Hoeijmakers, 1993; Vermeulen et al., 1994; Friedberg et al., 1996). This hypothesis suggests that the clinical features of CS are caused by subtle defects in gene expression during development. Despite intensive investigation in recent years it remains unclear if the primary defect in CS is related to repair or transcription or neither (Friedberg, 1996).

Clinically "pure" CS falls into 2 genetic complementation groups, CS-A and CS-B. Mutations in either the *CSA* or *CSB* genes result in UV radiation sensitivity and a defect in the preferential NER of transcriptionally-active genes. Preferential repair, also known as transcription coupled repair (TCR) or transcription-dependent repair, is a phenomenon first reported by Hanawalt and colleagues (Bohr et al., 1985) in which DNA lesions are more rapidly removed from the template strand of active genes than either the non-template strand or the genome overall. It has been proposed that the *CSA* and *CSB* gene products physically couple repair proteins to elongating RNAP II complexes stalled at sites of DNA damage (Hanawalt, 1994). In support of this model CSB protein has been shown to interact with both RNAP II and XPG protein (van Gool et al., 1997; Tantin et al., 1997) and CSA protein has been shown to interact with a subunit of TFIIH (Henning et al, 1995). Transcription-coupled repair has been observed in both prokaryotic and eukaryotic cells. While well understood in bacteria (Selby and Sancar, 1993), the molecular mechanism underlying this phenomenon in higher organisms has not been elucidated. The lack of a system that supports transcription-dependent repair *in vitro* has impeded progress in this area. The development of such a system may prove exceedingly difficult in light of the presence of a factor in crude extracts of human cells which promotes the release of RNAP II molecules stalled at lesions *in vitro* (Selby et al., 1997).

A yeast homolog of *CSB*, *RAD26* has been isolated (van Gool et al., 1994) and recently a yeast gene homologous to *CSA*, *RAD28*, has also been identified (Bhatia et al., 1996). The *RAD28* gene encodes a member of the WD-repeat family, a series of proteins involved in a variety of cellular processes including transcription (Neer et al., 1994). The WD domain is thought to provide a surface for protein/protein interactions, as many of these proteins are components of multisubunit complexes. Rad26, like Rad16 is a member of the Swi2/Snf2 superfamily of DNA-dependent ATPases. By analogy with the Swi/Snf complex perhaps the function of Rad26 (and possibly Rad28) is to overcome the action of inhibitors of RNAP II elongation. Such inhibitors could be DNA lesions or some element of DNA or chromatin structure. Indeed, cyclobutane pyrimidine dimers block RNAP II elongation (Donahue et al., 1994; Selby et al., 1997) and HMG2, a non histone chromosomal protein, inhibits transcription *in vitro* (Stelzer et al., 1994). A role for Rad26/CSB in elongation would be consistent with the transcription hypothesis of CS, however, involvement in initiation cannot be excluded. In support of this notion is the recent finding that CS-A, CS-B and XP-B/CS cells exhibit abnormal levels of RNAP II-directed transcription in crude extracts (Dianov et al., 1997) and *in vivo* (Balajee et al., 1997). In yeast we have also shown that Rad26 (but not Rad28) is required for the recovery of RNA synthesis following UV irradiation (Reagan and Friedberg, 1997).

Neither Rad26 or Rad28 is essential for viability and unlike their human counterparts, neither protein is required for UV radiation resistance (van Gool et al., 1994; Bhatia et al., 1996). Consistent with the lack of UV radiation sensitivity neither Rad26 nor Rad28 is required for NER *in vitro* (You et al., 1998). Similarly, *rad26* and *rad28* mutant extracts support wild-type levels of RNAP II transcription (You et al., 1998) unlike their human counterparts (Dianov et al., 1997). These findings suggest that there may be functional differences between the yeast and human proteins. Perhaps CSA and CSB have evolved additional functions not possessed by the yeast factors. Growth of *rad26* and *rad28* deletion mutants as well as *rad26/28* double mutants is indistinguishable from wild-type cells on various carbon sources (glucose, raffinose, sucrose, galactose and glycerol) at 15°C, 30°C and 37°C (W.J.F. and E.C.F., unpublished). Deletion of *RAD26* and/or *RAD28* does not result in sensitivity to caffeine, 6-azauracil, high concentrations of salt or sorbitol, or inositol auxotrophy (W.J.F. and E.C.F., unpublished). Analysis of *rad26* and *rad28* mutants has therefore proved largely refractory to genetic investigation.

Yeast strains lacking Rad26 (but not Rad28) are largely defective for TCR (van Gool et al., 1994; Bhatia et al., 1996). Interestingly, the low level of residual TCR in a *rad26* mutant is restricted to a region close to the promoter, prior to the point at which TFIIH is believed to dissociate from the elongating RNA polymerase (Tijsterman et al., 1997). These authors concluded that the role of Rad26 in TCR is to recruit TFIIH to stalled polymerase complexes, however, to date an interaction between Rad26 and TFIIH has not been reported. One might expect to see similar results if Rad26 acts as an elongation factor.

The study of TCR will hopefully provide insights into the relationship(s) between transcription and DNA repair. To further this end we have been investigating competition between NER and transcription *in vitro*. In yeast whole cell extracts which support both NER and transcription we have observed that ongoing NER significantly inhibits transcription by RNAP II (You et al., 1998). Conversely, increased levels of transcription had no effect on NER. Inhibition could be specifically relieved by the addition of a transcriptionally competent form of transcription factor IIH (TFIIH). This result was not unexpected, as TFIIH is required both for NER and transcription. Surprisingly, inhibition of transcription by NER also requires Rad26 protein (but not Rad28). Our current model to explain these results is that Rad26 is involved in the assembly and/or disassembly of TFIIH-containing transcription complexes.

REFERENCES

A. Aboussekhra, M. Biggerstaff, M.K.K.Shivji, J.A. Vilpo, V. Moncollin, V.N. Podust, M. Protic, U. Hubscher, J.M. Egly and R.D. Wood (1995) Mammalian DNA nucleotide excision repair reconstituted with purified protein components. *Cell* **80**, 859–868.

R. Ayyagari, K.J. Impellizzeri, B.L. Yoder, S.L. Gary and P. M. J. Burgers (1995) A mutational analysis of the yeast proliferating cell nuclear antigen indicates distinct roles in DNA replication and DNA repair. *Mol. Cell. Biol.* **15**, 4420–4429.

A.S. Balajee, A. May, G.L. Dianov, E.C. Friedberg and V.A. Bohr (1997) Reduced RNA polymerase II transcription in intact and permeabilized Cockayne syndrome B cells. *Proc.Natl. Acad. Sci. USA* **94**, 4306–4311.

D.D. Bang, R.A. Verhage, N. Goosen, J. Brouwer and P. van de Putte (1992) Molecular cloning of RAD16, a gene involved in differential repair in Saccharomyces cerevisiae. *Nucleic Acids Res.* **20**, 3925–3931.

A.J. Bardwell, L. Bardwell, D.K. Johnson and E.C. Friedberg (1993) Yeast DNA recombiantion and repair proteins constitute a complex in vivo mediated by localized hydrophobic domains. *Mol. Microbiol.* **8**, 1177–1188.

A.J. Bardwell, L. Bardwell, A.E. Tomkinson and E.C. Friedberg (1994a) Specific cleavage of model recombination and repair intermediates by the yeast Rad1-Rad10 DNA endonuclease. *Science* **265**, 2082–2085.

A.J. Bardwell, L. Bardwell, N. Iyer, J.Q. Svejstrup, W.J. Feaver, R.D. Kornberg and E.C. Friedberg (1994c) Yeast nucleotide excision repair proteins Rad2 and Rad4 interact with RNA polymerase II basal transcription factor b (TFIIH). *Mol. Cell Biol.* **14**, 3569–3576.

P.K. Bhatia, R.A. Verhage, J. Brouwer and E.C. Friedberg (1996) Molecular cloning and characterization of S. cerevisiae RAD28, homolog of the human Cockayne syndrome A (CSA) gene. *J. Bacteriol.* **128**, 5877–5987.

V.A. Bohr, C.A. Smith, D.S. Okkumoto and P.C. Hanawalt (1985) DNA repair in an acitve gene: Removal of pyrimidine dimers from the DHFR gene of CHO cells is much more efficient than in the genome overall. *Cell* **40**, 359–369.

D. Bootsma and J.H.J. Hoeijmakers (1993) Engagement with transcription. *Nature* **363**, 114–115.

S. Brill and B. Stillman (1989) Yeast replication factor-A functions in the unwinding of the SV40 origin of DNA replication. *Nature* **342**, 92–95.

S. Brill and B. Stillman (1991) Replication factor-A from Saccharomyces cerevisiae is encoded by three essential genes coordinately expressed at S phase. *Genes & Dev.* **5**, 1589–1600.

A.A. Davies, E.C. Friedberg, A.E. Tomkinson, R.D. Wood. and S.C. West. (1995) Role of the Rad1 and Rad10 proteins in nucleotide excision repair and recombination. *J. Biol. Chem.* **270**, 24638–24641.

G.L. Dianov, J.F. Houle, N. Iyer, V.A. Bohr and E.C. Friedberg (1997) Reduced RNA polymerase II trasncription in extracts of Cockayne syndrome and xeroderma pigmentosum/Cockayne syndrome cells. *Nucelic Acids Res.* **25**, 3636–3642.

R.J. Dohmen, P. Wu. and A. Varshavsky (1994) Heat-inducible degron: A method for constructing temperature-sensitive mutants. *Science* **263**, 1273–1276.

B.A. Donahue, S. Yin, J.S., Taylor, D. Rienes and P.C. Hanawalt (1994) Transcript cleavage by RNA polymerase II arrested by a cyclobutane pyrimidine dimer in the DNA template. *Proc. Natl. Acad. Sci. USA* **91**, 8502–8506.

M.P. Fairman and B. Stillman (1988) Cellular factors required for multiple stages of SV40 DNA replication in vitro. *EMBO J.* **7**, 1211–1218.

W.J. Feaver, O. Gileadi and R.D. Kornberg (1991a) Purification and characterization of yeast RNA polymerase II transcription factor b. *J. Biol. Chem.* **266**, 19000–19005.

W.J. Feaver, O. Gileadi, Y. Li and R.D. Kornberg (1991b) CTD kinase associated with yeast RNA polymerase II initiation factor b. *Cell* **67**, 1223–1230.

W.J. Feaver, J.Q. Svejstrup, L. Bardwell, A.J. Bardwell, S. Buratowski, K.D. Gulyas, T.F. Donahue, E.C. Friedberg and R.D. Kornberg (1993) Dual roles of a multiprotein complex from S. cerevisiae in transcription and DNA repair. *Cell* **75**, 1379–1387.

W.J. Feaver, J.Q. Svejstrup, N.L. Henry and R.D. Kornberg (1994) Relationship of CDK-activating kinase and RNA polymerase II CTD kinase TFIIH/TFIIK. *Cell* **79**, 1103–1109.

W.J. Feaver, N.L. Henry, Z.Wang, X. Wu, J.Q. Svejstrup, D.A. Bushnell, E.C. Friedberg and R.D. Kornberg (1997) Genes for Tfb2, Tfb3 and Tfb4 subunits of yeast transcription/repair factor IIH: Homology to human "CAK" and IIH subunits. *J. Biol. Chem.* **272**, 19319–19327.

E.C. Friedberg, G.C. Walker and W. Siede (1995) DNA Repair and Mutagenesis. ASM Press, Washington, DC.

E.C. Friedberg (1996) Cockayne syndrome - A primary defect in DNA repair, transcription, both or neither? *BioEssays* **18**, 731–738.

S.N. Guzder, P. Sung, L. Prakash and S. Prakash (1993) Yeast DNA-repair gene RAD14 encodes a zinc metalloprotein with affinity for ultraviolet-damaged DNA. *Proc. Natl. Acad. Sci. USA* **90**, 5433–5437.

S.N. Guzder, Y. Habraken, P. Sung, L. Prakash and S. Prakash (1995a) Reconstitution of yeast nucleotide excision repair with purified rad proteins, replication protein A, and transcription factor TFIIH. *J. Biol. Chem.* **270**, 12873–12976.

S.N. Guzder, V. Bailly, P. Sung, L. Prakash and S. Prakash (1995b) Yeast DNA repair protein RAD23 promotes complex formation between transcription factor TFIIH and DNA damage recognition factor RAD14. *J. Biol. Chem.* **270**, 8385–8388.

S.N. Guzder, P. Sung, L. Prakash and S. Prakash (1996) Nucleotide excision repair in yeast is mediated by sequential assembly of repair factors and not by a pre-assembled repairosome. *J. Biol. Chem.* **271**, 8903–8910.

S.N. Guzder, P. Sung, L. Prakash and S. Prakash (1997) Yeast Rad7–16 complex, specific for the nontranscribed DNA strand, is an ATP-dependent DNA damage sensor. *J. Biol. Chem.* **272**, 21665–21668.

Y. Habraken, P. Sung, L. Prakash and S. Prakash (1993) Yeast excision repair gene RAD2 encodes a single-stranded DNA endonuclease. *Nature* **366**, 365–368.

P.C. Hanawalt (1994) Transcription coupled repair and human disease. *Science* **266**, 1957–1958.

Z. He, L.A. Hendrickson, M.S. Wold and C.J. Ingles (1995) RPA involvement in the damage-recognition and incision steps of nucleotide excision repair. *Nature* **374**, 566–569.

Z. He and C.J. Ingles (1997) Isolation of human complexes proficient in nucleotide excision repair. *Nucleic Acids Res.* **25**, 1136–1141.

K.A. Henning, L. Li, N. Iyer, L.D. McDaniel, M.S. Regan, S. Legerski, R.A. Shultz, M. Stefanini, A.R. Leman, L.V. Mayne and E.C. Friedberg (1995) The Cockayne syndrome group A gene encodes a WD repeat protein that interacts with CSB protein and a subunit of RNA polymerase II transcription factor IIH. *Cell* **82**, 555–564.

J.H.J. Hoeijmakers (1993a) Nucleotide excision repair I: from E. coli to yeast. *Trends Genet.* **9**, 173–177.

J.H.J. Hoeijmakers (1993b) Nucleotide excision repair. II: From yeast to mammals., *Trends Genet*, **9**, 211–217.

W. Huang, W.J. Feaver, A.E. Tomkinson, and E.C. Friedberg (1998) The N-degron protein degradation strategy for investigating the function of essential genes: Requirement for RPA and PCNA proteins for nucleotide excision repair in yeast. *Mutation Res.* **408**, 183–194.

Z. Kelman (1997) PCNA: structure, functions and interactions. *Oncogene* **14**, 629–640.

B.C. Laurent, I. Treich and M. Carlson (1993) *Genes & Dev.* **7**, 583–592.

J.C. Marinoni, R. Roy, W. Vermeulen, P. Miniou, Y. Lutz, G. Weeda, T. Seroz, D.M. Gomez, J.M. Hoeijmakers and J.M. Egly (1997) Cloning and characterization of p52, the fifth subunit of the core of the transcription/DNA excision repair factor TFIIH. *EMBO J.* **16**, 1093–1102.

T. Matsunaga, D. Mu, C.H. Park, J.T. Reardon and A. Sancar (1995) Human DNA repair excision endonuclease: Analysis of the roles of the subunits involved in dual incisions by using anti-XPG and anti-ERCC1 antibodies. *J. Biol. Chem.* **270**, 20862–20869.

K.A. Nasmyth (1982) The regulation of yeast mating-type chromatin structure by SIR: an action at a distance affecting both transcription and transposition. *Cell* **30**, 567–578.

E.J. Neer, C.J. Schmidt, R. Nambudiripad and T.F. Smith (1994) The ancient regulatory-protein family of WD-repeat proteins. *Nature* **371**, 297–300.

M.S. Reagan and E.C. Friedberg (1997) Recovery of RNA polymerase II synthesis following DNA damage in mutants of Saccharomyces cerevisiae defective in nucleotide excision repair. *Nucleic Acids Res.*, **25**, 4257–4236.

S.H. Reed, Z. You, and E.C. Friedberg (1998) The yeast Rad7 and Rad16 Proteins are required for post-incision events during nucleotide excision repair: *in vitro* and *in vivo* studies with *rad7* and *rad16* mutants and purification of a Rad7/Rad16 protein complex. *J. Biol. Chem.* **273**, 29481–29488.

K. Rodriguez, K.J. Talamantez, W. Huang, S.H. Reed, L. Chen, W.J. Feaver, E.C. Friedberg, and A.E. Tomkinson (1999) Affinity purification and partial characterization of a yeast multiprotein complex for nucleotide excision repair using histidine-tagged Rad14 protein. *J. Biol. Chem.* (in press).

C.P. Selby and A. Sancar (1993) Molecular mechanism of transcription repair coupling. *Science* **260**, 53–58.

C.P. Selby, R. Drapkin, D. Reinberg and A. Sancar (1997) RNA polymerase II stalled at a thymine dimer: footprint and effect on excision repair. *Nucleic Acids Res.* **25**, 787–793.

M.K.K. Shivji, M.K. Kenny and R.D. Wood (1992) Proliferating cell nuclear antigen is required for DNA excision repair. *Cell* **69**, 367–374.

G. Stelzer, A. Goppelt, F. Lottspeich and M. Meisterernst (1994) Repression of basal transcription by HMG2 is counteracted by TFIIH-associated factors in an ATP-dependent process. *Mol. Cell Biol.* **14**, 4712–4721.

P. Sung, S.N. Guzder, L. Prakash and S. Prakash (1996) Reconstitution of TFIIH and requirement of its DNA helicase subunits, Rad3 and Rad25, in the incision step of nucleotide excision repair. *J. Biol. Chem.* **271**, 10821–10826.

J.Q. Svejstrup, W.J. Feaver, J. LaPoint and R.D. Kornberg (1994) RNA polymerase transcription factor IIH holoenzyme from yeast. *J. Biol. Chem.* **269**, 28044–28048.

J.Q. Svejstrup, Z. Wang, W.J. Feaver, X. Wu, D.A. Bushnell, T.F. Donahue, E.C. Friedberg and R.D. Kornberg (1995) Different forms of TFIIH for transcription and DNA repair: Holo-TFIIH and a nucleotide excision repairosome. *Cell* **80**, 21–28.

J.Q. Svejstrup, P. Vichi and J.M. Egly (1996a) The multiple role of transcription/repair factor TFIIH. *Trends Biochem. Sci.* **249**, 346–350.

J.Q. Svejstrup, W.J. Feaver and R.D. Kornberg (1996b) Subunits of yeast RNA polymerase II transcription factor TFIIH encoded by the CCL1 gene. *J. Biol. Chem.* **271**, 643–645.

D. Tantin, A. Kansal and M. Carey (1997) Recruitment of the putative transcription/repair coupling factor CSB/ERCC6 to RNA polymerase II elongation complexes. *Mol Cell. Biol.*, **17**, 6803–6814.

M. Tijsterman, R.A. Verhage, P. van de Putte, J.G. Tasseron-De Jong and J. Brouwer (1997). Transitions in the coupling of transcription and nucleotide excision repair within RNA polymerase II transcribed genes of Saccharomyces cerevisiae. *Proc. Natl. Acad. Sci. USA* **94**, 8027–8032.

A.E. Tomkinson, A.J. Bardwell, L. Bardwell, N.J. Tapppe and E.C. Friedberg (1993) Yeast DNA repair and recombination proteins Rad1 and Rad10 constitite a single-stranded DNA endonuclease. *Nature* **362**, 860–862.

J.G. Valay, M. Simon and G. Faye (1993) The Kin28 protein kinase is associated with a cyclin in Saccharomyces cerevisiae. *J. Mol. Biol.* **234**, 307–310.

A.J. van Gool, R. Verhage, S.M.A. Swagemakers, P. van de Putte, J. Brouwer, C. Troelstra, D. Bootsma and J.H.J. Hoeijmakers (1994) RAD26, the functional S. cerevisiae homolog of the Cockayne syndrome B gene ERCC6. *EMBO J.* **13**, 5361–5369.

A.J. van Gool, E. Citterio, S. Rademakers, R. van Os, W. Vermeulen, A. Constantinou, J.M. Egly, D. Bootsma and J.H.J. Hoeijmakers (1997) The Cockayne syndrome B protein, involved in transcription coupled DNA repair, resides in an RNA polymerase II containing complex. *EMBO J.*, **16**, 5955–5965.

R.A. Verhage, A.M. Zeeman, N. de Groot, F. Gleig, D.D. Bang, P. van de Putte and J. Brouwer (1994) The RAD7 and RAD16 genes, which are essential for pyrimidine dimer removal from the silent mating type loci, are also required for repair of the nontranscribed strand of an active gene in Saccharomyces cerevisiae. *Mol. Cell Biol.* **14**, 6135–6142.

W. Vermeulen, A.J. van Vuuren, A.J., Chipoulet, L. Schaeffer, E. Appledoorn, G. Weeda, N.G.J. Jaspers, A. Priestly, C.F. Arlett, A.R. Lehmann, M. Stefanini, M. Mezzina, A. Sarasin, D. Bootsma, J.M. Egly and J.H.J. Hoeijimakers (1994). Three unusual repair deficiencies associated with transcription factor BTF2 (TFIIH): Evidence for the existence of a transcription syndrome. *Cold Spring Harbor Symposium of Quant. Biol.* ***LIX***, 317–329.

Z. Wang, J.Q. Svejstrup, W.J. Feaver, X.Wu, R.D. Kornberg and E.C. Friedberg (1994) Transcription factor b (TFIIH) is required during nucleotide excision repair in yeast. *Nature* **368,** 74–76.

Z. Wang, S. Buratowski, J.Q. Svejstrup, W.J. Feaver, X.Wu, R.D. Kornberg, T.F. Donahue and E.C. Friedberg (1995a) The yeast TFB1 and SSL1 genes, which encode subunits of transcription factor IIH, are required for nucleotide excision repair and RNA polymerase II transcription. *Mol. Cell. Biol.* **15**, 2288–2293.

Z. Wang, X. Wu and E.C. Friedberg (1995b) The detection and measurement of base and nucleotide excision repair in cell-free extracts of the yeast Saccharomyces cerevisiae. *METHODS: A Companion to Methods in Enzymology* **7**, 177–186.

Z. Wang, S. Wei, S.H. Reed, X. Wu, J.Q. Svejstrup, W.J. Feaver, R.D. Kornberg and E.C. Friedberg (1997) The RAD7, RAD16 and RAD23 genes of Saccharomyces cerevisiae: Requirement for transcription-independent nucleotide excision repair in vitro and interactions between the gene products. *Mol. Cell Biol.* **17**, 635–643.

J. Watkins, P. Sung, L. Prakash and S. Prakash (1993) The Saccharomyces cerevisiae DNA repair gene RAD23 encodes a nuclear protein containing a ubiquitin-like domain required for biological function. *Mol. Cell. Biol.* **13**, 7757–7765.

D.R. Wilcox and L. Prakash (1981) Incision and post-incision steps of pyrimidine dimer removal in excision-defective mutants of Saccharomyces cerevisiae. *J. Bacteriol.* **148**, 618–623.

C.R. Wobbe, L. Weissbach, J.A. Borowiec, F.B. Dean, Y. Murakami, P. Bullock and J. Hurwitz (1987) Replication of simian virus 40 origin-containing DNA in vitro with purified proteins. *Proc. Natl. Acad. Sci. USA* **84**, 1834–1838.

M.S. Wold and T. Kelly (1988) Purification and characterization of replication protein A, a cellular protein required for in vitro replication of simian virus 40 DNA. *Proc. Natl. Acad. Sci. USA* **85**, 2523–2527.

Z. You, W.J. Feaver and E.C. Friedberg (1998) Yeast RNA polymerase II transcription in vitro is inhibited in the presence of nucleotide excision repair: Complementation of inhibition by holo-TFIIH and requirement for RAD26. *Mol. Cell. Biol.* **18**, 2668–2676.

11

PHYSIOLOGICAL CHEMISTRY OF SUPEROXIDE AND NITRIC OXIDE INTERACTIONS

Implications in DNA Damage and Repair

Matthew B. Grisham,[1] David Jourd'heuil,[1] and David A. Wink[2]

[1]Department of Molecular and Cellular Physiology
Louisiana State University Medical Center
Shreveport, Louisiana 71130
[2]Radiation Biology, National Cancer Institute
Bethesda, Maryland 20892

1. ABSTRACT

Chronic inflammation of the colon and rectum is known to be associated with enhanced production of both nitric oxide (NO) and reactive oxygen species such as superoxide (O_2^-) and hydrogen peroxide (H_2O_2). Patients with long-standing ulcerative colitis are also known to be at increased risk of developing colorectal cancer. Although NO and reactive oxygen intermediates have been shown to modify DNA bases and to promote a wide array of mutagenic reactions, there is increasing evidence to suggest that the interaction between O_2^- and NO may dictate the type of mutagenic reaction produced at sites where both free radicals are produced. In the absence of O_2^-, NO derived nitrosating agents will N-nitrosate a variety of primary and secondary amines and promote the nitrosative deamination of DNA bases. Furthermore, these same NO-derived nitrosating agents will S-nitrosate certain thiol-requiring repair enzymes thereby inhibiting certain DNA repair proteins. As the flux of O_2^- is increased, N- and S-nitrosation reactions are suppressed but oxidative chemistry is enhanced. Thus, depending upon the fluxes of each radical either nitrosation or oxidation chemistry may predominate. A fundamental understanding of the interaction between O_2^- and NO may provide new insight in the mechanisms responsible for inflammation-induced DNA repair and damage.

2. INTRODUCTION

Active episodes of ulcerative colitis are characterized by infiltration of large numbers of phagocytic leukocytes into the mucosal interstitium. This enhanced inflammatory

Advances in DNA Damage and Repair, edited by Dizdaroglu and
Karakaya. Kluwer Academic / Plenum Publishers, New York, 1999.

infiltrate is accompanied by extensive injury to the mucosa. A growing body of clinical and experimental data suggests that severe long-standing inflammation of the colon is associated with an increased risk of colorectal cancer (Weitzman and Gordon, 1990; Collins et al., 1987; Korelitz, 1983). In addition, investigators have shown that inflammation enhances the formation of colonic tumors in experimental animals given known carcinogens (Pozhariski, 1975; Chestre et al., 1989). Despite these studies, the mechanisms by which inflammation promotes tumor formation remain poorly understood. It has been suggested that certain leukocyte-derived products may act as endogenous carcinogens or tumor promotors *in vivo* (Weitzman and Gordon, 1990). A series of recent studies have demonstrated that chronic gut inflammation is associated with the upregulation of the inducible isoform of nitric oxide synthase (iNOs) and enhanced production of nitric oxide (NO) (Moncada et al., 1991; McCall et al., 1989; Salvemini et al., 1989). Nitric oxide is unstable in the presence of molecular oxygen and will auto-oxidize to yield nitrogen oxide intermediates, some of which are potent nitrosating agents that will N-nitrosate primary and secondary amines (Marletta, 1988). Secondary nitrosamines require metabolic activation (i.e. hydroxylation) to yield alkylating agents that have been shown to activate certain oncogenes via the covalent modification of certain DNA bases (Bartsh et al., 1989). Nitrosative deamination of primary aromatic amines has been another suggested pathway by which NO-derived N-nitrosating agents produce transition and transversion mutations (Wink et al., 1991). The objective of this review is to discuss the chemical interactions between superoxide (O_2^-) and NO and to examine how these interactions may be involved in inflammation-induced DNA damage and repair.

3. NITRIC OXIDE AND N- AND S-NITROSATION REACTIONS

Exposure of critical genes to mutagenic conditions increases the probability of tumor development. Chronic inflammation is one such environment which promotes malignant transformation. First, tissues neighboring inflammatory foci undergo increased cell division. Consequently, mutagenic effects associated with chronic inflammation can become multiplicative, as the chance of DNA modifications and mis-repair increase (Cohen et al., 1991). Secondly, certain leukocyte-and tissue derived metabolites may cause genomic damage, thereby increasing the probability of nicks, deletions, and point mutations (Shacter et al., 1988).

During chronic inflammation of the colon for example, epithelial cells may be exposed to large amounts of NO (as much as 10^4 molecules/cell/s) (Tamir et al., 1996). Nitric oxide will rapidly and spontaneously autooxidize to yield nitrogen oxide intermediates:

$$2NO + O_2^- \rightarrow 2NO_2$$

$$2NO + 2NO_2 \rightarrow 2N_2O_3$$

$$2N_2O_3 + 2H_2O \rightarrow 4NO_2^- + 4H^+$$

where $NO_2^{\cdot}$, N_2O_3, and NO_2^- represent nitrogen dioxide, dinitrogen trioxide, and nitrite, respectively. Of these, N_2O_3 has drawn particular interest due to its ability to N and

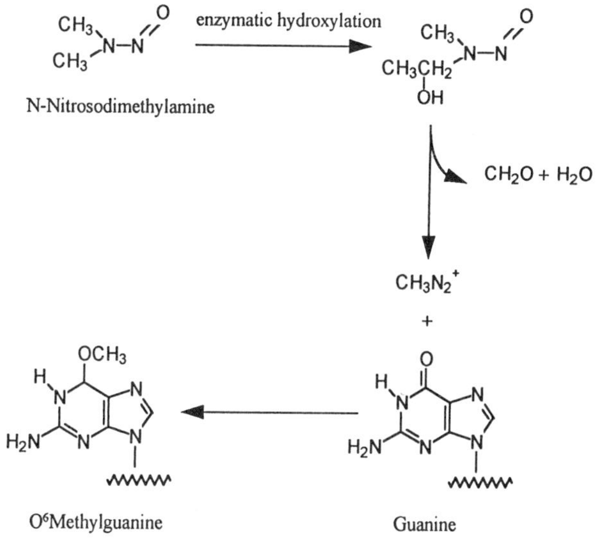

Figure 1. Nitrosamine-mediated alkylation of DNA bases. Secondary nitrosamines such as N-nitrosodimethylamine induce point mutations by alkylation of DNA bases such as guanine to form O^{6}methylguanine residues.

S-nitrosate certain nucleophilic substrates such as primary and secondary amines as well as thiol-containing proteins (Williams, 1988).

Such nitrosating species have been shown to promote the nitrosative deamination of primary aromatic amines, including purines and pyrimidines, via the formation of nitrosamine and diazonium ion intermediates. Deamination of cytosine, methyl cytosine, adenine, or guanine results in the formation of uracil, thymine, hypoxanthine, and xanthine respectively. Base conversion of cytosine and methyl cytosine can lead ultimately to a base pair substitution mutations, while deamination of adenine and guanine will result in transversion mutations. Moreover, the instability of hypoxanthine and xanthine in the DNA structure leads to rapid depurination and consequent single strand breaks. Even crosslinking with other nucleic acids or proteins have been suggested via reaction of a nucleophilic site on an adjacent macromolecule and the diazonium ion of the modified base (Tamir et al., 1996).

Nitrosation of secondary aliphatic and aromatic amines can also produce potentially carcinogenic nitrosamines. Secondary nitrosamines are more stable than their primary amine counterparts:

$$R_2NH + XNO \rightarrow R_2NNO + HX$$

Such nitrosamines, like many chemical carcinogens, are thought to promote mutagenesis and carcinogenesis via their ability to alkylate specific sites in DNA. For example, these types of nitrosamines undergoe enzymatic alpha-hydroxylation. The alpha-hydroxy nitrosamine decomposes to form the alkyl diaznoium ion and free alkyl carbocation (Fig 1). The alkyl diazonium salt or carbocations then can react with nucleophilic sites in DNA. To date, alkylation of DNA has been noted on the ring-nitrogen positions in the bases (adenine, guanine, cytosine, thymine), the oxygen atoms of hydroxyl or carbonyl groups (guanine, thymine, and cytosine) as well as on the phosphate groups (Williams, 1988).

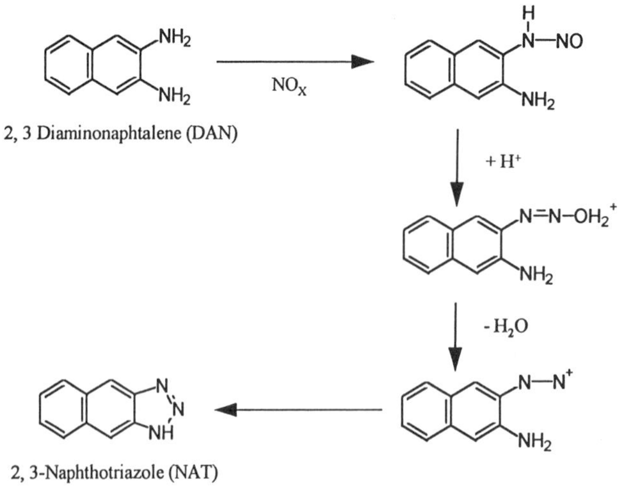

2, 3 Diaminonaphtalene (DAN)

NO_X

$+H^+$

$-H_2O$

2, 3-Naphthotriazole (NAT)

Figure 2. N-nitrosation of 2,3-diaminonaphthalene (DAN) to yield 2,3-naphthotriazole (NAT) by an NO-derived N-nitrosating agent (NO_X).

Thus, the limitation of NO-mediated genomic damage, rests primarily on the localized diffusion of the small molecule, the degree of reactivity and nitrosation, and ultimately the cells' replicative and DNA repair machinery. However, even in the latter case, DNA repair proteins such as O^6-methylguanine-DNA-methyltransferase and Fpg, have been shown to be inhibited by NO-derived nitrosating agents such as N_2O_3 *in vitro* and *in vivo* (Laval and Wink, 1995; Wink and Laval, 1994). These DNA-repair proteins are thought to be inhibited by the NO-dependent S-nitrosation of thiol-containing Zinc finger moiety in the case of Fpg or other cysteine residues in the case of the O^6-methylguanine-DNA-methyltransferase (Laval and Wink, 1995; Wink and Laval, 1994).

4. EFFECTS OF SUPEROXIDE ON NO-DEPENDENT N- AND S-NITROSATION

Coincident with the sustained overproduction of NO, inflammatory foci are also sites of enhanced production of reactive oxygen species, such as O_2^- and H_2O_2. Because O_2^- is known to rapidly react with NO, it was of interest to determine whether this reactive oxygen specie may modulate NO-dependent N- and S-nitrosation of primary aromatic amines and thiols. To address this possibility, we have used 2,3-diaminonaphthalene (DAN) and glutathione (GSH) as models to study the effect of O_2^- on the NO-dependent N and S-nitrosation of primary aromatic amines and thiols, respectively. DAN is N-nitrosated by N_2O_3 derived from NO to yield its highly fluorescent triazole derivative 1-naphtho(2,3)triazole (NAT) (Fig 2). A recent application of this sensitive technique may also be used to quantify nitrosothiol formation (Wink et al., 1997).

We have demonstrated that the addition of a O_2^- and H_2O_2 generator such as hypoxanthine/xanthine oxidase virtually eliminated the NO-dependent N-nitrosation of DAN and S-nitrosation of GSH (Wink et al., 1997; Miles et al., 1995). Inhibition was maximal when equimolar fluxes of NO and O_2^- were produced. We also noted that this inhibition was reversed by the addition of superoxide dismutase, but not catalase suggesting that O_2^- and not H_2O_2 was responsible for the inhibition. We proposed that at equimolar fluxes, O_2^- reacts rapidly with NO to generate products that possess only a limited ability to N- and S-nitrosate amino and thiol-containing compounds. Although we found that O_2^- inhibits the potentially mutagenic N-nitrosation of primary amines, interaction of O_2^- and NO may yield the potent oxidant peroxynitrite ($ONOO^-$) which could conceivably promote oxidative (and nitrative) modifications of DNA bases, thereby switching NO-mediated DNA damage from a nitrosative to a more oxidative pattern of mutagenic reactions (Miles et al., 1996).

5. SUPEROXIDE-NITRIC OXIDE INTERACTIONS: OXIDATIVE REACTIONS

Nitric oxide reacts with O_2^- to yield $ONOO^-$ and its conjugate acid peroxinitrous acid (ONOOH; pKa 6.6) with a second order rate constant of $6.7 \times 10^9\ M^{-1}.s^{-1}$ (Huie and Padmaja, 1993):

$$NO + O_2^- \rightarrow ONOO^- + H^+ \leftrightarrow ONOOH \rightarrow ONOOH^* \rightarrow NO_3^-$$

Peroxynitrous acid is very unstable and at physiological pH decomposes to yield nitrate (NO_3^-) via the intermediate formation of an excited form of ONOOH that possesses potent oxidizing activity. This may involve the homolysis of ONOOH to nitrogen dioxide radical ($NO_2^{\cdot}$) and hydroxyl radical ($OH^{\cdot}$) within a solvent cage, the two free radicals diffusing out of the solvent cage to mediate oxidation reactions (Pryor and Squadrito, 1995). Hydroxyl radical is an extremely reactive species, interacting with virtually all biomolecules at diffusion limited rates ($\sim 10^7 - 10^9\ M.sec^{-1}$) (Pryor, 1986). Nitrogen dioxide can initiate lipid peroxidaton and N-nitrosate certain amines to yield nitrosamines (Pryor, 1981). However, this mechanism is now known to be thermodinamically unfavorable. A more probable mechanism may involve the formation of an activated isomer of peroxynitrous acid, ONOOH* which would possess $NO_2^{\cdot}$ and $OH^{\cdot}$-like properties (Koppenol et al., 1992). From the standpoint of biologically relevant reactions, $ONOO^-$ is an oxidizing, hydroxylating, and nitrating agent. In regard to DNA modifications, $ONOO^-$ has been found to oxidize and nitrate isolated DNA resulting in DNA strand breaks (Tamir et al., 1996).

Little is known regarding the detailed reactions of $ONOO^-$ with DNA. Because of the multiplicity of DNA modifications produced during oxidative reactions, it has been difficult to establish the specificity of mutations engendered by individual oxidants such as $ONOO^-$. Oxidant-mediated DNA base modifications produce 8-hydroxydeoxyguanosine (8-OHdG) as a major product (Fig. 3) (Halliwell and Aruoma, 1991). The occurrence of this alteration has been associated with a number of conditions leading to increased oxidative stress including higher basal metabolic rate, gamma-irradiation, and hydrogen peroxide-mediated oxidative stress (Kasai et al., 1986). Nitric oxide and iron have also been implicated in the formation of 8-OHdG in asbestos-treated human lung epithelial cells (Chao et al., 1996). Peroxynitrite also mediates the oxidation of deoxyguanosine (King et al., 1993). In all of those conditions, the formation of 8-OHdG might lead to mutations by inducing misreading of the base itself and of the adjacent bases (Halliwell and Aruoma, 1991) which may represent an important source of mutations (Cheng et al., 1992). Peroxynitrite induces G:C toT:A mutations for the supF gene in *E. coli* and in human AD293 cells (Juedes and Wogan, 1996). In addition to oxidative reactions, recent data suggest that the interaction of $ONOO^-$ with DNA results in the nitration of guanine to form 8-nitroguanine (Fig. 4) (Yermilov et al., 1995a). This modification is potentially mutagenic,

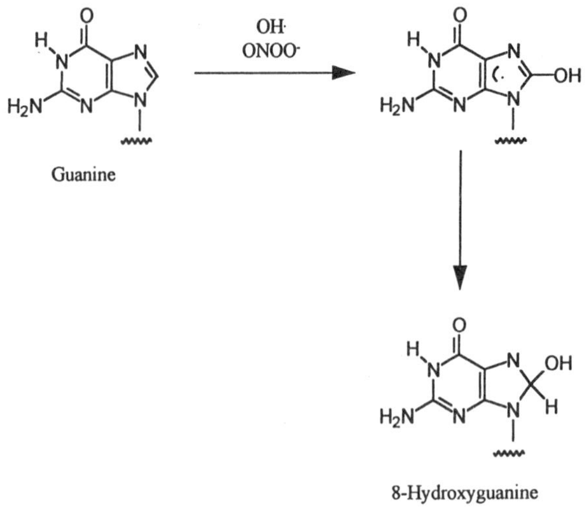

Figure 3. Hydroxyl radical ($OH^{\cdot}$) and peroxynitrite ($ONOO^-$)-mediated oxidation of guanine.

the depurination of 8-nitroguanine yielding apurinic sites with the resultant possibility of G:C to T:A transversions (Yermilov et al., 1995b). However, whether $ONOO^-$ mediates such DNA damage in cells and tissues is yet to be determined. Although the identification of 8-nitroguanine may be used as a marker of $ONOO^-$ -induced DNA modification, it should be noted that other NO-derived nitrating agents such as $NO_2^{\cdot}$ derived form myeloperoxidase-catalyzed oxidation of nitrite (NO_2^-) or OClNO produced from the interaction between HOCl and NO (Van der Vliet et al., 1997) would also be expected to nitrate DNA bases.

It is important to note that the evaluation of $ONOO^-$-mediated mutagenic properties has been assessed *in vitro* using bolus amounts of chemically generated oxidant. However, it is becoming increasingly evident that the formation of $ONOO^-$ at sites where both O_2^- and NO are produced may depend upon the relative fluxes of NO and O_2^-. Using the hypoxanthine/ xanthine oxidase system to generate both O_2^- and H_2O_2 and the spermine/NO adduct to generate varying fluxes of NO, we found that the simultaneous production of equimolar fluxes of O_2^- and NO dramatically increased the oxidation of the oxidant-sensitive probe dihydrorhodamine (DHR) (Miles et al., 1996). This oxidation was inhibited by superoxide dismutase but not catalase suggesting that O_2^- and not H_2O_2 interacted with NO to form $ONOO^-$/ONOOH. As the flux of one radical exceeded the other, oxidation of DHR was inhibited suggesting that excess production of either radical may act as an endogenous modulator of $ONOO^-$/ONOOH formation. Subsequent experiments by our labora-

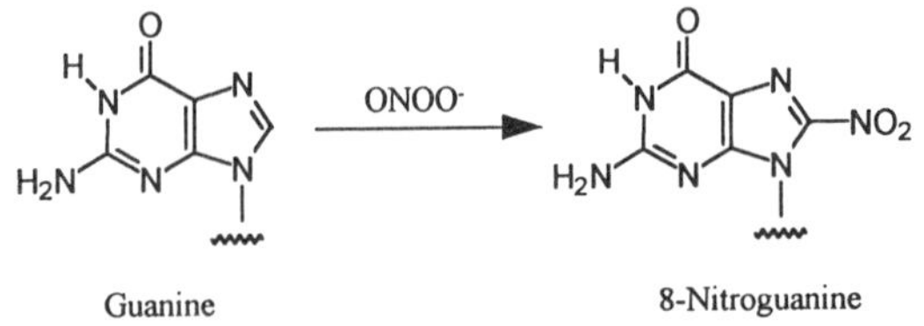

Figure 4. Peroxynitrite ($ONOO^-$)-mediated nitration of guanine to form 8-nitroguanine.

tory as well as others have demonstrated that NO (or O_2^-) interacts with and decomposes $ONOO^-/ONOOH$ (Miles et al., 1996; Pfieffer et al., 1997). Although this hypothesis suggests that NO and O_2^- may modulate steady state concentrations of $ONOO^-$, there has yet to be any direct evidence demonstrating such modulation of $ONOO^-$-mediated oxidative and/or nitrating reactions under physiological conditions.

6. SUPEROXIDE, FENTON CHEMISTRY AND NITRIC OXIDE

In general, O_2^- per se is not thought to be highly toxic to cells and tissues since it is a better reducing agent than an oxidant (Buettner, 1993). However, O_2^- will dismutate to form H_2O_2. Although H_2O_2 is an oxidizing agent, most of the H_2O_2-dependent oxidizing activity is mediated by secondary radicals such as $OH^·$ generated from metal (Mn)-catalyzed reactions. It is known that O_2^- and H_2O_2 interact with chelates of iron or copper to yield the potent oxidants $OH^·$ or $OH^·$-like species via the superoxide-driven Fenton reaction:

$$O_2^- + Mn^{+2} \rightarrow O_2 + Mn^{+1}$$

$$H_2O_2 + Mn^{+1} \rightarrow OH^· + OH^- + Mn^{+2}$$

The interaction of free radicals derived from superoxide-driven Fenton reactions with DNA has received considerable interest over the years (Imlay and Linn, 1988). Recent reports have focused on the ability of copper to participate in mutagenic reactions *in vivo* via Fenton-catalyzed reactions. Copper is an important structural metal in chromatin (Geierstanger et al., 1991) that in fact induces more DNA bases damage in the presence of H_2O_2 than does iron (Dizdaroglu et al., 1991).

There is now evidence to suggest that NO may modulate Fenton-driven oxidative reactions. We have recently investigated the ability of different fluxes of nitric oxide to modulate iron complex (Miles et al., 1996) and hemoprotein-catalyzed oxidative reactions (Jourd'heuil et al., 1998). We found that generation of O_2^- and H_2O_2 in the presence of 5 mM Fe^{+3}-EDTA stimulated dramatically the hydroxylation of benzoic acid. Catalase and superoxide dismutase were both effective at inhibiting this classic Fenton-driven reaction. Addition of NO inhibited this reaction in a concentration-dependent manner such that a ratio of $NO/O_2^-/H_2O_2$ of 1:1:1 inhibited hydroxylation of benzoic acid by 90% (Miles et al., 1996). Kanner *et al.* (1991) proposed that NO may inhibit iron-mediated oxidative reactions by forming nitrosyl complexes with ferrous iron:

$$Fe^{+3}\text{-EDTA} + O_2^- \rightarrow Fe^{+2}\text{-EDTA} + O_2$$

$$Fe^{+2}\text{-EDTA} + NO \rightarrow NO\text{-}Fe^{+2}\text{-EDTA}$$

$$NO\text{-}Fe^{+2}\text{-EDTA} + H_2O_2 \rightarrow Fe^{+3}\text{-EDTA} + HNO_2 + OH^-$$

We have recently assessed the ability of myoglobin to oxidize DHR in the presence or the absence of O_2^-, H_2O_2 and /or NO (Cheng et al., 1992). In the presence of equimolar fluxes of H_2O_2 and O_2^-, the addition of metmyoglobin (Mb-Fe^{+3}) dramatically enhanced DHR oxidation via the formation of ferryl myoglobin (Mb-Fe^{+4}). This oxidative reaction

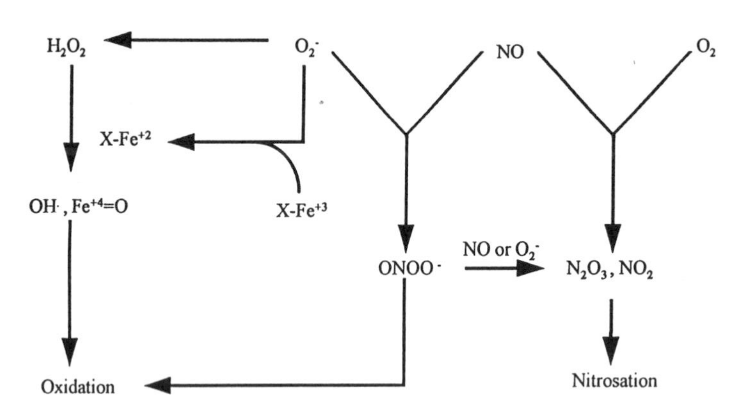

Figure 5. Modulation of oxidative and nitrosative reactions by NO and O_2^-.

was as expected inhibited by catalase but not superoxide dismutase. Addition of NO to this system further enhanced DHR oxidation which was inhibited by superoxide dismutase suggesting that O_2^- reacted with NO to yield $ONOO^-$/ONOOH in addition to Mb-Fe^{+4}. Further increases in NO flux dramatically inhibited DHR oxidation which was found to be due to the NO-mediated reduction of Mb-Fe^{+4} to Mb-Fe^{+3}. Taken together, these data suggest that NO may modulate iron complex or hemoprotein-catalyzed oxidative reactions depending upon the relative fluxes of O_2^-, H_2O_2 and NO. In accordance with our results, Pacelli *et al.* have shown that NO can inhibit DNA strand breaks induced by H_2O_2 and certain transition metals (Pacelli et al., 1994).

7. CONCLUSION

It is well known that chronic intestinal inflammation is associated with an increased risk of malignancy (Weitzman and Gordon, 1990; Colins et al., 1987; Korelitz, 1983). This pathological condition represents one example in which the chemistry we have described may play an important role. Indeed, the phagocytic leukocytes that accumulate within the chronically inflamed colon produce large amounts of reactive oxygen metabolites of oxygen and nitrogen which may mediate mutagenesis and possibly malignant transformation (Weitzman and Gordon, 1990). Thus, a fundamental understanding of the interplay between O_2^- and NO may provide new insight into the chemical role that these reactive species play in DNA damage and repair. In the absence of O_2^-, NO will N- and S-nitrosate amino and thiol-containing compounds to yield potentially mutagenic intermediates as well as S-nitroso derivatives of thiol-containing repair proteins (see Fig. 5). In the presence of both NO and O_2^-, N- and S-nitrosation chemistry may be suppressed but oxidation reactions may predominate.

ACKNOWLEDGMENTS

Some of the work reported in this manuscript was supported by grants from the National Institute of Health (DK 47663 and DK 43785).

REFERENCES

H.E. Bartsh, E. Hietanen and C. Malaveille: Carcinogenic nitrosamines: free radical apects of their action. *Free Rad. Bio.l Med.* **7**, 637–644 (1989)

Buettner, G.R.: The pecking order of free radicals and antioxidants: lipid peroxidation, alpha-tocopherol, and ascorbate. *Arch. Biochem. Biophys.* **300**, 535–543 (1993)

C.C. Chao, S.H. Park and A.E. Aust: Participation of nitric oxide and iron in the oxidation of DNA in abestos-treated human lung epithelial cells. *Arch. Biochem. Biophys.* **326**, 152–157 (1996)

K.C. Cheng, D.S. Cahill, H. Kasai, S. Nishimura and L.A. Loeb: 8-hydroxyguanine, an abundant form of oxidative DNA damage, causes G--->T and A--->C substitutions. *J. Biol. Chem.* **267**, 166–172 (1992)

J.F. Chester, H.A. Gaissert, J.S. Ross, R.A. Malt and S.A. Weitzman: Colonic cancer induced by 1,2-dimethylhydrazine: promotion by experimental colitis. *Br. J. Cancer* **59**, 704–705 (1989)

S.M. Cohen, D.T. Purilo and L.B. Ellwein: Pivotal role of increased cell proliferation in human carcinogenesis. *Mod. Pathol.* **4**, 371–382 (1991)

R.H. Collins, M. Feldman and J.S. Fordtran: Colon cancer, dysplasia and surveillance in patients with ulcerative colitis: a critical review. *New Engl. J. Med.* **316**, 1654–1658 (1987)

M. Dizdaroglu, G. Rao, B. Halliwell and E. Gajewski: Damage to the DNA bases in mammalian chromatin by hydrogen peroxide in the presence of ferric and cupric ions. *Arch. Biochem. Biophys.* **285**, 317–324 (1991)

B.H. Geierstanger, T.F. Kagawa, S. Chen, G.J. Quigley and P.S. Ho: Base-specific binding of copper (II) to Z-DNA. *J. Biol. Chem.* **266**, 20185–20191 (1991)

B. Halliwell and O.I. Aruoma: DNA damage by oxygen-derived species. *FEBS Lett.* **281**, 9–19 (1991)

R.E. Huie and S. Padmaja: The reaction of NO with superoxide. *Free Rad. Res.* **18**, 195–199 (1993)

J.A. Imlay and S. Linn: DNA damage and oxygen radical toxicity. *Science* **24**:1302–1309 (1988)

D. Jourd'heuil, L. Mills, A.M. Miles, and M.B. Grisham: Effect of nitric acid in hemoprotein-catalyzed oxidative reactions. *Nitric Oxide: Biology and Chemistry* **2**, 37–44 (1998)

M.J. Juedes and G.N. Wogan: Peroxinitrite-induced mutation spectra of pSP189 following replication in bacteria and human cells. *Mutat. Res.* **349**, 51–61 (1996)

J. Kanner, S. Harel and R. Granit: Nitric oxide as an antioxidant. *Arch. Biochem. Biophys.* **289**, 130–136 (1991)

H. Kasai, P.F. Crain, Y. Kuchino, S. Nishimura, A. Ootsuyama and H. Tanooka: Formation of 8-hydroxyguanine moiety in cellular DNA by agents producing oxygen radicals and evidence for its repair. *Carcinogenesis* **7**,1849–1851 (1986)

P.A. King, E. Jamison, D. Strahs, V.E. Anderson and M. Brenowitz: 'Footprinting' proteins on DNA with peroxynitrous acid. *Nucl. Acids Res.* **21**, 2473 (1993)

W.H. Koppenol, J.J. Moreno, W.A. Pryor, H. Ischiropoulos and J.S. Beckman. Peroxynitrite, a cloaked oxidant formed by nitric oxide and superoxide. *Chem. Res. Toxicol.* **5**, 834–842 (1992)

B.I. Korelitz. Carcinoma of the intestinal tract in Crohn's disease: results of a survey conducted by the National Foundation for Ileitis and Colitis. *Am. J. Gastroenterol.* **78**,44–46 (1983)

F. Laval and D.A. Wink: Inhibition by nitric oxide of the repair protein, O6-methylguanine-DNA-methyltransferase. *Carcinogenesis* **15**, 443–447 (1995)

M.A. Marletta: Mammalian synthesis of nitrite, nitrate, and N-nitrosating agents. *Chem. Res. Toxicol.* **1**, 249–257 (1988)

T.B. McCall, N.K. Boughton-Smith, R.M.J. Palmer, B.J.R. Whittle and S. Moncada: Synthesis of nitric oxide from L-arginine by neutrophils, release and interaction with superoxide anion. *Biochem J.* **261**, 293–296 (1989)

T.B. McCall, M. Feelisch, R.M.J. Palmer and A. Moncada: Identification of N-iminoethyl-L-ornithine as an irreversible inhibitor of nitric oxide synthase in phagocytic cells. *Br. J. Pharmacol.* **102**, 234–238 (1991)

A.M. Miles, M.F. Gibson, M. Kirshna, J.C. Cook, R. Parcelli, D. Wink and M.B. Grisham: Effects of superoxide on nitric oxide-dependent N-nitrosation reactions. *Free. Rad. Res.* **23**, 379–390 (1995)

A.M. Miles, D.S. Bohle, P.A. Glassbrenner, B. Hansert, D.A. Wink and M.B. Grisham: Modulation of superoxide-dependent oxidation and hydroxylation reactions by nitric oxide. *J. Biol. Chem.* **271**, 40–47 (1996)

S. Moncada, R.M.J. Palmer and E.A. Higgs: Nitric oxide: physiology, pathophysiology, and pharmacology. *J. Pharmacol. Exp. Ther.* **43**, 109–142 (1991)

R. Pacelli, M.C. Krishna, D.A. Wink and J.B. Mitchell: Nitric oxide protects DNA from hydrogen peroxide-induced double strand cleavage. *Proc. Am. Assoc. Cancer Res.* **35**, 540 (1994)

S. Pfieffer, A.C.F. Gorren, K. Scmidt, E.R. Werner, B. Hansert, D.S. Bohle and B. Mayer: Metabolic fate of peroxynitrite in aqueous solution. *J. Biol. Chem.* **272**, 3465–3470 (1997)

K.M. Pozharisski. The significance of non-specific injury for colon carcinogenesis in rats. *Cancer Res.* **35**, 3824 (1975)

W.A. Pryor and G.L. Squadrito: The chemistry of peroxynitrite: a product from the reaction of nitric oxide with superoxide. *Am. J. Physiol.* **268**, L699-L722 (1995)

W.A. Pryor: Oxy-radicals and related species: their formation, lifetimes, and reactions. *Ann. Rev. Physiol.* **48**, 657–667 (1986)

W.A. Pryor: Mechanisms of nitrogen dioxide reactions: initiation of lipid peroxidation and the production of nitrous acid. *Science* **214**, 435–437 (1981)

D. Salvemini, G. De Nucci, R.J. Gryglewsji and J.R. Vane: Human neutrophils and mononuclear cells inhibit platelet aggregation by releasing a nitric oxide-like factor. *Proc. Natl. Acad. Sci. USA* **86**,6328–6332 (1989)

E. Shacter, E.J. Beecham, J.M. Covey, K.W. Kohn and M. Potter: Activated neutrophils induce prolonged DNA damage in neighboring cells. *Carcinogenesis* **9**, 2297–2304 (1988)

S. Tamir, S. Burney and S.R. Tannenbaum: DNA damage by nitric oxide. *Chem. Res. Toxicol.* **9**, 821–827 (1996)

S. Tamir, T. Rojas-Walker, J.S. Wishnor and S.R. Tannenbaum: DNA damage and genotoxicity by nitric oxide. *Methods Enzymol.* **269**, 230–242 (1996)

S. Tamir, S. Burney and S.R. Tannenbaum: DNA damage by nitric oxide. *Chem. Res. Toxicol.* **9**, 821–827 (1996)

A. Van der Vliet, J.P. Eiserich, B. Halliwell, C.E. Cross: Formation of reactive nitrogen species during peroxidase-catalyzed oxidation of nitrite. *J. Biol. Chem.* **272**, 7617–7625 (1997)

S.A. Weitzman and L.I. Gordon: Inflammation and cancer: role of phagocyte-generated oxidants in carcinogenesis. *Blood* **76**, 655–663 (1990)

D.A. Wink, K.S. Kasprzak, C.M. Maragos, R.K. Elespuru, M. Misra, T.M. Dunams T.A. Cebula, W.H. Koch, A.W. Andrews, J.S. Allen and L. Keefer: DNA deamination ability and genotoxicity of nitric oxide and its progenitors. *Science* **254**, 1001–1003 (1991)

D.A. Wink, J.A. Cook, S. Kim, Y. Vodovotz, R. Pacelli, M.C. Krishna, A. Russo, J.B. Mitchell, D. Jourd'heuil, A.M. Miles and M.B. Grisham: Superoxide modulates the oxidation and nitrosation of thiols by nitric oxide-derived reactive intermediates. *J. Biol. Chem.* **272**, 11147–11151 (1997)

D.A. Wink and J. Laval: The Fpg protein, a DNA repair enzyme, is inhibited by the biomediator nitric oxide in vitro and in vivo. *Carcinogenesis* **15**, 2125–2129 (1994)

V. Yermilov, J. Rubio, J. Becchi, M.D. Friesen, B. Pignatelli and H. Ohshima: Formation of 8-nitroguanine by the reaction of guanine with peroxynitrite. *Carcinogenesis* **16**, 2045–2050 (1995a)

V. Yermilov, J. Rubio and H. Ohshima: Formation of 8-nitroguanine in DNA treated with peroxynitrite *in vitro* and its rapid removal from DNA by depurination. *FEBS Lett.* **376**, 207–210 (1995b)

12

RECOGNITION AND EXCISION OF BASES FROM OXIDATIVELY DAMAGED DNA BY Fpg, Ogg1, AND MutY PROTEINS

Arthur P. Grollman and Dmitry O. Zharkov

SUNY Stony Brook, Department of Pharmacological Sciences
Stony Brook, New York 11794-8651

1. ABSTRACT

Fpg and MutY proteins are DNA glycosylases involved in repair of oxidative DNA damage in bacteria. Ogg1 is a functional analog of Fpg found in eukaryotes. Fpg and Ogg1 catalyze excision of 8-oxoguanine paired with dC in duplex DNA while MutY is an adenine DNA glycosylase that acts preferentially on 8-oxodG:dA mispairs. By establishing structure-function relationships for these DNA repair enzymes, functional groups required for substrate recognition and catalysis have been identified. For efficient excision of damaged bases, Fpg and Ogg1 require the presence of 6-keto and 8-keto groups in the purine ring while MutY requires a dG(*syn*):dA(*anti*) mispair. Key structural motifs include the N-terminal proline and C-terminal zinc finger of Fpg and the helix-hairpin-helix motif and G/P...D loop of MutY and Ogg1. The mechanism for catalysis involves Schiff base formation between C1' and a nucleophillic group in the enzyme. Studies with substrates containing pre-formed extrahelical bulges suggest a requirement for an extrahelical base. The proposed mode of action involves rapid scanning of the grooves of DNA, formation of a stable hydrogen-bonded complex at the site of an oxidatively-damaged base, followed by base extrusion and cleavage of the glycosidic bond.

2. INTRODUCTION

Aerobic metabolism is utilized by many forms of life. In prokaryotes and eukaryotes, molecular oxygen is reduced to water in step-wise reactions generating superoxide anion radical ($O_2^{-\bullet}$), hydrogen peroxide (H_2O_2), and hydroxyl radical ($^{\bullet}OH$). These so-called reactive oxygen species also are produced in cells by ionizing radiation and

Advances in DNA Damage and Repair, edited by Dizdaroglu and
Karakaya. Kluwer Academic / Plenum Publishers, New York, 1999.

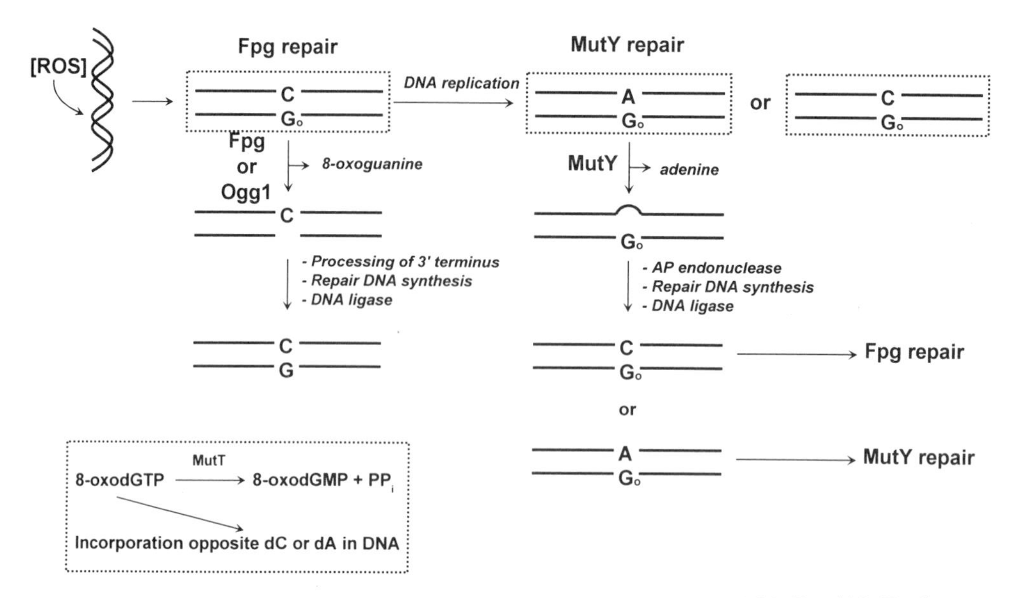

Figure 1. Scheme (modified after Grollman and Moriya, 1993) indicating how MutM, MutY and MutT reduce accumulation of 8-oxoguanine residues in DNA. 8-Oxoguanine is introduced into DNA by oxidative damage (unboxed pathway) or by incorporation of 8-oxodGTP during DNA synthesis (boxed pathway).

through Fenton chemistry, a non-enzymatic process in which Fe^{++} ion catalyzes the conversion of H_2O_2 to $^{\bullet}OH$ (Halliwell and Gutteridge, 1989).

A variety of mutagenic and/or cytotoxic lesions in DNA are formed by reactive oxygen species (von Sonntag, 1987), one of the most abundant being 8-oxoguanine (Shigenaga et al., 1989). During DNA replication, 8-oxodG pairs with dC and dA (Shibutani et al., 1991). The 8-oxodG•dA mispair leads to G:C→T:A transversions when 8-oxoguanine is in the template strand and to A:T→C:G transversions when the modified base is introduced into DNA via 8-oxodGTP, itself generated by reactive oxygen species (Grollman and Moriya, 1993). DNA polymerases catalyze incorporation of 8-oxodGTP into DNA (Cheng et al., 1992; Pavlov et al., 1994).

In *E. coli*, three DNA repair enzymes work in concert to minimize the mutagenic consequences of 8-oxodG (cf reviews by Grollman and Moriya, 1993; Michaels and Miller, 1992). MutM (Fpg) is a DNA glycosylase/AP lyase that acts preferentially on duplex DNA containing 8-oxodG:dC. MutY is an adenine DNA glycosylase that acts preferentially on 8-oxodG:dA. MutT is an 8-oxodGTPase. Fpg excises 8-oxoguanine from oxidatively-damaged DNA while MutY removes deoxyadenosine paired to 8-oxodG, presumably in the DNA replication intermediate, initiating a process by which the oxidized base can later be excised by Fpg. MutT cleanses the cellular deoxynucleoside triphosphate pool, preventing incorporation of 8-oxodGTP into DNA (Maki and Sekiguchi, 1992). These pathways for repair of 8-oxoguanine in *E. coli* are shown in Figure 1.

Bacterial enzymes involved in repair of 8-oxoguanine have functional homologs in eukaryotes. Genes coding for human and mouse homologs of MutT (Sakumi et al., 1993; Igarashi et al., 1997) and human MutY (Slupska et al., 1996) have been cloned. Yeast, human and mouse genes coding for an 8-oxoguanine DNA glycosylase that acts preferentially on 8-oxodG:dC (Ogg1), have been isolated by several laboratories (van der Kemp et al., 1996; Nash et al., 1996; Rosenquist et al., 1997; Radicella et al., 1997; Roldán-Arjona

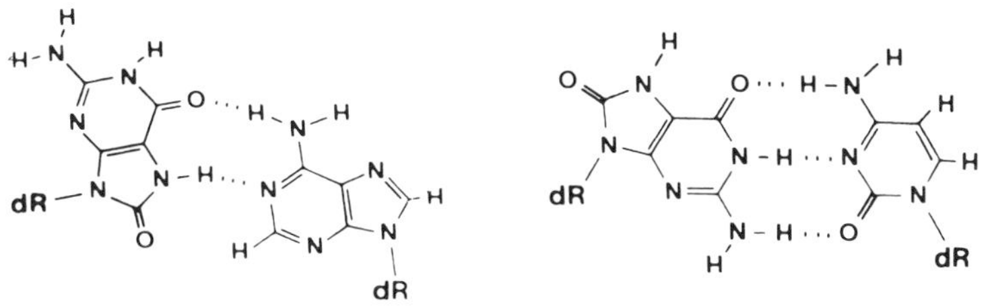

8-OxodG(*syn*)•dA(*anti*) **8-OxodG(*anti*)•dC(*anti*)**

Figure 2. Base pairs formed by 8-oxodG in *anti* and *syn* conformations.

et al., 1997; Arai et al., 1997; Aburatani et al., 1997; Lu et al., 1997; Bjørås et al., 1997). Eukaryotic MutT and MutY proteins are structurally similar to the corresponding *E. coli* enzymes. In contrast, Ogg1 shows no structural homologies with Fpg nor does it act efficiently on Fapy-containing substrates. However, Ogg1 does contain a conserved structural motif found in the active site of MutY and certain other DNA repair glycosylases (Thayer et al., 1995).

In this paper, we address two central questions: (i) how do the several DNA glycosylases involved in repair of 8-oxoguanine in bacteria and mammalian cells recognize their cognate lesions and (ii) what are the biochemical mechanisms by which 8-oxoguanine is efficiently excised from DNA? We analyze first those structural features of oxidatively-damaged DNA that confer recognition and binding of 8-oxoguanine by repair enzymes, then consider functional groups of DNA glycosylases involved in these processes. Finally, we discuss enzymatic mechanisms involved in excision of 8-oxoguanine from DNA.

3. RESULTS AND DISCUSSION

3.1. Substrate Recognition and Base Excision by Fpg

Electronic and conformational properties of deoxyguanosine are altered by oxidation at C-8. *Ab initio* calculations (Aida and Nishimura, 1987) and NMR studies (Culp et al., 1989) reveal that the 6,8-diketo form of 8-oxodG predominates under physiological conditions. *Syn* and *anti* forms of 8-oxodG exist in rapid equilibrium; the *syn* conformation is favored energetically (Uesugi and Ikehara, 1977; Culp et al., 1989). In duplex DNA, 8-oxoguanine (*syn*) forms a Hoogstein pair with dA (Kouchakdjian et al., 1991) while 8-oxoguanine (*anti*) pairs stably to dC (Oda et al., 1991) (Fig. 2). Apparently, steric interference between the 8-oxo and deoxyribose moieties in 8-oxodG:dC is offset by structural stabilization provided by the third hydrogen bond.

The steric disposition of hydrogen bond donors and acceptors in the major and minor grooves of duplex DNA (Seeman et al., 1976) contributes to recognition and binding of substrates by DNA repair enzymes. When 8-oxodG is in the an *anti* conformation, the C-8 carbonyl group is exposed in the major groove and the exocyclic amino group (N2) resides in the minor groove. When the modified base assumes the *syn* conformation, the relative position of these functional groups is reversed. In both conformations, the C-6 keto function lies in the major groove. These relationships are revealed in the molecular models shown in Figure 3.

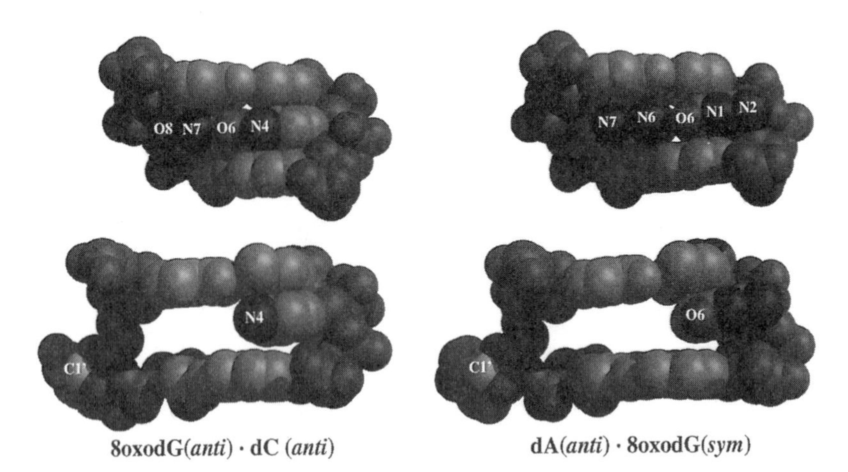

Figure 3. Top: molecular models showing central three bases in B-DNA for physiological substrates of Fpg (top left) and MutY (top right). The view is from the major groove. The C1' position of deoxyribose is inaccessible to solvent. Hydrogen bond donors and acceptors of the base pair containing 8-oxodG are labeled. Bottom: molecular models in which the base to be excised by MutM (lower left) and by MutY (lower right) has been rotated by 180°, leaving a gap in the duplex and exposing C1' to nucleophilic attack. These models were constructed by analogy to the crystal structure of the Hha DNA methyltransferase-DNA complex in which dC in a dG:dC pair is everted (Klimasauskas et al, 1994).

We have explored structure-function relationships for Fpg and its substrates using duplex oligodeoxynucleotides in which various functional groups are systematically modified. Specificity constants (k_{cat}/K_M) and apparent dissociation constants (K_d) were established for a variety of DNA structures (Tchou et al., 1994).

Duplex deoxynucleotides fall into three distinct classes—those that fail to bind Fpg, those that bind to the enzyme but are not efficiently cleaved, and those that bind and are cleaved (Table 1). For example, analogs lacking the 2-amino group (8-oxodI:dC) retain binding and cleavage activities. Replacement of the 6-keto function by hydrogen or substitution of carbon for oxygen in the deoxyribose ring creates molecules that bind Fpg but are resistant to cleavage. Substrates lacking the 8-keto function (dG:dC) are poorly bound and therefore not cleaved. Purines containing C8 substitutions other than a keto group (8-

Table 1. Kinetic parameters for binding and cleavage of duplex DNA substrates by Fpg[1]

Base	Modification	K_d (nM)	k_{cat}/K_M (min^{-1}/mM)
8-oxodG[2]	None	9	9
8-oxodI[2]	N2→H	10	18
8-oxodN[2]	O6,N2→H	7	0.024
8-oxodA[3]	O6→NH	n.d.	0.004
dG	O8→N	250	n.d
AP	loss of base	n.d.	210

[1]Tchou et al., 1994
[2]Opposite dC
[3]Opposite dT
n.d., not determined

Table 2. Effect of modifying bases opposite 8-oxodG[1]

Mispair	K_d (nM)	k_{cat}/K_M (min^{-1}/mM)
8-oxodG:dT	29	270
8-oxodG:dG	21	110
8-oxodG:AP	n.d.	30
8-oxodG:dC[1]	9	9.3
8-oxodG:dA[1]	340	0.5

[1]Tchou et al., 1994

methoxy-dG, 8-amino-dG) also are very poor substrates for Fpg (D.O.Z., unpublished observations). A duplex containing an abasic site, an intermediate structure in the AP lyase reaction, is rapidly cleaved.

Replacement of dC by T or dG on the strand opposite the lesion leads to 10–30 fold increase in specificity and 2-fold decrease in affinity for Fpg compared to 8-oxodG:dC (Table 2). Replacement of dC by dA decreases the specificity constant by 20-fold and reduces affinity for the enzyme 40-fold. Introduction of a mispair(s) adjacent to 8-oxodG:dC leads to a 10–25-fold increase in efficiency of cleavage, due mainly to lower K_M (Table 3).

DNA containing formamidopyrimidines (Fapy) is an excellent substrate for Fpg (Boiteux et al., 1990; Laval at al., 1990). Fapy are ring-open structures, existing in several rotameric forms (Boiteux et al., 1984). Exocyclic groups of these rotamers have considerable freedom of motion; in principle, different rotameric forms could be accommodated in the Fpg active site. Based on a structure-function analysis (Tchou et al., 1994), we proposed a model in which recognition and preferential binding of substrate by Fpg is guided by the C8-keto group in the major groove of DNA. The enzyme uses this structural feature *inter alia* to discriminate between sequences containing 8-oxoguanine and unmodified DNA. We suggest that the C-6 keto function is involved in catalysis by Fpg. Our model assumes that the rate of interconversion between *anti* and *syn* conformers of 8-oxopurines in duplex DNA is faster than glycosidic bond cleavage by the enzyme and slower when the modified base is stably hydrogen bonded to a base on the opposite strand (Grollman and Moriya, 1993). This model is generally consistent with published studies using randomly damaged DNA as substrate for Fpg (Boiteux et al., 1992); however, one pair of conflicting observations cannot easily be reconciled; namely, the facile excision of 4,6-

Table 3. Effect of neighboring mispairs on cleavage of duplex DNA substrates by Fpg[1]

Duplex	K_M nM	V_{max} nM/min	k_{cat} min^{-1}	k_{cat}/K_M min^{-1}/mM
-CXC- -GCG-	14	2.4	0.13	9.13
-CXC- -ACG-	2.0	0.84	0.50	250
-CXC- -GCA-	0.35	0.065	0.04	114
-CXC- -ACA-	2.1	0.40	0.24	114

[1]Tchou et al., 1994; X = 8-oxodG

diamino-5-formamidopyrimidine from DNA (Breimer, 1984; Karakaya et al., 1997) in light of the demonstrated resistance of 8-oxoadenine to this enzyme (Tchou et al., 1994). Another interesting, but as yet unconfirmed observation, is the reported cleavage of the 5-hydroxyl derivatives of dC and dU by Fpg (Hatahet et al., 1994).

The marked increase in biological activity, observed when hydrogen bonding to the opposite strand is impaired or when DNA containing 8-oxodG:dC is destabilized by a neighboring mispair, suggests that the base-flipping mechanism first reported for Hha methyltransferase (Klimasauskas et al., 1994) and later for several DNA glycosylases (Demple, 1995; Slupphaug et al., 1996) may be involved.

It is of interest to compare the preliminary structure-function analysis of Ogg1 with that of Fpg. Although structurally distinct, both enzymes show preference for substrates containing 8-oxodG:dC and appear functionally similar with the exception of the weak activity of Ogg1 towards substrates containing Fapy (van der Kemp et al., 1996; Rosenquist et al., 1997).

3.2. Substrate Recognition and Base Excision by MutY

MutY plays an essential role in repairing 8-oxoguanine in oxidatively-damaged DNA, accomplishing this function by excising adenosine paired in *syn* to the modified base (Michaels et al., 1993). MutY has relatively high apparent affinity (K_d = 21 nM) for unmodified duplex DNA. The apparent K_d for MutY bound to duplexes containing 8-oxodG:dA is 4-fold lower than for duplexes containing dG:dA, reflecting the generally tighter complex between the enzyme and various substrates oxidized at the C-8 position (Bulychev et al., 1996). These results parallel the enhanced thermodynamic stability of 8-oxodG:dA relative to the dG:dA duplex (Plum et al., 1995).

All base modifications tested in our studies of MutY are associated with reduced values for k_{cat}/K_M. However, for many of the substrates tested there is little apparent effect on binding to the enzyme even though the ability to recognize the mispair and/or performance of catalytic functions clearly is impaired (Bulychev et al., 1996 and Table 4).

The presence of the 8-keto function increased significantly the rate of removal of deoxyadenosine by MutY from all substrates tested. Replacement of dA by rA reduced the

Table 4. Kinetic parameters for binding and cleavage of duplex DNA substrates by MutY[1]

Modification	k_{cat}/K_M min^{-1}/mM	K_d nM
None	39,600	6
dA		
N→C	0	7
N6→O	35	8
H8→O	trace	11
2H'→OMe	trace	27
8-oxodG		
O8→H	383	26
O6→H;	824	25
N6→H		
N2→H	9280	8
2H'→OMe	81	11

[1]Bulychev et al., 1996

specificity constant from 39600 to 294 min^{-1}/mM, whereas replacement of dA by 2'-O-methyladenosine virtually abolished enzymatic activity. Modifications of dG generally were better tolerated than modifications of dA; however, introduction of a methyl ether at C-6 of dG produced a non-cleavable substrate and replacement of dG by 2'-O-methylguanosine generated a substrate with a low specificity constant. Rates of base excision from duplexes containing dA:dC and dA:tetrahydrofuran were three orders of magnitude lower than for the reference substrate. Duplexes containing a carbocyclic analog of dA were tightly bound by MutY but resistant to base excision (Bulychev et al., 1996). Porello et al. (1996) have shown that the dA analogs, 2'-deoxytubercidin and 2'-deoxyformycin A are also resistant to the glycosylase action of MutY.

In DNA containing 8-oxodG(*syn)*:dA(*anti*), N1 and N2 of guanine and N6 of adenine are potential hydrogen bond donors and O6 of guanine and N7 of adenine are hydrogen bond acceptors located in the major groove (Fig. 3). N3 of adenine and O8 of 8-oxodG are hydrogen bond acceptors exposed in the minor groove. For substrates in which N7, N6, or H8 of adenine have been replaced by C, O, and O, respectively, specificity constants for MutY are reduced by three orders of magnitude (Bulychev et al., 1996 and Table 4). The contribution of N1 of adenine was not directly explored although, in protonated form, this position plays an important role in stabilizing dG in the *syn* conformation. For each pair of substrates in which dG contains an 8-keto function, the specificity constant of the 8-oxo derivative is higher by approximately two orders of magnitude (Bulychev et al., 1996). This could reflect affinity of MutY for O8 of the modified base; more likely it represents the preference for dG in dG:dA pairs to assume the *anti* conformation at pH 7.5 (Bulychev et al., 1996). This partially stacked structure, stabilized by bifurcated hydrogen bonds, is a less effective substrate for MutY.

We conclude from our structure-function analysis of MutY that dG(*syn*):dA(*anti*) and 8-oxodG(*syn*):dA(*anti*) mispairs present unique configurations of hydrogen bond donors and acceptors which are recognized in the major groove by complementary groups in MutY. Specific binding is conferred by *simultaneous* interaction of MutY with donors and/or acceptors on each base involved in the mispair. The specificity constant for MutY with 8-oxodG:dA is 100-fold greater than the comparable value for dG:dA. This preference is consistent with the observation that, at pH 7.5, dG and dA tend to assume the *anti* conformation with protonation of dA required to form a Hoogstein pair and dG adopting the *syn* conformation. Neither of the hydrogen bond acceptors exposed in the minor groove (O8) of 8-oxodG(*syn*) and N3 of dA (*anti*), both having counterparts in the dA(*anti*):dT(*anti*) Watson Crick pair, appear to be involved in binding MutY. Methylation and ethylation experiments suggest that phosphate groups on either side of the mispair are involved in complex formation (Lu et al., 1995); the effect of nucleotide sequence context has not yet been systematically examined.

3.3. Protein Determinants of Substrate Binding

Amino acid sequence analysis of Fpg reveals a single zinc finger motif of the CC/CC type involving the Cys-244, -247, -264, and -267, residues located near the carboxyl terminus. Zinc finger motifs are found in a number of DNA-binding proteins; the importance of cysteine residues in Fpg is supported by the observation that hydroxymercury-phenylsulfonic acid, which specifically modifies sulfhydryl groups in proteins, inhibits the activity of this enzyme (O'Connor et al., 1993).

We have examined the role of the COOH terminal zinc finger of Fpg protein using a gel mobility shift assay to study binding of a duplex oligodeoxynucleotide containing a single tetrahydrofuran moiety. Selection of this abasic site analog was based on the observation that duplexes containing a natural abasic site are substrates for Fpg protein (Table 1), while those containing tetrahydrofuran are resistant to cleavage by AP-lyases (Tchou et al., 1991). Site-directed mutagenesis was used to dissect the role of cysteine residues in the zinc finger domain of Fpg protein (Tchou et al., 1993; O'Connor et al., 1993). Zinc finger mutants (C244S, C244H, C244A, C244amber, C247G, C264G, C267G, C244S/C247S) do not bind DNA. The loss of DNA binding activity appears not to affect secondary structure, as indicated by a circular dichroism study of the C244S/C247S mutant (Tchou et al., 1993), but rather reflects disruption of the zinc finger structure created by loss of zinc. Thus, the loss of enzymatic activity observed in these zinc finger mutants most likely is due to a failure to bind oxidatively damaged DNA. However, a role for the COOH-terminal zinc finger in the N-glycosylase and/or AP lyase activity of the protein cannot be excluded.

Fpg proteins found in various strains of *E. coli* contain several highly conserved sequences, among them the N-terminal PELPEVET. Addition of two amino acid residues (Gly-Met) to the amino terminus of Fpg or replacement of the N-terminal proline by glycine generates proteins that bind but fail to cleave duplexes containing 8-oxodG. These observations suggest that the N-terminal proline and/or position of the amino terminus is important for the catalytic function of the enzyme but does not play an essential role in the binding reaction.

3.4. DNA Binding at the Active Site

X-ray crystallographic analysis of *E. coli* endonuclease III (Endo III, Thayer et al., 1995) reveals a superfamily of DNA binding proteins defined by a helix-hairpin-helix

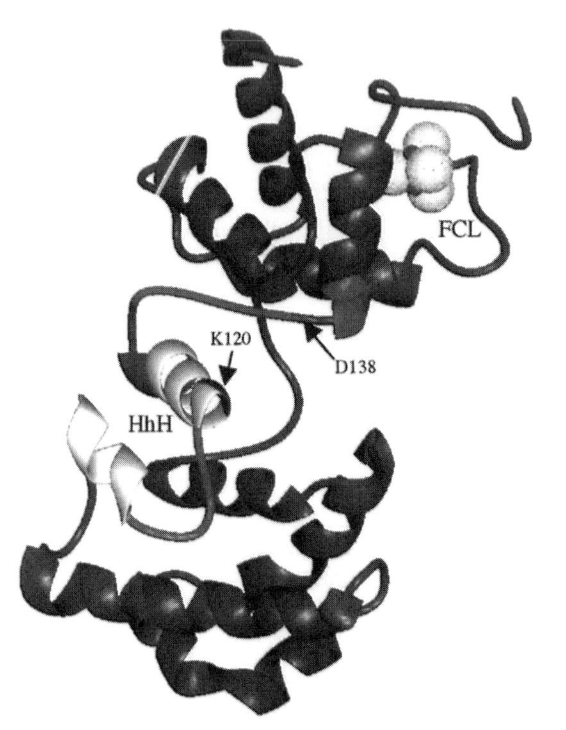

Figure 4. Ribbon model of E. coli endonuclease III derived from coordinates published by Thayer et al. (1995) showing position of potential active site residues, the helix-hairpin helix (HhH)-PVD loop, and the FCL cluster.

(HhH) structural motif (Fig. 4). This superfamily includes Endo III, 3-methyladenine DNA glycosylase, MutY and a G-T mismatch repair protein from *E. coli* (Thayer et al., 1995) along with several Ogg1 proteins recently identified in yeast (van der Kemp et al., 1996; Nash et al., 1996) and mammalian cells (Rosenquist et al., 1997; Radicella et al., 1997; Roldán-Arjona et al., 1997; Arai et al., 1997; Aburatani et al., 1997; Lu et al., 1997; Bjørås et al., 1997). Fpg does not contain the HhH motif and differs structurally from DNA glycosylases grouped in the superfamily. Endo III and MutY also share a [4Fe-4S] cluster (FCL) homology near the carboxy terminus of the enzyme (Cunningham et al., 1988; Michaels et al., 1990).

Endo III has two structural domains (Fig. 4). Between them and near a loop that binds thymine glycol is a deep solvent-filled pocket lined by polar side chains that lies in a shallow depression opposite the interdomain groove (Thayer et al., 1995). Within the groove, two amino acid residues, Lys-120 and Asp-138, bracket the solvent-filled pocket. Both of these residues are important for catalytic activity.

A general model has been proposed for the Endo III superfamily of DNA glycosylases based primarily on the common HhH motif and on kinetic evidence implicating Asp-138 and Lys-120 or their counterparts in the catalytic function of these enzymes (Thayer et al., 1995). The HhH motif, together with the Pro/Val rich sequence (PVD) that lies carboxy terminal to it, includes the active site of MutY and Ogg1. The PVD loop may be involved in recognition and binding of the modified base. Asp-138 is positioned to deprotonate Lys-120 (or water), a reaction required to activate a nucleophile that can attack at C1'. DNA can bind electrostatically to the positively-charged face of the protein with the backbone fitting into the interdomain groove of the protein near the active site residues. The nearby pocket can accommodate an extrahelical base.

Proteolysis of MutY with trypsin yields two distinct domains; a larger 26 kDa fragment representing the NH_2-terminal domain including the FCL cluster and a 13 kDa fragment containing the carboxy terminus. The p26 fragment binds and cleaves various substrates for MutY, retaining 15% of the DNA adenine glycosylase activity of the parent enzyme when tested with duplexes containing a G:A mispair. The activity of p26 against substrates containing 8-oxodG:dA is two orders of magnitude lower than intact MutY (Gogos et al., 1996). A DNA binding role has been proposed for the FCL cluster of MutY, based on the structural homology with Endo III (Thayer et al., 1995; Manuel et al., 1996); however, Fe does not contribute significantly to enzyme reactivity and may act instead to stabilize the enzyme structure.

3.5. Mechanism of Base Excision

Several mechanisms, all involving nucleophilic attack at C1' of deoxyribose, have been proposed to explain the action of DNA glycosylases on duplex DNA substrates. Nucleophilic attack could involve a basic group in the enzyme or an activated water molecule (Dodson et al., 1994); in itself, this does not assure efficient base release and must be accompanied by protonation of the base (Dodson et al., 1994; Tchou et al., 1994, 1995) or the deoxyribose oxygen. Structure-function analysis of Fpg (Tchou et al., 1994) suggests that this protonation may occur at the 6-keto group of 8-oxodG. Rabow and Kow (1997) suggest that Lys-155 functions as an acid catalyst by interacting with O8 in 8-oxodG.

Amine functions of proteins are capable of nucleophilic attack; for example, the N-terminal residue of T4 bacteriophage endonuclease V (Endo V, Dodson et al., 1993) and Lys- 120 of Endo III (R. Cunningham, personal communication) have been shown to be involved in Schiff base formation. For Fpg, we propose a mechanism similar to that reported for Endo V (Dodson et al., 1993). In this enzyme, the α-NH_2 group of Thr-1 serves as a nucleophile. Replacement of threonine by proline reduces enzymatic activity to negligible values (Schrock III and Lloyd, 1993). The fact that Pro-1 is conserved in all Fpg proteins reported to date and replacement of this residue by glycine abolishes cleavage while retaining binding properties (Tchou and Grollman, 1995) indicates that proline plays an important role in the catalytic mechanisms of Fpg.

An established procedure for stabilization of Schiff bases involves chemical reduction by borohydride. This approach has been used to prepare covalently-linked complexes between DNA glycosylases/lyases and their substrates (Sun et al., 1995). Fpg can be efficiently crosslinked to its active substrates (Tchou and Grollman, 1995); this reaction requires C1' of the modified base to be electrophilic since carba-8-oxodG does not form a crosslink (Zharkov et al., 1997). Mapping of Fpg (Tchou et al., 1995), combined with electrospray mass spectrometric analysis reveals that the Schiff base formed between Fpg and DNA involves the amino group of the N-terminal proline (Zharkov et al., 1997). The protein was resistant to Edman degradation following crosslinking, confirming that the N-terminus had been modified.

The N-terminal sequence of all Fpg proteins is highly conserved. Eight initial amino acids (PELPEVET), as well as Arg-12 and Leu/Ile-14, are almost absolutely conserved. The N-terminal sequence is rich in ionic and polar amino acids and can readily interact with substrate, water and/or other groups in the active site. Phosphate groups in the DNA backbone, the positively charged imino nitrogen of Pro-1, three glutamic acids and the guanidinium group of Arg-12 form a multipolar charge system that could stabilize the intermediate complex. A similar role was proposed for Glu-23 of Endo V, shown by X-ray analysis to lie in the proximity of Thr-2 (Morikawa et al., 1995). Glu-23 mutants bind but do not cleave substrates for Endo V (Manuel et al., 1995).

Ogg1 of yeast has been crosslinked to duplex DNA containing 8-oxodG (van der Kemp et al., 1996; Nash et al., 1996). The precise site of crosslinking for the human enzyme has been identified as Lys-249 (Nash et al., 1997), the nucleophilic group in *S. cerevisiae* therefore is likely to be Lys-241.

It has been proposed that all DNA glycosylases with AP lyase activity will form Schiff bases (Dodson et al., 1994); however, the literature contains conflicting data with respect to the presence of an AP/lyase function in MutY (cf Bulychev et al., 1996; Lu et al., 1996). MutY forms intermediates with DNA containing dA:8-oxodG and AP-8-oxodG (Lu et al., 1996), although the yield of covalently-bound complex derived is less than that obtained under similar conditions with Fpg. Analysis of the reaction products suggests that MutY does not possess true AP lyase activity (Zharkov, unpublished data); furthermore, MutY lacks Lys-120, the residue that characterizes members of the Endo III superfamily with AP lyase activity. MutY contains serine at a comparable position and therefore must utilize a different residue for the crosslinking reaction.

How do DNA Glycosylases Gain Access to C1'? Nucleophillic attack at C1' in duplex DNA involves a deeply buried deoxyribose residue. This sugar could be externalized by eversion, a mechanism known as base-flipping first reported for Hha methyltransferase (Klimasauskas et al., 1994). Base-flipping has been demonstrated for uracil DNA glycosylase and inferred for other DNA glycosylases (Demple, 1995). Eversion involves several steps: damage detection by the enzyme, binding to sugar phosphates to create backbone compression, and accommodation of the flipped out base in a pocket near the active site.

We have used a novel approach to test the possibility of an eversion mechanism for Fpg. Sterically, duplex DNA containing a bulge resembles DNA with a base everted 180°

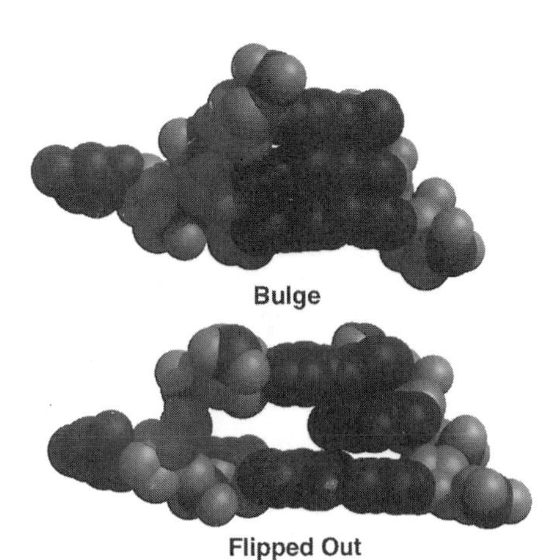

Figure 5. Molecular models showing a flipped-out base (see legend to Fig. 3) and a one-base bulge; the latter was derived from coordinates published by Joshua-Tor et al., 1992.

(Fig. 5). We have established kinetic parameters for Fpg using duplexes containing one-, two- or three-base bulges, one of the bulged bases being 8-oxodG. The most active substrate contained a two-base bulge; the specificity constant being only 4-fold lower than for an 8-oxodG:dC pair. The decrease reflects K_M rather than k_{cat}. Cleavage did not occur when dG replaced 8-oxodG. These experiments, together with the marked increase in k_{cat}/K_M, observed when 8-oxodG:dC is flanked by mispairs (Table 3), and the established nucleophilic attack by Fpg on C1' support the idea that the enzyme operates by a base-flipping mechanism. A molecular model for the proposed intermediate is shown in Fig. 3.

4. CONCLUSIONS

Based on these investigations, we propose the following model for recognition of 8-oxoguanine and excision of this base (or the adenosine opposite it) from duplex DNA: (i) the DNA glycosylases involved in repair (Fpg, Ogg1 and MutY) bind nonspecifically to DNA, an interaction mediated primarily by electrostatic interactions between the protein and phosphodiester backbone (von Hippel and Berg, 1989; Dowd and Lloyd, 1990); (ii) the enzyme scans the major groove of DNA for damage, a process known as facilitated linear diffusion (von Hippel and Berg, 1989). Damage recognition involves complementary hydrogen bond donor/acceptor pairs, in this case exposed in the major groove (Seeman et al., 1976); (iii) a specific complex is formed, stabilized by interactions with neighboring bases; (iv) eversion (base-flipping) occurs, facilitated by binding to a pocket near the active site and conformational changes in the protein filling the gap left by the flipped out base and DNA; (v) nucleophilic attack at C1' by the enzyme or by a water molecule, accompanied by protonation to displace the heterocyclic damaged base.

Many experimental questions remain to be answered. Atomic resolution structure of a DNA-Fpg complex would provide a more detailed understanding of substrate recognition by this enzyme. This study is underway. Substrate specificity of Ogg1 should be characterized, as was done for Fpg and MutY. The catalytic mechanism of MutY also needs to be elucidated. Issues relating to processivity and translocation of Fpg, MutY, and Ogg1 should be studied in more detail. Cloning of mammalian homologs for Fpg, MutY, and MutT now make it possible to examine the role of oxidative damage in aging and cancer;

these investigations would be enhanced by a mechanistic understanding of the DNA processes involved.

ACKNOWLEDGMENTS

Important contributions to the research discussed in this review were made by members of our research group, including Francis Johnson, Shinya Shibutani, Masaaki Moriya, Nick Bulychev and Julia Tchou.

REFERENCES

H. Aburatani, Y. Hippo, T. Ishida, R. Takashima, C. Matsuba, T. Kodama, M. Takao, A. Yasui, K. Yamamoto, M. Asano, K. Fukasawa, T. Yoshinari, H. Inoue, E. Ohtsuka, and S. Nishimura (1997) Cloning and characterization of mammalian 8-hydroxyguanine-specific DNA glycosylase/apurinic, apyrimidinic lyase, a functional mutM homologue. *Cancer Res.* **57**, 2151–2156.

M. Aida and S. Nishimura (1987) An ab initio molecular orbital study on the characteristics of 8-hydroxyguanine. *Mutat. Res.* **192**, 83–89.

K. Arai, K. Morishita, K. Shinmura, T. Kohno, S.-R. Kim, T. Nohmi, M. Taniwaki, S. Ohwada, and J. Yokota (1997) Cloning of a human homolog of the yeast *OGG1* gene that is involved in the repair of oxidative DNA damage. *Oncogene* **14**, 2857–2861.

M. Bjørås, L. Luna, B. Johnsen, E. Hoff, T. Haug, T. Rognes, and E. Seeberg (1997) Opposite base-dependent reactions of a human base excision repair enzyme on DNA containing 7,8-dihydro-8-oxoguanine and abasic sites. *EMBO J.* **16**, 6314–6322.

S. Boiteux, J. Belleney, B. P. Roques, and J. Laval (1984) Two rotameric forms of open ring 7-methylguanine are present in alkylated polynucleotides. *Nucl. Acids Res.* **12**, 5429–5439.

S. Boiteux, T.R. O'Connor, F. Lederer, A. Gouyette, and J. Laval (1990) Homogeneous *Escherichia coli* FPG protein. A DNA glycosylase which excises imidazole ring-opened purines and nicks DNA at apurinic/apyrimidinic sites. *J. Biol. Chem.* **265**, 3916–3922.

S. Boiteux, E. Gajewski, J. Laval, and M. Dizdaroglu (1992) Substrate specificity of the *Escherichia coli* Fpg protein (formamidopyrimidine-DNA glycosylase): excision of purine lesions in DNA produced by ionizing radiation or photosensitization. *Biochemistry* **31**, 106–110.

L.H. Breimer (1984) Enzymatic excision from γ-irradiated polydeoxyribonucleotides of adenine residues whose imidazole rings have been ruptured. *Nucl. Acids Res.* **12**, 6359–6367.

N.V. Bulychev, C.V. Varaprasad, G. Dormán, J.H. Miller, M. Eisenberg, M., A.P. Grollman, and F. Johnson (1996) Substrate specificity of *Escherichia coli* MutY protein. *Biochemistry* **35**, 13147–13156.

K.C. Cheng, D.S. Cahill, H. Kasai, S. Nishimura, and L.A. Loeb (1992) 8-Hydroxyguanine, an abundant form of oxidative DNA damage, causes G→T and A→C substitutions. *J. Biol. Chem.* **267**, 166–172.

S.J. Culp, B.P. Cho, F.F. Kadlubar, and F.E. Evans (1989) Structural and conformational analysis of 8-hydroxy-2'-deoxyguanosine. *Chem. Res. Toxicol.* **2**, 416–422.

R.P. Cunningham, H. Asahara, J.F. Bank, C.P. Scholes, J.C. Salerno, K. Surerus, E. Munck, J. McCracken, J. Peisach, and M.H. Emptage (1989) Endonuclease III is an iron-sulfur protein. *Biochemistry* **28**, 4450–4455.

B. Demple (1995) DNA repair flips out. *Current Biol.* **5**, 719–721.

M.L. Dodson, R.D. Schrock III, and R.S. Lloyd (1993) Evidence for an imino intermediate in the T4 endonuclease V reaction. *Biochemistry* **32**, 8284–8290.

M.L. Dodson, M.L. Michaels, and R.S. Lloyd (1994) Unified catalytic mechanism for DNA glycosylases. *J. Biol. Chem.* **269**, 32709–32712.

D.R. Dowd and R.S. Lloyd (1990) Biological significance of facilitated diffusion in protein-DNA interactions. Applications to T4 endonuclease V-initiated DNA repair. *J. Biol. Chem.* **265**, 3424–3431.

A. Gogos, J. Cillo, N.D. Clarke, and A-L. Lu (1996) Specific recognition of A/G and A/7,8-dihydro-8-oxoguanine (8-oxoG) mismatches by *Escherichia coli* MutY: removal of the C-terminal domain preferentially affects A/8-oxoG recognition. *Biochemistry* **35**, 16665–16671.

A.P. Grollman, and M. Moriya (1993) Mutagenesis by 8-oxoguanine: an enemy within. *Trends Gen.* **9**, 246–249.

B. Halliwell and J.M.C. Gutteridge (1989). *Free Radicals in Biology and Medicine*, 2nd ed., Clarendon Press, Oxford, U.K.

Z. Hatahet, Y.W. Kow, A.A. Purmal, R.P. Cunningham, and S.S. Wallace (1994) New substrates for old enzymes. 5-hydroxy-2'-deoxycytidine and 5-hydroxy-2'-deoxyuridine are substrates for *Escherichia coli* endonuclease III and formamidopyrimidine DNA *N*-glycosylase, while 5-hydroxy-2'-deoxyuridine is a substrate for uracil DNA *N*-glycosylase. *J. Biol. Chem.* **269**, 18814–18820

H. Igarashi, T. Tsuzuki, T. Kakuma, Y. Tominaga, and M. Sekiguchi (1997) Organization and expression of the mouse *MTH1* gene for preventing transversion mutation. *J. Biol. Chem.* **272**, 3766–3772.

L. Joshua-Tor, F. Frolow, E. Appella, H. Hope, D. Rabinovich, and J.L. Sussman (1992) Three-dimensional structures of bulge-containing DNA fragments. *J. Mol. Biol.* **225**, 397–431.

A. Karakaya, P. Jaruga, V.A. Bohr, A.P. Grollman, and M. Dizdaroglu (1997) Kinetics of excision of purine lesions from DNA by *Escherichia coli* Fpg protein. *Nucl. Acids Res.* **25**, 474–479.

S. Klimasauskas, S. Kumar, R.J. Roberts, and X. Cheng (1994) HhaI methyltransferase flips its target base out of the DNA helix. *Cell* **76**, 357–369.

M. Kouchakdjian, V. Bodepudi, S. Shibutani, M. Eisenberg, F. Johnson, A.P. Grollman, and D.J. Patel (1991) NMR structural studies of the ionizing radiation adduct 7-hydro-8-oxodeoxyguanosine (8-oxo-7*H*-dG) opposite deoxyadenosine in a DNA duplex. 8-Oxo-7*H*-dG(*syn*)•dA(*anti*) alignment at lesion site. *Biochemistry* **30**, 1403–1412.

J. Laval, S. Boiteux, and T.R. O'Connor (1990) Physiological properties and repair of apurinic/apyrimidinic sites and imidazole ring-opened guanines in DNA. *Mut. Res.* **233**, 73–79.

A-L. Lu, J.-J. Tsai-Wu, and J. Cillo (1995) DNA determinants and substrate specifities of *Escherichia coli* MutY. *J. Biol. Chem.* **270**, 23582–23588.

A-L. Lu, D.S. Yuen, and J. Cillo (1996) Catalytic mechanism and DNA substrate recognition of *Escherichia coli* MutY protein. *J. Biol. Chem.* **271**, 24138–24143.

R. Lu, H.M. Nash, and G.L. Verdine (1997) A mammalian DNA repair enzyme that excises oxidatively damaged guanines maps to a locus frequently lost in lung cancer. *Current Biol.* **7**, 397–407

H. Maki and M. Sekiguchi (1992) MutT protein specifically hydrolyses a potent mutagenic substrate for DNA synthesis. *Nature* **355**, 273–275.

R.C. Manuel, E.W. Czerwinski, and R.S. Lloyd (1996) Identification of the structural and functional domains of MutY, an *Escherichia coli* DNA mismatch repair enzyme. *J. Biol. Chem.* **271**, 16218–16226.

R.C. Manuel, K.A. Latham, M.L. Dodson, and R.S. Lloyd (1995) Involvement of glutamic acid 23 in the catalytic mechanism of T4 endonuclease V. *J. Biol. Chem.* **270**, 2652–2661.

M.L. Michaels, L. Pham, Y. Nghiem, C. Cruz, and J.H. Miller (1990) MutY, an adenine glycosylase active on G-A mispairs, has homology to endonuclease III. *Nucl. Acids Res.* **18**, 3841–3845.

M.L. Michaels and J.H. Miller (1992) The GO system protects organisms from the mutagenic effect of the spontaneous lesion 8-hydroxyguanine (7,8-dihydro-8-oxoguanine). *J. Bacteriol.* **174**, 6321–6325.

M.L. Michaels, J. Tchou, A.P. Grollman, and J.H. Miller (1993) A repair system for 8-oxo-7,8-dihydroguanine (8-hydroxyguanine). *Biochemistry* **31**, 10964–10968.

K. Morikawa, O. Matsumoto, M. Tsujimoto, K. Katayanagi, M. Ariyoshi, T. Doi, M. Ikehara, T. Inaoka, and E. Ohtsuka (1992) X-ray structure of T4 endonuclease V: an excision repair enzyme specific for a pyrimidine dimer. *Science* **256**, 523–526.

H.M. Nash, S.D. Bruner, O.D. Shärer, T. Kawate, T.A. Addona, E. Spooner, W.S. Lane, and G.L. Verdine (1996) Cloning of a yeast 8-oxoguanine DNA glycosylase reveals the existence of a base-excision DNA-repair protein superfamily. *Current Biol.* **6**, 968–980.

H.M. Nash, R. Lu, W.S. Lane, and G.L. Verdine (1997) The critical active-site amine of the human 8-oxoguanine DNA glycosylase, hOgg1: direct identification, ablation and chemical reconstitution. *Chemistry & Biology* **4**, 693–702

T.R. O'Connor, R.J. Graves, G. de Murcia, B. Castaing, and J. Laval (1993) Fpg protein of *Escherichia coli* is a zinc finger protein whose cysteine residues have a structural and/or functional role. *J. Biol. Chem.* **268**, 9063–9070.

Y. Oda, S. Uesugi, M. Ikehara, S. Nishimura, Y. Kawase, H. Ishikawa, H. Inoue, and E. Ohtsuka (1991) NMR studies of a DNA containing 8-hydroxydeoxyguanosine. *Nucl. Acids Res.* **19**, 1407–1412.

Y.I. Pavlov, D.T. Minnick, S. Izuta, and T.A. Kunkel (1994) DNA replication fidelity with 8-oxodeoxyguanosine triphosphate. *Biochemistry* **33**, 4695–4701.

G.E. Plum, A.P. Grollman, F. Johnson, and K.J. Breslauer (1995) Influence of the oxidatively damaged adduct 8-oxodeoxyguanosine on the conformation, energetics, and thermodynamic stability of a DNA duplex. *Biochemistry* **34**, 16418–16160.

S.L. Porello, S. D. Williams, H. Kuhn, M. L. Michaels and S. S. David (1996). Specific recognition of substrate analogs by the DNA mismatch repair enzyme MutY. *J. Amer. Chem. Soc.* **118,** 10684–10692.

L.E. Rabow and Y.W. Kow (1997) Mechanism of action of base release by *Escherichia coli* Fpg protein: role of lysine 155 in catalysis. *Biochemistry* **36**, 5084–5096.

J.P. Radicella, C. Dherin, C. Desmaze, M.S. Fox, and S. Boiteux (1997) Cloning and characterization of *hOGG1*, a human homolog of the *OGG1* gene of *Saccharomyces cerevisiae*. *Proc. Natl. Acad. Sci.* USA **94**, 8010–8015.

T. Roldán-Arjona, Y.-F. Wei, K.C. Carter, A. Klungland, C. Anselmino, R.-P. Wang, M. Augustus, and T. Lindahl (1997) Molecular cloning and functional expression of a human cDNA encoding the antimutator enzyme 8-hydroxyguanine-DNA glycosylase. *Proc. Natl. Acad. Sci.* USA **94**, 8016–8020.

T.A. Rosenquist, D.O. Zharkov, and A.P. Grollman (1997) Cloning and characterization of a mammalian 8-oxoguanine DNA glycosylase. *Proc. Natl. Acad. Sci.* USA **94**, 7429–7434.

K. Sakumi, M. Furuichi, T. Tsuzuki, T. Kakuma, S.-i. Kawabata, H. Maki, and M. Sekiguchi (1993) Cloning and expression of cDNA for a human enzyme that hydrolyzes 8-oxo-dGTP, a mutagenic substrate for DNA synthesis. *J. Biol. Chem.* **268**, 23524–23530.

R.D. Schrock III and R.S. Lloyd (1993) Site-directed mutagenesis of the NH_2 terminus of T4 endonuclease V. The position of the αNH_2 moiety affects catalytic activity. *J. Biol. Chem.* **268**, 880–886.

N.C. Seeman, J.M. Rosenberg, and A. Rich (1976) Sequence-specific recognition of double helical nucleic acids by proteins. *Proc. Natl. Acad. Sci.* USA **73**, 804–808.

S. Shibutani, M. Takeshita, and A.P. Grollman (1991) Insertion of specific bases during DNA synthesis past the oxidation-damaged base 8-oxodG. *Nature* **349**, 431–434.

M.K. Shigenaga, C.J. Gimeno, and B.N. Ames (1989) Urinary 8-hydroxy-2'-deoxyguanosine as a biological market of *in vivo* oxidative DNA damage. *Proc. Natl. Acad. Sci.* USA **86**, 9697–9701.

G. Slupphaug, C.D. Mol, B. Kavli, A.S. Arvai, H.E. Krokan, and J.A. Tainer (1996) A nucleotide-flipping mechanism from the structure of human uracil-DNA glycosylase bound to DNA. *Nature* **384**, 87–92.

M.M. Slupska, C. Baikalov, W.M. Luther, J.-H. Chiang, Y.-F. Wei, and J.H. Miller (1996) Cloning and sequencing a human homolog (*hMYH*) of the *Escherichia coli mutY* gene whose function is required for the repair of oxidative DNA damage. *J. Bacteriology* **178**, 3885–3892.

B.Sun, K.A. Latham, M.L. Dodson, and R.S. Lloyd (1995) Studies of the catalytic mechanism of five DNA glycosylases. Probing for enzyme-DNA imino intermediates. *J. Biol. Chem.* **270**, 19501–19508.

J. Tchou, V. Bodepudi, S. Shibutani, I. Antoshechkin, J. Miller, A.P. Grollman, and F. Johnson (1994) Substrate specificity of Fpg protein. Recognition and cleavage of oxidatively damaged DNA. *J. Biol. Chem.* **269**, 15318–15324.

J. Tchou and A.P. Grollman (1995) The catalytic mechanism of Fpg protein. Evidence for a Schiff base intermediate and amino terminus localization of the catalytic site. *J. Biol. Chem* **270**, 11671–11677.

J. Tchou, H. Kasai, S. Shibutani, M.-H. Chung, J. Laval, A.P. Grollman, and S. Nishimura (1991) 8-oxoguanine (8-hydroxyguanine) DNA glycosylase and its substrate specificity. *Proc. Natl. Acad. Sci.* USA **88**, 4690–4694.

J. Tchou, M.L. Michaels, J.H. Miller, and A.P. Grollman (1993) Function of the zinc finger in *Escherichia coli* Fpg protein. *J. Biol. Chem.* **268**, 26738–26744.

M.M. Thayer, H. Ahern, D. Xing, R.P. Cunningham, and J.A. Tainer (1995) Novel DNA binding motifs in the DNA repair enzyme endonuclease III crystal structure. *EMBO J.* **14**, 4108–4120.

S. Uesugi and M. Ikehara (1977) Carbon-13 magnetic resonance spectra of 8-substituted purine nucleosides. Characteristic shifts for the syn conformation. *J. Amer. Chem. Soc.* **99**, 3250–3253.

P.A. van der Kemp, D. Thomas, R. Barbey, R. de Oliveira, and S. Boiteux (1996) Cloning and expression in *Escherichia coli* of the *OGG1* gene of *Saccharomyces cerevisiae*, which codes for a DNA glycosylase that excises 7,8-dihydro-8-oxoguanine and 2,6-diamino-4-hydroxy-5-*N*-methylformamidopyrimidine. *Proc. Natl. Acad. Sci.* USA **93**, 5197–5202.

P.H. von Hippel and O.G. Berg (1989) Facilitated target location in biological systems. *J. Biol. Chem.* **264**, 675–678.

C. von Sonntag (1987) *The Chemical Basis of Radiation Biology*, Taylor & Francis, London, U.K.

D.O. Zharkov, R.A. Rieger, C.R. Iden, and A.P. Grollman (1997) NH_2-terminal proline acts as a nucleophile in the glycosylase/AP-lyase reaction catalyzed by *Escherichia coli* formamidopyrimidine-DNA glycosylase (Fpg) protein. *J. Biol. Chem.* **272**, 5335–5341.

13

DNA REPAIR AS A SUSCEPTIBILITY FACTOR IN CHRONIC DISEASES IN HUMAN POPULATIONS

Lawrence Grossman,[*,1] Genevieve Matanoski,[1] Evan Farmer,[2] Mohammad Hedyati,[1] Sugita Ray,[3] Bruce Trock,[3] John Hanfelt,[3] George Roush,[3] Marianne Berwick,[4] and Jennifer Hu[3]

[1]The Johns Hopkins University School of Hygiene and Public Health
[2]The Johns Hopkins University School of Medicine
Baltimore, Maryland 21205
[3]Georgetown University Medical Center
Washington, DC 20007
[4]New York University Medical Center
New York, New York 10010

1. ABSTRACT

Three case-control studies of DNA repair in the general population were conducted with: i. 88 primary basal cell carcinoma (BCC) cases and 135 cancer-free controls, ii 304 study subjects including 57 arsenical cancer patients and 247 noncancerous controls in Taiwan and iii 41 breast cancer patients and 73 controls. The host reactivation assay was used to measure cellular DRC capacity with cryopreserved peripheral lymphocytes from both the cases and their controls. In study i. reduced repair of UV-induced DNA damage contributed to the risk of sunlight-induced BCC. A family history of BCC is a predictor of low DNA repair. Repair of UV damaged DNA declines at a rate of about 0.6 per annum in in non-cancerous controls. In addition, reduced DNA repair is more likely seen in young BCC cases, indicating that BCC is a premature aging disease of the skin. The persistence of photochemical damage because of reduced repair, results in point mutations in the p53 gene and allelic loss of the nevoid BCC (Gorlin's syndrome) gene located on chromosome 9q. Xeroderma pigmentosum appears to be a valid paradigm for the role of DNA repair in BCC in the general population. An extension of these studies led to conclusions from Black Foot Disease (BDF) studies that DRC by itself is not a risk factor for arsenical skin cancer , but those individuals with low DRC,are at much greater risk when exposed to

* Fax: 410-955-2926, E-mail: lgrossman@jhsh.edu

high levels of arsenic in their drinking water or when they are on poor diets. DRC, therefore, appears to be a susceptibility factor in this disease. Further DRC is consistently lower in breast cancer cases than in controls and appears to be a susceptibility factor in breast cancer and DRC in lymphocytes may be employed as a biomarker for human breast cancer risk.

2. INTRODUCTION

Human populations typically display a range of inherent sensitivities to radiation and chemical carcinogens. Given a common carcinogenic insult, some individuals develop associated neoplasms, some may develop only an associated pre-neoplastic lesion, while others remain clinically-free from all related effects of the exposure. Even within specific cancer populations, age of onset, extent and severity of neoplasia often vary between patients. Such variability in host response, in part, may be due to inherent differences between individuals to monitor and repair damaged sites induced in their genetic material by exogenous and endogenous genotoxic agents. The manifestation of many cancers is associated with either activated oncogenes or deactivated tumor suppressor genes because of specific mutations. In most cases mutations at specific loci appear to provide the necessary signal releasing these genes from a cryptic or quiescent state eventuating in cancer formation. The linkage between persistent DNA damage and oncogene activity suggests that such long-lived DNA damage is a reflection of the diminished involvement of DNA repair or surveillance activities in tumorigenesis.

A human model supporting such assumptions exists with the rare, cancer-prone inherited disorder (Xeroderma pigmentosum) (Kraemer and Slor,1985 and Cleaver and Kraemer 1989). XP patients experience a greater than 2,000-fold excess frequency of sunlight related skin cancers. Coupled to this marked susceptibility is the consistent laboratory finding that all cells tested from the XP patients are defective in repairing DNA damage induced by ultraviolet UV and other UV-mimetic agents. Since the UV component of solar radiation exists as the predominant environmental risk factor for skin cancer, a causal association between UV exposure, defective repair of UV-induced DNA photoproducts, and skin cancer is inferred. The link is further strengthened by clinical reports showing the occurrence of skin tumors is practically ameliorated in XP patients afforded early and lifelong protection from sunlight exposure (Pausey et al 1979 and Lynch et al 1977).

Although the rarity of XP precludes any particular significance as a public health concern, the clear etiologic model of cancer susceptibility it provides may have relevance to the general population. This is made apparent through three basic observations. The first is that the defect in XP is not complete, but is expressed phenotypically as a range of diminished repair proficiencies estimated to encompass residual capabilities of < 2% up to approximately 80% of "normal". The second is that besides UV damage, XP cells are equally defective in repairing genetic damage induced by a variety of "bulky" genotoxic drugs and chemical carcinogens. Lastly, like many other human phenotypic traits, individual variability in DNA repair capacity DRC has been demonstrated (Setlow, 1985).

Taken together, these observations show that DRC could exist as an etiologic correlate of cancer risk outside of XP. If XP is considered to represent the lower range of repair capabilities in humans, those individuals expressing a somewhat reduced repair response within the upper quadrant of the shoulder of a dose-response curve (Fig. 1) may be at increased risk for skin cancer or internal neoplasms given an appropriate exposure. Because it can be anticipated that human populations may include individuals showing only marginal damage vulnerability, any assay must be able to detect the level of DRC found in the heterozygotes of autosomal recessive DNA repair diseases with extreme precision and minimal intra-assay variation.

The extent to which this hypothesis may be evaluated is dependent upon the availability of validated laboratory methodology with which to measure DNA repair proficiency within study populations of interest. Such laboratory methodology has been developed and substantiated with a pilot project (Athas et al, 1991) of 38 patients with BCC and 27 controls, which was further validated with 88 cases of this disease and 135 matched controls (Wei et al, 1993). It was the purpose of this program to extend these studies to examine the relationship of DRC to the risk of skin cancer patients.

3. QUANTITATIVE CONSIDERATIONS IN ESTABLISHING DRC ASSAYS

In developing a general population-oriented DNA repair assay it was our goal to be able to determine if human variational responses to environmental genotoxic agents were due, in part, to deviations in DRC. The preconditions for its laboratory development were guided by the consideration that approximately 5% of the cancer population might be carrying the genes for autosomal recessive DNA repair deficiency diseases such as XP, ataxia telangiectasia, etc. (Friedberg, 1985). Hence it was of importance to determine if the DRC of XP heterozygotes could be distinguished from matched normal subjects

When extrapolated to the size of a gene each one-percentage loss of repair reflects the persistence of 10 damaged nucleotides per thousand nucleotides in a cDNA. This is a significant amount of persistent damage, which is a direct reflection of the DRC. Figure 1 depicts a typical survival curve (exaggerated for purposes of explanation) for a biological system in which the extent of the shoulder delineates the saturation point of the repair system after which a first order rate of decline is observed. A 5% decrease along this "region of repair" is the guiding requirement in designing this assay procedure. The D_{37} is the theoretical dose required for a single inactivating hit according to random statistical considerations of dosimetry in photobiology (Jagger, 1976). This is further reflected in our ex-

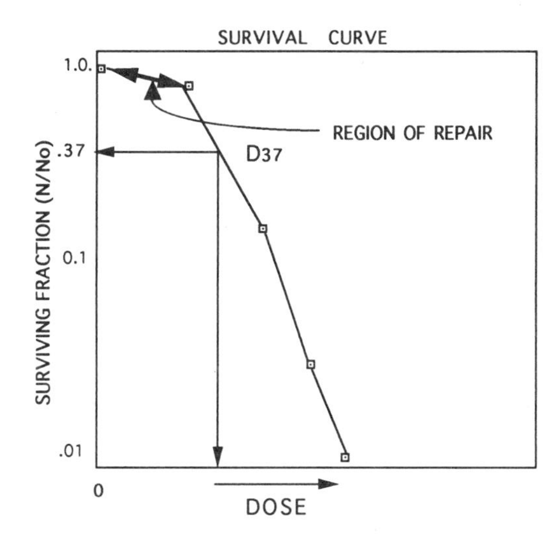

Figure 1. Theoretical survival curve of the cells exposed to genotoxic damaging agents. The shoulder of the curve represents the saturating dose which can be accommodated for effective repair. The survival rate diminishes dramatically in the first order rate. The D_{37} represents the dose that causes 37% survival after a single lethal hit. This is an effect which can be compared among different genotoxic agents.

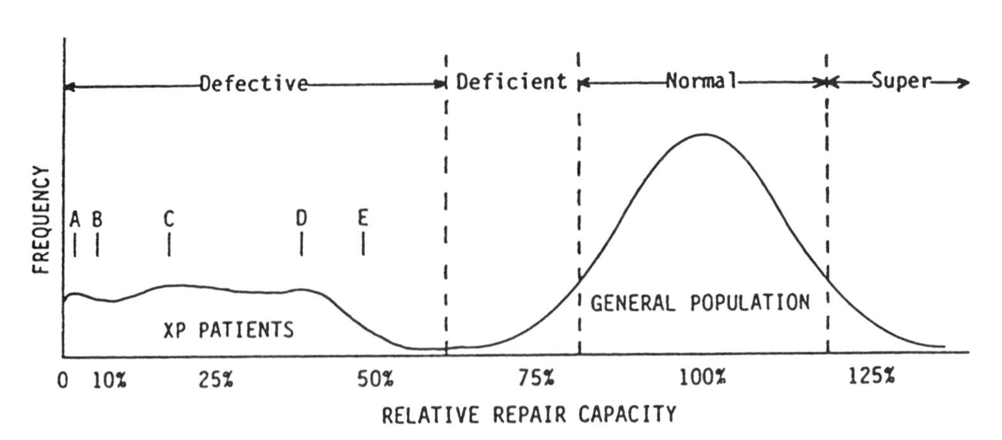

Figure 2. Theoretical distribution of relative DNA repair capacities among the population. The XP patients represent the subgroup with the most dramatic loss of repair capacity seen only in a small percentage of the general population. Relative to XP, however, there is a larger subgroup with reduced (deficient) repair capacity. This is the group that may have an increased risk of cancer associated with host susceptibility.

pectations of the distribution of DRC in various populations (Fig. 2). In this idealized distribution curve the differences between individuals would be reflected in the lower 5% of the normal distribution curve for the general population and larger reductions would be confined to autosomal recessive XP patients. Because different steps in the progression of DNA repair are affected in such diseases it was our objective to develop an assay which rather than measuring single or partial steps in repair would reflect the complete sequence of all repair steps. It was also considered desirable to use exogenously damaged genes or DNA as targets rather than damaging the cell since most genotoxic agents are also cytotoxic and might confound biochemical analyses. The comparative DNA repair capabilities of families who donated their cells to the NIGMS Human Genetic Mutant Cell Repository and who lived within 100 miles of Baltimore were examined so that the DRC of their established lymphoblasts could be compared with the repair capacity of circulating lymphocytes from the same donors.

4. PRINCIPLES OF ASSAY-HOST CELL REACTIVATION AS A MEASURE OF DRC

Exogenous DNA (Fig.3), appropriately engineered, can be transiently expressed when introduced into a receptive cell. If that DNA molecule is damaged, its expression will ultimately depend on the ability of the host cell to repair the damage to that DNA. In this case a non-replicating plasmid of 5 kb is engineered to contain a bacterial reporter gene that is novel to a mammalian cell. The gene is damaged in a controlled and quantitative manner, and the level of its expression is a direct measure of the repair capacity of that cell. The survival curves, the D_{37} and the ranking for repair of the various XP complementation groups are essentially identical. This assay does not distinguish nucleotide excision repair, DNA glycosylase nor direct reversal mechanisms because the assay measures the sum total of the repair capacity of the T lymphocytes. Biochemical analyses, however, can distinguish between these pathways. This assay is analogous to either the survival curves of intact cells exposed to damage or of damaged virus-infected cells. The repair ca-

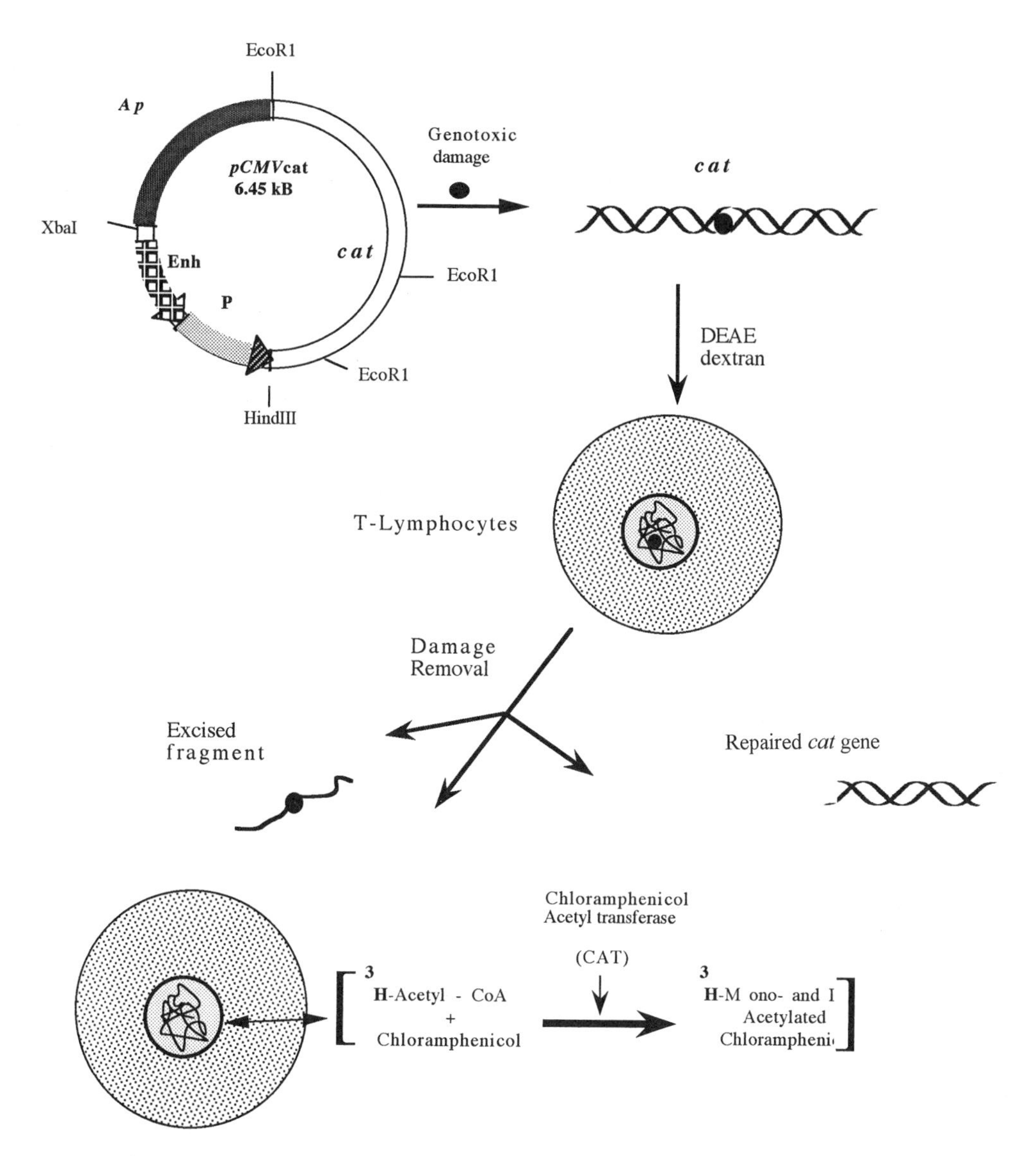

Figure 3. The schematic illustration of how the CAT assay measures host DRC. The non-replicating reporter gene, CAT, harbored in a recombinant DNA plasmid expression vector pCMVcat (5kb) is damaged before being introduced into the host cells (e.g., lymphocytes or fibroblasts) by transfection. The percent of reactivated CAT activity relative to undamaged CAT gene (100%) reflects host repair capacity.

pacity of the lymphocytes is, further, a reflection of the repair capacity of the donor. DRC levels in varying tissues within the same XP subject are virtually the same. Additionally, the level of repair in skin fibroblasts reflects the whole XP patient based on comparable repair as measured in situ, (UDS, unscheduled DNA synthesis) in skin cells, in other organs, and in lymphoid cell lines from peripheral blood.

The principle of this assay, projected for these studies, was originally used by Kraemer and his associates (Protic-Sabljic and Kraemer, 1985), to monitor mutagenic changes in rescued plasmid DNA . This experimental approach provided an important impetus for these population studies. The vector employed in these studies contains viral enhancer sequences permitting this bacterial gene to be expressed in human cells and, further, contains the elements of pBR322 for its perpetuation in bacteria (Foecking and Hofstter, 1986)

4.1. Plasmid DNA Substrates for Repair

(Foecking and Hofstter, 1986).The plasmid expression vector, pCMVcat, contains a bacterial gene coding for chloramphenicol acetyl transferase (cat), under the transcriptional control of the enhancer and promoter of the immediate early gene of human cytomegalovirus (Fig. 3). In comparison to other available cat expression vectors, this construct has demonstrated the highest efficiency for inducing cat enzyme activity under transient expression conditions in several human cell lines. For large-scale preparations, the plasmid is amplified with chloramphenicol in *E. coli* strain HB101, and isolated using standard molecular cloning procedures (Maniatis et al 1982), involving lysozyme/Triton X-100 lysis and double banding in cesium chloride/ethidium bromide equilibrium gradients. Greater than 95% of the plasmid is in the supercoiled, form I, configuration when checked by agarose gel electrophoresis. DNA concentrations are determined spectrophotometrically

4.2. Damage

It is because of the nature of the DRC assay in which stoichiometric damage removal is not measured that knowledge of the absolute adduct structure is not always necessary. However, when comparisons are made between the DRC of unrelated genotoxic agents then comparative concentrations per nucleotide should be made. For example, ionizing radiation results in the radiolysis of water generating a large variety of unstable and stable products which lead to a plethora of base damage primarily and strand breaks secondarily. It is the repair of the base damage which can be measured in the DRC assay but not strand breaks since the assay only measures expression of supercoiled DNA. Hence, it is necessary to generate very specific kinds of radiomimetic damage for DRC studies, whereas, methylene blue oxidation of DNA primarily generates 8-hydroxyguanine almost but not quite exclusively. The unresolved DNA oxidation products have been used extensively for substrate specificity studies by purified DNA glycosylases and are, as a consequence, valid substrates for DRC studies.

Typical DRC results are shown for one such family in Figure 4 in which the repair capacity of the homozygous XPA patient is compared with her siblings who are presumed heterozygotes for XP and with age-matched normal subjects. Because each dose-point represents a standard error of $\pm$ 0.02 of the DRC a variation of 2–5% is a statistically significant number providing sample sizes are sufficiently large in a population study. This becomes evident when these validation studies were extended to a larger population of subjects involved in examining the paradigm of photochemical (UV) damage and BCC (Wei et al, 1993). An example of the statistical evaluation is seen in the primary data in Figure 5 which is derived from individual dose curves carried out in triplicate including a standard curve using two different XP and one normal cell line with known levels of DRC. The data was subject to two-sided Student's t test (Table 1) and multiple linear regression analyses (Table 2) which created models for those factors contributing to DRC, generating p values = 0.00 to 0.003 for age effect on DRC which indicates the statistical significance of the data.

Table 1. Comparison of DRC between BCC patients and cancer-free controls

	N	Mean(%)±SD[a]	Percent difference[b]	t test[c]
Controls without FH[d] or actinic keratosis	106	8.00±2.2	0.0	reference
Controls with FH or actinic keratosis	29	7.28±2.2	-9.0	0.126
BCC case	88	7.35±2.0	-8.1	0.047

[a] Mean % CAT activity at UV dose to plasmid of 700 J/m^2 and its standard deviation.
[b] % reduction relative to the controls (reference group) which were labeled as 0% difference.
[c] Student's t test for the comparison of means to the reference group.
[d] FH = family history of BCC. (Reprinted with permission of the PNAS from Wei et al., 1993)

5. THE USE OF LYMPHOCYTES AS SURROGATE CELLS

T lymphocytes are useful as surrogate cells for the DNA repair-skin cancer studies for many reasons. (i). They are the most amenable nucleated cells for population studies because of the minimal invasiveness required for sampling; (ii). They are in equilibrium with virtually all cells in the body; (iii). They are able to pass through the blood-brain barrier; (iv). The DNA repair curves are the same as published repair data for established fibroblasts, lymphoblasts, and fresh lymphocytes from patients with DNA repair deficiencies (Athas et al, 1991 and Andrews et al, 1978); (v). In its landmark studies, the Nordic Study Group (includes Denmark, Finland, Norway and Sweden and the Cancer Registry) on the Health Risk of Chromosome Damage (Hagmar et al, 1994) it was found that chromosomal aberrations in T lymphocytes, as a biomarker, invariably reflect cancer of the GI tract (stomach, colon and rectum), respiratory tract (lung), breast, female genital organs, prostate, urinary system, skin, brain, lymphoma and leukemia. Additionally, lymphocytes also serve as surrogate cells for monitoring the damaging effects seen in chimney sweeps (Carstensen et al 1993); (vi). The effect of age of the donor on DRC is the same for both primary T lymphocytes as well as for skin fibroblasts (Moriwaki et al, 1996), which are corrected for passage number. However, there is no observed age related decline in DRC for transformed lines presumably because they have been immortalized by oncogenic vi-

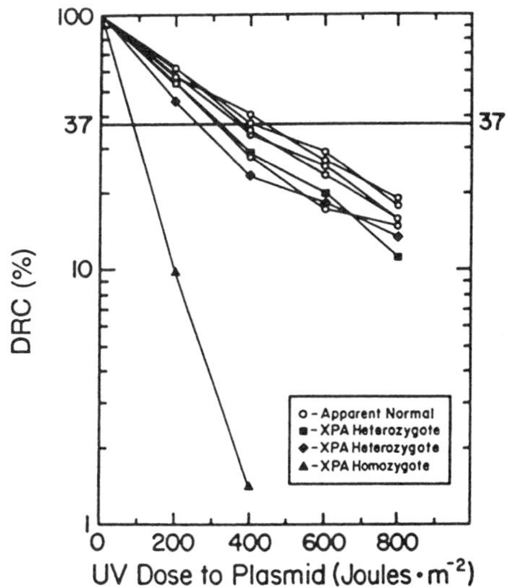

Figure 4. Dose-dependent decrease in transient *cat* gene expression in XP homozygotes, XP heterozygotes,and apparent normal peripheral blood lymphocytes transfected with UV-irradiated pCMV*cat*. All the cells were fresh circulating lymphocytes obtained from the subjects. (Reprinted with permission of the Cancer Research from Athas et al.1991).

Table 2. Multiple linear regression modelling of DRC related to risk factor among BCC and controls

Parameter	Estimate[a]	T-value [b]	p value[c]
Model 1			
(control only, n=135); F-test for the model: F-value=6.36; R2=0.088; p=0.002			
Intercept	11.725	10.23	0.000
Age in years	-0.071	-3.47	0.000
Sex	-0.409	-1.11	0.271
Model 2			
(BCC case only, n=88); F-test for the model: F-value=2.17; R2=0.049; p=0.120			
Intercept	10.445	6.96	0.000
Age in years	-0.047	-1.85	0.067
Sex	- 0.547	-1.26	0.211
Model 3			
(BCC case only, n=88); F-test for the model: F-value=3.51; R2=0.111; p=0.019			
Intercept	11.374	7.77	0.000
Age in years	-0.134	-3.09	0.003
Sex	-0.360	0.84	0.404
Age of onset	0.879	2.44	0.017
Model 4			
(All the subjects, n=223); F-test for the model: F-value=7.44; R2=0.092; p=0.0001			
Intercept	11.569	12.81	0.000
Age in years	-0.067	-4.24	0.000
Sex	-0.423	-1.51	0.133
BCC FH	-0.645	-1.93	0.055

[a] Least square estimate of regression coefficient from multiple linear regression.
[b] Student's t test for the null hypothesis that the estimate is equal to zero.
[c] Two-sided Student's t test.
Notes: All the models but model 2 are significantly different from a model with no risk according to the F-test. The variables included in these linear regression models are as follows:dependent variable is percent CAT activity at UV dose of 700 J/m2 (continuous variable); independent variables: Age (at assay, in years); Age of onset (age at first BCC, in group): 1=35, 2=35-44, 3=45-54, 4=55-60; Sex: 1=male, 2=female; BCC FH (family history): 1=without family history of BCC, 2=with BCC family history. (Reprinted with permission of the PNAS from Wei et al., 1993).

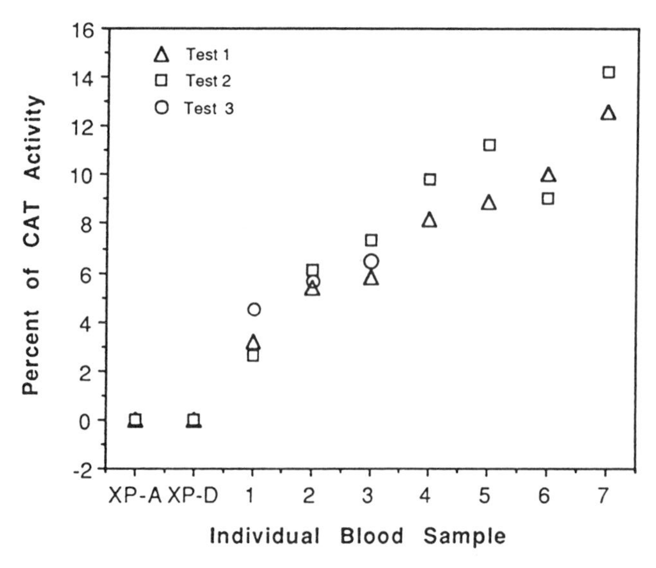

Figure 5. Relative ranking of percentage CAT activity of repeated assays with the same blood sample. The CAT activity at a UV dose of 700 J/m^2 measured by scintillation counting of radioactivity was standardized by CAT activity at zero UV dose. The assays were done at different times with one frozen 30-ml sample of peripheral blood from each individual. XP cell lines were used as the assay controls. (Reprinted with permission of the Cancer Bulletin from Wei et al., 1994).

ruses; (vii). One of the unexpected findings was the age-related decline in DRC as a function of aging. The mutation frequency for the hypoxanthine-guanine phosphoribosyl transferase (hprt) locus in circulating lymphocytes increases between 1.3–16% per annum, (Cole et al, 1988, Cole et al 1989 and Tates et al 1989), which reflects the same order of magnitude decline of DRC as a function of age (Wei et al, 1993). This same age-related mutational increase is also observed for the hemoglobin mutations in erythrocytes (Tates et al, 1989). The paradigm of persistent DNA damage arising from deficiencies in DRC with mutational frequencies is most easily measured in lymphocytes because of the ease of hprt mutant selection through 6-thioguanine (6TG) resistance. Hence, the selection process employing PCR of excised tumors or 6TG resistance allows for an examination of the mutational consequences of the persistence of DNA damage arising because of reduced DRC levels; (viii). Because of the genetic character of most repair deficiency diseases, DRC levels should be reduced in most cells of the patient.

6. RESULTS

The T lymphocytes isolated from subjects' peripheral blood were cryopreserved and assayed using aforementioned procedures. The assays used plasmid DNA containing photoproducts resulting from three UV doses and were carried out in triplicate for each dose from zero, 350, to 700 J/m^2. The results are reported as the percent of cat gene expression (% *cat* activity) following repair of damaged DNA compared to undamaged plasmid DNA. Lymphoblasts from patients with XP-A (Group A, most severe), XP-D (Group D, severe) and XP-C (Group C, classic form) provided the standard DNA repair of known levels of deficiency curves. Also lymphoblasts from normal individuals (GM0131 and GM1892) were included in the standard normal repair curves. The DRC measurements obtained at UV doses of 700 J/m^2 (about 26 pyrimidine dimers per DNA molecule) and 350 J/m^2 generated straight line functions. Under these circumstances, only the measurements at a dose of 700 J/m^2 were used for group comparisons.

The complete measurements of DRC were obtained from 88 cases with primary BCC as diagnosed through the dermatopathology laboratory of the Johns Hopkins Hospital, which serves multiple practicing dermatologists in Maryland. The 135 comparison controls had skin biopsies for diagnosis of mild skin disorders such as seborrheic keratosis, intradermal nevus, or subacute eczematous dermatitis. All subjects were Caucasians currently between 20–60 years of age, who lived in Baltimore City or its suburban area for most of their lives and who had skin biopsies in 1987–1990. The purpose of selecting only young subjects was to maximize the difference in risk of BCC due to genetic factors between cases and controls. At a clinical visit, dermatologists examined all participants to classify skin types and describe current skin conditions. The subjects then gave written informed consent, completed a structured questionnaire and provided blood. Control individuals with a self-reported history or clinical signs of skin cancers or other cancers were excluded.

6.1. Low DRC as a Risk Factor

In general, individual DRC values varied in this population. At a UV dose of 700 J/m^2, the residual DRCs for both cases and controls, relative to zero dose, were in the range of 3–15% which may be subject to the age effect (Figure 6) (Wei et al, 1993). The mean DRC of all BCC cases (n=88) is 5 percent lower than that of all control (n=135) and

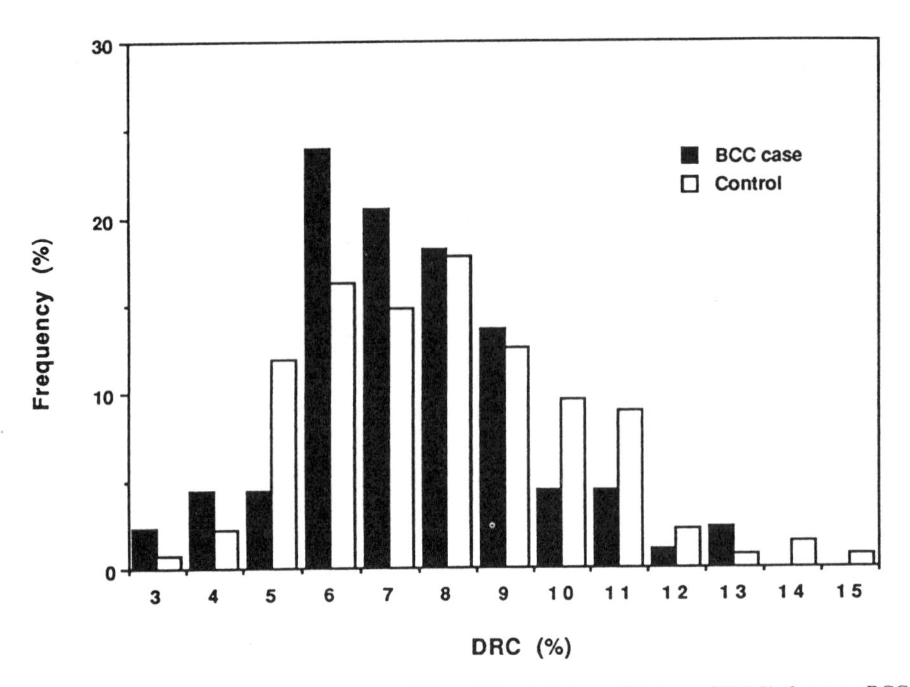

Figure 6. Percentage distribution of host cellular DRCs (CAT activity) at a UV dose of 700 J/m2 among BCC patients (n=88) and their controls (n=135). The DRCs of BCC patients were generally lower than their controls. (Reprinted with permission of the Cancer Research from Wei et al., 1994).

is of borderline significance (p=0.103). This difference is as high as 8 percent and statistically significant (p=0.047) when the controls with family history and premalignant skin lesions are removed from comparison. In addition, we found that individuals who had a DNA repair level below the 30th percentile of the controls had a greater than 2-fold increased risk for BCC (odds ratio, 2.3; 95% confidence interval, 1.2–4.5; adjusted for age) (Wei et al ,1994).

6.2. Age Effect on DRC

The DRC of subjects declined with age over the 20 to 60 year span studied. This age-related decline occurred in both cases and controls although it only reached significant decline in the controls (Fig. 7, Table 2, model 1 & 2). In the 106 controls, lacking any family history of skin cancer or presence of any premalignant skin lesions, the decline is 0.63 percent per year between 20–60 years of age (Table 2, model 1). This would amount to about 25 percent decrease in cumulative DRC over a 40 year period (Wei et al, 1993).

This age-related decline in DNA repair is accompanied by an increased accumulati on of persistent DNA damage affecting an increase in mutation fixation in structural and functional proteins (Moriwaki et al., 1996). The mutation rate accompanying aging in the *hprt* gene in human lymphocytes is reportedly between 1.3–1.6% (Cole et al, 1988, Cole et al 1989 and Tates et al 1989) which is twice the rate of decline of DNA repair reported here. Many known genetically-linked repair-deficiency diseases such as Cockayne's syndrome and xeroderma pigmentosum (XP) also manifest premature aging (Guzzeta et al, 1972 and Matin, 1978). In XP patients, the defect in DRC is associated with age-related

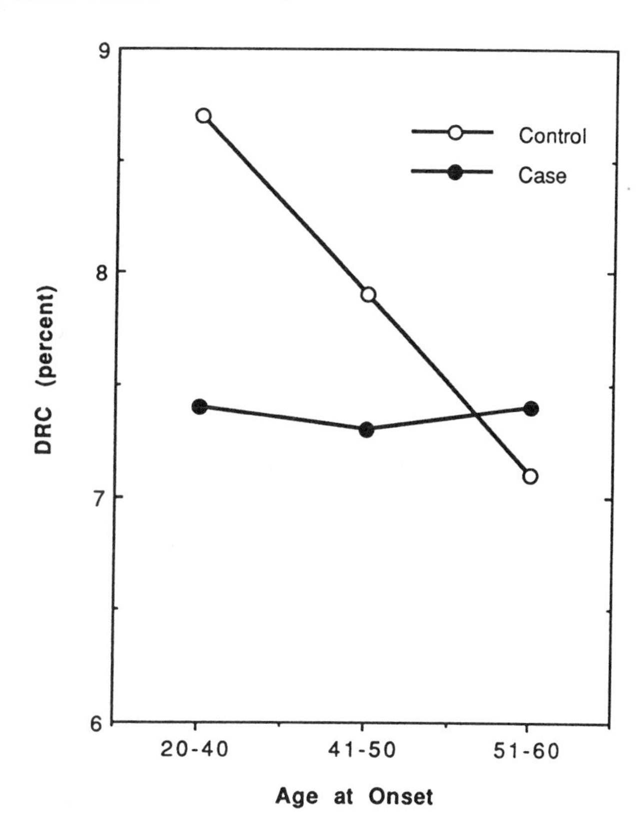

Figure 7. Relationship between age at first BCC and DRC. The age-related decline in DRC among controls in comparison with that of age-matched cases is displayed. The linear-regression modeling and statistical tests of these data are presented in Table 2. (Reprinted with permission of the PNAS from Wei et al., 1993).

skin changes and development of skin cancer at least 20 years earlier than the normal population (Cleaver and Kraemer, 1989). Other workers have reported age-related changes in unscheduled DNA synthesis (UDS) in lymphocytes (Lambert et al, 1979 and Singh et al, 1990) and epidermal cells (Nette et al, 1984)) from blood donors, especially in the aged. These findings are consistent with the age-related decline in DRC observed in this study.

Similar to the XP model, BCC cases with their first skin cancer at an early age repair DNA photoproducts poorly, when compared to controls or with those cases expressing BCC at a later age of onset. This suggests that poor DNA repair is associated with the precocious aging as manifested by an early age of the first BCC. After adjusting for age at onset, the age-related decline in the repair of UV damage amongst the cases was at least as sharp as that of controls and after controlling for current age, the age of onset of BCC was positively correlated with DNA repair (Table 2, model 3) (Wei et al, 1993).

6.3. Familial History

Several findings suggest that the early age of onset of skin cancer and the reduction of DRC have familial links. Those control subjects with a family history of BCC or with actinic keratosis had low DNA repair levels similar to those of the cases. After removing these "positive" controls, the overall difference in DNA repair between cases and controls was statistically significant (Table 1) (Wei et al,1993). Among those cases between the ages of 20 to 44 having BCC (n=38), about 45 percent had a family history of BCC,

Table 3. Relationship of family history of basal carcinoma (BCC) to the age of BCC onset

	Subject no.	Total with BCC family history no.	%	X2 test* for distribution
Controls	135	21	(15.6)	
BCC cases, age of onset				
20–44	38	17	(44.7)	
45–54	29	10	(34.5)	
55–60	21	2	(9.5)	p=0.022

*For BCC cases only. (Reprinted with permission from of the PNAS from Wei et al. 1993).

whereas only 10 percent of those cases who had BCC at ages 55–60 (n=21) had a similar history. This trend was statistically significant (Table 3) (7). In contrast, only 16 percent of controls (n=135) had a previous skin cancer history.

These findings support the influence of heredity in the early onset of BCC. In this study population, after adjustment for age and sex, prior family history of BCC is a statistically significant indicator of the individuals' DRC regardless of whether they were BCC cases or control subjects (Fig. 2 model 4.) (Wei et al, 1993). This suggests that an alteration in the genes regulating DNA repair may be responsible for the onset of BCC in younger subjects. The occurrence of BCC in the aged, however, may be due to the cumulative unrepaired photoproducts from excessive sunlight exposure.

6.4. Multiplicity of Tumors

To evaluate the relationship between DRC and multiplicity of BCCs, we used multiple linear regression models to correlate DRC with the number of skin cancers (Wei et al, 1994). It was found that the lower the DRC was, the greater number of skin tumors in individuals ($p<0.05$) after adjustment for age (Fig. 8). Further, family history of skin cancers

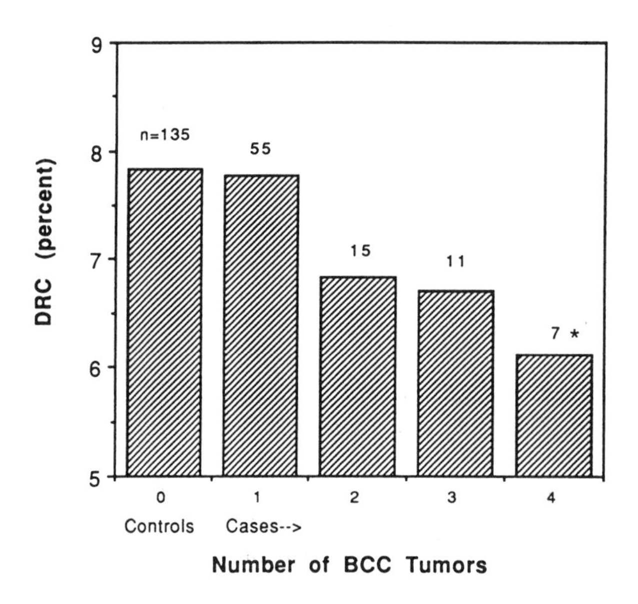

Figure 8. Relationship between DRC and the number of BCCs. The BCC patients with 4 or more tumors had a lower mean of DRC (* t test; $p<0.05$) compared with the controls. The means of DRC decreased significantly (trend test: $p<0.05$) as the number of tumors increased.

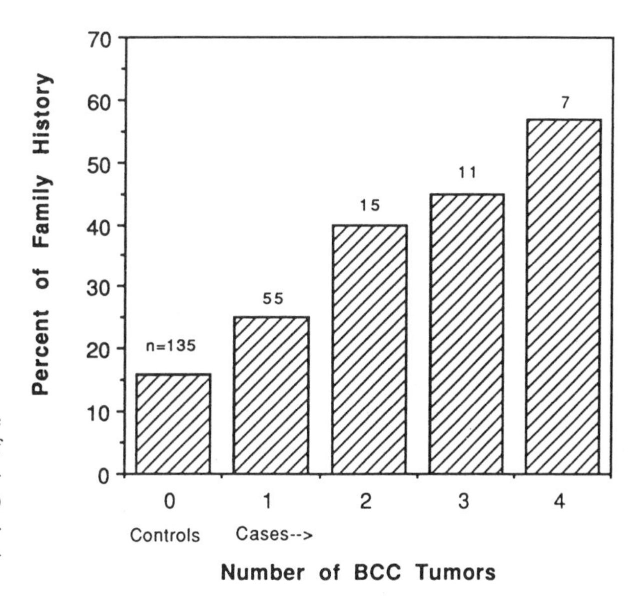

Figure 9. Relationship between the number of BCCs and a family history of skin cancer. The number of BCCs increased significantly (X2 test: $p<0.05$) as the percentage of subjects who reported a family history of skin cancer increased.

was more likely reported in those cases with multiple BCCs (Fig. 9). These findings are consistent with the fact that genetically determined low DRC tends to be associated with multiplicity of skin tumors as seen in XP patients (Cleaver and Kraemer, 1989).

6.5. Sunlight Exposure

A fundamental question is whether the genotoxicant-DNA repair paradigm of XP reflects the carcinogenic response of individuals who have relatively marginal reduction in DRC but who have over-saturated this capacity with excessive sunlight exposure. In those subjects who had 6 or more severe sunburns (blistering) in their lifetime, the DRC is 15 percent lower in cases than that of controls. Any of the measures of exposure affecting skin susceptibility to sunlight exposure such as skin type or fair complexion , poor tanning ability or burning tendency showed the same difference. The difference in the subjects with sunburns is not only due to the low DNA repair in BCC cases but also the higher DNA repair level in controls. These data suggest that if subjects are exposed to the genotoxicity of environmental UV light at doses which cause skin trauma then individuals with poor DRC are likely to develop skin cancer whereas those with high repair capacity are spared. This interactive effect between DNA repair levels and UV exposure is shown graphically in Figures 10 and 11. If the population is divided into high and low repair according to the median level of the controls, then the estimated risk of skin cancer after excessive sunlight exposure is almost 5 times higher in subjects with low repair and 1.9 times higher in those with high repair (Wei et al,1993).

Having been exposed to sufficient sunlight to have caused six or more severe sunburns in a lifetime appears to be associated with an increased incidence of BCC based on a large prospective study (Hunter et al, 1990). In this case-control study, the proportion of those cases (36%) having been overexposed in their lifetime is double that of the controls (17%). The development of BCC, therefore, probably reflects the mutation fixation as a consequence of the persistent DNA damage resulting from individuals having exceeded their DRC. After UV exposure, subjects with a family history of skin cancer, and possibly

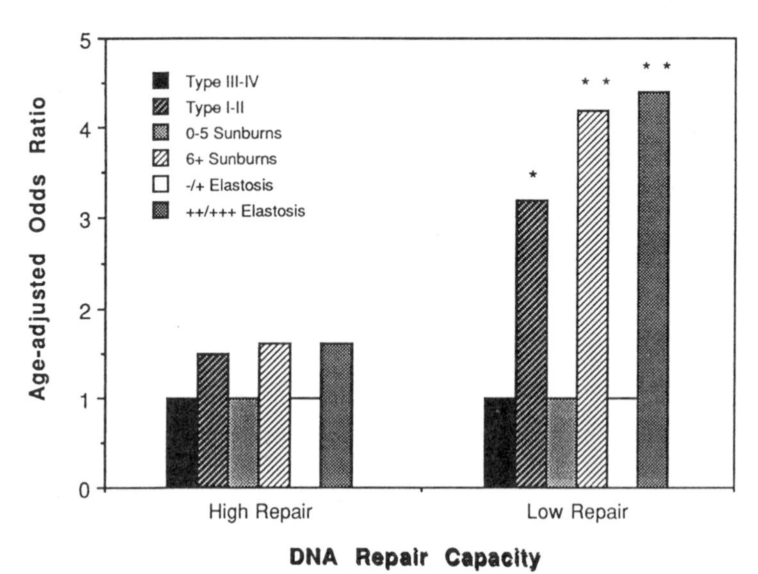

Figure 10. Age-adjusted odds ratios for BCC related to selected risk factors (skin type I to IV, sunlight exposure, and skin actinic elastosis) stratified by DRC. The median DRC at a UV dose of 700 J/m^2 of the controls was used as the cut-off value; values above it were considered high DRCs and values below it were considered as low DRCs. Significantly increased risk (* $p<0.05$; ** $p<0.01$) was observed only among subjects with low DRCs. (Reprinted with permission of the Cancer Bulletin from Wei et al., 1994).

having reduced repair compared to others of the same age, will develop this disease at an early age. It is implied in those subjects displaying a time-delayed onset of BCC that mutation fixation may also occur as a result of accumulated, excessive unrepaired DNA damage in those individuals possessing apparently normal DNA repair levels.

6.6. Gender Orientation

Stratification of the odds ratios according to gender (Fig. 11) revealed the vulnerability of women to BCC of the skin who have a history of overexposure to sunlight and who have reduced repair capacity. These results are in agreement with those obtained in the previous pilot study (Wei et al, 1993) in which young women who have a long standing history of sunbathing were predisposed to BCC of the skin. This gender orientation suggests either hormonal, genetic or combined factors to explain this gender orientation. The self-administered questionnaires included requests for information concerning stages of oestrous, hormonal supplementation as well as use of oral contraceptives. Surprisingly, post menopausal women receiving estrogen supplementation had a significant increase in their DRC (Table 4) (Wei et al, 1994). These data show that hormones may indirectly or directly,regulate DRC, although they may not account for the enhancement of DNA repair, but the data provide important clues concerning agents which can regulate DNA repair. These data are reflected in the cross-over of repair in Figure 7. When such age-stratified data is further stratified according to gender, there is a drop amongst the female BCC patients who are premenopausal and not on estrogen supplementation, whereas there is an increase in repair in postmenopausal female cases which appears to reflect a larger number of such women on estrogen supplementation (85).

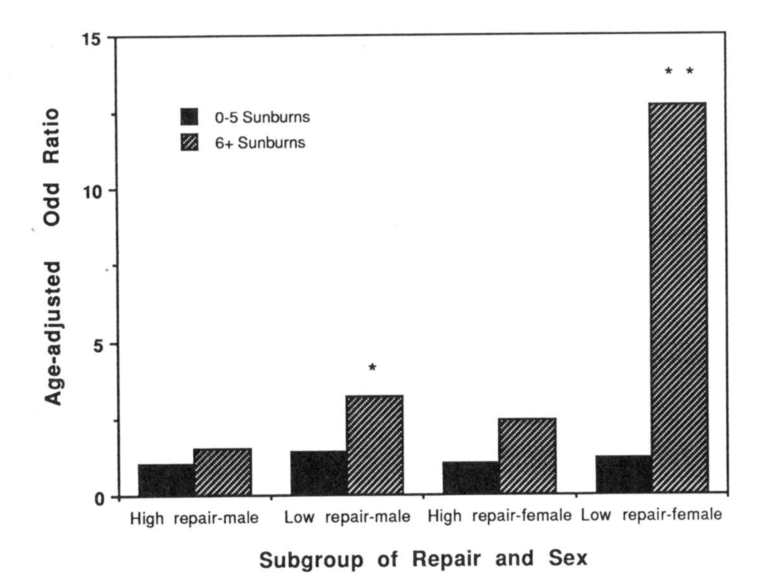

Figure 11. Effect of DRC on risk of BCC by gender: Relationship to the number of severe sunburns in a lifetime. The reference group had an odds ratio of 1. The significantly increased age-adjusted odds ratios were observed in both males (* $p<0.05$) and females (** $p<0.01$). (Reprinted with permission of the PNAS from Wei et al., 1993).

6.7. Mutational Consequences of Persistent DNA Damage

If DNA repair is a primary target for apoptosis, it can be anticipated that there is a corresponding increase in damage persistence as a function of aging (Moriwaki et al., 1996). These findings are also supported by the observations of Cole and her colleagues (Cole et al, 1988, Cole et al 1989 and Tates et al 1989) of a concomitant increase in mutation fixation of a number of genes at the rate corresponding to an increase of 1–2% per annum over a wide age range.

The persistence of UV-induced photoproducts, attributed to reduced DNA repair was also studied at the level of p53 genes in BCC. We analyzed 36 BCCs for p53 mutations, and a subset of these tumors for loss of chromosomes 17p and 9q (van der Riet, 1994). Sixty-nine percent of the BCCs had lost a 9q allele, with the common area of loss surrounding the putative gene for nevoid BCC, or Gorlin's Syndrome. Forty-four percent (16/36) of BCCs had a mutated 53 allele, usually opposite pyrimidine tracts, which is consistent with UV-induced pyrimidine dimer and 6–4 adduct generated mutations. Surpris-

Table 4. The effect of oral contraceptive (OC) and estrogen use on DRC

	User (n) mean±SD*	Non user (n) mean± SD	Diff[a] (%)	P[b]
Pre-menopausal OC use	(45) 7.9+2.1	(21) 7.5+2.4	+ 5.3	0.567
Post-menopausal estrogen use	(28) 7.5+2.1	(13) 6.0+2.0	+25.8	0.038

*The age-adjusted mean and its standard deviation of the percent CAT activity in response to UV dose to plasmid of 700 J/m^2.

[a] Relative difference was calculated as: [(user mean minus non-user mean) / non-user mean] x100%.

[b] Probability derived from two-sided t test for means of user vs non-user.

ingly, only one tumor lost a 17p allele, and in all BCCs only one p53 allele was inactivated. This is in direct contrast to other epithelial tumors, which usually progress by the inactivation of both p53 alleles. It is possible that the absence of clinical progression of BCC may be related to the lack of complete p53 inactivation in these tumors.

7. DISCUSSION

The impact of reduced DNA repair on risk of BCC is significant. Reduced host DRC serves as a susceptibility factor which is, additionally, linked to a family history of BCC. The risk of BCC, as a consequence of excessive sunlight exposure and low DNA repair, is greatly increased. A few other studies have suggested changes in DNA repair of skin cancer patients. In an early case-control study (Lehmann, 1985), a difference was observed in the UV-light induced unscheduled DNA synthesis (UDS) of peripheral blood lymphocytes of the cases (n=29, aged 25–83 years) and controls (n=25, aged 25–83). This difference was significant in patients with both BCC and SCC (n=10) and in those with BCC only (n=19). In a recent study, in which the DNA repair rate of pyrimidine dimers in fibroblasts from UV irradiated skin biopsies was tested, it was found that reduced nucleotide excision repair was observed in patients (n=22) with BCC compared to cancer-free subject (n=19) (Setlow, 1982), The small samples in these studies did not permit age adjustment or further stratification by known risk factors for skin cancer such as sunlight exposure.

The age-related decline in DNA repair activity of normal lymphocytes has been reported in assays using UV (Nette et al, 1984 and Roth et al, 1989), X-rays (Kovacs et al, 1984 and Harris et al, 1986) and gamma-irradiation (Licastro et al, 1982). A 30% reduction in DNA repair was observed in normal peripheral leucocytes from normal skin of 58 normal subjects aged 20 to 90 year (Singh et al, 1990). However, no age-related DNA repair changes were observed in other studies (Kutlaca et al, 1982 and Kovacs et al, 1984).

The present study represents an important advance. The use of the CAT assay to measure DNA repair does not require that the host cell be damaged which may compromise the endogenous repair mechanisms in which repair enzymes remain relatively constant (Hubscher et al, 1977). Given the programmed decline in DNA repair as a function of age, such tissues should be particularly vulnerable during the aging process because of their lack of regeneration. Hence, physiological differences such as these may serve to explain the elevated incidence of neurodegenerative diseases in many DNA repair deficiency diseases (Friedberg, 1985).

8. ARSENICAL CANCER PATIENTS

(Wu, 1997): The reduction in DRC predisposes Caucasians exposed to sunlight, and rather specifically leads to BCC, rarely seen in Oriental or Negro populations. Areas endemic for this Black foot disease, a form of BCC, have historically been over exposed to arsenic in the water taken from their artesian wells. Arsenic is a co-mutagen, co-carcinogenic and a co-clastogenic substance and has been shown recently to interfere with DNA repair (Okui and Fujiwara, 1986 and Li and Rossman, 1989). The objectives of this study, therefore, were, two-fold. First we wanted to determine if DRC plays a role in the development of arsenical skin cancer. Second we sought to determine the factors which are associated with DRC.

To address the objectives of this study, a community-based case-control study was conducted in three villages located in a BFD endemic area on the island of Taiwan. Cases

Table 5. DRC in breast cancer cases and controls

Variables	Study controls		Cancer cases		
	Mean	SD	Mean	SD	p value[1]
(A) New York University Medical center (NYU)	(n=25)		(n=25)		
Age	44	13	47	8	0.45
Baseline CAT(dpm)	25198	14180	22456	17992	0.55
%DRC	13.804	4.88	11.09	3.73	0.03*
(B) Georgetown University Medical Center (GT)	(n=48)		(n=16)		
Age	53	10	56	12	0.30
Baseline CAT(dpm)	29582	16253	26218	7020	0.48
%DRC	14.15	5.46	10.88	3.56	0.03*
(C) Combined (NYU+GT)	(n=73)		(n=41)		
Age	50	12	51	11	0.86
Baseline CAT(dpm)	28081	15620	23924	17500	0.19
% DRC	14.03	5.24	11.01	3.62	0.001

[1]p<0.05, Students t test, 2-sided.
*Ray, S. et al. 1997

were patients with clinically diagnosed skin cancer and controls were subjects without any cancers. Data collection included a questionnaire interview, a skin lesion examination and the drawing of blood from study subject, were performed at a screening clinic located in the study area. The blood samples were processed for lymphocytes and were frozen in Taiwan. The collective frozen lymphocytes were then delivered to the United States where the test of DRC were carried out. A total of 304 study subjects, including 57 arsenical skin cancer patients and 247 non-cancerous controls were included in the study to analyze the risk factors associated with the development of arsenical skin cancer as their interactions with DRC.In the analysis for the factors associated with DNA reapir capacity, 53 arsenical skin cancer patients and 227 noncancerous controls were included and analyzed separately.

It was found that poor living conditions (diet, education, etc.), cumulative arsenic exposure and family history of any cancer in siblings were significant risk factors for the high risk of arsenical skin cancer. The conclusions from the BFD studies are that DRC by itself is not a risk factor for arsenical skin cancer. However, those individuals with low DRC are at much greater risk for BFD when they are exposed to high levels of arsenic in their drinking water and or when they are on poor diets. DRC, therefore, appears to be a susceptibility factor in this disease

9. BREAST CANCER STUDIES

(Ray et al, 1997). In a pilot study the hypothesis that defective DNA repair is associated with breast cancer was tested. Lymphocytes from two breast cancer studies (39 breast cancer cases and 61 controls from two breast cancer case - control studies (Georgetown University Medical Center and New York University Medical Center) were used for DRC assays. The data show that the mean DRC activity is consistently lower in breast cancer cases (10.53–11.25% in the cases and 13.16–13.86 for the controls, respectively,Table 5). In a combined analysis there is a significant decrease in DRC (81% of controls) in cancer cases compared to controls ($p<0.05$) based on a weighted mean of the data from the two locations. These findings suggest that DNA repair may serve as a potential biomarker for breast cancer. However, a larger study is required to further confirm the association and mechanisms involved in the regulation of DNA repair.

ACKNOWLEDGMENTS

This work was supported by a MERIT AWARD from the National Institutes of Health (NIH; GM-22846) to L.G.

REFERENCES

A.D. Andrews and S.F. Barrett et al. Cockayne's Syndrome fibroblasts have increased sensitivity to ultraviolet light but normal rates of unscheduled DNA synthesis. *J Invest Dermatol* 1978; 70:237–239.

W.F. Athas and M.A. Hedayati et al. Development and field-test validation of an assay for DNA repair in circulating lymphocytes. *Cancer Res* 1991; 51:5786–579.

U. Carstensen, A.K. Alexandrie and B. Hogstedt et al. B- and T- lymphocytes macronuclei in chimney sweeps with respect to genetic polymorphism for CYPA1 and GST (class Mu). *Mutat Res* 1993; 289:187–195.

J.E. Cleaver and K. Kraemer, Xeroderma pigmentosum. In Scriver CR, Beauber AL, Sly WS, et al. ed: *The metabolic basis of inherited diseases II*, New York, NY; McGraw-Hill Inc., 1989:2949–2971.

J. Cole, C.F. Arlett and M.H.L. Green et al. *Measurement of mutant frequency to 6-thioguanine resistance in circulating T-lymphocytes for human population monitoring.* In: New Trends in Genetic Risk Assessment, Academic Press, London, 1989:175–203.

J. Cole, M.H.L. Greenand S.E. James et al. A further assessment of factors influencing measurements of 6-thioguanine-resistant mutant frequency in circulating T-lymphocytes. *Mutat Res* 1988; 204:493–507

M.K. Foecking and H. Hofstetter, Powerful and versatile enhancer-promoter unit for mammalian expression vectors. *Gene* 1986; 45:101–105.

E.C. Friedberg, *DNA Repair*. W.H. Freeman Inc. New York, 1985:505–574.

F.Guzzetta, Cockayne-Neill-Dingwall-Syndrome In: Vinken PJ, Bruyn W ed. Handbook of Clinic Neurology. Amsterdam, North-Holland Biomedical Press, 1972:431–440.

L. Hagmar, A. Brǐgger and I.L. Hansteen et al. Cancer risk in humans predicted by increased levels of chromosomal aberrations in lymphocytes: Nordic study group on the health risk of chromosomal damage. *Cancer Res* 1994; 54:2919–2922.

G. Harris, A. Holmes and S.A. Sabovljev et al. Sensitivity to X-irradiation of peripheral blood lymphocytes from aging donors. *Int J Radiat Biol* 1986; 50:685–694.

U. Hubscher, C.C. Kuenzle and S. Spadari, Variation of DNApolymerase during perinatal tissue growth and differentiation. *Nucl. Acid Res.* 1977; 4:2917–2929.

D.J. Colditz and MJ. Stampfer et al. Risk factors for BCC in a prospective cohort of women. *Ann Epidemiol* 1990; 1:13–23.

J. Jagger. Ultraviolet inactivation of biological systems In: Wang, S.Y, ed. Photo- chemistry and Photobiology of Nucleic Acids. vol II. *Academic Press*, 1976:147–185.

R. Kutlaca, R. Seshadri and A.A. Morley. Effect of age on sensitivity of human lymphocytes to radiation. A brief note Mech. Ageing Dev 1982; 19:97–101.

E. Kovacs, W. Weber and H.J. Muller. Age related variation in the DNA repair synthesis after UV-C irradiation in unstimulated lymphocytes of healthy blood donors. *Mutat Res* 1984; 131:231–237.

K.H. Kraemer and H. Slor. Xeroderma pigmentosum. *Clin Derm* 1985; 3:33–69.

B. Lambert, U. Ringborg and L. Skoog. Age-related decrease of ultraviolet light-induced DNA repair synthesis in human peripheral leukocytes. *Cancer Res* 1979; 39:2792–2795.

A.R. Lehmann. Ageing, DNA repair of radiation damage and carcinogenesis: Fact and fiction. In: Likhachev A et al. ed. Age-Related Factors in Carcinogenesis. WHO, IARC, Lyon, 1985:203–209.

J.H. Li and T.G. Rossman.Inhibition of DNA ligase activity by arsenic: a possible mechanism of co-mutagenesis. *Mol. Toxicol.*1989, 2; 1–9

F. Licastro, C. Fransechi and M. Chirocolo et al. DNA repair after gamma radiation and superoxide dismutase activity in lymphocytes from subjects of far advanced age. *Carcinogenesis* 1982; 3:45–48.

H.T. Lynch, B.C. Frichot and J.F. Lynch. Cancer control in xeroderma pigmentosum. *Arch Dermatol* 1977; 113:193–195.

T. Maniatis, E.F. Fritsch and J. Sambrook. Molecular Cloning: A Laboratory Manual. Cold Spring Harbor Laboratory 1982: 458–468.

M.G. Matin. Genetic syndromes in man with potential relevance to the pathology of ageing. In:Bergsma D, Harrison DH ed. Genetic Effects on Aging. New York, Liss., 1978:5–39.

S.I. Moriwaki, S. Ray, R.E. Tarone, K.H. Kraemer and L. Grossman. The effect of donor age on the processing of UV-damaged DNA by cultured human cell: Reduced DNA repair capacity and increased DNA mutability. *Mutation Res.* 1996; 364: 117–123

B. Munch-Petersen, G. Frenz and B. Squire et al. Abnormal lymphocyte response to UV radiation in multiple skin cancer. *Carcinogenesis* 1985; 6:843–845.

E.G. Nette, Y.P. Xi, Y.K. Sun et al. A correlation between aging and DNA repair in human epidermal cells. *Ageing Devel* 1984; 24:238–292.

T. Okui and Y. Fujiwara. Inhibition of human excision DNA repair by inorganic arsenic and the co-mutagenesis effect in V79 CHO cells. 1986; 172:69–76

S.A. Pawsey, I.A. Magnus and C.A. Ramsey et al. Clinical, genetic and DNA repair studies on a consecutive series of patients with xeroderma pigmentosum Quart *J Med* 1979; 48:179–210.

M. Protic-Sabljic and K.H. Kraemer. One pyrimidine dimer inactivated expression of a transfected gene in XP cells. *Proc Natl Acad Sci* USA 1985; 82:6622–6626.

M. Roth, L.R. Emmons and M. Haner et al. Age-related decrease in an early step of DNA repair of normal lymphocytes exposed to ultraviolet irradiation. *Exp Cell Res* 1989; 180:171–177.

S. Ray, L. Grossman, M. Berwick and J.J. Hu et al. DNA repair in human breast cancer. A molecular epidemiology study. 1997;Proc. *Am.Assoc. for Can. Res.* 38:215

R.B. Setlow. DNA repair, aging, and cancer. *Natl Cancer Inst. Monogr.* 60, (U.S. Department of Health and Human Services, Public Health Service, National Institute of Health, 1982:249–255.

R.B. Setlow. In: Castellani E (ed). Epidemiology and quantitation of environmental risk in humans from radiation and other agents. New York, Plenum Press, 1985:205–212.

N.P. Singh, D.B. Danner and R.R. Tice et al. DNA damage and repair with age in individual human lymphocytes. *Mutat Res* 1990; 237:123–130.

D. Tates, L.F. Bernini and A.T. Natarajan et al. Detection of somatic mutants in man: HPRT mutations in lymphocytes and hemoglobin mutations in erythrocytes. *Mutat Res* 1989; 213:73–82.

R.R. Tice and R.B. Setlow. DNA repair and replication in aging organisms and cells. In: Finch CE, Schneider EL ed. Handbook of the Biology of Aging. Van Nostrand Reinhold, New York, 1985:173–224.

P. van der Riet, D. Karp, E. Farmer, Q. Wei, L. Grossman, K. Tokino, J.M. Ruppert, and D. Sidransky. Progression of basal cell carcinoma through loss of chromasome 9q and inactivation of a single p53 allele. *Cancer Res. 1994;* 54: 25–27.

Q. Wei, M.A. Hedayati and G.M. Matanoski et al. DNA repair and aging in basal cell carcinoma: a molecular epidemiology study. *Proc Natl Acad Sci* USA 1993; 90:1614–1618.

Q. Wei, G.M. Matanoski and E.R. Farmer et al. DNA repair, multiple skin cancers, and drug use.*Cancer Res* 1994; 54:437–440.

M.M. Wu. DNA repair and arsenical skin cancer; a case-control study in a Blackfoot disease endemic area in Taiwan. 1997. A Ph.D. thesis in Epidemiology, The Johns Hopkins School of Hygiene and Public Health. Baltimore, MD 21205.

14

TRANSCRIPTION-COUPLED DNA REPAIR

Which Lesions? Which Diseases?

Philip C. Hanawalt and Graciela Spivak

Department of Biological Sciences, Stanford University
Stanford, California 94305-5020

1. ABSTRACT

Certain DNA lesions are removed preferentially from the transcribed strands of active genes in bacteria, yeast and mammalian cells. Initially it was thought that only lesions removed by nucleotide excision repair (NER) were subject to this pathway of transcription-coupled repair (TCR), but recent investigations have shown that some lesions caused by reactive oxygen species, that are recognized by glycosylases and are subject to base excision repair (BER), can be preferentially repaired in the transcribed strands of active genes in mammalian cells. We will discuss the pathways and the proteins involved in the repair of different lesions or groups of lesions, produced by environmental agents or endogenous metabolic activities. The victims of the human hereditary diseases, xeroderma pigmentosum (XP) and Cockayne syndrome (CS) are highly sensitive to DNA lesions induced by sunlight but only the former exhibit predisposition to cancer. While XP patients are generally deficient in NER, those with CS are specifically defective in TCR and are characterized by dwarfism and severe developmental abnormalities. The role of TCR in human health will be discussed with specific consideration of the problems of cancer and early development.

2. INTRODUCTION

DNA repair is not merely an extraordinary scheme needed by organisms that are exposed to the DNA damaging ultraviolet (UV) wavelengths in sunlight. Rather, it is an essential set of mechanisms required to maintain genomic stability in the face of a plethora of threats, deriving from endogenous metabolic events as well as from environmental exposures to radiations and noxious chemicals (Friedberg et al., 1995). Even the intrinsic chemical instability of the DNA molecule itself must be accommodated through appropri-

Advances in DNA Damage and Repair, edited by Dizdaroglu and
Karakaya. Kluwer Academic / Plenum Publishers, New York, 1999.

ate restoration schemes (Lindahl, 1993) and in addition the DNA polymerases occasionally introduce errors during replication that must be corrected (Echols and Goodman, 1991). We now appreciate that the inherent redundancy of the genetic message and the complementary base pairing in the double stranded DNA molecule are essential, not only for replication but also for the recovery of information through excision repair when one of those strands is damaged.

3. DNA REPAIR: SEVERAL PATHWAYS, TWO MODES

The simplest pathway for repair of damaged DNA involves direct reversal of the damaged base(s), exemplified by the photoreversal of UV induced photoproducts by photolyases. Otherwise the incorrect or damaged nucleotides must be removed from the DNA by an excision repair process. These "cut and patch" schemes include mismatch repair, an important pathway that corrects mistakes resulting from incorrect pairing of the bases in the otherwise complementary strands of DNA. Two pathways involve the removal and replacement of short stretches of DNA containing lesions, nucleotide excision repair (NER) and base excision repair (BER). NER is the most versatile of the repair mechanisms; it recognizes and removes a wide variety of DNA lesions, including those caused by diverse chemical agents, as well as the cyclobutane pyrimidine dimers (CPD) and 6–4 photoproducts (6–4PP) produced by UV. It employs multi-enzyme complexes rather than single proteins to recognize broad classes of structure-distorting lesions. Thus, in the *E. coli* system the polypeptides encoded by the *uvrA*, *uvrB* and *uvrC* genes are not independently operating nucleases. Lesion recognition requires a complex interaction between UvrB and a dimerized form of UvrA. UvrB is loaded onto the DNA at the lesion site, and then the UvrC protein (the limiting element with only a dozen copies per cell) joins the UvrB-DNA complex to unleash cryptic nuclease activities that initiate the excision repair process by producing dual incisions, 12–13 nucleotides apart, bracketing the lesion (Sancar, 1996). In eukaryotes the NER mechanism is essentially the same, but nearly 30 proteins are required to restore DNA to its undamaged form. A larger gap, of 28–30 nucleotides, is produced and then filled by the same machinery used for chromosomal replication (Wood, 1996). In the case of BER, the recognition and initiation steps are simpler and evidently more specific. The predominant mechanism in BER appears to require the swinging out of the altered or incorrect base from the DNA helix so that it's detailed dimensions and charge structure can be assessed in a "pocket" of the relevant enzyme. Different enzymes are required for recognition of each class of altered base. Thus, specific glycosylases remove the damaged bases, the resulting abasic sites are cut by AP endonucleases and the process is completed by the synthesis of a short repair patch (1 to 6 nucleotides) by DNA polymerase, that is then sealed by ligase.

3.1. Heterogeneity of DNA Repair

The DNA repair pathways described above are effective for the removal of lesions from the entire genomes of living cells, but it is now clear that some regions of the genome are repaired better and/or faster than others. An early finding was that chemical adducts in the highly repetitive alpha DNA sequences in African green monkey cells are repaired less than half as well as those in the overall genome (Zolan et al., 1982). This first example of deficient repair of a chromosomal domain was largely attributed to the unique chromatin structure of the centromere-associated alpha DNA species. Preferential repair of an expressed gene was then discovered in Chinese hamster ovary cells (CHO) in which the dihydrofolate reductase (*DHFR*) gene was amplified about 50-fold. The early results indicated that repair of UV-induced cyclobutane pyrimidine dimers (CPD) was much more efficient in the expressed *DHFR* gene than in a silent sequence downstream (Bohr et al., 1985). These findings were confirmed in the *DHFR* gene of human cells (Mellon et al., 1986), and in mouse cells (Madhani et al., 1986), in which the expressed *c-abl* protooncogene was much more efficiently repaired than the silent *c-mos* protooncogene. Isabel Mellon suggested that the efficient repair in expressed genes might be due to preferential repair of the transcribed DNA strands. That hypothesis turned out to be correct and universal as it was validated in CHO and human cells (Mellon et al., 1987) in *E. coli* (Mellon and Hanawalt, 1989), in yeast (Leadon and Lawrence, 1992; Smerdon and Thoma, 1990; Sweder and Hanawalt, 1992) and in the slime mold *Dictyostelium discoidium* (Mauldin and Deering, 1994).

3.2. Coupling of DNA Repair to Transcription

Transcription-coupled repair (TCR) is clearly targeted to expressed genes (which may or may not be essential genes) but it specifically deals with lesions only in the transcribed (template) strand. In eukaryotes only RNA polymerase II transcribed genes are subject to TCR and the process requires that the polymerase is in the actively elongating mode. Pol I transcribed ribosomal genes are not subject to TCR (Christians and Hanawalt, 1993; Vos and Wauthier, 1991) nor are Pol III transcribed genes (Dammann and Pfeifer, 1997). TCR is eliminated by alpha amanitin, an inhibitor of Pol II elongation (Carreau and Hunting, 1992; Christians and Hanawalt, 1992; Leadon and Lawrence, 1991) and abolished at the restrictive temperature in a temperature sensitive Pol II mutant in yeast (Leadon and Lawrence, 1992; Sweder and Hanawalt, 1992). It operates equally well on an active gene on a plasmid as when that gene is chromosomal in yeast (Sweder and Hanawalt, 1992). TCR does not exhibit cell cycle related variations in mammalian cells for the *DHFR* gene that is expressed continuously throughout the cycle (Lommel et al., 1995). The lengths of the repair patches are similar for TCR and GGR in UV-irradiated human cells (Bowman et al., 1997).

3.3. Which Lesions?

TCR appears to be specific for lesions that arrest transcription although there are possible exceptions, discussed below. The effect of transcription on repair of lesions in expressed genes must depend intimately upon how the elongating RNA polymerase complex responds upon encountering the lesion. Different lesions may be treated differently. Thus, methyl purines, such as 7-methylguanine that is thought not to block transcription and 3-methyladenine that does block transcription, are efficiently repaired by the glycosylase-initiated pathway of BER with no evidence of TCR (Scicchitano and Hanawalt, 1989); however ethyl purines, that are repaired mainly by BER, exhibit TCR but only in cells that are NER-proficient (Sitaram et al., 1997). This may be an example of a class of lesions that are normally repaired by BER; but when they pose a block to transcription, a "cross-over" to repair by the NER mode may occur through the transcription-repair coupling factor. However, some lesions normally repaired by BER follow a somewhat different pathway. Thymine glycols induced in DNA by IR and oxidizing agents, like hydrogen peroxide, are subject to TCR in the complete absence of most of the proteins required for the early steps of NER (Cooper et al., 1997). Interstrand crosslinks due to psoralen photoad-

Table 1. Lesions and the pathways for their repair

Agent	Lesion	Repair	Effect on transcription	TCR	Reference
UV 254 nm	CPD	NER	block	+	(Mellon, et al., 1987)
UV 254 nm	6-4PP	NER	block	+	(van Hoffen et al., 1995)
Psoralen + light	interstrand CL	NER	block	+	(Islas, et al., 1994)
Psoralen + light	monoadduct	NER	termination	–	(Islas, et al., 1994)
IR, H_2O_2	thymine glycol	BER	block	+	(Leadon and Cooper, 1993)
MNU, MMS	3-Me-A	BER	?	–	
MNNG	7-Me-G		block	–	(Scicchitano and Hanawalt, 1989)
N-ethyl-N-nitrosourea	ethylpurine	BER/NER	?	–/+	(Sitaram, et al., 1997)
NA-AAF	dG-C8-AF	NER	transient block	±	(van Oosterwijk, et al., 1996a)
Cisplatin	intrastrand CL	NER	?	+	(May et al., 1993)
Cisplatin	interstrand CL	NER	?	–	(May, et al., 1993)
Acridine orange + light	8-oxo-dG	BER	termination	–	(Taffe et al., 1996)

Abbreviations: CL: crosslink; BcPHDE: (+-)-3 alpha,4 beta-dihydroxy-1 alpha,2 alpha-epoxy-1,2,3,4-tetrahydrobenzo[c]phenanthrene; BPDE: benzo[a]pyrene diol epoxide; MNU: methyl nitrosourea; MNNG: methyl-N'-nitro-N-nitrosoguanidine; MMS: methyl methanesulfonate; 7-Me-G: 7-methylguanine; 3-Me-A: 3-methyladenine; NA-AAF: N-acetoxy-2-acetylaminofluorene; 8-oxo-dG: 8-oxodeoxyguanosine.

ducts are subject to TCR but, curiously, psoralen monoadducts are not (Islas et al., 1994; Vos and Hanawalt, 1987), possibly because the RNA polymerase is released from the DNA, terminating transcription before recruiting the repair enzymes (Wang and Rana, 1997). Another feature to be considered is the background level of overall genomic repair: sometimes the observation of TCR is masked by the operation of highly efficient NER; the dissection of the pathways used is then made possible by the use of repair mutants. For example, repair of dG-C8-AF in the *ada* gene in normal human cells and in Cockayne syndrome (CS) cells treated with N-acetoxy-2-acetylaminofluorene (NA-AAF) is efficient, with no evident strand bias; in XPC cells, deficient in global genomic repair (GGR) as elaborated below, repair is confined to the transcribed strand, it is lower than in normal cells, and it is abolished upon treatment with alpha amanitin. This indicates that AF lesions are capable of arresting Pol II and can be repaired by TCR in GGR deficient cells, but by GGR in repair proficient cells and in CS cells, deficient in TCR (van Oosterwijk et al., 1996a; van Oosterwijk et al., 1996b). dG-C8-AF lesions caused a transient arrest but were not absolute blocks to transcription by Pol II in an *in vitro* assay (Donahue et al., 1996). Table 1 lists some DNA lesions, the primary pathways used to repair them, their effect on transcription, and whether they are subject to TCR.

3.4. Which Diseases?

The fields of NER and transcription have now converged as a consequence of the discovery of TCR and the additional revelation that protein subunits of the basal transcription initiation factor, TFIIH, are essential for NER (Feaver et al., 1993; Qui et al., 1993; Schaeffer et al., 1993; Sweder and Hanawalt, 1994). Furthermore, the genes implicated in repair and/or transcription are defective in xeroderma pigmentosum (XP) and/or other rare human hereditary disorders: CS and trichothiodystrophy (TTD). XP patients belong to one of seven complementation groups designated XPA, XPB, etc.; to XPG. We will focus here on two of those groups: the most severely affected patients belong to XPA; their cells are deficient in the lesion recognition step and are completely deficient in NER. Some moderately affected patients belong to XPC, in which NER is confined to the transcribed strands of active genes, as noted above. While the victims of XP are highly predisposed to cancer in sun-exposed skin, CS and TTD patients are not. However, CS patients suffer severe developmental problems with growth retardation, neurological deficiencies, and skeletal abnormalities. TTD is characterized by brittle hair, short stature, scaly skin, and mental underdevelopment.

3.4.1. Could These Developmental Problems Be Attributed to Defects in Repair? Interestingly, the helicases encoded by the XPB and XPD genes are components of the transcription initiation factor, TFIIH, so they evidently serve roles in opening up the DNA both for NER and for initiation of transcription by RNA polymerase II. Mutations in other domains of one of these helicases in XPB or in XPD patients can also result in the symptoms of CS, and in some cases of TTD as well. These findings led to the suggestion that CS and TTD might be "transcription syndromes" due to malfunctioning of the XPB or XPD gene products, such that regulation of early development of neuronal or ectodermal tissues would be defective (Bootsma and Hoeijmakers, 1993). The *haywire* gene of *Drosophila* is a homolog of XPB; the viable mutants are UV sensitive, sterile, and they display neurological abnormalities (Mounkes et al., 1992). However, there have been no reports of specific defects in transcription of particular genes in XP, CS, TTD, or even *haywire* itself. An alternative model proposes that CS cells lack a factor that shuttles TFIIH from repair to transcription complexes, thus transcription in those cells is normal unless the cells are injured, in which case TFIIH becomes "trapped" in repair complexes inhibiting both transcription and repair (van Oosterwijk, et al., 1996b). Using immunological probes it has been shown that the amount of TFIIH is indeed significantly reduced in XPB and XPD cells. However, no deficiency in transcription was detected in an optimized cell-free transcription assay (Satoh and Hanawalt, 1997). In contrast, other studies have suggested a rather profound deficiency in overall transcription in CSB cells (Balajee et al., 1997). Of course a transcription defect in the CSB patient could not be very severe or it would be lethal early in embryogenesis.

The clinical features of CS might result from the DNA repair defect that is unique in CS, namely the deficiency in TCR (Hanawalt and Mellon, 1993; Venema et al., 1990). Two of the complementation groups of CS (CSA and CSB) exhibit no defect in overall genomic NER of UV induced CPD, but they are completely deficient in TCR. They do not overlap with XP and have no documented involvement in transcription, except that in an *in vitro* assay the addition of CSB stimulates Pol II transcription 3-fold (Selby and Sancar, 1997). Half of the XPG patients have CS symptoms, but the XPG gene product has not been directly implicated in transcription (Nouspikel et al., 1997; Vermeulen et al., 1993).

3.4.2. How Could a Defect in the Repair of Expressed Genes Result in a Rather Specialized Set of Developmental Abnormalities? A hint is provided by the discovery that CSA and CSB are deficient in TCR of some class(es) of lesions that are produced by IR (Leadon and Cooper, 1993). IR produces lesions in large part indirectly, through the generation of oxidative free radicals. The observation of TCR in γ-irradiated XPA cells has the important implication that repair pathways other than NER (such as those initiated by some glycosylases) may be coupled to transcription. One might wonder then why XP patients do not also have CS. The explanation could be found in that unique class of oxidative damage, thymine glycols, that is subject to TCR in XPA, XPF and XPG cells but not

Table 2. Clinical and biochemical characteristics of normal and repair deficient syndromes (adapted from Ljungman and Zhang, 1996)

	UV			Oxidative damage				
Syndrome	GGR	TCR	Sensitivity	GGR	TCR	Sensitivity	Cancer	Neural/ growth abnormalities
Normal	+	+	+	+	+	+	+	–
CS	+	–	++	+	–	++	+	++
XPA	–	–	++++	±	+	++	+++	+
XPC	–	+	++	+	+	?	+++	–

in XPB/CS, XPD/CS, XPG/CS cells nor in CS-A and CS-B (Cooper et al., 1997). The age related neurodegeneration that accompanies the most severe XPA cases could be due to the accumulation of a unique type of free radical damage in neurons, that is not repaired in XPA (Satoh et al., 1993). A summary of salient characteristics of normal and repair deficient syndromes in presented in Table 2.

Cells with unusually high metabolic activity, such as neurons and those cells that are proliferating rapidly during early development, may produce higher levels of free radicals of the same types as those generated by IR, causing oxidative damage to DNA. The nuclear stabilization of the p53 tumor supressor is correlated specifically with unrepaired damage in expressed genes (Yamaizumi and Sugano, 1994). Arrested DNA Pol II at a lesion may constitute a signal for p53 activation and the resultant apoptotic response unless the lesion is repaired. Thus, cells unable to remove lesions from genes transcribed by Pol II are much more sensitive to damage-induced apoptosis than are TCR proficient cells (Ljungman and Zhang, 1996). The demyelination of neurons that is characteristic of CS could then be the consequence of excessive neuronal cell death during early development. Likewise, the low cancer incidence in CS could also be the consequence of the p53 regulated apoptosis pathway, if cell death occurs before tumors can develop. Interestingly, p53 has been shown to have a regulatory role in GGR but not in TCR of CPD in UV-irradiated cells (Ford and Hanawalt, 1995; Ford and Hanawalt, 1997).

3.5. What Is the Mechanism of Transcription-Coupled Repair? Which Proteins Are Involved?

We originally suggested that arrest of transcription at lesions and release of RNA polymerase from the template could serve as a specific signal to accelerate repair in active domains (Mellon, et al., 1986). Selby et al. (1991) showed that a transcription-repair coupling factor from *E. coli* binds to and releases the RNA polymerase blocked at a lesion. This factor (the product of the *mfd* gene, discovered by Evelyn Witkin over 30 years ago) then interacts with the excision repair complex to remove the lesion. In human cells the product of the CSB gene, ERCC6, has been implicated in the coupling process (Troelstra et al., 1992) following upon the revelation that hamster UV61 cells (deficient in ERCC6) are deficient in TCR (Lommel and Hanawalt, 1991).

The template DNA strand is preferentially repaired in the region just upstream of the point at which TFIIH is released from the transcription initiation complex, in normal, CSB cells and in the yeast homolog of CSB, *rad26*; preferential repair of the transcribed strand beyond the point of TFIIH release depends on the presence of CSB or *rad26*. This suggests a role for CSB and *rad26* in re-recruiting TFIIH to the repair complex when the

RNA Pol II is arrested at a lesion while in elongation mode (Tijsterman et al., 1997; Tu et al., 1997).

The CSA gene product belongs to a class called WD-repeat proteins, implicated in chromatin remodeling but its role in TCR is not known; in yeast the putative CSA homolog, *rad28*, is not required for TCR.

The XPB protein and a region of the XPD protein are essential for TCR, in addition to their role in overall NER; their function in the TCR pathway has not been resolved.

Mutations in the mismatch repair genes *mutS* or *mutL* in *E. coli* or their homologs in human cells have been shown to inhibit TCR but the mechanism of this effect is not yet understood (Mellon and Champe, 1996a; Mellon et al., 1996b). In yeast, mismatch repair mutants do not appear to be defective in TCR (Sweder et al., 1996).

3.6. How Does the Repair Complex Gain Access to a Lesion that Is Obscured by an Arrested RNA Pol II?

A possible scenario is provided by the transcription elongation factor SII, based upon its activity that catalyzes nascent transcript cleavage by RNA polymerase II at a natural pause site, enabling the polymerase to "back off" and try again without aborting the incomplete transcript. A similar reaction has been demonstrated at the site of a CPD in a model DNA template *in vitro* (Donahue et al., 1994) so this might be an important component in the process of TCR. On the other hand a recent study has implicated CSA and CSB in the ubiquitination of RNA Pol II in UV-irradiated cells, in support of a model in which the ternary complex is disrupted and the arrested polymerase is marked for proteasomal degradation (Bregman et al., 1996; Ratner et al., 1998). Neither model addresses the question of how the repair enzymes are recruited to the lesion. Figure 1 outlines a model for NER in association with the nuclear matrix, and the consequences of faulty repair of lesions in transcribed and non-transcribed DNA domains.

3.7. TCR Affects Various Cellular Activities

An important question about the mechanism of TCR is whether the stalling of RNA polymerase II is sufficient to initiate a repair event or whether the repair complex still has the opportunity to distinguish between a natural, sequence dependent "pause site" and a *bona fide* lesion. If the former is true then the system might occasionally initiate a repair reaction at a pause site. This gratuitous TCR would result in the reiterative generation of repair patches at lesion-free sites. This could lead to higher levels of "spontaneous" mutagenesis in a frequently transcribed gene due to the natural error frequency of the DNA repair polymerase carrying out the patching, specially if such patching is not subject to mismatch repair. Long ago it was shown that a low but detectable level of repair replication occurred in *E. coli* that had not been UV irradiated (Grivell et al., 1975). Interestingly this DNA "turnover" was inhibited by rifampicin, a specific inhibitor of transcription. Evidence that spontaneous mutagenesis increases when the transcription rate is enhanced has been reported in yeast (Datta and Jinks-Robertson, 1995; Korogodin et al., 1991).

Other results consistent with the possibility of "gratuitous TCR" include the report that the hypervariable region promoter placed before the constant region in the immunoglobulin system leads to high mutation rates in the constant region (Peters and Storb, 1996). Also, Wand et al. (1996) have shown that antisense DNA oligomers that form triple helix structures are substrates for NER; they observed mutations in the relevant human

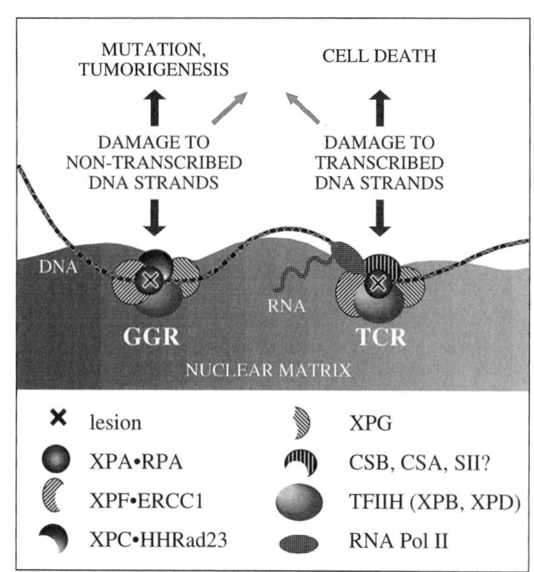

Figure 1. A model for global and transcription-coupled nucleotide excision repair and the cellular consequences of deficiencies in either or both repair pathways.The RNA polymerase II complex (presumably in association with the nuclear matrix) translocates as the DNA is pulled through it until a lesion is encountered in the template strand. In our favored model, RNA Pol II has backed off to allow access of the repair enzymes to the lesion at the site of transcription blockage; alternatively the polymerase may be released from the DNA. In either case the DNA strands at the site of the lesion reanneal so that NER may be initiated. Following removal of the lesion and the completion of repair replication, transcription can resume or recommence on the damage-free template. Lesions in non-transcribed DNA sequences must be brought to the nuclear compartment at the matrix where repair (and presumably most other DNA transactions) takes place (cf. Koehler and Hanawalt, 1996). Cells deficient in TCR are exquisitely sensitive to certain DNA damaging agents; cells deficient in GGR are more sensitive than normal cells but their survival is high enough that translesion replication in the unrepaired DNA may lead to mutations and cancer.

gene but this effect was not seen in CSA or CSB cells. Thus, TCR was implicated in the mutagenic mechanism.

TCR may also affect recombination. Deng and Nickoloff (1994) have found that increased transcription levels attenuated UV-induced intrachromosomal recombination involving an expressed gene in mouse cells. They suggested that the enhanced transcription resulted in more efficient TCR of photoproducts in the expressed DNA sequence that were responsible for UV-induced recombination. Many more complexities will surely be revealed before we fully understand the mechanism of TCR and its interactions with other cellular DNA transactions. Attempts to establish *in vitro* cell-free assays for TCR in our laboratory (G. Spivak, unpublished results) and elsewhere have not been successful to date.

3.8. Rationale for Existence of TCR

Clearly, TCR is necessary to sustain normal cell development. Various cellular activities that could be hindered by delays in the removal of lesions from template DNA strands include replication, expression of essential genes, and repair *per se*; perhaps the most important reason for the existence of TCR in mammalian cells is to eliminate situations that can trigger a cascade of reactions that irreversibly lead to apoptosis.

ACKNOWLEDGMENTS

Our research is supported by an Outstanding Investigator Grant (CA 44349) from the National Cancer Institute (NIH).

REFERENCES

A. S. Balajee, A. May, G. L. Dianov, E. C. Friedberg and V. A. Bohr (1997). Reduced RNA polymerase II transcription in intact and permeabilized Cockayne syndrome group B cells. *Proc. Natl. Acad. Sci. USA* **94**, 4306–4311.

V. A. Bohr, C. A. Smith, D. S. Okumoto, and P. C. Hanawalt (1985). DNA repair in an active gene: removal of pyrimidine dimers from the DHFR gene of CHO cells is much more efficient than in the genome overall. *Cell* **40**, 359–369.

D. Bootsma and J. H. J. Hoeijmakers (1993). DNA repair: Engagement with transcription. *Nature* **363**, 114–115.

K. K. Bowman, C. A. Smith and P. C. Hanawalt (1997) Excision-repair lengths are similar for transcription-coupled repair and global genome repair in UV-irradiated human cells. **Mut. Res. 385**, 95–105.

D. B. Bregman, R. Halaban, A. J. van Gool, K. A. Henning, E. C. Friedberg and S. L. Warren (1996) UV-induced ubiquitination of RNA polymerase II: a novel modification deficient in Cockayne syndrome cells. *Proc. Natl. Acad. Sci. USA* **93**, 11586–11590.

M. Carreau and D. Hunting (1992) Transcription-dependent and independent DNA excision repair pathways in human cells. *Mut. Res.* **274**, 57–64.

F. C. Christians and P. C. Hanawalt (1992) Inhibition of transcription and strand-specific DNA repair a-amanitin in Chinese hamster ovary cells. Mut. Res. **274**, 93–101.

F. C. Christians and P. C. Hanawalt (1993) Lack of transcription-coupled repair in mammalian ribosomal RNA genes. *Biochemistry* **32**, 10512–10518.

P. K. Cooper, T. Nouspikel, S. G. Clarkson, and S. A. Leadon (1997) Defective transcription-coupled repair of oxidative base damage in Cockayne syndrome patients from XP group G. *Science* **275**, 990–993.

R. Dammann and G. P. Pfeifer (1997) Lack of gene- and strand-specific DNA repair in RNA polymerase III-transcribed human tRNA genes. *Mol. Cell. Biol.* **17**, 219–229.

A. Datta and S. Jinks-Robertson (1995) Association of increased spontaneous mutation rates with high levels of transcription in yeast. *Science* **268**, 1616–1619.

W. P. Deng and J. A. Nickoloff (1994) Preferential repair of UV damage in highly transcribed DNA diminishes UV-induced intrachromosomal recombination in mammalian cells. *Mol. Cell. Biol.* **14**, 391–399.

B. A. Donahue, R. P. P. Fuchs, D. Reines, and P. C. Hanawalt (1996) Effects of aminofluorene and acetylaminofluorene DNA adducts on transcriptional elongation by RNA polymerase II. *J. Biol. Chem.* **271**, 10588–10594.

B. A. Donahue, S. Yin, J.-S. Taylor, D. Reines, and P. C. Hanawalt (1994) Transcript cleavage by RNA polymerase II arrested by a cyclobutane pyrimidine dimer in the DNA template. *Proc. Natl. Acad. Sci. USA* **91**, 8502–8506.

H. Echols and M. F. Goodman (1991) Fidelity mechanisms in DNA replication. Ann. Rev. *Biochemistry* **60**, 477–571.

J. W. Feaver, J. Q. Svejstrup, L. Bardwell, J. A. Bardwell, S. Buratowski, K. D. Gulyas, T. F. Donahue, E. C. Friedberg and R. D. Kornberg (1993) Dual roles of a multiprotein complex from S. cerevisiae in transcription and DNA repair. *Cell* **75**, 1379–1387.

J. M. Ford and P. C. Hanawalt (1995) Li-Fraumeni syndrome fibroblasts homozygous for p53 mutations are deficient in global DNA repair but exhibit normal transcription-coupled repair and enhanced UV-resistance. *Proc. Natl. Acad. Sci. USA* **92**, 8876–8880.

J. M. Ford and P. C. Hanawalt (1997) Expression of wild type p53 is required for efficient global genomic nucleotide excision repair in the non-transcribed strand of an active gene. *J. Biol. Chem.* **272**, 28073–28080.

E. C. Friedberg, G. C. Walker and W. Siede (1995) *DNA Repair and Mutagenesis*. American Society for Microbiology Press, Washington D.C.

A. R. Grivell, M. B. Grivell, and P. C. Hanawalt (1975) Turnover in bacterial DNA containing thymine or 5-bromouracil. *J. Mol. Biol.* **98**, 219–233.

P. C. Hanawalt and I. Mellon (1993) DNA repair: stranded in an active gene. *Current Biology* **3**, 67–69.

A. L. Islas, F. J. Baker and P. C. Hanawalt (1994) Transcription-coupled repair of psoralen crosslinks but not monoadducts in Chinese hamster ovary cells. *Biochemistry* **33**, 10794–10799.

D. R. Koehler and P. C. Hanawalt (1996) Recruitment of damaged DNA to the nuclear matrix in hamster cells following ultraviolet irradiation. *Nucl. Acids. Res.* **24**, 2877–2884.

V. I. Korogodin, V. L. Korogodina, C. Fajszi, A. I. Chepurnoy, N. Mikhova-Tsenova and N. V. Simonyan (1991) On the dependence of spontaneous mutation Rrates on the functional state of genes. *Yeast* **7**, 105–117.

S. A. Leadon and P. K. Cooper (1993) Preferential repair of ionizing radiation-induced damage in the transcribed strand of an active human gene is defective in Cockayne syndrome. *Proc. Natl. Acad. Sci. USA* **90**, 10499–10503.

S. A. Leadon and D. A. Lawrence (1991) Preferential repair of DNA damage on the transcribed strand of the human metallothionein genes requires RNA polymerase-ii. *Mut. Res.* **255**, 67–78.

S. A. Leadon and D. A. Lawrence (1992) Strand-selective repair of DNA damage in the yeast gal7-gene requires RNA polymerase-II. *J. Biol. Chem.* **267**, 23175–23182.

T. Lindahl (1993) Instability and decay of the primary structure of DNA. *Nature* **362**, 709–715.

M. Ljungman and F. Zhang (1996) Blockage of RNA polymerase as a possible trigger for u.v. light-induced apoptosis. *Oncogene* **13**, 823–831.

L. Lommel, C. Carswell-Crumpton and P. C. Hanawalt (1995) Preferential repair of the transcribed DNA strand in the dihydrofolate reductase gene throughout the cell cycle in UV-irradiated human cells. *Mut. Res.* **336**, 181–192.

L. Lommel and P. C. Hanawalt (1991) The genetic defect in the Chinese hamster ovary cell mutant UV61 permits moderate selective repair of cyclobutane pyrimidine dimers in an expressed gene. *Mut. Res.* **255**, 183–191.

D. H. Madhani, V. A. Bohr and P. C. Hanawalt (1986) Differential DNA repair in transcriptionally active and inactive proto-oncogenes: c-abl and c-mos. *Cell* **45**, 417–423.

S. K. Mauldin and R. A. Deering (1994) Differential repair of UV damage in developmentally regulated gene of Dictyostelium discoideum. *Mut. Res.* **314**, 187–198.

I. Mellon, V. A. Bohr, A. C. Smith and P. C. Hanawalt (1986) Preferential DNA repair of an active gene in human cells. *Proc. Natl. Acad. Sci. USA* **83**, 8878–8882.

I. Mellon and G. N. Champe (1996a) Products of DNA mismatch repair genes mutS and mutL are required for transcription-coupled nucleotide-excision repair of the lactose operon in *Escherichia coli*. *Proc. Natl. Acad. Sci. USA* **93**, 1292–1297.

I. Mellon and P. C. Hanawalt (1989) Induction of the *Escherichia coli* lactose operon selectively increases repair of its transcribed DNA strand. *Nature* **342**, 95–98.

I. Mellon, D. K. Rajpal, M. Koi, C. R. Boland and G. N. Champe (1996b) Transcription-coupled repair deficiency and mutations in human mismatch repair genes. *Science* **272**, 557–560.

I. Mellon, G. Spivak and P. C. Hanawalt (1987) Selective removal of transcription-blocking DNA damage from the transcribed strand of the mammalian DHFR gene. *Cell* **51**, 241–249.

L. C. Mounkes, R. S. Jones, B.-C. Liang, W. Gelbart and M. T. Fuller (1992) A Drosophila model for xeroderma pigmentosum and Cockayne's syndrome: *haywire* encodes the fly homolog of *ERCC3*, a human excision repair gene. *Cell* **71**, 925–937.

T. Nouspikel, P. Lalle, S. A. Leadon, P. K. Cooper and S. G. Clarkson (1997) A common mutational pattern in Cockayne syndrome patients from xeroderma pigmentosum group G: implications for a second XPG function. *Proc. Natl. Acad. Sci. USA* **94**, 3116–3121.

A. Peters and U. Storb (1996) Somatic hypermutation of immunoglobulin genes is linked to transcription initiation. *Immunity* **4**, 57–65.

H. Qui, E. Park, L. Prakash and S. Prakash (1993) The *Saccharomyces cerevisiae* DNA repair gene *RAD25* is required for transcription by RNA polymerase II. *Genes & Development* **7**, 2161–2171.

J. N. Ratner, B. Balasubramanian, J. Corden, S. L. Warren and D. B. Bregman (1998) Ultraviolet-induced ubiquitination and proteasomal degradation of the large subunit of RNA polymerase II. *J. Biol. Chem.* **273**, 5184–5189.

A. Sancar (1996) DNA excision repair. *Ann. Rev. Biochem.* **65**, 43–81.

M. S. Satoh and P. C. Hanawalt (1997) Competent transcription initiation by RNA polymerase II in cell-free extracts from xeroderma pigmentosum groups B and D in an optimized RNA transcription assay. *Biochim. Biophys. Acta* **1354**, 241–251.

M. S. Satoh, C. J. Jones, R. D. Wood and T. Lindahl (1993) DNA excision-repair defect of xeroderma pigmentosum prevents removal of a class of oxygen free-radical induced base lesions. *Proc. Natl. Acad. Sci. USA* **90**, 6335–6339.

L. Schaeffer, R. Roy, S. Humbert, V. Moncollin, W. Vermeulen, J. H. J. Hoeijmakers, P. Chambon and J.-M. Egly (1993) DNA repair helicase: A component of BTF2 (TFIIH) basic transcription factor. *Science* **260**, 58–63.

D. A. Scicchitano and P. C. Hanawalt (1989) Repair of N-methylpurines in specific DNA sequences in Chinese hamster ovary cells: Absence of strand specificity in the dihydrofolate reductase gene. *Proc. Natl. Acad. Sci. USA* **86**, 3050–3054.

C. B. Selby, E. M. Witkin and A. Sancar (1991) *Escherichia coli mfd* mutant deficient in "mutation frequency decline" lacks strand-specific repair: *In vitro* complementation with purified coupling factor. *Proc. Natl. Acad. Sci. USA* **88**, 11574–11578.

C. P. Selby and A. Sancar (1997) Cockayne syndrome group B protein enhances elongation by RNA polymerase II. *Proc. Natl. Acad. Sci. USA* **94**, 11205–11209.

A. Sitaram, G. Plitas, W. Wang and D. A. Scicchitano (1997) Functional nucleotide excision repair is required for the preferential removal of N-ethylpurines from the transcribed strand of the dihydrofolate reductase gene of Chinese hamster ovary cells. *Molec. Cell. Biol.* **17**, 564–570.

M. J. Smerdon and F. Thoma (1990) Site-specific DNA-repair at the nucleosome level in a yeast minichromosome. *Cell* **61**, 675–684.

K. S. Sweder and P. C. Hanawalt (1992) Preferential repair of cyclobutane pyrimidine dimers in the transcribed strand of a gene in yeast chromosomes and plasmids is dependent on transcription. *Proc. Natl. Acad. Sci. USA* **89**, 10696–10700.

K. S. Sweder and P. C. Hanawalt (1994) The COOH terminus of supressor of stem loop (SSL2) (RAD25) in yeast is essential for overall genomic excision repair and transcription-coupled repair. *J. Biol. Chem.* **269**, in press.

K. S. Sweder, R. A. Verhage, D. J. Crowley, G. F. Crouse, J. Brouwer and P. C. Hanawalt (1996) Mismatch repair mutants in yeast are not defective in transcription-coupled DNA repair of UV-induced DNA damage. *Genetics* **143**, 1127–1135.

B.G. Taffe, F. Larminat, J. Laval, D.L. Croteau, R.M. Anson, and V.A. Bohr (1996) Gene-specific nuclear and mitochondrial repair of formamidopyrimidine DNA glycosylase-sensitive sites in Chinese hamster ovary cells. *Mut. Res.* **364**, 183–192.

M. Tijsterman, R. A. Verhage, P. van de Putte, J. G. Tasseron-de Jong and J. Brouwer (1997) Transitions in the coupling of transcription and nucleotide excision repair within RNA polymerase II-transcribed genes of *Saccharomyces cerevisiae*. *Proc. Natl. Acad. Sci. USA* **94**, 8027–8032.

C. Troelstra, A. van Gool, J. D. Wit, W. Vermeulen, D. Bootsma and J. H. J. Hoeijmakers (1992) *ERCC6*, a member of a subfamily of putative helicases, is involved in Cockayne's syndrome and preferential repair of active genes. *Cell* **71**, 939–953.

Y. Tu, S. Bates and G. P. Pfeifer (1997) Sequence-specific and domain-specific DNA repair in xeroderma pigmentosum and Cockayne syndrome cells. *J. Biol. Chem.* **272**, 20747–20755.

M. F. van Oosterwijk, R. Filon, W. H. Kalle, L. H. Mullenders and A. A. van Zeeland (1996a) The sensitivity of human fibroblasts to N-acetoxy-2-acetylaminofluorene is determined by the extent of transcription-coupled repair, and/or their capability to counteract RNA synthesis inhibition. *Nucl. Acids Res.* **24**, 4653–4659.

M. F. van Oosterwijk, A. Versteeg, R. Filon, A. A. van Zeeland and L. H. Mullenders (1996b) The sensitivity of Cockayne's syndrome cells to DNA-damaging agents is not due to defective transcription-coupled repair of active genes. *Mol. Cell. Biol.* **16**, 4436–4444.

J. Venema, L. H. F. Mullenders, A. T. Natarajan, A. A. van Zeeland and L. V. Mayne (1990) The genetic defect in Cockayne syndrome is associated with a defect in repair of UV-induced DNA damage in transcriptionally active DNA. *Proc. Natl. Acad. Sci. USA* **87**, 4707–4711.

W. Vermeulen, J. Jaeken, N. G. J. Jaspers, D. Bootsma and J. H. J. Hoeijmakers (1993) Xeroderma-pigmentosum complementation group-G associated with Cockayne syndrome. *Am. J. Hum. Genet.* **53**, 185–192.

J. M. H. Vos and P. C. Hanawalt (1987) Processing of psoralen adducts in an active human gene: repair and replication of DNA containing monoadducts and interstrand crosslinks. *Cell* **50**, 789–799.

J. M. H. Vos and E. L. Wauthier (1991) Differential introduction of DNA damage and repair in mammalian genes transcribed by RNA polymerase-I and polymerase-II. *Mol. Cell. Biol.* **11**, 2245–2252.

G. Wand, M. M. Seidman and P. M. Glazer (1996) Mutagenesis in mammalian cells induced by triple helix formation and transcription-coupled repair. *Science* **271**, 802–805.

Z. Wang and T. M. Rana (1997) DNA damage-dependent transcriptional arrest and termination of RNA polymerase II elongation complexes in DNA template containing HIV-1 promoter. *Proc. Natl. Acad. Sci. USA* **94**, 6688–6693.

R. D. Wood (1996) DNA repair in eukaryotes. *Ann. Rev. Biochem.* **65**, 135–167.

M. Yamaizumi and T. Sugano (1994) UV-induced nuclear accumulation of p53 is evoked through DNA damage of actively transcribed genes independent of the cell cycle. *Oncogene* **9**, 2775–2784.

M. E. Zolan, G. A. Cortopassi C. A. Smith and P. C. Hanawalt (1982) Deficient repair of chemical adducts in alpha DNA of monkey cells. *Cell* **28**, 613–619.

15

GENOTOXICITY TESTS

Application to Occupational Exposure as Biomarkers

Ali E. Karakaya, Semra Sardas, and Sema Burgaz

Gazi University, Faculty of Pharmacy
Department of Toxicology
Hipodrom 06330, Ankara, Turkey

1. ABSTRACT

Using biomarkers for screening in workers who had known or suspected contact with genotoxic chemicals can be useful in quantifying exposure and assessing genotoxic risk. In this research the genotoxic risk of the following Turkish occupational groups (furniture workers, nurses, operating room personnel, hospital sterilising staff, engine repair workers, road-paving workers, car painting workers and hair colorists) were evaluated. Results of some markers of exposure and cytogenetic biomarkers (sister chromatid exchange, micronucleus assay and comet assay) has been assessed in the above mentioned occupational groups.

2. INTRODUCTION

Epidemiologists have tentatively attributed about 4 per cent of cancer deaths in industrialized countries such as the United States, to occupational causes (Doll and Peto, 1981). A total of 10 to 20 % of lung cancers and 21 to 27 % of bladder cancers are estimated to be related to occupational exposure (Ward, 1995). Obviously there are some difficulties in making such estimates, including the lack of accurate data on history of exposure and current exposures as well as confounding factors such as socioeconomic status and smoking.

Certain occupational groups are exposed at much higher concentrations than is the general population to potentially hazardous genotoxins. The possibility exists that these exposures may significantly increase the risk of cancer for some of the workers.

The most comprehensive source of information about both occupational and non-occupational carcinogens is a series of monographs published by the International Agency for Research on Cancer (IARC, 1987). Each monograph is the product of an individual

Advances in DNA Damage and Repair, edited by Dizdaroglu and Karakaya. Kluwer Academic / Plenum Publishers, New York, 1999.

working group of experts in chemical carcinogenesis and related fields. The evaluation is based on published information available at the time the working group has convened. IARC has published reviews on over 1000 substances. IARC Classifies agents (or exposure circumstances) according to their carcinogenicity as follows:

- Group 1: The agent is carcinogenic to humans
- Group 2A: The agent is probably carcinogenic to humans
- Group 2B: The agent is possibly carcinogenic to humans
- Group 3: The agent is not classifiable as to its carcinogenicity to humans.
- Group 4: The agent is probably not carcinogenic to humans.

There are 28 industrial chemicals or processes in group 1 that have been shown to cause human cancer. There are 23 industrial chemicals and processes in group 2A for which there is sufficient animal evidence but limited or inadequate human evidence of carcinogenicity, and about 70 agents in group 2B, for which there is less compelling evidence to suggest possible human carcinogenicity. There are an additional 147 substances for which there was sufficient evidence of animal carcinogenicity but for which there was either inadequate or nonexistent epidemiologic data (Fischman et al., 1990).

In industrialized countries, with strict governmental regulations of actual and potential industrial health hazards during the last two decades, it is likely that this figure will decrease even further in the future (Pitot and Dragan, 1996). But on the other hand while cancer deaths due to occupational exposure may be only a small portion of all cancer deaths in the developing countries , the risk may increase owing to rapid industrialization , uncontrolled transfer of technology and lack of regulations controlling hazardous substances (ILO, 1988).

The importance of occupational agents as risk factors for certain human cancers is indicated by two types of evidence, one deriving from the extensive data base on the carcinogenicity and mutagenicity of chemicals to which humans are exposed occupationally, and the second from epidemiological studies associating such exposures with elevated incidence of specific forms of malignant disease. Traditional epidemiologic studies have the potential for providing the strongest evidence for human carcinogenicity, because the subjects studied are human. These studies have some limitations. Failure to demonstrate a positive association in an epidemiologic study does not always indicate that there is no association between the agent and the effect studied. In some cases, a "false-negative" epidemiologic study may result because of a variety of shortcomings. Some of these limitations include difficulties in identifying exposure and effects, difficulties in choosing appropriate exposed and control populations, requirements for prolonged follow-up because of long induction-latency periods, and the relative lack of sensitivity of epidemiologic methods Recent advances in molecular biology and genetics have produced powerful new tools . Epidemiologic studies of populations at risk for exposure-induced malignant disease can now apply these new, powerful molecular tools to define biological markers associated with exposure and disease. Molecular epidemiology aims at detecting, understanding, and preventing these diseases (NIH, 1992; Fischman et al., 1990).

Possible genotoxic exposure and their detection in the occupational environment have been the object of much concern during recent years. Screening workers who have had known or suspected contact with genotoxic chemicals can be useful in quantifying exposures and assessing risks. As methods improve in analytical chemistry and molecular biology, direct biological monitoring of exposed populations is possible. Biomonitoring involves the use of biological or molecular markers as indicators signaling events in individuals exposed to occupational chemicals (Perera, 1993; Perera and Whyatt, 1994).

3. BIOLOGICAL MARKERS

A biomarker is a change in a biological system that can be related to an exposure to, or effect from, a specific xenobiotic or type of toxic material (Henderson et al., 1989). Biomarkers may serve as a bridge between experimental and epidemiologic studies of hazardous substances as they reflect biochemical or molecular changes associated with environmental exposure to these substances. Thus, biomarkers, together with the use of molecular epidemiology, enable risk analysis to increasingly focus on risk and associated interindividual variability resulting from exposure to hazardous substances (DeRosa et al., 1993). Biomarkers can be classified into a number of categories to the type of information obtained. *Markers of exposure* indicate whether exposure to an agent has taken place, and include measurement of specific metabolites and/or adduct formed by reaction of the compound or its metabolites with macromolecules. *Markers of susceptibility* can be used to identify specific individuals at greater risk than the general population as a result of a genetic and other predisposition effects of exposure. These might include the activity of specific enzymes involved in activation or detoxification of a specific chemical or DNA repair capacity for specific types of DNA damage. *Markers of effect* provide an indication of early events in development of toxicity, carcinogenesis and disease. Markers can be measured at various stages in progression from exposure to end effect and present a continuum of events involved (Fennell, 1990). Some biomarkers fall at the boundaries between these classifications. For example, a DNA adduct may be associated with a specific chemical, making it characteristic of a marker of exposure, but its formation may result from the metabolism, distribution, and reaction of the chemical with DNA which reflects cellular processes. Thus, to a degree, it is a marker of effect as well (Ward and Henderson, 1996).

Recently we have used some biomarkers to evaluate and assess some occupational genotoxic chemical exposure in different workplaces in Turkey. In our research the genotoxic risk of the following occupational groups were evaluated.

4. OCCUPATIONAL GROUPS AT GENOTOXIC RISK

4.1. Styrene Exposed Furniture Workers

Styrene is one of the monomers used worldwide, and its polymers and copolymers have an increasingly wide range of application. Global production of styrene in 1992 was 14 282 thousand tons. In Turkey, styrene-alkyd coating of furniture and production of glass-reinforced products, such as boats, storage tanks and wall panels are the most extensive sources of occupational exposure to styrene with respect to number of workers and levels of exposure. Most of this production is carried out in the small workshop in which an estimated that 20 000 workers are employed in these workshops in Turkey (Karakaya et al., 1997). Styrene was classified in 1987 within the IARC criteria as 2B (possibly carcinogenic to humans) (IARC, 1987). In our study workers involved in hand-spraying lamination processes using styrene-contained unsaturated polyester were investigated (Karakaya et al., 1997).

4.2. Nurses Handling Antineoplastic Drugs

Many anticancer agents have been shown to be mutagenic, teratogenic and carcinogenic in experimental systems and second malignancies are known to be associated with several specific therapeutic treatments. Anticancer agents thus represent a class of occupational carcinogens (Sorsa et al., 1985). In our study nurses working in the hematology and oncology departments of various university hospitals in Ankara were investigated. These nurses had been handling antineoplastic drugs: most frequently, cyclophosphamide, methotrexate, vincristine, adriamycin, cisplatinum, etoposide, 5-fluorouracil and bleomycin (Bayhan et al., 1987; Şardaş et al., 1991).

4.3. Operating Room Personnel

The possibility of a potential mutagenic and carcinogenic action of inhalation anesthetics has been subject of various studies (Baden and Simmon, 1980). Our study group consisted of operating room personnel employed at Ankara University, Faculty of Medicine Department of Anesthesiology exposed to waste anesthetic gases such as halothane, nitrous oxide and isoflurane (Şardaş et al., 1992).

4.4. Hospital Sterilising Staff Exposed to Ethylene Oxide

Ethylene oxide is widely used in the sterilization of medical supplies in hospitals and the pharmaceutical industry. In experimental animals ethylene oxide has been found to be carcinogenic, mutagenic and teratogenic (WHO, 1985). Ethylene oxide was classified in 1987 within the IARC criteria as 2A (probably carcinogenic to humans) (IARC, 1987). In our study hospital staff working in the sterilization facilities of five hospitals in Ankara were investigated (Burgaz et al., 1992a).

4.5. Engine Repair Workers Exposed to Polycyclic Aromatic Hydrocarbons

Polycyclic aromatic hydrocarbons (PAHs) are a complex group of chemical compounds. PAHs are produced by the incomplete combustion of organic matter and include compounds that are potent carcinogens in experimental animals. Humans are environmentally exposed to PAHs which occur in the atmosphere from such sources as coal burning, motor vehicle emissions and cigarette smoking. In addition, significant quantities of PAHs are produced industrially in the manufacture of coal tar, in iron foundries, in coke oven plants, and in automotive repair workshops and it is strongly suspected that PAHs plays and important role in the etiology of urban and industrially related human cancers. IARC stated that there is sufficient evidence that 11 PAHs are carcinogenic to experimental animals. Epidemiological studies have shown an increase in cancer incidence among workers exposed to PAHs. (IARC, 1983).

Apprentice practice is very common in Turkey. According to 1993 demographic statistics 17 % of the Turkish population is in the 12–19 age group and 34 % of this population (3 400 000) is working as apprentice. In our study child laborers (age: between 13–19) who work engine repair workshops were chosen as PAHs exposed group (Karahalil et al., 1998).

4.6. Road-Paving Workers Exposed to Bitumen Fumes

A total of 90 to 95% of bitumen is used hot (>100 °C) in road construction, roofing and flooring. Fumes from these operations contain PAHs. Although the PAH concentrations are small, they have been suggested as causative agents for the carcinogenic effects of bitumens observed in some animal experiments (IARC, 1985). It is estimated that the current annual use of bitumens in Turkey is approximately 0.35 million tons. As bitumens in our present study, workers involved in road paving operations were chosen as PAHs exposed group (Burgaz et al., 1992b).

4.7. Car Painting Workers

These workers are exposed not to a single agent but to a complex mixture of solvents and dye pigments which are potentially genotoxic agents. In small workshops painted cars are air dried on the work premises and in none of the studied areas workers were applying protective measures such as respiratory masks and appropriate ventilation , to prevent inhalation of solvents and dust during color mixing (Şardaş et al., 1994).

4.8. Hair Colorists Exposed to Oxidation Hair Dyes

Oxidation type of hair dyes are usually contain 7–12 aromatic substances. There is evidence that many hair dye constituents (including certain phenylenediamines, diaminotoluenes, diaminoanisoles) are mutagenic, and some are carcinogenic to rodents (Burnett, 1980). These compounds readily penetrate the human skin. Most of the hair dyes sold on Turkish market are oxidation type, and we have performed a genotoxicity study on professional male hair colorists selected from the busiest hairdressing saloons for dye application in Ankara. None of them were wearing gloves during applications (Şardaş et al., 1997; İlbars et al., 1997).

5. TESTS TO ASSESS GENOTOXIC CHEMICAL EXPOSURE

5.1. Metabolic Markers

Specific metabolites as exposure biomarkers were also used in some of our studies. 1-hydroxypyrene and mandelic acid+phenylglyoxylic acid were measured in urine samples to assess PAHs (Jongeneelen et al., 1987) and Styrene (DeRosa et al., 1993) exposure respectively. Such measurements are indicators of internal dose of the exposed specific chemical. Table 1 shows results of our studies.

5.2. Thioether Test

Electrophilic agents-a class of chemicals that includes most genotoxic compounds-can be inactivated by reaction with glutathione or other SH-bearing molecules. The conjugates so formed often appear in the urine as mercapturic acids or other thioether products.

Because the urinary thioether test may be potentially suitable as an index of exposure to a wide range of electrophilic chemicals that includes most genotoxic compounds and chemical mixtures, we used this test for some occupational groups in assessing the to electrophilic chemical exposure. Table 2 shows urinary thioether levels in various occupational groups.

As shown in Table 2 the results of the thioether test may be considered as a signal indicating that genotoxic chemical exposure may have taken place. In any case more precise and specific techniques should be performed. To estimate genotoxic exposure risks in the occupational environment we have used cytogenetic biomarkers as marker of effect.

Table 1. 1-Hydroxypyrene and mandelic acid+phenylglyoxylic acid levels in urine samples of exposed workers

Occupational group	Exposure biomarker	Exposed	Control	References
Road-paving workers	1-hydroxypyrene (μmol/mol creat.)	0.61±0.38	0.28±0.17[a]	Burgaz et al., 1992b.
		$p<0.01$		
Engine repair workers	1-hydroxypyrene (μmol/mol creat.)	4.71±0.53	1.55±0.28[b]	Karahalil et al., 1998.
		$p<0.0001$		
Furniture workers	Mandelic acid + Phenylglyoxylic acid (mg/g creat.)	207±42.8	12±1.70[b]	Karakaya et al., 1997.
		$p<0.01$		

[a] Mean±SD
[b] Mean±SE

5.3. Cytogenetic Biomarkers

We have especially focused on the following cytogenetic biomarkers to assess occupational genotoxic exposure risk: Sister chromatid exchange, micronucleus assay and comet assay in our studies. Recent review articles (Perera, 1993; Perera and Whyatt, 1994; Henderson et al., 1989) agree that the said tests are sensitive, cost-effective, reliable, require small amount of material and they are widely applicable in different fields of genetic toxicology.

5.3.1. Sister Chromatid Exchange (SCE). Although there is a great uncertainty about the mechanism by which SCE are formed in which DNA damage or DNA synthesis stimulates its formation, it can be defined as segment exchanges between two chromatids of a chromosome which is visible cytologically through differential staining of chromatids (Perry and Evans, 1975). Many mutagens induce SCE in cultured cells and in mammals in vivo.

Quantification of DNA damage by SCE assay. Slides for SCE should be coded and scored blindly. The coded slides should be scanned under low magnification (100–200x) and well differentiated metaphases should be accepted for scoring. Overlapping chromosomes should be disregarded; at least 43 chromosomes should be analyzed per cell. De-

Table 2. Urinary thioether levels in various occupational groups (mmol.SH-/mol creatinine)

	Urinary thioether		
Occupational group	exposed	control	References
Nurses	5.30±0. 53	12.93±2.89[a]	Bayhan et al., 1987
	$p<0.05$		
Road paving workers	4.61±2.59	7.76±4.70[b]	Burgaz et al., 1992b
	$p<0.01$		
Hospital sterilising staff	6.07±0.55	11.16±1.20[a]	Burgaz et al., 1992a
	$p<0.001$		
Furniture workers	2.75±1.78	4.43±3.42[b]	Karakaya et al., 1997
	$p<0.05$		

[a] Mean ± SE
[b] Mean ± SD

Table 3. Mean levels of SCE frequencies/cell in peripheral blood lymphocytes of various occupational groups

	SCE frequency/cell		
Occupational group	Control	Exposed	References
Nurses	5.22±0.24	6.48±0.25[a]	Şardaş et al., 1991.
	p<0.01		
Operating room personnel	5.22±1.70	7.66±1.81[b]	Şardaş et al., 1992
	p<0.001		
Car painting workers	4.92±0.10	7.81±1.50[b]	Şardaş et al., 1994
	p<0.001		
Professional hair colorist	4.78±0.24	4.05±1.2[b]	Şardaş et al., 1997.
	p>0.05		
Furniture workers	5.23±1.23	6.20±1.56[b]	Karakaya et al.,1997.
	p<0.01		
Engine repair workers	4.06±0.16	4.47±0.09[a]	Karahalil et al., 1998.
	p<0.05		

[a] Mean±SE
[b] Mean±SD

pending on the desired statistical sensitivity, a minimum of 50 cells should be scored, with 25 cells from each replicate culture (Abe and Sasaki, 1982). Table 3 shows mean values SCE frequencies /cell in peripheral blood lymphocytes of various occupational groups.

5.3.2. Micronucleus Assay (MN). Micronuclei are chromatin-containing bodies which represent chromosal fragments or whole chromosomes that were not incorporated into a nucleus during mitosis. Since micronuclei usually represent fragments, they are used as simple indicators of chromosomal damage (Heddle, 1973; Fenech and Morley, 1985). Micronuclei can be observed in almost any cell type and for this reason many variations of the assay exit. Peripheral blood lymphocytes and target cells, i.e exfoliated bladder cells, buccal and nasal mucosa cells have been recently used for human biomonitoring in order to assess exposure to mutagenic and carcinogenic chemicals. Lymphocyte cultures are set up and microscopic analyses are carried out on stained slides. The frequency of micronuclei is evaluated by scoring a total of 1000 cells/individual (Heddle et al., 1991). In the exfoliated cell micronucleus assay, the air-dried cell smears are stained; and for each sample depending on the exfoliated cell type; such as urinary bladder, buccal and nasal a minimum 500, 1000, and 3000 intact epithelial cell are examined for the presence of micronuclei respectively (Stich and Rosin, 1984).The results of lymphocyte cell micronuclei assay applied in various occupational situations are given in Table 4.

5.3.3. Comet Assay. The comet assay also called single cell gel electrophoresis (SCGE), is an electrophoretic technique for direct visualization of DNA damage in individual cells. The electric current pulls the broken or relaxed DNA fragments from nucleus and migrates to anode which appears an image as comet . In this technique cells were embedded in agarose gel on microscope slides. Lysed by detergents and high salt, an then electrophoressed a short period under neutral conditions. Cells with increased DNA damage display increased migration of DNA from nucleus toward the anode, giving the appearance of the tail of comet (the head being the residue of the nucleus). The comets are observed by florescence microscopy after staining with suitable dyes; ethidium bromide,acridine orange or propidium iodide (Ostling and Johanson, 1984).

Table 4. Mean levels of MN frequencies (‰) in peripheral blood lymphocytes of various occupational groups

Occupational group	MN frequencies (‰) Control		Exposed	References
Furniture workers	2.09±0.35		1.98±0.50[a]	Karakaya et al., 1997.
		p<0.01		
Engine repair workers	1.56±0.06		1.87±0.04[b]	Karahalil et al., 1998.
		p<0.05		
Road paving workers	1.79±0.31		2.25±0.42[a]	Erdem et al., 1997.
		p<0.0001		
Professional hair colorists	2.09±0.35		1.47±0.63[a]	İlbars et al., 1997.
		p<0.05		

[a] Mean±SD
[b] Mean±SE

However, neutral conditions for lysis and electrophoresis permit the detection of double-strand breaks, they do not allow for the detection of single stranded ones. Since many agents induce from 5- to 2000- fold more single stranded breaks than double stranded breaks, neutral conditions are not sensitive as alkaline conditions in detecting DNA damage. Modified method capable of detecting DNA single-strand breaks under alkali conditions was developed (Singh et al., 1988;). In our study (Aygün et al., 1997), operation room personnel currently employed at Ankara Hospitals, who had been continuously exposed to various anaesthetics were evaluated by comet assay. Results have been shown in Table 5.

It is known that there are wide interpersonal variation in the frequency of spontaneous DNA damages between individuals and therefore it is essential to take as many confounding factors as possible in to consideration in the planning of a study.Cigarette smoking is now generally agreed to be a confounding factor and numerous studies have published on the influence of smoking and DNA damage. The relationship between tobacco usage and variety of human cancers are already well documented. Tobacco smoke has been classified as Group 1 (Agent carcinogenic to humans) by IARC (IARC, 1987). Therefore in all the above mentioned studies we have divided both the exposed and control groups as smokers and nonsmokers and thus statistically compared and evaluated the results. The discrimination between smokers and nonsmokers seems to be a indicator of the sensitivity of these assays according to smoking habits (Şardaş et al., 1991; Burgaz et al., 1995; Burgaz et al., 1987).

6. CONCLUSION

Avoidance of carcinogens in the workplace and reduction of exposure to them are two main approaches to prevent occupational cancer (Vainio, 1995). As for the avoidance;

Table 5. Comet assay (SCGE) data in peripheral lymphocytes of operating room personnel

Occupational group		Mean(±SE)Grade of damage/100 cells Normal	Stretch	Comet	Reference
Operating room personnel	Exposed	26.79 ±3.67	28.47±2.69	44.72±4.06	Aygün et al., 1997.
	Control	91.27 ±1.03	6.22±0.77	2.56 ±0.44	
		p<0.001	p<0.001	p<0.001	

Table 6. Effect of smoking on some markers of exposure and cytogenetic markers studied

Assay	Target	Number of SCE cell		References
		Nonsmoker	Smoker	
SCE	Lymphocytes	4.11±0.18	6.52±0.30[a] p< 0.001	Şardaş et al., 1991.
MN	Lymphocytes (micronucleated cell(%o))	1.82±0.30	2.20± 0.31[b] p< 0.01	Karahalil et al., 1998
MN	Exfoliated urothelial cells (micronucleated cell (%))	0.66±0.05	1.93± 0.11[a] p< 0.001	Burgaz et al.,1995.
Thioether	Urine (mmol. SH- /mol creat.)	7.20± 2.15	17.84± 4.37[a] p< 0.001	Burgaz et al., 1987.
SCGE (Comet)	Lymphocytes			
	Normal	92.80±1.20	7.09±1.44[a] p< 0.05	Aygün et al., 1997
	Stretch	5.67±0.96	7.73±1.15[a] p> 0.05	
	Comet	1.55±0.38	5.36±0.78[a] p< 0.001	

[a] Mean±SE
[b] Mean±SD

known or suspected human carcinogens should be removed from commercial use and replaced with the materials that serve same or similar functions. However, the substitute is often not adequate or costs more. There are two reasons why these substitutions are delayed in developing countries and one is the reflection of scientific data to regulations are not as rapid is the case with developed countries. For example DDT was banned in the United States in 1972, but, the compound was widely used until 1984 in Mexico (Carrillo et al., 1996) and the same figure is also valid for Turkey and other developing countries. Occupational carcinogens are regulated by national occupational health authorities. Occupational Safety and Health Administration (OSHA) in the United States has lowered TWA value of carcinogenic ethylene oxide from 50 ppm to 1 ppm in 1983. However this value is still 50 ppm in most countries. Delays in substitution is also caused by the behavior of producers who in their search for cost-minimization, apply pressure on decision making bodies against legislation of stricter limits and/or for lax implementation of existing regulations.

Technological advances in engineering and the utilization of personal protective devices can reduce or eliminate workers exposure to hazardous substances. However, cost is a limiting factor this approach also and a major barrier for the implementation of such preventive measures in developing countries.

Interaction between cigarette smoking and many occupational carcinogens are well documented. Generally, it is accepted that the amount of cigarette consumption is increasing in developing countries compared to the developed ones.

Apprenticeship in small workshops is very common in developing countries. The International Labor Office (ILO) believes that children account for up to 11% of the workforce in some countries in Asia, up to 17% in Africa, and up to a quarter in Latin America. Children are the most easily exploited of all workers, and in the developing world, their numbers are ever increasing. The worldwide population of children under fourteen who work full-time is thought to exceed 200 million. (LaDou, 1996).When the long latency period in chemical carcinogenicity is taken into consideration, being exposed to chemical carcinogens starting at an early age can be particularly riskful.

The risk assessment and management of chemicals is the rational way in chemical usage. However, this approach is new even for developed countries and has not seriously entered the agenda in developing countries yet.

These type of cost effective occupational biomonitoring assays as presented here which can be used as indicators signaling genotoxic risk, also encourages the regulatory bodies in developing countries to take preventive actions against carcinogenic risk at workplaces.

REFERENCES

S.Abe and M.Sasaki (1982) SCE as an index of mutagenesis and /or carcinogenesis. In:A.Sandberg (Ed.),Sister Chromatid Exchange.Alan R.Liss.Inc.,New York, pp.461–464.

N.Aygün, S.Sardas, Y.Ünal, M.Gamli, N.Berk and A.E.Karakaya (1997) *Assessment of induced DNA damage by anaesthetic gases in operating room personnel by single-cell gel electrophoresis technique.*2nd National Congress of Toxicology, April 3–6,Antalya,Turkey.Abs.P-76.

J.M.Baden and V.F.Simmon (1980) Mutagenic effects of inhalation anaesthetics. *Mutation Res.*, **75**,169.

A.Bayhan, S.Burgaz and A.E.Karakaya (1987) Urinary thioether excretion in nurses at an oncologic department. *J.Clin.Pharm.Therap.*,**12**,303–306.

S.Burgaz, A.Bayhan and A.E.Karakaya (1987) Urinary thioethers in cigarette smokers. *J.Fac.Pharm.Gazi*, **4**,63–67.

S.Burgaz, R.Rezanko, S.Kara and A.E.Karakaya (1992) Thioethers in urine of sterilization personnel exposed to ethylene oxide, *J.Clin Pharm. Therap.*,**17**,169–172.

S.Burgaz, P.J.A.Borm and F.J.Jongeneelen (1992) Evaluation of urinary excretion of 1-hydroxypyrene and thioethers in workers exposed to bitumen fumes, *Int.Arch.Occup. Environ. Health*, **63**, 397–401.

S.Burgaz, A.Iscan, Z.K.Büyükbingöl, A.Bozkurt and A.E.Karakaya (1995) Evaluation of micronuclei in exfoliated urothelial cells and urinary thioether excretion of smokers, *Mutation Res.*, **335**,163–169.

S. Burgaz, O. Erdem, B. Karahalil, A.E. Karakaya (1998) Cytogenetic biomonitoring of workers exposed to bitumen fumes. *Mutation Res.*, in press.

C.Burnett (1980) Evaluation of toxicity and carcinogenecity of hair dyes. *J. Toxicol. Environ. Health*, **6**, 247–257.

L.P.Carrillo, L.T. Arreola, L.T.Sanchez, F.E.Torres, C. Jimenez, M.Cabrian, S.Waliszewski and O. Saldate (1996) Is DDT use a public health problem in Mexico *Environ. Health. Perspect.*,**104**, 584–588.

E.DeRosa, N.Cellini, G.Sessa, C.Saletti, G.Rausa, G.Marcuzzo and G.B.Bartolucci (1993) Biological monitoring of workers exposed to styrene and acetone. *Int.Arch.Occup.Environ.Health*, **65**,107–110.

C.T. DeRosa, Y.W. Stevens, J.W. Wilson, A.A. Ademeyero, S.D. Buchanan, W.Cibulas, P.J. Hughes, M.M. Mumtaz, R.E. Neft, H.R. Pohl and M.M. Johnson (1993) The agency for toxic substances and disease registry's role in development and application of biomarkers in public health service. *Toxicol. Industr. Health*, **9**, 979–994.

R. Doll and R. Peto (1981) The causes of cancer. Oxford University Press, Oxford.

M.Fenech and A.A.Morley (1985) Measurement of micronuclei in lymphocytes. *Mutation Res.*, **147**,29–36.

T.R. Fennell (1990) Biological markers of exposure to chemical carcinogens. *CIIT Activities* ,**10**(1), 1–7.

M.L. Fischman, E. C. Cadman and S. Desmond (1990) *Occupational cancer*. In Occupational Medicine (Ed. J.LaDou), Prentice-Hall, Connecticut, pp. 182–208.

J.A.Heddle (1973) A rapid in vivo test for chromosomal damage. *Mutation Res.*, **18**,187–190.

J.A. Heddle, M.C. Cimino, M. Hayashi, F. Romagna, M.D., Shelby, J.D. Tucker, Ph.Vanparys and J.T. Macgregor (1991) Micronuclei as an index of cytogenetic damage: Past, present, and future. *Environ Mol. Mutagen.* **18**, 277–291.

R.F. Henderson, W.E. Bechtold, J.A. Bond and J.D.Sun (1989) The use of biological markers in toxicology. CRC Critical Rev. *Toxicol.*, **20**, 65–82.

H.İlbars, A.E.Karakaya and S.Burgaz (1997) *Evaluation of possible genotoxic exposure of professional hair colorists by micronucleus assay.* 2nd National Congress of Toxicology, April 3–6, Antalya, Turkey.Abs.P-83.

International Agency for Research on Cancer (1983) IARC Monographs, Polynuclear Aromatic Compounds. Part 1. Chemical, environmental and experimental data.Vol.32,Lyon.

International Agency for Research on Cancer (1985) IARC Monographs, Polynuclear compounds,Bitumens,Coaltars and Derived Products, Shale oils and Soots. Vol.35.Lyon.

International Agency for Research on Cancer (1987) IARC Monographs on the Evaluation of Carcinogenic Risks to Humans. Overall Evaluations of Carcinogenicity. IARC Publication Supplement 7, Lyon.

International Labour Organisation (1988) Occupational Cancer:Prevention and Control. Occupational Safety and Health Series No. 39, Geneva.

F.J.Jongeneelen, R.B.M.Anzion, P.Th.Henderson (1987) Determination of hydroxylated metabolites of polycyclic aromatic hydrocarbons in urine. *J.Chromatog.*, **413**, 227–232.

B.Karahalil, S.Burgaz, G.Fişek and A.E.Karakaya (1998) Biological monitoring of young workers exposed to polycyclic aromatic hydrocarbons in engine repair workshops. *Mutation Res.*, **412**, 261–269.

A.E.Karakaya, B.Karahalil, M.Yýlmazer, N.Aygün, S.Sardas and S.Burgaz (1997) Evaluation of genotoxic potential of styrene in furniture workers using unsaturated polyester resins. *Mutation Res.*, **392**, 261–268.

La Dou (1996) The role of multinationale corporations in providing occupational health and safety in developing countries. Int.Arch.Occup.Environ. *Health*, **68**,363–366.

National Institute of Health(1992) Human health and the environment some research needs. NIH Publication No. 92–3344.

O.Östling and K.J. Johanson (1984) Microelectrophoretic study of radiation-induced DNA damages in individual cells. Biochem. Biophys. *Res. Commun.*, **123**, 291–293.

F. Perera (1993) Biomarkers and molecular epidemiology of occupational related cancer. *J. Toxicol. Environ. Health*, **40**, 203–215.

F. Perera and R.M. Whyatt (1994) Biomarkers and molecular epidemiology in mutation/cancer research. *Mutation Res.*, **313**, 117–129.

P.Perry and H.J.Evans (1975) Cytological detection of mutagen-carcinogen exposure by SCE. *Nature*, **258**,121–125.

H.C. Pitot and Y.P. Dragan (1996) Chemical carcinogenesis. In Casarett and Doull's Toxicology. Fifth Edition (Ed C.D. Klaassen), McGraw-Hill, New York, pp.201–267.

S.Şardaş, S.Gök and A.E.Karakaya (1991) Increased frequency of sister chromatid exchanges in the peripheral lymphocytes of cigarette smokers. *Toxic. In vitro*, **5**,263–265.

S.Şardaş, S. Gök and A.E. Karakaya (1991) Sister chromatid exchanges in lymphocytes of nurses handling antineoplastic drugs. *Toxicol. Lett.*, **55**, 331–335.

S.Şardaş, H.Cuhruk, A.E.Karakaya and Y.Atakurt (1992) Sister chromatid exchanges in operating room personnel, *Mutation Res.*, **279**, 117–120.

S.Şardaş, A.E. Karakaya and Y. Furtun (1994) Sister chromatid exchanges in workers employed in car painting workshops. Int. Arch. Occup. *Environ. Health*, **66,** 33–35.

S.Şardaş, N.Aygün and A.E.Karakaya (1997) Genotoxicity studies on professional hair colorists exposed to oxidation hair dyes. *Mutation Res.*, **394**, 153–161.

N.P. Singh, M.T.McCoy, R.R..Tice and E.L.Schneider (1988) A simple technique for quantitation of low levels of DNA damage in individual cells. Exp. *Cell Res.*, **17,** 184–191.

M. Sorsa, K. Hemminki and H. Vainio (1985) Occupational exposure to anticancer drugs-Potential and real hazards. *Mutation Res.*, **154**, 135–149.

H.F. Stich and M.P. Rosin (1984) Micronuclei in exfoliated human cells as a tool for studies in cancer risk and cancer intervention. *Cancer Lett.*, **22**,241–253.

H. Vainio (1995) Carcinogenesis and its prevention. In Occupational Toxicology (ed. N.H. Stacey), Taylor&Francis, London, pp. 149–162.

E.Ward (1995) Overview of preventable industrial causes of occupational cancer. *Environ. Health Perspect.*, **103,** 197–203.

J.B. Ward and R.E. Henderson (1996) Identification of needs in biomarker research. *Environ. Health. Perspect.*, **104**, 895–900.

World Health Organization (1985) Environmental Health Criteria for Ethylene Oxide. Vol: 55,Geneva.

16

STUDIES ON OXIDATIVE MECHANISMS OF METAL-INDUCED CARCINOGENESIS

Recent Developments

Kazimierz S. Kasprzak, Wojciech Bal, Dale W. Porter, and Karol Bialkowski

Laboratory of Comparative Carcinogenesis
National Cancer Institute, FCRDC
Frederick, Maryland 21702-1201

1. ABSTRACT

Two ways by which carcinogenic metals, such as Ni(II), Co(II), Cu(II), or Cd(II), may promote oxidative DNA damage, including direct effects consisting of activation of oxygen species and mediation of their attack on DNA, and indirect effects through suppression of cellular antimutagenic defenses, are discussed. The mechanisms of the direct attack may involve chelation of a metal by nuclear proteins, especially the histones and protamines, and activation of metabolic oxygen species by the resulting metal complexes at close proximity to DNA. We found that human protamine HP2 has a typical binding motif for Ni(II) and Cu(II), Arg-Thr-His-, at its N-terminus. A synthetic pentadecapeptide modeling this terminus formed strong chelates with these metals and, in addition, enhanced oxidative DNA damage by Ni(II) plus H_2O_2, but suppressed, though not completely, the damage by Cu(II) plus H_2O_2. Since protamines carry DNA in the sperm, the observed DNA damage may have spermicidal or transgenerational carcinogenic effects in man exposed to metals, as observed epidemiologically. The indirect effects of metals on DNA may involve inhibition of 8-oxo-dGTPases, a class of enzymes preventing incorporation of the 8-oxoguanine lesion from oxidatively-damaged deoxynucleotide pool into DNA. Cd(II) and Cu(II), and to a limited extent also Ni(II) and Co(II), were found to *in vitro* inhibit the enzymatic activity of a bacterial (MutT) and human (MTH1) 8-oxo-dGTPases. This may allow redox-inactive metals, such as Cd(II), to introduce the promutagenic 8-oxoguanine lesion from endogenously damaged 8-oxo-dGTP into DNA.

2. INTRODUCTION

Over the last several years, our research effort has been focused on the mechanisms of interaction of nickel and some other transition metals, carcinogenic to humans, with

Advances in DNA Damage and Repair, edited by Dizdaroglu and Karakaya. Kluwer Academic / Plenum Publishers, New York, 1999.

DNA and nuclear proteins. Special attention has been given to metal-mediated oxidative damage to those molecules, resulting in genotoxic and epigenetic effects (reviewed in Kasprzak 1995 a, b; 1996; 1997; Bal and Kasprzak 1997). In particular, we have been continuing studies on two major ways by which the metals may promote the occurrence of oxidatively modified DNA bases, many of which are mutagenic. These are: (a) mediation by transition metals of the generation of active oxygen species and their attack on DNA bases, and (b) inhibition by the metals of cellular antioxidant defenses.

Thus far, wide spectra of promutagenic oxidative DNA base damage were observed in Ni(II)- and Co(II)-treated animals (Kasprzak 1995 a, b; 1996; Kasprzak et al. 1990; 1992; 1994; 1997). In addition, a tissue-specific response to Ni(II) was revealed in rats, with apparently faster DNA base damage repair in liver than in kidney, the main target of Ni(II) carcinogenicity (Kasprzak et al. 1997). *In vitro*, several natural ligands, cysteine, glutathione, histidine, carnosine, homocarnosine, and anserine, were found to facilitate Ni(II) and Co(II) reactions with ambient O_2, H_2O_2, and/or lipid hydroperoxides and produce O-, C-, and S-centered DNA-damaging radicals at physiological pH (reviewed in Kasprzak 1995 a, b; 1996). We have also noticed that oxidative DNA damage in the presence of Ni(II) was promoted by nuclear proteins, most likely the histones (Kasprzak and Bare 1989; Nackerdien et al. 1991). The search for the role of histones in producing that effect resulted in the finding of amino acid binding motifs for Ni(II) in core histones H3 and H2A. These motifs have been investigated with the use of peptide models. The results for one such model for histone H3, acetyl-Cys-Ala-Ile-His-amide (CAIH), were discussed previously (Bal et al. 1995; 1996 a; Bal and Kasprzak 1997). Similar investigations on the histone H2A binding site, acetyl-Thr-Glu-Ser-His-His-Lys-amide (TESHHK), are under way. Most recently, we have found that not only the core histones, but also another class of DNA carrier proteins, the protamines, can offer a plethora of potential metal binding sites. The most important of them is, perhaps, the N-terminal Arg-Thr-His- motif of human and mouse protamine HP2, typical of other Ni(II) and Cu(II) carrier proteins and peptides (Sarkar 1997). Our investigations of the protamine models (Bal et al. 1997 a, b) are summarized and discussed in the present review.

Besides being generated directly in DNA, the damaged nucleobases found in metal-treated animals, predominantly 8-oxo-7,8-dihydroguanine (8-oxo-Gua), may be incorporated into DNA from the deoxynucleotide pool. In healthy cells, such incorporation is prevented by the "sanitizing" enzymes, including 8-oxo-dGTPases (Bessman et al. 1996). We hypothesized that the metals, while promoting oxidative damage to DNA and free nucleotides, could also inhibit 8-oxo-dGTPases. To test the latter possibility, we developed an HPLC assay for determination of the activity of these enzymes (Porter et al. 1996) and investigated the *in vitro* effects of several carcinogenic metals, such as Ni(II), Co(II), Cd(II), and Cu(II), on bacterial (MutT) and human (MTH1) 8-oxo-dGTPases (Porter et al. 1996; 1997). The results, presented in detail elsewhere (Porter et al. 1997), are reviewed below and discussed along with some novel, yet unreported, data from our laboratory.

3. POSSIBLE ROLE OF PROTAMINES IN METAL-MEDIATED OXIDATIVE DAMAGE IN THE SPERM. RELEVANCE TO TRANSGENERATIONAL CARCINOGENESIS

3.1. Rationale

As mentioned above, Ni(II) compounds are carcinogenic (IARC 1990), acting, at least in part, through oxidative damage mechanisms (Kasprzak 1995 a, b; 1996; 1997). Also, evidence is emerging that copper may be a powerful mutagen and an oxidative dam-

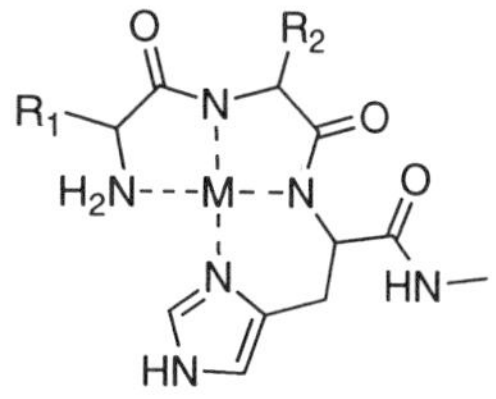

Figure 1. Structure of the four-nitrogen complexes of X-X-His- peptides: M, Cu(II) or Ni(II); R_1, R_2, side chains of amino acids at positions 1 and 2 (Camerman et al. 1967; Bal et al. 1994; 1997 a).

age catalyst *in vivo* (Aruoma et al. 1991; Tkeshelashvili et al. 1991). One of the major ways of exposure is inhalation of nickel-, copper-, and other metal-containing fumes and dusts, e.g., by welders and machinists (IARC 1990). Epidemiology indicates that pre-conceptional paternal exposure of this kind may contribute to higher incidence of cancer in the progeny (Bunin et al. 1992; Anderson et al. 1994; Buckley 1994). Metal-related sperm DNA damage seems to be a possibility for an explanation of this phenomenon.

In the mammalian sperm, the histones in chromatin are replaced by protamines that results in tighter compaction as well as suppression of transcription, replication, and the repair processes of DNA (Chandley and Kofman-Alfaro 1971; Sega 1974; Pogany et al. 1981; McKay et al. 1986). Most importantly, protamines might also be involved in the mechanisms of sperm damage by metals, e.g., nickel (Xie et al. 1995), owing to chelation of a metal cation close to DNA and facilitation of its redox activity that may eventually lead to pathogenic effects. The primary target for metals would be protamine HP2. HP2 is a 57-amino acid peptide that constitutes ca. 50–70% of human protamines (Ammer et al. 1986; Arkhis et al. 1991) and has N-terminal motif, Arg-Thr-His-, of the X-X-His- type found in various Cu(II)- and Ni(II)- binding proteins, including serum albumins (Camerman et al. 1967; Peters and Blumenstock 1967; Minamino et al. 1984; Hilgenfeldt 1988; Predki et al. 1992; Del Rio and De la Fuente 1994; Bal et al. 1994) (Figure 1). The above facts encouraged us to investigate the coordinative properties of HP2 toward Ni(II) and Cu(II) and, subsequently, DNA damage in the presence of the resulting complexes (Bal et al. 1997a,b).

A thorough investigation of metal binding to the whole HP2 molecule with its 9 histidine and 5 cysteine residues (Ammer et al. 1986) would be difficult. We therefore targeted the N-terminal sequence of this protamine represented by two synthetic peptides, RTHGQ-NH_2 (HP2$_{1-5}$) and RTHGQ-SHYRR-RHCSR-NH_2 (HP2$_{1-15}$). A third peptide, Ac-RHCSR-NH_2 (HP2$_{10-15}$) was also synthesized to test possible interference of the -SH involvement in metal binding by the HP2$_{1-15}$ peptide that emerged during experiments (Bal et al. 1997 a, b).

The methodology of our investigations has been described in detail elsewhere (Bal et al. 1997 a, b). Shortly, the protonation constants of the peptides, the stability constants of their Cu(II) and Ni(II) complexes, and the speciation of the complexes at pH 3 - 11, were determined using pH-metric titrations and the UV-vis and circular dichroism (CD) spectroscopies.

The oxidation-mediating capacity of the resulting metal complexes with HP2$_{1-15}$ was studied for H_2O_2 versus the guanine residues in 2'-deoxyguanosine (dG) and calf thymus DNA (Bal et al. 1997 b). The formation of 8-oxo-2'-deoxyguanosine (8-oxo-dG) as well as DNA strand scission was observed with the use of HPLC with electrochemical detection and agarose gel electrophoresis, respectively. To detect possible oxidative damage to HP2$_{1-15}$ during the DNA oxidation reactions, the reaction mixtures were also subjected to quantitative amino acid analyses, including an assay for 2-oxohistidine according to Lewisch and Levine (1995).

Table 1. Comparison of log β and log *K values for selected complexes of X-X-His-peptides with Cu(II) and Ni(II)[a]

	Cu(II)		Ni(II)	
Peptide	log β	log *K[b]	log β	log*K[b]
Gly-Gly-His[c]	-1.73	-16.43	-6.93	-21.81
Gly-Gly-His[d]	-1.55	-16.33	—	—
Gly-Gly-hist[e]	-2.48	-17.14	-7.9	-22.65
Val-Ile-His-Asn[f]	-	-	-5.39	-19.75
$HP2_{1\text{-}5}$[a]	-1.11	-14.24	-5.95	-19.23
$HP2_{1\text{-}15}$[a,g]	-0.96	-13.13	-5.95	-19.29

[a] From Bal et al. 1997 a

[b] log *K = log β ($MH_{n\text{-}j}L$)- log β (H_nL); M, metal; H, proton; L, ligand

[c] Hay et al. 1993

[d] Farkas et al. 1984

[e] Gajda et al. 1996; hist stands for histamine

[f] Bal et al. 1996 b

[g] CuH_2L and NiHL complexes, respectively

3.2. Results and Discussion

The results of pH-metric titrations and the analysis of UV/vis and CD spectra allowed for determination of the protonation and metal complex stability constants and the assignment of pK_a values to specific amino acid residues (Bal et al. 1997 a). As derived from these data, the complexes of either metal with $HP2_{1\text{–}15}$ predominating at physiological pH were neutral (doubly-deprotonated ligand), with charged complexes (one or more than two deprotonated nitrogen donors) constituting < 20% admixture; all of 1:1 metal:peptide stoichiometry. The data obtained for all three peptide complexes up to 1:1 metal:peptide molar ratios indicated that the $HP2_{10\text{–}15}$ amino acid motif did not participate in the formation of metal complexes in the complete $HP2_{1\text{–}15}$ peptide. Hence, in a biologically feasible situation where the excess of protamine over toxic metal is assured, there will be no binding to the -His[12]-Cys[13]- motif in HP2.

The protonation-corrected stability constant, log*K, obtained for the Ni(II)-$HP2_{1\text{–}15}$ complex was very high, equalling -19.29. Likewise, high was the stability constant found for the Cu(II) complex, log*K = -13.13. The comparison of *K constants, describing competition of the metal ions with protons for the peptide binding, $M(II) + H_nL \rightarrow M(II)H_{n\text{-}j}L + j\,H^+$ [where M(II) stands for the metal cation and H_nL is the fully protonated ligand], allows to conclude that stabilities of both Cu(II) and Ni(II) complexes are 2 - 3 orders of magnitude higher than those of the "generic" Gly-Gly-His- species (Table 1). The Ni(II) complexes of HP2 peptides are even more stable than the corresponding complex of Val-Ile-His-Asn, the most stable Ni(II)-peptide complex within the X-X-His- family known previously (Bal et al. 1996 b). This means that the protamine metal binding motif has the potential to sequester Ni(II) from other natural peptide and protein carriers, including albumin, and points at HP2 as a very likely target molecule for toxic metals Ni(II) and Cu(II).

The CD spectral pattern observed for free $HP2_{1\text{–}15}$ was typical for unordered peptides (Kortemme and Creighton 1995, Woody, 1994). Complexation of $HP2_{1\text{–}15}$ with Cu(II), together with partial deprotonation of non-binding histidine residues, induced a CD pattern strikingly resembling a mixture of α helix and parallel β sheet (Woody 1994; Perczel et al. 1992). A similar, but less pronounced change also resulted from Ni(II) coordination to $HP2_{1\text{–}15}$. These results indicate that both N-terminally bound Cu(II) and Ni(II) introduce partial ordering, possibly short stretches of α helix located in different parts of the $HP2_{1\text{–}15}$

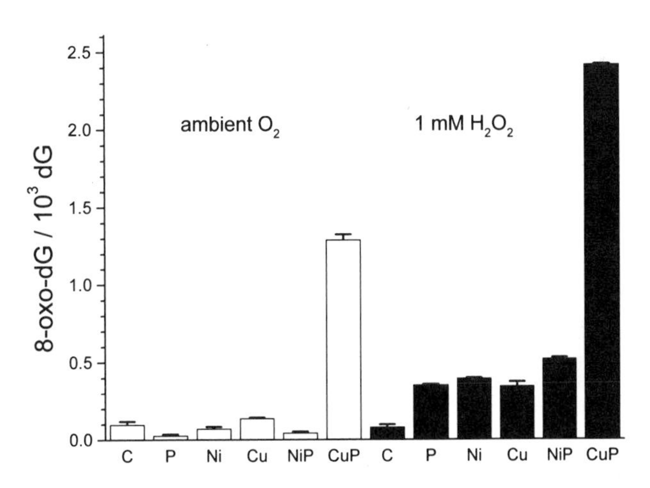

Figure 2. Formation of 8-oxo-dG from dG (0.1 mM) in the presence of combinations of $HP2_{1-15}$, Ni(II), and Cu(II) and/or 1mM H_2O_2 after 24 h incubations at 37°C in 100 mM phosphate buffer, pH 7.4. Means of two determinations ± range. C, control; P, 0.1 mM $HP2_{1-15}$; Ni, 0.1 mM $NiCl_2$; Cu, 0.1 mM $CuCl_2$; NiP, 0.1 mM $NiCl_2$ + 0.1 mM $HP2_{1-15}$; CuP, 0.1 mM $CuCl_2$ + 0.1 mM $HP2_{1-15}$ (Bal et al. 1997 b).

molecule. The formation of a β sheet would require peptide aggregation, which is improbable for a peptide carrying an electronic charge of +6 or more, and was in fact not observed. It is interesting to notice that the Cu(II)- and Ni(II)-imposed changes in CD spectra of HP_{1-15} are similar to those produced in the whole HP2 by Zn(II) (Gatewood et al. 1990), interpreted as formation of multiple β turns. Further studies are necessary to test whether there is a common phenomenon behind these spectral effects and how they affect the HP2-DNA interaction. Thus far, our studies revealed that Cu(II) and Ni(II) could increase the amount of DNA precipitated from solution by $HP2_{1-15}$.

The Cu(II)-$HP2_{1-15}$ complex was also found to be an effective promoter of the formation of 8-oxo-dG from both dG and DNA with ambient O_2 and H_2O_2 (Bal et al. 1997 b). The Ni(II)-$HP2_{1-15}$ complex was ineffective with O_2 versus 8-oxo-dG production from both substrates, but markedly enhanced generation of 8-oxo-dG by H_2O_2 in dG and DNA as compared with that in free (untreated) substrates (Figures 2 and 3). However, when compared with the results for DNA treated with H_2O_2 + uncomplexed metals, the effect of complexation has to be viewed as negligible for Ni(II) and strongly inhibitory for Cu(II) (Figure 3). Likewise, both Cu(II)- and Ni(II)-$HP2_{1-15}$ promoted strand scission by H_2O_2 of

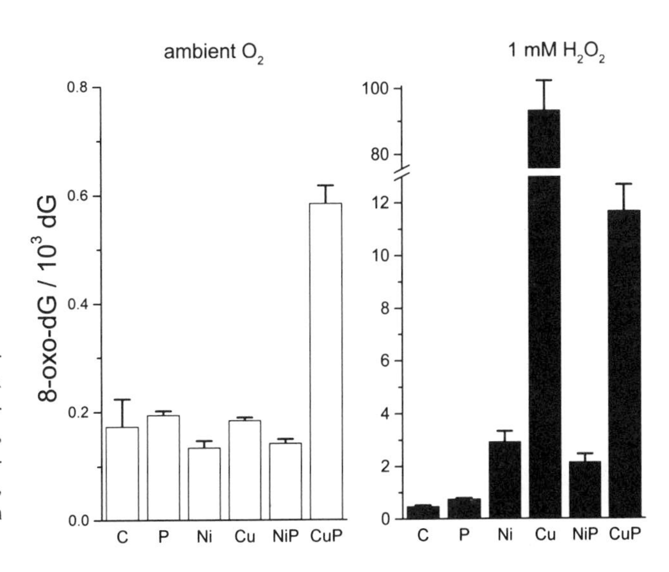

Figure 3. Formation of 8-oxoguanine (quantified as 8-oxo-dG) in DNA (0.5 mg/mL) in the presence of combinations of $HP2_{1-15}$, Ni(II), Cu(II), and/or H_2O_2. Means of two determinations ± range. Other conditions, concentrations, and symbols as in Figure 2 (Bal et al. 1997 b).

Table 2. Relative changes in Arg, His, Tyr and Asp contents in $HP2_{1-15}$ after treatment with Cu(II), Ni(II), and H_2O_2 [a, b]

Treatment	Arg[c]	His[c]	Tyr[c]	Asp[d]
None	100.0 ± 5.4	100.0 ± 1.4	100.0 ± 1.9	0.5 ± 0.5
Cu(II)	96.7 ± 1.7	94.6 ± 2.6	84.1 ± 11.1	3.1 ± 0.6
Ni(II)	102.6 ± 2.4	96.0 ± 6.9	96.0 ± 4.0	0.5 ± 0.5
H_2O_2	98.6 ± 3.4	95.2 ± 2.8	99.6 ± 4.0	0.8 ± 0.8
Cu(II) + H_2O_2	81.2 ± 1.2	50.8 ± 6.8	47.8 ± 14.8	21.7 ± 5.4
Ni(II) + H_2O_2	83.3 ± 1.7	51.1 ± 18.6	18.7 ± 4.3	9.5 ± 3.1

[a] From Bal et al. 1997 b.
[b] Samples were incubated for 16 h at 37^0C in 10 mM phosphate buffer, pH 7.4; concentrations: 0.1 mM $HP2_{1-15}$, Cu(II), or Ni(II); 1 mM H_2O_2. The contents of the remaining amino acid residues in $HP2_{1-15}$, i.e., Ser, Thr, Glu, and Gly were not affected by the incubations. Cys was not determined.
[c] In percentages of the respective amino acid content in $HP2_{1-15}$ incubated alone. The amino acid analysis of untreated $HP2_{1-15}$ yielded the following composition (based on the presence of 1Gly; means of two determinations ± range): 5.17 ± 0.28 Arg, 2.86 ± 0.04 His, 1.69 ± 0.04 Ser, 0.94 ± 0.01 Thr, 1.04 ± 0.02 Tyr, and 1.10 ± 0.02 Glu.
[d] In percentages of the Gly residue in a given incubation mixture.

calf thymus DNA; significantly more so for Ni(II)-$HP2_{1-15}$ compared with the free metal, but to a greatly lesser extent for Cu(II)-$HP2_{1-15}$ in comparison with free Cu(II) (Bal et al. 1997 b). Thus, chelation with $HP2_{1-15}$ had an opposite effect on the DNA oxidation mediating activity of the two metals.

All the above variations indicate different mechanisms of DNA damage, depending on the oxidant and the mediating agents. The present results do not allow for drawing any definite conclusions as to those mechanisms. Judging by the limited effect of Cu(II) alone on the oxidation of free dG, which does not form strong complexes with Cu(II) (Sigel 1993) versus a very strong effect in DNA, where Cu(II) is chelated by the bases (Tajmir-Riahi et al. 1993), metal-associated oxidants were more likely to predominate over "free" (diffusible) oxidants. Generation of oxidizing species requires redox cycling of the metal ions. Participation of Cu(I) in Cu(II)-catalyzed oxidation by H_2O_2 has been well documented. Cu(I) peroxide was proposed as a major intermediate (Gunther et al. 1995). In particular, Cu(II)-Gly-Gly-His complex was found to facilitate cleavage of DNA by H_2O_2 but only in the presence of ascorbate (Chiou 1983), thus suggesting that reduction of Cu(II) to Cu(I) could be involved in the mechanism of the cleavage (Mack et al. 1988; Harford et al. 1996; Chiou 1983). However, our observations are more consistent with the Cu(II)/Cu(III) redox couple in copper-Gly-Gly-His complex, as reported by others (McDonald et al. 1995). Involvement of transient Ni(III) in oxidations mediated by Ni(II) complexes of X-X-His peptides is firmly established (Nieboer et al. 1984; Bal et al. 1993, 1994), but the details of reaction mechanisms and the presence of free or metal-associated reactive oxygen species is disputed (Inoue and Kawanishi 1989; Torreilles et al. 1990; Torreilles and Guerin 1990; Nieboer et al. 1989; Cotelle et al. 1992). Poor dG oxidation by the Ni(II) complex of $HP2_{1-15}$ seems to point against diffusible oxidants being generated by this complex under the present experimental conditions.

The promotion by the metal-$HP2_{1-15}$ complexes of dG and DNA oxidation with H_2O_2 was accompanied in our investigation by oxidative damage to the complexes themselves, consisting of decreasing contents of their His and especially Tyr residues, as well as appearance of aspartic acid, the known oxidation product of His residues in peptides and proteins (Table 2) (Bal et al. 1997 b). Generation of 2-oxo-His, yet another marker of protein oxidation (Lewisch and Levine 1995; Uchida and Kawakishi 1993), was not observed. Under the same conditions, the incubation of $HP2_{1-15}$ alone with H_2O_2 had no effect whatsoever on the amino acid contents. Since all amino acids are susceptible to oxidation by "free" oxygen radicals

(Stadtman 1993), our findings might signify a mostly "site specific" generation of oxidants at the metal-binding site of the complexes. Nonetheless, the Arg and Thr residues of $HP2_{1-15}$, also located at the metal-binding site, were not markedly damaged (except, perhaps, for Arg). The difference might result from different susceptibility of the three amino acids to oxygen attack (Stadtman 1993). The demise of the Tyr residue, located away from the metal-binding site, provides, however, an argument for molecular electron transfer that could also be engaged in the mechanisms of mediation of the oxidative DNA damage by those complexes. The nature of this oxidation remains to be unveiled.

3.3. Biological Significance

The results of the present study show that the binding affinity of Ni(II) and Cu(II) for the N-terminal motif of HP2 is at least as high as that for serum albumin, the major metal carrier in blood. Therefore HP2 may be an important target for toxicity of both metals. One possible pathogenic effect of Ni(II) and Cu(II) on HP2 would be partial ordering of its molecule that may affect proper assembly of HP2 with DNA. Another one includes mediation by the bound metals of oxidative damage to both HP2 and associated DNA. Mature spermatids, loaded with mitochondria, have a very high rate of oxygen metabolism. Hydrogen peroxide is an inadvertent byproduct of such metabolism, and is known to leak from mitochondria (Sohal and Dubey 1994). A recent study showed that copper-zinc superoxide dismutase (SOD-1) is extensively expressed in mouse testis, and specifically in post-meiotic germ cells (Gu et al. 1995), as part of the protection of developing spermatids from oxidative injury. SOD-1 has been shown to be susceptible to oxidative damage at the copper site, and to lose copper in the process (Yim et al. 1990). This is just one mechanism that may make Cu(II) available inside a spermatid. Our studies (Bal et al. 1997 a, b), show that Cu(II) may be rapidly and strongly bound at the N-terminal motif of human protamine HP2, that the resulting complex can damage DNA even with ambient O_2, and that the presence of H_2O_2 highly potentiates the damage. We also presented evidence that the N-terminal Cu(II) binding may affect the DNA-protamine interaction. Taken together, our results provide a mechanism of sperm DNA damage, with all components already present in the sperm cell, even without additional exposure to copper.

Ni(II) is toxic and carcinogenic. HP2 may be an important target for Ni(II) toxicity. Indeed, as shown recently, intraperitoneal exposure of male mice to $NiCl_2$ resulted in transient accumulation of nickel in the testes, followed by testicular damage and loss of fertility (Xie et al. 1995). Reduced sperm counts and increased incidence of sperm chromosome aberrations were also observed (Li and Wang 1989). Additionally, nickel treatment slightly increased copper levels in the testes (Xie et al. 1995), thus providing another way of Cu(II) attack on protamine. Mouse protamine P2 shares the X-X-His- motif with HP2 (Chauviere et al. 1992), and so their binding properties should be similar. The present results indicate that Ni(II) binding to HP2 will likely result in DNA damage, and therefore offer a molecular mechanism that may be responsible for some toxic effects of Ni(II) in the testes and, eventually, transgenerational carcinogenesis (Anderson et al. 1997).

4. INHIBITION BY CARCINOGENIC METALS OF ANTIMUTAGENIC ENZYMES, 8-OXO-dGTPases

4.1. Rationale

The presence of 8-oxo-Gua in genomic DNA has become the focus of intensive studies because of its promutagenic properties (Grollman and Moriya 1993; Shibutani et al

1991; Cheng et al. 1992; Pavlov et al. 1994). Elevated levels of this damaged base have been observed in organs of rats exposed to Ni(II) and Co(II) (Kasprzak et al. 1990; 1992; 1994; 1997), or accumulating excessive amounts of Cu(II) (Yamamoto et al. 1993). Certain biocomplexes of these metals, including those with histones and protamines discussed above, are known to activate oxygen species under physiological conditions, and enhance oxidative damage to DNA and other molecules (Spear and Aust 1995; Kasprzak 1995 a). In addition, such damage may be sustained through inhibition by metals of antioxidant and DNA repair mechanisms (Kasprzak 1995 a; 1996; 1997; Stohs and Bagchi 1995; Rodriguez et al. 1996; Misra et al. 1991; Hartwig 1994). These "inhibitory pathways" might explain why Cd(II), which is also a potent carcinogen (IARC 1993) but not a redox active metal under physiological conditions, mediates oxidative DNA damage (Yang et al. 1996), including generation of 8-oxoguanine (Mikhailova et al. 1997) and inhibition of its repair (Hirano et al. 1997).

It is presumed that 8-oxoguanine can arise in DNA due to a direct attack of active oxygen species on chromatin, or through incorporation from an oxidatively damaged nucleotide pool, namely, from 8-oxo-7,8-dihydro-2'-deoxyguanosine-5'-triphosphate (8-oxo-dGTP), during DNA synthesis (Shibutani et al 1991; Cheng et al. 1992; Pavlov et al. 1994; Maki and Sekiguchi 1992; Mo et al. 1992). A class of enzymes, designated MutT in *Escherichia coli* (Treffers et al. 1954) and MTH1 in humans (Furuichi et al. 1994), was found to prevent such incorporation (Maki and Sekiguchi 1992; Mo et al. 1992; Grollman and Moriya 1993; Sekiguchi 1996). MutT and MTH1 act as 8-oxo-7,8-dihydro-2'-deoxyguanosine-5'-triphosphatases (8-oxo-dGTPases) hydrolyzing 8-oxo-dGTP to 8-oxo-7,8-dihydro-2'-deoxyguanosine-5'-monophosphate (8-oxo-dGMP) (Maki and Sekiguchi 1992; Mo et al. 1992). The latter can no longer be incorporated into DNA or phosphorylated back to 8-oxo-dGTP (Hayakawa et al. 1995). These enzymes are activated by Mg(II) which can make them sensitive to inhibition by other metals. If so, we hypothesized that inhibition of mammalian 8-oxo-dGTPases might, at least in part, account for the increased 8-oxoguanine levels found in DNA exposed *in vivo* to Ni(II), Co(II), Cd(II), or Cu(II) (Kasprzak et al. 1990; 1992; 1994; 1997; Mikhailova et al. 1997; Yamamoto et al. 1993). In order to test this hypothesis, effects of the above metals on the *in vitro* enzymatic activity of 8-oxo-dGTPases of two different species, bacterial MutT and human MTH1, were determined.

The methodology of investigations based on a HPLC assay of the enzyme kinetics developed earlier (Porter et al. 1996) has been described elsewhere (Porter et al. 1997). The syntheses of the reference compounds, 8-oxo-dGTP and 8-oxo-dGMP, as well as isolation and purification of the enzymes, were also published (Porter et al. 1996). The effects of Ni(II), Co(II), Cu(II), and Cd(II) chlorides on enzymatic activity of MTH1 and MutT were tested in the presence of 8 mM Mg(II) (assuring maximum enzymatic activity at pH 7.4) plus variable concentrations of the other metals. The metal concentrations that suppressed enzymatic activity of either enzyme to 50% of its original value (IC_{50}) were determined using the fractional inhibition method (Chou 1976). To test if the carcinogenic metals could substitute for Mg(II) as the metal cofactor, the MTH1 and MutT kinetic assays were also conducted in the absence of Mg(II). In control assays, the enzymes were omitted to determine the rate of 8-oxo-dGTP hydrolysis catalyzed by the metal alone.

4.2. Results and Discussion

The initial velocities of 8-oxo-dGTP hydrolysis by MTH1 and MutT as a function of the free Mg(II) concentration were fitted to the Michaelis-Menten equation. The results re-

Table 3. Estimates of $K_{m\ app}$ and $V_{max\ app}$ of MTH1 and MutT for Mg(II)[a,b]

Enzyme	$K_{m\ app}$ mM Mg(II)	$V_{max\ app}$ pmoles 8-oxo-dGMP produced/μg protein/min
MTH1	0.72 ± 0.16	575 ± 158
MutT	0.50 ± 0.09	1,187 ± 157

[a] From Porter et al. 1997
[b] Values represent means ± SE determined from three experiments.

vealed that both enzymes were activated by Mg(II), which is consistent with previous reports (Mo et al. 1992; Frick et al. 1994). The kinetic parameters $K_{m\ app}$ and $V_{max\ app}$ derived from the Michaelis-Menten equations are presented in Table 3 (Porter et al. 1997). A comparison of those parameters of the two enzyme preparations studied demonstrated that both had similar affinities for Mg(II), as indicated by their respective $K_{m\ app}$ values, but MTH1 had approximately 50% lower activity than MutT, as indicated by the values of $V_{max\ app}$.

The IC_{50} values for all four metals are given in Table 4 (Porter et al. 1997). They indicate that for MTH1, Cu(II) and Cd(II) were the strongest inhibitors, followed by Co(II) and Ni(II), the overall ranking being Cu(II) [1] > Cd(II) [1.8] >> Co(II) [22] > Ni(II) [47] [the numbers in brackets represent IC_{50} values relative to that of Cu(II)]. For MutT, the pattern was similar with the following ranking: Cd(II) [3.5] > Cu(II) [6.2] >> Ni(II) [86] >> Co(II) [517] [IC_{50} values relative to that of Cu(II) for MTH1, as above]. Generally, the human enzyme appeared to be more susceptible to inhibition by all the metals tested than the bacterial enzyme.

To determine whether the enzymatic activity remaining at and above certain concentration of each inhibitory metal is not due to this metal's ability to act as the enzyme cofactor in place of Mg(II), the enzymes were assayed with the four metals in the absence of Mg(II) (Porter et al. 1997). The concentrations of the metals chosen for the assays were based on previous determinations of concentrations in which the metals exerted maximum inhibition. Control assays to check for possible enhancement by the metals of 8-oxo-dGTP hydrolysis in the absence of enzymes were also performed. The results are summarized in Tables 5 and 6. They show that most of the metals could sustain some hydrolytic activity of the enzymes. Thus the activity of MTH1 ranged from a low, ca. 16% of that with Mg(II), in the presence of Cd(II), Cu(II), or Ni(II) to a maximum of 34% in the presence of Co(II) alone (Table 5). The activity of MutT ranged from an undetectable level in the

Table 4. IC_{50} values for MTH1 and MutT[a,b]

Metal	IC_{50} (μM) MTH1	MutT
Ni(II)	801 ± 97	1,459 ± 96
Co(II)	376 ± 71	8,788 ± 1,003
Cu(II)	17 ± 2	107 ± 7
Cd(II)	30 ± 8	60 ± 6

[a] From Porter et al. 1997
[b] IC_{50}, metal concentration suppressing the enzyme activity to 50% of its original value. The numbers represent means ± SE determined from three experiments.

Table 5. Comparison of MTH1 activity with various metals as cofactors [a, b]

Metal	Concentration (mM)	MTH1 activity (pmoles 8-oxo-dGMP produced/μg MTH1 protein/min)	Non-enzymatic hydrolysis (pmoles 8-oxo-dGMP produced/min)
Mg(II)	8	521.8 ± 124.7	not detectable
Ni(II)	2.4	89.5 ± 26.2	1.1 ± 1.2
Co(II)	6	177.7 ± 68.9	not detectable
Cu(II)	0.2	91.5 ± 32.9	3.4 ± 1.6
Cd(II)	0.1	77.2 ± 14.9	4.5 ± 1.7

[a] From Porter et al. 1997
[b] Values represent means ± SE determined from three experiments.

presence of Cd(II), through 11- 19% with Cu(II) or Ni(II), to 73% of that with Mg(II) in the presence of Co(II). Thus, Co(II) appeared to be a relatively efficient Mg(II) substitute for MutT. Although the activity of both enzymes could be sustained to a certain degree by the heavy metals substituting for Mg(II), this activity could not be accounted for by non-enzymatic metal-catalyzed hydrolysis of 8-oxo-dGTP, since the latter accounted only for 0 to at most 9% of the corresponding enzymatic assay with the same metal (Tables 5 and 6).

Under conditions of the HPLC assay used in this study (Porter et al. 1996, 1997), besides the substrate, 8-oxo-dGTP, and the enzymes' product, 8-oxo-dGMP, two other possible degradation products of 8-oxo-dGTP, 8-oxo-dGDP and 8-oxo-dG could be detected. However, the latter two were not observed in any of the tested mixtures. Thus, the inhibitory metals had no effect on the identity of the product of 8-oxo-dGTP hydrolysis in either the presence or absence of Mg(II).

Judging by the $K_{m\ app}$ value of MTH1 for Mg(II) and the IC_{50} values of the other metals, the affinity of Ni(II) for MTH1 was similar to that of Mg(II), while the affinities of Co(II) and especially Cd(II) and Cu(II) were higher. Therefore, it would seem possible that at least Cd(II) and Cu(II) might compete with Mg(II) for binding at the active site of MTH1 and their inhibition should have a competitive mechanism. The same might be true for MutT. However, the present results are more consistent with noncompetitive (or uncompetitive) inhibition, similar to that observed by us previously for Ni(II) and MutT (Porter et al. 1996). The transition metals studied have high affinity for histidine (His), and especially cysteine (Cys) ligands in peptides and proteins (Eidsness et al. 1988; Predki et al. 1992). Therefore, these two amino acid residues in MTH1 and MutT molecules may be considered as the most likely binding sites for these metals. MTH1 has 3 His and 2 Cys residues (Cai et al. 1995) whereas MutT has 3 His and no Cys residues (Bhatnagar and Bessman 1988; Akiyama et al. 1987). None of them is located in the evolutionarily con-

Table 6. Comparison of MutT activity with various metals as cofactors [a, b]

Metal	Concentration (mM)	MutT activity (pmoles 8-oxo-dGMP produced/μg MutT protein/min)	Non-enzymatic hydrolysis (pmoles 8-oxo-dGMP produced/min)
Mg(II)	8	1,204 ± 164	not detectable
Ni(II)	2.4	231 ± 82	1.1 ± 1.2
Co(II)	12	882 ± 238	0.9 ± 0.3
Cu(II)	0.5	138 ± 14	11.8 ± 1.3
Cd(II)	0.1	not detectable	not detectable

[a] From Porter et al. 1997
[b] Values represent means ± SE determined from three experiments.

served active regions of the enzymes, i.e., between amino acid residues 36 and 58 (Cai et al. 1995; Kakuma et al. 1995; Sakumi et al. 1993). The observed sensitivity of MTH1 to metal inhibition, generally higher than that of MutT, seems to reflect the potentially greater metal-binding capacity of MTH1 [and/or the possibility of metal-catalyzed damage through -SH oxidation (Cavallini et al. 1969; Hartman and Hartman 1992)] and argue for noncompetitive nature of the inhibitory effects. The exact mechanisms of these effects remain to be established.

4.3. Biological Significance

All four metals tested in the present study had been found in various *in vivo* experiments to increase the levels of 8-oxoguanine in DNA (Kasprzak et al. 1990; 1992; 1994; 1997; Mikhailova et al. 1997; Yamamoto et al. 1993). Whether or not inhibition of 8-oxo-dGTPase is involved in producing this effect remains to be established. Although the IC_{50} values found in the present study appear to be high and cytotoxic, we have to consider that both epidemiologic and experimental studies point at particulate, water-insoluble, metal compounds as the most potent carcinogens (IARC 1993; Hughes et al. 1995; Oller et al. 1997; Costa and Mollenhauer 1980), and phagocytosis as the way of their uptake by target cells (Costa and Mollenhauer 1980; Huang et al. 1994). Once phagocytized, such particles are fragmented and slowly solubilized (Hildebrand et al. 1990) thus producing gradients of released metal ions that may reach local concentrations high enough to damage cytosolic proteins, including 8-oxo-dGTPase. Further, the inhibition could be favored by accumulation of metals, also those administered systemically as promptly soluble salts, in specific cell organelles, e.g., in mitochondria (Kasprzak and Poirier 1985), having their own 8-oxo-dGTPase activity (Kang et al. 1995). Obviously, *in vivo* studies are necessary to validate these assumptions. They should include (a) tests for the inhibition in tissues, cells, and intracellular fractions (nuclei, mitochondria, cytosol) exposed to metals under conditions known to result in mutations and cancer, and (b) search for the A → C transversions that are characteristic for the incorporative mutagenicity of 8-oxo-dGTP, and for G → T transversions, typical for 8-oxoguanine generated in a DNA strand (Cheng et al. 1992), in metal-induced tumors. Distinguishing between these two types of mutations will be especially interesting for the redox active metals Cu(II), Ni(II), and Co(II), which can catalyze generation of 8-oxo-dGTP in the nucleotide pool and 8-oxoguanine directly in DNA, as opposed to redox-inactive Cd(II), which might be only capable of promoting incorporation of 8-oxoguanine from 8-oxo-dGTP produced in the cells by natural oxidants, such as H_2O_2 and lipid peroxides. Investigations of the *in vivo* effects of Cd(II) and Ni(II) on rat 8-oxo-dGTPase (83% homology with the human enzyme; Cai et al. 1995) are under way in our laboratory. The preliminary results of these studies seem to confirm the inhibitory effect of cadmium on 8-oxo-dGPTase activity in cultured Chinese hamster ovary (CHO) cells. As shown in Figure 4, the cells exposed for 2 or 24 hr to non-toxic concentrations of 0.3, 1, and 3 μM, had their 8-oxo-dGTPase activity significantly reduced to approximately 70% of the original value. More experiments are needed, however, to determine other parameters (e.g., dose, time, and cell type dependence) and biological significance of this inhibition.

In conclusion, the carcinogenic transition metal ions Ni(II), Cu(II), Co(II), and Cd(II) are able to inhibit *in vitro* both bacterial and human 8-oxo-dGTPases, Cu(II) and Cd(II) being stronger inhibitors than Ni(II) and Co(II), and the human MTH1 enzyme being relatively more sensitive to inhibition than the bacterial MutT enzyme. If this inhibition occurs *in vivo*, it may lead to mutagenesis associated with incorporation of the

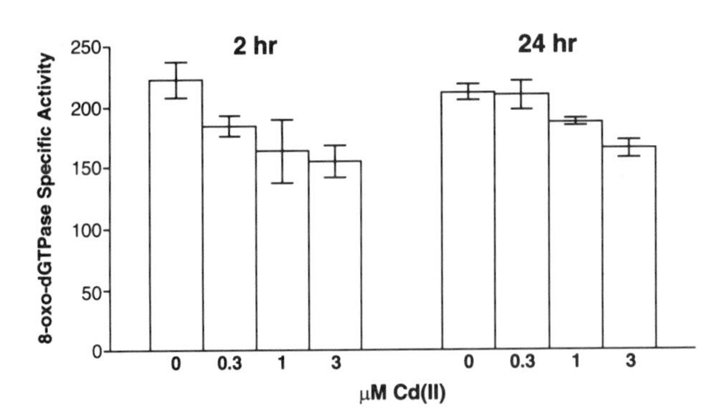

Figure 4. 8-Oxo-dGTPase activity (pmol 8-oxo-dGMP/min./mg of cytosolic proteins) in CHO cells incubated at 37°C for 2 or 24 hrs with 0 - 3 μM $CdCl_2$ in the DMEM culture medium supplemented with fetal calf serum, penicillin, and streptomycin. The activity was determined in ≤30 kDa cytosolic fraction using an HPLC assay (Bialkowski, unpublished; Porter et al. 1996).

8-oxoguanine lesion into genomic DNA, and thus be involved in the process of metal carcinogenicity; but this remains to be proven. Further studies are also needed to find the exact mechanisms of inhibitory action of the metals upon 8-oxo-dGTPases.

REFERENCES

M. Akiyama, T. Horiuchi and M. Sekiguchi (1987) Molecular cloning and nucleotide sequence of the *mutT* mutator of *Escherichia coli* that causes A:T to C:G transversion. *Mol. Gen. Genet.* **206**, 9–16.

H. Ammer, A. Henschen and C.-H. Lee (1986) Isolation and amino-acid sequence analysis of human sperm protamines P1 and P2. Occurrence of two forms of protamine P2. *Biol. Chem. Hoppe-Seyler* **367**, 515–522.

L.M. Anderson, K.S. Kasprzak and J.M. Rice (1994) Preconception exposure of males and neoplasia in their progeny: Effects of metals and consideration of mechanisms. In *Male-Mediated Developmental Toxicity* (eds. A.F. Olshan and D.R. Mattison), Plenum Press, New York, pp. 129–140.

A. Arkhis, A. Martinage, P. Sautiere and P. Chevaillier (1991) Molecular structure of human protamine P4 (HP4), a minor basic protein of human sperm nuclei. *Eur. J. Biochem.* **200**, 387–392.

O.I. Aruoma, B. Halliwell, E. Gajewski and M. Dizdaroglu (1991) Copper-ion-dependent damage to the bases in DNA in the presence of hydrogen peroxide. *Biochem. J.* **273**, 601–604.

W. Bal and K.S. Kasprzak (1997) Modeling the metal binding sites in core histones: Interactions of carcinogenic Ni(II) with the -CAIH- motif of histone H3. In *Cytotoxic, Mutagenic and Carcinogenic Potential of Heavy Metals Related to Human Environment* (ed. N.D. Hadjiliadis) Kluwer Academic Publishers, Dordrecht, The Netherlands, pp. 107–121.

W. Bal, H. Kozlowski, G. Kupryszewski, Z. Mackiewicz, L.D. Pettit and R. Robbins (1993) Complexes of Cu(II) with Asn-Ser-Phe-Arg-Tyr-NH_2; an example of metal ion-promoted conformational organization which results in exceptionally high complex stability. *J. Inorg. Biochem.* **52**, 79–87.

W. Bal, M.I. Djuran, D.W. Margerum, E.T. Gray, Jr., M.A. Mazid, R.T. Tom, E. Nieboer and P.J. Sadler (1994) Dioxygen-induced decarboxylation and hydroxylation of [Ni^{II}(Glycyl-Glycyl-l-Histidine)] occurs via Ni^{III}: X-ray crystal structure of [Ni^{II}(Glycyl-Glycyl-a-hydroxy-d,l-Histamine)]×$3H_2O$. *J. Chem. Soc., Chem. Comm.* 1889–1890.

W. Bal, J. Lukszo, M. Jezowska-Bojczuk and K.S. Kasprzak (1995) Interactions of nickel(II) with histones. Stability and solution structure of complexes with CH_3CO-Cys-Ala-Ile-His-NH_2, a putative metal binding sequence of histone H3. *Chem. Res. Toxicol.* **8**, 683–692.

W. Bal, J. Lukszo and K.S. Kasprzak (1996 a) Interactions of nickel(II) with histones: Enhancement of 2'-deoxyguanosine oxidation by Ni(II) complexes with CH_3CO-Cys-Ala-Ile-His-NH_2, a putative metal binding sequence of histone H3. *Chem. Res. Toxicol.* **9**, 535–540.

W. Bal, G.N. Chmurny, B.D. Hilton, P.J. Sadler and A. Tucker (1996 b) Axial hydrophobic fence in highly-stable Ni(II) complex of des-angiotensinogen N-terminal peptide. *J. Am. Chem. Soc.* **118**, 4727–4728.

W. Bal, M. Jezowska-Bojczuk and K.S. Kasprzak (1997 a) Binding of nickel(II) and copper(II) to the N-terminal sequence of human protamine HP2. *Chem. Res. Toxicol.* **10**, 906–914.

W. Bal, J. Lukszo and K.S. Kasprzak (1997 b) Mediation of oxidative DNA damage by nickel(II) and copper(II) complexes with the N-terminal sequence of human protamine HP2. *Chem. Res. Toxicol.* **10**, 914–921.

M.J. Bessman, D.N. Frick and S.F. O'Handley (1996) The MutT proteins or "nudix" hydrolases, a family of versatile, widely distributed, "housecleaning" enzymes. *J. Biol. Chem.* **271**, 25059–25062.

S.K. Bhatnagar and M.J. Bessman (1988) Studies on the mutator gene, mutT of *Escherichia coli*. *J. Biol. Chem.* **263**, 8953–8957.

J. Buckley (1994) Male-mediated developmental toxicity: Paternal exposures and childhood cancer. In *Male-Mediated Developmental Toxicity* (eds. A. F. Olshan and D. R. Mattison), Plenum Press, New York, pp. 169–175.

G.R. Bunin, K. Noller, P. Rose and E. Smith (1992) Carcinogenesis. In *Occupational and Environmental Reproductive Hazards: A Guide for Clinicians* (ed. M. Paul), Williams and Wilkins, Baltimore, pp. 76–88.

J.-P. Cai, T. Kakuma, T. Tsuzuki and M. Sekiguchi (1995) cDNA and genomic sequences for rat 8-oxo-dGTPase that prevents occurrence of spontaneous mutations due to oxidation of guanine nucleotides. *Carcinogenesis* **16**, 2343–2350.

N. Camerman, A. Camerman and B. Sarkar (1967) Molecular design to mimic the copper(II) transport site of human albumin. The crystal and molecular structure of copper(II)-glycylglycyl-l-histidine-*N*-methyl amide monoaquo complex. *Can. J. Chem.* **54**, 1309–1316.

A.C. Chandley and S. Kofman-Alfaro (1971) "Unscheduled" DNA synthesis in human germ cells following UV irradiation. *Exp. Cell Res.* **69**, 45–48.

M. Chauviere, A. Martinage, M. Debarle, P. Sautiere and P. Chevaillier (1992) Molecular characterization of six intermediate proteins in the processing of mouse protamine P2 precursor. *Eur. J. Biochem.* **204**, 759–765.

D. Cavallini, C. De Marco, S. Dupre and D. Rotilio. (1969) The copper catalyzed oxidation of cysteine to cystine. *Arch. Biochem. Biophys.* **130**, 354–361.

K.C. Cheng, D.S. Cahill, H. Kasai, S. Nishimura and L.A. Loeb (1992) 8-Hydroxy-guanine, an abundant form of oxidative DNA damage, causes G-T and A-C substitutions. *J. Biol. Chem.* **267**, 166–172.

S.-H. Chiou (1983) DNA- and protein-scission activities of ascorbate in the presence of copper ion and a copper-peptide complex. *J. Biochem.* **94**, 1259–1267.

N. Cotelle, E. Tremolieres, J.L. Bernier, J.P. Catteau and J.P. Henichart (1992) Redox chemistry of complexes of nickel(II) with some biologically important peptides in the presence of reduced oxygen species: An ESR study. *J. Inorg. Biochem.* **46**, 7–15.

T.-C. Chou (1976) Derivation and properties of Michaelis-Menten type and Hill type equations for reference ligands. *J. Theor. Biol.* **59**, 253–276.

M. Costa and A. Mollenhauer (1980) Carcinogenic activity of particulate metal compounds is proportional to their cellular uptake. *Science* **209**, 515–517.

M. Del Rio and M. De la Fuente (1994) Chemoattractant capacity of bombesin, gastrin-releasing peptide and neuromedin C is mediated through PKC activation in murine peritoneal leukocytes. *Regul. Peptides* **49**, 185–193.

M.K. Eidsness, R.J. Sullivan and R.A. Scott (1988) Electronic and molecular structure of bioinorganic nickel as studied by X-ray absorption spectroscopy. In *The Bioinorganic Chemistry of Nickel* (ed. J.R. Lancaster, Jr.), VCH Weinheim, New York, pp. 73–91.

E. Farkas, I. Sovago, T. Kiss and A. Gergely (1984) Studies on transition-metal-peptide complexes. Part 9. Copper(II) complexes of tripeptides containing histidine. *J. Chem. Soc. Dalton Trans.* 611–614.

D.N. Frick, D.J. Weber, J.R. Gillespie, M.J. Bessman and A.S. Mildvan (1994) Dual divalent cation requirement of the MutT dGTPase. Kinetic and magnetic resonance studies of the metal and substrate complexes. *J. Biol. Chem.* **269**, 1794–1803.

M. Furuichi, M.C. Yoshida, H. Oda, T. Tajiri, Y. Nakabeppu, T. Tsuzuki and M. Sekiguchi (1994) Genomic structure and chromosome location of the human *mutT* homologue gene *MTH1* encoding 8-oxo-dGTPase for prevention of A:T to C:G transversion. *Genomics* **24**, 485–490.

T. Gajda, B. Henry, A. Aubry and J.-J. Delpuech (1996) Proton and metal ion interactions with glycylglycylhistamine, a serum albumin mimicking pseudopeptide. *Inorg. Chem.* **35**, 586–593.

J.M. Gatewood, G.P. Schroth, C.W. Schmid and E.M. Bradbury (1990) Zinc-induced secondary structure transitions in human sperm protamines. *J. Biol. Chem.* **265**, 20667–20672.

A.P. Grollman and M. Moriya (1993) Mutagenesis by 8-oxoguanine: an enemy within. *Trends Genet.* **9**, 246–249.

W. Gu, C. Morales and N.B. Hecht (1995) In male mouse germ cells, copper-zinc superoxide dismutase utilizes alternative promoters that produce multiple transcripts with different translation potential. *J. Biol. Chem.* **270**, 236–243.

M.R. Gunther, P. Hanna, R.P. Mason and M.S. Cohen (1995) Hydroxyl radical formation from cuprous ion and hydrogen peroxide: a spin-trapping study. *Arch. Biochem. Biophys.* **316**, 515–522.

Z. Hartman and P.E. Hartman (1992) Copper and cobalt complexes of carnosine and anserine: production of active oxygen species and its enhancement by 2-mercaptoimidazoles. *Chem.-Biol. Interactions* **84**, 153–168.

C. Harford, S. Narindrasorasak and B. Sarkar (1996) The designed protein M(II)-Gly- Lys-His-Fos(138–211) specifically cleaves the AP-1 binding site containing DNA. *Biochemistry* **35**, 4271–4278.

A. Hartwig (1994) Role of DNA repair inhibition in lead- and cadmium-induced genotoxicity: a review. *Environ. Health. Perspect.* **102 (Suppl 3)**, 45–50.

R.W. Hay, M.M. Hassan and C. You-Quan (1993) Kinetic and thermodynamic studies of the copper(II) and nickel(II) complexes of glycylglycyl-l-histidine. *J. Inorg. Biochem.* **52**, 17–25.

H. Hayakawa, A. Taketomi, K. Sakumi, M. Kuwano and M. Sekiguchi (1995) Generation and elimination of 8-oxo-7,8-dihydro-2'-deoxyguanosine-5'-triphosphate, a mutagenic substrate for DNA synthesis, in human cells. *Biochemistry* **34**, 89–95.

H.F. Hildebrand, M. Collyn D'Hooghe, P. Shirali, C. Bailly and J.P. Kerckaert (1990) Uptake and biological transformation of βNiS and αNi_3S_2 by human embryonic pulmonary epithelial cells (L132) in culture. *Carcinogenesis* **11**, 1943–1950.

U. Hilgenfeldt (1988) Half-life of rat angiotensinogen: influence of nephrectomy and lipopolysaccharide stimulation. *Mol. Cell. Endocrinol.* **56**, 91–98.

T. Hirano, Y. Yamaguchi and H. Kasai, H. (1997) Inhibition of 8-hydroxyguanine repair in testes after administration of cadmium chloride to GSH-depleted rats. *Toxicol. Appl. Pharmacol.* **4**, 9–14.

X. Huang, C.B. Klein and M. Costa. (1994) Crystalline Ni_3S_2 specifically enhances the formation of oxidants in the nuclei of CHO cells as detected by dichlorofluorescein. *Carcinogenesis* **15**, 545–548.

K. Hughes, M.E. Meek, R. Newhook and P.K.L. Chan (1995) Speciation in health risk assessments of metals: Evaluation of effects associated with forms present in the environment. *Regul. Toxicol. Pharmacol.* **22**, 213–220.

S. Inoue and S. Kawanishi (1989) ESR evidence for superoxide, hydroxyl radicals and singlet oxygen produced from hydrogen peroxide and nickel(II) complex of glycylglycyl-L-histidine. *Biochem. Biophys. Res. Commun.* **159**, 445–451.

International Agency for Research on Cancer (1990) *IARC Monographs on the Evaluation of Carcinogenic Risk to Humans.* Vol. **49**. *Chromium, Nickel and Welding.* IARC, Lyon, France.

International Agency for Research on Cancer (1993) *IARC Monographs on the Evaluation of Carcinogenic Risk to Humans.* Vol. **58**. *Cadmium, Mercury, Beryllium and the Glass Industry.* IARC, Lyon, France.

T. Kakuma, J. Nishida, T. Tsuzuki and M. Sekiguchi (1995) Mouse MTH1 protein with 8-oxo-7,8-dihydro-2'-deoxyguanosine-5'-triphosphatase activity that prevents transversion mutation. cDNA cloning and tissue distribution. *J. Biol. Chem.*, **270**, 25942–25948.

D. Kang, J. Nishida, A. Iyama, Y. Nakabeppu, M. Furuichi, T. Fujiwara, M. Sekiguchi and K. Takeshige (1995) Intracellular localization of 8-oxo-dGTPase in human cells, with special reference to the role of the enzyme in mitochondria. *J. Biol. Chem.* **270**, 14659–14665.

K.S. Kasprzak (1995 a) Possible role of oxidative damage in metal-induced carcinogenesis. *Cancer Invest.* **13**, 411–430.

K.S. Kasprzak (1995 b) Oxidative mechanisms of nickel(II) and cobalt(II) genotoxicity. In *Genetic Response to Metals* (ed. B. Sarkar), M. Dekker, New York, pp. 69–85.

K.S. Kasprzak (1996) Oxidative DNA damage in metal-induced carcinogenesis. In *Toxicology of Metals* (ed. L. Chang), Lewis Publishers, Boca Raton, pp. 299–320.

K.S. Kasprzak (1997) The oxidative damage hypothesis of metal-induced genotoxicity and carcinogenesis. In *Cytotoxic, Mutagenic and Carcinogenic Potential of Heavy Metals Related to Human Environment* (ed. N.D. Hadjiliadis) Kluwer Academic Publishers, Dordrecht, The Netherlands, pp.73–92.

K.S. Kasprzak and R.M. Bare (1989) In vitro polymerization of histones by carcinogenic nickel compounds. *Carcinogenesis* **10**, 621–624.

K.S. Kasprzak and L.A. Poirier (1985) Effects of calcium(II) and magnesium(II) on nickel(II) uptake and stimulation of thymidine incorporation into DNA in the lungs of strain A mice. *Carcinogenesis* **6**, 1819–1821.

K.S. Kasprzak, B.A. Diwan, N. Konishi, M. Misra and J.M. Rice (1990) Initiation by nickel acetate and promotion by sodium barbital of renal cortical epithelial tumors in male F344 rats. *Carcinogenesis* **11**, 647–652.

K.S. Kasprzak, B.A. Diwan, J.M. Rice, M. Misra, C.W. Riggs, R. Olinski and M. Dizdaroglu (1992) Nickel(II)-mediated oxidative DNA base damage in renal and hepatic chromatin of pregnant rats and their fetuses. Possible relevance to carcinogenesis. *Chem. Res. Toxicol.* **5**, 809–815.

K.S. Kasprzak, T.H. Zastawny, S.L. North, C.W. Riggs, B.A. Diwan, J.M. Rice and M. Dizdaroglu (1994) Oxidative DNA base damage in renal, hepatic, and pulmonary chromatin of rats after intraperitoneal injection of cobalt(II) acetate. *Chem. Res. Toxicol.* **7**, 329–335.

K.S. Kasprzak, P. Jaruga, T.H. Zastawny, S.L. North, C.W. Riggs, R. Olinski and M. Dizdaroglu (1997) Oxidative DNA base damage and its repair in kidneys and livers of nickel(II)-treated male F344 rats. *Carcinogenesis* **18**, 271–277.

T. Kortemme and T.E. Creighton (1995) Ionisation of cysteine residues at the termini of model α-helical peptides. Relevance to unusual thiol pK_a values in proteins of the thioredoxin family. *J. Mol. Biol.* **253**, 799–812.

S.A. Lewisch and R.L. Levine (1995) Determination of 2-oxohistidine by amino acid analysis. *Anal. Biochem.* **231**, 440–446.

J. Li and X. Wang (1989) Toxic effects of nickel chloride on testis of mice. *Weisheng Dulixue Zazhi* **3**, 222–224; referred to by Xie et al. 1995

D.P. Mack, B.L. Iverson and P.B. Dervan (1988) Design and chemical synthesis of a sequence-specific DNA-cleaving protein. *J. Am. Chem. Soc.* **110**, 7572–7574.

H. Maki and M. Sekiguchi (1992) MutT protein specifically hydrolyses a potent mutagenic substrate for DNA synthesis. *Nature* **355**, 273–275.

M.R. McDonald, W.M. Scheper, H.D. Lee and D.W. Margerum (1995) Copper(III) complexes of tripeptides with histidine and histamine as the third residue. *Inorg. Chem.* **34**, 229–237.

D.J. McKay, B.S. Renaux and G.H. Dixon (1986) Human sperm protamines. Amino-acid sequences of two forms of protamine P2. *Eur. J. Biochem.* **156**, 5–8.

M.V. Mikhailova, N.A. Littlefield, B.S. Hass, L.A. Poirier and M.W. Chou (1997) Cadmium-induced 8-hydroxy-deoxyguanosine formation, DNA strand breaks and antioxidant enzyme activities in lymphoblastoid cells. *Cancer Lett.* **115**, 141–148.

N. Minamino, K. Kangawa and H. Matsuo (1984) Neuromedin B is a major bombesin-like peptide in rat brain: regional distribution of neuromedin B and neuromedin C in rat brain, pituitary and spinal cord. *Biochem. Biophys. Res. Commun.* **124**, 925–932.

M. Misra, R.E. Rodriguez and K.S. Kasprzak (1991) Nickel-induced renal lipid peroxidation in different strains of mice: Concurrence with nickel effects on antioxidant defense systems. *Toxicol. Lett.* **58**, 121–133.

J.-Y. Mo, H. Maki and M. Sekiguchi (1992) Hydrolytic elimination of a mutagenic nucleotide, 8-oxo-dGTP, by human 18-kilodalton protein: Sanitization of nucleotide pool. *Proc. Natl. Acad. Sci. USA* **89**, 11021–11025.

Z. Nackerdien, K.S. Kasprzak, G. Rao, B. Halliwell and M. Dizdaroglu (1991) Nickel(II)- and cobalt(II)-dependent damage by hydrogen peroxide to the DNA bases in isolated human chromatin. *Cancer Res.* **51**, 5837–5842.

E. Nieboer, P.I. Stetsko and P.Y. Hin (1984) Characterization of the Ni(III)/Ni(II) redox couple for the nickel(II) complex of human serum albumin. *Ann. Clin. Lab. Sci.* **14**, 409.

E. Nieboer, R.T. Tom and F.E. Rossetto (1989) Superoxide dismutase activity and novel reactions with hydrogen peroxide of histidine-containing nickel oligopeptide-complexes and nickel(II)-induced structural changes in synthetic DNA. *Biol. Trace Elem. Res.* **21**, 23–33.

A.R. Oller, M. Costa and G. Oberdorster (1997) Carcinogenicity assessment of selected nickel compounds. *Toxicol. Appl. Pharmacol.* **143**, 152–166.

Y.I. Pavlov, D.T. Minnick, S. Izuta and T.A. Kunkel (1994) DNA replication fidelity with 8-oxodeoxyguanosine triphosphate. *Biochemistry* **33**, 4695–4701.

A. Perczel, K. Park and G.D. Fasman (1992) Deconvolution of the circular dichroism spectra of proteins: the circular dichroism spectra of the antiparallel β-sheet in proteins. *Proteins* **13**, 57–69.

T. Peters, Jr. and F.A. Blumenstock (1967) Copper-binding properties of bovine serum albumin and its amino-terminal peptide fragment. *J. Biol. Chem.* **242**, 1574–1578.

G.C. Pogany, M. Corzett, S. Weston and R. Balhorn (1981) DNA and protein content of mouse sperm. *Exp. Cell Res.* **136**, 127–136.

D.W. Porter, V.C. Nelson, M.J. Fivash, Jr. and K.S. Kasprzak (1996) Mechanistic studies of the inhibition of MutT dGTPase by the carcinogenic metal Ni(II). *Chem. Res.Toxicol.* **9**, 1375–1381.

D.W. Porter, H. Yakushiji, Y. Nakabeppu, M. Sekiguchi, M.J. Fivash, Jr. and K.S. Kasprzak (1997) Sensitivity of *Escherichia coli* (MutT) and human (MTH1) 8-oxo-dGTPases to *in vitro* inhibition by the carcinogenic metals, nickel(II), copper(II), cobalt(II) and cadmium(II). *Carcinogenesis* **18**, 1785–1791.

P.F. Predki, C. Harford, P. Brar and B. Sarkar (1992) Further characterization of the N-terminal copper(II)- and nickel(II)-binding motif of proteins. *Biochem. J.* **287**, 211–215.

R.E. Rodriguez, M. Misra, B.A. Diwan, C.W. Riggs and K.S. Kasprzak (1996) Relative susceptibilities of C57BL/6, $(C57BL/6xC3H/He)F_1$, and C3H/He mice to acute toxicity and carcinogenicity of nickel subsulfide. *Toxicology* **107**, 131–140.

K. Sakumi, M. Furuichi, T. Tsuzuki, T. Kakuma, S. Kawabata, H. Maki and M. Sekiguchi (1993) Cloning and expression of cDNA for a human enzyme that hydrolyzes 8-oxo-dGTP, a mutagenic substrate for DNA synthesis. *J. Biol. Chem.* **268**, 23524–23530.

B. Sarkar (1997) Design of proteins with ATCUN motif which specifically cleave DNA. In *Cytotoxic, Mutagenic and Carcinogenic Potential of Heavy Metals Related to Human Environment* (ed. N.D. Hadjiliadis), Kluwer Academic Publishers, Dordrecht, The Netherlands, pp. 477–484.

G.A. Sega (1974) Unscheduled DNA synthesis in the germ cells of male mice exposed *in vivo* to the chemical mutagen ethyl methanesulfonate. *Proc. Natl. Acad. Sci. USA* **71**, 4955–4959.

M. Sekiguchi (1996) MutT-related error avoidance mechanism for DNA synthesis. *Genes Cells* **1**, 139–145.

S. Shibutani, M. Takeshita and A.P. Grollman (1991) Insertion of specific bases during DNA synthesis past the oxidation-damaged base 8-oxo-dG. *Nature* **349**, 431–434.

H. Sigel (1993) Interactions of metal ions with nucleotides and nucleic acids and their constituents. *Chem. Soc. Rev.* 255–267.

R.S. Sohal and A. Dubey (1994) Mitochondrial oxidative damage, hydrogen peroxide release, and ageing. *Free Radic. Biol. Med.* **16**, 621–626.

N. Spear and S.D. Aust (1995) Hydroxylation of deoxyguanosine in DNA by copper and thiols. *Arch. Biochem. Biophys.*, **317**, 142–148.

E.R. Stadtman (1993) Oxidation of free amino acids and amino acid residues in proteins by radiolysis and by metal-catalyzed reactions. *Annu. Rev. Biochem.* **62**, 797–821.

S.J. Stohs and D. Bagchi (1995) Oxidative mechanisms in the toxicity of metal ions. *Free Rad. Biol. Med.*, **18**, 321–336.

H.A. Tajmir-Riahi, M. Naoui and R. Ahmad (1993) The effect of Cu^{2+} and Pb^{2+} on the solution structure of calf thymus DNA: DNA condensation and denaturation studied by Fourier transformed IR difference spectroscopy. *Biopolymers* **33**, 1819–1827.

L.K. Tkeshelashvili, T. McBride, K. Spence and L.A. Loeb (1991) Mutation spectrum of copper-induced DNA damage. *J. Biol. Chem.* **266**, 6401–6406.

J. Torreilles and M.-C. Guerin (1990) Nickel(II) as a temporary catalyst for hydroxyl radical generation. *FEBS Lett.* **272**, 58–60.

J. Torreilles, M.-C. Guerin and A. Slaoui-Hasnaoui (1990) Nickel(II) complexes histidyl-peptides as Fenton-reaction catalysts. *Free Rad. Res. Commun.* **11**, 159–166.

H.P. Treffers, V. Spinelli and N.O. Belser (1954) A factor (or mutator gene) influencing mutation rates in *Escherichia coli*. *Proc. Natl. Acad. Sci. USA* **40**, 1064–1071.

K. Uchida and S. Kawakishi (1993) 2-Oxo-histidine as a novel biological marker for oxidatively modified proteins. *FEBS Lett.* **332**, 208–210.

R.W. Woody (1994) Circular dichroism of peptides and proteins. In *Circular Dichroism. Principles and Applications* (eds. K. Nakanishi, N. Berova and R.W. Woody), VCH Publishers, New York, pp. 473–496.

J. Xie, T. Funakoshi, H. Shimada and S. Kojima (1995) Effects of chelating agents on testicular toxicity in mice caused by acute exposure to nickel. *Toxicology* **103**, 147–155.

F. Yamamoto, H. Kasai, Y. Togashi, N. Takeichi, T. Hori and S. Nishimura (1993) Elevated level of 8-hydroxydeoxyguanosine in DNA of liver, kidneys, and brain of Long-Evans-Cinnamon rats. *Jpn. J. Cancer* **84**, 508–511.

J.-L. Yang, J.I. Chao and J.-G. Lin (1996) Reactive oxygen species may participate in the mutagenicity and mutational spectrum of cadmium in Chinese hamster ovary-K1 cells. *Chem. Res. Toxicol.* **9**, 1360–1367.

M.B. Yim, P.B. Chock and E.R. Stadtman (1990) Copper, zinc superoxide dismutase catalyzes hydroxyl radical production from hydrogen peroxide. *Proc. Natl. Acad. Sci. USA* **87**, 5006–5010.

17

MECHANISM OF ACTION OF *ESCHERICHIA COLI* FORMAMIDOPYRIMIDINE N-GLYCOSYLASE AND ENDONUCLEASE V

Yoke W. Kow and Lois E. Rabow

Division of Cancer Biology, Department of Radiation Oncology
Emory University School of Medicine
145 Edgewood Ave. Atlanta, Georgia 30335

1. ABSTRACT

DNA base damages generated either from oxidation or replication errors are repaired through the base excision and the mismatch repair pathways, respectively. In base excision repair, base damages are recognized specifically by N-glycosylases. Formamidopyrimidine N-glycosylase recognizes many oxidative products of purines including the formamidopyrimidines and 8-oxoguanine. The enzyme has an associated β,δ-lyase activity. The extent of β,δ-elimination was influenced by mutation in a conserved lysine in the fpg protein, sequence context surrounding the lesion as well as the location of the DNA lesion. These data suggest that the stability of the β-complex is critical in determining the extent of β,δ-elimination for fpg protein. Endonuclease V exhibit high affinity for deoxyinosine and might be important for the repair of this lesion *in vivo*. Mismatch recognition by endonuclease V is strand specific and depends on the presence of a 5' terminus, thus suggest that endonuclease V might be important in the repair of mismatches generated in the Okazaki fragment. It was recently found that mutation in endonuclease V lead to an increase frequency in the expansion of CAG repeats in DNA, a phenotype that was also shared by the yeast RAD27 mutant. These data suggest that structure specific endonucleases might play an important role in the maintenance of these triplet repeat sequences.

2. INTRODUCTION

Base damages in DNA can be readily formed from deamination or oxidation. In bacteria as well as in eukaryotes, most of these base damages are repaired through the base excision repair pathway (Friedberg et al., 1995). In the base excision repair pathway, the

Advances in DNA Damage and Repair, edited by Dizdaroglu and
Karakaya. Kluwer Academic / Plenum Publishers, New York, 1999.

damaged base is recognized by DNA N-glycosylases, which release the damaged base. The abasic (AP) site formed after the action of an N-glycosylase is then processed further by AP endonucleases, resulting in a one base gap in a duplex DNA. The repair process is then completed following the combine action of DNA polymerase and ligase.

While most of the base damages are repaired through the base excision repair pathway, mismatch bases generated from replication errors are repaired predominantly through the mismatch recognition system (Modrich, 1991). In *Escherichia coli*, base mismatches are repaired through a long patch repair pathway involving mutH, mutL and mutS proteins. Discrimination of the nascent DNA strand from the template strand is guided by the methylation status of the GATC sites within the newly replicated DNA (Kraemer et al., 1984; Jones et al., 1987). However, some specific base mismatches such as A/G mismatch can also be repaired through the base excision repair pathway involving base mismatch specific DNA N-glycosylases (Michaels et al, 1990; Tsai-Wu et al., 1992)

3. FORMAMIDOPYRIMIDINE N-GLYCOSYLASE

Oxidative purine base damages induced by reactive oxygen species are predominantly repaired through the base excision repair pathway. In *E. coli*, these oxidative purine damages are recognized by formamidopyrimidine N-glycosylase (fpg or mut M protein; Boiteux et al, 1987; Chetsanga and Lindahl, 1979; Tchou et al, 1991). The enzyme appears to be ubiquitous and has been found in most cells, from archaebacteria to human, suggesting that fpg protein and its homologs are fundamentally essential for the removal oxidative purine damages. 8-oxoG, a major oxidative damage of guanine is promutagenic. During replication, unrepaired 8-oxoG can pair with an A, leading to a G to T transversion mutation (Wood et al., 1990; Shibutani et, 1991; Moriya, 1993). Due to the ease of formation of 8-oxoG in DNA as well as in the nucleotide pool, cells have evolved a complicated repair system for the removal of 8-oxoG involving fpg (mutM), mutY (oxoG/A mismatch N-glycosylase) and mutT (8-oxo-dGTP pyrophosphatase) proteins (Friedberg, et al., 1995). The repair system involving these three enzymes is dealt with details by other investigators in this symposium and will not be described here.

3.1. Substrate Specificity

Formamidopyrimidine N-glycosylase (Fpg) recognizes a diverse purine base modifications including 7-hydro-8-oxoguanine (8-oxoG), formamidopyrimidines (FaPyG and FapyA, imidazole ring opened forms of guanine and adenine, respectively), and N_7-methyl-formamidopyrimidines. Interestingly, the enzyme is unable to recognize 8-oxoA, the hydroxylation damage of deoxyadenosine. In addition, fpg is unable to recognize most of the known oxidative products of pyrimidine. However, it is not clear whether fpg is able to recognize the 5-hydroxy derivatives of cytosine (5-hydroxycytosine and 5-hydroxyuracil; Hatehat et al., 1994; Karakaya et al, 1997). In addition to the N-glycosylase activity, Fpg has an associated AP lyase activity, catalyzing successive β- and δ-elimination reactions. The combine and concerted action of the N-glycosylase and AP lyase activity of Fpg will lead to strand sessions 3' and 5' to the base damage, resulting in the release of the damaged base, a monomeric five-carbon fragment derived from deoxyribose (4-oxo-2-pentenal) and a one base gapped DNA terminated by 3' and 5' phosphates (Bhagwat and Gerlt, 1996). Interestingly, fpg protein also recognizes urea residues as well as lesion homolog generated by the action of alkoxyamine on AP site (Purmal et al., 1996).

3.2. Mechanism of Action

Fpg is an N-glycosylase with an associated β,δ-lyase activity (Bhagwat and Gerlt, 1996; Zharkov et al, 1997; Rabow and Kow, 1997). It has been shown by many investigators that fpg forms a covalent imine intermediate with the DNA substrate that can be stabilized by sodium borohydride. The covalent intermediates was shown to involve the N-terminal proline residue and the C_1-aldehyde of the deoxyribose moiety (Zharkov et al, 1997). The N-glycosylase reaction has been proposed to be a direct nucleophilic attack at the $C_{1'}$-N_9-glycosylic bond by the N-terminal proline residue, leading to an imine intermediate. Based on our observations that fpg is able to recognize urea residue and alkoxyamine modified AP sites (Purmal et al, 1996), we suggested that the initial enzymatic step(s) preceeding the N-glycosylase reaction involved an enzyme assisted ring opening of the sugar moiety at which the damaged based is attached, generating an imine structure at the C1' of the deoxyribose ring (Figure 1). Transimminization reaction catalyzed by the N-terminal praline residue of fpg will lead to the release of the damaged base and the concomitant formation of the imine covalent intermediate. Similar reaction mechanism had been proposed by us earlier to describe the reaction pathway for endonuclease III (Kow and Wallace, 1987). The covalent enzyme complex then catalyzes the β-elimination reaction. Depending on the stability of the β-complex (the covalent intermediate after the β-lyase step), the enzyme complex can further catalyze a second β-elimination (or commonly called the δ-elimination), lead to the observed the β,δ-elimination product. Similar reaction pathway was also employed by other N-glycosylases/AP lyases including endonuclease III (Kow and Wallace, 1987) and T4 endonuclease V (Dodson et al, 1993). However, for both endonuclease III and T4 endonuclease V, the reaction pathway end at the β-elimination step, generating a one base gap terminated with 5' phosphate and 3'-modified sugar.

In order to further understand the reaction mechanism of fpg, we have generated several point mutants of fpg protein. Several conserved lysine residues were mutated individually and mutant proteins were purified and examined for changes in the kinetic parameters. K57A mutant was found to have a diminished activity in both the N-glycosylase and AP lyase activity. The decrease in enzymatic activity was accompanied by a concomitant change in the CD spectrum, suggesting that replacement of a lysine 55 residue by an alanine lead to changes in the conformation of the protein structures (Rabow and Kow, 1996).

However, the K155A mutant presented some interesting variation in the enzymatic properties as compared to the wild-type enzyme (Rabow and Kow, 1997). K155A showed a substantial decrease in the N-glycosylase activity. In contrast, the AP lyase activity of K155A was actually increased by more than 2-fold as compared to the wild-type enzyme. The decrease in the N-glycosylase activity was mainly due to a reduction in the kcat and an increase in the apparent Km values. Despite drastic reduction in the N-glycosylase activity of K155A, lysine 155 is not essential for covalent catalysis. This is shown by the fact that K155A mutant protein is fully capable of catalyzing the β,δ-elimination reaction, even at a much higher rate than the wild-type when DNA containing AP site was used as a substrate. Furthermore, in the presence of sodium borohydride, covalent intermediate formed between K155A fpg protein and the DNA substrate was stabilized and can be easily analyzed with polyacrylamide gel electrophoresis. The trapped covalent intermediate was shown to be mostly derived from reaction of fpg with the C1-aldehyde of the intermediary AP site generated after the N-glycosylase reaction; however, a small amount of covalent intermediate that was formed with the cleaved DNA product (β-complex) was also

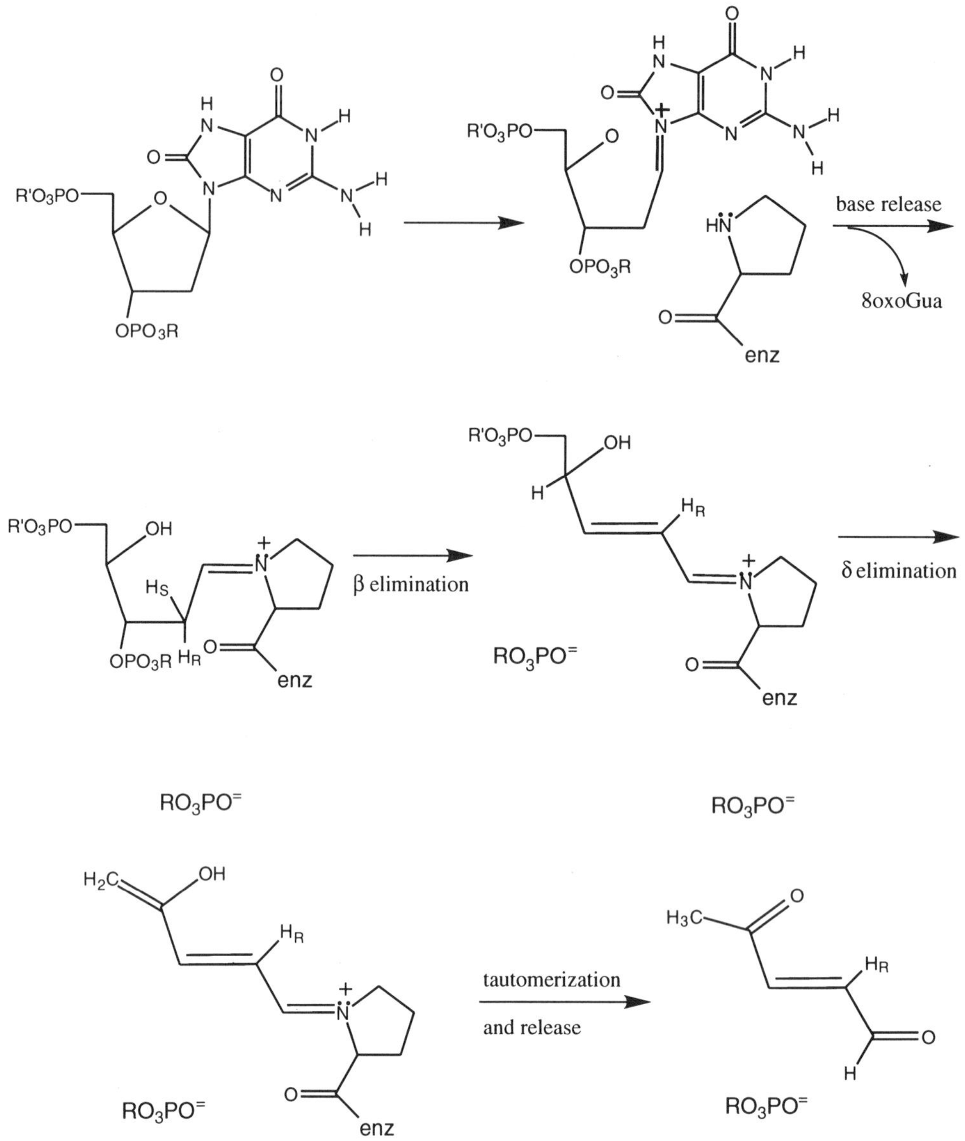

Figure 1. Reaction mechanism for Fpg protein.

observed. These data suggested that formation of β-complex is necessary before the β, δ-elimination reaction can occur.

Since lysine 155 is not involved in covalent catalysis, it is therefore interested to find out what functional role does K155 plays in the reaction catalyzed by fpg. In order to do this, a detail kinetic study was performed for the wild-type and the K155A fpg protein. It is interesting to note that when oxo-G containing DNA was used as DNA substrates, the rate of reaction for K155A mutant (kcat/Km = 0.003 min^{-1} nM^{-1}) is 50 fold lower than that for the wild-type enzyme (0.16 min^{-1} nM^{-1}). Similar reduction in enzymatic activity of K155A was also observed for 8-oxo-nebularine containing DNA. In contrast, when N_7-

Table 1. Distribution of products generated by wild-type and K155A Fpg proteins

	Wild-type		K155A	
Substrate	β-product	β,δ-product	β-product	β,δ-product
AP-27-mer/C[1]	0%	100%	0.65%	99.5%
Ap-27-mer/A[2]	0%	100%	51%	49%

[1] 5'-GGTCGACTSAGGAGGATCCCCGGGTAC
CCAGCTGACTCCTCCTAGGGGCC-5'
S = AP SITE
[2] 5'-GGTCGACTSAGGAGGATCCCCGGGTAC
CCAGCTGAATCCTCCTAGGGGCC-5'

methyl-FapyGua containing DNA was used as substrate, the kcat/Km values for the wild-type (0.01 min^{-1} nM^{-1}) was only 4-fold greater than the values for K155A protein (0.0025 min^{-1} nM^{-1}). These data suggested that lysine 155 might be important for interaction with the 8-oxo group in the imidazole ring of 8-oxoG (Rabow and Kow, 1997).

3.3. Effect of DNA Lesions on the Yield of β,δ-Product

As we have shown earlier, fpg has an associated β,δ-lyase activity. Under most reaction conditions, wild-type fpg proteins yield only β,δ-products, that is , cleaved DNA terminated with 3' phosphoryl ends. However, it is interesting to note that when DNA containing an AP site that is located directly opposite to a nick was used as a substrate, the reaction performed with K155A protein generated a mixture of β and β,δ-products. Similarly, fpg reaction with DNA substrate containing an AP site situated opposite an one-base gap yielded a mixtures of β and β,δ-products. However, when the gap size opposite the AP site increased to seventeen nucleotides wide or when fpg was incubated with single stranded DNA containing an AP site, only β,δ-product was observed. In order to explain differences in the product distribution, we suggest that the extent of β,δ-elimination is influence by the stability of the β-complex. When fpg was incubated with duplex DNA containing an AP site opposite a nick or a one base gap (or a small gap), the β-lyase activity of fpg will lead to a double strand break. The resulting double strand break will favor the hydrolysis of the covalent imine bond, destabilizes the β-complex and lead to the dissociation of fpg protein from the β-complex. The premature dissociation of the β-complex thus lead to the observed β-elimination product. However, some of the β-complex can continue to carry out the δ-elimination, yielding the observed β,δ-product.

The generation of β-product was also observed when mutant K155A was incubated with DNA substrate containing an AP site. The distribution of the β and β,δ-elimination products depended on the nature of the base opposite the AP site. When the AP site was opposite an C, the product generated by K155A fpg reaction was greater than 95% β, δ-elimination product. However, when the AP site was opposite an A, K155A fpg protein produced a mixture of β and β,δ-products (Table 1). For wild-type fpg, only β,δ-product was observed under similar conditions. These data is in agreement with the apparent binding constant (K_m) as well as the K_d of fpg towards these lesions by other investigators. It was shown that the Kd of fpg for oxoG/C or AP/C is 10 fold lower than the oxoG/A or AP/A pairs (Tchou et al, 1991; Castaing et al, 1993). A weaker K_d thus suggests a weaker β-complex formed by fpg with AP/A as compared to AP/C, thus promoting the dissociation of the β-complex.

3.4. Effect of N-Terminal Mutant on the Yield of B,Δ-Product

In order to further understand the role of N-terminal proline, N-terminal deletion fpg mutant proteins (Δ2–3 and Δ2–24) were prepared. These N-terminal deletion fpg proteins showed greatly diminished enzymatic activities (both proteins exhibit less than 0.5% of the wild-type N-glycosylase and AP lyase activities). Interestingly, the product generated by the N-terminal deletion mutants consist of only β,δ-product. Since β,δ-product is suggestive of the involvement of a covalent intermediate, it is expected that these N-terminal deletion proteins are still capable of forming covalent intermediates that can be trapped and analyzed using sodium borohydride. It is important to note that mutant K155A, protein which only exhibit less than 2% of the wild-type N-glycosylase activity, the covalent intermediate formed can be easily analyzed in the presence of sodium borohydride. However, under similar reaction conditions for the analysis of covalent intermediate, we were unable to demonstrate the presence of a covalent intermediate formed between the N-terminal deletion proteins and the DNA substrates. These data suggest that, in addition to the reaction pathway that lead to a covalent intermediate, other reaction pathway might also be possible for generating the observed β,δ-product (Bhagwat and Gerlt, 1996).

4. ENDONUCLEASE V

We have initially identified a deoxyinosine specific enzyme, deoxyinosine 3' endonuclease from *E. coli* (Yao et al, 1994a). In addition to deoxyinosine (Yao et al, 1994a), the enzyme also recognizes other lesions including AP sites (Yao et al, 1994a), urea residues (Yao et al, 1994a), and interestingly, single base and abnormal replicative DNA structures which including hairpin, loop and flap DNA structures (Yao and Kow, 1994b;1996; 1997). When the enzyme was cloned and sequenced, it was found to be identical to another *E. coli* repair enzyme, endonuclease V (Yao and Kow, 1996; 1997) . Endonuclease V which was recently cloned by Weiss' laboratory (Guo et al, 1997). Endonuclease V was partially characterized in the early 70's as an enzyme that recognized single stranded DNA, or DNA heavily treated with UV, osmium tetroxide, acid/alkali and interesting 7-bromomethyl-benz[a]anthracean modified DNA (Gates and Linn, 1977; Demple and Linn, 1982). Deoxyinosine 3' endonuclease, or better known as endonuclease V cleaves DNA substrates one nucleotide 3' away from the lesion, generating a 3'hydroxyl and a 5' phosphoryl group (Yao and Kow, 1994a; 1994b).

4.1. Repair of Deoxyinosine and Deoxyuridine

When the enzyme is interacting with deoxyinosine residue in DNA, the enzyme forms two protein-DNA complexes, with K_{d1} for complex I = 4 nM and K_{d2} for complex II around 40 nM (Yao and Kow, 1995; 1997). Based on the footprinting experiments, we showed that the enzyme interacts only with the DNA strand that contains the lesion. Interestingly, the enzyme binds to dI containing DNA with the same affinity with or without Mg^{++} . In fact, endonuclease V binds to both the substrate and the product (nicked dI containing DNA) with the same affinity. No turnover of endonuclease V was observed when the enzyme interacts with DNA containing dI. These data suggest that after endonuclease V cleaves the DNA containing dI, the enzyme remains bound to the DNA. Therefore, in order to complete the repair of deoxyinosine on DNA, another enzyme, possibly a 5' endonuclease that recognize the dI or a 5' to 3' exonuclease is required to remove the lesion,

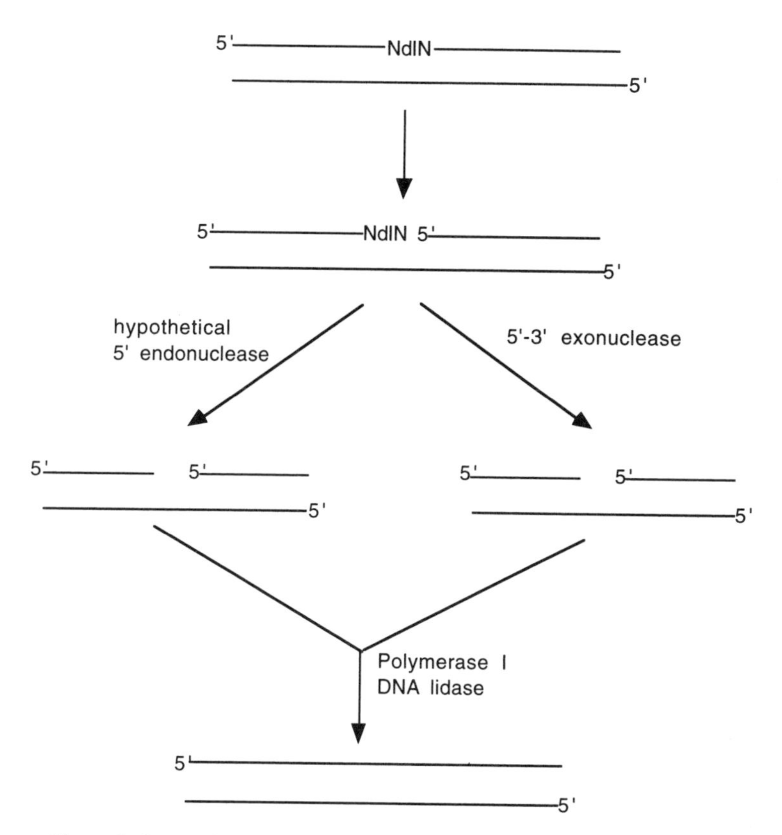

Figure 2. Proposed Pathway for the repair of deoxyinosine by endonuclease V.

generating a small gap ready for DNA polymerase and ligase to complete the repair process. This repair process is depicted in Figure 2. Since endonuclease V has a high affinity for deoxyinosine, it is expected that endonuclease V is important for the repair of deoxyinosine. The possible *in vivo* role of endonuclease V is supported by the fact that *E. coli nfi* mutant exhibit increased sensitivity towards sodium nitrite, an agent known to deaminates both cytosine and adenine (Guo and Weiss, 1998).

Endonuclease V also recognizes deoxyuridine (dU) residue, the deamination product of deoxycytosine (Yao and Kow, 1997). However, the ability to recognize dU is about two orders of magnitude lower. In contrast to DNA containing dI, endonuclease V does not remain bound on DNA containing dU. The complexes formed by endonuclease V with dU containing DNA is much less stable, with K_{d1} and K_{d2} in the range of 40 nM. The ability of endonuclease V to recognize dU suggest that under conditions when uracil N-glycosylase (*ung*) is absence, endonuclease V might be important for the repair of deoxyuracil. This is suggested from the recent study from Weiss (personal communication) demonstrating that *ung nfi* mutant shows a large increase in C-T mutation induction by nitrite.

4.2. Repair of AP Site

Endonuclease V has a very active AP endonuclease activity, cleaving on the second phosphodiester bond 3' to the AP site (Yao et al, 1994a). Since the major AP endonuclease

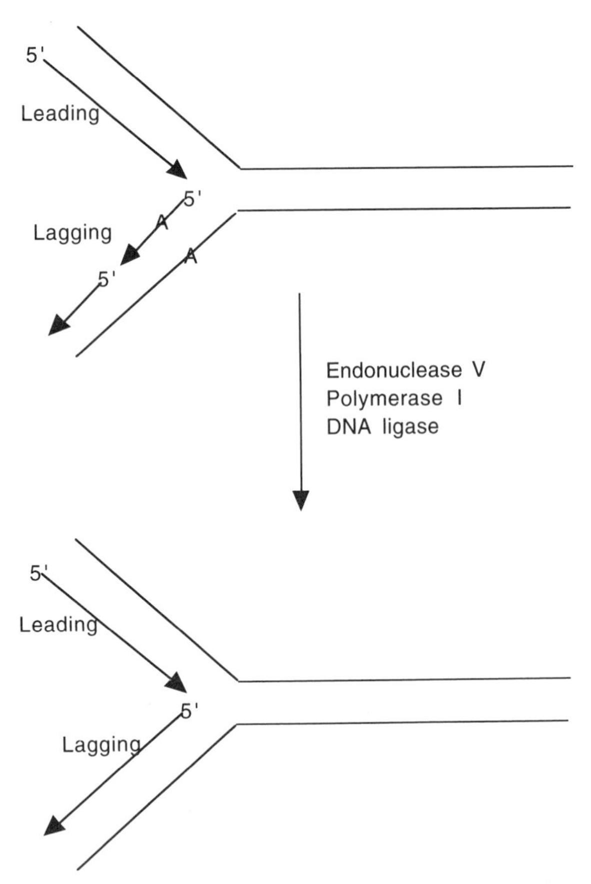

Figure 3. Proposed mismatch repair mediated through endonuclease V.

activities of *E. coli* (exonuclease III, *xth*; and endonuclease IV, *nfo*) cleave the phosphodiester bond 5' to an AP site, it is thus reasonable to think that endonuclease V might participate in the repair of AP site. The *in vivo* role of endonuclease in the repair of AP site is supported by the survival of an *nfi* mutant. Weiss showed that the survival of *ung(ts) xth* is 10^{-4} lower than the wild type *E. coli* cells. This is presumably due to excessive AP sites that are not be repaired in the absence of exonuclease III, the major AP endonuclease in *E. coli*. By inactivating the second major 5' AP endonuclease activity of *E. coli*, endonuclease IV, survival of *ung(ts) xth nfo* mutant cells decrease to 10^{-6}, two orders of magnitude lower than the *ung(ts)xth*. These data suggest that endonuclease IV is important for survival when exonuclease III is absent. Interestingly, *ung(ts) xth nfi* also showed a 5-fold decrease in survival as compared to *ung(ts) xth* (Guo and Weiss, 1998). This data suggest that when the major AP endonuclease is absent, *nfi* becomes important for the repair of AP sites, possibly working cooperatively with endonuclease IV.

4.3. Repair of Base Mismatches

In *E. coli* , base mismatches is repaired through a methyl-directed mutHLS system. Strand discrimination is achieved by determining the status of GATC methylation (Kraemer et al., 1984; Jones et al, 1987; Modrich, 1991). The A in the GATC sequence in the template strand is usually methylated, while the newly synthesized nascent strand is not. MutLS proteins bind to the mismatch base, while mutH binds to the hemi-methylated GATC site, making an incision at the unmethylated GATC site. The mismatch in the nascent strand (the nicked strand) can then be removed by a combination of either a 3' to 5' exonuclease or a 5' to 3' exonuclease, and the repair is completed following polymerization and a ligase reaction.

In is interesting to note that endonuclease V efficiently recognizes base mismatches in a strand specific manner (Yao and Kow, 1994b; 1997). The enzyme is able to recognize all 12 mismatches; however, the ability of endonuclease V to recognize C/C, C/T, C/A mismatch is poor. Interestingly, the substrate specificity is similar to the substrate specificity of the methyl-directed mismatch repair activity of the MutHLS system of *E. coli*. The ability of endonuclease V to recognize mismatch is reduced when the mismatch is flanked by a G/C pair, either 3' or 5' to the mismatch. The strand specificity of the enzyme is governed by the location of the base mismatch. Since endonuclease V interacts only with one strand of the DNA, we proposed that endonuclease V enters the DNA through the 5' terminus, and then move along the DNA in a 5' to 3' direction. The enzyme will make an incision 3' to the first mismatch pair it encounters. Based on this, we suggest that the enzyme might be important for the repair of mismatches on the lagging DNA strand. This is illustrated in Figure 3.

4.4. Repair of Abnormal DNA Structures

Instability of simple repetitive or microsatellite DNA sequences has been observed in many sporadic tumors. Defects in DNA mismatch repair in eukaryotic cells were shown to result in an increase in the instability of repetitive sequences. Mutations in hMSH2, hMLH1, hPMS1 and hPMS2 are associated with hereditary nonpolyposis colon cancer. In yeast, mutation in RTH1 (or RAD 27) also lead to increase instability in these simple repetitive sequences. RTH1 is a flap endonuclease, and its human homolog, FEN1, has been demonstrated to be important for the processing of Okazaki fragments (Harrington and Lieber, 1994a; 1994b; Sommers et al., 1995). One of the substrate that is recognized by RTH1 or FEN1 protein is the flap and pseudo-Y DNA structure. In *E. coli*, this activity was shown to be part of DNA polymerase I (Lyamichev et al, 1993). Interestingly, endonuclease V also exhibits a very active flap endonuclease activity (Yao and Kow, 1996). Similar to FEN1 or RTH1 nucleases, endonuclease V is unable to cleave 3' single stranded tail of pseudo-Y or 3'-flap DNA structure. In addition, endonuclease V also recognizes

Table 2. Comparative substrate specifities of RAD2, RTH1, FEN1 and Endonuclease V

Substrate	RAD2	RTH1	FEN1	Endonuclease V
Pseudo-Y	+	+	+	+
3'-flap	-	-	-	-
5'-flap	+	+	+	+
Hairpin	-	-	-	+
Loop	-	-	-	+
Mismatch	-	-	-	+
dI	-	-	-	+
AP	-	-	-	+
Urea	-	-	-	+

hairpin and DNA loop structures, DNA substrates that are not recognized by the RTH1 class endonucleases. Table 2 shows a comparison of the substrate specificity of RTH1, FEN1 and endonuclease V.

Recently, yeast mutant deficient in RTH1 activity also shows increase instability in the triplet repeat sequences (Schweitzer and Livingston, 1998; Freudenreich et al, 1998). Mutation in RTH1 not only lead to instability in the microsatellite sequences (mostly dinucleotide repeats; Johnson et al, 1995), however, also lead to large expansion of trinucleotide repeat. Trinucleotide repeat is found to be associated with many human disease including fragile X syndrome, muscular dystrophy, huntington disease and several others. Recently, we also showed that endonuclease V mutant lead to an expansion of the trinucleotide repeats (Mitas et al, 1998), suggesting that these structure specific endonucleases might be important for maintaining the stability of these triplet repeat sequences.

REFERENCES

Bhagwat, M and Gerlt, J.A. (1996) 3'- and 5'-Strand cleavage reactions catalyzed by the fpg protein from *Escherichia coli* occur via successive β- and δ-elimination mechanisms, respectively. *Biochemistry* **35**, 659–665.

Boiteux, S., O'Connor, T. R., and Laval, J. (1987) Formamidopyrimidine DNA glycosylase of *Escherichia coli*: cloning and sequencing of the fpg structural gene and overproduction of the protein. *EMBO J.* **6**, 3177–3183.

Castaing, B., Geiger, A., Seliger, H., Nehls, P., Laval , J., Zelwer, C., and Boiteux S . (1993) Cleavage and binding of a DNA fragment containing a single 8-oxoguanine by wild type and mutant FPG proteins. *Nucleic Acids Res.* **2**, 2899–2905.

Chetsanga, C. J., and Lindahl, T. (1979) Release of 7-methylguanine residues whose imidazole rings have been opened from damaged DNA by a DNA glycosylase from *Escherichia coli. Nucleic Acids Res.* **6**, 3673–3683.

Demple, B., and Linn, S. (1982) On the recognition and cleavage mechanism of *Escherichia coli* endodeoxyribonuclease V, a possible DNA repair enzyme. *J. Biol. Chem.* **257**, 2848–2855.

Dodson, M. L., Schrock, R. D., III, and Lloyd, R. S. (1993) Evidence for an imino intermediate in the T4 endonuclease V reaction. *Biochemistry* **32**, 8284–8290.

Freudenreich, C. H., Kantrow, S. M., and Zakian, V. A. (1998) Expansion and length-dependent fragility of CTG repeats in yeast. *Science* **279**, 853–856.

Friedberg, E. C., Walker, G. C., and Siede, W. (1995) DNA Repair and Mutagenesis, ASM Press, Washington, DC.

Gates, F. T., III, and Linn, S. (1977) Endonuclease V of *Escherichia coli.*. *J. Biol. Chem.* **252**, 1647–1653.

Guo, G., Ding, Y., and Weiss, B. (1997) Nfi, the gene for endonuclease V in *Escherichia coli* K-12. *J. Bacteriol.* **179**, 310–316.

Guo, G., and Weiss, B. (1998) Endonuclease V (*nfi*) mutant of *Escherichia coli* K-12. *J. Bacteriol.* **180**, 46–51.

Harrington, J. J., and Lieber, M. R. (1994a) Functional domains within FEN-1 and RAD2 define a family of structure-specific endonucleases: implications for nucleotide excision repair. *Genes Devel.* **8**, 1344–55

Harrington, J. J., and Lieber, M. R. (1994b) The characterization of a mammalian DNA structure-specific endonuclease. *EMBO Journal.* **13**, 1235–1246.

Hatehet, Z., Kow, Y. W., Purmal, A. A., Cunningham, R. P., and Wallace, S. S. (1994) New substrates for old enzymes: 5-hydroxy-2'-deoxycytidine and 5-hydroxy-2' deoxyuridine are substrates for *Escherichia coli* endonuclease III and formamidopyrimidine DNA glycosylase while 5-hydroxy-2'deoxyuridine is a substrate for uracil DNA N-glycosylase. *J. Biol. Chem.* **269**, 18814–18820.

Johnson, R.E., Kovvali, G. K., Prakash, L., and Prakash, S. (1995) Requirement of the yeast RTH1 5' to 3' exonuclease for the stability of simple repetitive DNA. *Science* **269,** 238–240.

Jones, M., Wagner, R., and Radman, M. (1987) Mismatch repair and recombination in *E. coli. Cell* **50**, 621–626.

Karakaya, A., Jaruga, P., Bohr, V. A., Grollman, A. P., and Dizdaroglu , M . (1997) Kinetics of excision of purine lesions from DNA by *Escherichia coli* Fpg protein. *Nucleic Acids Res.* **25**, 474–479.

Kow, Y. W., and Wallace, S. S. (1987) Mechanism of Action of *Escherichia coli* endonuclease III. *Biochemistry* **26**, 8200–8206

Kraemer, B., Kraemer, W., and Fritz, H. J. (1984) Different base/base mismatches are corrected with different efficiencies by the methyl-directed DNA mismatch-repair system of *E. coli*. *Cell* **38**, 879–887.

Lyamichev, V., Brow, M. A., and Dahlberg, J. E. (1993) Structure-specific endonucleolytic cleavage of nucleic acids by eubacterial DNA polymerases. *Science* **260,** 778–783.

Michaels, M. L., Pham, L., Nghiem, Y., Cruz, C., and Miller, J. H. (1990) MutY, an adenine glycosylase active on G-A mispairs, has homology to endonuclease III. *Nucleic Acids Res.* **18**, 3841–3843.

Mitas, M., Yao, M., McEwen, B., Blunt, C., DaSing, Rocha, J., Roger, J., Weiss, B., and Kow, Y. W. (1998) Precise Expansion of d(CAG.CTG) Triplet repeat sequences is enhanced in an *Escherichia coli* strain deficient in endonuclease V activity. Submitted to *Proc. Natl. Acad Sci. USA*

Modrich, P. (1991) Mechanisms and biological effects of mismatch repair. *Annu. Rev. Genet.* **25**, 229–253.

Moriya, M. (1993) Single-stranded shuttle phagemid for mutagenesis studies in mammalian cells: 8-oxoguanine in DNA induces targeted G.C-->T.A transversions in simian kidney cells *Proc.Natl. Acad. Sci. USA* **90**, 1122–1126.

Moriya, M., and Grollman, A. P. (1993) Mutations in the mutY gene of Escherichia coli enhance the frequency of targeted G:C-->T:A transversions induced by a single 8-oxoguanine residue in single-stranded DNA. *Mol. Gen. Genet.* **239**, 72–76.

Purmal, A. A., Rabow, L., Lampman, G. W., Cunningham, R. P., and Kow, Y. W. (1996). A common mechanism of action for the N-glycosylase activity of DNA N-glycosylase/AP lyases from *E. coli* and T4. *Mut. Res.* **364**, 193–207.

Rabow, L. E., and Kow, Y. W. (1996) . Properties of *Escherichia coli* fpg protein site-specific mutants. *FASEB J.* **10**, A966 (abstract)

Rabow, L. E. , and Kow, Y. W. (1997) Mechanism of Action of Base Release by *E. coli* Fpg Protein: Role of Lysine 155 in Catalysis. *Biochemistry* **36**, *5084–5096.*

Schweitzer, J. K., and Livingston, D. M. (1998) Expansion of CAG repeat tracts are frequent in a yeast mutant defective in Okazaki fragment maturation. *Human Mol. Genet.* **7**, 69–74.

Shibutani, S., Takeshita, M., and Grollman, A. P. (1991) Insertion of specific bases during DNA synthesis past the oxidation-damaged base 8-oxodG. *Nature* **349**, 431–434.

Sommers, C. H., Miller, E. J., Dujon, B., Prakash, S., and Prakash, L. (1995) Conditional lethality of null mutations in RTH1 that encodes the yeast counterpart of a mammalian 5'- to 3'-exonuclease required for lagging strand DNA synthesis in reconstituted systems. *J. Biol. Chem.* **270**, 4193–4196.

Tchou , J., and Grollman, A. P . (1995) The catalytic mechanism of Fpg protein. Evidence for a Schiff base intermediate and amino terminus localization of the catalytic site. *J. Biol. Chem.* **270**, 11671–11677.

Tchou, J., Kasai, H., Shibutani, S., Chung, M. H., Laval, J., Grollman, A. P., and Nishimura, S. (1991) 8-oxoGuanine (8-hydroxyguanine) DNA glycosylase and its substrate specificity. *Proc. Natl. Acad. Sci. USA* **88**, 4690–4694.

Tsai-Wu, J. J., Liu, H. F., and Lu, A. L. (1992) Escherichia coli MutY protein has both N-glycosylase and apurinic/apyrimidinic endonuclease activities on A-C and A-G mispairs. *Proc. Natl. Acad. Sci. USA* **89**, 8779–8783.

Wood, M. L., Dizdaroglu, M., Gajewski, E., and Essigmann, J. M. (1990) Mechanistic studies of ionizing radiation and oxidative mutagenesis: genetic effects of a single 8-hydroxyguanine (7-hydro-8-oxoguanine) residue inserted at a unique site in a viral genome. *Biochemistry* **29**, 7024–7032.

Yao, M., Hatahet, Z., Melamede, B., and Kow, Y. W. (1994a_) Purification and characterization of a novel *Escherichia coli* repair enzyme, deoxyinosine 3' endonuclease. *J. Biol.Chem.* **269**, 16260–16268.

Yao, M., and Kow, Y. W. (1994b) Strand-specific Cleavage of Mismatch-containing DNA by Deoxyinosine 3' Endonuclease from *Escherichia coli*. *J. Biol. Chem.* **269**, 31390–31396.

Yao, M., and Kow, Y. W. (1995) Interactions of deoxyinosine 3' endonuclease with DNA containing deoxyinosine. *J. Biol. Chem.* **270**, 28609–28616.

Yao, M., and Kow, Y. W. (1996) Cleavage of insertion/deletion mismatches, flap and pseudo Y DNA structures by deoxyinosine 3' endonuclease from *Escherichia coli*. *J. Biol. Chem.* **271**, 30672–30676.

Yao, M., and Kow, Y. W. (1997) Further characterization of *Escherichia coli* endonuclease V. Mechanism of recognition of deoxyinosine, deoxyuridine and base mismatches in DNA. *J. Biol. Chem.* **272**, *30774–30779.*

Zharkov, D. O., Rieger, R. A., Iden, C. R., and Grollman, A. P. (1997) NH2-terminal proline acts as a nucleophile in the glycosylase/AP-lyase reaction catalyzed by *Escherichia coli* formamidopyrimidine-DNA glycosylase (Fpg) protein. *J. Biol. Chem.* **272**, 5335–5341.

18

HUMAN URACIL-DNA GLYCOSYLASE

Gene Structure, Regulation, and Structural Basis for Catalysis

Hans E. Krokan, Frank Skorpen, Marit Otterlei, Sangeeta Bharati, Kristin Steinsbekk, Hilde Nilsen, Camilla Skjelbred, Bodil Kavli, Rune Standal, and Geir Slupphaug

UNIGEN Center for Molecular Biology
Norwegian University of Science and Technology, The Medical Faculty
N-7005 Trondheim, Norway

1. ABSTRACT

Uracil in DNA results from either misincorporation of dUMP residues during replication, or from deamination of cytosine residues. The latter process results in premutagenic U:G mispairs that, unless uracil is removed, will cause GC→AT transitions in the subsequent round of replication. The 13.8 kb gene for human uracil-DNA glycosylase, *UNG*, is highly conserved and comprises 7 exons. It encodes more than 98% of the total uracil-DNA glycosylase activity in the cell. The crystal structure of the catalytic domain of UNG in complex with target DNA has demonstrated that all essential contacts are with the uracil-containing strand. The structure also reveals the mechanism of enzyme-assisted flipping of the uracil-containing nucleotide into the deep catalytic pocket that specifically binds uracil. Nuclear (UNG2) and mitochondrial (UNG1) forms of the enzyme result from the use of two promoters, P_A and P_B, and alternative splicing. mRNA for UNG1 encodes 304 amino acids, the first 35 of which are unique to this form. mRNA for UNG2 encodes 313 amino acids, the first 44 of which are unique to UNG2. The unique N-terminal sequences in UNG1 and UNG2 are required for mitochondrial and nuclear sorting, respectively, but not for catalytic activity. The 269 amino acid residues common to the two forms include the compact catalytic domain of approximately 220 C-terminal residues and an N-terminal part that binds replication protein A (RPA), indicating a possible role for RPA in base excision repair.

2. INTRODUCTION

Uracil is not a normal constituent in DNA, but may still be present in low amounts either due to misincorporation of dUMP during replication, or due to spontaneous or in-

Advances in DNA Damage and Repair, edited by Dizdaroglu and Karakaya. Kluwer Academic / Plenum Publishers, New York, 1999.

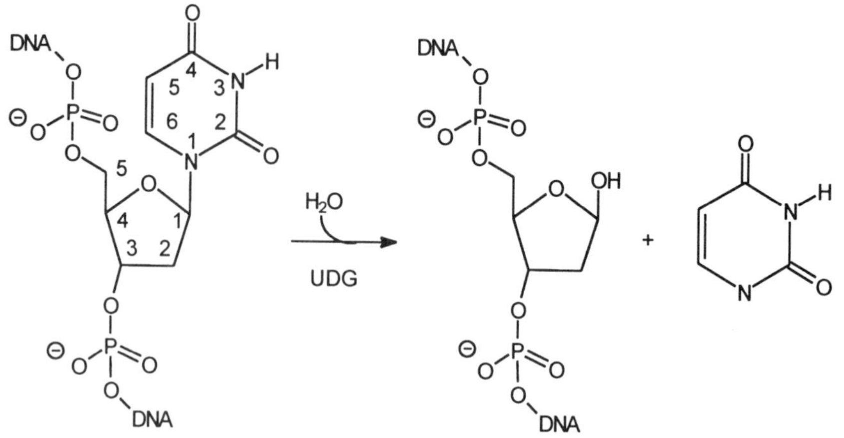

Figure 1. Removal of uracil from DNA by uracil-DNA glycosylase, resulting in an apyrimidinic site.

duced deamination of cytosine residues in DNA (reviewed in Krokan et al., 1997a; Krokan et al., 1997b). Uracil-DNA glycosylase (UDG) excises uracil from DNA by hydrolyzing the bond between N1 in uracil and C1' in deoxyribose, thereby releasing uracil as a free base (Fig.1). It was the first DNA glycosylase to be identified (Lindahl, 1974). UDG is a damage recognizing enzyme that initiates the base excision repair (BER) pathway for removal of uracil from DNA. Successful reconstitution of the BER pathway has recently been carried out with purified proteins (Sobol et al., 1996; Kubota et al., 1996; Nicholl et al., 1997) and is reviewed in detail in a separate chapter in this volume. The primary function of UDG is presumably to remove uracil from premutagenic U:G mispairs resulting from cytosine deamination. The significance of UDG in mammalian cells has not yet been established. However, *Escherichia coli* or yeast cells carrying mutations in the gene for UDG show a several-fold increased spontaneous mutation rate (Duncan et al., 1982; Impellizzeri et al., 1991). The mutations in UDG-deficient cells are predominantly GC→AT transitions indicating that cytosine deamination is the major premutagenic event (Fix et al., 1987). Even though UDG is a highly selective DNA glycosylase, it nevertheless recognizes and excises certain damaged cytosine-residues that are structurally related to uracil, but with low efficiencies. These include alloxan, isodialuric acid and 5-hydroxyuracil produced by oxidative stress such as γ-radiation (Dizdaroglu et al., 1996) and 5-fluorouracil in DNA resulting from chemotherapy (Mauro et al., 1993). These uracil-homologues are, however, only slowly removed and the possible significance of UDG in protection against damage due to oxidative stress is not known.

Genes for UDG from prokaryotes, eukaryotes and eukaryote viruses are highly conserved (Olsen et al., 1989). Crystal structures of UDGs from herpes simplex type-1 (HSV-1) and humans are also very similar and the sequence conservation of other UDGs indicate that their active sites are likely to be essentially identical (Savva et al., 1995; Mol et al., 1995b). These apparently simple enzymes have revealed very interesting properties that now can serve as a model for interactions between modified DNA and proteins. In the present chapter, we review the biochemical properties and structure of human UDG from the *UNG*-gene, as well as the structure and expression of the *UNG*-gene.

3. RESULTS AND DISCUSSION

3.1. Uracil in DNA

3.1.1. Misincorporation of dUMP during Replication. DNA polymerases from various sources generally do not discriminate between dUTP and dTTP, although rates of incorporation may be slightly lower for dUTP (Slupphaug et al., 1993a). dUTP is a normal intermediate in nucleotide metabolism, but the actual levels of dUTP in the nuclei of mammalian cells are not well established. Briefly, a cellular dUTPase rapidly hydrolyzes dUTP to dUMP, which is converted to dTMP by thymidylate synthase. Two kinases subsequently convert dTMP to dTTP (reviewed in Krokan et al., 1997a). Thymidylate synthase requires N^5,N^{10}-methylene tetrahydrofolate as a methyldonor. Factors that either reduce thymidylate synthase activity (such as the anticancer drug 5-fluorouracil), or reduce the level of N^5,N^{10}-methylene tetrahydrofolate (such as a folate-deficient diet, or the anticancer drug methotrexate) may therefore increase dUMP incorporation by increasing the ratio of dUTP to dTTP. Although misincorporated dUMP-residues are not mispaired, they may influence genetic processes in more indirect ways. In animal experiments, a folate-deficient diet has recently been shown to increase the content of uracil in DNA and lead to DNA fragmentation presumably due to removal of uracil by UDG and subsequent strand cleavage by an AP-endonuclease (Pogribny et al., 1997). Furthermore, folate deficiency has been demonstrated to cause massive incorporation of dUMP into human DNA (Blount et al., 1997), thus increasing the risk for two opposing nicks being formed and subsequent chromosome breakage. dUTPase is essential for viability of both *E. coli* and yeast cells, possibly in part due to fragmentation of DNA by repair processes in its absence (Gadsden et al., 1993; el Hajj et al., 1988; el Hajj et al., 1992). In addition, dUMP-incorporation may compromise transcription due to inefficient recognition of transcription factor elements by some transcription factors (Focher et al., 1992; Verri et al., 1990).

3.1.2. Deamination of Cytosine Residues. Cytosine in DNA is inherently unstable and is slowly deaminated to uracil in a hydrolytic process (Shapiro et al., 1966). The major route for deamination probably involves protonation at the N3 position, followed by an attack on the protonated base by H_2O at position 4 in a general acid-catalyzed reaction, leading to release of NH_4^+ (Shapiro, 1980, reviewed in Krokan et al., 1997a). Rates of deamination are enhanced by bisulfite, which is specific for cytosine, as well as by nitrous acid that also deaminates adenine and guanine (Shapiro, 1980). The rate of cytosine deamination under physiological conditions is 200–300-fold higher in single stranded DNA, as compared to double stranded. The rate constant for deamination of cytosine in single stranded DNA was estimated to be approximately 2×10^{-10} residues per second at 37°C (Shapiro, 1980) and this is close to results obtained using sensitive genetic assays for detection of deamination of cytosine in single stranded- and double stranded DNA (Frederico et al., 1990; Bockrath et al., 1986). Since a fraction of genomic DNA in cells is single stranded due to replication, transcription and repair processes, as well as breathing in AT-rich sequence contexts, the *in vivo* rates of deamination may be higher than those measured for double stranded DNA. Current estimates for mammalian cells are 100–500 deaminations of cytosine per genome per day (Lindahl, 1993).

In cyclobutyl cytosine dimers resulting from UV-irradiation, deamination is increased approximately 7–8 orders of magnitude as compared to other cytosine residues. It has therefore been suggested that replication across deaminated cytosine residues may contribute to mutagenesis after exposure to UV-light (Barak et al., 1995). Uracil in DNA may also result from enzymatic deamination of cytosine. Thus, (cytosine-5) methyltransferase, that normally methylates cytosine residues in a CpG context may cause deamination when the level of the methyldonor S-adenosylmethionine (SAM) is very low (Shen et al., 1992) or when the transferase is mutated so that it is unable to bind SAM (Shen et al., 1995). Interestingly, the level of (cytosine-5)methyltransferase is frequently increased several hundred-fold, and occasionally up to 3000-fold in tumor cells (Bandaru et al., 1996). This might lead to an imbalance between methyltransferase and SAM and it can not be ruled out that premutagenic U:G mispairs may be formed by an enzymatic process in some tumor cells, thereby contributing to a mutator phenotype.

3.2. Biochemical Properties of UDGs

All identified UDGs of the conserved family are relatively small, monomeric and basic proteins that do not require cofactors for their activity. In contrast to deoxyribonucleases they are fully active in the absence of divalent cations and in the presence of metal chelating agents such as EDTA (Krokan et al., 1997a). These properties are, with one possible exception, shared with other DNA glycosylases (Krokan et al., 1997b). Compared with other DNA glycosylases, UDGs have relatively high turnover numbers, typically in the range 500–1000 uracils per min, depending on assay conditions. UDG-activity can therefore easily and unambiguously be quantified in crude cell extracts of relatively few cells in the presence of EDTA, although UDGs are not abundant proteins. These properties also have a downside: The positive charge of UDGs and their high catalytic activity have probably resulted in questionable results where UDG-activity has been assigned to more abundant proteins with which small amounts of UDG copurify due to physical interaction or similarity in biochemical properties.

3.3. Biochemical Properties of Human Uracil-DNA Glycosylase

A catalytically active form of human UDG was originally obtained from human placenta after 131,000-fold purification in 7 chromatographic steps. A contaminating protein of approximately 26.5 kDa was separated from UDG by SDS-polyacrylamide gel electrophoresis and the N-terminal sequence of UDG determined (Wittwer et al., 1989). Based on this sequence information, cDNA for human UDG (named UNG15) was cloned from a cDNA library from human placenta and found to be homologous to UDG from *E. coli*, yeast and open reading frames in several animal viruses. The open reading frame in the human cDNA encoded 77 N-terminal amino acid residues not present in the UDG purified from human placenta (Olsen et al., 1989). The significance of N-terminal sequences in human UDG is described in more detail below. UDG purified from human placenta had an apparent molecular weight of 29 kDa and had properties essentially resembling those of bacterial and yeast UDGs. The enzyme was a strongly basic protein with a K_m value of approximately 2 μM and a turnover number of 600 uracils per min on dsDNA. It was some 2–3-fold more active on single stranded DNA, as compared with double stranded, and was stimulated 4-fold with monovalent salt up to 60 mM, whereas higher concentrations were inhibitory. Essentially similar properties were found later using a homogeneous form of human UDG expressed in *E. coli* (Slupphaug et al., 1995). This form lacked 7 N-terminal conserved amino acid residues present in the placental enzyme and was named UNGΔ84. The pH optimum for UNGΔ84 was 7.7–8.0, the activation energy 50.6 kJ/mol and the p*I* between 10.4 and 10.8. K_m values were 0.45 μM and 1.6 μM on ssDNA and dsDNA, respectively, but both the K_m and the V_{max} increased with increasing salt concentrations. The recombinant form of UDG displayed a striking sequence specificity in removal of U from U:A base pairs in M13 dsDNA, as previously found for bovine UDG (Eftedal et al., 1993) and UDG from *E. coli* (Nilsen et al., 1995). The sequence specificity for removal of U from U:G mispairs was essentially similar. In general U was removed faster from U:G mispairs, although there were several exceptions from this rule. The rates of removal of U differs as much as 20-fold between the «best» (A/T)UA(A/T) and «worst» (G/C)U(G/C/T) sequences. This may be biologically significant because, at least in *E. coli,* a low rate of removal tends to correlate with the occurrence of hot spots for mutation (Nilsen et al., 1995). Polyclonal antibodies raised against UNGΔ84 neutralized more than 98% of all detectable UDG-activity in crude extracts from HeLa cells, strongly indicating that human UDG corresponding to the UNG cDNA represents the major uracil-DNA glycosylase in human cells (Slupphaug et al., 1995). As described in more detail later, the essentially complete inhibition of UDG-activity is explained by the fact that the *UNG*-gene encodes both mitochondrial and nuclear forms of UDG and these forms have identical catalytic domains represented by the UNGΔ84 enzyme.

3.4. Structure of Human Uracil-DNA Glycosylase

The truncated form of human UDG (UNGΔ84) was efficiently expressed in *E. coli*, easy to purify and soluble at high concentrations, thus serving as an excellent candidate for crystallization and structural studies. The UNGΔ84 structure (Mol et al., 1995a) consists of a single α/β-domain containing eight α-helices and a central four-stranded parallel and twisted β-sheet (Fig. 2). A long groove containing many basic residues runs along the surface of the protein and has a width of 21 Å at the rim, equaling the diameter of double stranded DNA. The sides of the groove are made up by five sequence regions, as depicted in figure 3. At one end of the groove a deep pocket is located above the C-terminal end of the β-sheet. This is the pocket in which uracil binds (Fig 4). The amino acid residues forming this pocket are highly conserved and structurally identical in HSV-1 (Savva et al.,

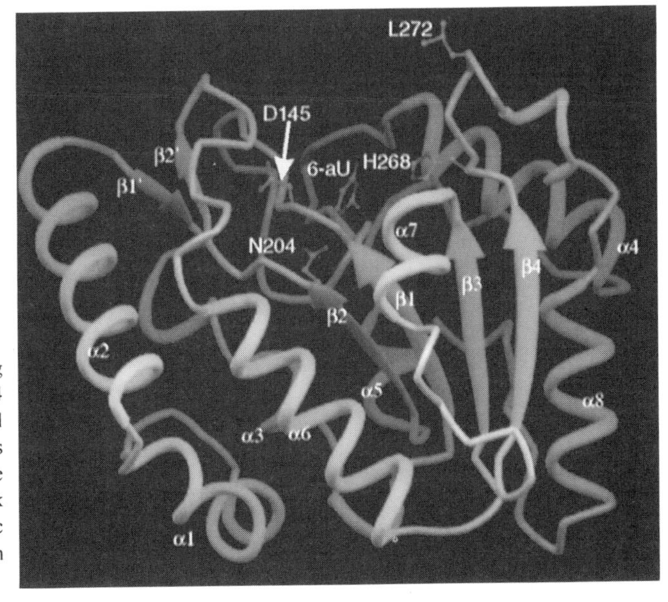

Figure 2. Ribbon diagram showing the secondary structure of UNGΔ84 in complex with 6-aminouracil (ball and stick model, top). The numbers of the β strands and the α helices are as shown in figure 3. Ball and stick models indicate important catalytic side chains, and Leu272 involved in nucleotide flipping.

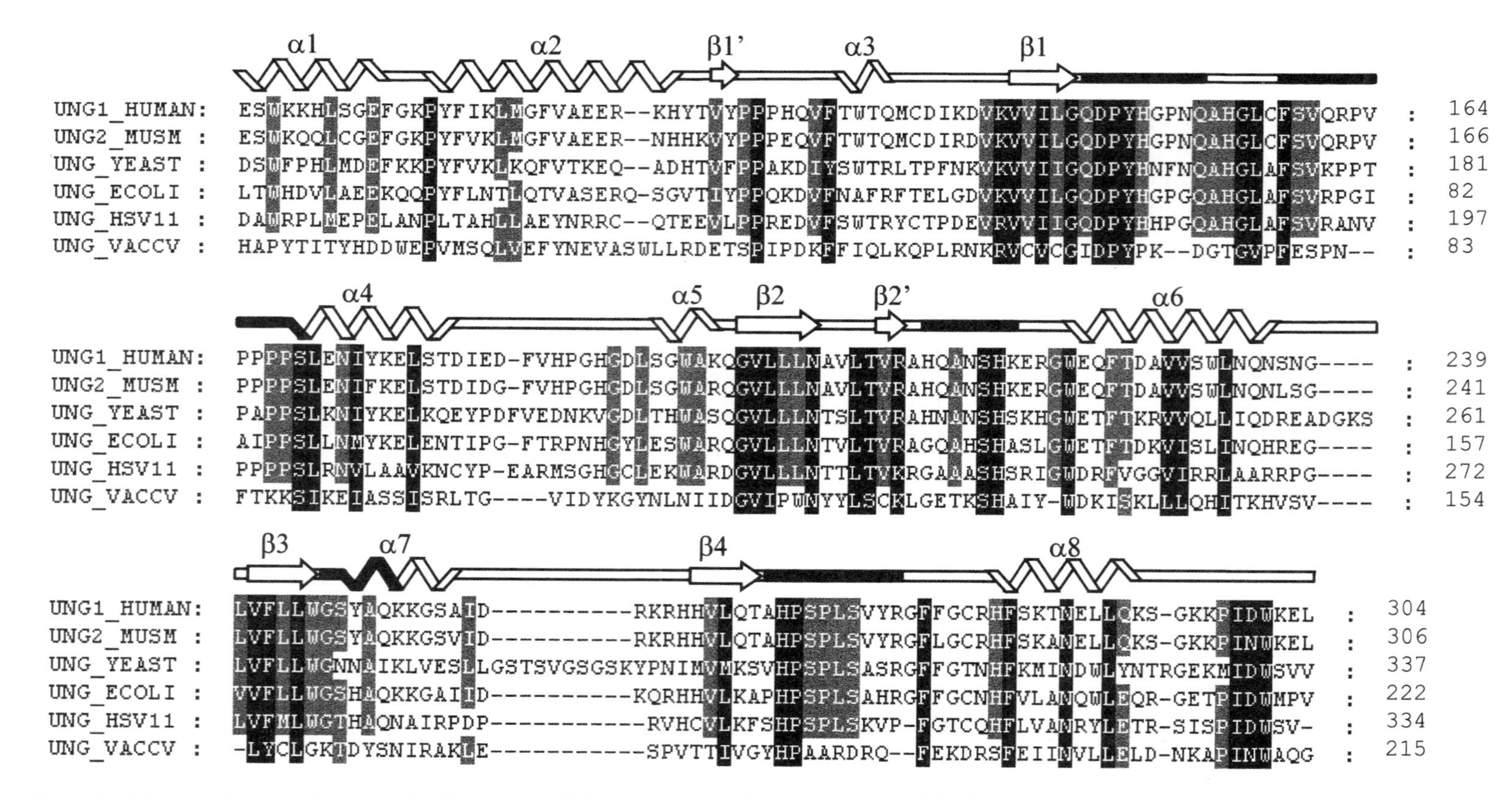

Figure 3. Alignment of conserved amino acid regions in some UDGs representing the highly conserved UDG family (human through Herpes simplex virus type 1), and in the more distantly related vaccinia UDG. Secondary structure elements observed in the crystal structure of human UDG are illustrated above the sequences, and the five regions known to make up the walls of the DNA-binding groove are illustrated by black boxes inside the secondary structure. The sequences with accession numbers are: UNG1_HUMAN, human, P13051; UNG2_MUSM, mouse, Y08975; UNG_YEAST, *Saccharomyces cerevisiae*, P12887; UNG_ECOLI, *Eschericia coli*, P12295; UNG_HSV11, Herpes simplex virus type 1, P10186; UNG_VACCV, Vaccinia virus, P20536.

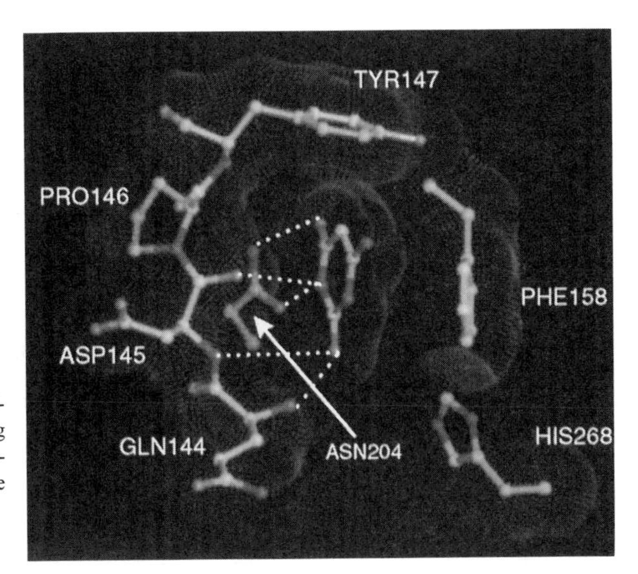

Figure 4. Ball and stick model of the 6-aminouracil bound in the uracil binding pocket. Main chain and side chain hydrogen bonds to conserved residues in the pocket are indicated by dotted lines.

1995; Pearl et al., 1995) and human UDGs (Mol et al., 1995a). All residues implicated in catalysis by site directed mutagenesis are located within this groove, and the largest effects of site directed mutagenesis are seen after changes of residues located in the pocket. The pocket is too narrow and too deep to accommodate double stranded DNA, thus it was suggested that UDG must excise uracil "flipped out" from the double helix. Asp145, the only acidic residue within the pocket, as well as Gln144, Pro146, Tyr147 and Asn204 line one side of the active site. The Tyr147 side chain OH forms a hydrogen bond with the Phe158 backbone carbonyl oxygen and these residues are conserved among known UDGs. The side chain of Phe158 lies along the bottom of the active site and interacts with uracil by stacking. His268, Ser169 and Ser270 line the other side of the active site.

The specificity of the active site for uracil is well established: The Asn204 side chain Nδ2 and Oδ1 atoms form hydrogen bonds with O4 and N3 in uracil, respectively, and exclude binding of cytosine. The N3 position also forms a hydrogen bond with the backbone carbonyl of Asp145. The backbone amides of Asp145 and Gln144 both form hydrogen bonds to the O2 atom of uracil, which also forms a hydrogen bond with the Nε2 atom of His-268 via a bridging water molecule. Thymine can not bind in the active site pocket because its 5-methyl group would sterically clash with the side chain of Tyr147. Consistent with these structural data, certain active site mutants not only have strongly reduced catalytic rates, but also reduced specificity. Thus, an Asn204Asp mutant, in addition to having reduced UDG-activity, has gained a cytosine-DNA glycosylase (CDG) activity because its carboxyl side chain may hydrogen-bond to the amino group in the 4-position. Furthermore, the mutants Tyr147Ala, Tyr147Ser and Tyr147Cys all have strongly reduced UDG activity, but have gained thymine-DNA glycosylase (TDG) activity (Kavli et al., 1996). Other mutants (Gln144Leu, Asp145Asn, Asn204Gln and His268Leu) have strongly reduced UDG activities, but do not recognize thymine or cytosine. All these biochemical results are entirely consistent with the structural data on the mechanism of uracil-binding.

Recently, the structure of a double mutant of UDG (Asp145Asn/Leu272Arg) in complex with a double stranded oligonucleotide containing uracil opposite of guanine was reported (Slupphaug et al., 1996). This has demonstrated the basis for enzyme-assisted nucleotide flipping. In the UDG-DNA complex the positively charged groove traversing the UDG surface orients the enzyme along the DNA (Fig. 5). A conserved leucine (Leu272 in the

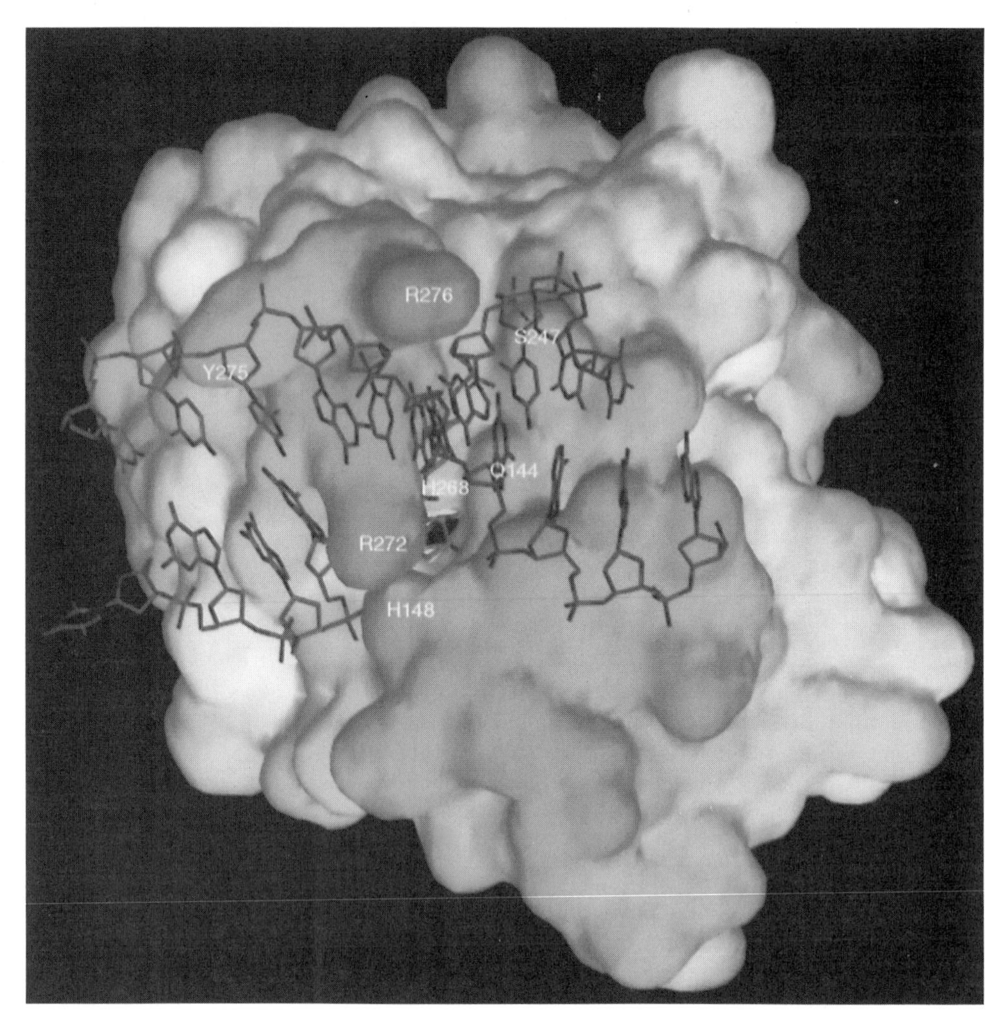

Figure 5. UDG-DNA complex viewed from above the DNA-binding groove and down into the uracil-binding pocket occupied by the uracil (black) and the abasic deoxyribose. Some of the residues lining the DNA-binding groove and uracil binding pocket are indicated.

wild type enzyme, but Arg272 in the mutant) located directly above the uracil-binding pocket aids in minor groove scanning and expulsion («push») of uracil from the dsDNA base stack via the major groove. A concomitant compression of the DNA backbone phosphates flanking the uracil and specific recognition of the 5'-phosphate, deoxyribose and uracil by UDG active site residues ("pull") stabilizes the extrahelical nucleotide conformation and promotes concerted condensation of the surrounding residues to form a productive complex specific for uracil excision. Hydrogen bonds between Ser169 and deoxyribose 5'-phosphate and between Ser270 and the deoxyribose 3'-phosphate also assist in the flipping of the uracil-containing nucleotide. The significance of interaction with 5'-phosphate and 3'-phosphate is in agreement with the observations that the minimal substrate for UDG is p(dU)p(dN)p (Varshney et al., 1991) and that uracil at the 3'-end is poorly removed (Krokan et al., 1981). A schematic presentation of interactions between UDG and uracil-containing target DNA is presented in

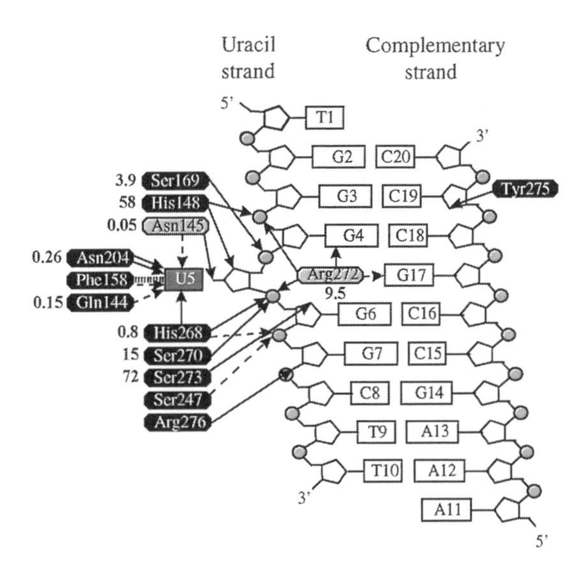

Figure 6. UDG-DNA interactions. Nucleotide numbering follows the 5'-3'-direction, starting from the uracil-containing strand. Amino acids interact with the DNA bases primarily along the sugar-phosphate backbone surrounding uracil. Numbers adjacent to amino acids in boxes indicate per cent specific UDG activity remaining when the respective residues are substituted with structurally similar non-polar residues. Numbers adjacent to Asn145 and Arg272 indicate specific activity of these mutants compared to the wild type Asp145 and Leu272, respectively. Arrows, side chain interactions; broken arrows, main chain interactions; vertical lines, stacking interactions.

figure 6. All interactions between UDG and DNA, except one, involve the uracil-containing strand, and most contacts are via the minor groove.

3.5. Mechanism of Catalysis

Catalysis requires specific binding of target DNA to UDG and nucleotide flipping, as described above, followed by cleavage of the bond between C1 in deoxyribose and N1 in uracil. Upon binding of target DNA, the active site undergoes concerted conformational changes and contraction of the active site. This includes movement of the Arg272 (Leu272 in the wild type) side chain such that it penetrates the DNA double helix, as well as an approximately 2 Å movement of His268, which is linked to Leu272 by the intervening rigid sequence Pro-Ser-Pro. The movement of His268 allows interaction with uracil O2, thereby stabilizing an oxyanion intermediate. Furthermore, in the buried active site, the side chain of Asn145 (Asp145 in the wild type) rotates towards bound uracil and may activate a water molecule, with the resultant hydroxyl nucleophile making a direct in-line attack on deoxyribose C1', as suggested previously (Savva et al., 1995). The residue 272 backbone carbonyl hydrogen bonds to the exocyclic amino group of the orphan guanine, and may thus contribute to the observed selectivity for U:G over U:A. The described mechanism for selective uracil-binding and catalysis does not explain the dramatic reduction in catalytic rate caused by the Gln144Leu (main chain interaction only) and Tyr147Ala (steric hindrance of thymine-binding). Possibly Gln144 may orient the water molecule or the catalytic Asp145, whereas both residues may contribute to bulk water exclusion from the active site. One key question is why UDG attacks uracil in DNA, but not in RNA. The selectivity is due to steric hindrance in His268 movement imposed by the ribose 2-hydroxy group and 3'-*endo* puckering.

3.6. Structure of the Human *UNG*-Gene and Mechanism for Formation of Mitochondrial and Nuclear Form of UDG

Cloning of genes or cDNAs for UDGs from bacteria, yeast, human cells, as well as homologous genes from several herpesviruses have demonstrated a striking similarity between these enzymes, ranging from 40.3 to 55.7% identity at the amino acid level (Olsen et al., 1989 and references therein). Unexpectedly, the proteins of human and bacterial origin were found to be most closely related, 73.3% similarity when conservative amino acid substitutions were included. The similarity is confined to several discrete boxes. The *UNG*-gene in mouse encodes proteins with approximately 90% identity to the human proteins (Nilsen et al., 1997). These findings demonstrate that UDGs from phylogenetically distant species are highly conserved. UDGs from various pox viruses are more distantly related, but the active site region is also highly conserved in these viruses.

The human uracil-DNA glycosylase gene (*UNG*) spans approximately 13.8 kb including the promoter regions and has been assigned to position 12q23-q24.1. (Haug et al., 1996). *UNG* comprises seven exons and exhibits typical features of housekeeping genes, including a 5' CpG island and two TATA-less promoters, P_A and P_B, required for formation of nuclear (UNG2) and mitochondrial (UNG1) forms of human UDG, respectively (Nilsen et al., 1997). Within the 15 kb sequence examined, 16 *Alu*-retroposons, 3 copies of medium reiteration (MER) frequency repeats and 1 copy of a mammalian-wide interspersed repetitive element (MIR) were detected. In addition, a 300 bp TA-dinucleotide repeat is located in the last intron. This repeat is hypervariable (M. Akbari, F. Skorpen and H.E. Krokan, unpublished results). *In vitro* methylation of the promoter region strongly reduces promoter activity, but methylation may not be involved in the regulation of *UNG in vivo* since the promoter region appears to be invariably methylation free.

Previously, similarity of the mitochondrial and nuclear form of UDG was indicated by similar biochemical properties (Wittwer and Krokan, 1985), inhibition of both forms by the highly selective protein inhibitor Ugi (Karran et al., 1981), transfection and immunocytochemistry experiments (Slupphaug et al., 1993b) as well as complete inhibition of UDG activity by antibodies in crude cellular extracts (Slupphaug et al., 1995). The mechanism for generation of the mitochondrial (UNG1) and nuclear (UNG2) forms from the *UNG*-gene has now been established (Nilsen et al., 1997). It basically involves the use of two promoters and alternative splicing (Fig. 7). P_B is located in intron 1 and contains a large number of putative transcription factor binding elements, mediating both negative and positive regulation of transcription (Haug et al., 1994). Transcription from this promoter results in an mRNA encoding a protein of 304 residues, the first 35 of which are unique to UNG1 and are required for sorting to mitochondria. Transcription from P_A, located upstream of exon 1A which encodes 44 amino acids, and splicing of this exon to a consensus splice acceptor site after codon 35 in exon 1B, results in an mRNA encoding the 313 amino acid residues of UNG2. The 44 N-terminal amino acid residues unique to UNG2 are required for transport of this form to nuclei. The sequence downstream of codon 35 in exon 1B, altogether 269 amino acid residues, are common to UNG1 and UNG2

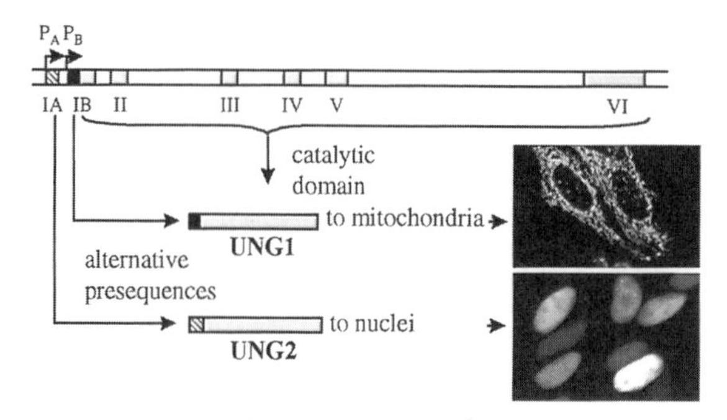

Figure 7. Formation of UNG1 and UNG2 by alternative promotor usage and alternative mRNA splicing. The unique N-terminal sequences thus formed are necessary for mitochondrial (UNG1) and nuclear translocation (UNG2), respectively. Upper micrograph, HeLa cells expressing an UNG1-EGFP fusion protein; lower micrograph, HeLa cells expressing an UNG2-EGFP fusion protein.

(Nilsen et al., 1997). We have recently shown that the 35 unique amino acid residues in UNG1 are required and sufficient for mitochondrial import. In contrast, the 44 unique residues in UNG2 are required, but not sufficient for nuclear import. The nuclear localization signal is in fact very complex and complete sorting to nuclei requires more than 100 N-terminal residues (Otterlei et al., 1998). This novel mechanism for generation of mitochondrial and nuclear species of a protein may not be unusual, at least not for DNA repair enzymes. Thus, closely related forms of mammalian 8-oxoG-DNA glycosylase are found in rat mitochondria (Croteau et al., 1997) and nuclei from human cells (Bjørås et al., 1997), and in addition closely related human forms with unknown subcellular localization (Roldan-Arjona et al., 1997; Radicella et al., 1997) have been identified. These apparently result from different splice forms of mRNAs from the same gene. Furthermore, cDNAs with different 5'-ends generated by alternative splicing and the use of different promoters have been described for human alkyl-DNA glycosylases (MPGs) (O'Connor et al., 1991; Samson et al., 1991; Chakravarti et al., 1991; Vickers et al., 1993) and the corresponding transcripts were found simultaneously in different cell lines and tissues (Pendlebury et al., 1994). A tissue-specific mode of expression was therefore ruled out, making it likely that the alternative N-terminal ends may be involved in subcellular targeting.

3.7. Expression of the *UNG*-Gene

UDG-activity is generally significantly higher in proliferating tissues, as compared to resting tissues and terminally differentiated cells (Krokan et al., 1997a; Krokan et al., 1997b). With the possible exception of rat neuronal cells (Focher et al., 1990), UDG-activity has been detected in all mammalian cells investigated. The actual levels apparently vary from individual to individual, the basis for which is not known (Krokan et al., 1983; Myrnes et al., 1983; Myrnes et al., 1984). UDG-activity is induced several-fold in rat liver after partial hepatectomy (Aprelikova et al., 1982) and in lymphocytes after treatment with phytohemagglutinin (Sirover, 1979). Both in human lymphocytes and fibroblasts, UDG activity starts to increase just prior to the S-phase and is elevated throughout the S-phase (Sirover, 1979; Gupta et al., 1981; Slupphaug et al., 1991). In fibroblasts, *UNG* transcript levels increase 8–10-fold in late G_1-phase, preceding an increase in enzyme activity in the S-phase (Slupphaug et al., 1991). Since assays for UDG-activity do not discriminate between nuclear and mitochondrial activities, the relative contributions of enzyme activities from UNG1 and UNG2 in different tissues and cell lines are not known. Furthermore, cDNAs, and presumably mRNAs, for UNG1 and UNG2 are of very similar sizes, 2058 and 2061 bp, respectively, and therefore they were not resolved in Northern blots in previous studies (Olsen et al., 1989; Slupphaug et al., 1991). Recent studies, using probes specific for exon 1A (UNG2) or the 5'-part of exon 1B (UNG1), have demonstrated that a) mRNAs for both forms are induced late in the G_1-phase, although UNG2 increases more than UNG1, b) the expression pattern in different human tissues is strikingly different with UNG1 mRNA being expressed in all tissues examined whereas UNG2 mRNA is essentially only expressed in proliferating tissues (Haug et al., 1998). Promoters P_A (UNG2) and P_B (UNG1) both contain a number of putative and verified transcription factor binding elements which have been studied by mutational analysis and transfection experiments. c-Myc and Sp1-binding elements present in both promoters are positive regulators, although overexpression of c-Myc is inhibitory. An E2F-1 binding element in P_B is a positive regulator when it binds free E2F-1, but a negative regulator of both promoters when E2F-1 is complexed to Rb. Somewhat surprisingly, expression from P_A, which does not contain a consensus E2F-binding site, is also suppressed by an E2f-1/Rb complex. The molecular basis for differential regulation of P_A and P_B thus remains unclear.

A region N-terminal to the catalytic domain, but common to UNG1 and UNG2 (residues 73–84 in UNG2) has been shown to interact with the 34kDa subunit (RPA2) of replication protein A, (RPA) (Nagelhus et al., 1997) a heterotrimeric protein required for DNA replication, nucleotide excision repair (NER) and recombination (Wold, 1997). Interestingly, the region in UNG-proteins required for interaction carries homology to the RPA2-interacting region in XPA, one of the damage recognizing proteins in NER. Furthermore, a second region that interacts with RPA has recently been identified in the very N-terminal region of UNG2 (residues 7–18) (Otterlei et al., 1998). These observations may indicate a role for RPA in base excision repair. UNG2 apparently has two subnuclear localizations; one fraction is homogeneously spread over nuclei, whereas another fraction is seen as intensively staining spots that morphologically resemble replication foci. UNG2 is, however, apparently not directly involved in DNA replication since polyclonal antibodies that essentially completely inhibit UNG2-activity in nuclei have no effect on DNA replication in isolated nuclei (Otterlei et al., 1998).

In conclusion, research on human uracil-DNA glycosylase has provided detailed information into the genetics, structure, cell biology and biochemistry of the enzyme. Still, many questions remain unanswered. Are mitochondrial and nuclear forms both essential, or is only one of them, or possibly neither form, essential? Are deficiencies or dysfunction of one of the forms of the enzyme linked to heritable or acquired human disease? May the enzyme have functions not yet considered? At the molecular level it still remains unclear how the enzyme is able to detect the relatively few uracil residues present among the large excess of closely related, but normal pyrimidines in a complex chromatin context. We have calculated that there are some 10^4 (fibroblasts) to 10^5 (HeLa cells) molecules of human uracil-DNA glycosylase per cell, some 70% of which are nuclear in transformed cells. This implies that each molecule somehow must scan some 10^4 to 10^5 DNA nucleotides per day in cells with a doubling time of 24 hrs in order to avoid mutations. These numbers are much higher than the estimated number (some 100–500) of deaminations per day. In fact, the numbers roughly correspond to the number of replication units per cell. Uracil-DNA glycosylase is upregulated late in the G_1/early S-phase and this might suggest that it may "prescan" DNA prior to replication to repair U:G mispairs, or "postscan" newly replicated DNA to remove misincorporated dUMP residues. It has been demonstrated that uracil-DNA glycosylase is preferentially associated with replicating SV40 chromatin (Krokan, 1981) and that it rapidly removes misincorporated uracil-residues (Wist et al., 1978), but there is no reason to assume that the function of the nuclear form is limited to the S-phase. One possibility is that the nuclear form may be organized in different subnuclear compartments, one of which specific for the S-phase and the other one independent of the S-phase, allowing repair of deaminated cytosines and misincorporated uracils throughout the cell cycle. Based upon an estimated turnover number of 600 uracils per second with a substrate having a high density of uracil in DNA, we have calculated that the 10^4–10^5 enzyme molecules may spend more than 97% of the time scanning the complete genome in one day and less than 3% of the time in actually excising the few uracil residues present. The "working conditions" for the enzyme may therefore not be all that unfavorable. As mentioned, uracil-DNA glycosylase also recognizes several analogs formed upon irradiation of cytosine-residues. These are removed 3–4 orders of magnitude less efficiently when compared with uracil and the biological significance of this is not clear. It should be noted, however, that the rate of catalysis of uracil-DNA glycosylase is also at least 2–3 orders of magnitude higher than other DNA glycosylases. The significance of different repair enzymes in various functions can therefore not easily be assessed solely from their biochemical properties.

ACKNOWLEDGMENTS

This work was supported by the Norwegian Cancer Society, the Research Council of Norway and the Cancer Foundation at the University Hospital, Norwegian University of Science and Technology.

REFERENCES

O.N. Aprelikova and N.V. Tomilin (1982) Activity of uracil-DNA glycosylase in different rat tissues and in regenerating rat liver. *FEBS Lett.* **137**, 193–195.

B. Bandaru, J. Gopal, and A.S. Bhagwat (1996) Overproduction of DNA cytosine methyltransferases causes methylation and C-->T mutations at non-canonical sites. *J. Biol. Chem.* **271**, 7851–7859.

Y. Barak, O. Cohen-Fox, and Z. Livneh (1995) Deamination of cytosine-containing photodimers in UV-irradiated DNA: significance for UV-light mutagenesis. *J. Biol. Chem.* **270**, 24174–24179.

M. Bjørås, L. Luna, B. Johnsen, E. Hoff, T. Haug, T. Rognes, and E. Seeberg (1997) Opposite base-dependent reactions of a human base excision repair enzyme on DNA containing 7,8-dihydro-8-oxoguanine and abasic sites. *EMBO J.* **16**, 6314–6322.

B.C. Blount, M.M. Mack, C.M. Wehr, J.T. MacGregor, R.A. Hiatt, G. Wang, S.N. Wickramasinghe, R.B. Everson, and B.N. Ames (1997) Folate deficiency causes uracil misincorporation into human DNA and chromosome breakage: implications for cancer and neuronal damage. *Proc. Natl. Acad. Sci.* U.S.A. **94**, 3290–3295.

R. Bockrath and P. Mosbaugh (1986) Mutation probe of gene structure in *E. coli:* suppressor mutations in the seven-tRNA operon. *Mol. Gen. Genet.* **204**, 457–462.

D. Chakravarti, G.C. Ibeanu, K. Tano, and S. Mitra (1991) Cloning and expression in *Escherichia coli* of a human cDNA encoding the DNA repair protein N-methylpurine-DNA glycosylase. *J. Biol. Chem.* **266**, 15710–15715.

D.L. Croteau, C.M.J. ap Rhys, E.K. Hudson, G.L. Dianov, R.G. Hansford, and V.A. Bohr (1997) An oxidative damage-specific endonuclease from rat liver mitochondria. *J. Biol. Chem.* **272**, 27338–27344.

M. Dizdaroglu, A. Karakaya, P. Jaruga, G. Slupphaug, and H.E. Krokan (1996) Novel activities of human uracil DNA N-glycosylase for cytosine-derived products of oxidative DNA damage. *Nucl. Acids Res.* **24**, 418–422.

B.K. Duncan and B. Weiss (1982) Specific mutator effects of ung (uracil-DNA glycosylase) mutations in *Escherichia coli. J. Bacteriol.* **151**, 750–755.

I. Eftedal, P.H. Guddal, G. Slupphaug, G. Volden, and H.E. Krokan (1993) Consensus sequences for good and poor removal of uracil from double stranded DNA by uracil-DNA glycosylase. *Nucleic Acids Res.* **21**, 2095–2101.

H.H. el Hajj, H. Zhang, and B. Weiss (1988) Lethality of a dut (deoxyuridine triphosphatase) mutation in *Escherichia coli. J. Bacteriol.* **170**, 1069–1075.

H.H. el Hajj, L. Wang, and B. Weiss (1992) Multiple mutant of *Escherichia coli* synthesizing virtually thymineless DNA during limited growth. *J. Bacteriol.* **174**, 4450–4456.

D.F. Fix and B.W. Glickman (1987) Asymmetric cytosine deamination revealed by spontaneous mutational specificity in an Ung- strain of *Escherichia coli. Mol. Gen. Genet.* **209**, 78–82.

F. Focher, P. Mazzarello, A. Verri, U. Hubscher, and S. Spadari (1990) Activity profiles of enzymes that control the uracil incorporation into DNA during neuronal development. *Mutat. Res.* **237**, 65–73.

F. Focher, A. Verri, S. Verzeletti, P. Mazzarello, and S. Spadari (1992) Uracil in OriS of herpes simplex 1 alters its specific recognition by origin binding protein (OBP): does virus induced uracil-DNA glycosylase play a key role in viral reactivation and replication? *Chromosoma* **102**, S67–71.

L.A. Frederico, T.A. Kunkel, and B.R. Shaw (1990) A sensitive genetic assay for the detection of cytosine deamination: determination of rate constants and the activation energy. *Biochemistry* **29**, 2532–2537.

M.H. Gadsden, E.M. McIntosh, J.C. Game, P.J. Wilson, and R.H. Haynes (1993) dUTP pyrophosphatase is an essential enzyme in *Saccharomyces cerevisiae. EMBO J.* **12**, 4425–4431.

P.K. Gupta and M.A. Sirover (1981) Stimulation of the nuclear uracil DNA glycosylase in proliferating human fibroblasts. *Cancer Res.* **41**, 3133–3136.

T. Haug, F. Skorpen, P.A. Aas, V. Malm, C. Skjelbred, and H.E Krokan (1998) Regulation of expression of nuclear and mitochondrial forms of human uracil-DNA glycosylase. *Nucleic Acids Res.* **26**, 1449–1457.

T. Haug, F. Skorpen, H. Lund, and H.E. Krokan (1994) Structure of the gene for human uracil-DNA glycosylase and analysis of the promoter function. *FEBS Lett.* **353**, 180–184.

T. Haug, F. Skorpen, K. Kvaloy, I. Eftedal, H. Lund, and H.E. Krokan (1996) Human uracil-DNA glycosylase gene: Sequence organization, methylation pattern, and mapping to chromosome 12q23-q24.1. *Genomics* **36**, 408–416.

K.J. Impellizzeri, B. Anderson, and P.M. Burgers (1991) The spectrum of spontaneous mutations in a *Saccharomyces cerevisiae* uracil-DNA-glycosylase mutant limits the function of this enzyme to cytosine deamination repair. *J. Bacteriol.* **173**, 6807–6810.

P. Karran, R. Cone, and E.C. Friedberg (1981) Specificity of the bacteriophage PBS2 induced inhibitor of uracil-DNA glycosylase. *Biochemistry* **20**, 6092–6096.

B. Kavli, G. Slupphaug, C.D. Mol, A.S. Arvai, S.B. Petersen, J.A. Tainer, and H.E. Krokan (1996) Excision of cytosine and thymine from DNA by mutants of human uracil-DNA glycosylase. *EMBO J.* **15,** 3442–3447.

H. Krokan (1981) Preferential association of uracil-DNA glycosylase activity with replicating SV40 minichromosomes. *FEBS Lett.* **133**, 89–91.

H. Krokan and C.U. Wittwer (1981) Uracil DNA-glycosylase from HeLa cells: general properties, substrate specificity and effect of uracil analogs. *Nucl. Acids Res.* **9**, 2599–2613.

H. Krokan, A. Haugen, B. Myrnes, and P.H. Guddal (1983) Repair of premutagenic DNA lesions in human fetal tissues: evidence for low levels of O6-methylguanine-DNA methyltransferase and uracil-DNA glycosylase activity in some tissues. *Carcinogenesis.* **4**, 1559–1564.

H.E. Krokan, R. Standal, S. Bharati, M. Otterlei, T. Haug, G. Slupphaug and F. Skorpen (1997a) *Uracil in DNA and the family of conserved DNA glycosylases. In Base Excision Repair of DNA damage* (ed. I.D. Hickson), Landes Bioscience, Austin, Texas, pp.7–30.

H.E. Krokan, R. Standal, and G. Slupphaug (1997b) DNA glycosylases in the base excision repair of DNA. *Biochem.* J. **325**, 1–16.

Y. Kubota, R.A. Nash, A. Klungland, P. Schär, D.E. Barnes, and T. Lindahl (1996) Reconstitution of DNA base excision-repair with purified human proteins: interaction between DNA polymerase β and the XRCC1 protein. *EMBO J.* **15**, 6662–6670.

T. Lindahl (1974) An N-glycosidase from *Escherichia coli* that releases free uracil from DNA containing deaminated cytosine residues. *Proc. Natl. Acad. Sci.* U.S.A. **71**, 3649–3653.

T. Lindahl (1993) Instability and decay of the primary structure of DNA. *Nature* **362**, 709–715.

D.J. Mauro, J.K. de Riel, R.J. Tallarida, and M.A. Sirover (1993) Mechanisms of excision of 5-fluorouracil by uracil DNA glycosylase in normal human cells. *Mol. Pharmacol.* **43**, 854–857.

C.D. Mol, A.S. Arvai, R.J. Sanderson, G. Slupphaug, B. Kavli, H.E. Krokan, D.W. Mosbaugh, and J.A. Tainer (1995a) Crystal structure of human uracil-DNA glycosylase in complex with a protein inhibitor: protein mimicry of DNA. *Cell* **82**, 701–708.

C.D. Mol, A.S. Arvai, G. Slupphaug, B. Kavli, I. Alseth, H.E. Krokan, and J.A. Tainer (1995b) Crystal structure and mutational analysis of human uracil-DNA glycosylase: structural basis for specificity and catalysis. *Cell* **80**, 869–878.

B. Myrnes, K.E. Giercksky, and H. Krokan (1983) Interindividual variation in the activity of O6-methyl guanine-DNA methyltransferase and uracil-DNA glycosylase in human organs. *Carcinogenesis* **4**, 1565–1568.

B. Myrnes, K. Norstrand, K.E. Giercksky, C. Sjunneskog, and H. Krokan (1984) A simplified assay for O6-methylguanine-DNA methyltransferase activity and its application to human neoplastic and non-neoplastic tissues. *Carcinogenesis* **5**, 1061–1064.

T. Nagelhus, T. Haug, K.K. Singh, K.F. Keshav, F. Skorpen, M. Otterlei, S. Bharati, T. Lindmo, S. Benichou, R. Benarous, and H.E. Krokan (1997) A sequence in the N-terminal region of human uracil-DNA glycosylase with homology to XPA interacts with the C-terminal part of the 34 kDa subunit of replication protein A. *J. Biol. Chem.* **272**, 6561–6566.

I.D. Nicholl, K. Nealon, and M.K. Kenny (1997) Reconstitution of human base excision repair with purified proteins. *Biochemistry* **36**, 7557–7566.

H. Nilsen, S.P. Yazdankhah, I. Eftedal, and H.E. Krokan (1995) Sequence specificity for removal of uracil from U.A pairs and U.G mismatches by uracil-DNA glycosylase from *Escherichia coli*, and correlation with mutational hotspots. *FEBS Lett.* **362**, 205–209.

H. Nilsen, M. Otterlei, T. Haug, K. Solum, T.A. Nagelhus, F. Skorpen and H.E. Krokan (1997) Nuclear and mitochondrial uracil-DNA glycosylase are generated by alternative splicing and transcription from different positions in the *UNG* gene. *Nucleic Acids Res.* **25**, 750–755.

T.R. O'Connor and J. Laval (1991) Human cDNA expressing a functional DNA glycosylase excising 3-methyladenine and 7-methylguanine. *Biochem. Biophys. Res. Commun.* **176**, 1170–1177.

L.C. Olsen, R. Aasland, C.U. Wittwer, H.E. Krokan, and D.E. Helland (1989) Molecular cloning of human uracil-DNA glycosylase, a highly conserved DNA repair enzyme. *EMBO J.* **8**, 3121–3125.

H. Otterlei, T. Haug, T.A. Nagelhus, G. Slupphaug, T. Lindmo, and H.E. Krokan (1998) Nuclear and mitochondrial splice forms of human uracil-DNA glycosylase contain a complex nuclear localisation signal and a strong classical mitochondrial localisation signal, respectively. *Nucleic Acids Res.* **26**, 4611–4617.

L.H. Pearl and R. Savva (1995) DNA repair in three dimensions. *Trends Biochem. Sci.* **20**, 421–426.

A. Pendlebury, I.M. Frayling, M.F.S. Koref, G.P. Margison, and J.A. Rafferty (1994) Evidence for the Simultaneous Expression of Alternatively Spliced Alkylpurine N-glycosylase Transcripts in Human Tissues and Cells. *Carcinogenesis* **15**, 2957–2960.

I.P. Pogribny, L. Muskhelishvili, B.J. Miller, and S.J. James (1997) Presence and consequence of uracil in preneoplastic DNA from folate/methyl-deficient rats. *Carcinogenesis* **18**, 2071–2076.

J.P. Radicella, C. Dherin, C. Desmaze, M.S. Fox, and S. Boiteux (1997) Cloning and characterization of hOGG1, a human homolog of the OGG1 gene of *Saccharomyces cerevisiae*. *Proc. Natl. Acad. Sci.* U. S. A. **94**, 8010–8015.

T. Roldan-Arjona, Y.F. Wei, K.C. Carter, A. Klungland, C. Anselmino, R.P. Wang, M. Augustus, and T. Lindahl (1997) Molecular cloning and functional expression of a human cDNA encoding the antimutator enzyme 8-hydroxyguanine-DNA glycosylase. *Proc. Natl. Acad. Sci.* U.S.A. **94**, 8016–8020.

L. Samson, B. Derfler, M. Boosalis, and K. Call (1991) Cloning and characterization of a 3-methyladenine DNA glycosylase cDNA from human cells whose gene maps to chromosome 16. *Proc. Natl. Acad. Sci.* U.S.A. **88**, 9127–9131.

R. Savva, K. McAuley Hecht, T. Brown, and L. Pearl (1995) The structural basis of specific base-excision repair by uracil-DNA glycosylase. *Nature* **373**, 487–493.

R. Shapiro and R.S. Klein (1966) The deamination of cytidine and cytosine by acidic buffer solutions. Mutagenic implications. *Biochemistry* **5**, 2358–2362.

R. Shapiro. (1980) Damage to DNA caused by hydrolysis. *In Chromosome Damage and Repair* (eds. E. Seeberg and K. Kleppe), Plenum Press, New York, pp.3–18.

J.C. Shen, W.M. Rideout, and P.A. Jones (1992) High frequency mutagenesis by a DNA methyltransferase. *Cell* **71**, 1073–1080.

J.C. Shen, J.M. Zingg, A.S. Yang, C. Schmutte, and P.A. Jones (1995) A mutant HpaII methyltransferase functions as a mutator enzyme. *Nucl. Acids Res.* **23**, 4275–4282.

M.A. Sirover (1979) Induction of the DNA repair enzyme uracil-DNA glycosylase in stimulated human lymphocytes. *Cancer Res.* **39**, 2090–2095.

G. Slupphaug, L.C. Olsen, D. Helland, R. Aasland, and H.E. Krokan (1991) Cell cycle regulation and in vitro hybrid arrest analysis of the major human uracil-DNA glycosylase. *Nucl. Acids Res.* **19**, 5131–5137.

G. Slupphaug, I. Alseth, I. Eftedal, G. Volden, and H.E. Krokan (1993a) Low incorporation of dUMP by some thermostable DNA polymerases may limit their use in PCR amplifications. *Anal. Biochem.* **211**, 164–169.

G. Slupphaug, F.H. Markussen, L.C. Olsen, R. Aasland, N. Aarsaether, O. Bakke, H.E. Krokan, and D.E. Helland (1993b) Nuclear and mitochondrial forms of human uracil-DNA glycosylase are encoded by the same gene. *Nucl. Acids Res.* **21**, 2579–2584.

G. Slupphaug, I. Eftedal, B. Kavli, S. Bharati, N.M. Helle, T. Haug, D.W. Levine, and H.E. Krokan (1995) Properties of a recombinant human uracil-DNA glycosylase from the UNG gene and evidence that UNG encodes the major uracil-DNA glycosylase. *Biochemistry* **34**, 128–138.

G. Slupphaug, C.D. Mol, B. Kavli, A.S. Arvai, H.E. Krokan, and J.A. Tainer (1996) A nucleotide-flipping mechanism from the structure of human uracil-DNA glycosylase bound to DNA. *Nature* **384**, 87–92.

R.W. Sobol, J.K. Horton, R. Kuhn, H. Gu, R.K. Singhal, R. Prasad, K. Rajewsky, and S.H. Wilson (1996) Requirement of mammalian DNA polymerase-beta in base-excision repair. *Nature* **379**, 183–186.

U. Varshney and J.H. van de Sande (1991) Specificities and kinetics of uracil excision from uracil-containing DNA oligomers by *Escherichia coli* uracil DNA glycosylase. *Biochemistry* **30**, 4055–4061.

A. Verri, P. Mazzarello, G. Biamonti, S. Spadari, and F. Focher (1990) The specific binding of nuclear protein(s) to the cAMP responsive element (CRE) sequence (TGACGTCA) is reduced by the misincorporation of U and increased by the deamination of C. *Nucleic. Acids. Res.* **18**, 5775–5780.

M.A. Vickers, P. Vyas, P.C. Harris, D.L. Simmons, and D.R. Higgs (1993) Structure of the human 3-methyladenine DNA glycosylase gene and localization close to the 16p telomere. *Proc. Natl. Acad. Sci.* U.S.A. **90**, 3437–3441.

E. Wist, O. Unhjem and H.E. Krokan (1978) Accumulation of small fragments of DNA in isolated HeLa cell nuclei due to transient incorporation of dUMP. *Biochim. Biophys. Acta.* **250**, 253–270.

C.U. Wittwer and H. Krokan (1985) Uracil-DNA glycosylase in HeLa S3 cells: interconvertibility of 50 and 20 kDa forms and similarity of the nuclear and mitochondrial form of the enzyme. *Biochim. Biophys. Acta.* **832**, 308–318.

C.U. Wittwer, G. Bauw, and H.E. Krokan (1989) Purification and determination of the NH2-terminal amino acid sequence of uracil-DNA glycosylase from human placenta. *Biochemistry* **28**, 780–784.

M.S. Wold (1997) Replication protein A: a heterotrimeric, single-stranded DNA-binding protein required for eukaryotic DNA metabolism. *Ann. Rev. Biochem.* **66**, 61–92.

19

REPAIR OF DNA DAMAGED BY FREE RADICALS

Jacques Laval, Cécile Bauche, Juan Jurado, Franck Paillard,
Murat Saparbaev, and Olga Sidorkina

Groupe "Réparation des lésions Radio- et Chimio-Induites"
UMR 1772 CNRS, Institut Gustave Roussy
94805 Villejuif Cedex, France

1. ABSTRACT

Reactive oxygen species (ROS) are generated in cells either by aerobic metabolism or by exposure to ionising radiations and other oxidizing agents. They can cause damage to DNA bases and strand breaks. Unrepaired oxidative damage to DNA has been suggested to play a role in cancer, aging, and the other degenerative processes. In most organisms, the repair of oxidative DNA base damages is thought to be primarily mediated via the base excision repair (BER) pathway (combined action of a glycosylase, AP endonuclease, polymerase and ligase).

In *Escherichia coli*, three glycosylases are known to recognize oxidized purines and pyrimidines. The Fpg protein recognizes and removes from DNA damaged purines such as Formamidopyrimidine and 8-oxoGuanine residues. Oxidized pyrimidines (like thymine glycols) are excised from DNA either by the Nth or the Nei proteins, which share a common range of substrates. In this bacteria, the BER pathway is completed by the action of endonucleases such as Endo IV and Exo III proteins. This pathway has also been identified in the radio-resistant bacteria *Deinococcus radiodurans*, with the characterization of a Fpg a Nei-like proteins.

In both yeast and mammalian, homologs of *E. coli* glycosylases and AP endonucleases have been cloned and sequenced (Ogg1, hOgg1, hNth...), demonstrating the existence of a BER pathway in eukaryotes. In mammalian cells, a short and long-patch BER have been identified, depending on the size of the gap generated at the base damage site. The short BER pathway leads to the resynthesis of a single nucleotide, and involves the DNA polymerase β. When a gap greater than one nucleotide is generated, the completion of repair (long patch BER) is PCNA-dependent and requires the nuclease FEN 1 and the polymerase δ/ε.

Advances in DNA Damage and Repair, edited by Dizdaroglu and
Karakaya. Kluwer Academic / Plenum Publishers, New York, 1999.

The availability of mammalian cDNA coding for BER enzymes is expected to help the evaluation of the pathological and genotoxic effects of ROS *in vivo*.

2. INTRODUCTION

Reactive oxygen species (ROS) are generated by cellular metabolism, cell injury (MacCord, 1987) and exposure to physical and chemical oxygen radical-forming agents (Sies, 1991). They are believed to be major contributors to aging, cancerogenesis (Trus and Kensler, 1991) and degenerative processes (Halliwell and Gutteridge, 1989). All the cellular macromolecules are targets for reactive oxygen radicals (Sies, 1991; Jaruga *et al.*, 1994).

Oxygen radicals $O_2\bullet^-$, H_2O_2 and OH• are formed *in vivo* during cellular metabolism in all aerobic organisms. Exogenous factors like γ-rays, UV-radiations (Doetsch *et al.*, 1995) and chemical carcinogens can also generate oxygen radicals. Oxidative damages generated by ROS to DNA, lipids, carbohydrates and proteins have been implicated in aging, ischemia (McCord, 1987), cancer (Trus and Kensler, 1991), autoimmune diseases (Halliwell, 1982), neural cell death (Reiter, 1995). Evidence has accumulated that lack of protection against free radicals and lack of repair of oxidative damage in biological macromolecules have significant role in mutagenesis and carcinogenesis.

Free radicals can damage nucleobases and deoxyribose in DNA either directly or indirectly, by products of membrane lipids peroxidation (Park and Floyd, 1992). About 100 different base and sugar damages have been identified. Although cells have efficient repair mechanisms, an accumulation of oxidised DNA damages products occurs in genome during aging (Kaneko *et al.*, 1996). The most active species, hydroxyl radicals, predominantly react with C_8 of purines, forming 7,8-dihydro-8-oxo-2'-deoxyguanosine or imidazol ring-opened products, and also with C_5-C_6 double bond of pyrimidines forming thymine glycol (Imlay and Linn, 1988; Kasai and Nishimura, 1986; Cadet and Berger, 1985). The abstraction of hydrogen atom from deoxyribose at C1' and C4' generates DNA-strand breaks with 3'-phosphoglycolate esters and 3'-phosphates (Dizdaroglu *et al.*, 1977; Henner *et al.*, 1983). Reactive aldehydes, a products of membrane lipids peroxidation, react with DNA bases forming the cyclic adducts 1,N^6-ethenoadenine, 1,N^2-ethenoguanine, N^2, 3-ethenoguanine, and 3,N^4-ethenocytosine (Marnett, 1994; El-Ghissassi *et al.*, 1995).

Base damages like 8-oxoG, 5-hydroxycytosine, hypoxanthine and ethenoadducts have miscoding properties and, if not repaired, could lead to mutations upon replication. Other such as thymine glycol, formamidopyrimidine, fragmented forms of thymine and oxidized forms of deoxyribose cause replication block and therefore are believed to be lethal lesions. In DNA, ROS generate DNA lesions, like oxidized bases, that does not cause major helix distorsions. These lesions are substrates for DNA glycosylases, that remove several structurally different modified bases. These enzymes are involved in the base excision repair (BER) pathway. First N-DNA glycosylases cleave the N-glycosylic bond between the damaged base and the deoxyribose, leaving an abasic (AP) site in DNA. AP-sites are non informative and must be eliminated, so the DNA backbone next to the abasic site is cleaved by an AP-endonuclease or AP-lyase (Krokan *et al.*, 1997). In mammalian cells two BER pathways have been described: a 'short-patch' one involving replacement of one nucleotide (Dianov *et al.*, 1992) and 'long-patch' pathway with gap-filling of several nucleotides (Frosina *et al.*, 1996; Klungland and Lindahl, 1997).

3. REPAIR OF OXIDIZED PURINES AND PYRIMIDINES BY *E. COLI* ENZYMES

3.1. The Fpg Protein (Formamidopyrimidine-DNA Glycosylase)

The *fpg* gene of *E.coli* coding for the Fpg protein was cloned (Boiteux *et al.*, 1987) and the physical and enzymatic properties of the protein established (Laval *et* al., 1990; Boiteux, 1993). It is a globular monomer of 30.2 Kda (269 amino acids) having a stoke radius of 2.5 nm and an isoelectric point of 8.6. It contains one zinc atom per molecule of enzyme (Boiteux *et al*, 1990). The Fpg protein contains in the COOH-terminus one zinc-finger motif (Cys-X_2-Cys-X_{16}-Cys-X_2-Cys-X_2-COOH) that is mandatory for Fpg binding to DNA and to its enzymatic activities (O'Connor *et al.*, 1993). *In vitro*, the Fpg protein excises a broad spectrum of modified purines, in particular, 2,6-diamino-4-hydroxy-5N-methylformamidopyrimidine (Fapy) and 8-oxoG residues (Laval, 1996). Fpg protein is also endowed of an AP lyase activity that incises DNA at abasic sites by a β-δ elimination mechanism (O'Connor and Laval, 1989; Bailly *et al.*, 1989) and an activity excising 5'-terminal deoxyribose phosphate (dRPase, Graves *et al.*, 1992). It is inhibited by NO (Wink *et al.*, 1994). *In vivo*, the Fpg protein has an antimutator effect preventing G/C→T/A spontaneous transversion. It acts in concert with the MutY protein (Tajiri *et al.*, 1995). The *fpg mutY* double mutant CC104 has an extreme mutator phenotype that can be reversed by plasmids carrying the *fpg* gene (Demple *et al.*, 1983). In contrast, the active site of Fpg protein, as judged by mapping analysis, could be located within the first 73 amino acid residues of the amino terminus (Tchou and Grollman, 1995). The N-terminal proline can be trapped with the substrate (Zharkov *et al.*, 1997), whereas the zinc finger motive is at the COOH-terminus.

For the N-glycosylases endowed of an AP-lyase activity, an unified catalytic mechanism involving a Shiff's base intermediate between the enzyme and the substrate DNA has been proposed (Dosdon *et al.*, 1994). In this model, the N-glycosylase/AP lyase utilizes an amino group as the attacking nucleophile that react with the C_1 of the deoxyribose to displace the damaged base and produces an imino (Schiff's base) intermediate. Possible candidate for such nucleophilic attack could be the α-NH_2 group of the amino terminal amino acid or the ε-amino group of lysine. Among two highly conserved lysine residues - at positions 57 and 155- no one is responsible for producing Schiff's base complex with C_8-oxoG containing DNA (Sidorkina and Laval, 1998; Rabow and Kow, 1997). In contrast, the N-terminal proline, that belongs to highly conserved sequence PELPEVE in NH_2-ends of the Fpg protein and its procariotic homologs, can be linked to DNA in reduced Shiff base complex (Zharkov *et al.*, 1997). Using site-directed mutagenesis, it was shown that Pro-2 is mandatory for AP lyase activity and production of a Shiff's base with C_8-oxoG containing DNA(Sidorkina and Laval, unpublished data). Pro-2 is not of paramount importance in the organization of the conformation of the active center (Kouznetsov *et al.*, 1998).

3.2. The Nth Protein (Endonuclease III)

Endonuclease III does not require divalent cations for its activity, and is EDTA-resistant (Radman, 1976; Gates and Linn, 1977). It's a DNA glycosylase having a broad substrate specificity, excising ring-saturated, ring-opened, and ring-fragmented pyrimidines, such as thymine glycol, 5,6-dihydrothymine, 5-hydroxy-6-hydrothymine, 5,6-dihydrouracil, alloxan, 5-hydroxy-6-hydrouracil, 5-hydroxycytosine, 5-hydroxyuracil (Hatahet *et al.*, 1994), β-ureidoisobutiric acid (Mazuder *et al.*, 1991), uracil glycol (Dizdaroglu *et al.*, 1993), and recently α-R-hydroxy-β-ureidoisobutiric acid a fragmentation product of the 5R-thymidine C5-hydrate (Jurado *et al.*, 1998).

However, oxidized pyrimidines can be substrate for other DNA glycosylases. It has been shown that cytosine derived products of oxidative damages are substrates for the human uracyl DNA-glycosylase (Dizdaroglu *et al.*, 1996) and isodialuric acid for the *E. coli* uracil DNA-glycosylase (Zastawny *et al.*, 1995) and αRT is excellent substrate for the Fpg protein (Jurado *et al.*, 1998).

The nicking activity at abasic sites of endonuclease III is due to its AP lyase function. The phosphodiester cleavage occurs via β-elimination (the protein does not cleave at reduced AP sites) generating 5' ends bearing 5'-phosphate and 3' ends with 2,3-unsaturated abasic residue 4-hydroxy-2-pentanal (Bailly and Verly, 1987; Kim and Linn, 1988).

The gene coding for endonuclease III activity, the *nth* gene, has been cloned (Cunningham and Weiss, 1985). It maps at 36 min. The Nth protein has been purified and its properties investigated (Ashara *et al.*, 1989). The Nth protein is a monomeric protein of 23.4kDa (211 aa). It contains an iron sulfur motive and its three dimensional structure is established (Kuo *et al.*, 1992). It is an elongated molecule with a cleft separating two similar size domains: a continuous domain formed by six-α-helices and a second domain formed by three C-terminal α-helices and the N-terminal helix. The C terminal loop contains an iron-sulphur center which is anchored to the protein by a Cys-X6-Cys-X2-Cys-X5-Cys sequence. This cluster has a structural function in positioning basic residues for DNA-binding (Thayer *et al.*, 1995) and it is also present in the MutY protein (Michaels *et al.*, 1990).

E. coli nth mutant deficient in Nth protein does not show apparent phenotype (Cunningham and Weiss, 1985) They are not sensitive to X-rays, H_2O_2 or other agents which produce ring saturation or fragmentation products in DNA. However, the double mutant *nth nei* exhibits a strong spontaneous mutator phenotype and is hypersensitive to ionizing radiation and H_2O_2 (Jiang *et al.*, 1997).

3.3. The Nei Protein (Endonuclease VIII)

Beside the Nth protein, endonuclease VIII (Nei protein) also excises thymine glycol and oxidized cytosine (Melamede *et al.*, 1994). This protein was purified from *E.coli* extract lacking the Nth protein (Melamede *et al.*, 1994) and presents many enzymatic similarities with the Nth protein. It excises DNA containing thymine glycol, dihydrothymine, β-ureido-isobutyric acid and urea residues. The Nei protein has been recently sequenced and it showed a striking similarity in amino acid composition with the Fpg protein (Jiang *et al.*, 1997), including a zinc finger motive.

3.4. The Hypoxanthine-DNA Glycosylase Activity of AlkA Protein

Hypoxanthine-DNA glycosylase releases hypoxanthine (HX) from DNA containing dIMP (Karran and Lindahl, 1978). This lesion is generated in DNA by deamination of adenine and also by nitric oxide (Wink *et al.*, 1994). HX residues in DNA are mutagenic, if unrepaired, since they can pair with cytosine, generating AT to GC transitions after DNA replication (Karran and Lindahl, 1980; Hill-Perkins *et al.*, 1986). The gene coding for the HX-DNA glycosylase activity in *E. coli* is the *alkA* gene (Nakabeppu *et al.*, 1984), coding for the 3-methyladenine DNA glycosylase (Saparbaev and Laval, 1994). The ANPG, APDG and MAG proteins, the human, rat, and yeast homologues of AlkA release HX residues from DNA, the mammalian enzymes being by far the most efficient (Saparbaev and Laval, 1994). The AlkA protein also catalyses the excision of ethenobases such as N^2,3-ethenoguanine (Habraken *et al.*, 1991) and 1,N^6-ethenoadenine (Saparbaev *et al.*, 1995). These lesions are generated in DNA by products of lipid peroxidations. (Marnett, 1994; El-Ghissassi *et al.*, 1995). The oxidation of vinyl halides by cytochrome P450 also generates etheno adducts in DNA (Guengerich, 1992).

The 5-formyluracil (5-foU) residue, an oxidized thymine lesion in DNA and a major lesion induced by ionizing radiation that induces AT to GC transitions (Kasai *et al.*, 1990), is also repared in *E. coli* by the AlkA protein (Bjelland *et al.*, 1994). These oxidized lesions repaired by the AlkA protein are added to the broad list of substrates of this enzyme. It should be recalled that purines and pyrimidines alkylated which alkyl groups protrude either into the major groove (7meG) or into the minor groove (3meA and other alkyl groups are substrates for the AlkA protein, Seeberg *et al.*, 1995). The three dimentional structure of AlkA has been established. It is a compact globular protein with a prominent hydrophobic cleft on its surface and three equal-sized domains (Labahn *et al.*, 1996; Yamagata *et al.*, 1996).

3.5. Endonuclease IV (Nfo Protein)

Endonuclease IV is an EDTA-resistant AP endonuclease that represents only 5% of the total AP endonuclease activity in *E. coli* wild type (Ljungquist *et al.*, 1976). It is inducible by oxidative stress and is under the control of the soxRS system (Chan and Weiss, 1987; Nunoshiba *et al.*, 1993; Tsaneva and Weiss, 1990; Demple, 1991). The gene coding in *E. coli* for EndoIV, *nfo*, has been cloned (Cunninghamn *et al.*, 1986), and characterised (Levin *et al.*, 1988; Ljungquist, 1977). Nfo protein is a monomer of 30kDa having several activities: an AP endonuclease (Ljungquist, 1977), a 3'-phosphatase and 3'-phosphoglycoaldehyde diesterase activities (Levin *et al.*, 1988). These 3' repair activities probably clean the 3' ends, a prerequisite for DNA synthesis.

Nfo protein hydrolyses the phosphodiester bond 5' to α-deoxyadenosine, the α-anomer of deoxyadenosine (Ide *et al.*, 1994). It also incises an oligonucleotide containing 3,N4-benzetheno-2'-deoxycytidine (pBQ-dC). In addition, exonuclease III and human AP-endonuclease incises oligonucleotides containing pBQ-dC (Hang *et al.*, 1996).

Mutant deficient in the product of the *nfo* gene are extremely sensitive to the lethal effects of oxidative agents such as bleomycin and *t*-butyl-hydroperoxide and this mutation enhances the sensitivity of *xth* mutants to H_2O_2, and alkylating agents. The double mutants *xth nfo* are also sensitive to γ-radiation (Cunningham *et al.*, 1986)

3.6. Exonuclease III (Xth Protein)

Exonuclease III, coded for by the *xth* gene, is a 3'→5' exonuclease active on double-stranded DNA and a 3'-phosphatase (Richarson *et al.*, 1964; Richardson and Kornberg, 1964). It is the major AP endonuclease of *E. coli*, accounting for more than 80% of the total AP-endonuclease activity in wild type strains(Weiss and Grossman, 1987; Levin *et al.*, 1988: Weiss, 1976; Grossard and Verly, 1978). This protein is also active on single-stranded DNA and resistant to EDTA (Rogers and Weiss, 1980). The 3'-phosphatase activity of ExoIII represents about 99% of the total activity in *E. coli.* and also has 3'-repair diesterase and ribonuclease H activities. The 3' termini generated by ExoIII are normal nucleotides with 3'hydroxyl group that are effective primers for DNA polymerases (Rogers and Weiss, 1980; Warner *et al.*, 1980; Demple *et al.*, 1986).

The *xth* mutants are extremely sensitive to H_2O_2, and also to oxidative damage generated by near-UV light (Sammartano and Tuveson, 1983). The cristal structure of the Exo III protein has been solved (Mol *et al.*, 1995).

4. MECHANISM OF ACTION

In order to understand the base excision repair pathway, inhibitors that form stable complexes with DNA glycosylases have been synthesized. Double-standed DNA containing a single pyrrolidine residue is a potent inhibitor of the AlkA protein (Schärer *et al.*, 1995). Further investigations of the interaction of this inhibitor with the following DNA glycosylases (*E. coli* Nth, Fpg, MutY, uracil glycosylase and Tag proteins, human ANPG) using binding/or inhibition assays have shown that with the exception of uracil DNA-glycosylase, they bind specifically to the inhibitor (Schärer *et al.*, 1998). By comparing the interaction of the pyrrolidine- and tetrahydrofuran-containing DNA with the various DNA glycosylases, the importance of the positive charge of the pyrrolidine in binding to these enzymes was investigated. One can expect that such a general inhibitor of DNA glycosylases will provide a valuable tool to study stable protein-DNA complexes of these enzymes (Schärer *et al.*, 1998).

The structural basis for the recognition and removal of damaged bases from DNA by repair enzymes is not yet fully understood. Endonuclease III forms a hydrophilic pocket that can recognise oxidised pyrimidine DNA modifications (Thayer *et al.*, 1995). Two active site residues, Lys-120 and Asp-138, are at the edge of this pocket (Thayer *et al.*,1995). It has been proposed that Asp-138 deprotonates the ε-NH_3^+ group of Lys-120, which attacks the glycosidic bond of the damaged base to form a covalent enzyme-substrate intermediate (Labahn *et al.*,1996). The three-dimensional structure of the Fpg protein is not yet available. It has been shown that the N-terminal proline of Fpg protein is linked to DNA (Zharkov *et al.*, 1997). It was supposed that Pro-1 is the nucleophile that initiates the catalytic excision of oxidised bases from DNA (Zharkov *et al.*, 1997). From our and others studies by site-directed mutagenesis (Sidorkina and Laval, 1998; Rabow and Kow, 1997; Sidorkina and Laval, unpublished data; Kouznetsov *et al.*, 1998) we propose the following mode of action for the Fpg protein: positively charged Lys-57 and Lys-155 facilitate specifically the disruption of the N-glycosilic bond of DNA contaiing 8-oxoG. Pro-2 is indispensable for AP-lyase activity, but not for Schiff base production with preformed AP-site (Sidorkina and Laval, 1998).

5. REPAIR OF OXIDIZED BASES IN *DEINOCOCCUS RADIODURANS*

Deinococcus radiodurans is a bacterium extremely resistant to the lethal and mutagenic effects of ionizing radiations (Anderson, *et al.*, 1956), as well as other physical and chemical DNA-damaging agents (including mitomycin and UV, Moseley, 1983; Murray, *et al.*, 1986; Smith *et al.*, 1992). It has been suggested that this resistance is due to unusually efficient DNA repair mechanisms (Moseley, 1983; Minton, 1994), however these mechanisms have not been investigated in details.

In an effort to characterize the repair of oxidized purines in *Deinococcus radiodurans*, we purified from this bacterium two proteins presenting a Fapy-DNA glycosylase activity. One of these proteins shows a strong 8-oxoG DNA glycosylase and AP-lyase associated activities. It presents kinetic parameters (Km and Kcat) for Fapy and 8-oxoG residues similar to the *E. coli* FPG protein ones. The mechanisms of action of this protein is comparable to the FPG protein : it forms a transient Schiff base intermediate with 8-oxoG or AP-site containing DNA, and the AP-lyase activity processes by a β,δ-elimination. Anti *E. coli* Fpg antibodies recognize this enzyme, showing that they share common structural motives. An apparent molecular weight of ~32 kD has also been determined. We have shown that the *D. radiodurans* enzyme is able to recognize and repair C_8-oxoG/A mismatch although at a slow but detectable rate. This result may suggest that *Deinococcus radiodurans* may have a different 8-oxoG repair system (Bauche and Laval, 1999).

The second protein shows a weaker affinity for Fapy residues, and a low 8-oxoG/C and 8-oxoG/T DNA glycosylase activity, but the velocity of a reaction for these substrates is comparable to the *E. coli* Fpg protein. This enzyme presents an AP-lyase activity, processing by a β,δ-elimination mechanism. Like Fpg, it also forms a transient Schiff base intermediate with 8-oxoG or AP-site containing DNA. An associated thymine glycol ADN glycosylase has been found. Anti *E. coli* Fpg antibodies inhibit all those activities, indicating structural similarities between the two proteins. An apparent molecular weight of ~35 kD has also been determined. This enzyme could be an homolog to the *E. coli* endonuclease VIII protein.

Deinococcus radiodurans presents two proteins showing sequence homologies with the *E. coli* FPG protein, and sharing a common range of substrates, which confirms the existence of a base excision repair system in this bacteria. Further characterization of these two proteins and of their corresponding genes will be necessary to determine their exact roles in the unusual resistance of *D. radiodurans* to ionizing radiations and to other agents that damage DNA.

6. DNA GLYCOSYLASES EXCISING OXIDIZED BASES IN EUKARYOTES

Molecular cloning of *Saccharomyces cerevisiae* 8-oxoG DNA glycosylase, the counterpart of the *E. coli* Fpg protein, designated as OGG1, demonstrated that eukaryotes use the DNA glycosylase pathway to repair oxidized purines in DNA as shown in *E. coli* (Auffret van der Kemp, 1996; Nash *et al.*, 1996).

Comparison of amino acid sequences of the Ogg1 protein and of the *E. coli* Fpg protein does not reveal any obvious homology. Ogg1 protein has associated AP-lyase activity and cleaves DNA at abasic sites through β-elimination and δ-elimination was not observed. In contrast to the bacterial enzyme, the Ogg1 protein repairs 8-oxoG only when paired with pyrimidines. The removal of Fapy residues was less efficient as compared to 8oxoG. Nash et al. purified and characterized a second *S. cerevisiae* Ogg2 protein which preferentially acts on 8-oxoG/G base pair (Nash *et al.*,1996).

Recent results on molecular cloning of the human Nth-homolog and 8-oxoguanine glycosylase have extended eukaryotic family of base excision repair enzymes. Several groups have reported molecular cloning of the cDNA of the human 8-oxoG DNA glycosylase (hOGG1) (Rosenquist *et al.*, 1997; Radicella *et al.*, 1997; Roldan-Arjona *et al.*, 1997; Lu *et al.*, 1997;Arai *et al.*, 1997; Aburatani *et al.*, 1997). The amino acid sequence of hOGG1 showed 33% identity and 54% similarity to *S. cerevisiae* OGG1, and conserved HhH/PVD motif which is present in endoIII/MutY/AlkA family of DNA glycosylases. Purified hOGG1 protein, similarly to *E. coli* Fpg protein, acts as a DNA glycosylase towards duplex DNA containing 8-oxoG/C base pair and has an associated AP-lyase activity. We have purified from calf thymus extracts three different DNA glycosylases excising Fapy resisdues. Two of them show properties similar to hOGG1 and hNth. The third one has not been discribed yet (Paillard and Laval, manuscipt in preparation).

Homologs of the *E. coli* endonuclease III in yeast strains and human have been identified (Eide *et al.*, 1996; Roldan-Arjona *et al.*, 1996; Aspinwall *et al.*, 1997; Hilbert *et al.*, 1997). Eukaryotic homologs of *E. coli* Nth protein act as DNA glycosylases with associated AP-lyase activity, and they have similar substrate specificity as the *E. coli* enzyme towards damaged pyrimidine derivatives that result from ring saturation, ring fragmentation or ring contraction. Most of eukaryotic endoIII-homologs contained the conserved HhH - motif and the [4Fe-4S] cluster loop motif except *S. cerevisiae* NTG1 protein which has lost iron-binding motif. Quite unexpectedly it was found that *S. cerevisiae* NTG1 protein also releases formamidopyrimidines residues from DNA with a high efficiency comparable to the *E. coli* Fpg protein.

Eukaryotic AP-endonucleases with enzymatic properties similar to the *E. coli* endo IV and exo III proteins have been cloned from yeast and mammals. Yeast AP endonuclease I (40,5 kDa) resembles *E. coli* endoIV (Popoff *et al.*,1990). Like the *E. coli* enzyme, it is a metalloenzyme excising 3'-phosphoglycoaldehyde, 3'-phosphoryl groups, and 3'-α,β unsaturated aldehydes (Johnson and Demple,1988). Amino acids sequences of AP-endonucleases cloned from Drosophila (RRP1, Lowenhaupt *et al.*, 1989), bovine (BAP1, Robson and Hickson, 1991), mouse (APEX, Seki *et al.*, 1991) and human (HAP1) (Demple *et al.*, 1991; Robson *et al.*, 1991; Seki *et al.*, 1992) share significant similarity towards the *E. coli* exonuclease III. It should be noted that HAP1 and probably most eukaryotic AP-endonucleases, involved in direct repair of some bulky exocyclic adducts in DNA generated by benzene metabolites (Hang *et al.*, 1996). However, enzymatic properties of the mammalian enzymes differ from those of the bacterial protein. The human enzyme has a kcat for hydrolytic AP-endonuclease activity 10-fold higher than *E. coli* exo III. Among mammalian exo III homologs, the levels of 3'-phosphatase, 3'-phosphoglycolate aldehyde diesterase and 3'$\rightarrow$5' exonuclease activity are much lower than their AP-endonuclease function.

ANPG protein is the human counterpart of the *E.coli* AlkA-protein. A human cDNA expressing a 3-methyladenine-DNA glycosylase has been isolated (O'Connor and Laval, 1991) and purified (O'Connor, 1993), the gene maps to chromosome 16 (Samson *et al.*, 1991; Vickers *et al.*, 1993). This protein releases 3-meA and 7-meG from methylated DNA (O'Connor and Laval, 1991), but it is also able to excise oxidized bases from DNA such as hypoxantine and 1,N6-ethenoadenine, even more efficiently than the AlkA-protein (Saparbaev and Laval, 1994; Saparbaev *et al.*, 1995). In contrast to the AlkA protein, the ANPG-protein does not release 5-formyluracil residues from DNA. Nevertheless, there is an activity in human cell free extracts excising this lesion (Bjelland *et al.*, 1995). It should be noticed that the human protein does not share significant aminoacid sequence homology with the bacterial AlkA.

7. THE SHORT- AND LONG-PATCH BER IN MAMMALIAN CELLS

The proteins involved, in mammalian cells, in the repair of uracil residue in the G/U mispair originating from the deamination of the cytosine residue in a G/C pair has been investigated. As expected from the reconstitution, using pure *E.coli* proteins, of the BER pathway (Dianov and Lindahl 1994), the combined action of human uracil-DNA glycosylase, HAP1, DNA polymerase β and either DNA ligase I or DNA ligase III lead to the successful repair of the uracil residue (Kubota et al 1996). This pathway is called the short-patch base excision repair and leads to the resynthesis of a single nucleotide.

However, a gap greater than one nucleotide can be generated at the base damage site and the completion of repair in this case is PCNA-dependent (Stucki *et al.*, 1998). *In vitro* reconstitution of this long-patch base excision repair has confirmed that PCNA is required for this process (Kim *et al.*, 1998) and has shown that the structure-specific nuclease FEN 1 is essential for cleavage of the reaction intermediate produced by the template strand displacement (Klungland and Lindahl, 1997). The DNA polymerase δ/ε (Pol δ/ε) is implicated in this pathway (Stucki *et al.*, 1998; Fortini *et al.*, 1998).

8. CONCLUSIONS

Several lines of evidence suggest that ROS are important contributors to aging and degeneratives diseases (cancer, cardiovascular diseases, rhumatoid arthrities...), and play a role in some pathologic states (reoxygenation injury, post traumatic degeneration...). It is the role of DNA repair mechanisms to ensure the stability of the genetic information.

Important breakthrough were obtained using mutants of *E. coli* deficient in base excision repair pathway. They demonstrated the crucial role of DNA glycosylases and AP-endonucleases in protecting cells from mutagenic and cytotoxic effects of free radicals (Michaels *et al.*, 1992; Duwat *et al.*, 1995; Demple *et al.*, 1983; Cunningham *et al.*, 1986). The availability of mammalian cDNA coding for base excision repair enzymes makes possible the generation of mutants animals deficient in the repair of oxidative damage in DNA. These mutants will help the evaluation of the pathological and genotoxic effects of ROS *in vivo*.

ACKNOWLEDGMENTS

The experiments of the authors reported in this work were supported by grants from European Community, Association pour la Recherche sur le Cancer, Electricité de France, Fondation Franco-Norvegienne, fellowships from ESF and from Fondation pour la Recherche Médicale.

REFERENCES

Aburatani, H., Y. Hippo, T. Ishida, R. Takashima, C. Matsuba, T. Kodama, M. Takao, A. Yasui, K. Yamamoto, M. Asano, K. Fukasawa, T. Yoshinari, H. Inoue, E. Ohtsuka, and S. Nishimura (1997) Cloning and characterisation of mammalian 8-hydroxyguanine-specific DNA glycosylase/apurinic, apyrimidinic lyase, a functional mutM homologue, *Cancer Res.*, **57**, 2151–2156.

Anderson, A.W., H.C. Norden, R.F. Cain, G. Parrish, and D. Duggan. (1956). Studies on a radio-resistant micrococcus. I. Isolation, morphology, cultural characteristics, and resistance to gamma radiation. *Food Technol.* **10 :** 575–578.

Arai, K., K. Morishita, K. Shinmura, T. Kohno, S-R. Kim, T. Nohmi, M. Taniwaki, S. Ohwada, and J. Yokota (1997) Cloning of a human homolog of the yeast OGG1 gene that is involved in the repair of oxidative DNA damage, *Oncogene*, **14**, 2857–2861.

Asahara, H., P.M. Wistort, J.F. Bank, R.H. Bakerian and R.P. Cunningham (1989) Purification and characterization of *Escherichia coli* endonuclease III from the cloned *nth* gene, *Biochemistry*, **28**, 4444–4449.

Aspinwall, R., D.G. Rothwell, T. Roldan-Arjona, C. Anselmino, C.J. Ward, J.P. Cheadle, J.R. Sampson, T. Lindahl, P.C. Harris, and I.D. Hickson (1997) Cloning and characterisation of a functional human homolog of *Escherichia coli endonuclease* III, *Proc. Natl. Acad. Sci. USA*, **94**, 109–114.

Auffret van der Kemp, P., Thomas, D., Barbey, R., de Oliveira, R., and Boiteux, S. (1996) Cloning and expression in *Escherichia coli* of the *OGG1* gene of *Saccharomyces cerevisiae*, which codes for a DNA glycosylase that excises 7,8-dihydro-8-oxoguanine and 2,6-diamino-4-hydroxy-5-N-methylformamidopyrimi-dine, *Proc. Natl. Acad. Sci. USA*, **93**, 5197–5202.

Bailly, V., and W.G. Verly (1987) *Escherichia coli* endonuclease III is not an endonuclease but a β-elimination catalyst. *Biochem. J.,* **242**, 565–572.

Bailly,V., W.G.Verly, T.O'Connor and J.Laval. (1989) Mechanism of DNA strand nicking at apurinic/apyrimidinic sites by *Escherichia coli* [formamidopyrimidineDNA glycosylase], B*iochemistry* **262**, 581–589

Bauche, C., and Laval, J., (1999) Repair of the extremely radiation resistant bacterium *Deinococcus Radiodiorans*, *J. Bacteriol.*, in press.

Bjelland, S., N.K. Birkeland, T. Benneche, G. Volden, and E. Seeberg (1994) DNA glycosylase activities for thymine residues oxidized in the methyl group are functions of the AlkA enzyme in *Escherichia coli. J. Biol. Chem.* **269**, 30489–30495.

Bjelland, S., L. Eide, R.W. Time, R. Stote, I. Eftedal, G. Volden, and E. Seeberg (1995) Oxidation of thymine to 5-formyluracil in DNA: mechanisms of formation, structural implications, and base excision by human cell free extracts. *Biochemistry* **34**, 14758–14764.

Boiteux,S., T.O'Connor and J.Laval. (1987) Formamidopyrimidine-DNA glycosylase of *Escherichia* coli : cloning and sequencing of the *fpg* structural gene and overproduction of the protein. *EMBO J.* **6**, 3177–3183

Boiteux,S., T.R. O'Connor, F.Lederer, A.Gouyette, J.Laval. (1990) Homogeneous *Escherichia coli* FPG protein. A DNA glycosylase which excises imidazole ring-opened purines and nicks DNA at apurinic/apyrimidinic sites. *J. Biol. Chem.*, **265**, 3916–3922.

Boiteux,S. (1993) Properties and biological functions of the NTH and FPG proteins of *Escherichia coli*: two DNA glycosylases that repair oxidative damage in DNA. *J. Photochem. Photobiol. B: Biol.* **19**, 87–96.

Cadet, J., M. Berger (1985) Radiation-induced decomposition of the purine bases within DNA and related model compounds, *Int. J. Radiat. Biol.*, **47**, 127–143.

Chan, E., and B. Weiss (1987) Endonuclease IV of *Escherichia coli* is induced by paraquat. *Proc. Natl. Acad. Sci. USA*, **84**, 3189–3193

Cunningham, R.P., and B. Weiss (1985) Endonuclease III (nth) mutants of *Escherichia coli*, *Proc. Natl. Acad. Sci. USA*, **82**, 474–478

Cunningham, R.P., S. Saporito, S.G. Spitzer, and B. Weiss, B. (1986) Endonuclease IV (nfo) mutant of *Escherichia coli*, *J. Bacteriol.*, **168**, 1120–1127.

Demple B., J.H. Halbrook, and S. Linn (1983) *Escherichia coli* xth mutants are hypersensitive to hydrogen peroxide, *J. Bacteriol.*, **153**, 1079–1082.

Demple, B., A. Johnson, and D. Fung. (1986) Exonuclease III and endonuclease IV remove 3' blocks from DNA synthesis primers in H2O2-damaged *Escherichia coli*. *Proc. Natl. Acad. Sci. USA*, **83**, 7731–7735.

Demple, B. (1991) Regulation of bacterial oxidative stress genes. *Annu. Rev. Genet.* **25**, 315–337

Demple, B., T. Herman, and D.S. Chen (1991) Cloning and expression of APE, the cDNA encoding the major human apurinic endonuclease: definition of a family of DNA repair enzymes, *Proc. Natl. Acad. Sci. U.S.A.*, **88**, 10450–10454.

Dianov, G., A. Price, and T. Lindahl. (1992) Generation of single-nucleotide repair patches following excision of uracil residues from DNA. *Mol. Cell. Biol.*, **12**, 1605–1612.

Dianov G, Lindahl T (1994). Reconstitution of the DNA base excision-repair pathway. *Current Biol.*, 1994, **4** : 1069–1076

Dizdaroglu, M., D. Schulte-Frohlinde, C. von Sonntag (1977) Isolation of 2-deoxy-D-erythro-pentonic acid from an alkali-labile site in gamma-irradiated DNA, *Int. J. Radiat. Biol.*, **32**, 481–483.

Dizdaroglu M., J. Laval, and S. Boiteux (1993) Substrate specificity of the Escherichia coli endonuclease III: excision of thymine- and cytosine-derived lesions in DNA produced by radiation-generated free radicals, *Biochemistry*, **32**, 12105–12111.

Dizdaroglu, M., Karakaya, A., Jaruga, P., Slupphaug, G. and Krokan, H.E. (1996) Novel activities of human uracil DNA-glycosylase for cytosine-derived products of oxidative DNA damage. *Nucl. Acids Res.* **24** 418–422.

Doetsch, P.W., Zastawny, T.H., Martin, A.M. and Dizdaroglu, M. (1995) Monomeric base damage products from Adenine, Guanine and Thymine induced by exposure of DNA to Ultraviolet radiation, *Biochemistry* **34**, 737–742.

Dodson,M.L., M.L.Michaels and R.S.Lloyd. (1994) Unified catalytic mechanism for DNA glycosylases. *J. Biol. Chem.* **269**, 32709–32712

Duwat, P., R. de Oliveira, S.D. Ehrlich, and S. Boiteux (1995) Repair of oxidative DNA damage in gram-positive bacteria: the Lactococcus lactis Fpg protein, *Microbiology*, **141**, 411–417.

Eide, L., M. Björas, M. Pirovano, I. Alseth, K.G. Berdal, and E. Seeberg (1996) Base excision of oxidative purine and pyrimidine DNA damage in *Saccharomyces cerevisiae* by a DNA glycosylase with sequence similarity to endonuclease III from *Escherichia coli, Proc. Natl. Acad. Sci USA* **93**, 10735–10740.

El-Ghissassi, F.E., A. Barbin, J. Nair, and H. Bartsch (1995) Formation of 1,N6-ethenoadenine and 3,N4-ethenocytosine by lipid peroxidation products and nucleic acid bases, *Chem. Res. Toxicol.*, **8**, 278–283.

Fortini, P., Pascucci, B., Parlanti, E., Sobol, R., Wilson, S.H. and Dogliotti E. (1998) Different DNA polymerases are involved in the short- and long-patch base excision repair in mammalian cells. *Biochemistry*, submitted.

Frosina, G., P. Fortini, O. Rossi, F. Carrozino, G. Raspaglio, I.S. Cox, D.P. Lane, A. Abbondandolo, and E. Dogliotti. (1996) Two pathways for base excision repair in mammalian cells. *J. Biol. Chem.* **271**, 9573–9578.

Gates III, F.T. and S. Linn (1977) Endonuclease from *Escherichia coli* that acts specifically upon duplex DNA damaged by ultraviolet light, osmium tetroxide, acid, or X-rays, *J. Biol. Chem.*, **252**, 2802–2807.

Graves,R.J., I.Felzenszwabb, J.Laval and T.O'Connor. (1992) Excision of 5'-terminal deoxyribose phosphate from damaged DNA is catalysed by the Fpg protein of *Escherichia coli. J. Biol. Chem.* **267**, 14429–14435

Grossard, F., and W.G. Verly (1978) Properties of the main endonuclease specific for apurinic sites of Escherichia coli (endonuclease VI). Mechanism of apurinic site excision from DNA, *Eur. J. Biochem.*, **82**, 321–332.

Guengerich, F.P. (1992) Roles of the vinyl chloride oxidation products 1-chlorooxirane and 2-chloroacetaldehyde in the in vitro formation of etheno adducts of nucleic acid bases, *Chem. Res. Toxicol.* **5**, 2–5

Habraken, Y., C.A. Carter, M. Sekiguchi, and D.B. Ludlum (1991) Release of N2,3-ethanoguanine from haloethylnitrosourea-treated DNA by *Escherichia coli* 3-methyladenine DNA glycosylase II. *Carcinogenesis,* **12**, 1971–1973.

Halliwell, B. (1982) Production of superoxide, hydrogen peroxide and hydroxyl radicals by phagocytic cells: a cause of chronic inflammatory disease? *Cell Biol. Intl. Rep.*, **6**, 529–542.

Halliwell, B. and Gutteridge J. M. C. (1989). Iron toxicity and oxygen radicals. *Baillieres Clin. Haematol.* **2** : 195–256

Hang, B., A. Chenna, H. Fraenkel-Conrat, and B. Singer. (1996) An unusual mechanism for the major human apurinic/apyrimidinic (AP) endonuclease involving 5' cleavage of DNA containing a benzene-derived exocyclic adduct in the absence of an AP-site. *Proc. Natl. Acad. Sci. USA* **93**, 13737–13741.

Hatahet, Z., Kow, Y. W., Purmal, A. A., Cunningham, R. P. and S. Wallace. (1994) New substrates for old enzymes. 5-Hydroxy-2'-deoxycytidine and 5-hydroxy-2'-deoxyuridine are substrates for *Escherichia coli* endonuclease III and formamidopyrimidine DNA N-glycosylase, while 5-hydroxy-2'-deoxyuridine is a substrate for uracil DNA N-glycosylase. *J. Biol. Chem.* **269**, 18814–18820.

Henner, W.D., L.O. Rodriguez, S.M. Hecht, W.A. Haseltine (1983) gamma Ray induced deoxyribonucleic acid strand breaks. 3' glycolate termini, *J. Biol. Chem.*, **258**, 711–713.

Hilbert, T.P., W. Chaung, R.J. Boorstein, R.P. Cunningham, and G.W. Teebor (1997) Cloning and expression of the cDNA encoding the human homologue of the DNA repair enzyme, *Escherichia coli* endonuclease III, *J. Biol. Chem.*, **272**, 6733–6740.

Hill-Perkins, M., M.D. Jones, and P. Karran. (1986) Site-specific mutagenesis in vivo by single methylated or deaminated purine bases. *Mutat. Res.* **162**, 153–163.

Ide, H., K. Tedzuka, H. Shimzu, Y. Kimura, A.A. Purmal, S.S. Wallace, and Y.W. Kow. (1994) α-Deoxyadenosine, a major anoxic radiolysis product of adenine in DNA, is a substrate for *Escherichia coli* endonuclease IV. *Biochemistry,* **33**, 7842–7847.

Imlay, J.M., and S. Linn (1988) DNA damage and oxygen radical toxicity, *Science*, **240**, 1302–1308.

Jaruga, P., Zastawny, T.H., Skokowski, J., Dizdaroglu, M. and Olinski, R. (1994) Oxidative DNA base damage and antioxidant enzyme activities in human lung cancer, *FEBS Lett.* **341**, 59–64.

Jiang,D., Z.Hatahet, J.O.Blaisdell, R.J.Melamede and S.S.Wallace. (1997) *Escherichia coli* endonuclease VIII: cloning, sequencing, and overexpression of the *nei* structural gene and characterization of *nei* and *nei nth* mutants. *J. Bacteriol.* **179** , 3773–3782

Johnson, A.W., and B. Demple (1988) Yeast DNA 3'-repair diesterase is the major cellular apurinic/apyrimidinic endonuclease: substrate specificity and kinetics, *J. Biol. Chem.*, **263**, 18017–18022.

Jurado, J., Saparbaev, M., Matray, M.J., Greenberg, M.M., and J. Laval (1998) The Ring Fragmentation product of Thymidine C5-hydrate when present in DNA is repaired by the *Escherichia coli* Fpg and Nth proteins. *Biochemistry.* 37, 7757–7763.

Kaneko, T., S. Tahara, and M. Matsuo (1996) Non-linear accumulation of 8-hydroxy-2'-deoxyguanosine, a marker of oxidized DNA damage, during aging. *Mutat. Res.* **316**, 277–285.

Karran, P., and T. Lindahl (1978) Enzymatic excision of free hypoxanthine from polydeoxynucleotides and DNA containing deoxyinosine monophosphate residues. *J Biol. Chem.* **253**, 5877–5879.

Karran P., and T. Lindahl (1980) Hypoxanthine in deoxyribonucleic acid: generation by heat-induced hydrolysis of adenine residues and release in free form by a deoxyribonucleic acid glycosylase from calf thymus. *Biochemistry* **19**, 6005–6011.

Kasai, H., and S. Nishimura (1986) Hydroxylation of guanine in nucleosides and DNA at the C-8 position by heated glucose and oxygen radical-forming agents. *Envir. Health Pers.* **67**, 111–116.

Kasai, H., A. Iida, Z. Yamaizumi, S. Nishimura, and H.Tanooka (1990) 5-Formyldeoxyuridine: a new type of DNA damage induced by ionizing radiation and its mutagenicity to salmonella strain TA102. *Mutat. Res.* **243**, 249–253.

Kim, J., and S. Linn, (1988) The mechanism of action of *E.coli* endonuclease III and T4UV endonuclease (endonuclease V) at AP sites. *Nucl. Acids Res.*, **16**, 1135–1141

Kim, K., Biade, S., and Matsumoto Y. (1998) Involvment of Flap endonuclease 1 in base excision DNA repair. *J. Biol. Chem.* **273** 8842–8848.

Klungland, A., and T. Lindahl. (1997) Second pathway for completion of human DNA base excision-repair: reconstitution with purified proteins and requirement for DNase IV (FEN1). *EMBO J.* **16**, 3341–3348.

Kouznetsov,S.V., O.M.Sidorkina, J.Jurado, M.Bazin, P.Deanjean, J.C.Brochon, J.Laval, R.Santus. (1998) Effect of the single mutations: P2G, P2E and K57G on the fluorescent properties of *Escherichia coli* Fpg protein. *Eur. J. Biochem.* **253**, 413–420.

Krokan H.E., Standal R., Slupphaug G. (1997). DNA glycosylases in the base excision repair of DNA. *Biochem. J.* **325** : 1–16

Kubota Y., Nash R.A., Klungland A., Schar P., Barnes D.E., Lindahl T. (1996). Reconstitution of DNA base excision-repair with purified human proteins: interaction between DNA polymerase beta and the XRCC1 protein. *EMBO J.*, **15**, 6662–6670

Kuo, C.F., D.E. McRee, C.L. Fisher, S.F. O'Handley, R.P. Cunningham, and J.A. Tainer (1992) Atomic structure of the DNA repair [4Fe-4S] enzyme endonuclease III. *Science*, **258**, 434–440

Labahn, J., O.D. Scharer, A. Long, K. Ezaz-Nikpay, G.L. Verdine, and T.E. Ellenberger (1996) Structural basis for the excision repair of alkylation-damaged DNA. *Cell* **86**, 321–329.

Laval, J., S.Boiteux and T.R.O'Connor. (1990) Physiological properties and repair of apurinic/apyrimidinic sites and imidazole ring-opened guanines in DNA. *Mutat. Res.*, 1990, **233**, 73–79.

Laval, J. (1996) Role of DNA repair enzymes in the cellular resistance to oxidative stress. *Path. Biol.* **44**, 14–24

Levin, J.D., A.W. Johnson, B. Demple (1988) Homogeneous Escherichia coli endonuclease IV. Characterization of an enzyme that recognizes oxidative damage in DNA. *J. Biol. Chem.* **263**, 8066–8071.

Ljungquist, S., T. Lindahl, and P. Howard-Flanders (1976) Methyl methane sulfonate-sensitive mutant of *Escherichia coli* deficient in an endonuclease specific for apurinic sites in deoxyribonucleic acid. *J. Bacteriol.* **126**, 646–653.

Ljungquist, S., (1977) A new endonuclease from *Escherichia coli* acting at apurinic sites in DNA. *J. Biol. Chem.* **252**, 2808–2814.

Lowenhaupt, K., M. Sander, C. Hauser, and A. Rich (1989) Drosophila melanogaster strand transferase. A protein that forms heteroduplex DNA in the absence of both ATP and single-strand DNA binding protein, *J. Biol. Chem.*, **264**, 20568–20575.

Lu, R., H.M. Nash, and G.L. Verdine (1997) A mammalian DNA repair enzyme that excises oxidatively damaged guanines maps to a locus frequently lost in lung cancer, *Curr. Biol.*, **7**, 397–407.

McCord, J.M. (1987) Oxygen-derived radicals: a link between reperfusion injury and inflammation, *Fed. Proc.* **46**, 2402–2406

Marnett, L.J. (1994) DNA adducts of α,β-unsaturated aldehydes and dicarbonyl compounds. In *DNA Adducts: Identification and Biological Significance* (Hemminki, K., Dipple, A., Shuker, D.E.G., Kadlubar, F.F., Segerbäck, D., and Bartsch, H., Eds.), pp. 151–163, LARC Scientific Publications (No. 125), Lyon, France.

Mazumder, A., Gerlt, J.A., Absalon, M.J., Stubbe, J., Cunningham, R.P., Withka, J.M., and P.H. Bolton, (1991) Stereochemical studies of the beta-elimination reactions at aldehydic abasic sites in DNA: endonuclease III from Escherichia coli, sodium hydroxide, and Lys-Trp-Lys. *Biochemistry*, **30**, 1116–1126.

Melamede,R.J., Z.Hatahet, Y.W.Kow, H.Ide, S.S.Wallace. (1994) Isolation and characterization of endonuclease VIII from *Escherichia coli*. *Biochemistry* **33**, 1255–1264

Michaels, M.L., L. Pham, Y. Ngheim, C. Cruz and J.H. Miller (1990) MutY, an adenine glycosylase active on G-A mispairs, has homology to endonuclease III. *Nucl. Acids Res.*, **18**, 3841–3845.

Michaels, L.M., C. Cruz, A.P. Grollman, and J.H. Miller (1992) Evidence that MutY and MutM combine to prevent mutations by an oxidatively damaged form of guanine in DNA, *Proc. Natl. Acad. Sci. USA* **89**, 7022–7025.

Minton, K.W. (1994) DNA repair in the extremely radioresistant bacterium *Deinococcus radiodurans*. *Molec. Microbiol.* **13** : 9–15.

Mol, C.D., C.F. Kuo, M.M. Thayer, R.P. Cunningham, and J.A. Tainer (1995) Structure and function of the multifunctional DNA-repair enzyme exonuclease III. *Nature* **374**, 381–386. 7.

Moseley, B.E.B. 1983. Photobiology and radiobiology of *Micrococcus (Deinococcus) radiodurans*. *Photochem. Photobiol. Rev.* **7** : 223–274.

Murray, R.G.E., and B.W. Brooks. (1986). *Deinococcus*, p. 1.35–1043. *In* P. Sneath, N. Mair, M. Sharpe, and J. Holt (ed.), Bergey's manual of systematic bacteriology, vol. 2. Williams & Wilkins, Baltimore.

Nakabeppu, Y., H. Kondo, and M. Sekiguchi (1984) Cloning and characterization of the alkA gene of *Escherichia coli* that encodes 3-methyladenine DNA glycosylase II. *J. Biol. Chem.* **259**, 13723–13729

Nash, H.M., S.D. Bruner, O.D. Schärer, T. Kawate, T.A. Addona, E. Spooner, W.S. Lane, and G.L. Verdine (1996) Cloning of a yeast 8-oxoguanine DNA glycosylase reveals the existence of a base-excision DNA-repair protein superfamily, *Curr. Biol.*, **6**, 968–980.

Nunoshiba, T., E. Hidalgo, Z. Li, B. Demple (1993) Negative autoregulation by the *Escherichia coli* SoxS protein: a dampening mechanism for the soxRS redox stress response. *J. Bacteriol.* **175**, 7492–7494

O'Connor,T.R. and J.Laval. (1989) Physical association of the 2,6-diamino-4-hydroxy-5N-formamidopyrimidine-DNA glycosylase of *Escherichia coli* and an activity nicking DNA at apurinic/apyrimidinic sites. *Proc. Natl. Acad. Sci. USA* **86**, 5222–5226

O'Connor, T.R., and J. Laval J (1991) Human cDNA expressing a functional DNA glycosylase excising 3-methyladenine and 7-methylguanine. *Biochem. Biophys. Res. Commun.* **176**, 1170–1177

O'Connor T.R., (1993) Purification and characterization of human 3-methyladenine-DNA glycosylase. *Nucl. Acids Res.* **21**, 5561–5569.

O'Connor,T.R., R.J.Graves, G.de Murcia, B.Castaing and J.Laval. (1993) Fpg protein of *Escherichia coli* is a zinc finger protein whose cysteine residues have a structural and/or functional role. *J. Biol. Chem.* **268**, 9063–9070

Park J.W., and Floyd R.A. (1992). Lipid peroxidation products mediate the formation of 8-hydroxydeoxyguanosine in DNA. *Free Radic. Biol. Med.* **12,** 245–250 .

Popoff, S.C., A.I. Spira, A.W. Johnson, and B. Demple (1990) Yeast structural gene (APN1) for the major apurinic endonuclease: homology to Escherichia coli endonuclease IV, *Proc. Natl. Acad. Sci. USA* **87**, 4193–4197.

Rabow,L.E. and Y.W.Kow. (1997) Mechanism of action of base release by *Escherichia coli* Fpg protein: role of lysine 155 in catalysis. *Biochemistry* **36**, 5084–5096.

Radicella, J.P., C. Dherin, C. Desmaze, M.S. Fox, and S. Boiteux (1997) Cloning and characterization of *hOGG1*, a human homolog of the *OGG1* gene of *Saccharomyces cerevisiae*, *Proc. Natl. Acad. Sci. USA*, **94**, 8010–8015.

Radman M. (1976) An endonuclease from *Escherichia coli* that introduces single polynucleotide chain scissions in ultraviolet-irradiated DNA. *J. Biol.Chem.* **251**, 1438–1445.

Reiter, R.J. (1995) Oxidative processes and antioxidative defence mechanisms in the aging brain, *FASEB J.* **9**, 526–533.

Richardson, C.C., I.R. Lehman and A. Kornberg (1964) A deoxiribonucleic acid phosphatase-exonuclease from *Escherichia coli*. II. Characterization of the exonuclease activity, *J. Biol. Chem.*, **239**, 251–258.

Richardson, C.C., and A. Kornberg (1964)) A deoxiribonucleic acid phosphatase-exonuclease from *Escherichia coli*. I. Purification and Characterization of the phosphatase activity, *J. Biol. Chem.*, **239**, 242–250.

Robson, C.N., and I.D. Hickson (1991) Isolation of cDNA clones encoding an enzyme from bovine cells that repairs oxidative DNA damage in vitro: homology with bacterial repair enzymes, *Nucl. Acids Res.* **19**, 1087–1092.

Robson, C.N., A.M. Milne, D.J.C. Pappin, and I.D. Hickson (1991) Isolation of cDNA clones encoding a human apurinic/apyrimidinic endonuclease that corrects DNA repair and mutagenesis defects in *E. coli* xth (exonuclease III) mutants, *Nucl. Acids Res.*, **19**, 5519–5523.

Rogers, S.G., and B. Weiss (1980) Cloning of the exonuclease III gene of Escherichia coli. *Gene* **11**, 187–195.

Roldan-Arjona, T., C. Anselmino, and T. Lindahl (1996) Molecular cloning and functional analysis of a *Schizosaccharomyces pombe* homologue of *Escherichia coli* endonuclease III, *Nucl. Acids Res.* **24**, 3307–3312.

Roldan-Arjona, T., Y-F. Wei, K.C. Carter, A. Klungland, C. Anselmino, R-P. Wang, M. Augustus, and T. Lindahl (1997) Molecular cloning and functional expression of a human cDNA encoding the antimutator enzyme 8-hydroxyguanine-DNA glycosylase, *Proc. Natl. Acad. Sci. USA* **94**, 8016–8020.

Rosenquist, T.A., D.O. Zharkov, and A.P. Grollman (1997) Cloning and characterization of a mammalian 8-oxoguanine DNA glycosylase, *Proc. Natl. Acad. Sci. USA* **94**, 7429–7434.

Sammartano, L.J., and R.W. Tuveson (1983) *Escherichia coli* xthA mutants are sensitive to inactivation by broad-spectrum near-UV (300- to 400-nm) radiation. *J. Bacteriol.* **156**, 904–906

Samson, L., B. Derfler, M. Boosalis,and K. Call (1991) Cloning and characterization of a 3-methyladenine DNA glycosylase cDNA from human cells whose gene maps to chromosome 16. *Proc. Natl. Acad. Sci. USA* **88**, 9127–9131.

Saparbaev, M., and J. Laval (1994) Excision of hypoxanthine from DNA containing dIMP residues by the *Escherichia coli*, yeast, rat, and human alkylpurine DNA glycosylases. *Proc. Natl. Acad. Sci. USA* **91**, 5873–5877.

Saparbaev, M., K. Kleibl, and J. Laval (1995) *Escherichia coli, Saccharomyces cerevisiae*, rat and human 3-methyladenine DNA glycosylases repair 1,N6-ethenoadenine when present in DNA. *Nucl. Acids Res.* **23,** 3750–3755.

Schärer, O.D, Ortholand, J.Y., Ganesan, A., Ezaz-Nikpay, K. and Verdine, G.L. (1995) Specific binding of the DNA repair enzyme AlkA to a pyrrolidine-based inhibitor. *J. Am. Chem. Soc.* **117**, 6623–6624

Schärer, O.D., Nash, H.M., Jiricny, J., Laval, J. ang Verdine, G.L. (1998) Specific binding of a designed pyrrolidine abasic site analog to multiple DNA glycosylases. J. *Biol. Chem.* **15,** 8592–8597

Seeberg, E., L. Eide, and M. Bjoras (1995) The base excision repair pathway. *Trends Biochem. Sci.* **20**, 391–397

Seki, S., K. Akiyama, S. Watanabe, M. Hatsushika, S. Ikeda, and K. Tsutsui (1991) cDNA and deduced amino acid sequence of a mouse DNA repair enzyme (APEX nuclease) with significant homology to Escherichia coli exonuclease III, *J. Biol. Chem.*, **266**, 20797–20802.

Seki, S., M. Hatsushika, S. Watanabe, K. Akiyama, K. Nagao, and K. Tsutsui (1992) cDNA cloning, sequencing, expression and possible domain structure of human nuclease homologous to Escherichia coli exonuclease III, *Biochim. Biophys. Acta*, **1131**, 287–299.

Sidorkina,O.M. and J.Laval. (1998) Role of lysine-57 in the catalytic activities of *Escherichia coli* formamidopyrimidine-DNA glycosylase (Fpg protein) *Nucl. Acids Res.* **26**, in press.

Smith, M.D., C.I. Masters, and B.E.B. Moseley. (1992) Molecular biology of radiation resistant bacteria, p. 258–280. *In* R. A. Herbert and R. J. Sharp (ed.), Molecular biology and biotechnology of extremophiles. Chapman & Hall, New York.

Stucki, M., Pascuccu, B., Parlanti, E., Fortini, P., Wilson, S.H., Hübsher, U. and Dogliotti, E. (1998) Mammalian base excision repair by DNA polymerases δ and ε. Oncogene. In press.

Tajiri,T., H.Maki, M.Sekiguchi. (1995) Functional cooperation of MutT, MutM and MutY proteins in preventing mutations caused by spontaneous oxidation of guanine nucleotides in *Escherichia coli. Mutat. Res.*, **336**, 257–267.

Tchou,J. A.P.and Grollman. (1995) The catalytic mechanism of Fpg protein. Evidence for a Schiff base intermediate and amino terminus localization of the catalytic site. *J. Biol. Chem.* **270**, 11671–11677

Thayer,M.M., H.Ahern, D.Xing, R.P.Cunningham and J.A.Tainer. (1995) Novel DNA binding moyifs in the DNA repair enzyme endonuclease III crystal structure. *EMBO J.* **14**,4108–4120

Trus, M.A. and Kensler, T.W. (1991) An overreview of the relationship between oxidative stress and chemical carcinogenesis, *Free Radical Biol. Med.* **10**, 201–209.

Tsaneva, I.R., and B. Weiss (1990) *soxR*, a locus governing a superoxide response regulon in *Escherichia coli* K-12. *J. Bacteriol.* **172**, 4197–4205.

Vickers, M.A., P. Vyas, P.C. Harris, D.L. Simmons, and D.R. Higgs (1993) Structure of the human 3-methyladenine DNA glycosylase gene and localization close to the 16p telomere. *Proc. Natl. Acad. Sci. USA* **90**, 3437–3441.

Warner, H.R., B.F. Demple, W.A. Deutsch, C.M. Kane, and S. Linn (1980) Apurinic/apyrimidinic endonucleases in repair of pyrimidine dimers and other lesions in DNA. *Proc. Natl. Acad. Sci. USA* **77**, 4602–4606.

Weiss, B., (1976) Endonuclease II of Escherichia coli is exonuclease III. *J. Biol. Chem.* **251**, 1896–1901.

Weiss, B., and L. Grossman (1987) Phosphodiesterases involved in DNA repair. *Adv. Enzymol. Relat. Areas Mol. Biol.* **60**, 1–34.

Wink D.A., I. Hanbauer , F. Laval, J.A. Cook, M.C. Krishna, J.B. Mitchell (1994) Nitric oxide protects against the cytotoxic effects of reactive oxygen species. *Ann. N. Y. Acad. Sci.* **738**, 265–278.

Yamagata, Y., M. Kato, K. Odawara, Y. Tokuno, Y. Nakashima, N. Matsushima, K. Yasumura, K. Tomita, K. Ihara, Y. Fujii, Y. Nakabeppu, M. Sekiguchi, and S. Fujii (1996) Three-dimensional structure of a DNA repair enzyme, 3-methyladenine DNA glycosylase II, from Escherichia coli. *Cell* **86**, 311–319

Zastawny, T.H., Doetsch, P.W., and Dizdaroglu, M. (1995) A novel activity of Uracil DNA N-glycosylase : excision of isodialuric acid (5,6-Dihydroxyuracil) from DNA, a major product of oxidative DNA damage. *FEBS Lett.* **364**, 255–260.

Zharkov,D.O., R.A.Rieger, C.R.Iden and A.P.Grollman. (1997) NH_2-terminal proline acts as a nucleophile in the glycosylase/AP lyase reaction catalyzed by *Escherichia coli* formamidopyrimidine-DNA glycosylase(Fpg) protein. *J. Biol. Chem.* **272**, 5335–5341.

20

DNA LESIONS GENERATED *IN VIVO* BY REACTIVE OXYGEN SPECIES, THEIR ACCUMULATION AND REPAIR

Tomas Lindahl

Imperial Cancer Research Fund
Clare Hall Laboratories, South Mimms
Hertfordshire EN6 3LD, U.K.

1. ABSTRACT

Endogenous DNA lesions are generated continuously by exposure to water, reactive small molecules such as certain coenzymes and metabolites, and reactive oxygen species. The proportion of total spontaneous DNA lesions *in vivo* due to oxidative damage is not known, but the occurrence of specific repair enzymes for oxidized bases indicates that this is one of the most significant forms of damage. The major human DNA glycosylases that remove base lesions generated by oxygen free-radicals are known, and corresponding cDNAs have been cloned and overexpressed to yield functional products, allowing for detailed biochemical studies. Due to efficient DNA repair, the steady state levels of oxidized bases in the DNA of normal mammalian cells are likely to be so low that they cannot be detected by current methodologies.

2. INTRODUCTION

DNA is an unstable chemical entity under *in vivo* conditions, because it is a target of a variety of reactive small molecules. Thus, DNA undergoes hydrolysis resulting in depurination and deamination, degradative reactions caused by active oxygen, and base adduct formation by reactions with metabolites and coenzymes (for reviews, see Lindahl, 1993; Marnett and Burcham, 1993; Lindahl, 1996).

The relative importance of the different modes of degradation has been subject to much debate. This is because only some of the reaction rates for different types of DNA decay can be estimated with any degree of accuracy. However, the DNA depurination rate at 37 °C and pH7.4 has been precisely determined, and corresponds to the loss of 10,000

Advances in DNA Damage and Repair, edited by Dizdaroglu and Karakaya. Kluwer Academic / Plenum Publishers, New York, 1999.

(Lindahl and Nyberg, 1972) or 9000 (Nakamura *et al.*, 1998) bases per day by non-enzymatic hydrolytic cleavage of glycosyl bonds in a mammalian cell. The remaining uncertainty here is whether DNA in chromatin is depurinated 2–5 times more slowly than naked DNA, although DNA remains fully hydrated in chromatin. The reason for suspecting a possible moderate decrease in depurination rate *in vivo* is that DNA charge neutralization by histones is effective in nucleosomes, and depurination rates are slower in high-salt than in low-salt buffers. On the other hand, the presence or absence of divalent cations such as Mg^{2+} in buffers of physiological ionic strength does not significantly affect depurination rates, in spite of the improved charge shielding of the DNA phosphates achieved in the presence of Mg^{2+}.

The rate of hydrolytic cytosine deamination in DNA *in vivo* is somewhat more difficult to estimate, because of the much higher rate of deamination of single-stranded DNA than double-stranded DNA. However, even 100% double-stranded DNA is deaminated at a significant rate. Most likely, the range is within 100–500 deaminated cytosines per day in a mammalian cell (Lindahl and Nyberg, 1974; Frederico *et al.*, 1990; Shen *et al.*, 1994), with the higher value dependent on temporary exposure of a small but significant proportion of single-stranded DNA due to active transcription and replication processes. The spontaneous mutator phenotype of yeast and *E.coli ung* mutants defective in removal of deaminated cytosine strongly indicates that the reaction is of relevance *in vivo*. Hydrolytic deamination of 5-methylcytosine is 2- to 3-fold faster than that of unsubstituted cytosine, and this moderate rate increase may partly account for 5-methylcytosine residues being mutational hot spots in mammalian cells, although the slower rate of repair of the deaminated form of 5-methylcytosine than that of cytosine is likely to be more important in this regard (Shen *et al.*, 1994).

S-Adenosylmethionine (SAM) is the reactive methyl donor in most cellular transmethylation reactions, and acts like a weak alkylating agent of the S_N2 type in a nonenzymatic reaction. The most important DNA product generated by SAM is the cytotoxic lesion 3-methyladenine. SAM is present both in the cell nucleus and cytoplasm, typically at a concentration of 30μM, with reported values for different types of cells generally being 10–80μM. The rate of reaction of SAM with DNA under physiological solvent conditions can be determined conveniently and accurately by using ^{3}H-SAM, and corresponds to the formation in DNA of 600 3-methyladenine residues per mammalian cell in 24h (Rydberg and Lindahl, 1982). There is no significant protection of DNA in chromatin against alkylating agents, as determined by studies on the effects of dimethyl sulfate and methyl methanesulfonate treatment, so this value for DNA alkylation by SAM should be correct under *in vivo* conditions. It is 10–20 times lower than the rate of DNA depurination, but still high enough to make it advantageous to living cells to have the specific repair enzyme 3-methyladenine-DNA glycosylase.

Bacteria and eukaryotic cells also have specific repair enzymes to remove the various major types of oxidative damage to DNA. Moreover, a defect in DNA repair of the most mutagenic base lesion, 8-hydroxyguanine (8-oxoG), leads to a spontaneous weak mutator phenotype in bacteria and yeast. These data strongly suggest that repair of endogenous DNA damage generated by reactive oxygen species is relevant *in vivo*, and that significant endogenous oxidative damage to DNA occurs continuously. However, the relative importance of this form of DNA decay is not known in comparison with other types of intrinsic DNA damage, in spite of much debate in the literature. This is because the intracellular, and especially intranuclear, concentrations of the relevant reactive oxygen species can not be determined, whereas the intracellular concentration of water that accounts for hydrolytic degradation is known to be 55M and, as already noted, the SAM concentration is about 30 μM. Reactive oxygen species that damage DNA are generated by iron-mediated Fenton reactions (Henle and Linn, 1997), and the frequency of such reactions in the cell nucleus is not known. In eukaryotic cells, most oxygen metabolism has been delegated to mitochondria, so the cell nucleus is practically anoxic (Joenje, 1989). Processes such as lipid peroxidation may still occur to some extent in nuclei and cytoplasm. Mammalian mutant cell lines with perturbed or inflated lipid peroxidation are now required to elucidate the problem. A gene knockout mouse defective in *klotho*, which may encode an enzyme involved in sphingolipid metabolism, interestingly displays several symptoms reminiscent of human senescence (Kuro-o *et al.*, 1997), but it is not yet known whether lipid peroxidation and subsequent DNA oxidation are perturbed in *klotho*$^{-}$ cells.

In order to evaluate the relative contribution to endogenous DNA damage by reactive oxygen species *in vivo*, it is now important to elucidate the key DNA repair enzymes and pathways involved in the correction of such damage, and then provide mouse gene knockouts that allow relevant biochemical and pathological studies.

3. RESULTS AND DISCUSSION

3.1. Cloning of Human cDNAs Encoding DNA Glycosylases Active on Oxidative Damage

DNA repair enzymes active in base excision-repair often have been highly conserved during evolution. This may reflect the continuous occurrence of significant endogenous DNA decay in all living cells; the problem is the same for *E.coli* as for man, so the strategy to solve it has been conserved. *E. coli* endonuclease III (Nth) is a DNA glycosylase that catalyzes the release of cytosine glycol, thymine glycol, and several other ring-saturated and ring-fragmented pyrimidine derivatives from DNA. These lesions can be generated by exposure of DNA to oxygen free-radicals.

By database homology searches, an open reading frame encoding a protein similar to *E. coli* Nth was detected in the genome of both *Schizosaccharomyces pombe* (Roldán-Arjona *et al.*, 1996) and *Saccharomyces cerevisiae* (Eide *et al.*, 1996). The expressed protein shows DNA glycosylase and AP lyase activities similar to those of *E. coli* Nth. Moreover, an *S. cerevisiae* enzyme-deficient mutant obtained by targeted gene disruption is anomalously sensitive to H_2O_2 and menadione, indicating that the wild-type yeast enzyme is required for repair of oxidative damage (Eide *et al.*, 1996).

Attempts to detect a human cDNA, or expressed sequence tags, homologous to *E. coli* Nth in available data banks initially were unsuccessful. Instead, the relevant human cDNA was cloned by a fortuitous event. In mapping, cloning and sequencing studies on the human tuberous sclerosis gene in the chromosome 16p 13.3 region, an adjacent smaller gene in a head-to-head orientation was also sequenced. This turned out to encode the human equivalent of *E. coli* Nth (Aspinwall *et al.*, 1997). Expression of the cloned cDNA yielded a protein with partial homology to Nth which possessed the expected DNA glycosylase activities against oxidized pyrimidine residues. Unusually, the purified human enzyme was yellow/brown-colored, as is also the case for *E. coli* Nth and its *S. pombe* counterpart; this reflects the presence of an iron/sulfur cluster in the C-terminal region of the protein, which is involved in DNA binding by the enzyme. The same cDNA encoding the human homolog of Nth has also been isolated by the more conventional and general approach of purifying the enzyme to homogeneity from bovine tissue, obtaining several peptide sequences, and using this sequence information to identify a cDNA clone (Hilbert *et al.*, 1997). The enzyme is present both in cell nuclei and mitochondria.

Later steps of repair by the base excision-repair pathway are not discussed here, but oxidative damage of DNA can also result in the generation of an altered oxidized form of deoxyribose with simultaneous non-enzymatic release of the attached base residue. Such abasic sites may not be susceptible to β-elimination events and require repair by the alternative, long-patch (2–6 nucleotides) form of base excision-repair (Klungland and Lindahl, 1997).

The most mutagenic, and therefore probably most relevant, of the major base lesions generated by reactive oxygen species is 8-oxoG. Bacteria excise this lesion employing the *fpg/mutM*-encoded DNA glycosylase. However, attempts in several laboratories to find a counterpart in eukaryotic cells were unsuccessful. This problem was solved by Boiteux and coworkers (van der Kemp *et al.*, 1996), who isolated the relevant yeast cDNA by a functional assay, measuring reduction of the mutator phenotype of an *E. coli* strain defective in 8-oxoG repair by expression of the *S. cerevisiae* cDNA. There is little or no homology between the bacterial and yeast DNA glycosylases that excise 8-oxoG, but enzyme assays showed that the two proteins perform similar functions. The yeast enzyme was called OGG1. The same cDNA was also cloned by Verdine's group, using the alternative strategy of relying on the AP lyase activity of the enzyme in a sodium borohydride trapping assay with an 8-oxoG containing oligonucleotide, to isolate sufficient amounts of the protein for peptide microsequencing (Nash *et al.*, 1996). The determination of the yeast OGG1 sequence allowed for the rapid isolation of the corresponding, partly homologous human cDNA by several groups (Arai *et al.*, 1997; Aburatani *et al.*, 1997; Radicella *et al.*, 1997; Roldán-Arjona *et al.*, 1997; Bjoras *et al.*, 1997; Lu *et al.*, 1997; Rosenquist *et al.*, 1997; for a recent review see Cunningham, 1997). Intriguingly, hOGG1 has also been cloned from a B-cell subtractive cDNA library representing genes overexpressed or induced in the zone of lymph node germinal centers where somatic hypermutation of antibody V gene segments takes place (Kuo and Sklar, 1997). The human enzyme efficiently excises 8-oxoG opposite a C residue in double-stranded DNA, but acts very slowly and inefficiently on 8-oxoG opposite A, G, or T. As for the yeast OGG1, the human enzyme (hOGG1) partly suppresses the spontaneous mutator phenotype of an *E. coli* mutant deficient in 8-oxoG excision. The *hOgg1* gene is located to chromosome 3p25.

The hOGG1 enzyme occurs in at least two different forms, due to alternative splicing (Aburatani *et al.*, 1997; Tani *et al.*, 1998). Most groups isolated the cDNA encoding the smaller but more abundant form of hOGG1; this is a nuclear enzyme (Bjoras *et al.*, 1997). In our work, expressed sequence tags were used to isolate several hOGG1 cDNAs, and the two longest ones were selected for DNA sequencing in order to avoid putative problems with incomplete sequences at the 5' end, a common problem with cDNAs. This approach resulted in the isolation of a distinct splice variant of hOGG1 with a longer C-terminal sequence (Roldán-Arjona *et al.*, 1997). This form is also specific for 8-oxoG opposite C, and is antimutagenic in *E. coli* just like the shorter, major form of the enzyme. It is not yet known if the longer splice variant of hOGG1 is a nuclear or mitochondrial protein.

Several of the groups involved in cloning of *hOgg*1 now are taking the next logical step, attempting to construct gene knockout mice deficient in the activity. The pathology of such mice should be of considerable interest : If endogenous 8-oxoG formation is as frequent and important as has often been claimed in the literature, the mice might be expected to exhibit phenotypes of early senescence and high cancer frequency, which could perhaps be suppressed if the mice are given diets containing antioxidants. Results on

OGG1-deficient mice, and cell strains derived from such mice, should be available in the near future, and such data will allow a better assessment of the amounts and physiological effects of this particular DNA lesion.

3.2. Levels of 8-OxoG in Human DNA

Since endogenous damage to DNA is significant, a certain number of endogenous lesions might be expected to be present in mammalian DNA *in vivo*. However, the chemical instability of some lesions, and more importantly the evolution of efficient DNA repair mechanisms to remove damaged residues ensure that the number is kept very low and does not result in a mutagenic "error catastrophe" in normal cells. The presence of abasic sites in DNA is a good example. The major source of such lesions is likely to be nonenzymatic depurination, with additional contributions from release of damaged DNA bases by DNA glycosylases, and slower nonenzymatic depyrimidination events. It is very likely that the final number of such sites generated in a mammalian cell per day is close to 10,000 sites, and within 5000–20,000 sites. If no DNA repair mechanism would exist, a steady state situation in the range of this level of abasic sites in DNA would soon be achieved, because nonenzymatic cleavage of 3' phosphodiester bonds at abasic sites by a β-elimination reaction may be estimated to occur in 1–2 days at 37°C under *in vivo* conditions. It seems highly unlikely that this equilibrium process would change with time, or with the age of an organism. In normal mammalian cells, an abasic site in DNA is removed by an excision-repair process in less than 4 min (Moran and Ebisuzaki, 1987). This means that the expected steady state level of abasic sites in a normal, repair-competent cell would be 100 sites per cell, *or less*. A repair-deficient cell lacking the AP endonuclease HAP1/APE might have a steady state level of abasic sites 10–100 times higher than wild-type cells, assuming that some slow repair of abasic sites by the nucleotide excision-repair pathway would still occur. However, attempts to construct gene knockout mice with this deficiency have resulted in an early embryonic lethal phenotype, indicating that a high steady-state level of abasic sites in the mammalian genome is incompatible with cellular proliferation.

Deamination of cytosine to uracil occurs more slowly than depurination, and again such residues are removed quickly by the ubiquitous uracil-DNA glycosylase, which has a high turnover number. Thus, the expected steady state level of uracil in mammalian DNA *in vivo* would only be 1–10 residues per genome. Present experimental techniques do not allow for the direct analytical detection of lesions at this extremely low level. A number of early reports were published on the apparent presence of detectable levels of uracil in isolated human DNA, in the vicinity of 0.01%, but these results are now generally ascribed to artificial acid-catalyzed deamination of DNA during the hydrolysis procedure preceding base analysis.

Similar reasoning with regard to the expected steady state level of 8-oxoG in the DNA of a mammalian cell again would yield low estimates. The level of hOGG1 is lower than that of HAP1 or uracil-DNA glycosylase in mammalian cells, but experiments on the rates of introduction and subsequent removal of 8-oxoG residues after exposure of mammalian cells to ionizing radiation strongly indicate that an 8-oxoG residue is repaired in less than 20 min *in vivo*. The amount of oxidative damage to nuclear DNA is quenched about 50-fold by histones, which largely prevent iron binding to DNA and thereby suppress local formation of Fenton oxidants. The absence of histones in mitochondria, and the active oxygen metabolism in this organelle, suggest that mitochondria should have a steady state level of 8-oxoG 100–1000 times higher than nuclear DNA, per 10^6 nucleotide residues. Interestingly, the careful investigations by Higuchi and Linn (1995) show that DNA isolated from mitochondria of HeLa cells contains only barely detectable levels of 8-oxoG, less than 1 residue per 10^6 G residues. In view of this result, and by comparison with the data on abasic sites and uracil in DNA, a likely estimate is that there are 5–100 8-oxoG residues in the DNA of a normal mammalian cell at any specific time. This is below the detection level of current methodologies. Obviously, exposure to substantial doses of ionizing rediation might greatly increase these numbers. While the ratio of 8-oxoG to G should be much lower in nuclei than in mitochondria, there is much more DNA in the nucleus, so the total amount of 8-oxoG in DNA from a mammalian cell may be expected to be similar in the nucleus *vs* the mitochondria of a cell.

This reasoning implies that all recent determinations of 8-oxoG in the DNA from normal mammalian cells may be artifacts due to 8-oxoG formation during the isolation procedure. In spite of the sensitive detection methods developed, such as mass spectrometry analysis and electrochemical detection techniques after HPLC, and a variety of precautions taken to attempt to reduce DNA oxidation *in vitro*, formation of 8-oxoG residues during DNA isolation remains a major problem in this field (Collins *et al.*, 1996). The fact that some of the earlier data on very high 8-oxoG levels in DNA now have been retracted and replaced with lower estimates (Helbock *et al.*, 1998) does not necessarily mean that the new numbers are correct. Construction of OGG1-deficient mice should allow for better estimates of the rate of formation of 8-oxoG residues in DNA under *in vivo* conditions.

REFERENCES

H. Aburatani, Y. Hippo, T. Ishida *et al.* (1997) Cloning and characterization of mammalian 8-hydroxyguanine-specific DNA glycosylase/apurinic, apyrimidinic lyase, a functional mutM homologue. Cancer Res. 57, 2151–2156.

K. Arai, K. Morishita, K. Shinmura *et al.* (1997) Cloning of a human homolog of the yeast *OGG1* gene that is involved in the repair of oxidative DNA damage. Oncogene 14, 2857–2861.

R. Aspinwall, D. G. Rothwell, T. Roldán-Arjona *et al.* (1997) Cloning and characterization of a functional human homolog of *Escherichia coli* endonuclease III. Proc. Natl. Acad. Sci. USA 94, 109–114.

M. Bjørås, L. Luna, B. Johnsen, E. Hoff, T. Haug, T. Rognes, and E. Seeberg (1997) Opposite base-dependent reactions of a human base excision repair enzyme on DNA containing 7,8-dihydro-8-oxoguanine and abasic sites. EMBO J. 16, 101–109.

A. R. Collins, M. Dusinská, C. M. Gedik, and R. Stetina (1996) Oxidative damage to DNA: Do we have a reliable biomarker? Env. Health Perspec. 104, 465–469.

R. P. Cunningham (1997) DNA repair: Caretakers of the genome? Current Biol. 7, R576-R579.

L. Eide, M. Bjørås, M. Pirovano, I. Alseth, K. G. Berdal, and E. Seeberg (1996) Base excision of oxidative purine and pyrimidine DNA damage in *Saccharomyces cerevisiae* by a DNA glycosylase with sequence similarity to endonuclease III from *Escherichia coli*. Proc. Natl. Acad. Sci. USA 93, 10735–10740.

L.A. Frederico, T.A. Kunkel and B.R. Shaw (1990) A sensitive genetic assay for the detection of cytosine deamination: determination of rate constants and activation energy. Biochemistry 29, 2532–2537.

H. J. Helbock, K. B. Beckman, M. K. Shigenaga, P. B. Walter, A. A. Woodall, H. C. Yeo, and B. N. Ames (1998) DNA oxidation matters: The HPLC- electrochemical detection assay of 8-oxo-deoxyguanosine and 8-oxo-guanine. Proc. Natl. Acad. Sci. USA 95, 288–293.

E. S. Henle and S. Linn (1997) Formation, prevention, and repair of DNA damage by iron-hydrogen peroxide. J. Biol. Chem. 272, 19095–19098.

Y. Higuchi and S. Linn (1995) Purification of all forms of HeLa cell mitochondrial DNA and assessment of damage to it caused by hydrogen peroxide treatment of mitochondria or cells. J. Biol. Chem. 270, 7950–7956.

T. P. Hilbert, W. Chaung, R. J. Boorstein, R. P. Cunningham, and G. W. Teebor (1997) Cloning and expression of the cDNA encoding the human homologue of the DNA repair enzyme, *Escherichia coli* endonuclease III. J. Biol. Chem. 272, 6733–6740.

H. Joenje (1989) Genetic toxicology of oxygen. Mut. Res. 219, 193–208.

Klungland and T. Lindahl (1997) Second pathway for completion of human DNA base excision-repair: reconstitution with purified proteins and requirement for DNaseIV (FEN1). EMBO J. 16, 3341–3348.

F. C. Kuo and J. Sklar (1997) Augmented expression of a human gene for 8-oxoguanine DNA glycosylase (MutM) in B lymphocytes of the dark zone in lymph node germinal centers. J. Exp. Med. 186, 1547–1556.

M. Kuro-o, Y. Matsumura, H. Aizawa *et al.* (1997) Mutation of the mouse *klotho* gene leads to a syndrome resembling ageing. Nature 390, 45–51.

T. Lindahl (1996) The Croonian Lecture: Endogenous damage to DNA. Phil. Trans. R. Soc. Lond. B 351, 1529–1538.

T. Lindahl (1993) Instability and decay of the primary structure of DNA. Nature 362, 709–715.

T. Lindahl and B. Nyberg (1972) Rate of depurination of native DNA. Biochemistry 11, 3610–3618.

T. Lindahl and B. Nyberg (1974) Heat-induced deamination of cytosine residues in deoxyribonucleic acid. Biochemistry 13, 3405–3410.

R. Lu, H.M. Nash and G.L. Verdine (1997) A mammalian DNA repair enzyme that excises oxidatively damaged guanines maps to a locus frequently lost in lung cancer. Current Biol. 7, 397–407.

L. J. Marnett and P. C. Burcham (1993) Endogenous DNA adducts: Potential and paradox. Chem. Res. Toxicol. 6, 771–785.

M. F. Moran and K. Ebisuzaki (1987) Base excision repair of DNA in γ-irradiated human cells. Carcinogenesis 8, 607–609.

J. Nakamura, V. E. Walker, P. B. Upton, S-Y. Chiang, Y. W. Kow, and J. A. Swenberg (1998) Highly sensitive apurinic/apyrimidinic site assay can detect spontaneous and chemically induced depurination under physiological conditions. Cancer Res. 58, 222–225.

H. M. Nash, S. D. Bruner, O. D. Schärer, T. Kawate, T. A. Addona, E. Spooner, W. S. Lane, and G. L. Verdine (1996) Cloning of a yeast 8-oxoguanine DNA glycosylase reveals the existence of a base-excision DNA-repair protein superfamily. Current Biol. 6, 968–980.

J. P. Radicella, C Dherin, C. Desmaze, M. S. Fox, and S. Boiteux (1997) Cloning and characterization of *hOGG1*, a human homolog of the *OGG1* gene of *Saccharomyces cerevisiae*. Proc. Natl. Acad. Sci. USA 94, 8010–8015.

T. Roldán-Arjona, Y-F. Wei, K. C. Carter, A. Klungland, C. Anselmino, R-P. Wang, M. Augustus, and T. Lindahl (1997) Molecular cloning and functional expression of a human cDNA encoding the antimutator enzyme 8-hydroxyguanine-DNA glycosylase. Proc. Natl. Acad. Sci. USA 94, 8016–8020.

T. Roldán-Arjona, C. Anselmino, and T. Lindahl (1996) Molecular cloning and functional analysis of a *Schizosaccharomyces pombe* homologue of *Escherichia coli* endonuclease III. Nucl. Acids Res. 24, 3307–3312.

T. A. Rosenquist, D. O. Zharkov, and A. P. Grollman (1997) Cloning and characterization of a mammalian 8-oxoguanine DNA glycosylase. Proc. Natl. Acad. Sci. USA 94, 7429–7434.

B. Rydberg and T. Lindahl (1982) Nonenzymatic methylation of DNA by the intracellular methyl group donor S-adenosyl-L-methionine is a potentially mutagenic reaction. EMBO J. 1, 211–216.

J-C. Shen, W. M. Rideout III, and P. A. Jones (1994) The rate of hydrolytic deamination of 5-methylcytosine in double-stranded DNA. Nucl. Acids Res. 22, 972–976.

M. Tani, K. Shinmura, T. Kohno *et al.* (1998) Genomic structure and chromosomal localization of the mouse *Ogg1* gene that is involved in the repair of 8-hydroxyguanine in DNA damage. Mammalian Genome 9, 32–37.

P.A. van der Kemp, D. Thomas, R. Barbey, R. de Oliveira and S. Boiteux (1996) Cloning and expression in *Escherichia coli* of the *OGG1* gene of *Saccharomyces cerevisiae*, which codes for a DNA glycosylase that excises 7,8-dihydro-8-oxoguanine and 2,6-diamino-4-hydroxyl-5-N- methylformamidopyrimidine. Proc. Natl. Acad. Sci. USA 93, 5197–5202.

21

DNA DAMAGE BY IRON AND HYDROGEN PEROXIDE

Stuart Linn

Division of Biochemistry and Molecular Biology
Barker Hall, University of California
Berkeley, California 94720–3202

1. ABSTRACT

DNA is nicked by Fe^{2+}-mediated Fenton reactions with 0.5 mM H_2O_2 by oxidants which are sensitive to higher concentrations of H_2O_2. Above 5 mM H_2O_2, however, the level of DNA nicking is independent of H_2O_2 concentration and is mediated by other oxidants. This reaction can be driven by NADH which regenerates Fe^{2+}, but not by NADPH and *E. coli* responds by increasing its NADPH pools at the expense of its NADH pools. We have proposed that the different kinetics of nicking by Fe^{2+} in 0.5 mM- *vs.* 50 mM H_2O_2 arise because of differences in which Fe^{2+} associates with DNA (Luo *et al.*, PNAS **91**, 12438 (1994)).

To explore the nature of these associations, the locations of DNA strand breaks in 0.5 mM and 50 mM H_2O_2 have been analyzed using terminally-labeled duplex oligonucleotides and Maxam-Gilbert sequence analysis. With 50 mM H_2O_2, preferential cleavages occurred only at the nucleoside 5' to each of the dG moieties in the sequence **RGG**G, and these cleavages were resistant to inhibition by 100 mM ethanol. (**Underscored bold** nucleotides are cleaved.) However, cleavage with 0.5 mM H_2O_2 was sensitive to ethanol and occurred at dT in the sequences RTGR, YATTY, YTTA. The sequence RGGG is noteworthy because it is commonly found in telomere repeats that are known to diminish in number during cellular aging and, indeed, a telomer inserted into a plasmid is preferentially cleaved in high concentrations of H_2O_2. The sequence RTGR is a common and necessary motif in upstream transcription elements of genes coding for proteins responding to oxidative and electrophilic stresses and genes coding for proteins involved in iron uptake and base excision repair. Energy minimization computer modeling of the interaction of Fe^{2+} with RTGR in B-DNA suggests that Fe might interact with this sequence by associating with the three purine N^7's in an octahedrally-oriented coordination with the dT flipped

Advances in DNA Damage and Repair, edited by Dizdaroglu and
Karakaya. Kluwer Academic / Plenum Publishers, New York, 1999.

$$(1)\quad O_2 \xrightarrow[(-0.16\ V)]{e^-} O_2^{\bullet -} \xrightarrow[(0.94\ V)]{e^-,\ 2H^+} H_2O_2$$

$$\xrightarrow[(0.32\ V)]{e^-,\ H^+} \bullet OH + H_2O \xrightarrow[(2.31\ V)]{e^-,\ H^+} 2\,H_2O$$

$$(2)\quad 2O_2^{\bullet -} + 2H^+ \rightarrow O_2 + H_2O_2$$

$$(3)\quad O_2^{\bullet -} + Fe^{3+} \rightarrow Fe^{2+} + O_2$$

$$(4)\quad [4Fe\text{-}4S]^{2+} + O_2^{\bullet -} + 2H^+ \rightarrow [3Fe\text{-}4S]^{+} + H_2O_2 + Fe^{2+}$$

$$(5)\quad Fe^{2+} + H^+ + H_2O_2 \rightarrow Fe^{3+} + \bullet OH + H_2O$$

$$(6)\quad Fe^{2+} + H_2O_2 \rightarrow FeO^{2+} + H_2O$$

$$(7)\quad FeO^{2+} + H^+ \rightarrow FeOH^{3+} \rightarrow Fe^{3+} + \bullet OH$$

$$(8)\quad NADH + H^+ + R \rightarrow NAD^+ + RH_2$$

Figure 1. Reactions cited in the text. Reaction 1 shows the univalent reductions of aqueous oxygen to water at pH 7. Potentials are for 1M aqueous oxygen at pH 7 (Koppenol, 1994). The large potential for •OH reduction and its radical nature allows it to oxidize organic molecules almost indiscriminately.

out. NMR studies of Fe^{2+} bound to RTGR in a 16-bp duplex are consistent with a specific interaction of the iron ion with the sequence.

2. INTRODUCTION

Oxygen is a powerful oxidant, but the triplet ground state of dioxygen constitutes a kinetic barrier for oxidation of biological molecules which are mostly singlet state (Koppenol, 1994). However, the unpaired orbitals of dioxygen can sequentially accommodate single electrons to yield $O_2^{\bullet -}$, H_2O_2, •OH, and water (Reaction 1, Fig. 1). Aerobic life is based upon harnessing energy *via* the catalytic spin pairing of triplet oxygen by the electron transport chain (Babcock and Wikström, 1992), but that process also gives rise to $O_2^{\bullet -}$ and other reactive oxygen species (Imlay and Fridovich, 1991) that cause genetic damage (Imlay and Linn, 1988; Lindahl, 1993; Stadtman, 1993; Dix and Aikens, 1993). In addition, catabolic oxidases such as xanthine oxidase, anabolic processes such as nucleoside reduction, and defense processes such as phagocytosis also produce oxygen radicals.

Although DNA is a biologically important target for reactive oxygen species, $O_2^{\bullet -}$ is relatively unreactive with DNA (Bielski *et al.*, 1985). However, $O_2^{\bullet -}$ dismutates (*via* spontaneous- or enzyme-catalyzed reactions) to produce H_2O_2 (Reaction 2). $O_2^{\bullet -}$ can also reduce and liberate Fe^{3+} from ferritin (Reif, 1992) (Reaction 3), or liberate Fe^{2+} from iron-sulphur clusters (Flint, *et al.*, 1993) (Reaction 4). Subsequently reactive oxygen species can form from Fe^{2+} and H_2O_2 *via* the Fenton reaction (Reaction 5). Thus, the cytotoxic effects of $O_2^{\bullet -}$ (as well as of iron and H_2O_2) have been linked to DNA damage by way of the Fenton reaction (Imlay and Linn, 1988; Keyer and Imlay, 1996; Mello-Filho and Meneghini, 1991; Henle and Linn, 1997).

Iron has 5 oxidation states in aqueous solution, Fe(II) - Fe(VI). Fe(II) and Fe(III) are the most common, and their reactions with oxygen and its reduced forms are well-documented (Koppenol, 1994; Walling, 1975). More recently, however, reactions with Fe(IV) have been implicated in biological processes and proposed to be involved in damage to cellular components (Koppenol, 1994; Goldstein *et al.,* 1993; Yamazaki and Piette, 1991; Wink *et al.,* 1994; Wardman and Candeias, 1996). In the case of Fe^{2+} chelates with ADP, *ortho*-phosphate or EDTA, the oxidant formed from H_2O_2 behaves differently than expected for •OH, and it has been proposed to be the ferryl radical (Reaction 6). A caged- or bound •OH, often denoted as $[Fe\ H_2O_2]^{2+}$ or $[FeOOH]^{+}$, could alternatively account for the noted differences (Yamazaki and Piette, 1991), but this distinction might be arbitrary, as this bound •OH might be an intermediate of Reactions 5 or 6 (Goldstein *et al.*, 1993) and the ferryl radical itself could give rise to •OH *via* Reaction 7 (Koppenol, 1994; Yamazaki and Piette, 1991).

It thus appears that the chemical properties of iron-ligand complexes are strongly determined by the particular ligands (Koppenol, 1994; Goldstein *et al.*, 1993; Yamazaki and Piette, 1991). Luo *et al.,* (1994) found that in the presence of the very complex ligand, DNA, there are 3 kinetically-distinguishable oxidants formed from H_2O_2 and Fe^{2+} that cause DNA strand breakage. One of these is easily scavengable, consistent with it being a freely diffusible •OH, whereas the other two vary in their scavenging susceptibilities. One or both of these might be an iron(IV) species. This article will describe studies of some of these $DNA/Fe^{2+}/H_2O_2$ reactions in detail.

3. RESULTS

3.1. Killing and Mutagenesis by H_2O_2

Using *E. coli* and H_2O_2 as a model system to study oxyradical damage, we noted that 1–2 mM H_2O_2 is more toxic than concentrations in the range of 2–20 mM (Imlay and Linn, 1986, 1987a, 1987b). It was also observed that strains that are deficient in recombinational DNA repair (*recA*), excision repair (*xth*) or both types of repair (*polA*) are extremely sensitive to such killing, but the same peculiar shaped dose response is observed. This characteristic dose response was observed for mutagenesis or other indicators of DNA damage- in each instance the rate of DNA damage was 2- to 3-fold greater at 1–2 mM than at 2–20 mM H_2O_2 and essentially unchanged between 5 and 20 mM H_2O_2. This unusual and characteristic dose response provided us with a distinct hallmark with which to proceed further in the characterization of the mechanism of toxicity.

Among the first observations we made, was that DNA damage did not occur if cells were exposed to H_2O_2 in buffer lacking a carbon source (Imlay and Linn, 1986). Evidently respiration was required for the H_2O_2 toxicity to be observable. Conversely, DNA damage was enhanced in cells grown anaerobically (Imlay and Linn, 1987a, 1987b). A regulon for the response of *E. coli* to anaerobic growth is *fnr* (*f*umarate *n*itrate *r*eductase), a regulon which positively controls the synthesis of fumarate, nitrate and nitrite reductases, several cytochromes, etc., but negatively regulates *ndh*, the gene for NADH dehydrogenase II (Spiro *et al.*, 1989). A genetic analysis of mutants of these genes allowed us to implicate the negative regulation of NADH dehydrogenase as the sensitizing step (Imlay and Linn, 1987a; Imlay *et al.,* 1988). NADH dehydrogenase, or diaphorase catalyzes Reaction 8 (Fig. 1).

3.2. The Role of NADH in H_2O_2 Toxicity

NADH levels were then raised by blocking respiration with KCN, and the same sensitization to H_2O_2 ensued. Based on these results, it was proposed (Imlay and Linn, 1987b,

1988) that NADH is normally utilized in aerobic respiration ultimately to donate two electron pairs to molecular oxygen to form water. It is also a potential source of two electrons to reduce H_2O_2 to water. However, by some means, one electron can be transferred from NADH to H_2O_2 to generate •OH. The rate of this transfer is limited by electron flow to form NADH, however, so as to govern a maximum possible rate of •OH formation. On the other hand, •OH can oxidize H_2O_2 to $•O_2^-$ so that at high H_2O_2 concentrations the available •OH for cell damage is lowered.

What is the agent that mediates one-electron transfers from NADH to H_2O_2? One possibility is the iron-catalyzed Fenton reaction (Reaction 5). Indeed, iron chelators such as *o*-phentanthroline or dipyridyl protect against killing or mutagenesis by H_2O_2 in the 1–20 mM range (Imlay *et al.,* 1988). However, various alcohol scavengers of free •OH do not protect against H_2O_2 toxicity, implying that freely diffusible •OH was not the damaging agent (Imlay and Linn, 1988; Imlay *et al.,* 1988).

We then turned to studying DNA damage by Fe2+ and H_2O_2 *in vitro.* When DNA nicking is generated by H_2O_2 and $FeSO_4$, nicking occurs rapidly (probably within several seconds) and then ceases as the Fe^{2+} is oxidized. With NADH present, however, nicking continues as the Fe^{3+} product is reduced back to Fe^{2+}. As expected, when Fe^{3+} rather than Fe^{2+} is added with the H_2O_2, DNA nicking occurs only when NADH is present. In summary, NADH can drive the Fenton reaction - and the DNA damaging reactions - by reducing Fe^{3+} to Fe^{2+}. While NADH reduction of Fe^{3+} *via* a charge transfer complex was described by Gutman and Eisenbach (1973) and Gutman *et al.* (1968), the intermediate species involved, especially in the presence of DNA, remain undetermined.

In view of the above, one might predict that a cell could alter its pyridine nucleotide pools in response to oxygen stress, and this is indeed the case. In *E. coli*, the *soxRS* regulon is induced by superoxide (Chan and Weiss, 1987; Kogoma *et al.,* 1988; Greenberg and Demple, 1989; Wu and Weiss, 1991). Among the enzymes induced is a diaphorase (NADH dehydrogenase), that would presumably result in decreased levels of NADH and also act to reduce redox cycling chemicals such as paraquat so as to eliminate the production of free radicals. Another enzyme induced by the *soxRS* response, glucose 6-phosphate dehydrogenase, catalyzes the oxidation of glucose 6-phosphate with the concomitant formation of NADPH. The increase of NADPH levels as a result of the induction of glucose 6-phosphate dehydrogenase during the *soxRS* response can be rationalized by the fact that NADPH is the cofactor for glutathione reductase and both NADPH and reduced glutathione are cofactors for peroxidases. Glutathione is also important for maintaining proteins and iron in the reduced state - certainly in red blood cells. But, if the cell depletes NADH so as to prevent the driving of Fenton reactions, cannot NADPH assume that role? The answer is no - NADPH is at least an order of magnitude slower in reducing Fe^{3+} and competes very effectively with NADH for iron binding (Luo 1993). Finally, *E. coli* aconitase is inactivated by superoxide (Gardner and Fridovich, 1991), thus shutting down the Krebs cycle and NADH production.

As predicted from the above responses, we have noted that the relative pools of pyridine nucleotides do change significantly after exposure of *E. coli* to H_2O_2 (Table 1). In particular, the NADPH/NADH ratio doubles. However, these were measurements of total pyridine nucleotide pools; it would be of interest to measure the non-enzyme-bound pools which might be more active in Fe^{3+} reduction.

An interesting response to strand breaks caused by oxygen radicals exists in animal cells: nuclear NAD^+ is depleted by its polymerization to poly(ADP-ribose) (Simbulan-Rosenthal *et al.,* 1996). This reaction would also result in nuclear NADH depletion and presumably avoid the NADH-mediated reduction of Fe^{3+}.

Table 1. Total pyridine nucleotide pools in aerobic, log phase *E. coli**

	10^6 molecules/cell	
	Before challenge	15 min after 75 μM H_2O_2 challenge
NAD^+	15.3 ± 3.6	14.6 ± 2.9
NADH	1.9 ± 0.7	1.4 ± 0.7
NADP	3.4 ± 0.9	3.9 ± 1.3
NADPH	2.1 ± 0.6	3.0 ± 0.4
Total	22.7	22.9
NADPH/NADH ratio	1.1	2.2

*F Li, Y. and Linn, S. unpublished observations.
Extraction of nucleotides, resolution by thin layer chromatography, and quantitation were by a combination of procedures described in Linquist and Olivera (1971) and Wimpenny and Firthe (1972).

3.3. Sequence Specificity of the DNA Damage

As noted above, a large portion of H_2O_2-dependent DNA damage appears not to be due to diffusible hydroxyl radicals *in vitro* and none might be so-caused *in vivo* (Mello-Filho and Meneghini, 1984; Imlay and Linn, 1988; Luo *et al.*, 1994;). Instead, DNA-damaging Fenton oxidants are produced on Fe^{2+} atoms associated with DNA and it would appear that the location of iron binding may determine the substrate and nature of attack (Luo *et al.*, 1994). There appears to be at least two distinguishable classes of iron-mediated Fenton oxidants of DNA (Imlay and Linn, 1988; Luo *et al.*, 1994): Type I oxidants are moderately sensitive to H_2O_2 and ethanol, whereas Type II oxidants are resistant to ethanol and to higher concentrations of H_2O_2. To test whether the distinguishing characteristics of these oxidants might be predominantly due to localization of the iron which gives rise to them, we asked whether nicks caused by Type I oxidants occur at different nucleotide sequences than those caused by Type II oxidants.* (The former were tested by utilizing 50 μM H_2O_2 where they predominate, whereas the latter were tested at 50 mM H_2O_2 at which concentration the Type I oxidants are totally quenched.)

DNA nicks occur predominantly due to H-abstraction from one of the deoxyribose carbons for which the predominant consequence is eventual strand breakage and base release (von Sonntag, 1987; Henle *et al.*, 1995). In approximately half of these alterations, a 5'-phosphate end group is located 3' to the cleavage, a 3'-phosphoglycolate is located 5' to the cleavage, and a base propenal is released which subsequently decomposes to the free base and malondialdehyde (Janicek *et al.*, 1985; Bertoncini and Meneghini, 1995). The majority of other sugar damages yield 5'- and 3'- phosphomonoesters flanking a one-nucleoside gap. Some sugar alterations, such as the gamma-lactone, do not give this product immediately, but do so after adequate time or treatment (von Sonntag, 1987). In general, attack by reactive oxygen species on the base moieties of DNA do not give rise to altered sugars or strand breaks except when base modifications labilize the N-glycosyl bond, allowing the formation of baseless sites which are subject to β-elimination (Suzuki *et al.*, 1994). Thus, in the following studies, we presumably were studying nucleotide sequence specificity of attack at sugar residues.

With 50 μM H_2O_2, DNA nicks occurred within the sequences R**T**GR, YA**T**TY, and YT**T**A (the bold, underscored nucleotides are the sites of cleavage). However, with 50 mM H_2O_2 present, preferential cleavage occurred in the sequence **RGG**G.*(Double strand breaks were not observed to occur.)

In the case of RGGR, breaks occurred at any residue that was 5' to a dG residue. RGGG is a consensus sequence in a majority of telomer repeats, and indeed we have noted that when a *Tetrahymena* telomer, $(TTAGGG)_{81}$, was inserted into a plasmid, it was preferentially cleaved within the plasmid in 50 mM- but not 50 μM H_2O_2.† However, it may be that the sites of nicking are not necessarily the iron binding sites. RGGG sites in particular may be sinks for radical electrons which are formed elsewhere on the helix and travel through the base stack (Hall *et al.*, 1996). This phenomenon appears not to apply here, however, because the preferential cleavage occurred only if the Fe^{2+} was first mixed with the DNA for roughly one hour prior to the addition of H_2O_2 implying a time-dependent association of Fe^{2+} with the RGGG sequence.† Whether this result implies that Fe^{2+}/H_2O_2 damage might engender age-related telomer shortening (Blackburn and Greider, 1990) is an intriguing question.

In the case of 50 μM H_2O_2, to date we have studied in depth only the sequence RTGR. Computer modeling for energy minimization predicts what Fe^{2+} imbeds itself into the sequence by displacing (flipping out) the thymidine residue, then interacting with the 3 purine N^7 atoms, the guanine O_2, the thymidine phosphate, and the thymidine dribose through a water bridge.‡ The thymidine is displaced (or held displaced) due to steric hindrance of the 5-methyl group. NMR analysis has confirmed this structure.‡

RTGR is of biological importance: it is a necessary element in the promotor region of genes responding either to iron stress or oxygen stress (Primiano *et al.*, 1997). Whether this fact is related to the ability of Fe^{2+} to interact with this sequence and whether this interaction is relevant to its necessity as a promotor regulatory element is to be investigated.

4. CONCLUSIONS

Reactive oxygen species and the damage which they cause are implicated in arthritis, heart disease, cancer, and degenerative phenomena during aging processes. It appears that DNA damage is a major target of these reactive oxygen species, that such damage could be important in these diseased states and that iron plays a significant role in the processes which generate this damage in both procaryotes and eucaryotes. Hence, it is very likely that molecular biologists, cell biologists, biochemists and pharmacologists will turn toward monitoring DNA damages caused by oxygen radicals as a dosimeter of damage by these species. DNA damages are being better defined but reliable assays for them are just being developed and validated, as are reliable methods for preparing DNA samples. We ought also to turn our attention to iron-binding agents (in addition to free radical scavengers) as antagonists of the generation of DNA damage and other forms of cellular damage by iron and oxygen species. Finally, with our understanding of how DNA damage is repaired, we will be in a position to understand whether-and to what extent-deficiencies in these repair processes might lead to predispositions to the various diseases with which damage by reactive oxygen species is implicated.

---

* Han, Z. X., Falk, M. S., Henle, E. S., Luo, Y., and Linn, S. unpublished observations

† Tang, N. and Linn, S. unpublished observations

‡ Henle, E. S. and Linn, S. unpublished observations

ACKNOWLEDGMENTS

Research cited from our laboratory was supported by Grants GM19020 and P30E08196 from the National Institutes of Health. I am grateful to James Imlay, S. M. Chin, Ernst Henle, Yongzhang Luo, Ying Li, Xan-Zhu Han and Ning Tang who not only did much of the work from our laboratory which is cited in this review, but also provided much of the intellectual stimulus for it.

REFERENCES

G. T. Babcock and M. Wikstrom (1992) Oxygen activation and the conservation of energy in cell respiration. *Nature* **356**, 301–309.

C. R. Bertoncini and R. Meneghini (1995) DNA strand breaks produced by oxidative stress in mammalian cells exhibit 3'-phosphoglycolate termini. *Nucl. Acids Res.* **23**, 2995–3002.

B. H. J. Bielski, D. E. Cabelli, R. L. Arudi and A. B. Ross (1985) Reactivity of $HO_2/O_2{\bullet}^-$ radicals in aqueous solution. *J. Phys. Chem. Ref. Data* **14**, 1041–1100.

E. H. Blackburn and C. W. Greider (1990) *Telomeres*. Cold Spring Harbor Laboratory Press, New York.

E. Chan and B. Weiss (1987) Endonuclease IV of *Escherichia coli* is induced by paraquat. *Proc. Natl. Acad. Sci. USA* **84**, 3189–3192.

T. A. Dix and J. Aikens (1993) Mechanisms and biological relevance of lipid peroxidation initiation. *Chem. Res. Tox.* **6**, 2–18.

D. H. Flint, J. F. Tuminello and M. H. Emptage (1993) The inactivation of Fe-S cluster containing hydro-lyases by superoxide. *J. Biol. Chem.* **268**, 22369–22376.

P. R. Gardner and I. Fridovich (1991) Superoxide sensitivity of the *Escherichia coli* aconitase. *J. Biol. Chem.* **266**, 19328–19333.

S. Goldstein, D. Meyerstein and G. Czapski (1993) The Fenton reagents. *Free Rad. Biol. Med.* **15**, 435–445.

J. T. Greenberg and B. Demple (1989) A global response induced in *Escherichia coli* by redox-cycling agents overlaps with that induced by peroxide stress. *J. Bacteriol.* **171**, 3933–3939.

M. Gutman and M. Eisenbach (1973) On the complexation of ferric ions by reduced nicotinamide adenine dinucleotide. *Biochemistry* **12**, 2314–2317.

M. Gutman, R. Margalit and A. Schejter (1968) A charge-transfer intermediate in the mechanism of reduced diphosphopyridine nucleotide oxidation by ferric ions. *Biochemistry* **7**, 2778–2785.

D. B. Hall, R. E. Holmlin and J. K. Barton (1996) Oxidative DNA damage through long-range electron transfer. *Nature* **382**, 731–735.

E. S. Henle and S. Linn (1997) Formation, prevention, and repair of DNA damage by iron/hydrogen peroxide. *J. Biol. Chem.* **272**, 19095–19098.

E. S. Henle, R. Roots, W. R. Holley and A. Chatterjee (1995) DNA strand breakage is correlated with unaltered base release after gamma irradiation. *Radiat. Res.* **143**, 144–150.

J. A. Imlay, S. M. Chin and S. Linn (1988) Toxic DNA damage by hydrogen peroxide through the Fenton reaction *in vivo* and *in vitro*. *Science* **240**, 640–642.

J. A. Imlay and I. Fridovich (1991) Assay of metabolic superoxide production in *Escherichia coli*. *J. Biol. Chem.* **266**, 6957–6965.

J. A. Imlay and S. Linn (1986) Bimodal pattern of killing of DNA-repair-defective or anoxically grown *Escherichia coli* by hydrogen peroxide. *J. Bacteriol.* **166**, 519–527.

J. A. Imlay and S. Linn (1987a) Mutagenesis and stress responses induced in *Escherichia coli* by hydrogen peroxide. *J. Bacteriol.* **169**, 2967–2976.

J. A. Imlay and S. Linn (1987b) Toxicity, mutagenesis and stress responses induced in *Escherichia coli* by hydrogen peroxide. *J. Cell Sci. Suppl.* **6**, 289–301.

J. A. Imlay and S. Linn (1988) DNA damage and oxygen radical toxicity. *Science* **240**, 1302–1309.

M. R. Janicek, W. A. Haseltine and W. D. Henner (1985) Malondialdehyde presursors in gamma-irradiated DNA, deoxynucleotides and deoxynucleosides. *Nucl. Acids Res.* **13**, 9011–9029.

K. Keyer and J. A. Imlay (1996) Superoxide accelerates DNA damage by elevating free-iron levels. *Proc. Natl. Acad. Sci. USA* **93**, 13635–13640.

T. Kogoma, S. B. Farr, K. M. Joyce and D. O. Natvig (1988) Isolation of gene fusions (*soi::lacZ*) inducible by oxidative stress in *Escherichia coli*. *Proc. Natl. Acad. Sci. USA* **85**, 4799–4802.

W. H. Koppenol (1994) In: *Free Radical Damage and its Control* (eds. C. A. Rice-Evans and R. H. Burdon), Elsevier Science, New York, pp. 3–24.

T. Lindahl (1993) Instability and decay of the primary structure of DNA. *Nature* **362**, 709–715.

R. Lundquist and B. Olivera (1971) Pyridine nucleotide metabolism in *Escherichia coli* I. Exponential Growth. *J. Biol. Chem.* **246**, 1107–1116.

Y. Luo (1993) Characterization of Fenton oxidants and DNA damage. Ph.D. thesis, Univ. of California, Berkeley.

Y. Luo, Z.-X. Han, S. M. Chin and S. Linn (1994) Three chemically distinct types of oxidants formed by iron-mediated Fenton reactions in the presence of DNA. *Proc. Natl. Acad. Sci. USA* **91**, 12438–12442.

A. C. Mello-Filho and R. Meneghini (1984) *In vivo* formation of single-strand breaks in DNA by hydrogen peroxide is mediated by the Haber-Weiss reaction. *Biochim. Biophys. Acta* **781**, 56–63.

A. C. Mello-Filho and R. Meneghini (1991) Iron is the intracellular metal involved in the production of DNA damage by oxygen radicals. *Mut. Res.* **251**, 109–113.

T. Primiano, T. R. Sutter and T. W. Kensler (1997) Antioxidant-inducible genes. *Adv. Pharmacol.* **38**, 293–328.

D. W. Reif (1992) Ferritin as a source of iron for oxidative damage. *Free Rad. Biol. Med.* **12**, 417–427.

C. M. G. Simbulan-Rosenthal, D. S. Rosenthal, R. Ding, J. Jackman and M. E. Smulson (1996) Depletion of nuclear poly(ADP-ribose) polymerase by antisense RNA expression: Influence on genomic stability, chromatin organization, DNA repair and DNA replication. *Prog. Nuc. Acid Res. Mol. Biol.* **55**, 135–156.

S. Spiro, R. E. Roberts and J. R. Guest (1989). FNR-dependent repression of the *ndh* gene of *Escherichia coli* and metal ion requirement for FNR-regulated gene expression. *Molec. Microbiol.* **3**, 601–608.

E. R. Stadtman (1993) Oxidation of free amino acids and amino acid residues in proteins by radiolysis and by metal-catalyzed reactions. *Annu. Rev. Biochem.* **62**, 797–821.

T. Suzuki, S. Ohsumi and K. Makino (1994) Mechanistic studies on depurination and apurinic site chain breakage in oligodeoxyribonucleotides. *Nucl. Acids Res.* **22**, 4997–5003.

C. von Sonntag (1987) *The Chemical Basis of Radiation Biology*, Taylor and Francis, New York.

C. Walling (1975) Fenton's reagent revisited. *Accts. Chem. Res.* **8**, 125–131.

P. Wardman and L. P. Candeias (1996) Fenton chemistry: An introduction. *Radiat. Res.* **145**, 523–531.

J. W. T. Wimpenny and A. Firthe (1972) Levels of nicotinamide adenine dinucleotide and reduced nicotinamide adenine dinucleotide in facultative bacteria and the effect of oxygen. *J. Bacteriol.* **111**, 24–32.

D. A. Wink, R. W. Nims, J. E. Saavedra, W. E. Utermahlen, Jr. and P. C. Ford (1994) The Fenton oxidation mechanism: Reactivities of biologically relevant substrates with two oxidizing intermediates differ from those predicted for the hydroxyl radical. *Proc. Natl. Acad. Sci.* USA **91**, 6604–6608.

J. Wu and B. Weiss (1991) Two divergently transcribed genes, *soxR* and *soxS*, control a superoxide response regulon of *Escherichia coli*. *J. Bacteriol.* **173**, 2864–2871.

I. Yamazaki and L. H. Piette (1991) EPR spin-trapping on the oxidizing species formed in the reaction of the ferrous ion with hydrogen peroxide. *J. Am. Chem. Soc.* **113**, 7588–7593.

22

MEASUREMENT OF OXIDATIVE DAMAGE TO DNA NUCLEOBASES IN VIVO

Interpretation of Nuclear Levels and Urinary Excretion of Repair Products

Steffen Loft[1] and Henrik Enghusen Poulsen[2]

[1]Department of Pharmacology
University of Copenhagen, Denmark
[2]Department of Clinical Pharmacology
University Hospital Copenhagen, Denmark

1. ABSTRACT

Oxidatively modified nucleobases can be measured in cells although with large inter-methodological variation. The excretion rates of corresponding repair products correspond to 10^4 oxidative DNA modifications per cell per day indicating >99% repair. The urine and nuclear measurements represent two fundamentally different estimates, i.e. the average rate of damage and the local balance between damage and repair, respectively. The important determinants of oxidative damage in humans include smoking, air pollution, oxygen consumption, cancer therapy, inflammation and neurodegenerative diseases, whereas diet and antioxidant supplements have minimal influence. The oxidative DNA damage biomarkers may provide proof of causal relationships with cancer and aging and improve prevention.

2. BIOMARKERS OF OXIDATIVE DNA DAMAGE

Oxidative damage to DNA has been proposed to be an important factor in carcinogenesis, supported by experimental studies in animals and in vitro (Ames et al. 1995, Loft and Poulsen 1996, Wiseman and Halliwell 1996). With respect to aging, mainly the oxidative modifications of DNA in both nucleus and mitochondria are thought to be involved (Ames et al. 1993). Indeed cells are constantly exposed to oxidants from both physiological processes, such as mitochondrial respiration (Chance et al. 1979), and pathophysi-

Advances in DNA Damage and Repair, edited by Dizdaroglu and
Karakaya. Kluwer Academic / Plenum Publishers, New York, 1999.

ological conditions such as inflammation, ischemia/reperfusion, foreign compound metabolism and radiation (Ames et al. 1995). The continously ongoing damage to DNA is constantly repaired with high efficiency in the cells in the body (Demple and Harrison 1994). Many of the repair products are excreted in quantifiable form in the urine.

The use of biomarkers of oxidative DNA damage may provide further proof of a causal relationship with cancer and aging as well as serve as intermediate endpoints in human intervention studies which may target the optimum intervention strategy for the large scale intervention (Schulte and Perera 1993). Moreover, the biomarker approach is applicable in mechanistic animal experiments and cancer bioassays as well as in vitro.

In DNA more than 100 different oxidative modifications have been observed (Dizdaroglu 1991, Cadet et al. 1994). However, so far only a few of the base modifications have been used as biomarkers and of these the oxidative C-8 adduct of guanine is by far the most studied as either the nucleoside or base. In principle, the level in nuclear or mitochondrial DNA from target or surrogate tissues or cells or the excretion of repair products into the urine can be measured. Under the usual steady state conditions the latter will reflect the rate of damage whereas the former will reflect the balance between damage and repair.

3. ANALYSIS OF OXIDATIVELY MODIFIED NUCLEOBASES AND NUCLEOSIDES

3.1. Cellular DNA

In tissue or cell samples the level of oxidatively modified nucleobases can be measured by various techniques, including HPLC-EC (or MS or UV), GC/MS-SIM, TLC with ^{32}P-postlabelling and various immunoassays. Except in the slot blot technique and immunohistochemistry (Musarrat and Wani 1994, Yarborough et al. 1996), DNA or chromatin is isolated and hydrolyzed by enzymes or acid at high temperature. In all the assays the abundant unmodified nucleobases may be oxidized and thus cause artificially high values. Particularly, the derivatisation with silyl-groups for the GC/MS is prone to give rise oxidation and could be carried out after removal of the unmodified base (Ravanat et al. 1995, Douki et al. 1996) or under controlled temperature and other conditions (Hamberg and Zhang 1995). Similarly, the gamma-radiation from the ^{32}P-phosphate used for post-labeling could oxidize guanine and thus explain the rather high values measured in human lymphocytes and rat organs by that method (Wilson et al. 1993, Devanaboyina and Gupta 1996, Collins et al. 1997a). Even with the HPLC-EC method the reported values for 8-oxodG in leukocyte DNA varies from 0.3 to 13 per 10^5 dG (Loft and Poulsen 1996). There is no doubt that oxidation may occur during DNA extraction, particularly with the use of impure phenol and during drying the DNA after ethanol precipitation (Floyd et al. 1990, Claycamp 1992, Adachi et al. 1995). The lowest values have been obtained with anaerobic DNA extraction (Collins et al. 1996, Nakajima et al. 1996a). Some of the immunoassays are calibrated by HPLC-EC values (Degan et al. 1991, Musarrat and Wani 1994) whereas others appear to yield much higher 8-oxodG values than the HPLC-EC assays (Yin et al. 1995) but whether that is due to insufficient specificity of the antibodies is unknown.

Oxidative damage to nucleobases can be assessed indirectly as strand breaks, DNA unwinding or relaxation of supercoiling induced by treatment of nuclear material by the relevant repair enzymes, i.e. Fpg and endonuclease III for purines, inluding 8-oxodG, and pyrimidine lesions, respectively (Epe 1995, Collins et al. 1997b). This approach has been applied in alkaline elution and alkaline unwinding of DNA as well as in alkaline single cell gel electrophoresis (Epe 1995, Hartwig et al. 1996, Collins et al. 1997b). However, the 8-oxodG values estimated by those assays are around 0.3 per 10^6 dG, i.e. 10 times lower than the lowest values obtained by HPLC-EC. So far, it is unknown whether the HPLC-EC assay still gives artificially high values or the enzyme based assays are missing 8-oxodG in some of the DNA as suggested by a completeness of only 50% of FPG repair of damaged DNA in vitro (Karakaya et al. 1997). Nevertheless, there was a strong correlation ($r=0.89$, $p<0.01$) between 8-oxodG concentration and comet tail length after 2-nitropropane induced DNA damage in rat bone marrow cells (Deng et al. 1997).

3.2. Urine

The repair products from oxidative DNA damage, i.e. oxidized bases and nucleosides, are poor substrates for the enzymes involved in nucleotide synthesis, fairly water soluble, and generally excreted into the urine without further metabolism (Shigenaga et al. 1989, Loft et al. 1995a). Indeed, animal experiments have shown that injected 8-oxodG is readily excreted unchanged into the urine whereas 8-oxodG in the diet or oxidation of dG during excretion does not contribute (Shigenaga et al. 1989, Park et al. 1992, Loft et al. 1995a). Among the possible repair products from oxidative DNA modifications 8-oxodG, 8-oxoGua, Tg, dTg and 5-OHmU have so far been identified in urine (Fig. 1, (Cathcart et al. 1984, Shigenaga et al. 1989, Simic and Bergtold 1991, Faure et al. 1993, Suzuki et al. 1995, Teixeira et al. 1995, Loft and Poulsen 1996). Of these 8-oxodG and the thymine derivatives are the most intensively studied ones. The levels of concentration and excretion of the oxidized bases and nucleosides obtained in different laboratories are in the same range (Loft and Poulsen 1996).

The assays for the urinary DNA repair products include HPLC with detection by electrochemistry for 8-oxodG and 8-oxoGua and by UV absorbance for dTg and Tg, whereas all the repair products can potentially be measured by GC/MS (Simic and Ber-

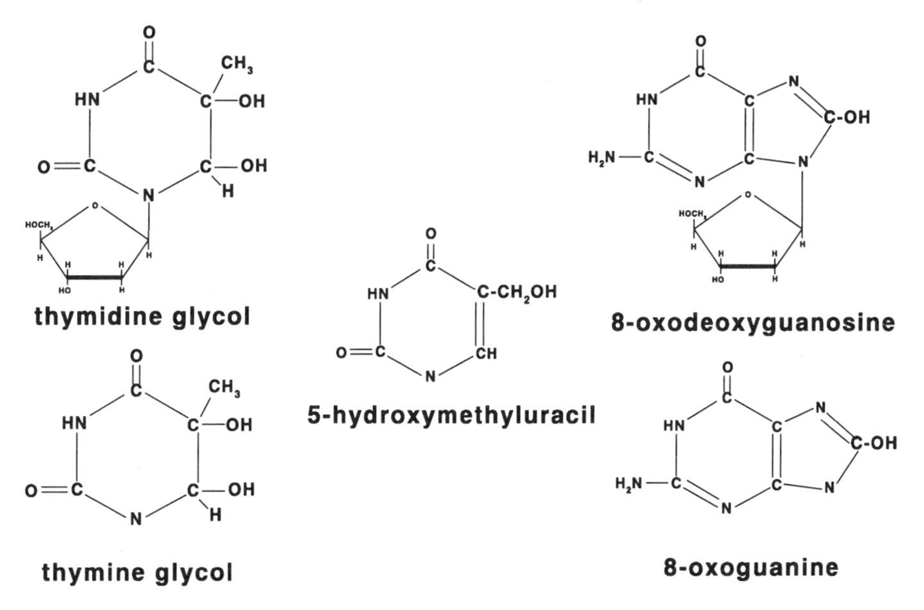

Figure 1. Repair products of oxidative DNA damage identified in urine and used as biomarkers.

gtold 1991, Faure et al. 1993, Dizdaroglu 1994, Teixeira et al. 1995). The major problem with all these assays involves separation of the very small amounts of analyte from urine that is a very complicated matrix. Thus, although several of the products are electrochemically active and high sensitivity is achievable the HPLC methods require extensive clean up procedures such as multiple solid phase extractions, HPLC column switching techniques or immunoaffinity columns (Cathcart et al. 1984, Shigenaga et al. 1989, 1994, Loft et al. 1992, Park et al. 1992, Tagesson et al. 1992, 1995, Brown et al. 1995, Germadnik et al. 1997). The complicated extraction procedures cause recovery problems in both HPLC and GC/MS-SIM methods and may require labeled internal standards. Moreover, the complicated procedures limit the analytical capacity. In the future HPLC/MS-MS is likely to solve many of the problems. An ELISA assay based on monoclonal antibodies has been developed for estimation of 8-oxodG in urine samples (Osawa et al. 1995). However, the values obtained in rat urine were 3–5 times higher than other published values. Similarly, in 4 smokers studied before and after smoking cessation the urinary 8-oxodG excretion values estimated by the ELISA method were 8 times higher than and showed only a weak correlation (r=0.42) with the values obtained by HPLC (Priemé et al. 1996). In another study the ELISA method yielded an 8-oxodG to creatinine ratio of around 8 nmol/mmol in healthy subjects (Erhola et al. 1997), as compared with 1 to 2 nmol/mmol in a large number of reports using HPLC-EC and GC/MS assays (Loft and Poulsen 1996).

Collection of urine for 24 hours or longer is quite straight forward, however, it may present some practical problems. For convenience the use of spot urine samples corrected for creatinine would be simpler than 24-h collection of urine. However, in 74 healthy subjects we collected urine for 24 h and a spot urine sample from the subsequent voiding (unpublished data). The correlation between the 8-oxodG to creatinine ratio in the spot samples and the 24 h excretion of 8-oxodG per kg lean body mass was rather poor (r=0.50). The insufficiency of creatinine corrected spot urine samples for estimation of the 8-oxodG excretion rates may partly explain some of the apparent differing data obtained in various human studies as discussed below.

4. INTERPRETATION OF NUCLEAR DNA LEVELS AND URINARY EXCRETION OF OXIDIZED NUCLEOBASES AND NUCLEOSIDES

The level of oxidized bases in DNA from tissues or cells is in a steady state determined by a simple balance of the influx and efflux of oxidized bases as outlined for 8-oxodG in Fig. 2. The major part of the oxidized bases in DNA arises from oxidation of bases within the DNA whereas incorporation of oxidized nucleotides for the cellular pool is probably of minor quantitative importance although highly mutagenic and thus of large qualitative importance (Tajiri et al. 1995). The efflux is determined mainly by the repair of the modified bases that for 8-oxodG in DNA results in 8-oxodG or 8-oxoGua by nucleotide excision and base excision, respectively (Bessho et al. 1993). Recently, the human 8-oxoGua glycosylase was cloned by several groups (Radicella et al. 1997, Roldan-Arjona et al. 1997) whereas nucleotide excision repair was shown to contribute to the repair of 8-oxodG in DNA. (Reardon et al. 1997). 8-OxodG will also come from the highly specific 8-oxodGTP phosphatase (MutT) and 8-oxodGMP nucleotidase enzymes sanitizing the nucleotide pool (Mo et al. 1992, Hayakawa, H. et al. 1995). Digestion of DNA from apoptotic cells and turnover of mitochondria could also be a source of 8-oxodG. A few of the

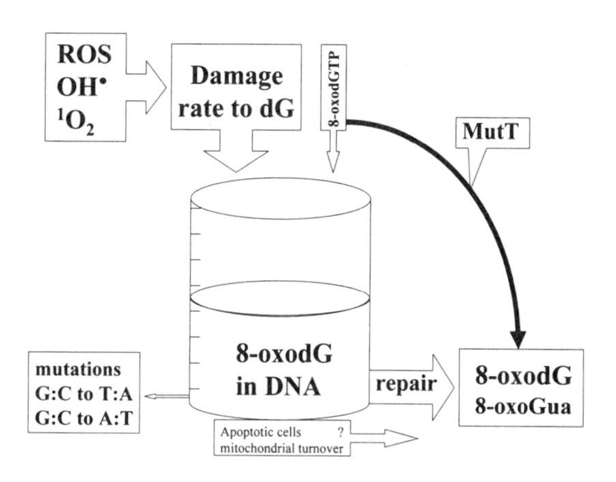

Figure 2. Mass balance of 8-oxodG formation in DNA and nucelotide pool and fates in terms of repair, cell and mitochondrial turnover and mutations. ROS are reactive oxygen species. MutT are 8-oxodGTP phosphatase and 8-oxodGMP nucleotidase sanitizing the nucleotide pool (Mo et al. 1992, Hayakawa, H. et al. 1995). Repair result in 8-oxodG or 8-oxo-Gua by nucleotide excision and base excision, respectively (Bessho et al. 1993).

8-oxodG lesions in DNA will after replication without repair or misrepair lead to mutations, 8-oxodG formed within DNA can lead to G to T transversions whereas incoporation of 8-oxodGTP can result in both G to T and G to A transversions (Kuchino et al. 1987, Shibutani et al. 1991, Tajiri et al. 1995).

If the rate of damage to DNA, e.g. 8-oxodG formation, is increased by oxidative stress, the nuclear level will increase until an increased repair rate matches the 8-oxodG-influx rate and a new steady state is achieved. An increased enzymatic repair rate can result just from the increased substrate availability (i.e. the 8-oxodG level) as well as from increased enzyme activity as shown in lymphocytes from smokers (Asami et al. 1996). The concept of steady state in the 8-oxodG level is supported by the limited accumulation with age shown in various human and animal cells and tissues investigated so far (Table 1) (Loft and Poulsen 1996). Moreover, in cultured human cells induced levels of oxidized nucleobases return to initial values within some hour's (Jaruga and Dizdaroglu 1996). In human and experimental animals exposed to radiation the increase in oxidized bases in DNA from leukocytes or liver as well as urinary excretion of repair products is temporary (Kasai et al. 1986, Bergtold et al. 1990, Blount et al. 1991, Olinski et al. 1996). The effect of the environmental factors on the level of modified bases in DNA in humans studied so far have been in the order of a factor two or less (Loft and Poulsen 1996). Further support for the concept of a steady state is the relative low variation in individual urinary excretion of 8-oxodG in the control as well as intervention groups in controlled trials (Loft et al. 1995b, Verhagen et al. 1995, Priemé et al. 1997).

The use of urinary excretion of oxidized nucleosides and bases as biomarkers requires almost complete repair and minimal accumulation as argued above. In a study of 8-oxodG in human brain the accumulation rate in the nuclear DNA corresponded 2 lesions per cell per day and possible less in other cells (Mecocci et al. 1993). In humans the reported values of the urinary excretion of the repair products, 8-oxodG, are in the range 15–50 nmol per 24 h (Loft and Poulsen 1996) and the alternative repair product 8-oxoGua appears to be excreted in similar amounts (Suzuki et al. 1995). The sum of these products thus corresponds to an average of 300–1000 lesions per day for each of the assumed 5×10^{13} cells in the body per day (Loft et al. 1992, 1995b). Accordingly, the calculated repair efficiency under these assumptions ranges from 99.4% to 99.8%. Each cell contains around 2×10^9 dG residues and assuming that 1 per 10^5 is oxidized, the cellular burden is 20,000 8-oxodG's. If repair suddenly stopped a doubling of the number of oxidized nucleobases would take about 20 to 66 days

Table 1. Factors studied with regard to the level of oxidative modifications nucleobases in DNA from humans

Factor	lesion(s)	Assay(s)	Cells or tissues showing a significant increase (or decrease) related to factor	Cells or tissues not showing a significant effect
Sex	8-oxodG	HPLC-EC		leukocytes (Degan et al. 1995)
Age	8-oxodG	HPLC-EC	mtDNA: brain (Mecocci et al. 1993), diaphragm (Hayakawa, M. et al. 1991), heart (Hayakawa, M. et al. 1992), nuDNA: brain (Mecocci et al. 1993), leukocytes (Degan et al. 1995)	leukocytes (Takeuchi et al. 1994), MN+PMN (Bashir et al. 1993, Nakajima et al. 1996b)
Smoking	8-oxodG	HPLC-EC Immunoassay	leukocytes (Kiyosawa et al. 1990, Degan et al. 1995, Asami et al. 1996), sperm (Fraga et al. 1996) oral mucosa cells (Yarborough et al. 1996), placenta (Yin et al. 1995)	leukocytes (Hanaoka et al. 1993, Takeuchi et al. 1994), MN+PMN (Nakajima et al. 1996b), placenta (Daube et al. 1997)
Exercise	8-oxodG	HPLC-EC	lymphocytes (decrease) (Inoue et al. 1993)	
Vitamin C deficiency	8-oxodG	HPLC-EC	sperm cells (Fraga et al. 1991)	lymphocytes (Jacob et al. 1991)
Energy restriction	8-oxodG	HPLC-EC		lymphocytes (Velthuis-te Wierik et al. 1995)
Low fat diet	5-OHmU	GC/MS	leukocytes (decrease) (Djuric et al. 1991)	
Asbestos exposure	8-oxodG	HPLC-EC		leukocytes (Hanaoka et al. 1993)
Radiation	8-oxodG	HPLC-EC+TLC-^{32}P	leukocytes (Wilson et al. 1993)	
	7 bases[o]	GC/MS-SIM	lymphocytes (Olinski et al. 1996)	
Autoimmune diseases[a]	8-oxodG	HPLC-EC	lymphocytes+PMN (Bashir et al. 1993)	
Liver diseases (liver from)	8-oxodG	HPLC-EC	chronic hepatitis (Shimoda et al. 1994), Wilson's disease (decrease) (Carmichael et al. 1995),	cirrhosis (Shimoda et al. 1994), haemochromatosis (Carmichael et al. 1995)
Helicobacter pylori infection	8-oxodG	HPLC-EC	infected gastric mucosa (Baik et al. 1996)	
Cancers of the	8-oxodG	HPLC-EC	kidney (Okamoto et al. 1994), colon (Oliva et al. 1997)	breast (Nagashima et al. 1995), liver (Shimoda et al. 1994)
	8-oxodG	immunoassay	breast (Musarrat et al. 1996)	
	up to 12 bases[b]	GC/MS-SIM	breast (Malins and Haimanot 1991, Malins et al. 1993), lung (Olinski et al. 1992, Jaruga et al. 1994), other sites (Olinski et al. 1992)	
	5-OHmdUridine	GC/MS	whole blood from breast cancer patients (Djuric et al. 1996)	
Neurodegenerative diseases of (various regions of the brain)	8-oxodG [b]	HPLC-EC	Huntington (Browne et al. 1997), Alzheimer (Lyras et al. 1997), Parkinson (Sanchez-Ramon et al. 1994)	
	6 bases	GC/MS-SIM	Alzheimer (Lyras et al. 1997)	
Diabetes mellitus	8-oxodG	HPLC-EC	MN (Dandona et al. 1996)	
Fanconi's anaemia	8-oxodG	HPLC-EC	leukocytes (Degan et al. 1995)	
Prostatic hyperplasia	7 bases[b]	GC/MS-SIM	hyperplastic tissue (Olinski et al. 1995)	

mt: mitochondrial; nu: nuclear; PMN: polymorphonuclear granulocytes; MN: mononuclear leukocytes; [a]SLE, RA, vasculitis and Behçet's disease; [b]of 8-oxoGua, 8-oxoAde, 5-OHmU, Tg, 5-OH-Cyt, 2-OH-Ade, 5,6-diOH-Ura, 5-OH-Ura, 5-OH-Hyd, 5-OH-5-MeHyd, FapyGua and/or FapyAde: not all were increased related to factors.

Several points can be inferred from the concept of a steady state outlined in Fig. 2. As efflux must equal influx, the rate of damage is estimated from the rate of excretion of the repair product, i.e. 8-oxodG in this case. The unknowns in that equation are the contribution of the glycosylase pathway to the repair and the contribution from cellular and mitochondrial turnover to 8-oxodG formation. However, these contributions would also be expected to be in a steady state and the latter actually a part of oxidative damage to DNA. It should be noted that the urinary excretion represent the cumulated body burden and that it usually cannot be determined if this originated from an impact to all body cells or to much higher insult to one or several organs. In contrast, tissue or cellular levels represent the measure of a concentration, reflecting the balance between rate of oxidation and the rate of repair. Moreover, in human studies the cellular levels are frequently measured in surrogate cells, such as lymphocytes, rather than in true target tissues. The general interpretation of a change in the urinary excretion rate of oxidized nucleosides and bases is a change in the rate of damage inflicted by oxidative stress. A change in tissue concentration of oxidized nucleobases/nucleosides cannot with certainty be related to a change in oxidative stress, a change in repair or a combination. Accordingly, the two groups of biomarkers are supplementary. However, much more knowledge regarding repair pathways and kinetics are warranted for the optimum interpretation of these biomarkers.

In a recent experimental rat study we compared target tissue levels and urinary excretion of 8-oxodG. A temporary excess of around 3 8-oxodG per 10^5 dG was induced in the liver by administration of the hepatocarcinogen, 2-nitropropane. Assuming that a rat liver contains 6 x 10^8 cells and that each cell contains 2 x 10^9 dG residues this excess corresponds 3.6 x 10^{13} molecules or 60 pmoles of 8-oxodG. In the same study period the excess urinary excretion of 8-oxodG was 40 pmoles showing a correspondence between the target tissue level and the urinary excretion of repair product. The remaining 8-oxodG may have been excreted as 8-oxoGua that is difficult to quantify in rats because dietary purines are the major source (Park et al. 1992). The present calculations are of course subject to substantial variation and assumptions, however the consistent numbers appear to support that urinary 8-oxodG is a biomarker of 8-oxodG formation in target tissues.

5. FACTORS DETERMINING OXIDATIVE DNA DAMAGE IN HUMANS

A large number of (patho)physiological and environmental factors have been studied in humans with regard to influence on the tissue or cell level of oxidized bases/nucleosides in DNA and the urinary excretion of repair products, i.e. the damage rate (Table 1 and 2). Most of the exact data were recently summarized (Loft and Poulsen 1996).

Sex and age are probably not important determinants of the urinary excretion of 8-oxodG (Table 1 and 2). The excretion may be slightly higher in men and it may decrease some with age (Loft and Poulsen 1996). According to the steady state concept discussed above such a decrease is most likely due to a decreasing damage rate following a decreased metabolic rate (Loft et al. 1994, Loft and Poulsen 1996). The possible accumulation of nuclear 8-oxodG in brain tissue (Mecocci et al. 1993) and in leukocytes showed in one study (Degan et al. 1995) but not in other studies (Bashir et al. 1993, Takeuchi et al. 1994, Nakajima et al. 1996b) is probably caused by failing repair. However, related to neurodegenerative diseases also associated with aging the increased accumulation of oxidized nucleobases could be related to an increased rate of damage as other markers of oxidative stress are increased (Sanchez-Ramon et al. 1994, Browne et al. 1997, Lyras et al.

Table 2. Factors studied with regard to the urinary excretion of oxidatively modified nucleobases and deoxynucleosides in humans

Factor	Lesion(s)	Assay	Excretion parameters showing a significant increase (or decrease) related to factor	Excretion parameters not showing a significant effect
Sex	8-oxodG	HPLC-EC	male>female: 24 h excretion (Loft et al. 1992),	ratio to creatinine (Tagesson et al. 1995)
	dTg and Tg	HPLC-UV	concentration (Tagesson et al. 1996)	24 h excretion (Lunec et al. 1994)
Age	8-oxodG	HPLC-EC	ratio to creatinine (Lagorio et al. 1994), concentration (Tagesson et al. 1996)	24 h excretion (Loft et al. 1992), ratio to creatinine (Lunec et al. 1994)
Smoking	8-oxodG	HPLC-EC	24-h excretion (Loft et al. 1992, 1994, Prieme et al. 1998), ratio to creatinine or concentration (Tagesson et al. 1992, 1996)	ratio to creatinine (Lagorio et al. 1994)
	8-oxoGua	HPLC-EC	ratio to creatinine (Suzuki et al. 1995)	
Oxygen consumption	8-oxodG	HPLC-EC	24 h excretion (Loft et al. 1994)	
Exercise	8-oxodG	HPLC-EC	ratio to creatinine after extensive exercise for 30 d (Poulsen et al. 1996)	ratio to creatinine after swimming/running (Inoue et al. 1993) or rowing (Nielsen et al. 1995)
Antioxidants: beta-carotene, vitamins C and E and coenzyme Q	8-oxodG 8-oxoG	HPLC-EC HPLC-EC		24 h excretion (van Poppel et al. 1995, Priemé et al. 1997), 24 h excretion (Witt et al. 1992)
Brussels sprouts rich diet	8-oxodG	HPLC-EC	decreased 24 excretion in non-smokers (Verhagen et al. 1995)	24 h excretion (Verhagen et al. 1997)
Vegetable and fruit rich diet	8-oxodG	HPLC-EC		24 h excretion (Hertog et al. 1997)
Energy restriction	8-oxodG	HPLC-EC		24 h excretion in 16 non-smokers (Loft et al. 1995b)
	8-oxodG + dTg	GC/MS	decreased 24 h excretion in 1 subject (Simic and Bergtold 1991)	
Benzene exposure	8-oxodG	HPLC-EC	ratio to creatinine (Lagorio et al. 1994), pre-postshift (Nilsson et al. 1996)	
glass work fumes	8-oxodG	HPLC-EC	exposed smokers-unexposed non-smokers (Tagesson et al. 1996)	exposed-unexposed smokers or non-smokers (Tagesson et al. 1996)
rubber + asbestos + azo dye	8-oxodG	HPLC-EC	ratio to creatinine (Tagesson et al. 1993)	
polluted urban air	8-oxoGua	HPLC-EC	ratio to creatinine (Suzuki et al. 1995)	
Various cancer chemotherapy	8-oxodG	HPLC-EC	ratio to creatinine (Tagesson et al. 1995)	
Adriamycin therapy	5-OhmU	GC/MS	24 h excretion (Faure et al. 1996)	
Radiation	8-oxodG+dTg	GC/MS, HPLC-EC, ELISA	24 h excretion (Bergtold et al. 1990), ratio to creatinine (Blount et al. 1991, Tagesson et al. 1995, Erhola et al. 1997)	
Autoimmune diseases	8-oxodG	HPLC-EC	ratio to creatinine increased in RA, decreased in SLE (Lunec et al. 1994)	
Cystic fibrosis	8-oxodG	HPLC-EC	ratio to creatinine in children (Brown et al. 1995)	
Various cancers	8-oxodG	HPLC-EC	ratio to creatinine (Tagesson et al. 1992, 1995)	24 h excretion (Tagesson et al. 1995)
	8-oxodG	ELISA	ratio to creatinine (Erhola et al. 1997)	ratio to creatinine (Erhola et al. 1997)
	dTg and Tg	HPLC-UV		24 h excretion (Cao and Wang 1993)

1997). Possibly the increased levels of 8-oxodG in leukocytes from patients with diabetes mellitus could be attributed to either increased oxidative stress or reduced repair if urinary excretion of 8-oxodG was measured (Dandona et al. 1996).

The urinary excretion of 8-oxodG was closely correlated with oxygen consumption or metabolic rate within a group of young women (Loft et al. 1994) and in men after energy restriction (Loft et al. 1995b). Similar correlations also including dTg excretion have been found across species (Adelman et al. 1988, Cutler 1991, Simic and Bergtold 1991, Loft et al. 1993). This relationship is in keeping with the concept of mitochondrial respiration as an important source of ROS (Chance et al. 1979). Exercise would thus be expected to increase the rate of oxidative DNA damage. Indeed, the 8-oxodG to creatinine ratio was increased after 30 days of intense exercise although short term running, swimming or rowing had no effect in this respect (Inoue et al. 1993, Nielsen et al. 1995). The decrease in the 8-oxodG level in lymphocytes seen after running could be due to recruitment of young lymphocytes as supported by increased counts after exercise (Inoue et al. 1993). In a controlled human study energy restriction had no beneficial effect on lymphocyte DNA level and urinary excretion of 8-oxodG (Loft et al. 1995b, Velthuis-te Wierik et al. 1995) although such intervention can reduce oxidative DNA damage in animal experiments (Kaneko et al. 1997).

Tobacco smoke is a major source of ROS per se and can induce endogenous production of ROS in leukocytes and via an increased metabolic rate. In accordance, the urinary excretion of 8-oxodG and 8-oxoGua was consistently elevated in smokers (Loft et al. 1992, 1994, Tagesson et al. 1992, 1996, Suzuki et al. 1995, Prieme et al. 1998) with the exception of one study with creatinine corrected spot samples (Lagorio et al. 1994). Moreover, in a recent smoking cessation study there was a dose-response relationship and the relevant decrease in 8-oxodG excretion that mirrored an increase in plasma vitamin C (unpublished observations). Similarly, exposure to air pollution from urban air and related to glasswork, benzene and other occupational exposures appear to cause the expected increases in 8-oxodG and 8-oxoGua excretion. However, half of the studies of 8-oxodG in DNA from leukocytes and placenta as well as a study of asbestos exposed workers have failed to show a difference between smoker and non-smokers (Hanaoka et al. 1993, Takeuchi et al. 1994, Nakajima et al. 1996b), whereas the other half and studies of DNA and oral mucosal cells have showed increased levels in smokers (Kiyosawa et al. 1990, Degan et al. 1995, Asami et al. 1996, Fraga et al. 1996). There is no explanation for this apparent discrepancy as several of the studies with both outcomes were reasonably sized and employed similar methods. The level of oxidized pyrimidines measured by the enzyme based comet assay in lymphocytes was increased in smokers (Duthie et al. 1996)

Antioxidants would be expected to decrease oxidative DNA damage, however, intervention studies have generally failed to show effects of the traditional nutritional antioxidants, vitamins C and E, beta-carotene and coenzyme Q, on nuclear levels in leukocytes and urinary excretion of 8-oxodG (Jacob et al. 1991, Witt et al. 1992, van Poppel et al. 1995, Priemé et al. 1997). The only exception relates to the increase in 8-oxodG levels in sperm after depletion of vitamin C (Fraga et al. 1991). Similarly, in a comparison of subjects with a habitual diet rich and poor in fruit and vegetables there was no difference in 8-oxodG excretion (Hertog et al. 1997) whereas intervention with a diet rich in Brussels sprouts reduced the excretion (Verhagen et al. 1995), although this effect was only partly reproduced in men in a later study (Verhagen et al. 1997). In contrast, intervention with a combination of vitamins C and E and beta-carotene reduced the level of oxidized pyrimidines measured by the enzyme based comet assay in lymphocytes from smokers (Duthie et al. 1996)

In most (but not all (Shimoda et al. 1994, Nagashima et al. 1995)) of the studies of malignant (and a benign) tumors increased levels of oxidized nucleobases/nucleosides have been found in the tumor DNA as compared with the surrounding tissue (Malins and Haimanot 1991, Olinski et al. 1992, 1995, Malins et al. 1993, Jaruga et al. 1994, Okamoto et al. 1994, Musarrat et al. 1996, Oliva et al. 1997). Whether this is a causal relationship or an effect of ongoing inflammation in the tumor is not known. Indeed, inflammation, such as related to autoimmune diseases, hepatitis and Helicobacter pylori infections cause increases in the nuclear levels of 8-oxodG and the urinary excretion of 8-oxodG in rheumatoid arthritis (Bashir et al. 1993, Lunec et al. 1994, Shimoda et al. 1994, Baik et al. 1996). In some cancer patients the ratio of 8-oxodG to creatinine was reported to be increased (Tagesson et al. 1992, 1995, Erhola et al. 1997), however, the 24 h excretion was not (Tagesson et al. 1995) and the 24 excretion of dTg and Tg was not increased in cancer patients from another study (Cao and Wang 1993). Accordingly, an apparent effect of cancer in this context may be related to a low creatinine production and the use of creatinine corrected spot urine samples is probably not prudent. In contrast, an expected increasing effect on oxidative DNA damage of cancer chemotherapy and radiation appears quite consistent from all reported studies of both leukocytes and urinary excretion of repair products (Bergtold et al. 1990, Blount et al. 1991, Wilson et al. 1993, Tagesson et al. 1995, Faure et al. 1996, Olinski et al. 1996, Erhola et al. 1997).

6. CONCLUSIONS

A number of different oxidatively modified nucleobases in cellular and tissue DNA and repair products excreted in urine can be measured with a variety of methods. However, problems remain with respect to the true values and inter-method and inter-laboratory variation in particular regarding the levels in nuclear DNA.

It should be emphasized that the levels of oxidized nucleobases in tissue/cells and the excretion of the repair products represent two fundamentally different estimates that are supplementary. The urine measurement represents the number of repaired bases summed from all organs and cells during a given time period, i.e. the rate of damage. The tissue measurement is a concentration measurement in the specific tissue/cells in the moment of sampling dependent on the balance between the rates of damage and repair.

The rate and levels of oxidative DNA modifications in humans have been studied extensively. The data obtained so far indicate that the important determinants of the oxidative damage rate include tobacco smoking, air pollution, oxygen consumption, cancer therapy and some inflammatory and neurodegenerative diseases, whereas diet composition, energy restriction and antioxidant supplements have minimal influence, possibly with the exception of yet unidentified phytochemicals, e.g. from brassica vegetables. Generally these effects are in the order of a factor 2 or less. The data support that oxidative DNA damage caused by endogenous or exogenous oxidative stress is involved in the pathogenesis of cancer and degenerative diseases of aging. In the future the use of the biomarkers may provide further proof of a causal relationship in this respect as well as elucidate possible preventive measures. However, much more knowledge regarding the true sources, pathways and kinetics of repair of the oxidized nucleobases and nucleosides as well as their true levels in DNA by accurate assays is warranted.

REFERENCES

S. Adachi, M. Zeisig, L. Moller (1995) Improvements in the analytical method for 8-hydroxydeoxyguanosine in nuclear DNA. *Carcinogenesis* **16**, 253–258.

R. Adelman, R.L. Saul, B.N. Ames (1988) Oxidative damage to DNA: Relation to species metabolic rate and life span. *Proc. Natl. Acad. Sci. USA* **85**, 2706–2708.

B.N. Ames, M.K. Shigenaga, T.M. Hagen (1993) Oxidants, antioxidants, and the degenerative diseases of aging. *Proc. Natl. Acad. Sci. USA* **90**, 7915–7922.

B.N. Ames, L.S. Gold, W.C. Willett (1995) The causes and prevention of cancer. *Proc. Natl. Acad. Sci.* USA **92**, 5258–5265.

S. Asami, T. Hirano, R. Yamaguchi, Y. Tomioka, H. Itoh, H. Kasai (1996) Increase of a type of oxidative DNA damage, 8-hydroxyguanine, and its repair activity in human leukocytes by cigarette smoking. *Cancer Res.* **56**, 2546–2549.

S.C. Baik, H.S. Youn, M.H. Chung, W.K. Lee, M.J. Cho, G.H. Ko, C.K. Park, H. Kasai, K.H. Rhee (1996) Increased oxidative DNA damage in Helicobacter pylori-infected human gastric mucosa. *Cancer Res.* **56**, 1279–1282.

S. Bashir, G. Harris, M.A. Denman, D.R. Blake, P.G. Winyard (1993) Oxidative DNA damage and cellular sensitivity to oxidative stress in human autoimmune diseases. *Ann. Rheum. Dis.* **52**, 659–666.

D.S. Bergtold, C.D. Berg, Simic;MG (1990) Urinary biomarkers in radiation therapy of cancer. *Adv. Exp. Med. Biol.* **264**, 311–316.

T. Bessho, K. Tano, H. Kasai, E. Ohtsuka, S. Nishimura (1993) Evidence for two DNA repair enzymes for 8-hydroxyguanine (7,8-dihydro-8-oxoguanine) in human cells. *J. Biol. Chem.* **268**, 19416–19421.

S. Blount, H.R. Griffiths, J. Lunec (1991) Reactive oxygen species damage to DNA and its role in systemic lupus erythematosus. *Molec. Aspects Med.* **12**, 93–105.

R.K. Brown, A. McBurney, J. Lunec, F.J. Kelly (1995) Oxidative damage to DNA in patients with cystic fibrosis. *Free Rad. Biol. Med.* **18**, 801–806.

S.E. Browne, A.C. Bowling, U. MacGarvey, M.J. Baik, S.C. Berger, M.M. Muqit, E.D. Bird, M.F. Beal (1997) Oxidative damage and metabolic dysfunction in Huntington's disease: selective vulnerability of the basal ganglia. *Ann. Neurol.* **41**, 646–653.

J. Cadet, J.L. Ravanat, G.W. Buchko, H.C. Yeo, B.N. Ames (1994) Singlet oxygen DNA damage: chromatographic and mass spectrometric analysis of damage products. *Meth. Enzymol.* **234**, 79–88.

E.-H. Cao, J.-J. Wang (1993) Oxidative damage to DNA: levels of thymine glycol and thymidine glycol in neoplastic human urines. *Carcinogenesis* **14**, 1359–1362.

P.L. Carmichael, A. Hewer, M.R. Osborne, A.J. Strain, D.H. Phillips (1995) Detection of bulky DNA lesions in the liver of patients with Wilson's disease and primary haemochromatosis. *Mutation Res.* **326**, 235–243.

R. Cathcart, E. Schwiers, R.L. Saul, B.N. Ames (1984) Thymine glycol and thymidine glycol in human and rat urine: a possible assay for oxidative DNA damage. *Proc. Natl. Acad. Sci. USA* **81**, 5633–5637.

B. Chance, H. Sies, A. Boveris (1979) Hydroperoxide metabolism in mammalian organs. *Physiol.* Rev. **59**, 527–605.

H.G. Claycamp (1992) Phenol sensitization of DNA to subsequent oxidative damage in 8-hydroxyguanine assays. *Carcinogenesis* **13**, 1289–1292.

A.R. Collins, M. Dusinska, C.M. Gedik, R. Stetina (1996) Oxidative damage to DNA: do we have a reliable biomarker? *Environ. Health Perspec.* **104 Suppl 3**, 465–469.

A.R. Collins, J. Cadet, B. Epe, C. Gedik (1997a) Problems in the measurement of 8-oxoguanine in human DNA. Report of a workshop, DNA Oxidation, held in Aberdeen, UK, 19–21 January, 1997. *Carcinogenesis* **18**, 1833–1836.

A.R. Collins, V.L. Dobson, M. Dusinska, G. Kennedy, R. Stetina (1997b) The comet assay: what can it really tell us? *Mutation Res.* **375**, 183–193.

R.G. Cutler (1991) Human longevity and aging: possible role of reactive oxygen species. *Ann. N. Y. Acad. Sci.* **621**, 1–28.

P. Dandona, K. Thusu, S. Cook, B. Snyder, J. Makowski, D. Armstrong, T. Nicotera (1996) Oxidative damage to DNA in diabetes mellitus. *Lancet* **347**, 444–445.

H. Daube, G. Scherer, K. Riedel, T. Ruppert, A.R. Tricker, P. Rosenbaum, F. Adlkofer (1997) DNA adducts in human placenta in relation to tobacco smoke exposure and plasma antioxidant status. *J. Cancer. Res. Clin. Oncol.* **123**, 141–151.

P. Degan, M.K. Shigenaga, E.-M. Park, P.E. Alperin, B.N. Ames (1991) Immunoaffinity isolation of urinary 8-hydroxy-2'deoxyguanosine and 8-hydroxyguanine and quantitation 8-hydroxy-2'-deoxyguanosine in DNA by polyclonal antibodies. *Carcinogenesis* **12**, 865–871.

P. Degan, S. Bonassi, M. De Caterina, L.G. Korkina, L. Pinto, F. Scopacasa, A. Zatterale, R. Calzone, G. Pagano (1995) In vivo accumulation of 8-hydroxy-2'-deoxyguanosine in DNA correlates with release of reactive oxygen species in Fanconi's anaemia families. *Carcinogenesis* **16**, 735–741.

B. Demple, L. Harrison (1994) Repair of oxidative damage to DNA: enzymology and biology. *Ann. Rev. Biochem.* **63**, 915–948.

X.-S. Deng, J.-S. Tuo, H.E. Poulsen, S. Loft (1997) 2-Nitropropane induced DNA damage in rat bone marrow. *Mutation Res.* **391**, 165–169.

U. Devanaboyina, R.C. Gupta (1996) Sensitive detection of 8-hydroxy-2'deoxyguanosine in DNA by 32p-postlabeling assay and the basal levels in rat tissues. *Carcinogenesis* **17**, 917–924.

M. Dizdaroglu (1991) Chemical determination of free radical-induced damage to DNA. *Free Rad. Biol. Med.* **10**, 225–242.

M. Dizdaroglu (1994) Chemical determination of oxidative DNA damage by gas chromatography-mass spectrometry. *Meth. Enzymol.* **234**, 3–16.

Z. Djuric, L.K. Heilbrun, B.A. Reading, A. Boomer, F.A. Valeriote, S. Martino (1991) Effects of low-fat diet on levels of oxidative damage to DNA in human peripheral nucleated blood cells. *J. Natl. Cancer. Inst.* **83**, 766–769.

Z. Djuric, L.K. Heilbrun, M.S. Simon, D. Smith, D.A. Luongo, P.M. LoRusso, S. Martino (1996) Levels of 5-hydroxymethyl-2'-deoxyuridine in DNA from blood as a marker of breast cancer. *Cancer.* **77**, 691–696.

T. Douki, T. Delatour, F. Bianchini, J. Cadet (1996) Observation and prevention of an artefactual formation of oxidized DNA bases and nucleosides in the GC-EIMS method. *Carcinogenesis* **17**, 347–353.

S.J. Duthie, A. Ma, M.A. Ross, A.R. Collins (1996) Antioxidant supplementation decreases oxidative DNA damage in human lymphocytes. *Cancer Res.* **56**, 1291–1295.

B. Epe (1995) DNA damage profiles induced by oxidizing agents. *Rev. Physiol. Biochem. Pharmacol.* **127**, 223–249.

M. Erhola, S. Toyokuni, K. Okada, T. Tanaka, H. Hiai, H. Ochi, K. Uchida, T. Osawa, M.M. Nieminen, H. Alho, P. Kellokumpu-Lehtinen (1997) Biomarker evidence of DNA oxidation in lung cancer patients: association of urinary 8-hydroxy-2'-deoxyguanosine excretion with radiotherapy, chemotherapy, and response to treatment. *FEBS. Letters.* **409**, 287–291.

H. Faure, M.F. Incardona, C. Boujet, J. Cadet, V. Ducros, A. Favier (1993) Gas chromatographic-mass spectrometric determination of 5-hydroxymethyluracil in human urine by stable isotope dilution. *J. Chromatogr.* **616**, 1–7.

H. Faure, C. Coudray, M. Mousseau, V. Ducros, T. Douki, F. Bianchini, J. Cadet, A. Favier (1996) 5-Hydroxymethyluracil excretion, plasma TBARS and plasma antioxidant vitamins in adriamycin-treated patients. *Free Rad. Biol. Med.* **20**, 979–983.

R.A. Floyd, M.S. West, K.L. Eneff, J.E. Schneider, P.K. Wong, D.T. Tingey, W.E. Hogsett (1990) Conditions influencing yield and analysis of 8-hydroxy-2'deoxyguanosine in oxidatively damaged DNA. *Anal. Biochem.* **188**, 155–158.

C.G. Fraga, P.A. Motchnik, M.K. Shigenaga, H.J. Helbock, R.A. Jacob, B.N. Ames (1991) Ascorbic acid protects against endogenous oxidative DNA damage in human sperm. *Proc. Natl. Acad. Sci. USA* **88**, 11003–11006.

C.G. Fraga, P.A. Motchnik, A.J. Wyrobek, D.M. Rempel, B.N. Ames (1996) Smoking and low antioxidant levels increase oxidative damage to sperm DNA. *Mutation Res.* **351**, 199–203.

D. Germadnik, A. Pilger, H.W. Rudiger (1997) Assay for the determination of urinary 8-hydroxy-2'-deoxyguanosine by high-performance liquid chromatography with electrochemical detection. *J. Chromatogr.* **689**, 399–403.

M. Hamberg, L.Y. Zhang (1995) Quantitative determination of 8-hydroxyguanine and guanine by isotope dilution masss spectrometry. *Anal. Biochem.* **229**, 336–344.

T. Hanaoka, S. Tsugane, Y. Yamano, T. Takahashi, H. Kasai, Y. Natori, S. Watanabe (1993) Quantitative analysis of 8-hydroxyguanine in peripheral blood cells: an application for asbestosis patients. *Int. Arch. Occup. Environ. Health.* **65**, S215-S217.

Hartwig, H. Dally, R. Schlepegrell (1996) Sensitive analysis of oxidative DNA damage in mammalian cells: use of the bacterial Fpg protein in combination with alkaline unwinding. *Toxicol. Lett.* **88**, 85–90.

H. Hayakawa, A. Taketomi, K. Sakumi, M. Kuwano, M. Sekiguchi (1995) Generation and elimination of 8-oxo-7,8-dihydro-2'-deoxyguanosine 5'-triphosphate, a mutagenic substrate for DNA synthesis, in human cells. *Biochemistry.* **34**, 89–95.

M. Hayakawa, K. Torii, S. Sugiyama, M. Tanaka, T. Ozawa (1991) Age-associated accumulation of 8-hydroxydeoxyguanosine in mitochondrial DNA of human diaphragm. *Biochem. Biophys. Res. Commun.* **179**, 1023–1029.

M. Hayakawa, K. Hattori, S. Sugiyama, T. Ozawa (1992) Age-associated oxygen damage and mutations in mitochondrial DNA in human hearts. *Biochem. Biophys. Res. Commun.* **189**, 979–985.

M.G.L. Hertog, A. de Vries, M.C. Ocké, A. Schouten, H. Bas Bueno-de-Mesquita, H. Verhagen (1997) Oxidative DNA damage in humans: comparison between high and low habitual fruit and vegetable consumption. *Biomarkers* **2**, 259–262.

T. Inoue, Z. Mu, K. Sumikawa, K. Adachi, T. Okochi (1993) Effect of physical exercise on the content of 8-hydroxydeoxyguanosine in nuclear DNA prepared from human lymphocytes. *Jpn. J. Cancer. Res.* **84**, 720–725.

R.A. Jacob, D.S. Kelley, F.S. Pianalto, M.E. Swendseid, S.M. Henning, J.Z. Zhang, B.N. Ames, C.G. Fraga, J.H. Peters (1991) Immunocompetence and oxidant defence during ascorbate depletion of healthy men. *Am. J. Clin. Nutr.* **54 suppl 6**, 1302S-1309S.

P. Jaruga, M. Dizdaroglu (1996) Repair of products of oxidative DNA base damage in human cells. *Nucleic. Acids. Res.* **24**, 1389–1394.

P. Jaruga, T.H. Zastawny, J. Skokowski, M. Dizdaroglu, R. Olinski (1994) Oxidative DNA base damage and antioxidant enzyme activities in human lung cancer. *FEBS. Lett.* **341**, 59–64.

T. Kaneko, S. Tahara, M. Matsuo (1997) Retarding effect of dietary restriction on the accumulation of 8-hydroxy-2'-deoxyguanosine in organs of Fischer 344 rats during aging. *Free Rad. Biol. Med.* **23**, 76–81.

Karakaya, P. Jaruga, V.A. Bohr, A.P. Grollman, M. Dizdaroglu (1997) Kinetics of excision of purine lesions from DNA by Escherichia coli Fpg protein. *Nucleic Acids Research* **25**, 474–479.

H. Kasai, P.F. Crain, Y. Kuchino, S. Nishimura, A. Ootsuyama, H. Tanooka (1986) Formation of 8-hydroxyguanine moiety in cellular DNA by agents producing oxygen radicals and evidence for its repair. *Carcinogenesis* **7**, 1849–1851.

H. Kiyosawa, M. Suko, H. Okudaira, K. Murata, T. Miyamoto, M.H. Chung, H. Kasai, S. Nishimura (1990) Cigarette smokeing induces formation of 8-hydroxydeoxyguanosine, one of the oxidative DNA damages in human peripheral leucocytes. *Free. Rad. Res. Comms.* **11**, 23–27.

Y. Kuchino, F. Mori, H. Kasai, H. Inoue, S. Iwai, K. Miura, E. Ohtsuka, S. Nishimura (1987) Misreading of DNA templates containing 8-hydroxydeoxyguanosine at the modified base and at adjacent residues. *Nature.* **327**, 77–79.

S. Lagorio, C. Tagesson, F. Forastiere, I. Iavarone, O. Axelson, A. Carere (1994) Exposure to benzene and urinary concentrations of 8-hydroxydeoxyguanosine, a biological marker of oxidative damage to DNA. *Occup. Environ. Med.* **51**, 739–743.

S. Loft, H.E. Poulsen (1996) Cancer risk and oxidative DNA damage in man. *J. Mol. Med.* **74**, 297–312.

S. Loft, K. Vistisen, M. Ewertz, A. Tjønneland, K. Overvad, H.E. Poulsen (1992) Oxidative DNA-damage estimated by 8-hydroxydeoxyguanosine excretion in humans: influence of smoking, gender and body mass index. *Carcinogenesis* **13**, 2241–2247.

S. Loft, A. Fischer-Nielsen, I.B. Jeding, K. Vistisen, H.E. Poulsen (1993) 8-Hydroxydeoxyguanosine as a biomarker of oxidative DNA damage. *J. Toxicol. Environ. Health.* **40**, 391–404.

S. Loft, A. Astrup, B. Buemann, H.E. Poulsen (1994) Oxidative DNA damage correlates with oxygen consumption in humans. *FASEB J.* **8**, 534–537.

S. Loft, P.N. Larsen, A. Rasmussen, A. Fischer-Nielsen, S. Bondesen, P. Kirkegaard, L.S. Rasmussen, E. Ejlersen, K. Tornøe, R. Bergholdt, H.E. Poulsen (1995a) Oxidative DNA damage after transplantation of the liver and small intestine in pigs. *Transplantation.* **59**, 16–20.

S. Loft, E.J.M.V. Velthuis-te Wierik, H. van den Berg, H.E. Poulsen (1995b) Energy restriction and oxidative DNA damage in humans. *Cancer Epidemiol. Biomarkers Prev.* **4**, 515–519.

J. Lunec, K. Herbert, S. Blount, H.R. Griffiths, P. Emery (1994) 8-Hydroxydeoxyguanosine. A marker of oxidative DNA damage in systemic lupus erythematosus. *FEBS. Lett.* **348**, 131–138.

L. Lyras, N.J. Cairns, A. Jenner, P. Jenner, B. Halliwell (1997) An assessment of oxidative damage to proteins, lipids, and DNA in brain from patients with Alzheimer's disease. *Journal of Neurochemistry* **68**, 2061–2069.

D.C. Malins, R. Haimanot (1991) Major alterations in the nucleotide structure of DNA in cancer of the female breast. *Cancer Res.* **51**, 5430–5432.

D.C. Malins, E.H. Holmes, N.L. Polissar, S. Gunselman (1993) The etiology of breast cancer. Characteristic alterations in hydroxyl radical-induced DNA base lesions during oncogenesis with potential for evaluating incidence risk. *Cancer Lett.* **71**, 3036–3042.

P. Mecocci, U. MacGarvey, A.E. Kaufman, D. Koontz, J.M. Shoffner, D.C. Wallace, M.F. Beal (1993) Oxidative damage to mitochondrial DNA shows marked age-dependent increases in human brain. *Ann. Neurol.* **34**, 609–616.

J.Y. Mo, H. Maki, M. Sekiguchi (1992) Hydrolytic elimination of a mutagenic nucleotide, 8-oxodGTP, by human 18-kilodalton protein: sanitization of nucleotide pool. *Proc. Natl. Acad. Sci. USA* **89**, 11021–11025.

J. Musarrat, A.A. Wani (1994) Quantitative immunoanalysis of promutagenic 8-hydroxy-2'-deoxyguanosine in oxidized DNA. *Carcinogenesis* **15**, 2037–2043.

J. Musarrat, J. Arezina-Wilson, A.A. Wani (1996) Prognostic and aetiological relevance of 8-hydroxyguanosine in human breast carcinogenesis. Eur. *J. Cancer.* **32A**, 1209–1214.

M. Nagashima, H. Tsuda, S. Takenoshita, Y. Nagamachi, S. Hirohashi, J. Yokota, H. Kasai (1995) 8-hydroxydeoxyguanosine levels in DNA of human breast cancer are not significantly different from those of non-cancerous breast tissues by the HPLC-ECD method. *Cancer Lett.* **90**, 157–162.

M. Nakajima, T. Takeuchi, K. Morimoto (1996a) Determination of 8-hydroxydeoxyguanosine in human cells under oxygen-free conditions. *Carcinogenesis* **17**, 787–791.

M. Nakajima, T. Takeuchi, T. Takeshita, K. Morimoto (1996b) 8-Hydroxydeoxyguanosine in human leukocyte DNA and daily health practice factors: effects of individual alcohol sensitivity. *Environ. Health Perspec.* **104**, 1336–1338.

H.B. Nielsen, B. Hanel, S. Loft, H.E. Poulsen, B.K. Pedersen, M. Diamant, K. Vistisen, N.H. Secher (1995) Restricted pulmonary diffusion capacity after exercise is not an ARDS-like injury. *J. Sport Sci.* **13**, 109–113.

R.I. Nilsson, R.G. Nordlinder, C. Tagesson, S. Walles, B.G. Jarvholm (1996) Genotoxic effects in workers exposed to low levels of benzene from gasoline. *Am. J. of Ind. Med.* **30**, 317–324.

K. Okamoto, S. Toyokuni, K. Uchida, O. Ogawa, J. Takenewa, Y. Kahehi, H. Kinoshita, Y. Hattori-Nakakuki, H. Hiai, O. Yoshida (1994) Formation of 8-hydroxy-2'deoxyguanosine and 4-hydroxy-2-nonenal-modified proteins in human renal-cell carcinoma. *Int. J. Cancer.* **58**, 825–829.

R. Olinski, T. Zastawny, J. Budzbon, J. Skokowski, W. Zegarski, M. Dizdaroglu (1992) DNA base modifications in chromatin of human cancerous tissues. *FEBS. Lett.* **309**, 193–198.

R. Olinski, T.H. Zastawny, M. Foksinski, A. Barecki, M. Dizdaroglu (1995) DNA base modifications and antioxidant enzyme activities in human benign prostatic hyperplasia. *Free Rad. Biol. Med.* **18**, 807–813.

R. Olinski, T.H. Zastawny, M. Foksinski, W. Windorbska, P. Jaruga, M. Dizdaroglu (1996) DNA base damage in lymphocytes of cancer patients undergoing radiation therapy. *Cancer Lett.* **106**, 207–215.

M.R. Oliva, F. Ripoll, P. Muniz, A. Iradi, R. Trullenque, V. Valls, E. Drehmer, G.T. Saez (1997) Genetic alterations and oxidative metabolism in sporadic colorectal tumors from a Spanish community. *Mol. Carcinogenesis* **18**, 232–243.

T. Osawa, A. Yoshida, S. Kawakishi, K. Yamashita, H. Ochi (1995) Protective role of dietary antioxidants in oxidative stress. *In Oxidative stress and aging* (eds R.G. Cutler, L. Packer, J. Bertram, A. Mori), Birkhauser Verlag, Basel, Switzerland, pp. 367–378.

E.-M. Park, M.K. Shigenaga, P. Degan, T.S. Korn, J.W. Kitzler, C.M. Wehr, P. Kolachana, B.N. Ames (1992) Assay of excised oxidative DNA lesions: Isolation of 8-oxoguanine and its nucleoside derivatives from biological fluids with a monoclonal antibody column. *Proc. Natl. Acad. Sci. USA* **89**, 3375–3379.

H.E. Poulsen, S. Loft, K. Vistisen (1996) Extreme exercise and oxidative DNA modification. *J. Sports. Sci.* **14**, 343–346.

H. Priemé, S. Loft, R.G. Cutler, H.E. Poulsen (1996) Measurement of oxidative DNA injury in humans: evaluation of a commercially available ELISA assay. *In Natural antioxidants and food quality in atherosclerosis and cancer prevention* (eds J.T. Kumpulainen, J.T. Salonen), The Royal Society of Chemistry,, pp. 78–82.

H. Prieme, S. Loft, M. Klarlund, K. Grønbæk, P. Tønnesen, H.E. Poulsen (1998) Effect of smoking cessation on oxidative DNA damage estimated by 8-oxo-7,8-dihydro-2'-deoxyguanosine. *Carcinogenesis* **19**, 347–351 .

H. Priemé, S. Loft, K. Nyyssönen, J.T. Salonen, H.E. Poulsen (1997) No effect of supplementation with vitamin E, ascorbic acid, or coenzyme Q10 on oxidative DNA damage estimated by 8-oxo-7,8-dihydro-2'-deoxyguanosine excretion in smokers. *Am. J. Clin. Nutr.* **65**, 503–507.

J.P. Radicella, C. Dherin, C. Desmaze, M.S. Fox, S. Boiteux (1997) Cloning and characterization of hOGG1, a human homolog of the OGG1 gene of Saccharomyces cerevisiae. *Proc. Natl. Acad. Sci. USA* **94**, 8010–8015.

J.-L. Ravanat, R.J. Turesky, E. Gremaud, L.J. Trudel, R.H. Stadler (1995) Determination of 8-oxoguanine in DNA by gas chromatography - mass spectrometry and HPLC - electrochemical detection: overestimation of the background level of the oxidized base by the gas chromatography - mass spectrometry assay. *Chem. Res. Toxicol.* **8**, 1039–1045.

J.T. Reardon, T. Bessho, H.C. Kung, P.H. Bolton, A. Sancar (1997) In vitro repair of oxidative DNA damage by human nucleotide excision repair system: possible explanation for neurodegeneration in Xeroderma pigmentosum patients. *Proc. Natl. Acad. Sci. USA* **94**, 9463–9468.

T. Roldan-Arjona, Y.F. Wei, K.C. Carter, A. Klungland, C. Anselmino, R.P. Wang, M. Augustus, T. Lindahl (1997) Molecular cloning and functional expression of a human cDNA encoding the antimutator enzyme 8-hydroxyguanine-DNA glycosylase. *Proc. Natl. Acad. Sci. USA* **94**, 8016–8020.

J.R. Sanchez-Ramon, E. Overvik, B.N. Ames (1994) A marker of oxy-radical-mediated DNA damage (8-hydroxy-2'-deoxyguanosine) is increased in nigro-striatum of Parkinson's disease brain. *Neurodegeration* **3**, 197–204.

P.A. Schulte, F.P. Perera (1993) *Molecular epidemiology*. Principles and practices. Academic Press, Inc., London.

S. Shibutani, M. Takeshita, A.P. Grollman (1991) Insertion of specific bases during DNA synthesis past the oxidation-damaged base 8-oxodG. *Nature.* **349**, 431–434.

M.K. Shigenaga, C.J. Gimeno, B.N. Ames (1989) Urinary 8-hydroxy-2'-deoxyguanosine as a biological marker of in vivo oxidative DNA damage. *Proc. Natl. Acad. Sci. USA* **86**, 9697–9701.

M.K. Shigenaga, E.N. Aboujaoude, Q. Chen, B.N. Ames (1994) Assays of oxidative DNA damage biomarkers 8-oxo-2'-deoxyguanosine and 8-oxoguanine in nuclear DNA and biological fluids by high-performance liquid chromatography with electrochemical detection. *Meth. Enzymol.* **234**, 16–33.

R. Shimoda, M. Nagashima, M. Sakamoto, N. Yamaguchi, S. Hirohashi, J. Yokota, H. Kasai (1994) Increased formation of oxidative DNA damage, 8-hydroxydeoxyguanosine, in human livers with chronic hepatitis'. *Cancer Res.* **54**, 3171–3172.

M.G. Simic, D.S. Bergtold (1991) Dietary modulation of DNA damage in human. *Mutation Res.* **250**, 17–24.

J. Suzuki, Y. Inoue, S. Suzuki (1995) Changes in urinary excretion level of 8-hydroxyguanine by exposure to reactive oxygen-genrating substances. *Free Rad. Biol. Med.* **18**, 431–436.

C. Tagesson, M. Källberg, P. Leanderson (1992) Determination of urinary 8-hydroxydeoxyguanosine by coupled-column high-perfomance liquid chromatography with electrochemical detection: a noninvasive assay for in vivo oxidative DNA damage in humans. *Toxicol. Meth.* **1**, 242–251.

C. Tagesson, D. Chabiuk, O. Axelson, B. Baranski, J. Palus, K. Wyszynska (1993) Increased urinary excretion of the oxidative DNA adduct, 8-hydroxydeoxyguanosine, as a possible early indicator of occupational cancer hazards in the asbestos, rubber, and azo-dye industries. *Polish J. of Occup. Med. Environ. Health* **6**, 357–368.

C. Tagesson, M. Kallberg, C. Klintenberg, H. Starkhammar (1995) Determination of urinary 8-hydroxydeoxyguanosine by automated coupled-column high performance liquid chromatography: a powerful technique for assaying in vivo oxidative DNA damage in cancer patients. *Eur. J. Cancer.* **31A**, 934–940.

C. Tagesson, M. Kallberg, G. Wingren (1996) Urinary malondialdehyde and 8-hydroxydeoxyguanosine as potential markers of oxidative stress in industrial art glass workers. *Int. Arch. Occup. Environ. Health.* **69**, 5–13.

T. Tajiri, H. Maki, M. Sekiguchi (1995) Functional cooperation of MutT, MutM and MutY proteins in preventing mutations caused by spontaneous oxidation of guanine nucleotide in Escherichia coli. *Mutation Res.* **336**, 257–267.

T. Takeuchi, M. Nakajima, Y. Ohta, K. Mure, T. Takeshita, K. Morimot (1994) Evaluation of 8-hydroxydeoxyguanosine, a typical oxidative DNA damage, in human leukocytes. *Carcinogenesis* **15**, 1519–1523.

A.J. Teixeira, M.R. Ferreira, W.J. van Dijk, G. van de Werken, A.P. de Jong (1995) Analysis of 8-hydroxy-2'-deoxyguanosine in rat urine and liver DNA by stable isotope dilution gas chromatography/mass spectrometry. *Anal. Biochem.* **226**, 307–319.

G. van Poppel, H. Poulsen, S. Loft, H. Verhagen (1995) No influence of beta carotene on oxidative DNA damage in male smokers. *J. Natl. Cancer. Inst.* **87**, 310–311.

E.J.M.V. Velthuis-te Wierik, R.E.W.van Leeuwen, H.F.J. Hendriks, H. Verhagen, S. Loft, H.E. Poulsen, H.van den Berg (1995) Short-term moderate energy restriction does not affect parameters of oxidative stress and genotoxicity in humans. *J. Nutr.* **125**, 2631–2639.

H. Verhagen, H.E. Poulsen, S. Loft, G. van Poppel, M.I. Willems, P.J. van Bladeren (1995) Reduction of oxidative DNA-damage in humans by Brussels sprouts. *Carcinogenesis* **16**, 969–970.

H. Verhagen, A. de Vries, W.A. Nijhoff, A. Schouten, G. van Poppel, W.H.M. Peters, H. van den Berg (1997) Effect of Brussels sprouts on oxidative DNA-damage in man. *Cancer Lett.* **114**, 127–130.

V.L. Wilson, B.G. Taffe, P.G. Shields, A.C. Povey, C.C. Harris (1993) Detection and quantification of 8-hydroxydeoxyguanosine adducts in peripheral blood of people exposed to ionizing radiation. *Environ. Health Perspec.* **99**, 261–263.

H. Wiseman, B. Halliwell (1996) Damage to DNA by reactive oxygen and nitrogen species: role on inflammatory disease and progression to cancer. *Biochem. J.* **313**, 17–29.

E.H. Witt, A.Z. Reznick, C.A. Viguie, P. Starke-Reed, L. Packer (1992) Exercise, oxidative damage and effects of antioxidant manipulation. *J. Nutr.* **122**, 766–773.

Yarborough, Y.J. Zhang, T.M. Hsu, R.M. Santella (1996) Immunoperoxidase detection of 8-hydroxydeoxyguanosine in aflatoxin B1-treated rat liver and human oral mucosal cells. *Cancer Res.* **56**, 683–688.

B. Yin, R.M. Whyatt, F.P. Perera, M.C. Randall, T.B. Cooper, R.M. Santella (1995) Determination of 8-hydroxydeoxyguanosine by an immunoaffinity chromatography-monoclonal antibody-based ELISA. *Free Rad. Biol. Med.* **18**, 1023–1032.

23

EFFECTS OF VITAMIN E SUPPLEMENTATION ON *IN VIVO* OXIDATIVE DNA DAMAGE

J. Lunec, I. D. Podmore, H. R. Griffiths, K. E. Herbert, N. Mistry, and P. Mistry

Division of Chemical Pathology, University of Leicester
Centre for Mechanisms of Human Toxicity
Leicester, UK

1. ABSTRACT

The aim of this study was to investigate the effect of vitamin E (400 i.u./day) supplementation in normal individuals in terms of lymphocyte levels of the base-lesion 7,8-dihydro 8-oxo-2'deoxyguanosine (8-oxoG) which is recognised as a specific marker of ROS induced damage, *in vivo*. During the study two techniques, high performance liquid chromatography (HPLC) and (GCMS), were established for determining oxidative DNA lesions in peripheral blood lymphocytes. In a placebo-controlled supplementation study 30 normal healthy individuals received six weeks placebo followed by six weeks of vitamin E (400 i.u./day). Blood was taken at intervals for analysis of vitamin E and C and 8-oxodG by the two methods described above. 8-oxo adenine (8 oxo A) a corresponding oxidative damage product for the base adenine was also measured for comparison. The sample were taken as follows:- baseline, week 3 (placebo), week 6 (placebo), week 9 (supplementation) and week 12 supplementation. At week 27 a washout sample was analysed to determine any post supplementation effect.

Results showed that the HPLC method detected a significant reduction in 8-oxoG (>50%; $p<0.01$); the GCMS showed no change in either 8-oxoG or 8-oxoA through the supplementation. The discrepancy was probably related to the superior sensitivity and reproducibility of the HPLC procedure over GCMS. This work demonstrates that vitamin E supplementation can effect *in vivo* DNA damage in a positive protective way. This could have major implications for the idea that a vitamin E rich diet influences risk for cancer in humans.

2. INTRODUCTION

Reactive oxygen species (ROS) are believed to be involved in the pathogenesis of several, major human diseases including; atherosclerosis, chronic inflammatory disorders such as rheumatoid arthritis (RA), systemic lupus erythematosus (SLE) and cancer (Halli-

Advances in DNA Damage and Repair, edited by Dizdaroglu and Karakaya. Kluwer Academic / Plenum Publishers, New York, 1999.

well and Gutteridge, 1990; Lunec, 1990; Lunec *et al.* 1994). Whilst the precise mechanism in each case is far from clear there is indirect evidence to suggest that DNA damage by oxygen radicals, in particular the hydroxyl radical (·OH), may be related to pathogenesis, particularly of both cancer and SLE.

Although DNA appears well protected against ROS by a multilevel defence system (extra-nuclear antioxidants, compartmentalisation, shielding by histones and polyamines, efficient enzyme repair systems), it has been calculated that in normal individuals the total number of oxidative events occurring at DNA per cell, per day, may be more than 10,000 (Ames 1989). Numerous base modifications have been identified as a result of ROS-induced damage to DNA, chromatin and cultured cells (Gajewski *et al.*, 1990; Dizdaroglu *et al.*, 1991). ROS-mediated DNA damage is believed to occur in human cells, *in vivo,* typically at extremely low levels (e.g. approximately 0.02 nmol/mg DNA to 2 nmol/mg DNA, depending on analytical procedure). These low levels of modified bases in the presence of "overwhelming" concentration of normal bases necessitates exquisitely sensitive techniques for the biomonitoring of such adducts in human cells. 8-oxodeoxyguanosine, or its base 8-oxoG, have become popular markers of oxygen radical mediated damage to DNA because of the relative ease of their detection and because of their ubiquitous production in DNA exposed to all types of oxidative free radical challenge. Kasai and Nishimura (1991) discovered the formation of the oxidative DNA damage product 8-oxoG during a study on DNA modification caused *in vitro* by heating carbohydrates. Various oxygen radical forming agents such as ionising radiation, cigarette smoke condensate and asbestos have been found to be effective in the formation of 8-oxodG in DNA *in vitro*. It has also been established that 8-oxodG is an important mutagenic lesion in DNA which could predispose to carcinogenesis (Kamiya *et al.*, 1995). Two basic procedures are available (for the measurement of 8-oxoguanine): HPLC and GC-MS. In 1986, Floyd *et al.* developed and improved the sensitivity of the HPLC assay of 8-oxodG by three orders of magnitude over the UV detection (to femtomole levels) by using an electrochemical detection system to monitor elution from a reversed-phase HPLC column (HPLC-EC). Because HPLC is readily available in many laboratories this method became very popular and is routinely used to mark evidence of oxidative damage to DNA *in vivo* (Shigenaga *et al.*, 1994; Nagashima *et al.*, 1995). Validation of the use of 8-oxodG as a marker of ROS damage to DNA has been confirmed by a number of groups, but in particular, using a different type of technique, that of GC-MS with selective ion monitoring (GC-MS-SIM) (Dizdaroglu, 1994). The latter technique measures 8-oxodG as the corresponding base product 8-oxoG, generated after hydrolysis and detection facilitated by trimethylsilyation. In this report we have utilised two separate procedures: a) GC-MS and b) HPLC-EC (Herbert *et al.,* 1996) to measure 8-oxoG in human peripheral blood lymphocytes. These techniques have been used subsequently to investigate 8-oxoG levels following *in vivo* administration of vitamin E. The major purpose of this research being to establish any potential efficacy that this vitamin may have in reducing oxidative DNA damage and, by implication, cancer risk in the human population.

3. METHODS

3.1. Supplementation Protocol

30 healthy volunteers consisting of 16 females and 14 males, aged between 17 and 49, were recruited from Leicester University at three sites: Leicester Royal Infirmary

(LRI), Glenfield General Hospital (GGH) and Division of Chemical Pathology (Centre for Mechanisms of Human Toxicity, CMHT). Written informed consent was obtained and each subject completed a short questionnaire to record details of drinking and dietary habits, and also medication being taken. Smokers, people taking vitamin supplements and/or salicylates were all excluded from the study.

All volunteers were bled for initial baseline values and then given a 6 week course of placebo (soya bean oil) and vitamin E (400 iu/ day, RRR-α-tocopherol, Henkel, Dusseldorf, Germany). Samples were collected at 3 weekly intervals: one at baseline, two during placebo, two during vitamin E and one following a 15 week washout period.

3.2. Sample Collection

Approximately 50 ml of blood was collected from each volunteer at every visit. 34 ml of blood was collected in lithium heparin tubes; the majority (28 ml) required for DNA extraction and subsequent analysis of oxidative damage, whilst the remainder (6 ml) was required for vitamin E analysis. 4 ml blood was collected in tubes containing EDTA for ascorbic acid analysis. A 9 ml portion of the blood was left untreated and allowed to clot. Following centrifugation (1500 g/15 mins/10 ºC) the serum was collected and stored at -80 ºC. Finally, at the beginning and end of the supplementation trial, a 3 ml portion of blood was set aside for leucocyte counts.

3.3. Isolation of DNA from Peripheral Blood Lymphocytes (Figure 1)

3.3.1. Materials. All chemicals were purchased as the highest purity. Histopaque 1077 and 1119 and sodium hydroxide pellets were purchased from Aldrich Chemical Co. (Gillingham, Dorset, UK). Trisodium citrate, sodium chloride (NaCl), phosphate buffered saline (PBS), Trizma base [Tris(hydroxymethyl)aminomethane], EDTA, N-lauroyl-sarcosine (sarkosyl), Trizma-HCl {Tris-HCl or [Tris(hydroxymethyl)aminomethane-HCl]}, orcinol (5-methylresorcinol; 3,5-dihydroxytoluene), ferric chloride, ammonium acetate, pronase E (protease, EC 3.4.24.31, type XIV from *Streptomyces griseus*) and ribonuclease A (RNase A, type 1-A from bovine pancreas; EC 3.1.27.5) were all obtained from Sigma

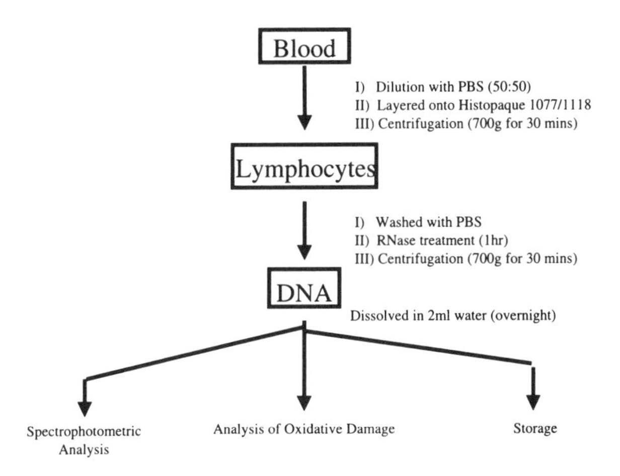

Figure 1. Protocol for the storage and analysis of DNA isolated from normal human peripheral blood lymphocytes.

Chemical Co. (Poole, Dorset, UK). Ethanol was acquired from Fisher Scientific (Loughborough, UK). Ultra-pure water (Milli-Q grade) was used throughout.

3.3.2. Extraction of Lymphocytes from Peripheral Blood. Extraction was carried out using a double density centrifugation technique which produced a relatively high yield of lymphocytes whilst minimising the yield of granulocytes (<5%). 1.5ml of Histopaque 1077 was carefully layered onto 1.5 ml of Histopaque 1119 followed by 7ml of whole blood which had previously been diluted in an equal volume of PBS. Following centrifugation at 700 g, 18 °C for 30 minutes, the blood components were separated into 5 layers. The uppermost plasma/platelet layer was removed and the second layer (lymphocytes) collected. The cells were then washed twice in PBS (1400 g /10 °C/15 mins) prior to extraction of DNA.

3.3.3. Extraction of DNA from Peripheral Blood Lymphocytes (PBL). The method for extraction of DNA from PBL is outlined in figure 1. Firstly, cells were lysed by suspension in 1.75 ml of ice cold buffer 1 (5 mM trisodium citrate, 20 mM NaCl, pH 6.5) followed by 2ml of buffer 2 (20 mM Trizma base, 20 mM EDTA, 1.5% w/v sarkosyl, pH 8.5). After vigorous mixing the samples were incubated with 250 μg of RNase A in RNase Buffer (10 mM Tris-HCl, 10 mM EDTA, 10 mM NaCl, pH 6.0) for 1 hour at 37 °C. N.B. the RNase solution (i.e. RNase A + buffer) was heated in a boiling water bath for 15 minutes to destroy any contaminating DNase activity, then allowed to cool at room temperature prior to addition to cells. Pronase E (4mg/ml in buffer 1) was subsequently added and the digestion mixture left overnight at 37 °C. The following day 2 ml of Buffer 3 (10 mM Tris-HCl, 10 mM EDTA, 10 mM NaCl, pH 6.0) and 0.5 ml 7.5 M ammonium acetate were added and samples mixed by inversion. DNA was precipitated with 18 ml ethanol and the resultant strands spooled into 70% ethanol solution. After 30 minutes the DNA was rinsed using 100% and 70% ethanol solutions for 10 minutes. Finally, DNA samples were evaporated to dryness using nitrogen gas and dissolved in 2 ml water overnight prior to assessment for nucleic acid content.

3.3.4. Measurement of DNA and RNA Content. DNA concentration was calculated using UV absorbance at 260 nm. Strictly, this absorbance measurement corresponds to total nucleic acid concentration (i.e. [DNA] + [RNA]). It was considered important, therefore, to assess batches of DNA solutions for RNA content. The method used for this purpose is based on the compound orcinol which, in the presence of ferric chloride, forms a coloured complex with the deoxyribose moiety of RNA. The complex has a maximum UV absorbance at 665 nm. In all batches of DNA samples examined by this method the RNA content was found to be <5%.

The method was as follows: 100 ml of each DNA sample (1 mg/ml) was added to 2 ml of orcinol reagent (0.3 g in 100 ml concentrated HCl to which 4 drops of a 10% solution of ferric chloride are added) and heated for 10 minutes in a boiling water bath. The samples were then cooled and the UV absorbance measured at 665 nm. Standards for a calibration curve were prepared in the range of 10 - 200 mg/ml.

3.4. Measurement of Oxidative DNA Damage

Two assays have been used to measure oxidative damage to DNA: the first involves the use of the technique gas chromatography-mass spectrometry (GC-MS) to measure the base lesions 8-oxoguanine (8-oxoG) and 8-oxoadenine (8-oxoA). The other requires high performance liquid chromatography with highly sensitive electrochemical detection

(HPLC-EC) to measure 8-oxoguanine. In order to measure levels of oxidative damage within DNA, it is necessary to break the DNA down into its constituents. This was achieved by the use of a chemical hydrolysis step.

3.4.1. Measurement of 8-Oxoguanine and 8-Oxoadenine by Gas Chromatography-Mass Spectrometry(GC-MS).

3.4.1.1. Materials. N,O- bis(trimethylsilyl)trifluoroacetamide (BSTFA) containing 1% trimethylchlorosilane (TMCS), silylation grade acetonitrile and dimethyldichlorosilane were purchased from Pierce and Warriner (Chester, Cheshire, UK). Formic acid (99%), 2-Amino-6,8-dihydroxypurine (8-oxoguanine), 2,6-diaminopurine (DAP) and trifluoroacetic acid (TFA) were from Aldrich. HPLC grade methanol, sulfuric acid and toluene were obtained from Fisher Scientific Ltd. 8-oxoA was synthesised in the laboratory by treatment of 8-bromoadenine (Sigma) with concentrated formic acid (99%) at 150 ºC for 45 minutes and purified by crystallisation from water. All other chemicals were obtained as described earlier.

3.4.1.2. GC-MS Equipment. The equipment consisted of a Perkin-Elmer Autosystem GC interfaced with a Perkin-Elmer Q-mass 910 quadrupole mass spectrometer. The GC-MS conditions were as follows: The injection port and GC-MS transfer line were kept at 250 ºC and 280 ºC respectively. Separations were carried out on a fused silica capillary column (15 m in length, 0.25 mm internal diameter) coated with cross-linked 5% phenylmethylsiloxane (film thickness 0.25 μm) [Rtx-5, Restek Corporation, purchased from Thames Chromatography, Maidenhead, Berks., UK]. Helium (CP grade) was the carrier gas with a constant flow rate of 1 ml/min. The column temperature was increased from 120 ºC to 250 ºC at 10 º/min after an initial 2 minutes at 120 ºC, subsequently, the temperature was raised to 290 ºC at 30 º/min and then kept at 290 ºC for 2 minutes.

3.4.1.3. Generation of Stock Solutions of Analytes and Internal Standards. Fresh stocks of 8-oxoguanine were made up before each analysis. 8-oxoguanine (100 μM) was dissolved in nitrogen-purged 10 mM NaOH and concentration determined by UV spectrophotometry (λ_{max} = 283 nm, ε = 11600 $M^{-1}cm^{-1}$). 8-oxoadenine was dissolved in water (40 μM) and, after determining its concentration (λ_{max} = 270 nm, ε = 12764 $M^{-1}cm^{-1}$) (Cavalieri and Bendich, 1950), 1 ml aliquots were frozen and stored at -80 ºC. Similarly, 1 ml aliquots from stocks (100 μM) of internal standard 2,6-diaminopurine (Spencer *et al.*, 1996) were frozen and stored at -80 ºC.

3.4.1.4. Formic Acid Hydrolysis of DNA and Derivatisation of Bases. Acid hydrolysis using formic acid cleaves the glycosidic bonds between the deoxyribose moieties and the bases in DNA, releasing both modified and unmodified bases. To each sample of approximately 80 μg of DNA was added 50 μl of 10 μM DAP (≡ 500 pmoles), 150 μl of H_2O and 350 μl of 99 % formic acid (giving 550 μl of approx. 60 % formic acid). Samples were heated at 140 °C for 45 minutes in evacuated and sealed hydrolysis tubes. N.B. All vacuum hydrolysis tubes were pre-silanised using 5% dimethylchlorosilane in toluene. Samples were allowed to cool for 1–2 hrs at room temperature before being transferred into presilanised tubes (CV2, Chromacol Ltd., Welwyn Garden City, Herts., UK). Following approximately 30 minutes at -80 °C the hydrolysates were lyophilised overnight in a freeze-drier (Lyoprep-3000, IEC, Dunstable Beds. UK). Standards for the generation of calibration curves were also treated with 60% formic acid in the same manner.

It has been shown recently that derivatisation at elevated temperatures in the presence of oxygen can lead to artifactual production of oxidised DNA bases (Hamberg and Zhang, 1995; Ravanat *et al.*, 1995; Douki *et al.*, 1996). To minimise the possibility of artifactual formation of 8-oxopurines a derivatisation method using low temperature was adopted (Hamberg and Zhang, 1995). 10 μl of trifluoroacetic acid (TFA) was added to lyophilised mixtures of 8-oxoG, 8-oxoA, DAP on ice. After 5 minutes, 100 μl of a mixture of BSTFA/acetonitrile (4:1 v/v) was added and samples allowed to warm to room temperature under a nitrogen atmosphere. 1 μl of each sample was then injected into the GC injection port using a splitless mode and calibration curves were produced over a range 50 fmol to 1 pmol (on GC column). It should be noted that all analytes and internal standards were treated using the same formic acid conditions as required to hydrolyse DNA samples.

3.4.1.5. Analysis of DNA. Analysis of both standards and DNA samples was carried out with GC-MS in selected ion monitoring (SIM) mode for increased sensitivity. The following ions were measured for each compound: 440.2 for 8-oxoG, 352.1 for 8-oxoA, 351.1 for DAP). Duplicate injections were carried out for the analysis of both 8-oxopurines in each sample. 1 μl of each derivatised sample of DNA hydrolysate was auto-injected with the GC injection port in splitless mode.

3.4.2. The Measurement of 8-Oxoguanine by High Performance Liquid Chromatography with Electrochemical Detection (HPLC-EC). Under the HPLC conditions used for the measurement of 8-oxo-2'-deoxyguanosine the chromatographic peak for 8-oxoguanine is found to co-elute with that for guanine. In order to solve this problem DNA hydrolysates are treated with the enzyme guanase (guanine deaminase; guanine aminohydrolase; EC 3.5.4.3; Sigma), which converts guanine base to xanthine thereby shifting the chromatographic peak. 8-oxoguanine is not found to be a substrate for this enzyme (Herbert *et al.*, 1996).

3.4.2.1. Chemical Hydrolysis of DNA. The hydrolysis procedure for the analysis of 8-oxoG in DNA by HPLC-ECD was essentially the same as that used by the GC-MS assay To 20 μg of DNA was added 80 μl of 200 μM 8-oxo-2,6-diaminopurine (8-oxoDAP; Sigma) (internal standard), 120 μl of H_20 and 350 μl of 99% formic acid.

3.4.2.2. Treatment with Guanase (Conversion of Guanine to Xanthine). DNA hydrolysates were reconstituted in ultra-pure water (500 μl) and separated into portions; 300 μl for treatment with guanase and 60 μl for guanine measurement. A stock solution of guanase (1 mg/ml protein, 0.1 U/ml) was prepared from the original guanase suspension supplied by Sigma. 15 μl of this stock was added to the 300 μl portion of DNA hydrolysate (0.2 mU protein/mg DNA) and the pH adjusted with approx. 2 ml sodium hydroxide solution (100 mM) to 8.0 - 8.5 for optimal activity of the enzyme. The untreated 60 μl portion of hydrolysate was diluted further with 240 μl H_2O. Both guanase-treated and -untreated samples were placed in an air incubator at 37 ºC for 1 hour prior to analysis by HPLC.

3.4.2.3. Analysis of DNA. DNA bases were separated by reversed-phase HPLC using a 3μm ODS-Hypersil column (150mm in length, 0.46 mm internal diameter) and elution was achieved isocratically at a flow rate of 1 ml/min with a mobile phase consisting of 40 mM di-potassium hydrogen phosphate, 1 mM EDTA and 1% methanol (v/v) (pH 5.1 using 6 M HCl). Duplicate analyses were performed with, in each case, 50 μl of sample injected into 100 μl loop. DNA bases and xanthine (Sigma) were detected by UV absorbance at 254 nm. 8-oxoguanine was detected electrochemically using a glassy carbon working electrode at an applied potential of +600 mV versus a Ag/AgCl reference electrode. Calibration curves for 8-oxoG were generated over a range 0 -5 pmol (on column) using 8-oxoDAP as the internal standard. 8-oxoDAP has been shown previously to be ideal as an internal standard for this procedure (Ravanat *et al.*, 1995).

3.5. Measurement of Plasma Ascorbic Acid by HPLC

The method is based upon that of Lunec and Blake (1985). Plasma was obtained from fasting blood samples (12 hours overnight), collected in EDTA tubes, by centrifugation at 1400 g for 15 minutes at 10 °C. The plasma was immediately stabilised by vigorous mixing with an equal volume of cold 10% (w/v) metaphosphoric acid (MPA), followed by centrifugation at 1500 g for 15 minutes at 10 °C. The supernatant was removed and samples stored at -80 °C for several days pending analysis.

MPA-precipitated plasma samples were thawed rapidly and EDTA solution was added to a final concentration of 1 mM, in order to prevent oxidation. An equal volume of 0.5 M citric acid/ sodium citrate (both from Sigma) buffer, pH 3.6 was added, and 20 ml of each sample injected onto a Lichrosorb NH_2 column (Merck, Ltd.). Elution was achieved at using acetonitrile, 5 mM citric acid/ sodium citrate buffer, pH 3.2, glacial acetic acid (Fisher Scientific) at a ratio of 85: 15: 0.1 (v/v) respectively, delivered at 1 ml/min. Ascorbic acid was detected by UV absorbance at 254 nm. Quantitation of plasma ascorbic acid was achieved using a standard curve in the range of 0 - 100 μM ascorbic acid.

3.6. Measurement of Plasma Vitamin E by HPLC

Plasma concentrations of vitamin E in all samples were measured using an HPLC method based upon that of Bieri (1979). Plasma was obtained from fasting blood samples, collected in lithium heparin tubes, by centrifugation at 1400 g for 15 mins at 10 ºC. The intra- and inter-batch precision of this procedure was 5% and 6% respectively.

3.7. Statistical Methods

Data were analysed by General Linear Model Analysis of Variance (ANOVA) with subsequent comparison between means using either Tukey One-way ANOVA, or Fishers Least Significant Difference test where there was a significant subject effect. Statistics were performed using Minitab software.

4. VITAMIN E SUPPLEMENTATION STUDY

A placebo controlled study of oxidative DNA damage in healthy human volunteers was carried out involving supplementation with 400IU/day RRR α-tocopherol. The variables investigated were: plasma concentrations of vitamins C and E, blood leucocyte counts and levels of purine base damage products in peripheral blood lymphocyte DNA. The purines were measured using GC-MS (8-oxoG and 8-oxoA) and HPLC (8-oxoG).

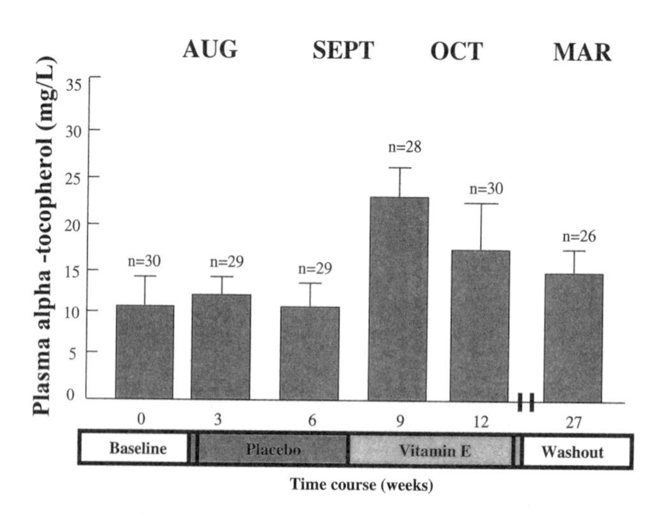

Figure 2. Changes in the plasma concentration of tocopherol in normal human volunteers undertaking supplementation of α tocopherol (400 i.u./day).

4.1. Effect of Vitamin E Supplementation on Plasma α-Tocopherol Concentrations in Healthy Volunteers

Supplementation of diets with 400IU/day α-tocopherol resulted in a highly significant increase ($p<0.01$ by ANOVA) in plasma vitamin E levels compared with the baseline and placebo values (Figure 2); the concentration increased from an initial mean of 11.7 (SD=3.3) mg/l for the baseline value to 24.1 (SD=4.9) mg/l and 18.1 (SD=5.8) mg/l for three and six weeks of supplementation respectively. There was a smaller but statistically significant ($p<0.05$) placebo effect; the concentration of vitamin E increased from baseline to the first placebo value - 13.3 (SD=2.4) mg/l. However there was no significant difference between baseline and second placebo (week 6) values. The levels of vitamin E had returned to the pre-trial (baseline) values by the washout (week 27; mean=16.2, SD=2.7mg/l).

4.2. Effect of Vitamin E Supplementation on Plasma Ascorbic Acid Concentrations in Healthy Volunteers

Supplementation of diets with 400IU/day vitamin E did not affect plasma ascorbic acid levels compared to both pre-supplementation and to placebo as determined by ANOVA (data not shown). The concentrations ranged from an initial mean of 54.2 (SD=17.0) μM to 54.4 (SD=16.0) μM following three weeks and to 49.6 (SD=15.7) μM after six weeks supplementation. Similarly there was no change in ascorbate level during oral placebo in comparison to the baseline value - the mean ascorbate concentrations were 54.6 (SD=14.4) μM and 53.7 (SD=15.0) μm after three and six weeks of the placebo. However, following a 15 week washout period the vitamin C levels (41.5; SD=16.5μM) were significantly reduced compared to the previous 'baseline' value ($p<0.05$) (data not shown).

4.3. Effect of Vitamin E Supplementation on Blood Leucocyte Counts in Healthy Volunteers

Dietary supplementation with 400IU/day vitamin E did not result in any significant changes in either circulating lymphocyte or neutrophil counts (Figure 3).

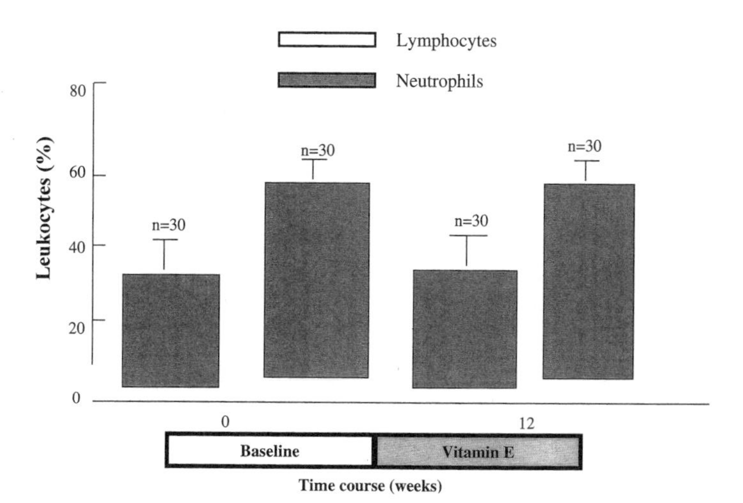

Figure 3. Changes in the relative proportion of lymphocytes and neutrophils in peripheral blood of human volunteers undertaking supplementation with α-tocopherol (400 i.u./day).

4.4. Effect of Vitamin E Supplementation on DNA Base Damage in Peripheral Blood Lymphocytes from Healthy Volunteers

4.4.1. 8-Oxoguanine. Supplementation of diets with 400IU/day vitamin E did not affect 8-oxoguanine levels, measured by GC-MS, in the DNA isolated from lymphocytes (Figure 4). The mean values were: 0.22 (SD=0.12) nmol 8-oxoG/mg DNA at week 0, or baseline, and 0.16 (SD=0.10) and 0.23 (SD=0.05) nmol/mg DNA at weeks 9 and 12 respectively. Similarly, no significant differences from baseline levels were observed during the placebo treatment; mean levels of 8-oxoG were similar to the baseline value (0.18, SD=0.05 nmol/mg DNA at week 3; and 0.23, SD=0.06 nmol/mg DNA at week 6). Follow-

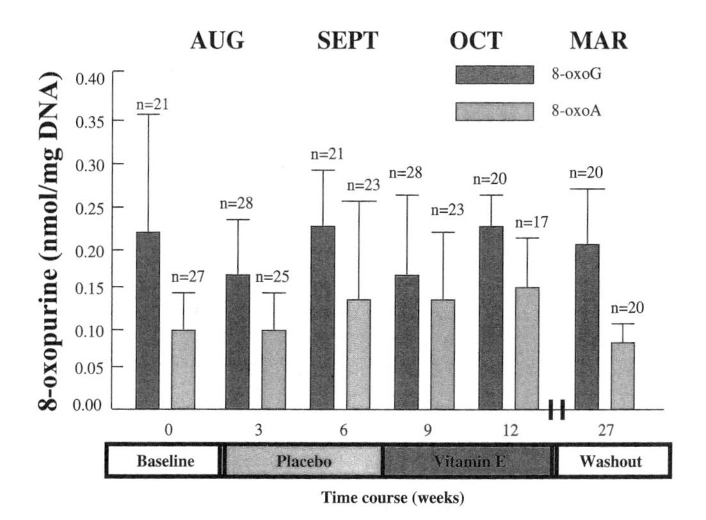

Figure 4. Changes in the concentration of 8-oxopurines, measured by GCMS-SIM, in lymphocyte DNA from normal human volunteers undertaking supplementation with α-tocopherol (400 i.u./day).

ing a 15 week 'washout' period, the mean level of 8-oxoG (0.20, SD=0.09 nmol/mg DNA) was also similar to values obtained during baseline/placebo.

When analyses were performed using HPLC-EC significant changes in 8-oxoG were observed subsequent to supplementation of diets with vitamin E (Figure 5). The mean values decreased significantly ($p<0.01$) from 0.21 (SD=0.12) nmol 8-oxoG/mg DNA at week 0, or baseline, to means of 0.13 (SD=0.06) and 0.12 (SD=0.06) nmol/mg at weeks 9 and 12 respectively. Although a significant *increase* in 8-oxoG was observed for the first placebo sample (0.29, SD=0.10 nmol/mg DNA at week 3), the second placebo value (0.26 SD=0.12 nmol/mg DNA at week 6) was not significantly different from the baseline value. Furthermore, compared with these values obtained during placebo, the decrease in 8-oxoG levels resulting from vitamin E supplementation was also statistically significant ($p<0.01$ by ANOVA). Following a 15 week 'washout' period, levels of 8-oxoG determined by HPLC (0.26, SD=0.08 nmol/mg DNA) had returned to those observed at 'baseline' and during the placebo period; that is, the only values which differed significantly ($p<0.01$) from the values obtained during the washout were those corresponding to the vitamin E supplementation phase.

4.4.2. 8-Oxoadenine. 8-oxoadenine (8-oxoA) levels were also determined by GC-MS analysis of DNA hydrolysates. Significant ($p<0.01$) increases in 8-oxoA levels in the DNA isolated from lymphocytes were observed both following placebo administration (second placebo sample only) and during part of the dietary supplementation with vitamin E (second vitamin E sample only) (Figure 4); this suggests a classical placebo effect. The mean values were: 0.09 (SD=0.05) nmol 8-oxoA/mg DNA at week 0, 0.14 (SD=0.07) and 0.15 (SD=0.06) nmol/mg at weeks 6 (placebo) and 12 respectively. No significant differences from the baseline (and subsequently from the washout) were observed for either the first placebo sample or the first vitamin E sample (Figure 4); for the placebo sample the mean level of 8-oxoA was similar to the baseline value (0.10, SD=0.04 nmol/mg DNA) whereas the vitamin E sample showed a high degree of variation - mean=0.14, SD=0.11. The same pattern of statistical significance was observed for the placebo and vitamin E phases when comparisons were made with the mean level of 8-oxoA obtained during the

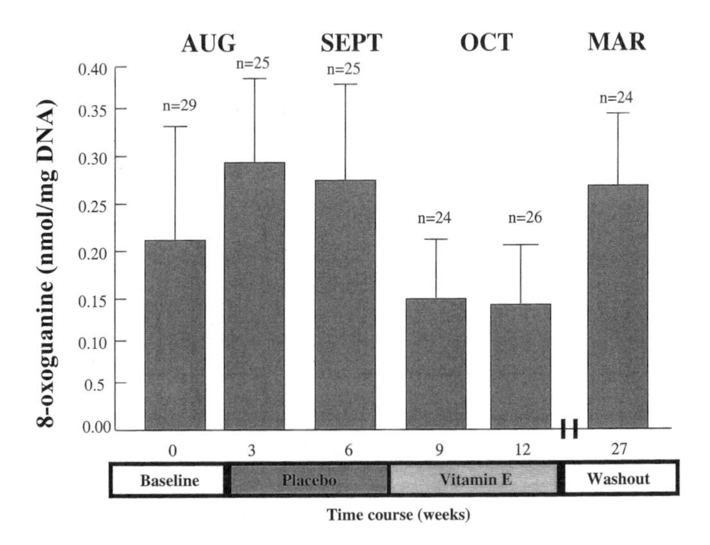

Figure 5. Changes in the concentration of 8-oxoguanine measured by HPLC-EC in lymphocyte DNA from normal human volunteers undertaking supplementation with α-tocopherol (400 i.u./day).

washout (mean=0.08, SD=0.02 nmol/mg DNA). This result also suggested that the levels of 8-oxoA had returned to baseline concentrations by the end of the washout period.

5. DISCUSSION

The data described relating to the changes in oxopurine following vitamin E supplements are less compelling than those following ascorbate (Podmore *et al.*, 1998). This is primarily due to a lack of agreement between the GC-MS and HPLC assays for 8-oxoG. Although the HPLC data strongly suggest that vitamin E decreases 8-oxoG levels (in a manner similar to that observed for vitamin C), the GC-MS data did not concur. Although these assays have similar limits of detection for this DNA lesion, the GC-MS assay consistently displays greater imprecision, particularly between batches and at levels of 8-oxoG close to the limits of detection of the method. It is therefore possible that in certain instances the GC-MS analysis may be presently incapable of detecting small changes in 8-oxoG, whereas the HPLC assay may be adequate (Inter-batch CVs for HPLC, 13.9% versus GC-MS 32.5%). This situation can be improved following the generation of isotopically-labelled 8-oxoG as an internal standard for GC-MS. The incorporation of such a standard will compensate for losses during the various manipulations, improving accuracy and reproducibility of the determination.

Elevations in 8-oxoA levels following vitamin E were accounted for by a similar change at the second placebo sample; that is a "placebo effect". Clearly improvements to the GC-MS analysis of 8-oxoA, primarily by using heavy isotope internal standards would also aid this determination. Nevertheless it would be highly beneficial if an alternative, perhaps HPLC, methodology could be employed in the future for this lesion. A recent study measuring genetic damage (micronucleus assay) showed no attenuation by vitamin E albeit at doses ten times less than that administered in the present study (Fenech *et al.*, 1997).

The data in this phase of the study suggest that vitamin E levels had been increased by the supplementation, and that this was fairly rapid, that is an increase was observed by three weeks. However even following 15 weeks of washout the vitamin E levels had not fallen to baseline values, although this may have been influenced by seasonal effects on vitamin E levels - the vitamin E supplements were given in autumn whereas the blood samples were obtained for the washout in mid-winter.

6. SUMMARY CONCLUSIONS

Although technically demanding and time consuming the assays were shown to be specific, sensitive and fairly robust through their use in a vitamin supplementation trial.

By measuring levels of 8-oxoguanine in a sample of commercially available calf thymus the assays were found to be comparable. However variations in between-batch analysis were found to be higher for GC-MS than HPLC-EC.

6.1. Vitamin E Supplementation Study

- Supplementation of diets with 400 IU/day vitamin E resulted in a highly significant increase ($p<0.01$ by least significant difference) in plasma α-tocopherol levels compared to both pre-supplementation and placebo.
- No systematic change in either circulating lymphocyte or neutrophil counts were found during the supplementation trial.

- Supplementation with 400 IU/day vitamin E resulted in a significant decrease ($p < 0.01$ by ANOVA) in 8-oxoG levels in DNA isolated from peripheral blood lymphocytes, compared to both pre-supplementation and placebo, but only when measured by the HPLC-EC assay.

REFERENCES

B.N. Ames (1989), Endogenous DNA damage as related to cancer and ageing. *Mut. Res.* **214,** 41–46.

J.G. Bieri (1979), Simultaneous determination of a-tocopherol and retinol in plasma or red cells by HPLC. *Am. J. Clin. Nutr.,* **32,** 2143 -2149.

L.F. Cavalieri and A. Bendich (1950), The ultraviolet absorption spectraof pyrimidines and purines. *J. Am Chem. Soc.* **72,** 2587–2594.

M. Dizdaroglu (1991), Chemical determination of free radical induced damage to DNA, *Free. Rad. Biol. Med* **10,** 225–242.

M. Dizdaroglu (1994), Chemical determination of oxidative DNA damage by gas chromatography-mass spectrometry, *Methods in Enzymology* **234**, 3 - 16.

T. Douki, T. Delatour, F. Bianchini and J. Cadet (1996), Observation and prevèntion of an artefactual formation of oxidised DNA bases and nucleosides in the GC-EIMS method, *Carcinogenesis,* **17**, 347 - 353.

M. Fenech, I. Dreosti and C. Aitken (1997), Vitamin-E supplements and their effect on vitamin-E status in blood and genetic damage rate in peripheral blood lymphocytes, *Carcinogenesis* **18, No.2.,** 359–364.

R.A. Floyd, J.J. Watson, P.K. Wong, D.H. Altmiller and R.C. Rickard (1986), Hydroxyl free radical adduct of deoxyguanosine: Sensitive detection and mechanisms of formation, *Free Rad. Res. Commun.* **1**, 163 - 172.

E. Gajewski, G. Rao, Z. Nackerdien and M. Dizdaroglu (1990), Modification of DNA bases in mammalian chromatin by radiation-generated free radicals, Biochemistry **29**, 7876 - 7882.

B. Halliwell and J.M.C. Gutteridge (1990), Role of free radicals and catalytic metal ions in human disease: An overview, *Methods in Enzymology* **186**, 1 - 85.

M. Hamberg and L.-Y. Zhang (1995), Quantitative determination of 8-hydroxyguanine and guanine by isotope dilution mass spectrometry, *Anal. Biochem.* **229**, 336 - 344.

K.E. Herbert, M.D. Evans, M.T. Finnegan, S. Farooq, N. Mistry, I.D. Podmore, P. Farmer and J. Lunec (1996), A novel HPLC procedure for the analysis of 8-oxoguanine in DNA, *Free Rad. Biol. Med.* **20**, 247 - 473.

H. Kamiya, H. Miura, N. Murata-Kamiya, H. Ishikawa, T. Sakaguchi, H. Inoue, T. Sasaki, C. Masutani, F. Hanaoka, S. Nishimura and E. Ohtsuka (1995), 8-Hydroxyguanine (7,8-dihydro-8-oxoadenine) induces misincorporation in *in vitro* DNA synthesis and mutations in HIH 3T3 cells, *Nucl. Acids Res.* **23**, 2893 - 2899.

H. Kasai and S. Nishimura (1991), Formation of 8-hydroxydeoxyguanosine in DNA by oxygen radicals and its biological significance, *In: Sies. H., ed. Oxidative Stress: Oxidants and antioxidants. London Academic Press* 99–116.

J. Lunec (1990), Free radicals, their involvement in the disease process, *Ann. Clin. Biochem.* **22,** 173–182.

J. Lunec and D.R. Blake (1985), The determination of dehydroascorbic acid and ascorbic acid in the serum and synovial fluid of patients with rheumatoid arthritis (RA), *Free Rad. Res. Comms.* **1**, 31 - 39.

J. Lunec, K. Herbert, S. Blount, H. Griffiths and P. Emery (1994), 8-Hydroxyguanosine: A marker of oxidative DNA damage in systemic lupus erythematosus, *FEBS Letters* **348**, 131–138.

Z. Nackerdien, R. Olinski and M. Dizdaroglu (1992), DNA base damage in chromatin of gamma irradiated cultured human cells, *Free Rad. Res. Comms.* **16**, 259 - 273.

M. Nagashima, H. Tsuda, S. Takenoshita, Y. Nagamachi, S. Hirohashi, J.Yokota and H. Kasai (1995), 8-hydroxyguanosine levels in DNA of human breast cancers are not significantly different from those of non-cancerous breast tissues by the HPLC-ECD method, *Cancer Lett.* **90**, 157 - 262.

I.D. Podmore, H.R. Griffiths, K.E. Herbert, N. Mistry, P. Mistry and J. Lunec (1998), *Nature* **392**, *559.*

J.-L. Ravanat, R.J. Turesky, E. Gremaud, L.J. Trudel and R.H. Stadler (1995), Determination of 8-oxoguanine in DNA by gas chromatography-mass spectrometry and HPLC-electrochemical detection: Overestimation of the background level of the oxidised base by the gas chromatography-mass spectrometry assay, *Chem. Res. Toxicol.* **8**, 1039 - 1045.

M.K. Shigenaga, E.N. Aboujaoude, Q. Chen and B.N. Ames (1994), Assays of oxidative DNA biomarkers 8-oxo-2'-deoxyguanosine and 8-oxoguanine in nuclear DNA and biological fluids by high performance liquid chromatography with electrochemical detection, *Methods Enzymol.* **234**, 16 - 33.

J.P.E. Spencer, A. Jenner, O.I. Aruoma, C.E. Cross, R. Wu and B. Halliwell (1996), Oxidative DNA damage in human respiratory tract epithelial cells. Time course in relation to DNA strand breakage, *Biochem. Biophys. Res. Commun.* **224**, 17 - 22.

24

REPAIR OF OXIDATIVE DNA DAMAGE AND AGING

Central Role of AP-Endonuclease

Sankar Mitra, Tadahide Izumi, Istvan Boldogh, Chilakamarti V. Ramana,*
Ching-Chyuan Hsieh, Hiroshi Saito, Julie Lock, and John Papaconstantinou

Sealy Center for Molecular Science and Department of Human Biological Chemistry and Genetics
University of Texas Medical Branch
Galveston, Texas 77555

1. ABSTRACT

Reactive oxygen species (ROS) generate many types of damage in cellular genomes, including base lesions, base loss, and nonligatable DNA strand breaks. These lesions are repaired predominantly via the base excision repair (BER) pathway, which is initiated by the removal of damaged bases from DNA by specific DNA glycosylases. There is strong evidence for the presence of two BER pathways, named BER I and BER II, in mammalian cells. AP-endonuclease (APE) functions as an endonuclease in BER I, which is utilized when abasic (AP) sites are generated by simple DNA glycosylases. APE functions as a DNA 3' phosphoesterase/exonuclease in the BER II pathway which is used when 3' blocked ends are generated in DNA either by ROS or during removal of oxidized base lesions by complex DNA glycosylase/AP lyases. In addition to its central role in the BER pathways, the major human APE (hAPE-1) possesses other unrelated activities as an activator of several transcription factors, and as a Ca^{2+}-dependent repressor of several genes including its own. APE-1 is activated transiently in human and rodent cells by a variety of sublethal levels of ROS, but not by other genotoxic agents. Similar activation was observed in the livers of mice treated with bacterial lipopolysaccharide, an ROS inducer. This effect was less pronounced in the aged (24 mo-old) than in the young (4 mo-old) ani-

---

* Present address: Cleveland Clinic Research Institute, Cleveland, Ohio 44195

Advances in DNA Damage and Repair, edited by Dizdaroglu and Karakaya. Kluwer Academic / Plenum Publishers, New York, 1999.

mals. These results indicate complex regulation of the APE level during age-dependent stress response of mammalian cells.

2. INTRODUCTION

Reactive oxygen species play a profound role in cellular physiology and homeostasis because these insidious agents are generated continuously in all living organisms. These agents, including partially reduced forms of oxygen, i.e. the superoxide anion O_2, H_2O_2, and the hydroxyl radical OH•, are generated endogenously as byproducts of respiration, and also during the inflammatory response which includes synthesis of reactive oxygen radicals and HOCl (Grisham and McCord, 1986; Gotz *et al.*, 1994; Parkins *et al.*, 1995; Ward, 1994; Tyrrell and Keyse, 1990). Furthermore, such exogenous agents as ionizing radiation, heavy metals and cytokines also induce formation of these ROS (Suzuki *et al.*, 1997).

2.1. Base Excision Repair of ROS-Induced DNA Damage

ROS may react with all cellular components. However, their reaction with DNA is genotoxic. It is interesting that, unlike other genotoxic agents, ROS induce a plethora of lesions in DNA (Dizdaroglu, 1992; Breen and Murphy, 1995). These include a multitude of oxidized bases, abasic (AP) sites due to base glycosylic bond cleavage, and DNA strand breaks caused by fragmentation and oxidation of deoxyribose in the DNA backbone so that 3' phosphoglycolate (glycolaldehyde) termini are generated. While the relative abundance of any particular lesion depends on the type of ROS and the presence of modifying factors, all of the lesions, including DNA single-strand breaks, are repaired primarily via the base excision repair (BER) pathway. BER is distinct from other excision repair systems, i.e., the nucleotide excision repair and mismatch repair pathways, both in regard to the component enzymes and the size of the DNA repair patch that replaces the lesion-containing region (Friedberg *et al.*, 1995). In the basic mechanism of BER (BER I in Fig. 1), the base lesion is removed by a specific DNA glycosylase, leaving an AP site in the DNA. AP-endonuclease, (APE; originally called class II enzymes), then cleaves the DNA strand 5' to the AP site and generates 3'-OH and 5' phosphodeoxyribose termini.

Repair synthesis is subsequently initiated at the 3' end; removal of the 5' blocking group with a 5' exonuclease or a specific deoxyribose phosphate hydrolase (lyase; dRPase) allows the refilling of the resulting single nucleotide gap (Fig. 1). Recently, *in vitro* repair systems with extracts of mammalian cells and tissues have been developed in the laboratories of Wilson and Lindahl (Dianov and Lindahl, 1994; Singhal *et al.*, 1995; Kubota *et al.*, 1996). In both cases, analysis of repair of uracil (U)-containing DNA (oligonucleotide) confirmed the pathway described above and outlined in Fig. 1 (BER I). A single nucleotide repair patch was observed after removal of U by uracil-DNA glycosylase (UDG) and cleavage of the resulting AP site with the major APE (Singhal et al., 1994). The DNA polymerase β (β-pol) is responsible for repair synthesis and was also shown to possess dRPase (AP lyase) activity via a βδ-elimination reaction on the AP site (Matsumoto and Kim, 1995). Because β-pol makes contact with the upstream 5' phosphate terminus incorporating a single nucleotide at the 3' terminus (Dianov *et al.*, 1992; Singhal *et al.*, 1995), the overall reaction can be explained by the following simple scenario. Once the U residue in DNA is cleaved by UDG, APE, complexed with β-pol, cleaves the AP site. There is recent evidence for *in vivo* interaction between human β-pol and human APE

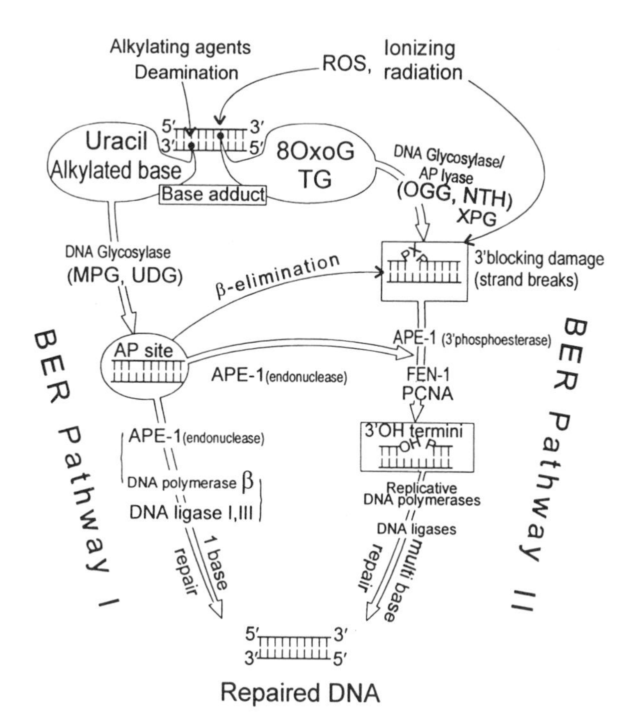

Figure 1. A schematic diagram describing the model of two DNA BER pathways in mammalian cells (Frosina *et al.*, 1996; Klungland and Lindahl, 1997; Mitra *et al.*, 1997). BER I is specific for uracil and alkylated bases (Dianov *et al.*, 1992; Piersen *et al.*, 1996; Singhal *et al.*, 1994) and BER II for oxidized bases and DNA strand breaks (Mitra *et al.*, 1997). APE-1 functions as an endonuclease in BER I and as a 3'-DNA phosphoesterase in the BER II pathway. Interaction between APE-1 and β-pol and between β-pol and DNA ligases I/III are indicated by brackets (Kubota *et al.*, 1996; Prasad *et al.*, 1996). FEN-1 and PCNA are required for efficient repair in BER II (Klungland and Lindahl, 1997; Kelman, 1997). The replicative DNA polymerases (Satoh *et al.*, 1993) include α, β, ε enzymes. The site of XPG action in BER II is not known (Cooper *et al.*, 1997).

hereafter referred to as hAPE-1(Bennett *et al.*, 1997). The polymerase then removes the deoxyribose phosphate residue in a concerted reaction via the dRPase activity located in its 8 kDa domain (Piersen *et al.*, 1996), and thus makes contact with the exposed 5' phosphate. The filling of the single nucleotide then occurs efficiently; DNA ligase I, which was also shown to complex with β-pol *in vitro* (Prasad *et al.*, 1996), then completes the repair process by sealing the nick. We should note that DNA ligase III, shown to interact with XRCC-1, has also been implicated in the sealing of the nick (Kubota *et al.*, 1996). Thus this repair pathway is simple, and requires only a few proteins. Although there is significant evidence that other accessory proteins, e.g., poly(ADP-ribose) polymerase (PARP), are also involved in this repair process (Satoh *et al.*, 1993), PARP only stimulates the repair reaction several fold and was found not to be essential for the repair process (Satoh *et al.*, 1994).

2.2. Two Pathways of BER

Repair of U has been commonly used as the paradigm for BER. However, several independent studies suggested that the BER process may be more complex (Klungland

and Lindahl, 1997). Repair of an oligonucleotide containing a reduced AP site (tetrahydrofuran, an AP analog) with Xenopus oocyte extract suggested that repair of the lesion involves proliferating cell nuclear antigen (PCNA), which is associated only with replicative DNA polymerases such as δ and ε, but not β-pol (Matsumoto *et al.*, 1994; Kelman, 1997). The involvement of the replicative DNA polymerases in BER of ionizing radiation-induced DNA strand breaks, rather than β-pol, was supported by studies with specific inhibitors (Satoh and Lindahl, 1993). Even experiments on U repair, independently performed by Frosina *et al.* (1996), indicated a repair patch containing multiple nucleotides. More recently, Klungland and Lindahl (1997) confirmed this observation by showing that repair of U can occur via two BER pathways.

The evidence for the presence of a second BER pathway was provided independently by studies on β-pol-knockout mutant cells. Although inactivation of the β-pol gene in the mouse causes embryonic lethality of the homozygous mutant animals, fibroblasts derived from the embryo grow normally, even though they lack β-pol completely (Sobol *et al.*, 1996). More interestingly, these cells are sensitive to alkylating agents; however, ROS and IR show no detectable effect on their viability (Sobol *et al.*, 1996). It is important to note in this context that while the BER pathway is normally involved in the repair of alkylated and oxidized base lesions and uracil in DNA, initiated by their removal by DNA glycosylases (Friedberg *et al.*, 1995), there is a fundamental difference in the activity of those glycosylases that are involved in the removal of oxidized base lesions vs. other DNA lesions, including U (Mol *et al.*, 1995; Roy *et al.*, 1995 and 1996). Uracil and alkylated bases are removed from DNA by simple DNA glycosylases, which appear to be initiate the cleavage reaction by nucleophilic activation of water which attacks the deoxyribose C'1 and thus labilizes the glycosylic bond. In contrast, oxidized bases are removed by the multifunctional DNA glycosylase/AP lyases. These enzymes cleave the *N*-glycosylic bond, and thus remove the base by attacking the C'1 residue of deoxyribose with an NH_2 group after its nucleophilic activation by an Asp (Dodson *et al.*, 1994; Nash *et al.*, 1996). The fundamental distinction between the two types of DNA glycosylases is the formation of a transient, Schiff base covalent intermediate with AP lyases. Such an intermediate carries out β or βδ elimination resulting in cleavage of the downstream phosphodiester bond with partial or complete removal of the sugar. Thus this AP lyase activity leads to a 3' phospho α,β-unsaturated aldehyde or a 3'-phosphate due to β or βδ elimination at the site of strand cleavage, respectively. The 5' terminus in all cases contains the phosphate residue. It is interesting that although ROS induces many types of base damage, only two DNA glycosylase/AP lyases, namely endonuclease III (NTH) and 8-oxoguanine-DNA glycosylase (OGG), with very broad substrate range, have been identified so far as repairing these lesions in mammalian cells.

The unique feature of the DNA glycosylase/AP lyase reaction is the formation of 3' blocked ends at strand breaks, which are, therefore, not available as primers for DNA polymerases during repair synthesis. As already mentioned, similar 3' blocked termini are also generated during ROS-induced DNA strand breakage.

2.3. Role of AP-Endonucleases in BER

AP-endonucleases possess a second activity, a rather non-specific phosphoesterase, that removes the 3' blocking ends in DNA. This activity was first discovered in *E. coli*, and was in fact identified and characterized as an exonuclease (exonuclease III or Xth) with DNA 3' phosphatase activity (Richardson *et al.*, 1963). This enzyme was later shown to be the major APE in *E. coli* (White *et al.*, 1976). *E. coli* has a second APE, named endonuclease IV (Nfo), also with 3' phosphoesterase like the Xth protein, but without the exonuclease activity (Doetsch and Cunningham, 1990). Both human and mouse APE (variously named as APEX, HAP1 as well as Ref-1) cDNAs have been cloned and recombinant proteins purified from *E. coli*. This major APE is a homolog of the Xth protein, and they share significant sequence identity (Demple *et al*., 1991; Seki *et al*., 1992). However, the mammalian endonucleases have weak 3' phosphoesterase and 3' exonuclease activities relative to the Xth protein (Seki *et al*., 1991; Suh *et al*., 1997). Thus an interesting dichotomy appears to exist in the dual activities of APE; because of its low level, the 3' end-removing activity may be rate-limiting for repair of oxidative lesions in DNA in mammalian cells but not in *E. coli*. Indeed, we have recently shown that APE-1 is rate-limiting during *in vitro* repair of both ROS- and 3' AP lyase-induced DNA strand breaks in repair assays *in vitro* using whole cell extracts of HeLa cells (Izumi *et al*., manuscript in preparation). Our results also appear to preclude the possibility that a second 3' DNA phosphoesterase activity identified earlier (Chen *et al*., 1991), contributes significantly to DNA strand break repair.

The various observations described so far have led us to propose the hypothesis that mammalian cells possess at least two distinct pathways of base excision repair, which we named BERI and BERII (Fig. 1; Mitra *et al*., 1997). BER II is generic, may be responsible for all types of repair of DNA damage, and utilizes replicative DNA polymerases and PCNA (Matsumoto *et al*., 1994; Frosina *et al*., 1996; Klungland and Lindahl, 1997). In contrast, the specialized BER pathway (BER I) is involved in the repair only of U and alkylated bases that are removed by monofunctional DNA glycosylases. This pathway utilizes β-pol. While the evolutionary significance of the presence of AP lyase activity in NTH and OGG is not clear, the functional significance is evident. Although the DNA glycosylase and AP lyase activities of NTH and OGG can be examined separately, the two activities act always in concert, at least during *in vitro* reaction. Thus NTH and OGG, acting on lesion-containing DNA oligonucleotides, cleave the DNA strands rather than leaving an abasic site as the reaction product throughout the course of reaction. While the key distinct feature of the two BER pathways is the presence of the blocking group at the 3' vs. 5' termini, it is tempting to speculate that APE is present *in vivo* as a complex, either with β-pol (Bennett *et al*., 1997) or with other proteins. The free AP site, after reaction with UDG, or *N*-methylpurine-DNA glycosylase (MPG) which is responsible for removal of *N*-alkylpurines from DNA, provides a target for successive and efficient action of APE and β-pol, and the conformation of APE is optimal for the endonuclease action. Interestingly, our preliminary observation that *in vitro* repair of ROS-induced strand breaks in plasmid DNA is inhibited by an excess amount of β-pol supports this notion (Izumi *et al*., manuscript in preparation). Although there is no evidence for distinct active sites for the endonuclease and 3' phosphoesterase activities in the APE polypeptide, it is possible that APE can assume a slightly altered structure as a 3' phosphoesterase for removing the 3' blocked ends generated by AP lyase, and which is inhibited when the protein is complexed with β-pol.

The rate-limiting activity of APE as a DNA 3'-phosphoesterase would predict that a change in the intracellular concentration of APE will have a profound impact on the repair of nonligatable 3' strand breaks in DNA. By the same token, the strong endonuclease activity of mammalian APE would suggest a lack of persistent AP sites in DNA. In contrast, it seems more likely that strand breaks with 3' blocked ends will persist in DNA in a cumulative fashion in cells of organisms as a function of age or chronic oxidative stress. There is some evidence in support of an age-dependent increase in the frequency of single-strand breaks in mammalian tissues (King *et al*., 1997; Rao & Loeb, 1992).

3. RESULTS AND DISCUSSION

Because APE appears to be rate-limiting in the repair of ROS-induced DNA damage, particularly single-strand breaks, regulation of this enzyme has a profound impact on the cellular responses to ROS.

3.1. The Major Human APE (hAPE-1) Is a Multifunctional Protein

While the predominant amount of AP-endonuclease activity *in vivo* in mammalian cells is contributed by one enzyme, namely APE-1, discovered more than fifteen years ago (Kane and Linn, 1981), its subsequent cloning showed its homology to *E. coli* Xth (Demple *et al.*, 1991; Seki *et al.*, 1992). This 36 kDa human enzyme turned out to be a remarkable, multifunctional protein. Although several groups independently cloned the cDNA of this enzyme based on its endonuclease activity, it was also independently cloned and named Ref-1 on the basis of its two unrelated activities. During *in vitro* studies of oxidative inactivation of several transcription factors, Curran and his collaborators observed that some of these factors, particularly c-Jun, can be reactivated by a second cellular protein, which they named Ref-1 (Xanthoudakis *et al.*, 1992). On the basis of this activity, the protein was purified to homogeneity, and its *N*-terminal sequence determined, leading to its cloning and identification as APE-1. Reductive activation of c-Jun by reduced hAPE-1 occurred via a thiol exchange reaction in which the Cys 272 of c-Jun in monomeric form is oxidized to SOH and is reduced back to the active form by the Cys 65 residue of APE-1. The resulting oxidized form of APE-1 is then reduced by thioredoxin. The rest of the thiol exchange cycle may include sequential reduction of oxidized thioredoxin by thioredoxin reductase (Abate *et al.*, 1990). While there is still some skepticism about the physiological relevance of the role of APE-1/Ref-1 as a reductive activator of transcription factors, observed only in *in vitro* reactions so far, several recent papers provide supporting evidence for the overall scheme. Reduction of oxidized hAPE-1 by thioredoxin requires specific interaction between these proteins. Recent NMR studies and experiments using 2-hybrid system provide evidence for thioredoxin interaction with hAPE-1 (Qin *et al.*, 1996; Hirota *et al.*, 1997). More recently, hAPE-1/Ref-1 has been shown to reductively activate p53 in a complex fashion (Jayaraman *et al.*, 1997). This protein is a critical tumor suppressor and transcriptional activator that contributes to the maintenance of genomic stability, thus dubbed as the "guardian of the genome" (Levine, 1997). It was shown that p53, activated by hAPE-1/Ref-1, migrates to the nucleus and thus activates transcription of p53-responsive genes (Jayaraman *et al.*, 1997). Another recent study showed that APE-1 deficiency in APE-1 heterozygous mutant mice promotes tumor growth in animals also deficient in p53 (Meira *et al.*, 1997). These results have been interpreted to confirm activation of p53 by APE-1 *in vivo* (Jayaraman *et al.*, 1997). That homozygous APE-1 knock-out mutant mouse embryos die 5.5 days after implantation is consistent with the lack of reductive activation of transcription factors (including C-Jun, p53 and others) being responsible for embryonic lethality.

3.2. APE-1/Ref-1 Is a Ca^{2+}-Dependent (Co)Repressor

A second and apparently unconnected function of hAPE-1 was discovered by Okazaki and his collaborators (Okazaki *et al.*, 1994). In their pursuit of identifying the *trans*-acting factor that recognizes the Ca^{2+}-dependent *cis*-element (nCaRE) in the parathyroid hormone gene which is subject to feedback negative regulation by Ca^{2+}, they cloned a corepressor of the gene and identified it as APE-1/Ref-1. This gene promoter has two types of nCaRE sequences, nCaRE-A and nCaRE-B, both of which bind to hAPE-1 (Okazaki *et al.*, 1994). However, a second protein, along with hAPE-1, is needed to bind these nCaRE sequences, presumably because the repressor functions as a heteroligomer. Later studies by Okazaki's group showed that the nCaRE-A sequence binds to a complex of hAPE-1 and Ku70 (or Ku86), an autoantigen that was later shown to be involved in the rejoining of DNA double strand breaks, which are intermediates in V(D)J recombination of immunoglobulin genes and may also be induced by ROS. Such heterodimer formation between APE-1 and Ku70 (Ku86) requires specific and stable interaction (Chung *et al.*, 1996). A recent report provided evidence for such interaction between hAPE-1 and Ku70 but not Ku86, using the yeast 2-hybrid assay system (B. Demple, personal communication).

Chung *et al.* showed that hAPE-1 and Ku70 binds only to nCaRE-A but not nCaRE-B (Chung *et al.*, 1996). Thus it is likely that a distinct protein interacts with the APE/Ref-1 in order to bind to the nCaRE-B sequence. Our preliminary studies using affinity chromatography of HeLa extract on an nCaRE-B oligonucleotide, followed by SDS/PAGE of the tightly bound fraction, showed the presence of two major bands. One of the bands was hAPE-1 itself as expected. Identification of the second, 60 kDa band, is in progress (D. Kuninger, T. Izumi, W. Washington, J. Papaconstantinou, and S. Mitra, unpublished experiment).

3.3. Multiple Partners of hAPE-1

The promiscuity of hAPE-1 in binding to different partners is indeed remarkable. Table 1 summarizes the interactions that have been experimentally demonstrated or predicted from indirect evidence. The role of APE-1 in DNA repair, reductive activation of unaltered transcription factors of diverse primary sequences, and in repression of nCaRE containing genes necessitates its recognition by distinct proteins in different fashions.

A GenBank data base search showed the presence of nCaRE-A and -B type sequences in more than 100 genes, in addition to the parathyroid hormone gene (McHaffie and Ralston, 1995). Whether these elements are biologically functional in suppressing the expression of these genes remains to be seen. However, it is interesting to note that some of these genes (including calmodulin, β-myosin, and erythropoietin) are involved in Ca^{2+}-dependent regulation or in Ca^{2+}-metabolism, far beyond the specific involvement in repair of oxidative DNA damage.

Table 1. Interacting partners of hAPE-1

Partner	Function	Evidence	Reference
DNA Polymerase β	DNA base excision repair I (repair of U)	*In vitro* interaction and yeast 2-hybrid studies	Bennet *et al.*, 1997
Thioredoxin	Redox signal transduction	NMR spectroscopy and yeast 2-hybrid studies	Qin *et al.*, 1996; Hirota *et al.*, 1997
Transcription factors	Reductive activation of transcription factors	*In vitro* interaction	Xanthoudakis *et al.*, 1992
Ku70	Binding to nCaRE-A	*In vitro* interaction and yeast 2-hybrid studies	Chung *et al.*, 1996; B. Demple (Personal commun.)
Protein X	Binding to nCaRE-B	Affinity chromatography	D. Kuninger *et al.*, (Unpublished experiment)

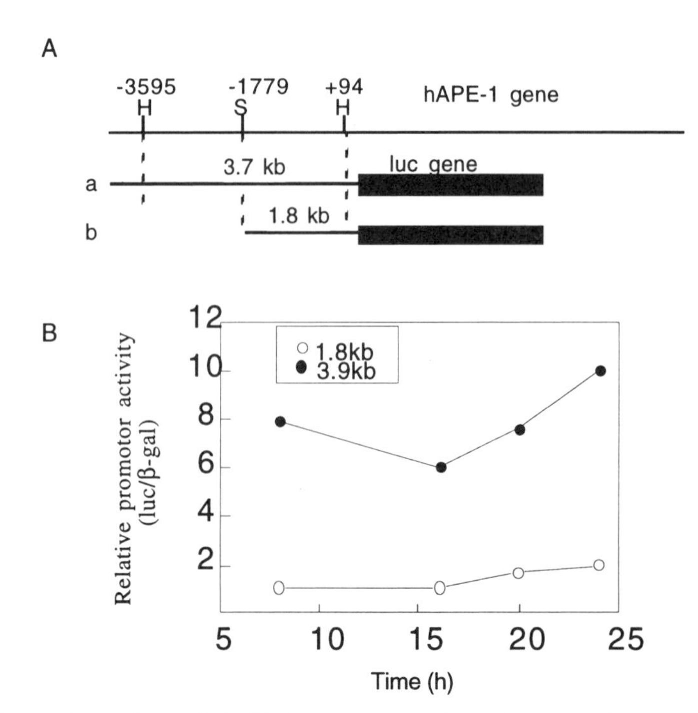

Figure 2. Negative regulatory element in hAPE-1 promoter. (A) Schematic diagram of the hAPE-1 gene (upper line). Luciferase (luc) coding sequence was fused to 3.9 kb (a) or 1.8 kb promoter fragment (b). The nucleotide numbers are relative to transcription start site (+1). H, *Hind* III; S, *Sma* I. (B) Transient luc expression from hAPE-1 promoter. The reporter plasmid a (o) or b (·), along with a β-galactosidase (β-gal) expression plasmid, was electroporated into TK6 lymphoblastoid cells, and luciferase activity normalized to β-gal expression was measured at various times after transfection.

3.4. Autoregulation of hAPE-1

During a detailed investigation of identifying and characterizing of the promoter region of the hAPE-1 gene, using reporter gene expression assay with extracts of transiently transfected human cells, we consistently noticed about an 8-fold lower level of expression of the luciferase reporter when the hAPE-1 promoter contained a 2 kb upstream segment (Fig. 2). We sequenced the 2 kb region and identified 3 putative nCaRE elements, including one nCaRE-A and two nCaRE-B types. We also showed that the 2 kb element functions as a negative enhancer because its repressor activity was independent of its orientation in the reporter plasmid construct (Izumi *et al.*, 1996). We then constructed a series of luciferase expression plasmids with sequential deletion of the 2 kb region and tested the level of luciferase expression in transiently transfected cells (Izumi *et al.*, 1996). Although the results were somewhat ambiguous, it was clear that the presence of one nCaRE-B (nCaRE-B2) sequence led to significant down-regulation of luciferase expression (Izumi *et al.*, 1996). Furthermore, this repression was eliminated by deletional mutation of the sequence (Izumi *et al.*, 1996). Based on these results, we tentatively concluded that hAPE-1 binds to its own promoter at the nCaRE-B2 site and down-regulates its own expression. We provided additional evidence in support of our conclusion by using electrophoretic gel mobility shift assay (EMSA) showing that HeLa nuclear extract binds to nCaRE-B sequence of the hAPE-1 gene; this binding can be competed with the nCaRE-B

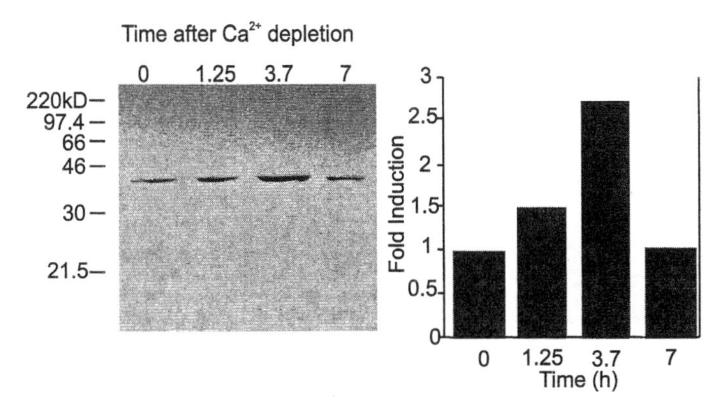

Figure 3. Transient induction of hAPE-1 in HeLa cells by Ca^{2+} depletion. Western blot analysis for APE-1 was carried out with HeLa cell extracts pretreated with 0.2 mM EGTA (left panel). The intensity of the APE band was plotted relative to the control (right panel).

sequence of parathyroid hormone gene, and preincubation of the extract with anti-hAPE-1 antibody inhibited the gel mobility shift (Izumi *et al.*, 1996). The fact that we could not show binding of purified, recombinant hAPE-1 to the nCaRE-B sequence is consistent with the requirement that APE-1 may be present as a heterodimer, in complex with another protein, in order to function as a repressor (see above).

Although we have not yet pursued the function of the nCaRE-A and nCaRE-B1 sequences in the hAPE-1 promoter, our EMSA studies showed that these sequences also bind protein(s) in nuclear extract containing hAPE-1. In order to explain the anomalous expression of luciferase from plasmids containing these nCaRE sequences, we speculate that autoregulation of hAPE-1 is rather complex and involves multiple positive and negative regulatory elements which are present in the 2 kb fragment (Izumi *et al.*, 1996). We could show the function of the nCaRE-B2 in negative regulation unambiguously only after eliminating other upstream positive and negative regulatory sequences. It should be possible to test this possibility by identifying positive regulatory *cis* elements and the proteins that bind to them in subsequent site-specific mutation studies. A prediction of the Ca^{2+}-dependent negative regulation of hAPE-1 is that a reduced level of intracellular free Ca^{2+} should enhance APE expression, but only transiently, due to autoregulation. This prediction was confirmed, as shown in Fig. 3. Although we have not yet measured the intracellular Ca^{2+} level in HeLa cells after depletion of Ca^{2+} from the medium, this result and these of Okazaki's earlier study can be explained by our model that intracellular free Ca^{2+} acts as the second messenger in affecting the hAPE-1 level, which in turn may affect expression of other nCaRE-containing genes. Robertson *et al.* (1997) recently showed that progression to apoptosis in the differentiating HL60 promyelocytic line is associated with a sharp decrease in the hAPE-1 level, below the level of immunohistochemical detection. This observation may be explained by our model, because apoptosis is associated with a large influx of free Ca^{2+}_i into the cell (McConkey and Orrenius, 1996).

3.5. Activation of hAPE-1 by ROS

Because of the central role of hAPE-1 in the repair of ROS-induced DNA damage, we expected that regulation of hAPE-1 is linked to the redox state of the cellular environ-

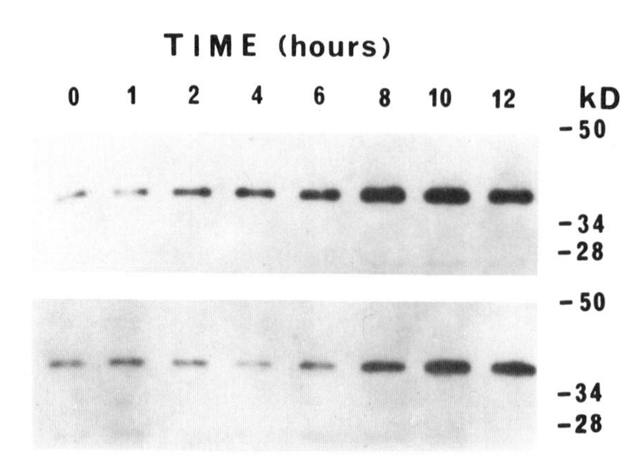

Figure 4. Western blot analysis of hAPE-1 expression in response to oxidative stress. Log-phase HeLa S (upper panel) and WI-38 (lower panel) cells were exposed to 50 μM H_2O_2 and harvested at various times as indicated.

ment. Although the Ca^{2+}-dependent modulation of nCaRE-containing genes may be directly affected by the level of hAPE-1, we hypothesized that such a level is responsive to oxidative stress. Therefore, we tested whether ROS and ROS generators alter the level of hAPE-1 expression in cultured human cells. Our results so far indicate that all varieties of ROS and ROS generators including ionizing radiation activate hAPE-1 expression (Ramana *et al*, 1998). Fig. 4 shows a representative example in which Western blot analysis demonstrated a time-dependent and dose-dependent increased level of hAPE-1 polypeptide after treatment with H_2O_2. Northern blot analysis indicated that the increased hAPE-1 polypeptide level was associated with a corresponding increase in its mRNA level (Fig. 5). We have further shown that activation of APE was associated with its translocation to the nucleus (Ramana *et al.*, 1998) Similar activation of APE after ROS treatment has also been observed by several other investigators (B. Mossman, B. Kaina, personal communications).

However, several points should be made regarding this activation. First, at least in our experiments with cultured cells, the activation was invariably transient, and represented a delayed response. Typically the increase in the hAPE-1 polypeptide level was observed 9–12 h after ROS exposure, and the amount of hAPE-1 polypeptide returned to the

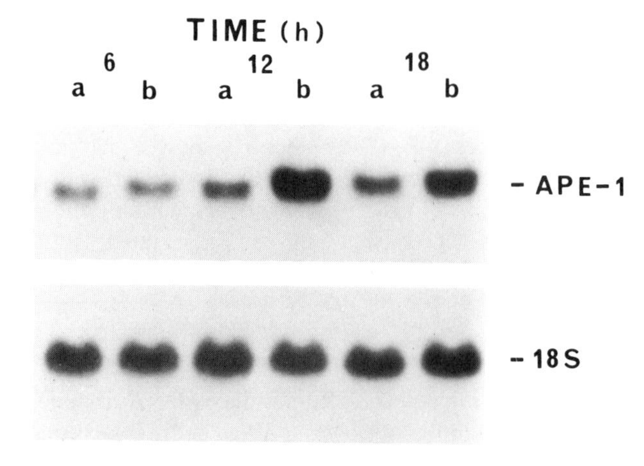

Figure 5. Activation of the hAPE-1 gene in HeLa S cells by H_2O_2. Cells were treated with H_2O_2 and total RNA is extracted at various times indicated for Northern blot analysis. a, mock-treated; b, treated with 50 μM H_2O_2.

Table 2. Activation of hAPE-1 by reactive oxygen species (ROS)

Treatment	Dose	Cell type	Fold increase at maximum activation*	
			RNA	Protein
A:				
HOCl	130 nM	HeLa S	4.2	4.5
	260 nM		6.3	6.5
	260 nM	WI-38	5.5	6.5
	260 nM	V79	NT	5.5
H_2O_2	50 μM	HeLa S	4.8	5
	100 μM		7.5	6.5
		V79	NT	5.5
γ–rays	2 Gy		5.5	NT
Bleomycin	5 μg/ml	HeLa S	4.5	5
PMS	5 μM	HeLa S	NT	4.5
B:				
MMS	2 mM	HeLa S		<1.0NT
UV-C	10 J/m^2	HeLa S		<1.0NT

*Average of at least 3 independent experiments; NT, not tested; PMS, phenazine methosulfate; MMS, methyl methanesulfonate; A, ROS or ROS generators; B, Non-ROS agents.

basal level in about 24 h. Although there is some variability in this temporal relationship depending on the oxidizing species, in general the APE-1 mRNA level peaked at 6–9 h.

Secondly, no APE activation was observed in some experiments. In fact, Harrison and Demple did not observe any ROS-induced activation (Harrison *et al.*, 1995). This discrepancy can be explained by a variable response of cells in culture to deliberate oxidative stress because the cells may already be oxidatively stressed from the culture medium and cellular manipulation, among other things. Table 2 summarizes our results on the extent of activation of hAPE-1 in HeLa tumor line, WI38 primary human fibroblast cells, and V79 Chinese hamster line after exposure to various oxidizing agents. Although a significant variation in the extent of induction was observed in different experiments, this variation may not be agent-specific.

While genotoxic agents induce a variety of similar genes in both bacteria and mammals, the response of these organisms differ in several significant ways. First, the bacterial response is rather agent-specific, i.e. only a limited set of genes is activated in response to a specific genotoxic agent, and there is no large overlap in the sets of genes that are responsive to multiple types of agents. Thus, UV light (UVC) induces a certain set of genes (or regulons), and most of these genes may not be responsive to other types of genotoxic agents such as ROS and alkylating agents (Walker, 1984). In contrast, mammalian cells' response to these variety of agents is generally more indiscriminate. In other words, distinct types of agents such as UV light and ROS and ionizing radiation activate many of the same mammalian genes. This basic difference between bacteria (*E. coli*) and mammalian cells may reflect the difference in the molecular mechanisms and cellular signalling in their gene activation. Unlike *E. coli*, in which an activator protein often regulates gene expression by specific interaction with the *cis* element and the transcriptional machinery, regulation of mammalian genes is mediated by families of *trans*-acting factors, e.g., NKκβ, AP-1, c/EBP among others. These proteins bind to their cognate *cis* elements in a combinatorial fashion for regulation of diverse genes and their fine tuning (Hill and Treisman, 1995). It is, therefore, surprising that only ROS was found to induce APE-1 activation in human cells. Another unusual feature of APE-1 activation is the level of ROS needed for optimum effect. In general, the level of genotoxic agents required for activation of a variety of genes in mammalian cells is often toxic. For example, many genes for alkylation damage repair such as *N*-methylpurine-DNA glycosylase and O^6-methylguanine-DNA methyltransferase and β-pol are induced in human and rodent cells by alkylating agents at concentration levels that would kill a significant fraction of the treated cells (Kaina *et al.*, 1998). In contrast, the concentration of ROS needed to activate APE-1 causes little or undetectable cytotoxicity. One of the best examples of such ROS is HOCl, a powerful oxidant that is produced by and released from activated neutrophils during the inflammatory response, and generated by the successive action of two inducible enzymes—NADH oxidase and myeloperoxidase (Thelen *et al.*, 1993). Clonogenic assays for cytotoxicity to HeLa and Chinese hamster V79 cells showed significant toxicity of the compound in the micromolar range. However, we observed that HOCl is one of the best inducers of APE-1 at a concentration of less than 300 nM. No effect on cellular survival was observed after treatment with this concentration of HOCl (Ramana *et al.*, 1998).

3.6. Activation of APE-1 by ROS in the Mouse and the Effect of Aging on Oxidative Stress Response

Because of the difficulty of investigating low level oxidative stress *in vitro* (as discussed earlier), we investigated the effect of such stress in whole animals. To induce ROS we used lipopolysaccharide (LPS), a component of the cell wall of gram negative bacteria, which induces an inflammatory response and oxidative stress, and is commonly used to investigate the acute phase response in experimental animals (Post *et al.*, 1991). Because aging is believed to be a manifestation of chronic oxidative stress which may be generated both endogenously and due to environmental agents (Papaconstantinou, 1994), we investigated whether oxidative stress due to LPS upregulates APE-1 expression, and then went on to test whether aging affects such stress responses. We examined the level of APE-1 mRNA in liver, the target organ for the LPS-induced acute phase response (Papaconstantinou, 1994) in young vs. old mice. Fig. 6 shows the kinetics of APE-1 activation in the liver of 4 mo- and 4 mo-old Balb/C mice. The unusual features of the results are immediately obvious. Based on the *in vitro* culture cell studies, we expected LPS-induced activation of APE-1. However, it was surprising that the induction occurred much earlier in the mouse liver than in cultured human cells, although the APE-1 mRNA returned to the basal level by 24 h in both *in vivo* and cellular studies. Furthermore, the extent of the increase in APE-1 mRNA level in the mouse liver was much higher than that observed in cultured human cells. We believe that this difference in cellular response is not due to their rodent or human origin but may be organ specific. Further experiments are needed to confirm this possibility.

In order to assess the physiological significance or consequence of such age-dependent variability in response to oxidative stress, we checked whether similar LPS-induced response also occurs in another mouse strain, C57 Bl/6. Although there was some difference in the level and kinetics of APE-1 activation, the young animals again showed a more robust and earlier response to LPS treatment (data not shown). Thus it appears that ROS activation of APE-1 is adversely affected by aging.

3.7. Biological Consequences of ROS-Induced APE-1 Activation

The teleological reason for APE-1 being activated exclusively by ROS and not other genotoxic agents is not clear. One obvious explanation is that a small inducing level of ROS forewarns the cell to be prepared for a possible onslaught of future acute oxidative

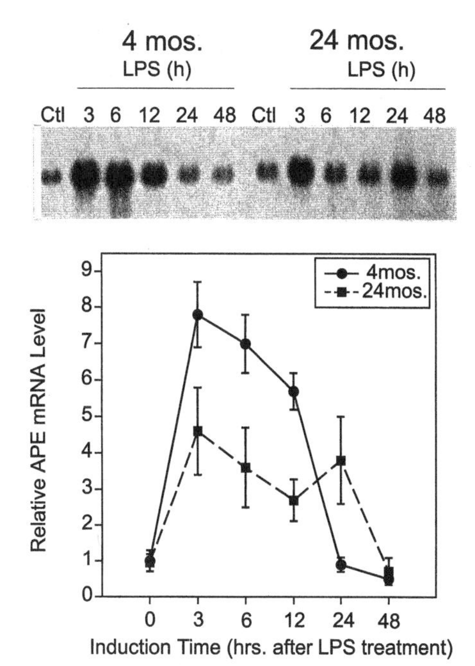

Figure 6. Lipopolysaccharide (LPS)-induced activation of the APE-1 gene in the livers of young (4 mos.) and aged (24 mos.) Balb/C mice. Northern blot analysis was carried out with total RNA (15 μg/lane) isolated from the livers of mice at various times after IP injection of 10 μg LPS (upper panel). Relative levels of APE-1 mRNA (averaged from 5 independent experiments), as determined by densitometric analysis of the autoradiograms, were plotted as a function of time after LPS treatment (lower panel).

stress (Cairns, 1980). The cell responds by activating multiple genes whose products counter the deleterious affects of ROS, including DNA damage. Among these genes, the APE-1 gene holds a unique place because of its additional, critical functions beyond its role in repair of oxidative DNA damage. It is also likely that the basal level of APE-1 is limiting in at least many of these processes in order for its transient increase to be effective. We have shown such to be the case during the "adaptive response" of ROS-treated cells when HeLa or V79 cells pretreated with nontoxic level (130 NM) of HOCl became significantly more resistant to toxic doses of such ROS as H_2O_2 and bleomycin (Ramana *et al.*, 1998). As discussed earlier, we have shown in other experiments that activated APE-1 may be responsible for this adaptive response because its basal level is limiting in repair of DNA single strand breaks, which may be critical genotoxic lesions induced by ROS (Izumi *et al.*, manuscript in preparation).

4. CONCLUDING REMARKS

The results described in this review and the relevant literature should make it clear that we have only a superficial understanding of the complex regulatory circuits involving APE-1, a highly unusual protein because of its multiple unrelated activities. Many facets of this regulation, and their *in vivo* implications, need to be explored. Some of these issues include (1) identification and characterization of the *cis* element(s), and their cognate *trans*-acting factors that are involved in ROS-induced activation of APE-1; (2) age-dependent modulation of these *trans*-acting factors; (3) the impact of APE-1 on Ca^{2+} homeostasis and the *in vivo* function of AP-1, p53 and other transcription factors. Fig. 7 summarizes our working model of the complex and interlocking roles of APE-1, and how aging may affect such regulation.

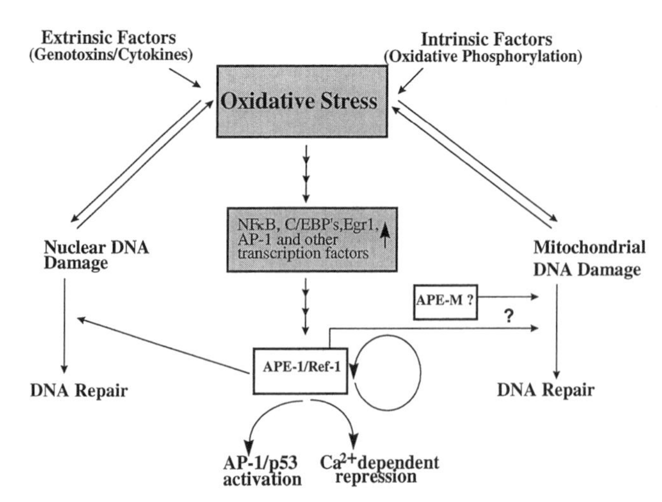

Figure 7. A schematic representation of established and proposed functions of hAPE-1 in DNA repair and gene regulation. ROS generated both intrinsically, and after exposure to external agents, are expected to activate transcription factors, one or more of which activate APE. APE, in turn, functions both as an activator and as a repressor. Furthermore it may be autoregulated (indicated by the circular direction). Mitochondrial DNA is more susceptible to oxidative damage than nuclear DNA (Yakes and Van Houten, 1997), and is subject to repair (Driggers *et al.*, 1997). Whether APE-1 or a mitochondria-specific APE (APE-m) is involved in mitochondrial DNA repair is not known. Aging is likely to be caused by chronic stress (Papaconstantinou, 1994) and may affect many of the regulatory pathways shown here.

ACKNOWLEDGMENT

The research described in this article was supported by U.S. Public Health Service Grants ES08457, and CA53791, and by the Publication No. 97c supported by U.S.P.H.S. grant PO1 AG10514 and the National Institute of Environmental Health Sciences Center grant ES06676. The expert technical assistance of Ms. G. Roy is gratefully acknowledged. We thank Dr. D. Konkel for critical reading of the manuscript and Ms. W. Smith for competent secretarial assistance.

REFERENCES

C. Abate, L. Patel, F. J. Rauscher III and T. Curran (1990) Redox regulation of fos and jun DNA-binding activity *in vitro*. *Science* **249**, 1157–1161.

R. A. O. Bennett, D. M. 3rd Wilson, D. Wong and B. Demple (1997) Interaction of human apurinic endonuclease and DNA polymerase beta in the base excision repair pathway. *Proc. Natl. Acad. Sci. USA*. **94**, 7166–7169.

A. P. Breen and J. A. Murphy (1995) Reactions of oxyl radicals with DNA. *Free Radic. Biol. Med.* **18**, 1033–1077.

J. Cairns (1980) Efficiency of the adaptive response of Escherichia coli to alkylating agents. *Nature* **286**, 176–178.

D. S.Chen, T. Herman and B. Demple (1991) Two distinct human DNA diesterases that hydrolyze 3'-blocking deoxyribose fragments from oxidized DNA. *Nucl. Acids Res*. **19**, 5907–5914.

U. Chung, T. Igarashi, T. Nishishita, H. Iwanari, A. Iwamatsu, A. Suwa, T. Mimori, K. Hata, S. Ebisu, E. Ogata, T. Fujita and T. Okazaki (1996) The interaction between Ku antigen and REF1 protein mediates negative gene regulation by extracellular calcium. *J. Biol. Chem*. **271**, 8593–8598.

P. K. Cooper, T. Nouspikel, S. G. Clarkson and S. A. Leadon (1997) Defective transcription-coupled repair of oxidative base damage in Cockayne syndrome patients from XP group G. *Science* **275**, 990–993.

B. Demple, T. Herman and D. S. Chen (1991) Cloning and expression of APE, the cDNA encoding the major human apurinic endonuclease: definition of a family of DNA repair enzymes. *Proc. Natl. Acad. Sci. USA* **88**, 11450–11454.

G. Dianov, A. Price anf T. Lindahl (1992) Generation of single-nucleotide repair patches following excision of uracil residues from DNA. *Mol. Cell Biol.* **12**, 1605–12.

G. Dianov and T. Lindahl (1994) Reconstitution of the DNA base excision-repair pathway. *Curr. Biol.* **4**, 1069–1076.

M. Dizdaroglu (1992) Oxidative damage to DNA in mammalian chromatin. *Mutat. Res.* **275**, 331–342.

M. L. Dodson, M. L. Michaels and R. S. Lloyd (1994) Unified catalytic mechanism for DNA glycosylases. *J. Biol. Chem.* **269**, 32709–32712.

P. W. Doetsch and R. P. Cunningham (1990) The enzymology of apurinic/apyrimidinic endonucleases. *Mutat. Res.* **236**, 173–201.

W. J. Driggers, V. I. Grishko, S. P. LeDoux and G. L. Wilson (1996) Defective repair of oxidative damage in the mitochondrial DNA of a xeroderma pigmentosum group A cell line. *Cancer Res.* **56**, 1262–1266.

E. C. Friedberg, G. C. Walker and W. Siede. (1995) Chapter 4: *Base Excision Repair in DNA Repair and Mutagenesis*, ASM Press, Washington, D.C.

G. Frosina, P. Fortini, O. Rossi, F. Carrozzino, G. Raspaglio, L. S. Cox, D. P. Lane, A. Abbondandolo snf E. Dogliotti (1996) Two pathways for base excision repair in mammalian cells. *J. Biol. Chem.* **271**, 9573–9578.

M. E. Gotz, G. Kunig, P. Riederer, and M. B. Youdim (1994) Oxidative stress: free radical production in neural degeneration. *Pharmacol. Ther.* **63**, 37–122.

M. B., Grisham and J. M. McCord (1986) *Physiology of Oxygen Radicals*, eds. Taylor, A. E., Matalon, S. and Ward, P. A (Waverly Press, Baltimore), pp. 1–18.

L. Harrison, A. G. Ascione, D. M. 3rd Wilson and B. Demple (1995) Characterization of the promoter region of the human apurinic endonuclease gene (APE). *J. Biol. Chem.* **270**, 5556–5564.

C. S. Hill and R. Treisman (1995) Transcriptional regulation by extracellular signals: mechanisms and specificity. *Cell* **80**, 199–211.

K. Hirota, M. Matsui, S. Iwata, A. Nishiyama, K. Mori and J. Yodoi (1997) AP-1 transcriptional activity is regulated by a direct association between thioredoxin and Ref-1. *Proc. Natl. Acad. Sci. USA* **94**, 3633–3638.

T. Izumi, W. D. Henner and S. Mitra (1996) Negative regulation of the major human AP-endonuclease, a multifunctional protein. *Biochemistry* **35**, 14679–14683.

L. Jayaraman, K. G. Murthy, C. Zhu, T. Curran, S. Xanthoudakis and C. Prives (1997) Identification of redox/repair protein Ref-1 as a potent activator of p53. *Genes. Dev.* **11**, 558–570.

B. Kaina, S. Haas, S. Grosch, T. Grombacher, J. Dosch, I. Boldogh, T. Biswas and S. Mitra. (1998) *Inducible responses and protective functions of mammalian cells upon exposure to UV light and ionizing radiation.* NATO Proceedings on Radiation Effects on Living Matter. (In Press)

C. M. Kane and S. Linn (1981) Purification and characterization of an apurinic/apyrimidinic endonuclease from HeLa cells. *J. Biol. Chem.* **256**, 3405–3414.

Z. Kelman (1997) PCNA: structure, functions and interactions. *Oncogene.* **14**, 629–640.

C. M. King, H. E. Bristow-Craig, E. S. Gillespie and Y. A. Barnett (1997) In vivo antioxidant status, DNA damage, mutation and DNA repair capacity in cultured lymphocytes from healthy 75- to 80-year-old humans. *Mutat. Res.* **377**, 137–147.

A. Klungland and T. Lindahl (1997) Second pathway for completion of human DNA base excision-repair: reconstitution with purified proteins and requirement for DNase IV (FEN1). *EMBO. J.* **16**, 3341–3348.

Y. Kubota, R. A. Nash, A. Klungland, P. Schar, D. E. Barnes and T. Lindahl (1996) Reconstitution of DNA base excision-repair with purified human proteins: interaction between DNA polymerase beta and the XRCC1 protein. *EMBO. J.* **15**, 6662–6670.

A. J. Levine (1997) p53, the cellular gatekeeper for growth and division. *Cell* **88**, 323–331.

L. B. Meira, D. L. Cheo, R. E. Hammer, D. K. Burns, A. Reis and E. C. Friedberg (1997) Genetic interaction between HAP1/REF-1 and p53. *Nature Genet.* **17**, 145.

Y. Matsumoto, K. Kim, and D. F. Bogenhagen (1994) Proliferating cell nuclear antigen-dependent abasic site repair in Xenopus laevis oocytes: An alternative pathway of base excision DNA repair. *Mol. Cell Biol.* **14**, 6187–6197.

Y. Matsumoto and K. Kim (1995) Excision of deoxyribose phosphate residues by DNA polymerase beta during DNA repair. *Science* **269**, 699–702.

D. J. McConkey and S. Orrenius (1996) Signal transduction pathways in apoptosis. *Stem Cells* **14**, 619–631.

G. S. McHaffie and S. H. Ralston (1995) Origin of a negative calcium response element in an ALU-repeat: implications for regulation of gene expression by extracellular calcium. *Bone* **17**, 11–14.

S. Mitra, T. K. Hazra, R. Roy, S. Ikeda, T. Biswas, J. Lock, I. Boldogh and T. Izumi (1997) Complexities of DNA base excision repair in mammalian cells. *Mol. Cells.* **7**, 305–312.

C. D. Mol, A. S. Arvai, G. Slupphaug, B. Kavli, I. Alseth, H. E. Krokan and J. A. Tainer (1995) Crystal structure and mutational analysis of human uracil-DNA glycosylase: structural basis for specificity and catalysis. *Cell* **80**, 869–878.

H. M. Nash, S. D. Bruner, O. D. Scharer, T. Kawate. T. A. Addona, E. Spooner, W. S. Lane and G. L. Verdine (1996) Cloning of a yeast 8-oxoguanine DNA glycosylase reveals the existence of a base-excision DNA-repair protein superfamily. *Curr. Biol.* **6**, 968–980.

T. Okazaki, U. Chung, T. Nishishita, S. Ebisu, S. Usuda, S. Mishiro, S. Xanthoudakis, T. Igarashi and E. Ogata (1994) A redox factor protein, ref1, is involved in negative gene regulation by extracellular calcium. *J. Biol. Chem.* **269**, 27855–27862.

J. Papaconstantinou (1994) Unifying model of the programmed (intrinsic) and stochastic (extrinsic) theories of aging. The stress response genes, signal transduction-redox pathways and aging. *Ann. NY. Acad. Sci.* **719**, 195–211.

C. S. Parkins, M. F. Dennis, M. R. Stratford, S. Hill, and D. J. Chaplin (1995) Ischemia reperfusion injury in tumors: the role of oxygen radicals and nitric oxide. *Cancer Res.* **55**, 6026–6029.

C. E. Piersen, R. Prasad, S. H. Wilson and R. S. Lloyd (1996) Evidence for an imino intermediate in the DNA polymerase beta deoxyribose phosphate excision reaction. *J. Biol. Chem.* **271**, 17811–17815.

D. J. Post, K. C. Carter and J. Papaconstantinou (1991) The effect of aging on constitutive mRNA levels and lipopolysaccharide inducibility of acute phase genes. *Ann. NY Acad. Sci.* **621**, 66–77.

R. Prasad, R. K. Singhal, D. K. Srivastava, J. T. Molina , A. E. Tomkinson and S. H. Wilson (1996) Specific interaction of DNA polymerase beta and DNA ligase I in a multiprotein base excision repair complex from bovine testis. *J. Biol. Chem.* **271**, 16000–16007.

J. Qin, G. M. Clore, W. P. Kennedy, J. Kuszewski and A. M. Gronenborn (1996) The solution structure of human thioredoxin complexed with its target from Ref-1 reveals peptide chain reversal. *Structure* **4**, 613–620.

C. V. Ramana, I. Boldogh, T. Izumi and S. Mitra (1998) Activation of AP-endonuclease in human cells by reactive oxygen species and its correlation with their adaptive response to genotoxicity of free radicals. *Proc. Natl. Acad. Sci. USA* (in press).

K. S. Rao and L. A. Loeb (1992) DNA damage and repair in brain: relationship to aging. *Mutat. Res.* **275**, 317–329.

C. C. Richardson, I. R. Lehman and A. Kornberg (1964) A deoxyribonucleic acid phosphatase-exonuclease from Escherichia coli II. Characterization of the exonuclease activity. *J. Biol. Chem.* **239**, 251–258.

K. A. Robertson, D. P. Hill, Y. Xu, L. Liu, S. Van Epps, D. M. Hockenbery, J. R. Park, T. M. Wilson and M. R. Kelley (1997) Down-regulation of apurinic/apyrimidinic endonuclease expression is associated with the induction of apoptosis in differentiating myeloid leukemia cells. *Cell Growth Differ.* **8**, 443–449.

R. Roy, C. Brooks and S. Mitra (1994) Purification and biochemical characterization of recombinant N-methylpurine-DNA glycosylase of the mouse. *Biochem.* **33**, 15131–15140.

R. Roy, A. Kumar, J. C. Lee and S. Mitra (1996) The domains of mammalian base excision repair enzyme N-methylpurine-DNA glycosylase - interaction, conformational change, and role in DNA binding and damage recognition. *J. Biol. Chem.* **271**, 23690–23697.

M. S. Satoh, G. G. Poirier and T. Lindahl (1993) NAD(+)-dependent repair of damaged DNA by human cell extracts. *J. Biol. Chem.* **268**, 5480–5487.

M. S. Satoh, G. G. Poirier and T. Lindahl (1994) Dual function for poly(ADP-ribose) synthesis in response to DNA strand breakage. *Biochemistry* **33**, 7099–7106.

S. Seki, S. Ikeda, Watanabe, S., M. Hatsushika, K. Tsutsui, K. Akiyama and F. Zhang (1991) A mouse DNA repair enzyme (APEX nuclease) having exonuclease and apurinic/apyrimidinic endonuclease activities: purification and characterization. *Biochim. Biophys. Acta* **1079**, 57–64.

S. Seki, M. Hatsushika, S. Watanabe, K. Akiyama, K. Nagao and K. Tsutsui (1992) cDNA cloning, sequencing, expression and possible domain structure of human APEX nuclease homologous to Escherichia coli exonuclease III. *Biochim. Biophys. Acta.* **1131**, 287–299.

R.K. Singhal, R. Prasad and S. H. Wilson (1995) DNA polymerase beta conducts the gap-filling step in uracil-initiated base excision repair in a bovine testis nuclear extract. *J. Biol. Chem.* **270**, 949–957.

R. W. Sobol, J. K. Horton, R. Kuhn, H. Gu, R. K. Singhal, R. Prasad, K. Rajewsky and S. H. Wilson (1996) Requirement of mammalian DNA polymerase-beta in base-excision repair. *Nature* **379**, 183–186.

D. Suh, D. M. 3rd Wilson and L. F. Povirk (1997) 3'-phosphodiesterase activity of human apurinic/apyrimidinic endonuclease at DNA double-strand break ends. *Nucleic Acids Res.* **25**, 2495–2500.

Y. J. Suzuki, H. J. Forman and A. Sevanian (1997) Oxidants as stimulators of signal transduction. *Free. Rad. Biol. Med.* **22**, 287–306.

M. Thelen, B. Dewald and M. Baggiolini (1993) Neutrophil signal transduction and activation of the respiratory burst. *Physiol. Rev.* **73**, 797–821.

R. Tyrrell and S. M. Keyse (1990) New trends in photobiology. The interaction of UVA radiation with cultured cells. *J. Photochem. Photobiol.* B. **4,** 349–361.

G. C. Walker (1984) Mutagenesis and inducible responses to deoxyribonucleic acid damage in Escherichia coli. *Microbiol. Rev.* **48**, 60–93.

J. F. Ward (1994) The complexity of DNA damage: relevance to biological consequences. *Int. J. Radiat. Biol.* **66**, 427–432.

B. J. White, S. J. Hochhauser, N. M. Cintron and B. Weiss (1976) Genetic mapping of xthA, the structural gene for exonuclease III in Escherichia coli K-12. *J. Bacteriol.* **126**, 1082–1088.

F. M. Yakes and B. Van Houten (1997) Mitochondrial DNA damage is more extensive and persists longer than nuclear DNA damage in human cells following oxidative stress. *Proc. Natl. Acad. Sci. USA* **94**, 514–519.

S. Xanthoudakis, G. Miao, F. Wang, Y. C. Pan and T. Curran (1992) Redox activation of Fos-Jun DNA binding activity is mediated by a DNA repair enzyme. *EMBO. J.* **11**, 3323–3335.

25

ACTION OF ANTIOXIDANTS AGAINST OXIDATIVE STRESS

Etsuo Niki

Research Center for Advanced Science and Technology
University of Tokyo, Japan

1. INTRODUCTION

The active oxygen species and free radicals are formed *in vivo* endogenously by several normal metabolic processes and abnormal events. In addition, exogenous sources of oxidants such as cigarette smoke and UV light may also increase oxidative stress *in vivo*. These oxidants attack DNA, lipids, sugars and proteins to cause oxidative damage, which eventually leads to various diseases, cancer and aging. It has been estimated that the number of oxidative hits to DNA per cell per day is about 105 in the rat and roughly 10 times fewer in the human (Ames et al. 1993). Although there is no definitive evidence that free radical involvement is obligatory in carcinogenisis, mutation and transformation, it is clear that free radicals in biological systems could lead to mutation, transformation, and ultimately cancer. Free radicals may attack DNA directly and DNA may also be modified through oxidation of lipids, sugars and proteins. The aerobic organisms are protected from such oxidative stress by an array of defense system. Various kinds of antioxidants with different functions act in the defense system *in vivo* (Table 1). From the type of functions, the antioxidants may be classified into the following three groups: (1) preventive antioxidants which suppress the formation of free radicals and active oxygens species, (2) radical-scavenging antioxidants which inhibit chain initiation and break chain propagation, and (3) repair and de-novo antioxidant enzymes. These antioxidants act as the first, second and third defense line, respectively. Furthermore, there is another function in which the appropriate antioxidant is formed at the right time and transfer it to the right place in a right concentration. This "adaptation mechanism" may be regarded as the fourth defense line.

The decomposition of hydroperoxides or hydrogen peroxide by transition metal ions is assumed to be one of the important pathways of free radical generation *in vivo*. Glutathione peroxidases, glutathione-S-transferase and catalase reduce hydroperoxides and hydrogen peroxide to corresponding alcohols and water, while some proteins such as transferrin and ceruloplasmin sequester metal ions. Thus the formation of free radicals by

Advances in DNA Damage and Repair, edited by Dizdaroglu and Karakaya. Kluwer Academic / Plenum Publishers, New York, 1999.

Table 1. Defense systems *in vivo* against oxidative damage

1. Preventive antioxidants: suppress the formation of free radicals	
a. Non-radical decomposition of hydroperoxides and hydrogen peroxide	
Catalase	decomposition of hydrogen peroxide $2H_2O_2 \rightarrow 2H_2O+O_2$
Glutathione peroxidase (cellular)	decomposition of hydrogen peroxide and free fatty acid hydroperoxides $H_2O_2+2GSH \rightarrow 2H_2O+GSSG$ $LOOH+2GSH \rightarrow LOH+H_2O+GSSG$
Glutathione peroxidase (plasma)	decomposition of hydrogen peroxide and phospholipid hydroperoxides $PLOOH+2GSH \rightarrow PLOH+H_2O+GSSG$
Phospholipid hydroperoxide glutathione peroxidase	decomposition of phospholipid hydroperoxides
Peroxidase	decomposition of hydrogen peroxide and lipid hydroperoxides $LOOH+AH_2 \rightarrow LOH+H_2O+A$ $H_2O_2+AH_2 \rightarrow 2H_2O+A$
Glutathione-S-transferase	decomposition of lipid hydroperoxides
b. Sequestration of metal by chelation	
Transferrin, lactoferrin	sequestration of iron
Haptoglobin	sequestration of hemoglobin
Hemopexin	stabilization of heme
Ceruloplasmin, albumin	sequestration of copper
c. Quenching of active oxygens	
Superoxide dismutase (SOD)	disproportionation of superoxide $2O_2^{\cdot -}+2H^+ \rightarrow H_2O_2+O_2$
Carotenoids, vitamin E	quenching of singlet oxygen
2. Radical-scavenging antioxidants: scavenge radicals to inhibit chain initiation and break chain propagation hydrophilic: vitamin C, uric acid, bilirubin, albumin lipophilic: vitamin E, ubiquinol, carotenoids, flavonoids	
3. Repair and *de novo* enzymes: repair the damage and reconstitute membranes: lipase, protease, DNA repair enzymes, transferase	
4. Adaptation: generate appropriate antioxidant enzymes and transfer them to the right site at the right time and in the right concentration	

the metal-induced decomposition of peroxides is inhibited or at least minimized. Neither superoxide nor nitric oxide (NO) is reactive enough per se to induce oxidative damage, but they react very rapidly to give peroxynitrite which is capable of causing various oxidative damage. Superoxide dismutase (SOD) quenches superoxide and inhibits the formation of peroxynitrite and superoxide-driven oxidative damage. UV light is another source of active oxygen species and free radicals and various carotenoids play an important role as a quencher of UV light. These antioxidants having different functions and working at different sites suppress the formation of active oxygen species and act as a preventive antioxidant.

The second line defense is radical-scavenging antioxidants. In spite of the efficient action of the above-mentioned preventive antioxidants, free radicals are still formed *in vivo*. The radical-scavenging antioxidants are responsible to scavenge and stabilize them before the radicals attack target molecules. There are both hydrophilic and lipophilic radical-scavenging antioxidants. Vitamin C, uric acid and bilirubin are hydrophilic antioxidants, while vitamin E, ubiquinol and carotenoids are lipophilic antioxidants. The actions of these antioxidants will be considered in more detail below.

The third defense line is composed of repair and *de novo* enzymes such as lipases, proteases, and DNA repair enzymes, which are covered in great detail in other chapters of this book. Transferases reconstitute membrane.

These antioxidants described above construct a defense system to cope with the oxidative stress challenged by active oxygen species and free radicals.

2. ACTION OF RADICAL SCAVENGING ANTIOXIDANTS

The potency of radical-scavenging antioxidant is determined not only by its reactivity toward radical but also by many other factors. Apparently, the reactivity of the antioxidant toward radical which is determined by its chemical structure is important, and the faster it scavenges radical, the stronger its activity. However, the rate of scavenging radical is dependent on various physical factors as well. For example, the apparent rate constant for scavenging peroxyl radical by α-tocopherol (vitamin E) decreases quite markedly in the membranes as compared with that in homogeneous organic solution (Niki et al, 1986: Barclay, 1992). The experiments using a spin label clearly show that the efficiency of scavenging radicals in the membrane by α-tocopherol decreases as the radical goes deeper into the interior of the membrane (Fig. 1, Takahashi et al, 1989). This is also true in the low density lipoprotein (Gotoh et al, 1996). The efficacy of collision of active phenolic hydrogen of α-tocopherol which is located at the surface of the membrane and the peroxyl radical present within the lipophilic domain of the membrane must be lower than in homogeneous solution. Interestingly, the drop in the efficacy is much less for 2,2,5,7,8-

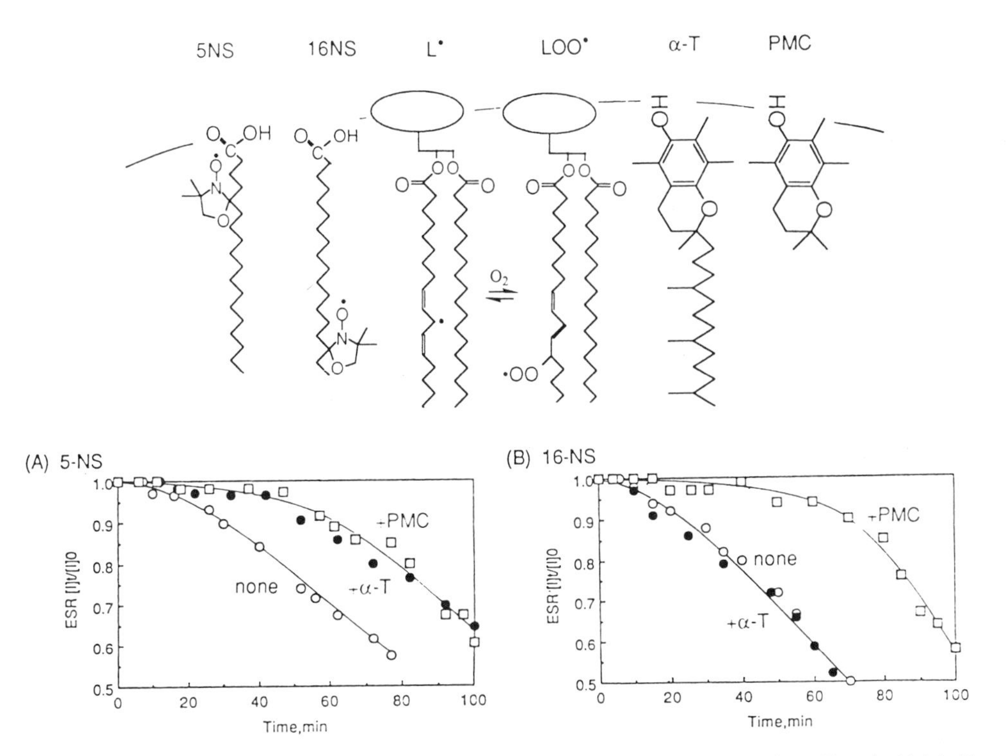

Figure 1. Efficacy of scavenging radicals by vitamin E as studied by spin labeling technique. Phosphatidylcholine liposomal membranes were oxidized by free radical initiator in the presence of spin label (5NS or 15NS) and either vitamin E or 2,2,5,7,8-pentamethyl-6-chromanol (PMC). Both vitamin E and PMC suppress 5-NS consumption, but vitamin E can not spare 16-NS while PMC can.

pentamethyl-6-chromanol, a model compound for α-tocopherol without phytyl side chain, than that of α-tocopherol, suggesting that the long side chain of α-tocopherol, although necessary biologically for incorporation and retainment in the membrane, reduces its activity in the membranes. In accordance with this observation, 2,6-di-tert-butyl-4-methylphenol, a well known synthetic antioxidant as BHT, is chemically much less reactive than α-tocopherol toward peroxyl radical in solution, but the difference between the activities of the two antioxidants is much smaller in the membrane.

It may be also noteworthy that hydrophilic antioxidants such as vitamin C (ascorbic acid) and uric acid are capable of scavenging aqueous radicals but that they can not scavenge radicals within the membranes or lipoproteins efficiently. For example, ascorbic acid acts as a potent radical-scavenging antioxidant and suppresses the oxidation of membranes induced by aqueous radicals, but it does not suppress the oxidation of membranes when it is induced by radicals generated within the lipophilic compartment of the membranes (Niki et al, 1985; Doba et al, 1985). This is confirmed more directly by the experiments using spin probe, doxyl stearic acid. Ascorbic acid reduces the nitroxide radical quite rapidly in solution, but the rate of reduction decreases as the nitroxide radical goes deeper from the membrane surface into the interior (Takahashi et al, 1988; Gotoh et al, 1996).

However, vitamin C is capable of exerting antioxidant action against the oxidation taking place within the membranes in conjunction with vitamin E, that is, although vitamin C can not scavenge the radicals within the membrane directly, it reduces vitamin E radical which is formed when vitamin E scavenges radical. Thus, when both vitamin C and vitamin E are present, vitamin C is consumed predominantly while vitamin E is spared even if the oxidation is taking place within the membranes. Such an efficient synergistic effect is not observed for uric acid.

It is also important to understand that the total antioxidant potency is determined by the relative importance of many competing reactions. For example, β-carotene radical may undergo several reactions as shown below (βC denotes β-carotene). β-Carotene scavenges peroxyl radical by addition reaction to give resonance-stabilized conjugated carbon-centered radical (Burton and Ingold, 1984). It may scavenge another peroxyl radical (reaction 1) to terminate chain reaction. It may react with oxygen molecule to give β-carotene peroxyl radical, which may react with β-carotene to induce autoxidation of β-carotene (reaction 2) or continue oxidation (reaction 3). The β-carotene radical may undergo β-scission reaction to give epoxide and alkoxyl radical, which is capable of continuing oxidation (reaction 4). Some of the reactions are chain-terminating, but others are not. The efficacy of β-carotene as an antioxidant is determined by the relative importance of these competing reactions. This is also true with other antioxidants, that is, the fate of antioxidant-derived radical is important in determining antioxidant potency.

$$LO_2^{\bullet} + \beta C \longrightarrow LOO\beta C^{\bullet}$$

$$LOO\beta C^{\bullet} \xrightarrow{LO_2^{\bullet}} LOO\beta COOL \quad (1)$$

$$LOO\beta C^{\bullet} \xrightarrow{O_2} LOO\beta COO^{\bullet} \longrightarrow \text{autoxidation of } \beta C \quad (2)$$

$$LOO\beta COO^{\bullet} \longrightarrow \text{continue oxidation} \quad (3)$$

$$LOO\beta C^{\bullet} \longrightarrow \text{epoxide} + LO^{\bullet} \longrightarrow \text{continue oxidation} \quad (4)$$

3. PROOXIDANT ACTION OF ANTIOXIDANT

It has been reported several times that antioxidant may act as a prooxidant under certain conditions. As described above, if the radical derived from the antioxidant is stable enough, it may act as an antioxidant. However, some antioxidant-derived radicals may attack substrate to induce oxidation. For example, Stocker and his colleagues (Bowry et al, 1992) reported that α-tocopherol acts as a prooxidant by phase transfer and chain transfer mechanisms: α-tocopherol scavenges aqueous radicals to give α-tocopheroxyl radical, which attacks lipid in the membrane to initiate chain oxidation, thus carrying aqueous radials into the membrane. Furthermore, they propose that the chain is propagated by α-tocopheroxyl radical, not by lipid peroxyl radical (Waldeck and Stocker, 1996). Such an effect may be important under certain circumstances. However, α-tocopheroxyl radical may well be reduced rapidly by reducing agents such as ascorbate and ubiquinol, and therefore the prooxidant action may probably be unimportant *in vivo*.

Another possible prooxidant action may arise from the reduction of metal ions. Both ascorbate and α-tocopherol are capable of reducing ferric ion Fe(III) and cupric ion Cu(II) to corresponding lower valency state, Fe(II) and Cu(I), which decompose hydroperoxide and hydrogen peroxide much faster than Fe(III) and Cu(II) respectively. Therefore, under such conditions, ascorbate and α-tocopherol act as a prooxidant. In fact, the combination of ascorbate and iron has been often used as an oxidant in the *in vitro* systems and such prooxidant action of α-tocopherol has been also reported. However, this does not necessarily mean that such prooxidant action takes place *in vivo* since substantially all of metal ions are sequestered by protein and are not readily reduced by ascorbate or α-tocopherol.

Polyphenols, which have received much attention recently as natural antioxidants contained in wine, green tea and plants, may also exert prooxidant effect under certain conditions *in vitro* by their autoxidation to give superoxide and hydrogen peroxide.

4. CONCLUDING REMARKS

As briefly reviewed above, we are protected from the oxidative stress by quite an efficient defense system. This is why we human being can survive as long as 100 years, although salad dressing composed of substrates with similar oxidizabilities as those of human being is oxidized and deteriorated quite easily even stored in the refrigerator. Nevertheless, the oxidative stress may overwhelm the defense capacity sometimes and cause damage. It is essential to understand the defense mechanism and try to enhance the defense capacity. The function of the antioxidants beyond the above-mentioned mechanism, such as in signal transduction and cell proliferation, may also be important and should be explored in the future.

REFERENCES

B. N. Ames, N. K. Shigenaga and T. M. Hagen (1993) Oxidants, antioxidants and the degenerative disease of aging. *Proc. Natl. Acad. Sci.* USA **90**, 7915–7922.

L. R. C. Barclay (1993) Model biomembranes: quantitative studies of peroxidation, antioxidant action, partitioning, and oxidative stress. *Can. J. Chem.* **71**, 1–16.

V. W. Bowry, K. U. Ingold and R. Stocker (1992) Vitamin E in human low-density lipoprotein. *Biochem. J.* **28**, 341–344.

G. W. Burotn and K. U. Ingold (1984) β-Carotene: An unusual type of lipid antioxidant. since **224**, 569–573.

T. Doba, G. W. Burton and K. U. Ingold (1985) Antioxidant and co-antioxidant activity of vitamin C. The effect of vitamin C, either alone or in the presence of vitamin E or a water-soluble vitamin E analogue, upon the peroxidation of aqueous multilammellar phospholipid liposome. *Biochim. Biophys.* Acta, **835**, 298–303.

N. Gotoh, N. Noguchi, J. Tsuchiya, K. Morita, H. Sakai, H. Shimasaki and E. Niki (1996) Inhibition of oxidation of low density lipoprotein by vitamin E and related compounds. *Free Rad. Res.* **24**, 123–134.

E. Niki, A. Kawakami, Y. Yamamoto and Y. Kamiya (1985) Oxidation of lipids. VIII. Synergistic inhibition of oxidation of phosphatidylcholine liposome in aqueous dispersion by vitamin E and vitamin C. Bull. *Chem. Soc.* Jpn. **58**, 1971–1975.

E. Niki, M. Takahashi and E. Komuro (1986) Antioxidant activity of vitamin E in liposomal membranes. *Chem. Lett.* **6**, 1573–1576.

M. Takahashi, J. Tsuchiya and E. Niki (1989) Scavenging of radicals by vitamin E in the membranes as studied by spin labeling. *J. Am. Chem. Soc.* **111**, 6350–6353.

A. R. Waldeck and R. Stocker (1996) Radical-initiated lipid peroxidation in low density lipoproteins: insights obtained from kinetic modeling. *Chem. Res. Toxicol.* **9**, 954–964.

26

8-HYDROXYGUANINE, DNA ADDUCT FORMED BY OXYGEN RADICALS

From Its Discovery to Current Studies on Its Involvement in Mutagenesis and Repair

Susumu Nishimura

Banyu Tsukuba Research Institute, in Collaboration with Merck Research Laboratories
3 Okubo, Tsukuba 300-2611, Japan

1. ABSTRACT

8-Hydroxyguanine (7,8-dihydro-8-oxoguanine, abbreviated as 8-OH-G or 8-oxo-G) was discovered in 1983 during the course of study to isolate mutagens/carcinogens present in broiled foods. 8-OH-G was produced in DNA by various oxygen radical forming agents not only in *in vitro* reaction, but in *in vivo* systems. 8-OH-G became a very important modified base, because it in DNA induces G to T transvertion. Since oxidative DNA damage is a common type of genomic damage, 8-OH-G may play a critical role in a broad range of pathophysiological process, such as carcinogenesis, aging and degenerative diseases. Studies by many investigators in late 1980 to early 1990 showed that the three genes, *MutM* (a glycosylase/AP-lyase), *MutY* (a monofunctional DNA glycosylase that cleaves the misincorporated A residue paired with 8-OH-G) and *MutT* (8-OH-dGTPase which hydrolyzed 8-OH-dGTP in nucleotide pool) are involved in repair of 8-OH-G in E. coli, indicating importance of 8-OH-G in living organisms. An important question raised is whether similar systems for 8-OH-G exist in mammalian cells. In the case of *mutY* and *mutT*, similar genes have been identified. On the contrary, mammalian *mutM* homologues have not been characterized until recently. In June, 1997, six groups independently obtained a human or mouse homologue of yeast OGG1 (*MutM* homologue). A single gene in human cells (*hMMH* or *hOGG1*) produces four isoforms (type Ia, Ib, Ic and II) by alternative splicing. Type 1a expressed in E. coli showed both glycosylase and lyase activity. The isoform, type Ia contained a nuclear localization signal, but others did not, indicating that type Ia is involved in repair of 8-OH-G in nuclear DNA, and others in mitochondrial DNA

Advances in DNA Damage and Repair, edited by Dizdaroglu and Karakaya. Kluwer Academic / Plenum Publishers, New York, 1999.

repair. Whether *hMMH* is relevant to incidence of certain type of human cancer is an important question answered in future study.

2. DISCOVERY OF 8-HYDROXYGUANINE

Discovery of 8-hydroxyguanine (the same as 7,8-dihydro-8-oxoguanine; abbreviated as 8-OH-G or 8-oxo-G) in DNA by us (Kasai and Nishimura, 1983, 1984a) in 1983 was not the outcome initially intended, but accidental during the course of study to isolate new mutagens/carcinogens present in broiled foods. In late 1970, Takashi Sugimura, (Director of National Cancer Center Research Institute, Tokyo at that time, and presently President Emeritus of National Cancer Center) and his group discovered the production of large amounts of mutagenicity, when fish or meat was broiled for cooking. Hiroshi Kasai in my group (at that time, I was in Biology Division, National Cancer Center Research Institute) collaborated with them to isolate and characterize structure of mutagenic principles present in broiled food. Subsequently we were able to identify three active mutagens, namely 1Q and Methyl 1Q from broiled sardine, and Methyl 1Qx from broiled beef meat (Kasai et al., 1980a, b, 1981). They were quite potent mutagens, when assayed by Ames's *Salmonella* test, and later all of them were found to be carcinogenic in animal models. However, during the course of isolation of those mutagens, we noticed that large portion of mutagenic activity without microsomal activation were lost, due to its instability. Therefore, Kasai adopted a different method to attack this problem. Since most of mutagens/carcinogens are known to produce guanine adduct in DNA, isopropyridene derivative of guanosine was first incubated with a heated glucose (as a model compound of broiled food), extracted by ethylacetate, and then the extract was analyzed by HPLC. The two additional peaks appeared. Subsequently their structures were determined by UV, Mass and nmr. The adduct 1 turned out to be the reaction product with glyoxal. Interestingly, the adduct 2 was the product simply having oxygen at 8 position of guanine residue. This observation immediately indicated us that oxygen radicals can react with guanine to form guanine adduct.[*] Subsequently, deoxyguanosine and single-stranded and double-stranded DNA were incubated with various oxygen radical forming agents, including Udenfriends system, asbestos plus hydrogen peroxide, and X-ray, and in all cases, 8-hydroxyguanine was found to be formed (Kasai and Nishimura, 1984a,b; Kasai et al., 1984b). It is interest to note that 8-hydroxydeoxyguanine is a new compound at that time; even no body previously chemically synthesized, nor isolated from natural sources.[†]

3. DETECTION OF 8-OH-G *IN VIVO* BY OXYGEN RADICAL FORMING AGENTS

Even though the amount of 8-hydroxydeoxyguanosine produced by *in vitro* reaction was more than one per cent of deoxyguanosine residues added in the reaction mixture, it was expected to be quite a small amount *in vivo* reaction, if any produced, thus difficult to

[*] Later active principle present in heated glucose to generate oxygen radials were identified as reductic acid and methyl reductic acid (Kasai et al., 1989)

[†] 8-Hydroxyaguanosine was previously chemically synthesized by M. Ikehara and hiscolleagues (Ikehara et al., 1965).

be detected. A breakthrough to solve this problem was made by R. Floyd and his colleagues in 1986. They showed that electrochemical detection coupled with HPLC is able to analyze 8-OH-dG with more than 1000 times sensitive than that by conventional UV detection (Floyd et al., 1986). We immediately adopted their method in order to detect the presence of 8-OH-G in DNA *in vivo*. A whole body of mice was irradiated with different amount of doses of γ-ray. Immediately after the irradiation, the mice were killed, and DNA from their liver was isolated. DNA was converted to deoxynucleoside mixture by incubating it by nuclease P_1, followed by acid phosphatase, and then analyzed by electrochemical detector coupled with HPLC. It was clearly shown that the amounts of 8-OH-G in liver DNA were increased proportionally with increased does of γ -ray, approximately 1 molecule per 10^5 guanine residues (Kasai et al., 1986). Another important finding at that time was quick decrease of 8-OH-G in liver DNA, when the mice were kept after the irradiation, indicating that there is a repair system for 8-OH-G in mice. The presence of the repair system for 8-OH-G is a strong indication for the importance of 8-OH-G in living organisms. The amount of 8-OH-G in DNA *in vivo* by the irradiation was the same magnitude as that of thymine glycol. It is noteworthy to comment that 8-OH-G was not detected previously even many studies were done for many years in radiation biology field; perhaps because pyrimidine adduct, thymine glycol was a central topic.

8-OH-G was also formed *in vivo* by chemical carcinogens which are known to generate oxygen radicals. In 1985, we first showed that administration of renal carcinogen, potassium bromate into rats increased amounts of 8-OH-G in DNA in kidney, but not in liver DNA, suggesting that 8-OH-G is a good marker for identification of potential of chemical carcinogens (Kasai et al., 1987a). In order to get meaningful data, it is essential to reduce the background level of 8-OH-G by eliminating artificial formation of 8-OH-G in DNA during isolation of DNA from tissue and subsequent conversion of DNA to deoxynucleosides mixture. The following are the current procedures used now by H. Kasai and his colleagues (Asami et al., 1996); Namely, 1) DNA is extracted from the cell homogenate by WB kit (Nal method). 2) DNA is mixed with 1mM EDTA. 3) Add 2M sodium acetate, nuclease P1 and acid phosphatase, and incubate. 4) Mix with anion-exchange resin and filtrate to get the supernatant. 5) Inject into HPLC for electro-chemical analysis. The background level of 8-OH-G in DNA should be around 1–2 8-OH-G residue per 10^6 guanine residues in DNA.

4. CHEMICAL NATURE OF 8-OH-G

In earlier days, we showed that a favorable conformation of 8-OH-G is tautomeric 8-oxo-form rather than 8-hydroxy-form by Ab intitio calculation (Aida and Nishimura, 1987), or x-ray crystallographic analysis of 9-ethyl-8-hydroxyguanine (Kasai et al., 1987b). Thus, 8-OH-G can be named either as 8-hydroxyguanine or 7,8-dihydro-8-oxoguanine. Perhaps, 8-hydroxyguanine is chemically oriented name, and 7,8-dehydro-8-oxoguanine is biologically oriented name.*

An unique chemical nature of 8-OH-G is its ability to form base pair either with cytosine or adenine on opposite strands. This conclusion was obtained by nmr analysis of double-stranded oligonucleotides containing 8-OH-G in one strand and cytosine or adenine at the paired position in other strands (Oda et al., 1991; Kouchakdfian et al., 1991).

---

* 8-oxoguanine is not a formal name. It is abbreviated name of 7,8-dihydro-8-oxoguanine.

Recently, the same conclusion was obtained by x-ray crystallographic analysis (Lipscomb et al., 1995).

5. MISREPLICATION BY 8-OH-G *IN VITRO* AND *IN VIVO*

In 1987, we first reported that 8-OH-G was misread during DNA replication in *in vitro* reaction. When the deoxyoligonucleotide containing 8-OH-G residue in a specific position was used as a template for DNA synthesis by using Klenow fragment, 8-OH-G was read to incorporate not only C, but also A, C and G. In addition neibouring base next to 8-OH-G was also misread (Kuchino et al., 1987). This observation may be exaggerated, because of the artifact caused by the use of dideoxynucleoside triphosphates in DNA sequencing reaction. The more elegant study was done later by A. P. Grollman and his colleagues (Shibutani et al., 1991). By using polyacrylamide gel electrophoresis for separation of the reaction products, they clearly showed that 8-OH-G is only misread to incorporate adenine in the opposite strand. It is noteworthy that extent of misincorporation is much greater in the case of polymerase α as compared with Klenow fragment. In addition, misrecognition of 8-OH-G by adenine in *E. coli* was proved in *in vivo* system by several investigators (Wood et al., 1990; Cheng et al., 1992; Moriya et al., 1991). In the case of mammalian systems, 8-OH-G was misread to recognize A but also other bases with a significant extent. In addition, neibouring bases next to 8-OH-G was also misread. Those data were obtained by analyzing the sequences of the codon 12 or 16 of c-Ha-*ras* of the transformants after transfection of c-Ha-*ras* genes containing 8-OH-G at codon 12 or 61 (Kamiya et al., 1992, 1995). The further study is needed to solve the discrepancy of the data between prokaryote and eukaryote.

6. REPAIR SYSTEMS OF 8-OH-G IN *E. COLI*

If the presence of 8-OH-G is crucial for living organisms, there must be a mechanism to repair it in the cells. Thus we tried to isolate an enzyme from *E. coli* which cleaves DNA specifically at the site of 8-OH-G residue. For this purpose, double-stranded oligonucleotides containing 8-OH-G in a specific position were used as substrate for identification of a specific endonuclease (Chung et al., 1991). The enzyme thus isolated was found to be specific for 8-OH-G. The enzyme was only active for double-stranded DNA containing 8-OH-G, but not active against mispaired DNA with normal bases. It was later turned out that the specific endonuclease isolated from *E. coli* is the same as FPG protein (Tchou et al., 1991). Since then so much works were carried out to clarify mechanisms of repair of 8-OH-G in DNA in *E. coli* by many investigators including my group (Fig.1). Namely, the three genes (the gene products) are involved in repair of 8-OH-G. FPG protein (*MutM* product) is the specific glycosylase and endonuclease for 8-OH-G. With its inactivation, G to A transversion is enhanced (Tchou et al., 1991; Michaels et al., 1991). *MutY* is the gene for the specific mismatch repair enzyme to remove adenine residue paired with 8-OH-G (Michaels et al., 1992). Mutation of this gene also induces G to A transversion. Another interesting gene is *MutT* which codes for the specific phosphatase to cleave 8-OH-dGTP to 8-OH-dGMP (Maki and Sekiguchi, 1992). Oxygen radicals produce not only 8-OH-G in DNA but also 8-OH-dGTP from dGTP in nucleotide pool. 8-OH-dGTP generated is incorporated into DNA recognizing A instead of C, thus induces mutation. Once 8-OH-dGMP is formed, it is not converted again to 8-OH-dGTP. Thus

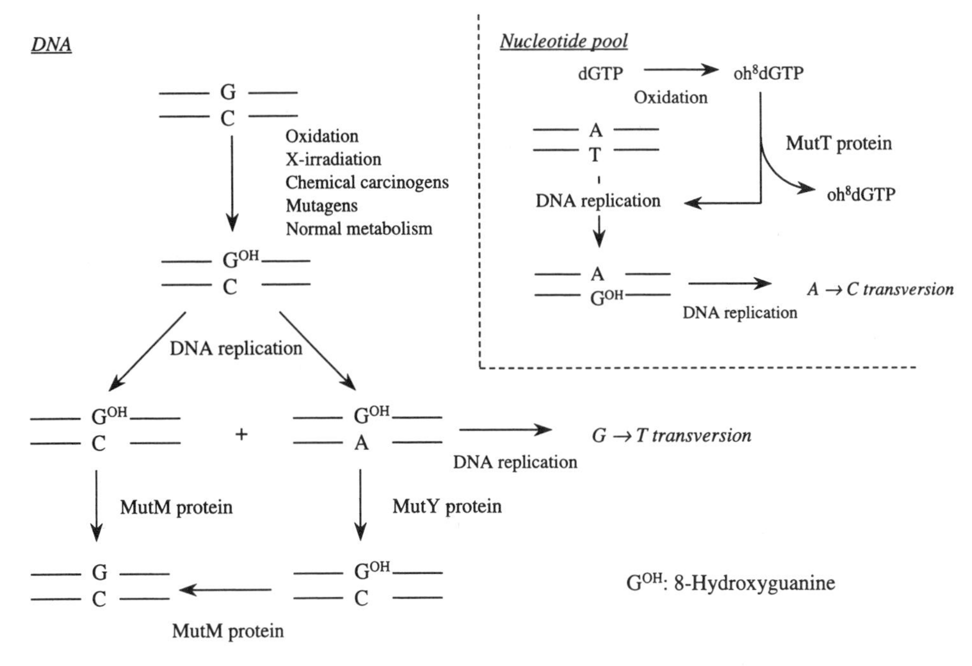

Figure 1. DNA repair for 8-hydroxyguanine in *E. coli.*

inactivation of *MutT* induces A to C transversion. The existence of the three genes for repair of 8-OH-G in DNA which beautifully cooperate to function, called as GO system by J. H. Miller (Michaels and Miller, 1992) is really an indication of the importance 8-OH-G in living organisms.

7. REPAIR SYSTEMS OF 8-OH-G IN MAMMALIAN CELLS

An important question raised is whether or not a similar mechanism for repair of 8-OH-G exists in mammalian cells. It has been reported that homologues of *MutY* and *MutT* exist in mammalian cells. The gene and its product for *MutT* were identified by M. Sekiguchi, and his coworkers (Sakumi et al., 1993). Cloning of a human homologue of *MutY* was also reported by J. H. Miller and his colleagues (Slupska et al., 1996). On the other hand, full characterization of mammalian *MutM* homologue has not been done until this year. We previously reported the presence of 8-OH-G specific endonuclease in crude extracts from various tissues from mouse, rat and cow and human leukocyte (Yamamoto et al., 1992; Chung et al., 1991; Besso et al., 1993). However, it was difficult to fully characterize the enzyme, due to its instability and divergence of the enzymatic activity during the course of the purification. Meanwhile, a gene of *MutM* homologue, called *OGG1* was isolated from yeast *Sacharomyces cerevisiae* in 1996 by S. Boiteux and his colleagues, and G. L. Verdine and his colleagues, independently (Van der Kemp et al., 1996; Nash et al., 1996). Recently six groups independently obtained a human or mouse homologue of *OGG1* by similarity search of human EST data base with yeast *OGG1* sequence (Aburantant et al., 1997; Rosenquist et al., 1997, Roldan-Arjona et al., 1997; Radicella et al., 1997; Arai et al., 1997; Lu et al., 1997). It is called *hOGGA* or *hMMH*. Y. Aburatani and

```
hMMH-1a :   1 MPARALLPRR MGHRTLASTP ALWASIPCPR SELRLDLVLP SGQSFRWREQ
hMMH-1b :   1 MPARALLPRR MGHRTLASTP ALWASIPCPR SELRLDLVLP SGQSFRWREQ
hMMH-1c :   1 MPARALLPRR MGHRTLASTP ALWASIPCPR SELRLDLVLP SGQSFRWREQ
hMMH-2  :   1 MPARALLPRR MGHRTLASTP ALWASIPCPR SELRLDLVLP SGQSFRWREQ
hOGG1#G :   1 MPARALLPRR MGHRTLASTP ALWASIPCPR SELRLDLVLP SGQSFRWREQ

hMMH-1a :  51 SPAHWSGVLA DQVWTLTQTE EQLHCTVYRG DKSQASRPTP DELEAVRKYF
hMMH-1b :  51 SPAHWSGVLA DQVWTLTQTE EQLHCTVYRG DKSQASRPTP DELEAVRKYF
hMMH-1c :  51 SPAHWSGVLA DQVWTLTQTE EQLHCTVYRG DKSQASRPTP DELEAVRKYF
hMMH-2  :  51 SPAHWSGVLA DQVWTLTQTE EQLHCTVYRG DKSQASRPTP DELEAVRKYF
hOGG1#G :  51 SPAHWSGVLA DQVWTLTQTE EQLHCTVYRG DKSQASRPTP DELEAVRKYF

hMMH-1a : 101 QLDVTLAQLY HHWGSVDSHF QEVAQKFQGV RLLRQDPIEC LFSFICSSNN
hMMH-1b : 101 QLDVTLAQLY HHWGSVDSHF QEVAQKFQGV RLLRQDPIEC LFSFICSSNN
hMMH-1c : 101 QLDVTLAQLY HHWGSVDSHF QEVAQKFQGV RLLRQDPIEC LFSFICSSNN
hMMH-2  : 101 QLDVTLAQLY HHWGSVDSHF QEVAQKFQGV RLLRQDPIEC LFSFICSSNN
hOGG1#G : 101 QLDVTLAQLY HHWGSVDSHF QEVAQKFQGV RLLRQDPIEC LFSFICSSNN

hMMH-1a : 151 NIARITGMVE RLCQAFGPRL IQLDDVTYHG FPSLQALAGP EVEAHLRKLG
hMMH-1b : 151 NIARITGMVE RLCQAFGPRL IQLDDVTYHG FPSLQALAGP EVEAHLRKLG
hMMH-1c : 151 NIARITGMVE RLCQAFGPRL IQLDDVTYHG FPSLQALAGP EVEAHLRKLG
hMMH-2  : 151 NIARITGMVE RLCQAFGPRL IQLDDVTYHG FPSLQALAGP EVEAHLRKLG
hOGG1#G : 151 NIARITGMVE RLCQAFGPRL IQLDDVTYHG FPSLQALAGP EVEAHLRKLG

hMMH-1a : 201 LGYRARYVSA SARAILEEQG GLAWLQQLRE SSYEEAHKAL CILPGVGTKV
hMMH-1b : 201 LGYRARYVSA SARAILEEQG GLAWLQQLRE SSYEEAHKAL CILPGVGTKV
hMMH-1c : 201 LGYRARYVSA SARAILEEQG GLAWLQQLRE SSYEEAHKAL CILPGVGTKV
hMMH-2  : 201 LGYRARYVSA SARAILEEQG GLAWLQQLRE SSYEEAHKAL CILPGVGTKV
hOGG1#G : 201 LGYRARYVSA SARAILEEQG GLAWLQQLRE SSYEEAHKAL CILPGVGTKV

hMMH-1a : 251 ADCICLMALD KPQAVPVDVH MWHIAQRDYS WHPTTSQAKG PSPQTNKELG
hMMH-1b : 251 ADCICLMALD KPQAVPVDVH MWHIAQRDYS WHPTTSQAKG PSPQTNKELG
hMMH-1c : 251 ADCICLMALD KPQAVPVDVH MWHIAQRDYS WHPTTSQAKG PSPQTNKELG
hMMH-2  : 251 ADCICLMALD KPQAVPVDVH MWHIAQRDYS WHPTTSQAKG PSPQTNKELG
hOGG1#G : 251 ADCICLMALD KEQAVPVDVH MWHIAQRDYS WHPTTSQAKG PSPQTNKELG
                                 ▼
hMMH-1a : 301 NFFRSLWGPY AGWAQAVLFS ADLRQSRHAQ EPPAKRRKGS KGPEG*(345 a.a.)
hMMH-1b : 301 NFFRSLWGPY AGWAQAVSVP RCPP*(324 a.a.)
hMMH-1c : 301 NFFRSLWGPY AGWAQATPPS YRCCSVPTCA NPAMLRSHQQ SAERVPKGRK
hMMH-2  : 301 NFFRSLWGPY AGWAQAGLLG NAFDGHQLLR PLIFCQDHLR EGPPIGRGDS
hOGG1#G : 301 NFFRSLWGPY AGWAQATPPS LQVLFSADLR QSRHAQEPPA KRRKGSKGPE


hMMH-1c : 351 ARWGTLDKEI PQAPSPPFPT SLSPSPPSLM LGRGLPVTTS KARHPQIKQS
hMMH-2  : 351 QGEELEPQLP SSLSSIPYGF CDHCWTKDVD DPPLVTHPSP GSRDGHMTQA
hOGG1#G : 351 G*(351 a.a.)


hMMH-1c : 401 VCTTRWGGGY*(410 a.a.)
hMMH-2  : 401 WPVKVVSPLA TVIGHVMQAS LLAL*(424 a.a.)
```

Figure 2. Amino acid sequence of isoforms of hMMH (hOGG1) i)hOGG1#G is an isoform identified by T.A. Rosenquist *et al* (Rosenquist, Zharkov and Grollman, 1997) ii)The closed triangle indicates the alternative site by which the four isoforms 1a, 1b,1c and z are produced.

his group in collaboration with us and others isolated four isoforms of the cDNAs produced by alternative splicing as well as genomic DNA (Aburatani et al., 1997). One of the human isoforms, type 1a was expressed in *E. coli* and its gene product was purified to homogeneity. It showed the both glycosylase and lyase activity, specific for 8-OH-G residue, similar to yeast OGG1. (Fig. 2, Table 1) Transfection of each four isoforms of cDNAs to *MutM* and *MutY* minus *E. coli* mutant reduced mutation frequency of G to A transvertion,

Table 1. Publications on *hMMH* (*hOOG1*)

Authors	Reported Clones	Activity
H. Aburatani *et al.* *Cancer Res.* (1997) (ref.38)	genomic DNA, hMMH-type 1a, 1b, 1c, 2	Complementation of *mutM mutY*/type 1a, 1b, 1c, 2 Nicking activity, Glycosylase activity/type 1a
T. A. Rosenquist *et al.* *Proc. Natl. Acad. Sci., USA* (1997) (ref.39)	hOGG1 (hMMH-type 1a+18bp addition)	Complementation of *mutM mutY*, Nicking activity, Glycosylase activity/ for mouseOGG1
T. Roldan-Arjona *et al.* *Proc. Natl. Acad. Sci., USA* (1997) (ref.40)	hOGG1 (hMMH-type 2)	Complementation of *mutM mutY*, Nicking activity, Glycosylase activity/type 2
J. P. Radicella *et al.* *Proc. Natl. Acad. Sci., USA* (1997) (ref.41)	hOGG1 (hMMH-type 1a)	Complementation of *mutM mutY*, Nicking activity/type 1a
K. Arai *et al.* *Oncogene* (1997) (ref.42)	hOOG1 (hMMH-type 1a)	Complementation of *mutM mutY*/type 1a
R. Lu *et al.* *Current Biology* (1997) (ref.43)	hOGG1 (hMMH-type 1a and hMMH-type 1c)	Nicking activity, Glycosylase activity/type 1a

indicating that all four gene products are active at least in *E. coli*. The isoform, type 1a contained a nuclear localization signal at its 3'-end, but others did not. It is likely that type 1a is involved in repair of 8-OH-G in nuclear DNA, and others in mitochondrial DNA repair.

In further studies, it is important to analyze quantitatively amounts of the proteins corresponding to each isoform in different tissues, together with their cellular localization. The human homologue of *MutM*, *hMMH* (called also as *hOGG1*) is localized in chromosome 3p25. This reageon is known to have frequently LOH (loss of heterozygosity) in various human cancers. Whether *hMMH* is relevant to incidence of certain type of human cancers in another important question answered in future study.

REFERENCES

H. Aburatani. Y. Hippo, T. Ishida, R. Takashima, C. Matsuba, T. Kodama, M. Takao, A. Yasushi, K. Yamamoto, M. Asano, K. Fukasawa, T. Yoshinari, H. Inoue, E. Ohtsuka and S. Nishimura (1997) Cloning and characterization of mammalian 8-hydroxyguanine-specific DNA glycosylase/apurinic, apyrimidinic lyase, a functional *mutM* homologue. *Cancer Res.* **57**, 2151–2156

M. Aida and S. Nishimura (1987) An ab inito molecular orbital study on the characterization of 8-hydroxyguanine. *Mutation Res.* **192**, 83–89

S. Asami, T. Hirano, R. Yamaguchi, Y. Tomioka, H. Itoh and H. Kasai (1996) Increase of a type of oxidative DNA damage, 8-hydroxyguanine, and its repair activity in human leukocytes by cigarette smoking. *Cancer Res.* **56**, 2546–2549

K. Arai, K. Morishita, K. Shinmura, T. Kohno, S.-R. Kim., T. Nohmi, M. Taniwaki, S. Ohwada and J. Yokota (1997) Cloning of a human homologue of the yeast *OGG1* gene that is involved in the repair of oxidatie DNA damage. *Oncogene* **14**, 2857–2861

T. Besso, K. Tano, H. Kasai, E. Ohtsuka and S. Nishimura (1993) Evidence for two DNA repair enzymes for 8-hydroxyguanine (7,8-dihydro-8-oxoguanine) in human cells. *J. Biol. Chem.* **268**, 19416–19421

K.C. Cheng, D.S. Cahill, H. Kasai, S. Nishimura and L.A. Loeb (1992) 8-Hydroxyguanine, abundant form of oxidative DNA damage, causes G-T and A-C substitution. *J. Biol. Chem.* **267**, 166–172

M.H. Chung, H. Kasai, D.S. Jones, H. Inoue, H. Ishikawa, H. Ohtsuka and S. Nishimura (1991) An endonuclease activity of *Escherichia coli* that specifically remove 8-hydroxyguanine residues from DNA. *Mutation Res.* **254**, 1–12

M.-H. Chung, S.-H. Kim, E. Ohtsuka., H. Kasai, F. Yamamoto and S. Nishimura (1991) An endonuclease activity in human polymorphonuclear neutrophils that removes 8-hydroxyguanine residues from DNA. *Biochem. Biophys. Res. Communin.* **178**, 1472–1478

R.A. Floyd, J.J. Watson, P.K. Wong, D.H. Altmiler and R.C. Richard (1986) Hydroxy free radical adduct of deoxyguanosine: sensitive detection and mechanism of formation. *Free Radical Res.*, Commun. **1**, 163–172

M. Ikehara, H. Tada and K. Muneyama (1965) Synthesis of 8-hydroxypurine nucleoside. *Chem. Pharm. Bull.* **13**, 1140–1142

H. Kamiya, K. Miura, H. Ishikawa, H. Inowe, S. Nishimura and E. Ohtsuka (1992) c-Ha-*ras* containing 8-hydroxyguanine at codon 12 induces point mutation at the modified and adjacent positions. *Cancer Res.* **52**, 3483–3485

H. Kamiya, N. Murata-Kamiya, S. Kaizumi, H. Inoue, S. Nishimura and E. Ohtsuka (1995) 8-Hydroxyguanine (7,8-dihydro-8-oxoguanine) in hot spots of the c-Ha-*ras* gene: effect of sequence contexts on mutation spectra. *Carcinogenesis* **16**, 883–889

H. Kasai and S. Nishimura (1983) Hydroxylation of the C-8 position of deoxyguanine by reducing agents in the presence of oxygen. *Nucl. Acids Res. Symp. Ser.* **No. 12**, s165–167

H. Kasai and S. Nishimura (1984a) Hydroxylation of deoxyguanosine at the C-8 position by ascorbic acid and other reducing agents. *Nucl. Acids Res.* **12**, 2137–2145

H. Kasai, Z. Yamaizumi, K. Wakabayashi, M. Nagao, T. Sugimura, S. Yokoyama, T. Miyagawa, N.E. Spingarn, J.H. Weisburger and S. Nishimura (1980a) Potent novel mutagens produced by broiling fish under normal conditions. *Proc. Japan Acid.* (Ser.B) **56**, 278–283

H. Kasai, Z. Yamaizumi, K. Wakabayashi, M. Nagao, T. Sugimura, S. Yokoyama, T. Miyazawa and S. Nishimura (1980) Structure and chemical synthesis of Me-1Q, a potent mutagen isolated from broiled fish. *Chemistry Lett.* 1391–1394

H. Kasai, Z. Yamaizumi, T. Shiomi, S. Yokoyama, T. Miyazawa, K. Wakabayashi, M. Nagao, T. Sugimura and S. Nishimura (1981) Structure of a potent mutagen isolated from fried beef. *Chemistry Lett.* 485–488

H. Kasai and S. Nishimura (1984b) Hydroxylation of deoxyguanosine at the C-8 position by polyphenols and aminophenol in the presence of hydrogen peroxide and ferric ion. *Gann* **75**, 565–566

H. Kasai, H. Tanooka and S. Nishimura (1984b) Formation of 8-hydroxyguanine residue in DNA by X-irradiation. *Gann* **75**,1037–1039

H. Kasai and S. Nishimura (1984) DNA damage induced by asbestos in the presence of hydrogen peroxide. *Gann* **75**, 841–843

H. Kasai, P.F. Crain, Y. Kuchino, S. Nishimura, A. Ootsuyama and H. Tanooka (1986) Formation of 8-hydroxyguanine moiety in cellular DNA by agents producing oxygen radicals and evidence for its repair. *Carcinogenesis* **7**, 1849–1851

H. Kasai, S. Nishimura, Y. Kurokawa and Y. Hayashi (1987) Oral administration of the renal carcinogen, potassium bromate, specifically produces 8-hydroxydeoxyguanosine in rat target organ DNA. *Carcinogenesis* **8**, 1959–1961

H. Kasai, M. Nakayama, N. Toda, Z. Yamaizumi, J. Oikawa and S. Nishimura (1989) Methylreductic and hydroxymethylreductic acid: oxygen radical-forming agents in heated starch. *Mutation Res.* **214**, 159–164

H. Kasai, S. Nishimura, Y. Toriumi, A. Itai and Y. Iitaka (1987) The crystal structure of 9-ethyl-8-hydroxyguanine. *Bull. Chem. Soc. Jpn.* **60**, 3799–3800

M. Kouchakdfian, V. Bodepudi, S. Shibutani, M. Eisenberg, F. Johnson, A.P. Grollman and D.J. Patal (1991) NMR structural studies of the ionizing radiation adduct 7-hydro-8-oxodeoxyguanosine (8-oxo-7H-dG) opposite deoxyadenosine in a DNA duplex. 8-Oxo-7H-dG(syn).dA(anti) alignment at lesion site. *Biochemistry* **30**, 1403–1412

Y. Kuchino, F. Mori, H. Kasai, S. Inoue, K. Iwai, K. Miura, E. Ohtsuka and S. Nishimura (1987) Misreading of DNA template containing 8-hydroxyguanine at the modified base and at adjacent residue. *Nature* **327**, 77–79

L.A. Lipscomb, M.E. Peek, M.L. Morningstar, S.M. Verghis, E.M. Miller, A. Ric, J.H. Essigmann and L.D. Williams (1995) X-ray structure of a DNA decamer containing 7,8-dihydro-8-oxoguanine. *Proc. Natl. Acad. Sci.*, USA **92**, 719–723

R. Lu, H.M. Nash and G.L. Verdine (1997) A mammalian DNA repair enzyme that excises oxidatively damaged guanines maps to a locus frequently lost in lung cancer. *Current Biology* **7**, 397–407

H. Maki and M. Sekiguchi (1992) *MutT* protein specifically hydrolyzes a potent mutagenic substrate for DNA synthesis. *Nature* **355**, 273–275

M.L. Michaels, L. Pham, C. Cruz, J.H. Miller (1991) MutM, a protein that prevents G.C→T.A transversions, is formamidopyrimidine-DNA glycosylase. *Nucl. Acids Res.* **19**, 3629–3632

M.L. Michaels and J.H. Miller (1992) The GO system protects organisms from the mutgaenic effect of the spontaneous lesin 8-hydroxyguanine (7,8-dihydro-8-oxoguanine). *J. Bacteriol.* **174**, 6321–6325

M.L. Michaels, C. Cruz, A.P. Grollman and J.H. Miller (1992) Evidence that *MutY* and *MutM* combine to prevent mutations by an oxidatively damaged form of guanine in DNA. *Proc. Natl. Acad. Sci.*, USA **89**, 7022–7025

M. Moriya, C. Cu, V. Bodepudi, F. Johnson, M. Takeshita and A.P. Grollman (1991) Site-specific mutagenesis using a gapped duplex vector: a study of translesion synthesis past 8-oxodeoxyguanosine in *E. coli*. Mutation Res. **254**, 281–288

H.M. Nash, S.D. Bruner, O.D. Scharer, T. Kawate, T.A. Addona, E. Spooner, W.S. Lane and G.L. Verdine (1996) Cloning of a yeast 8-oxoguanine DNA glycosylase reveals the existanve of a base-excision DNA repair protein superfamily. *Curr. Biol.* **6**, 968–980

Y. Oda, S. Uesugi, M. Ikehara, S. Nishimura, Y. Kawase, H. Ishikawa, H. Inoue and E. Ohtsuka (1991) NMR studies of a DNA containing 8-hydroxydeoxyguanosine. *Nucl. Acids. Res.* **19**, 1407–1412

J.P. Radicella, C. Dherin, C. Desmaze, M.S. Fox, and S. Boiteux (1997) Cloning and characterization of *hOGGA*, a human homologue of the *OGG1* gene of *Saccharomyces cerevisiaee. Proc. Natl. Acad. Sci.*, USA **94**, 8010–8015

T. Rodlan-Arjona, Y.-F. Wei, K.C. Carter, A. Klungland, C. Anselmino, R.-P. Wang, M. Augstutand T. Lindahl (1997) Molecular cloning and functional expression of a human cDNA encoding the antimutator enzyme 8-hydroxyauanine-DNA glycosylase. *Proc. Natl. Acad. Sci.*, USA **94**, 7429–7434

T.A. Rosenquist, D.I. Zharkov and A.P. Grollman (1997) Cloning and characterization of a 8-oxoguanine DNA glycosylase. *Proc. Natl. Acad. Sci.*, USA **94**, 7429–7434

K. Sakumi, M. Furuichi, T. Tsuzuki, T. Kakuma, S. Kawabata, H. Maki and M. Sekiguchi (1993) Cloning and expression of cDNA for a human enzyme that hydrolyzes 8-oxo-dGTP, a mutagenic substrate for DNA synthesis. *J. Biol. Chem.* **268**, 23524–23530

S. Shibutani, M. Takeshita and A.P. Grollman (1991) Insertion of specific bases during DNA synthesis part the oxidation-damaged base 8-oxodG. *Nature* **327**, 77–79

M.M. Slupska, C. Baikalov, W.M. Luther, J.H. Chiang, Y.F. Wei and J.H. Miller (1996) Cloning and sequencing a human homologue (*hMYH*) of the *Escherichia coli mutY* gene whose function is required for the repair of oxidative DNA damage. *J. Bacterial* **178**, 3885–3892

J. Tchou, H. Kasai, S. Shibutani, M.-H. Chug, J. Laval, A.P. Grollman and S. Nishimura (1991) 8-Oxoguanine (8-hydroxyguanine) DNA glycosylase and its substrate specificity. *Proc. Natl. Acad. Sci. USA* **88**, 4690–4694

P.A. van der Kemp, D. Thomas, R. Barkey, R. de Oliveira and S. Boitex (1996) Cloning and expression in *Escherichia coli* of the *Oggl* gene of *Saccharomyces cerevisiae* which codes for a DNA glycosylase that excises 7,8-dihydro-8-oxoguanine and 2,6-diamino-4-hydroxy-5-N-methylformamidopyrimidine. *Proc. Natl. Acad. Sci., USA* **93**, 5197–5202

M.L. Wood, M. Dizdaroglu, E. Gajewski and J.M. Essigmann (1990) Mechanistic studies of ionizing radiation and oxidative mutagenesis: genetic effect of a single 8-hydroxyguanine (7-hydro-8-oxoguanine) residue inserted at a unique site in a viral genome. *Biochemistry* **29**, 7024–7032

F. Yamamoto, H. Kasai, T. besso, M.H. Chung, H. Inoue, E. Ohtsuka, T. Hori and S. Nishimura (1992) Ubiquitous presence in mammalian cells of enzymatic activity specifically cleaving 8-hydroxyguanine-containing DNA. *Jpn. J. cancer Res.* **83**, 351–357

27

DNA DAMAGE INDUCED BY REACTIVE NITROGEN SPECIES

Hiroshi Ohshima,* Vladimir Yermilov,† Yumiko Yoshie,‡ and Julieta Rubio°

International Agency for Research on Cancer
Unit of Endogenous Cancer Risk Factors
150 Cours Albert-Thomas
69372, Lyon, Cedex 08, France

1. ABSTRACT

Chronic infection by bacteria, parasites or viruses and tissue inflammation such as gastritis and colitis are recognized risk factors for human cancers at various sites. Reactive oxygen and nitrogen species produced by activated inflammatory cells play an important role in the multistage carcinogenesis process, by inducing DNA and tissue damage and mutations. Nitric oxide (NO), generated enzymatically from L-arginine by constitutive and inducible NO synthases, acts not only as a signal molecule mediating various physiological functions, such as vasodilation and neurotransmission, but also acts as a mediator of the cytotoxic activity of macrophages, playing an important role in inflammation processes. Excess NO in the presence of oxygen can be converted to nitrosating agents, which can damage DNA through various mechanisms including deamination, induction of strand breaks and alkylation by nitrosamines formed by reaction of secondary amines with nitrosating agent(s) derived from NO. Superoxide also reacts rapidly with NO to yield a potent oxidant, peroxynitrite, which can induce DNA strand breaks and causes various DNA base modifications through nitration, oxidation and deamination. On the other hand, NO has been shown to protect against DNA damage or lipid peroxidation mediated by the

* Address for correspondence: International Agency for Research on Cancer, 150 Cours Albert-Thomas, 69372 Lyon, Cedex 08, France, Phone: 33 4 72 73 84 85; Fax: 33 4 72 73 85 75; E-mail: Ohshima@IARC.FR

† Present address: N. N. Petrov Research Institute of Oncology, Leningradskaya Street 68, Pesochny 2, 189646 St Petersburg, Russia

‡ Present address: Tokyo University of Fisheries, 5-7 Konan 4, Minato-ku, Tokyo 108, Japan

° Present address: Instituto de Investigaciones Biomédicas, UNAM, Apdo Postal 70228, Mexico 04510 D. F., Mexico

Advances in DNA Damage and Repair, edited by Dizdaroglu and Karakaya. Kluwer Academic / Plenum Publishers, New York, 1999.

Fenton-type reactions (H_2O_2 plus metal ions). In these cases NO may form a complex with a metal ion and/or reactive species or terminate chain reaction of lipid peroxydation, inhibiting formation of reactive species. The balance between production of NO and reactive oxygen species and interactions of NO with metal ions are important in inflammation-associated human diseases including cancer.

2. INTRODUCTION

Chronic infection by bacteria, parasites or viruses and tissue inflammation such as gastritis and colitis are recognized risk factors for human cancers at various sites (Ohshima and Bartsch, 1994). Reactive oxygen species such as superoxide anion, hydrogen peroxide and hydroxyl radical generated in inflamed tissues can cause injury to target cells and also damage DNA, which could contribute to cancer development. There is now increasing evidence to suggest that nitric oxide (NO) and its derivatives (peroxynitrite, NOx) produced by activated inflammatory cells also play an important role in the multistage carcinogenesis process (Ohshima and Bartsch, 1994; Liu and Hotchkiss, 1995; Tamir and Tannenbaum, 1996). NO, a potentially toxic gas with free radical properties, can be generated enzymatically from L-arginine by constitutive and inducible NO synthases. NO acts not only as a signal molecule mediating various physiological functions, such as vasodilation and neurotransmission, but also as a mediator of the cytotoxic activity of macrophages, playing an important role in inflammation processes. In this article, we review the various DNA modifications induced by exposure to NO and its derivatives.

3. DNA DAMAGE INDUCED BY NO AND NOx

Excess NO can damage DNA through various mechanisms such as deamination, induction of strand breaks and alkylation by nitrosamines formed by reaction of secondary amines with nitrosating agent(s) derived from NO (Fig. 1). NO gas and various NO-releasing compounds such as diethylamine-NONOate have been reported to be mutagenic and genotoxic in *Salmonella typhimurium* as well as in human lymphoblastoid cells (Wink et al. 1991; Nguyen et al. 1992). NO can be converted in the presence of oxygen to a strong nitrosating agent N_2O_3, which can deaminate various DNA bases (e.g. guanine to xanthine, adenine to hypoxanthine, cytosine to uracil, 5-methylcytosine to thymine). These DNA base modifications lead to a variety of mutations (Tannenbaum et al. 1994). Routledge et al. (1993, 1994) studied mutagenicity of NO gas or NO-releasing compounds in the *supF* gene of the pSP189 shuttle vector. Aerobic gaseous NO induced primarily A:T to G:C transitions (Routledge et al. 1993), whereas aqueous NO, derived from NO-releasing compounds, generated largely G:C to A:T transitions (Routledge et al. 1994). Deamination of adenine to hypoxanthine, which pairs with cytosine rather than thymine in DNA, could account for A:T to G:C transitions. G:C to A:T transitions could result from deamination of cytosine to uracil, of 5-methylcytosine to thymine and/or of guanine to xanthine (Routledge et al. 1993, 1994).

In several tumor suppressor genes such as *p53* and the retinoblatoma gene *RB*, G:C to A:T transitions are frequently observed at CpG sites. For example, about 80% of mutations found in human colon tumors contains these transitions at CpG sites (Greenblatt et al. 1994). Deamination by NO of 5-methylcytosine to thymine at CpGs has been proposed as being responsible for spontaneous C to T mutations in vivo (Wink et al. 1991; Nguyen

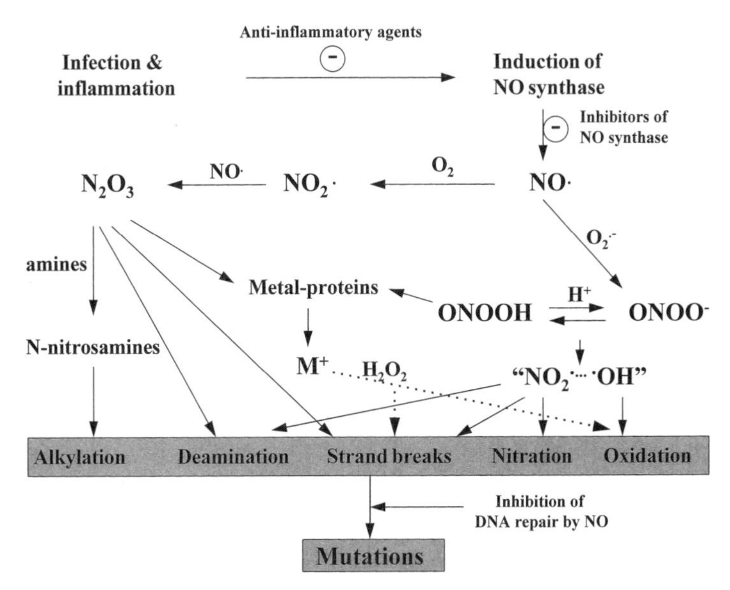

Figure 1. DNA damage induced by NO and its derivatives (NOx and peroxynitrite).

et al. 1992). However, recent studies have reported findings that the genotoxicity of NO is not caused by deamination of cytosine or 5-methylcytosine to uracil or thymine, respectively, in double-stranded DNA (Schmutte et al. 1994). It has been also reported that 5-methylcytosine in codon 248 of the *p53* gene, which is one of the hot spots for G:C to A:T mutations at CpG sites, is not deaminated in cultured human bronchial epithelial cells exposed to high concentrations of an NO-releasing compound (Felley Bosco et al. 1995).

DNA base deamination by nitrous acid could also lead to crosslinking between amino groups of deoxyguanosine to form a dG-to-dG interstrand cross-link at the duplex sequnce 5'-CG (Kirchner et al. 1992) (Fig. 2). Suzuki et al. (1996) recently have shown that 2'-deoxyoxanosine is produced as a major product in the reaction of deoxyguanosine with nitrous acid, together with 2'-deoxyxanthosine (Fig. 2). The same compounds were formed in calf thymus DNA treated with either acidic sodium nitrite or NO at neutral pH.

Carcinogenic *N*-nitrosamines are formed in macrophage culture on addition of secondary amines and incubation with immunostimulators such as lipopolysaccharide and interferon-γ for 24–72 h (Iyengar et al. 1987; Miwa et al. 1987; Ohshima et al. 1991). In addition to macrophages, hepatocytes and neutrophils can yield *N*-nitrosamines upon incubation with amines in culture (Liu et al. 1992; Grisham et al. 1992). Increased formation of nitrosamines have been reported to occur in vivo in experimental animals with acute and chronic inflammation (Leaf et al. 1991; Wu et al. 1993; Ohshima et al. 1994).

4. DNA BASE MODIFICATIONS CAUSED BY PEROXYNITRITE

Both NO and superoxide are free radicals. They undergo a rapid radical-radical reaction (reaction rate, $6.7 \times 10^9\ M^{-1} S^{-1}$) to form peroxynitrite anion ($ONOO^-$), which is a highly reactive species, causing rapid oxidation of sulfhydryl groups and thioethers, as well as nitration and hydroxylation of aromatic compounds, such as tyrosine and tryptophan (Pryor and Squadrito, 1995; Beckman and Koppenol, 1996). The ratio of superoxide

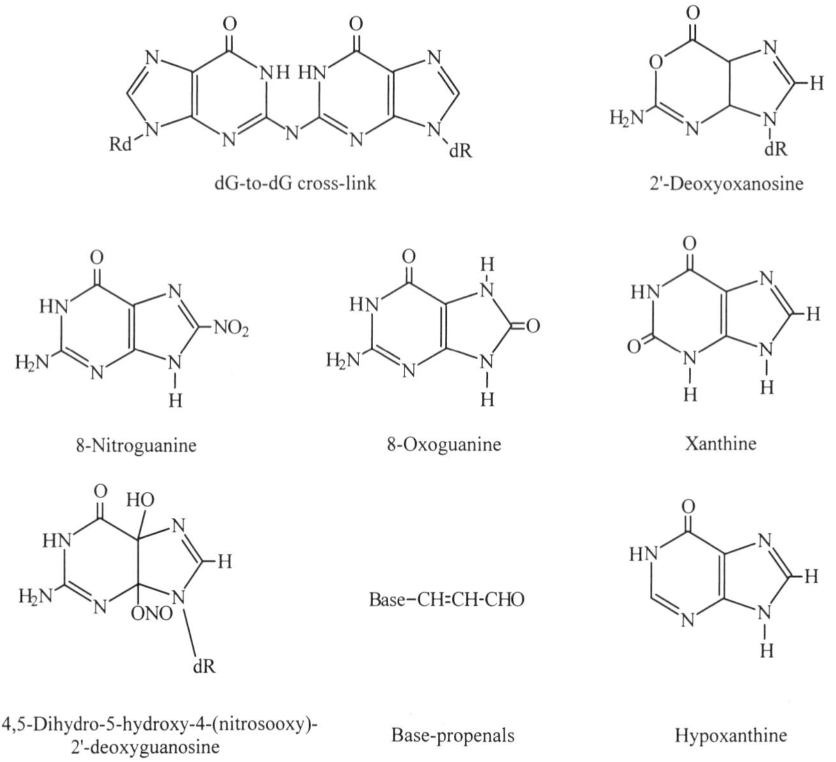

dG-to-dG cross-link

2'-Deoxyoxanosine

8-Nitroguanine

8-Oxoguanine

Xanthine

4,5-Dihydro-5-hydroxy-4-(nitrosooxy)-2'-deoxyguanosine

Base−CH=CH-CHO

Base-propenals

Hypoxanthine

Figure 2. DNA base modifications induced by reactive nitrogen species (NO, NOx and peroxynitrite). In addition to the modifications shown in this figure, increased production of various oxidative base modifications have also been reported (see text).

to NO is important in determining the reactivity of peroxynitrite: excess NO or excess superoxide affects the oxidation and nitration reactions elicited by peroxynitrite (Rubbo et al. 1994; Miles et al. 1996). The oxidant reactivity of peroxynitrite is mediated by an intermediate with the biological activity of the hydroxyl radical. However, this product appears not to be the free hydroxyl radical, but peroxynitrous acid (ONOOH) or its activated isomer (ONOOH*) (Pryor and Squadrito, 1995).

As shown in Fig. 2, peroxynitrite can initiate a number of DNA modifications. Yermilov et al. (1995a) studied the reaction of various nucleobases and nucleosides with peroxynitrite in vitro. The major yellow compound formed by the reaction between guanine and peroxynitrite has been identified as 8-nitroguanine (Yermilov et al. 1995a). Reactions of isolated DNA with authentic peroxynitrite also formed 8-nitroguanine dose-dependently (Yermilov et al. 1995b). Only peroxynitrite, but not nitrous acid, tetranitromethane or NO-releasing compounds, formed 8-nitroguanine. Therefore 8-nitroguanine in DNA could be measured as a specific marker for peroxynitrite-mediated DNA damage. Desferrioxamine and antioxidants such as uric acid, but not hydroxyl radical scavengers inhibited the reaction (Yermilov et al. 1995b). Bicarbonate (0–10 mM) caused a dose-dependent increase of up to 6-fold in the formation of 8-nitroguanine in DNA. 8-Nitroguanine was found to be depurinated rapidly from DNA incubated at pH 7.4, 37°C ($t_{1/2}$ = ~4 h), suggesting that 8-nitroguanine formed in DNA is potentially mutagenic, because its depurination yields apurinic sites, which can induce G:C to T:A transversions (Yermilov et al. 1995b).

Considerable controversy exists regarding whether synthesized peroxynitrite reacts with guanine to form 8-oxoguanine in DNA. Yermilov et al. (1995b), Douki and Cadet (1996) and Uppu et al. (1996) reported no significant increases in 8-oxoguanine in calf thymus DNA treated with synthesized peroxynitrite, compared to non-treated DNA or DNA treated with decomposed peroxynitrite. On the other hand, Inoue and Kawanishi (1995), Fiala et al. (1996) and Spencer et al. (1996) found significant increases in 8-oxoguanine in calf thymus DNA treated with peroxynitrite, the levels formed with peroxynitrite being 10–40 times higher than controls. Epe et al. (1996) recently reported that peroxynitrite could induce a large number of base modifications, which were sensitive to Fpg protein, suggesting that 8-oxoguanine was a major modification. More recently Kennedy et al. (1997) also reported that low doses (<50 μM) of peroxynitrite formed 8-oxoguanine in plasmid DNA dose-dependently, its level being decreased with a high dose (100 μM) of peroxynitrite. A similar plateau or decrease in 8-oxoguanine formation at high concentrations of peroxynitrite was also reported (Inoue and Kawanishi, 1995; Yermilov et al. 1995b). Possible reasons for this controversy could be that (i) formation of 8-oxoguanine is mediated by contaminants (e.g. hydrogen peroxide and metal ions), the concentrations of which may vary in different preparations of peroxynitrite; (ii) 8-oxoguanine may be formed artificially during isolation, hydrolysis and analyses of DNA (e.g. conversion of 8-nitroguanine to 8-oxoguanine); and (iii) under certain conditions, especially with high concentrations of peroxynitrite, the 8-oxoguanine that is produced may be further oxidized into the ring cleavage product by peroxynitrite (Uppu et al. 1996; Kennedy et al. 1997).

3-Morpholinosydnonimine (SIN-1), that simultaneously generates NO and superoxide, thus possibly forming peroxynitrite, has also been used to study DNA damage. Inoue and Kawanishi (1995) reported increased production of 8-oxoguanine in DNA treated with SIN-1. Yermilov et al. (1995b) also found that SIN-1 increased dose-dependently the level of 8-oxoguanine, but not that of 8-nitroguanine, in DNA, in contrast with the fact that authentic peroxynitrite formed 8-nitroguanine, but not 8-oxoguanine, in DNA. One possible reason for this observation could be that SIN-1 may produce other oxidants (such as the peroxynitrite radical $ONOO^{\cdot}$), which may be responsible for the formation of 8-oxoguanine (Yermilov et al. 1995b; Uppu et al. 1996).

Exposure of human skin epidermal keratinocytes to preformed peroxynitrite or SIN-1 led to extensive DNA base modification (Spencer et al. 1996). With both compounds, large increases in xanthine and hypoxanthine (deamination products of guanine and adenine, respectively) and 8-nitroguanine were observed, whereas only small increases in some oxidized bases including 8-oxoguanine and FAPy-guanine were found in the DNA from keratinocytes. Similarly, small increases in 8-oxoadenine and oxazolone were found in calf-thymus DNA treated with peroxynitrite (Douki and Cadet, 1996). Increased levels of xanthine, 5-(hydroxymethyl)uracil and 8-oxoguanine were also detected in the DNA of macrophages activated with lipopolysaccharide and interferon-γ, which were indicative of both oxidative and deaminative DNA damage (deRojas-Walker et al. 1995). Formation of both xanthine and 8-oxoguanine was inhibited by an NO synthase inhibitor, suggesting that NO plays a role in both deamination and oxidation reactions. The authors postulated that peroxynitrite or metal ions released from metal-proteins by NO, which could participate in the Fenton reaction, may be responsible for the oxidative damage (deRojas-Walker et al. 1995).

Juedes and Wogan (1996) studied mutagenicity of peroxynitrite in the *supF* gene of the pSP189 shuttle vector. The plasmid was exposed to peroxynitrite in vitro, then replicated in *Escherichia coli* and in human AD293 cells. Mutation frequency increased 21-fold in pSP189 replicated in *E. coli* and 9-fold in plasmid replicated in human AD293 cells. In both systems, G:C to T:A transversions (~65%) were predominantly detected. The G:C to T:A transversions could result from 8-oxoguanine or apurinic/apyrimidinic sites (Juedes and Wogan, 1996). On the other hand, qualitatively very different mutation spectra have been reported for NO using similar test systems (see above).

In addition to the above DNA damage, the reaction of 2'-deoxyguanosine with peroxynitrite was shown to yield several compounds, two of which were identified as 4,5-dihydro-5-hydroxy-4-(nitrosooxy)-2'-deoxyguanosine and 8-nitroguanine (Douki et al. 1996) (Fig, 2). Kinetics on the formation of 4,5-dihydro-5-hydroxy-4-(nitrosooxy)-2'-deoxyguanosine however have not been studied. The reaction of various deoxynucleosides with peroxynitrite was also shown to yield highly cytotoxic base-propenals (base-CH=CH-CHO) (Rubio et al. 1996). The reaction mechanism for base-propenal formation could involve radical abstraction of a hydrogen from the C4' position of deoxyribose by hydroxyl radical-like intermediate(s) (ONOOH*) or peroxynitrous acid (ONOOH). Base-propenals can be derived from bases with carbon atoms C1', C2' and C3' of the deoxyribose molecule after cleavage of C3'-C4' and C1'-(ring-O) bonds (Giloni et al. 1981; Burger et al. 1994).

5. DNA STRAND BREAKAGE INDUCED BY PEROXYNITRITE AND BY NO PLUS SUPEROXIDE

Several studies have shown that NO itself dose not cause strand breakage in vitro in plasmid DNA (Tamir et al. 1996; Yoshie and Ohshima, 1997a), whereas exposure of plasmid DNA to preformed peroxynitrite (Salgo et al. 1995b; Epe et al. 1996; Yermilov et al. 1996), SIN-1 (Inoue and Kawanishi, 1995; Yoshie and Ohshima, 1997a) or NO plus superoxide generated concurrently (Yoshie and Ohshima, 1997a, 1997c) can induce strand breakage. Single strand breakage can be induced in pBR322 plasmid by treatment with concentrations of peroxynitrite as low as 1 μM, whereas much higher concentrations of peroxynitrite (> 1 mM) (Rubio et al. 1996) or the presence of a catalyst such as manganese porphyrin (Groves and Marla, 1995) were needed to induce double strand breakage. DNA cleavage caused by peroxynitrite was observed at almost every nucleotide with a small dominance at guanine residues (Inoue and Kawanishi, 1995). Peroxynitrite induced significantly more single strand breaks at acidic pH than at neutral or alkaline pH, suggesting that hydroxyl radical-like intermediate(s) (ONOOH*) or peroxynitrous acid (ONOOH) are responsible for the damage (Yermilov et al. 1996). Salgo et al. (1995b) reported that mannitol failed to protect DNA from damage by peroxynitrite, while benzoate and dimethylsulfoxide amplified the breakage, indicating that the free hydroxyl radical is not involved in the damage. On the other hand, various compounds can inhibit peroxynitrite-mediated strand breakage, including selenomethionine, selenocystine, desferrioxamine, antioxidants such as ascorbate and urate, carbon dioxide/bicarbonate and carboxy-PTIO (Epe et al. 1996; Roussyn et al. 1996; Yermilov et al. 1996; Yoshie and Ohshima, 1997a, 1997c). A possible mechanism for peroxynitrite-mediated strand breakage has been proposed on the basis of available data in the literature (Szabo and Ohshima, 1997). The initial reaction could involve hydrogen abstraction and O_2 attack at either deoxyribose C4' or C5' by hydroxy-radical like intermediate(s) (ONOOH*) or peroxynitrous acid (ONOOH). After the modification at C4', the strand breakage can be induced by either C3'-(phosphate-O) cleavage or C3'-C4' plus C1'-(ring-O) bond cleavages. The

Table 1. Phenolic and related compounds that induce DNA strand breaks in the presence of NO*

Positive	Negative
Catechol	Phenol
1,4-Hydroquinone	Resorcinol
Pyrogallol	Guaiacol
DOPA	3-*O*-methyldopa
Dopamine	Tyrosine
Epinephrine	Tyramine
2-Hydroxy-estradiol or -estrone	β-Estradiol
4-Hydroxyestradiol or -estrone	Estron
Epigallocatechin	Catechin
Cigarette smoke (tar)	
Hypoxanthine/xanthine oxidase	

Positive and negative: compounds which either induce or do not induce DNA strand breakage in the presence of NO. NO alone does not induce the breakage.
* Yoshie and Ohshima, 1997a, 1997b, 1997c ,1998 and unpublished data.

strand breakage could also be induced by cleavage between C4' and C5', following the damage at C5'.

Concurrent generation of NO and superoxide can induce strand breakage under a variety of conditions. Yoshie and Ohshima (1997a) have recently found that concurrent incubation of plasmid DNA with an NO-releasing compound and a polyhydroxyaromatic compound, such as catechol or 1,4-hydroquinone, leads to a synergistic induction of DNA strand breaks. As shown in Table 1, a variety of catechol-type compounds, including dopamines and catechol-estrogens, can induce DNA strand breaks in the presence of NO-releasing compounds (Yoshie and Ohshima, 1997a, 1997b, 1997c, 1998). The single strand breakage caused by SIN-1 or catechol-type compounds plus NO was inhibited by superoxide dismutase as well as NO-trapping agents such as oxyhemoglobin and carboxy-PTIO, suggesting that concurrent generation of NO and superoxide is necessary to cause strand breaks. These results support the notion that NO and superoxide generated from SIN-1 or the quinone redox system react with each other to form peroxynitrite or other oxidant(s), which is responsible for DNA damage.

DNA single strand breakage has been also reported in intact cells exposed to peroxynitrite (Salgo et al. 1995a; Spencer et al. 1996; Szabo et al. 1996), indicating that extracellular peroxynitrite has the ability to enter the cells and reach the nucleus. During immunostimulation of various cells, in addition to DNA base modifications (see above), DNA single strand breakage also occurs (Zingarelli et al. 1996).

6. PROTECTIVE EFFECTS OF NO ON DNA DAMAGE MEDIATED BY THE FENTON-TYPE REACTION

Several studies have demonstrated protective effects of NO on oxidative stress, especially those mediated by the Fenton-type reactions (Kanner et al. 1991; Wink et al. 1993). The DNA strand breakage induced by H_2O_2 plus Fe^{2+} or Fe^{3+} or by 1,4-hydroquinone plus Cu^{2+} was dose-dependently inhibited by NO-releasing compounds or sodium ni-

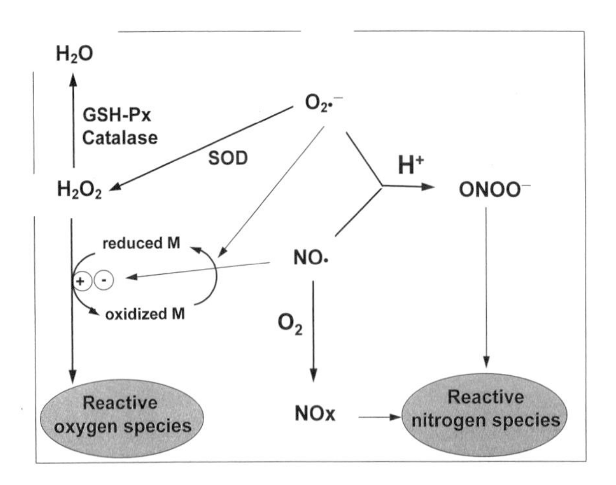

Figure 3. Interactions between NO, rective oxygen species and metal ions in relation to oxidative damage. GSH-Px, glutathione peroxidase: SOD, superoxide dismutase.

trite, but not by sodium nitrate (Yoshie and Ohshima, 1997a). Similarly, NO has been shown to inhibit lipid peroxidation (Rubbo et al. 1994; Goss et al. 1995). Possible mechanisms for NO/NO_2^--mediated inhibition of the Fenton reaction have been proposed, including (i) scavenging of reactive species formed by the Fenton reaction by NO and/or NO_2^-, (ii) formation of a complex between transition metals and NO and/or NO_2^-, inhibiting generation of reactive species from H_2O_2, or (iii) rapid reaction of NO with superoxide, which inhibits superoxide-mediated reduction of metal ions in the Haber-Weiss reaction and the subsequent production of hydroxyl radical (Yoshie and Ohshima, 1997a). NO has been also shown to act as an antioxidant to prevent Fenton-type lipid peroxidation by terminating lipid radical chain propagation reactions (Rubbo et al. 1994; Goss et al. 1995). These results indicate that the balance and interactions between production of reactive oxygen and nitrogen species and the presence of metal ions may play a crucial role in inflammation-associated human diseases including cancer (Fig. 3).

7. ROLE OF DNA DAMAGE INDUCED BY REACTIVE NITROGEN SPECIES IN CARCINOGENESIS

Since large amounts of NO and other reactive oxygen species could be formed in inflamed tissues, they may play an important role in inflammation-associated carcinogenesis. As described above, reactive nitrogen species (NO, NOx and peroxynitrite) cause various DNA modifications and induce strand breakage. NO has been also shown to inhibit several DNA repair enzymes, such as O^6-methylguanine-DNA-methyltransferase and the Fpg protein (Laval and Wink, 1994; Wink and Laval, 1994). Predominant mutations induced by NO and peroxynitrite are G:C to A:T and G:C to T:A, respectively. In particular, G:C to A:T transitions at CpG sites are very common in a variety of genes from all types of human cancers (Greenblatt et al. 1994). Similarly, G:C to T:A transversions account for 30% of all *p53* gene mutations in lung cancer, and are also found in a large number of liver and breast cancers. It has been proposed that G:C to T:A transversions are predominantly caused by polycyclic aromatic hydrocarbons such as benzo(a)pyrene present in tobacco smoke. However, the same G:C to T:A transversions can also be induced by peroxynitrite. There is now increasing evidence that suggests that peroxynitrite could be formed in the lung of cigarette smokers (Yoshie and Ohshima, 1997b). Similarly, the reaction between catechol-estrogens and NO can produce oxidants which are similar to peroxynitrite (Yoshie and Ohshima, 1998). Several recent studies have shown the importance of catechol-estrogens in breast cancer. Because NO is also produced by constitutive and inducible types of NO synthases in human breast tissues, it may react with catechol-estrogens to produce peroxynitrite, which can induce the G:C to T:A transversions observed frequently in breast tumors. Other sites, where inflammation, peroxynitrite production and DNA injury have been linked to carcinogenesis, include *Helicobacter pylori* infection, gastritis and gastric cancer, as well as ulcerative colitis and colon cancer. In precancerous lesions of the stomach of human subjects infected with *H. pylori*, increased expression of inducible NO synthase and decreased expression of Mn-superoxide dismutase have been detected (Pignatelli et al. 1995), suggesting that the production of both NO and superoxide is elevated. Under such conditions, an increased amount of peroxynitrite may also be formed.

In conclusion, NO and its derivatives, especially peroxynitrite, can induce multiple forms of DNA damage (various base modifications, single strand breakage etc). NO could inhibit several DNA repair enzymes. Persistent DNA damage, during chronic inflammation, may contribute to increased mutagenesis and risk of cancer. However, NO may act as an antioxidant under certain conditions. The balance between production of NO and reactive oxygen species and interactions of NO with metal ions may play crucial roles in inflammation-associated human diseases including cancer. Therefore compounds that limit production of and damage by NO, superoxide or peroxynitrite should be further explored as possible chemopreventive agents for cancer. These can include inhibitors of iNOS induction, NOS enzyme inhibitors (preferably with selectivity for iNOS) as well as scavengers of NO, superoxide, and peroxynitrite.

ACKNOWLEDGMENT

The authors thank Ms. P. Collard for secretarial assistance.

REFERENCES

J.S. Beckman and W.H. Koppenol (1996) Nitric oxide, superoxide, and peroxynitrite: the good, the bad, and ugly. *Am. J. Physiol.* 271, C1424-C1437.

R.M. Burger, K. Drlica and B. Birdsall (1994) The DNA cleavage pathway of iron bleomycin. Strand scission precedes deoxyribose 3-phosphate bond cleavage. *J. Biol. Chem.* 269, 25978–25985.

T. deRojas-Walker, S. Tamir, H. Ji, J.S. Wishnok and S.R. Tannenbaum (1995) Nitric oxide induces oxidative damage in addition to deamination in macrophage DNA. *Chem. Res. Toxicol.* 8, 473–477.

T. Douki, J. Cadet and B.N. Ames (1996) An adduct between peroxynitrite and 2'-deoxyguanosine: 4,5-dihydro-5-hydroxy-4-(nitrosooxy)-2'-deoxyguanosine. *Chem. Res. Toxicol.* 9, 3–7.

T. Douki and J. Cadet (1996) Peroxynitrite mediated oxidation of purine bases of nucleosides and isolated DNA. *Free Radic. Res.* 24, 369–380.

B. Epe, D. Ballmaier, I. Roussyn, K. Briviba and H. Sies (1996) DNA damage by peroxynitrite characterized with DNA repair enzymes. *Nucleic Acids Res.* 24, 4105–4110.

E. Felley Bosco, J. Mirkovitch, S. Ambs, K. Mace, A. Pfeifer, L.K. Keefer and C.C. Harris (1995) Nitric oxide and ethylnitrosourea: relative mutagenicity in the p53 tumor suppressor and hypoxanthine-phosphoribosyltransferase genes. *Carcinogenesis* 16, 2069–2074.

E.S. Fiala, R.S. Sodum, M. Bhattacharya and H. Li (1996) (-)-Epigallocatechin gallate, a polyphenolic tea antioxidant, inhibits peroxynitrite-mediated formation of 8-oxodeoxyguanosine and 3-nitrotyrosine. *Experientia* 52, 922–926.

L. Giloni, M. Takeshita, F. Johnson, C. Iden and A.P. Grollman (1981) Bleomycine-induced strand-scission of DNA. Mechanism of deoxyribose cleavage. *J. Biol. Chem.* 256, 8608–8615.

S.P. Goss, N. Hogg and B. Kalyanaraman (1995) The antioxidant effect of spermine NONOate in human low-density lipoprotein. *Chem. Res. Toxicol.* 8, 800–806.

M.S. Greenblatt, W.P. Bennett, M. Hollstein and C.C. Harris (1994) Mutations in the p53 tumor suppressor gene: clues to cancer etiology and molecular pathogenesis. *Cancer Res.* 54, 4855–4878.

M.B. Grisham, K. Ware, H.E. Gilleland, Jr., L.B. Gilleland, C.L. Abell and T. Yamada (1992) Neutrophil-mediated nitrosamine formation: role of nitric oxide in rats. *Gastroenterology* 103, 1260–1266.

J.T. Groves and S.S. Marla (1995) Peroxynitrite-induced DNA strand scission mediated by a manganese porphyrin. *J. Am. Chem. Soc.* 117, 9578–9579.

S. Inoue and S. Kawanishi (1995) Oxidative DNA damage induced by simultaneous generation of nitric oxide and superoxide. *FEBS Lett.* 371, 86–88.

R. Iyengar, D.J. Stuehr and M.A. Marletta (1987) Macrophage synthesis of nitrite, nitrate, and N-nitrosamines: precursors and role of the respiratory burst. *Proc. Natl. Acad. Sci.* U. S. A. 84, 6369–6373.

M.J. Juedes and G.N. Wogan (1996) Peroxynitrite-induced mutation spectra of pSP189 following replication in bacteria and in human cells. *Mutat. Res.* 349, 51–61.

J. Kanner, S. Harel and R. Granit (1991) Nitric oxide as an antioxidant. Arch. *Biochem. Biophys.* 289, 130–136.

L.J. Kennedy, K. Moore Jr., J.L. Caulfield, S.R. Tannenbaum and P.C. Dedon (1997) Quantitation of 8-oxoguanine and strand breaks produced by four oxidizing agents. *Chem. Res. Toxicol.* 10, 386–392.

J.J. Kirchner, S.T. Sigurdsson and P.B. Hopkins (1992) Interstrand cross-linking of duplex DNA by nitrous acid: covalent structure of the dG-to-dG cross-link at the sequence of 5'-CG. *J. Am. Chem. Soc.* 114, 4021–4027.

F. Laval and D.A. Wink (1994) Inhibition by nitric oxide of the repair protein, O^6-methylguanine-DNA-methyltransferase. *Carcinogenesis* 15, 443–447.

C.D. Leaf, J.S. Wishnok and S.R. Tannenbaum (1991) Endogenous incorporation of nitric oxide from L-arginine into N-nitrosomorpholine stimulated by Escherichia coli lipopolysaccharide in the rat. *Carcinogenesis* 12, 537–539.

R.H. Liu, J.R. Jacob, B.C. Tennant and J.H. Hotchkiss (1992) Nitrite and nitrosamine synthesis by hepatocytes isolated from normal woodchucks (Marmota monax) and woodchucks chronically infected with woodchuck hepatitis virus. *Cancer Res.* 52, 4139–4143.

R.H. Liu and J.H. Hotchkiss (1995) Potential genotoxicity of chronically elevated nitric oxide: a review. *Mutat. Res.* 339, 73–89.

A.M. Miles, D.S. Bohle, P.A. Glassbrenner, B. Hansert, D.A. Wink and M.B. Grisham (1996) Modulation of superoxide-dependent oxidation and hydroxylation reactions by nitric oxide. *J. Biol. Chem.* 271, 40–47.

M. Miwa, D.J. Stuehr, M.A. Marletta, J.S. Wishnok and S.R. Tannenbaum (1987) Nitrosation of amines by stimulated macrophages. *Carcinogenesis* 8, 955–958.

T. Nguyen, D. Brunson, C.L. Crespi, B.W. Penman, J.S. Wishnok and S.R. Tannenbaum (1992) DNA damage and mutation in human cells exposed to nitric oxide in vitro. *Proc. Natl. Acad. Sci.* U. S. A. 89, 3030–3034.

H. Ohshima, M. Tsuda, H. Adachi, T. Ogura, T. Sugimura and H. Esumi (1991) L-arginine-dependent formation of N-nitrosamines by the cytosol of macrophages activated with lipopolysaccharide and interferon-gamma. *Carcinogenesis* 12, 1217–1220.

H. Ohshima, T.Y. Bandaletova, I. Brouet, H. Bartsch, G. Kirby, F. Ogunbiyi, V. Vatanasapt and V. Pipitgool (1994) Increased nitrosamine and nitrate biosynthesis mediated by nitric oxide synthase induced in hamsters infected with liver fluke (Opisthorchis viverrini). *Carcinogenesis* 15, 271–275.

H. Ohshima and H. Bartsch (1994) Chronic infections and inflammatory processes as cancer risk factors: possible role of nitric oxide in carcinogenesis. *Mutat. Res.* 305, 253–264.

B. Pignatelli, B. Bancel, C. Malaveille, S. Calmels, P. Correa and H. Ohshima (1995) Defence against oxidative stress in relation to Helicobacter pylori infection and precancerous conditions of the stomach. *Eur. J. Cancer Prev.* 3, 108–109.

W.A. Pryor and G.L. Squadrito (1995) The chemistry of peroxynitrite: a product from the reaction of nitric oxide with superoxide. *Am. J. Physiol.* 268, L699-L722.

I. Roussyn, K. Briviba, H. Masumoto and H. Sies (1996) Selenium-containing compounds protect DNA from single-strand breaks caused by peroxynitrite. *Arch. Biochem. Biophys.* 330, 216–218.

M.N. Routledge, D.A. Wink, L.K. Keefer and A. Dipple (1993) Mutations induced by saturated aqueous nitric oxide in the pSP189 supF gene in human Ad293 and E. coli MBM7070 cells. *Carcinogenesis* 14, 1251–1254.

M.N. Routledge, D.A. Wink, L.K. Keefer and A. Dipple (1994) DNA sequence changes induced by two nitric oxide donor drugs in the supF assay. *Chem. Res. Toxicol.* 7, 628–632.

H. Rubbo, R. Radi, M. Trujillo, R. Telleri, B. Kalyanaraman, S. Barnes, M. Kirk and B.A. Freeman (1994) Nitric oxide regulation of superoxide and peroxynitrite-dependent lipid peroxidation. Formation of novel nitrogen-containing oxidized lipid derivatives. *J. Biol. Chem.* 269, 26066–26075.

J. Rubio, V. Yermilov and H. Ohshima (1996) DNA damage induced by peroxynitrite: Formation of 8-nitroguanine and base propenals. In The biology of nitric oxide, part 5 (eds. S. Moncada, J. Stamler, S. Gross and E.A. Higgs) *Portland press, London*, p. 34.

M.G. Salgo, E. Bermudez, G.L. Squadrito and W.A. Pryor (1995a) DNA damage and oxidation of thiols peroxynitrite causes in rat thymocytes. *Arch. Biochem. Biophys.* 322, 500–505.

M.G. Salgo, K. Stone, G.L. Squadrito, J.R. Battista and W.A. Pryor (1995b) Peroxynitrite causes DNA nicks in plasmid pBR322. *Biochem. Biophys. Res. Commun.* 210, 1025–1030.

C. Schmutte, W.M. Rideout, J.C. Shen and P.A. Jones (1994) Mutagenicity of nitric oxide is not caused by deamination of cytosine or 5-methylcytosine in double-stranded DNA. *Carcinogenesis* 15, 2899–2903.

J.P. Spencer, J. Wong, A. Jenner, O.I. Aruoma, C.E. Cross and B. Halliwell (1996) Base modification and strand breakage in isolated calf thymus DNA and in DNA from human skin epidermal keratinocytes exposed to peroxynitrite or 3-morpholinosydnonimine. *Chem. Res. Toxicol.* 9, 1152–1158.

T. Suzuki, R. Yamaoka, M. Nishi, H. Ide and K. Makino (1996) Isolation and characterization of a novel product, 2'-deoxyoxanosine, from 2'-deoxyguanosine, oligodeoxynucleotide, and calf thymus DNA treated by nitrous acid and nitric oxide. *J. Am. Chem. Soc.* 118, 2515–2516.

C. Szabo, B. Zingarelli, M. O'Connor and A. Salzman (1996) DNA strand breakage, activation of poly(ADP-ribose) synthetase, and cellular energy depletion are involved in the cytotoxicity in macrophages and smooth muscle cells exposed to peroxynitrite. *Proc. Natl. Acad. Sci.* U. S. A. 93, 1753–1758.

C. Szabo and H. Ohshima (1997) DNA damage induced by peroxynitrite: subsequent biological effects. *Nitric Oxide Biol. Chem.*, 1, 373–385.

S. Tamir, S. Burney and S.R. Tannenbaum (1996) DNA damage by nitric oxide. *Chem. Res. Toxicol.* 9, 821–827.

S. Tamir and S.R. Tannenbaum (1996) The role of nitric oxide (NO.) in the carcinogenic process. *Biochim. Biophys. Acta* 1288, F31-F36.

S.R. Tannenbaum, S. Tamir, T.D. Rojas-Walker and J.S. Wishnok (1994) DNA damage and cytotoxicity caused by nitric oxide. *ACS Symp. Ser.* 553, 120–135.

R.M. Uppu, R. Cueto, G.L. Squadrito, M.G. Salgo and W.A. Pryor (1996) Competitive reactions of peroxynitrite with 2'-deoxyguanosine and 7,8-dihydro-8-oxo-2'-deoxyguanosine (8-oxodG): relevance to the formation of 8-oxodG in DNA exposed to peroxynitrite. *Free Radic. Biol. Med.* 21, 407–411.

D.A. Wink, K.S. Kasprzak, C.M. Maragos, R.K. Elespuru, M. Misra, T.M. Dunams, T.A. Cebula, W.H. Koch, A.W. Andrews, J.S. Allen anf L.K. Keefer (1991) DNA deaminating ability and genotoxicity of nitric oxide and its progenitors. *Science* 254, 1001–1003.

D.A. Wink, I. Hanbauer, M.C. Krishna, W. DeGraff, J. Gamson and J.B. Mitchell (1993) Nitric oxide protects against cellular damage and cytotoxicity from reactive oxygen species. *Proc. Natl. Acad. Sci.* U. S. A. 90, 9813–9817.

D.A. Wink and J. Laval (1994) The Fpg protein, a DNA repair enzyme, is inhibited by the biomediator nitric oxide in vitro and in vivo. *Carcinogenesis* 15, 2125–2129.

Y. Wu, I. Brouet, S. Calmels, H. Bartsch and H. Ohshima (1993) Increased endogenous N-nitrosamine and nitrate formation by induction of nitric oxide synthase in rats with acute hepatic injury caused by Propionibacterium acnes and lipopolysaccharide administration. *Carcinogenesis* 14, 7–10.

V. Yermilov, J. Rubio, M. Becchi, M.D. Friesen, B. Pignatelli and H. Ohshima (1995a) Formation of 8-nitroguanine by the reaction of guanine with peroxynitrite in vitro. *Carcinogenesis* 16, 2045–2050.

V. Yermilov, J. Rubio and H. Ohshima (1995b) Formation of 8-nitroguanine in DNA treated with peroxynitrite in vitro and its rapid removal from DNA by depurination. *FEBS Lett.* 376, 207–210.

V. Yermilov, Y. Yoshie, J. Rubio and H. Ohshima (1996) Effects of carbon dioxide/bicarbonate on induction of DNA single-strand breaks and formation of 8-nitroguanine, 8-oxoguanine and base-propenal mediated by peroxynitrite. *FEBS Lett.* 399, 67–70.

Y. Yoshie and H. Ohshima (1997a) Nitric oxide synergistically enhances DNA strand breakage induced by polyhydroxyaromatic compounds, but inhibits that induced by the Fenton-reaction. *Arch. Biochem. Biophys.* 342, 13–21.

Y. Yoshie and H. Ohshima (1997b) Synergistic induction of DNA strand breakage by cigarette tar and nitric oxide. *Carcinogenesis* 18, 1359–1363.

Y. Yoshie and H. Ohshima (1997c) Synergistic induction of DNA strand breakage caused by nitric oxide together with catecholamine: implications for neurodegenerative disease. *Chem. Res. Toxicol.* 10, 1015–1022.

Y. Yoshie and H. Ohshima (1998) Synergistic induction of DNA strand breakage by catechol-estrogen and nitric oxide: implications for hormonal carcinogenesis. *Free Radic. Biol. Med.* 24, 341–348.

B. Zingarelli, M. O'Connor, H. Wong, A.L. Salzman and C. Szabo (1996) Peroxynitrite-mediated DNA strand breakage activates poly-adenosine diphosphate ribosyl synthetase and causes cellular energy depletion in macrophages stimulated with bacterial lipopolysaccharide. *J. Immunol.* 156, 350–358.

28

MODIFICATION OF IONIZING RADIATION DAMAGE TO CELLULAR DNA BY FACTORS AFFECTING CHROMATIN STRUCTURE

Effects of Histone Deacetylase Inhibitors Trichostatin A and Sodium Butyrate

Nancy L. Oleinick, Song-mao Chiu, Liang-yan Xue, and Karl J. Mann

Department of Radiation Oncology
Case Western Reserve University School of Medicine
Cleveland, Ohio

1. ABSTRACT

The radiosensitivity of DNA within the cell nucleus is greatly affected by its environment. Histones are major radioprotectors of chromatin against formation of single- or double-strand breaks (DSB), but not against DNA-protein crosslinks (DPC), lesions that form preferentially between nuclear matrix-associated DNA and matrix proteins rather than histones. Copper ion, which strengthens associations between matrix proteins and attached DNA, enhances the formation of DPC by radiation and, through matrix site-specific radical generation, cuts DNA to give chromosomal loop-sized DNA fragments in isolated nuclei or nucleoids. DNA radiosensitivity can also be modified by the polyamine spermine, which compacts DNA and thereby protects DNA and exposed regions of histone-depleted chromatin, but not histone-replete chromatin, against DSB formation. In contrast, DPC formation is suppressed by very low concentrations of spermine or putrescine, suggesting preferential protection of nuclear matrix sites. Transcriptionally active gene domains are transiently relaxed through core histone hyperacetylation by histone acetylases and deacetylases. Earlier reports by Bohm and Leith demonstrated radiosensitization of mammalian cells and DNA by the histone deacetylase inhibitor, sodium butyrate. Using a new, more specific histone deacetylase inhibitor, trichostatin A (TSA), we have revisited the question of radiosensitization through histone hyperacetylation. Treatment of cells with butyrate (5 or 10 mM for 16 h) or TSA (1 μM for 4–6 h) resulted in hyperacetylation of core histones. With 10 mM butyrate, radiosensitization was confirmed; with

Advances in DNA Damage and Repair, edited by Dizdaroglu and
Karakaya. Kluwer Academic / Plenum Publishers, New York, 1999.

TSA, however, radiosensitization required exposure of cells for >8 h. Butyrate, but not TSA, markedly elevated the yield of DSB. However, rejoining of DSB induced by 60 Gy was accelerated in cells treated with either of the inhibitors, and this occurred mainly during the early phase of the rejoining process. We hypothesize that the accelerated repair may result in misrejoining of DSB and contribute to the radiosensitization of cell killing in the presence of histone hyperacetylation.

2. INTRODUCTION

In contrast to random energy deposition in cell nuclei exposed to low-LET ionizing radiation, the radiosensitivity of DNA within the cell nucleus is not uniform. DNA damage distribution is greatly affected by the interaction with histones, non-histone proteins, and small molecules in the formation of chromatin, the degree of condensation of chromatin, and the interaction of chromatin with the nuclear matrix. This paper reviews the contribution of some of the modifying factors, as studied in various sub-cellular preparations, and describes one means of modifying DNA damage and repair of cellular chromatin.

2.1. Nucleosomes and Chromatin Compaction

Histones not only participate in the organization of eukaryotic DNA into nucleosomes and higher order structures, but they also play an important role in the protection of DNA against damage induced by ionizing radiation. The importance of histones has been demonstrated by stripping them from DNA with high ionic strength solutions. Removal of histones, especially core histones, from chromatin greatly enhances the susceptibility of the DNA to damage by radiation (Ljungman, 1991; Elia and Bradley, 1992; Xue et al., 1994).

The compaction or conformation of chromatin is one factor determining the radiosensitivity of the DNA housed therein. For example, regions of chromatin DNA active in transcription and having a less compact structure are several times more sensitive to radiation-induced strand breakage (Chiu et al., 1982; Bunch et al., 1992) and DNA-protein crosslink (DPC) formation (Chiu et al., 1986) than bulk DNA. Expansion of isolated chromatin in hypotonic buffers also results in a 5–10-fold increase in DNA damage production upon exposure to radiation (Heussen et al., 1987; Ljungman, 1991; Chiu et al., 1992; Warters and Lyons, 1992), and the enhancement in radiosensitivity can be reversed by the addition of a low concentration of Mg^{++} which promotes chromatin condensation.

2.2. Nuclear Matrix

Within nuclei, the proteinaceous nuclear skeleton anchors chromosomal loops, creates independent topological domains, and provides sites for DNA replication and transcription (Pienta and Coffey, 1984; Berezney, 1991; Davie, 1995). The nuclear matrix has unusual properties with respect to radiation damage, due to the binding of metal ions (Lewis and Laemmli, 1982; Chiu et al., 1993) and matrix-attachment DNA regions (MARs). MARs, consisting of some 300+ bp of DNA and containing AT-rich and topoisomerase II consensus sequences, exist in or near many genes (reviewed by Boulikas, 1995). Radiation-induced DPC are formed preferentially between matrix-associated DNA and proteins of the nuclear matrix rather than histones (Chiu et al., 1986). Study of the interaction of cloned MARs with isolated nuclear matrices has revealed the specific associations to be hypersensitive to DNA-nuclear matrix crosslinking upon γ-irradiation (Balasubramaniam and Oleinick, 1995). Roti Roti and colleagues (Kapiszewska et al., 1989; Roti Roti et al., 1993) have shown that the ability of anchored chromosomal loops, once dehistonized to form nucleoids, to unwind and rewind in the presence of intercalating agents is dependent upon (a) the absence of radiation-induced strand breaks, which interrupt supercoiling, and (b) the strength of DNA-matrix protein anchorage sites. Nucleoids from several types of radiosensitive cells, deficient in repair of double-strand breaks (DSB), appear to be depleted in specific nuclear matrix proteins, as compared to their radioresistant counterparts (Malyapa et al., 1994; 1995; 1996). Protein spectra of nuclear matrices isolated from various types of tumors have also been reported to differ from their normal cell counterparts (Woudstra et al., 1996). Furthermore, a role for the nuclear matrix in the repair of UV-induced DNA damage has also been demonstrated (McCready and Cook, 1984; Mullenders et al., 1987; Koehler and Hanawalt, 1996).

2.3. Metal Ions

Copper ion strengthens associations between matrix proteins and attached DNA (Lewis and Laemmli, 1982), enhances the formation of DPC by radiation (Chiu et al., 1993) and, with ascorbate and H_2O_2, produces 0.1–0.2 Mbp "chromosomal loop-sized" DNA fragments in isolated nuclei or nucleoids through site-specific radical generation (Chiu et al., 1995). A subset of matrix proteins has a high affinity for copper ion and is bound to the matrix in the presence of this metal ion. When agarose-embedded nuclei were exposed to Fe^{++}-EDTA- or Cu^{++} and assayed for DNA damage, we found that like radiation, Fe^{++}-EDTA induced DSB randomly and in enhanced yields after chromatin expansion or histone removal. In contrast, DNA fragments from Cu^{++}-treated nuclei were derived from a limited portion of the genome and, similar to DPC induction, were little affected by prior removal of histones, since they occurred at or near the nuclear matrix (Chiu et al., 1995). Thus, the nuclear matrix appears to be an important site for DNA damage in irradiated cells.

2.4. Polyamines

The possible role of polyamines in radioprotecting chromosomal DNA has also been investigated. Spermine is an efficient radioprotector of plasmid or viral DNA and of viral minichromosomes by a mechanism involving site-specific radical scavenging and the induction of DNA compaction and aggregation (Spotheim-Maurizot et al., 1995). Based on radioprotection of SV40 minichromosomes at a lower spermine concentration than needed for SV40 DNA, Newton *et al.* (1996) proposed that the differential concentration dependence could account for the greater radiosensitivity of open regions of cellular chromatin as compared to bulk inactive chromatin at physiological levels of spermine. However, whereas the effects of spermine on DNA DSB formation in dehistonized V79 DNA (nucleoids) are consistent with spermine-induced DNA compaction, spermine provides no radioprotection to native chromatin and only modest radioprotection of histone H1-depleted chromatin (Chiu and Oleinick, 1997). Thus, spermine is much less effective in protecting cellular chromatin from DSB formation than are histones (Chiu and Oleinick, 1997). In contrast to the relatively inefficient radioprotection of V79 chromatin against DSB formation, low concentrations (<0.1 mM) of spermine or putrescine (which is incapable of causing DNA compaction) provided partial radioprotection against DPC formation in both native and histone H1-depleted chromatin (Chiu and Oleinick, 1998). Whereas all DPC generated by the irradiation of chromatin, above the level generated in intact cells, could be blocked by 5 mM spermine, less than half could be blocked by 5 mM putrescine. The difference in efficiency of radioprotection between the two polyamines could not be accounted for simply by correcting for the number of charged amino groups per molecule. It appears that both spermine and putrescine bind preferentially and with high affinity at matrix-associated sites of DPC formation, disrupting the associations between DNA and protein that are essential for DPC formation and/or scavenging hydroxyl radicals at these sites; that some sites are susceptible to spermine but not to putrescine; and that endogenous spermine is a major cellular radioprotector against DPC formation (Chiu and Oleinick, 1998).

2.5. Histone Acetylation

2.5.1. Function and Regulation. Core histones are reversibly acetylated and deacetylated at specific lysine residues within the N-terminal portion. Acetylation and deacetylation of histones are catalyzed by nuclear matrix-associated acetyltransferases and deacetylases, respectively (reviewed by Davie, 1995). Histone acetylation has been suggested to be important in the modulation of the interactions between histones and DNA, since acetylation of histones neutralizes positive histone charges and inhibits the interaction with DNA (Perry and Chalkley, 1982; Matthews and Waterburg, 1985). The acetylation may free the N-terminal domain of the histone from core DNA (Cary et al., 1978, 1982). Histone acetylation is also essential in the control of transcription, and histones in chromatin regions engaged in transcription have a higher level of acetylation than those in non-transcribed regions. It has been suggested that acetylation of histones may open up chromatin conformation (Vidali et al., 1978) so that it becomes accessible to the transcription machinery (Gross and Garrard, 1988). Recent studies have identified certain transcription factors as histone acetylases, and some repressors are associated with histone deacetylases (reviewed by Grunstein, 1997). Rather than acting by simply turning overall histone acetylation on or off, these enzymes demonstrate some specificity in the lysine residues subject to modification. Therefore, through specific modifications in response to steroids or protein growth factors, histones may actively participate in gene regulation rather than simply serve a structural function in chromatin packing. Furthermore, because of their association with gene regulation, histones are believed to serve as a link to the control of cell proliferation and differentiation.

2.5.2. Inhibitors of Histone Deacetylases: Sodium Butyrate. The pleiotropic effects of histones on chromatin organization and cellular functions may explain the diversity of effects observed in cells treated with inhibitors of deacetylases. For example, treatment of cells with sodium butyrate (NaB) to produce hyperacetylated histones results in an increase in the size of nucleosome particles and an increase in their accessibility to nucleases, including UV endonuclease (Perry and Chalkley, 1982; Rezek et al., 1982; Smith, 1986; Gross and Garrard, 1987; Riehm and Harrington, 1987; Lopez-Larraza and Bianchi, 1993), without evident alteration of the higher order chromatin structure (Imai et al., 1986; Riehm and Harrington, 1987). The chromatin of NaB-treated cells is also more susceptible to ionizing radiation-induced strand breakage (Nackerdien et al., 1989) as well as to DNA damage by bleomycin (Lopez-Larraza and Bianchi, 1993). These phenomena have been interpreted as resulting from the chromatin decondensation accompanying histone hyperacetylation. Exposure of cells to NaB also results in a 1.2–1.4-fold increase in the sensitivity of V79 (Nackerdien et al., 1989) and CHO HA-1 cells (Leith, 1988) to the lethal effects of ionizing radiation. Heussen et al. (1987) found a similar increase in the yield of radiation-induced SSB in chromatin DNA isolated from NaB-treated cells and in decondensed chromatin depleted of histone H1. This finding led Nackerdien et al. (1989) to conclude that chromatin compaction and accessibility are important factors in determining DNA radiosensitivity.

2.5.3. Inhibitors of Histone Deacetylases: Trichostatin A. In the experiments with NaB, a high concentration was used, which can produce non-specific effects, such as hypermethylation of cytosine residues (Parker et al., 1986) and changes in other enzymes, the cytoskeleton, and cell membranes (Prasad and Sinha, 1976; Kruh, 1982). More recently, two microbial metabolites have been found to be potent inhibitors of histone deacetylases: trichostatin A (TSA) and trapoxin (reviewed by Yoshida et al., 1995). Whereas trapoxin is an irreversible inhibitor, TSA is both specific and reversible, making it an attractive agent to explore effects of histone hyperacetylation on cellular and chromatin responses to ionizing radiation. However, in addition to the primary effects of TSA on histone deacetylases, secondary effects include induction of expression of certain genes, cell differentiation, and arrest of cell cycle progression at G1 and G2 phases of the cell cycle (Yoshida et al., 1995). In the present study, we have begun to explore the influence of histone hyperacetylation, comparing TSA to NaB. Histone acetylation, radiation-induced DNA strand break production, and the kinetics of DSB repair have been studied.

3. MATERIALS AND METHODS

3.1. Cell Culture and Labeling

Exponentially growing monolayer cultures of Chinese hamster V79–379 lung fibroblasts were grown in McCoy's 5A medium supplemented with 10% calf serum. Cultures were incubated in medium containing 0.1 μCi/ml of ^{3}H-thymidine for 18–24 h to label nuclear DNA.

3.2. DNA DSB Measurement by Pulsed Field Gel Electrophoresis

For measurement of DSB induction, cells recovered from the monolayers with trypsin were washed and embedded in agarose plugs (0.5–1.0 x 10^{6} cells in each plug of 0.1 cm^{3}). The plugs were irradiated in culture medium on ice in a ^{137}Cs irradiator at a dose rate of 4.5 Gy/min. Two methods were used to study the repair of DSB. In the first method, monolayer cultures were irradiated on ice, the cold medium was replaced with pre-warmed medium, and the dishes were incubated at 37°C for various time intervals. The cells were collected by trypsinization on ice (5–10 min), and plugs were prepared and processed as described above. In the second method, trypsinized cells were maintained in suspension throughout irradiation and post-irradiation incubation. For measurement of DSB, the plugs were incubated in buffer (0.1 M EDTA, 10 mM Tris, 50 mM NaCl, pH 7.8, 0.5 mg/ml proteinase K) at 50°C overnight, then washed three times in TE buffer. Electrophoresis was performed using a CHEF DR III system (Bio-Rad) in 1% agarose in 0.5 x TBE with the following parameters: 14°C, single run of 20 h, 6 volts/cm, 120° angle, switch times linearly ramped from 50 to 90 seconds. The gels were stained with ethidium bromide and photographed. Each lane was sliced into 2.5-cm segments and radioactivities determined as previously described (Chiu and Oleinick, 1997).

3.3. Analysis of Histones

Nuclei were prepared by lysis of cells in solution containing 0.14 M NaCl, 10 mM Tris-HCl, pH 7.4, 2 mM $MgCl_2$, 0.5% Triton X-100, and the histones were extracted with 0.4 N H_2SO_4. After precipitation with TCA and washing with acetone, the histones were dissolved in 8 M urea, 0.9 N acetic acid and analyzed on 15% acrylamide minigels containing 2.5 M urea and running in 0.9% acetic acid at a constant current of 15 mA/gel for about 1.5 h. The gels were then stained with Coomassie blue.

4. RESULTS

4.1. Histone Hyperacetylation with TSA or NaB

Fig. 1 represents a gel analysis of histones extracted from untreated cells or cells treated with an inhibitor of histone deacetylases. As shown in the figure, the unacetylated isoform of histone H4 and the isoforms containing 1–4 acetyl groups can be clearly distinguished as the fastest migrating histone bands. Changes in the modification of other histones cannot be readily distinguished in this gel system. Therefore, our analysis of the effects of the inhibitors on histone acetylation relies upon histone H4. In untreated, exponentially growing cells, histone H4 exists predominantly in the unacetylated isoform (lanes 2 and 13); upon introduction of NaB or TSA, the level of acetylation increases, as revealed by a shift in the distribution of bands. For example, the majority of histone H4 is hyperacetylated in cells treated with 5 mM NaB for 16 hr or TSA at a concentration as low as 0.1 μM for 4 hr (lanes 4 and 5). Higher concentrations of TSA appear not to produce greater histone acetylation (compare lanes 4–7). Histone H4 acetylation reached an apparent maximum level very rapidly after addition of TSA to the culture, since the level after a

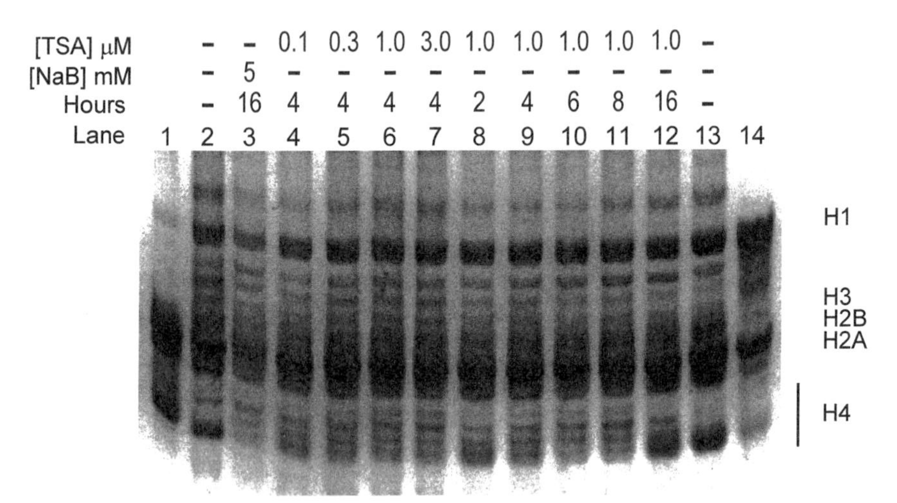

Figure 1. Acid-urea gel analysis of histones from untreated V79 cells or those treated with either NaB or TSA. Histones were extracted from nuclei and analyzed by gel electrophoresis as described in Materials and Methods. Lanes 1 and 14, histone standards; lanes 2 and 13, middle and late exponential cells, respectively, not treated with inhibitors; lane 3, cells exposed to 5 mM NaB for 16 hr; lanes 4–7, cells exposed to TSA (0.1, 0.3, 1.0 and 3.0 μM, respectively) for 4 hr; lanes 8–12, cells exposed to 1 μM TSA for 2, 4, 6, 8 and 16 hr, respectively.

2-hr treatment was similar to those seen after longer exposure times (lanes 8–11). However, histone modification by TSA was reduced after 16 hr of treatment (lane 12) due to the reversible nature of this inhibitor. In the studies to follow, unless otherwise specified, cells were exposed to 1 μM TSA for 4 hr or 5–10 mM NaB for 16–20 hr.

4.2. Double-Strand Break Formation

DSB are believed to be the primary lesions leading to cell killing, and the yield of radiation-induced DSB can be enhanced when nuclear chromatin is expanded. Therefore, the effect of the inhibitors on the production and rejoining of DSB by radiation was investigated. As measured by pulsed-field gel electrophoresis, addition of NaB to the cultures resulted in a doubling of the background level of DSB as well as a slight enhancement in the yield of DSB by radiation (Fig. 2). A smaller effect on both the background and the radiation-induced DSB was found for TSA.

4.3. Rejoining of Double-Strand Breaks

The rejoining of DSB has been determined both in cells that remained in monolayer culture during the post-irradiation repair period and in cells that were trypsinized and kept in suspension culture after irradiation, in order to determine whether or not accurate measurement of DSB level, particularly at early time points, is compromised by trypsinization. However, similar results were obtained for the two situations, so the data have been combined. As shown in Fig. 3, 70–80% of the radiation-induced DSB were rejoined within the first hour post-irradiation. Interestingly, in the presence of either TSA or NaB, the early phase of DSB rejoining was markedly accelerated. In addition, in spite of the higher level of background and radiation-induced DSB in the treated cells, by two hours post-irradiation, the level of residual DSB was lower in the inhibitor-treated cells than in the control cells.

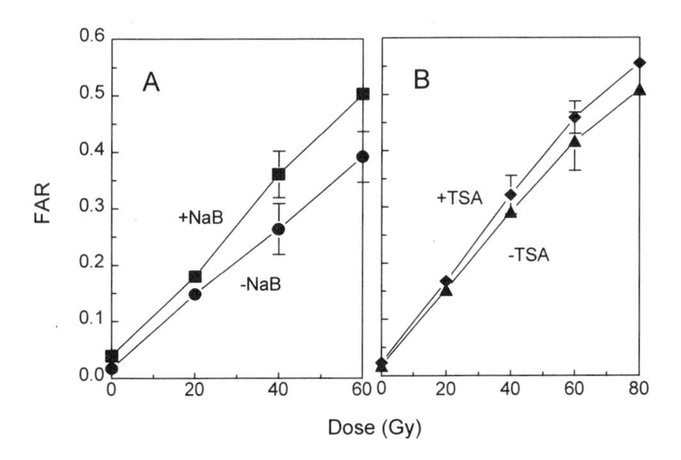

Figure 2. Determination of the yield of γ-radiation-induced DSB in V79 cells treated with NaB (A) or TSA (B). V79 cells were exposed to 5 mM NaB for 16–20 hr or 1 μM TSA for 4 hr or were untreated. The cells were embedded in agarose plugs and irradiated (0–80 Gy) in culture medium on ice. After irradiation, the samples were subjected to deproteinization and PFGE as described in Materials and Methods. Values of 0, 40 and 60 Gy represent mean ± SD of 4–5 determinations. The error bars for the 0-Gy points were smaller than the symbols. Data points for 20 and 80 Gy are the average of 2 determinations.

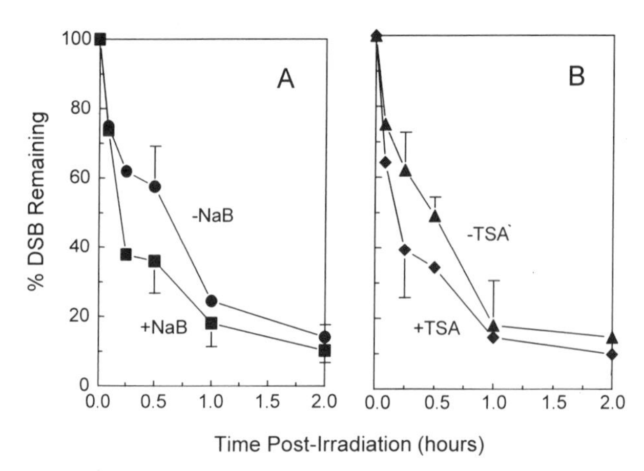

Figure 3. Kinetics of the rejoining of DSB in control cells and in cells treated with NaB (A) or TSA (B). Attached or suspension V79 cells from cultures treated with 5 mM NaB (16 hr) or 1 μM TSA (4 hr) or untreated were irradiated (60 Gy) on ice. Irradiated cells were then incubated in pre-warmed culture medium for 5 min to 2 hr. After various post-irradiation incubation periods, samples were collected, and FAR (fraction of activity released from the plug during electrophoresis) was determined by PFGE as described in Materials and Methods. The percentage of DSB remaining after a repair period was calculated from the ratio of net FAR at that time point to the net FAR at 0 min for each treatment. The average net FAR was 34.9 and 43.5 for control and NaB-treated cells, respectively (A), and 36.9 and 39.0 for control and TSA-treated cells, respectively (B), immediately after 60 Gy. Each datum represents the mean ± SD of 3–7 determinations.

5. DISCUSSION

Our results confirm previous reports that NaB and TSA induce histone hyperacetylation through the inhibition of histone deacetylases, and that TSA is a potent inhibitor of these enzymes. Using antibodies against acetylated histone H4 and immunofluorescence, Jeppesen et al. (1992) were able to distinguish chromatin regions according to their level of acetylation. It was observed that the labeling-positive regions were associated with coding regions, whereas heterochromatin, which is infrequently acetylated, showed a weak or negative reaction. Exposure of cells to NaB resulted in an increase in the intensity of fluorescence in those regions that normally were labeled but failed to detect an increase in labeling in most of the negative regions (Jeppesen and Turner, 1993). The authors concluded that underacetylation of heterochromatin is a result of reduced acetylase activity rather than an increase in deacetylase activity. Accordingly, even though our gel analysis indicates a virtual absence of unacetylated histone H4 in cells exposed to the inhibitors, the level of histone hyperacetylation, and therefore of decondensation of genomic chromatin, would be expected to be limited as well as heterogeneous.

Histone hyperacetylation induced by NaB renders chromatin more accessible to nucleases and UV endonuclease (Perry and Chalkley, 1982; Rezek et al., 1982; Smith, 1986; Lopez-Larraza and Bianchi, 1993). The chromatin of the treated cells is also more susceptible to DNA damage by radiation (Nackerdien et al., 1989) and by bleomycin (Lopez-Larraza and Bianchi, 1993). The increased accessibility of chromatin containing hyperacetylated histones has been interpreted as a result of chromatin decondensation. Although we observed a consistent, small increase in the yield of DSB when the cells were irradiated following treatment with inhibitors of histone deacetylase, the increase was not statistically significant due to large variations in FAR determinations. Thus, the decondensation of chromatin that results from histone hyperacetylation in cells did not radiosensitize chromatin as much as did histone H1 depletion of isolated chromatin, which led to a 1.3–1.5 fold increase in the yield of SSB (Heussen et al., 1987) or DSB (Xue et al., 1994).

It is interesting that the rejoining of DNA DSB is accelerated in cells treated with either of the histone deacetylase inhibitors. A similar acceleration of repair by NaB has been reported for the rejoining of strand breaks induced by bleomycin (Lopez-Larraza and Bianchi, 1993) and for repair synthesis after UV irradiation (Smerdon et al., 1982). Since accelerated rejoining occurred in cells treated with TSA for only 4 hr, which would be sufficient to induce histone hyperacetylation but would arrest only a small fraction of cells in the G2 phase, we speculate that histone hyperacetylation per se rather than secondary cell cycle arrest is responsible for this effect. Alternatively, histone hyperacetylation may induce the expression of certain repair genes, and thus enhance one or more repair mechanisms. Induction of gene expression by the inhibitors through histone hyperacetylation has been demonstrated (Hoshikawa et al., 1994; Miyashita et al., 1994; Arts et al., 1995). The fidelity of repair may or may not be compromised by this acceleration.

In summary, these initial studies of DSB formation by radiation in intact cells treated with the histone deacetylase inhibitors, NaB or TSA, indicate that the extent of chromatin expansion that can be achieved in cells is sufficient to cause only a minor increase in the initial yield of DSB. Even in chromatin regions that have extensively hyperacetylated histones, the presence of the histones may be sufficient to be radioprotective. Thus, histones remain the best radioprotectors identified thus far against DSB formation. A greater effect of the presence of the inhibitors in cells is the acceleration of DSB rejoining, which may or may not include compromised fidelity of repair. Preliminary data suggest that the inhibitors cause cell cycle progression delays and an increase in the lethal effects of radiation. How these effects may relate to the modified formation and rejoining of DSB in the presence of the inhibitors remains to be determined.

ACKNOWLEDGMENTS

The authors thank Dr. Miral Dizdaroglu for the opportunity to participate in the NATO/ASI meeting on DNA Damage and Repair and Dr. Minoru Yoshida (University of Tokyo) for generously providing TSA. This research was supported by USPHS Grants R37 CA15378 and P30 CA43703 awarded by the U.S. National Cancer Institute, DHHS.

REFERENCES

J. Arts, M. Lansink, J. Grimbergen, K. H. Toet, T. Kooistra (1995) Stimulation of tissue-type plasminogen activator gene expression by sodium butyrate and trichostatin A in human endothelial cells involves histone acetylation. *Biochem. J.* **310**, 171–176.

U. Balasubramaniam, N. L. Oleinick (1995) Preferential cross-linking of matrix-attachment region (MAR) containing DNA fragments to the isolated nuclear matrix by ionizing radiation. *Biochemistry* **34**, 12790–12802.

R. Berezney (1991) The nuclear matrix: A heuristic model for investigating genomic organization and function in the cell nucleus. *J. Cell Biochem.* **47**, 109–123.

T. Boulikas (1995) Chromatin domains and prediction of MAR sequences. *International Rev. of Cytology.* **162A**, 279–388.

R. T. Bunch, D. A. Gewirtz, L. F. Povirk (1995) Ionizing radiation-induced DNA strand breakage and rejoining in specific genomic regions as determined by an alkaline unwinding/Southern blotting method. *Int. J. Radiat. Biol.* **68**, 553–562.

P. D. Cary, C. Crane-Robinson, E.M. Bradbury, G. H. Dixon (1982) Effect of acetylation on the binding of N-terminal peptides of histone H4 to DNA. *Eur. J. Biochem.* **127**, 137–143.

P. D. Cary, T. Moss, E. M. Bradbury (1978) High resolution proton magnetic resonance studies of chromatin core particles. *Eur. J. Biochem.* **89**, 475–482.

S.-m. Chiu, N. L. Oleinick (1998) Radioprotection of cellular chromatin by the polyamines spermine and putrescine: Preferential action against formation of DNA-protein crosslinks. *Radiat. Res.* 149, 543–549.

S.-m. Chiu, N. L. Oleinick (1997) Radioprotection against the formation of DNA double-strand breaks in cellular DNA but not native cellular chromatin by the polyamine spermine. *Radiat. Res.* **148**, 188–192.

S.-m. Chiu, L.-y. Xue, L. R. Friedman, N. L. Oleinick (1995) Differential dependence on chromatin structure for copper and iron ion induction of DNA double-strand breaks. *Biochemistry* **34**, 2653–2661.

S.-m. Chiu, L.-y. Xue, L. R. Friedman, N. L. Oleinick (1993) Copper ion-mediated sensitization of nuclear matrix attachment sites to ionizing radiation. *Biochemistry* **32**, 6214–6219.

S.-m. Chiu, L.-y. Xue, L. R. Friedman, N. L. Oleinick (1992) Chromatin compaction and the efficiency of formation of DNA-protein crosslinks in γ-irradiated mammalian cells. *Radiat. Res.* **129**, 184–191.

S.-m. Chiu, L. R. Friedman, N. M. Sokany, L.-y. Xue, N. L. Oleinick (1986) Nuclear matrix proteins are crosslinked to transcriptionally active gene sequences by ionizing radiation. *Radiat. Res.* **107**, 24–38.

S.-m. Chiu, N. L. Oleinick, L. R. Friedman, P. J. Stambrook (1982)Hypersensitivity of DNA in transcriptionally active chromatin to ionizing radiation. *Biochim. Biophys. Acta* **699**, 15–21.

J. R. Davie (1995) The nuclear matrix and the regulation of chromatin organization and function. *Int. Rev. Cytol.* **162A**, 191–250.

M. C. Elia, M. O. Bradley (1992) Influence of chromatin structure on the induction of DNA double strand breaks by ionizing radiation. *Cancer Res.* **52**, 1580–1586.

D. S. Gross, W. T. Garrard (1988) Nucleosome hypersensitive sites in chromatin. *Annu. Rev. Biochem.* **57**, 159–197.

D. S. Gross, W. T. Garrard (1987) Poising chromatin for transcription. *TIBS* **12**, 293–297.

M. Grunstein (1997) Molecular acetylation in chromatin structure and transcription. *Nature* **389**, 349–352.

C. Heussen, Z. Nackerdien, B. J. Smit, L. Bohm (1987) Irradiation damage in chromatin isolated from V-79 Chinese hamster lung fibroblasts. *Radiat. Res.* **110**, 84–94.

Y. Hoshikawa, H. J. Kwon, M. Yoshida, S. Horinouchi, T. Beppu (1994) Trichostatin A induces morphological changes and gelsolin expression by inhibiting histone deacetylase in human carcinoma cell lines. *Exp. Cell Res.* **214**, 189–197.

B. S. Imai, P. Yau, K. Balwin, K. Ibel, R. P. May, E. M. Bradbury (1986) Hyperacetylation of core histones does not cause unfolding of nucleosomes. *J. Biol. Chem.* **26**, 8784–8792.

P. Jeppesen, B. M. Turner (1993) The inactive X chromosome in female mammals is distinguished by a lack of histone H4 acetylation, a cytogenetic marker for gene expression. *Cell* **74**, 281–289.

P. Jeppesen, A. Mitchell, B. Turner, P. Perry (1992) Antibodies to defined histone epitopes reveal variations in chromatin conformation and underacetylation of centric heterochromatin in human metaphase chromosomes. *Chromosoma* **101**, 322–332.

M. Kapiszewska, W. D. Wright, C. S. Lange, J. L. Roti-Roti (1989) DNA supercoiling changes in nucleoids from irradiated L5178Y-S and -R cells. *Radiat. Res.* **119**, 569–575.

D. R. Koehler, P. C. Hanawalt (1996) Recruitment of damaged DNA to the nuclear matrix in hamster cells following ultraviolet irradiation. *Nucleic Acids Res.* **24**, 2877–2884.

J. Kruh (1982) Effects of sodium butyrate, a new pharmacological agent, on cells in culture. *Mol. Cell Biochem.* **42**, 65–82.

J. T. Leith (1988) Effects of sodium butyrate and 3-aminobenzamide on survival of Chinese hamster HA-1 cells after X irradiation. *Radiat. Res.* **114**, 186–191.

C. D. Lewis, U. K. Laemmli (1982) Higher order metaphase chromosome structure: evidence for metalloprotein interactions. *Cell* **29**, 171–181.

M. Ljungman (1991) The influence of chromatin structure on the frequency of radiation-induced DNA strand breaks: A study using nuclear and nucleoid monolayers. *Radiat. Res.* **126**, 58–64.

D. M. Lopez-Larraza, N. O. Bianchi (1993) DNA response to bleomycin in mammalian cells with variable degrees of chromatin condensation. *Environ. Mol. Mutagen* **21**, 258–264.

R. S. Malyapa, W. D. Wright, J. L. Roti-Roti (1996) DNA supercoiling changes and nucleoid protein composition in a group of L5178Y cells of varying radiosensitivity. *Radiat. Res.* **145**, 239–242.

R. S. Malyapa, W. D. Wright, J. L. Roti-Roti (1995) Radiation Damage in DNA: Structure/Function Relationships at Early Times (eds. A. F. Fuciarelli, J. D. Zimbrick) Battelle Press, Columbus, Ohio, pp. 409–418.

R. S. Malyapa, W. D. Wright, J. L. Roti-Roti (1994) Radiation sensitivity correlates with changes in DNA supercoiling and nucleoid protein content in cells of three Chinese hamster cell lines. *Radiat. Res.* **140**, 312–320.

H. R. Matthews, J. H. Waterborg, in The Enzymology of Post-translational Modification of Proteins, R. Freedman, H. C. Hasskins, Eds. (Academic Press, New York, London, 1985) vol. 2, pp. 125–185.

S. J. McCready, P. R. Cook (1984) Lesions induced in DNA by ultraviolet light are repaired at the nuclear cage. *J. Cell Sci.* **70**, 189–196.

T. Miyashita, H. Yamamoto, Y. Nishimune, M. Nozaki, T. Morita, A. Matsushiro (1994) Activation of the mouse cytokeratin A (endo A) gene in tetratocarcinoma F9 cells by the histone deacetylase inhibitor Trichostatin A. *FEBS Lett.* **353**, 225–229.

L. H. F. Mullenders, A. A. v. Zeeland, A. T. Natarajan (1987) The localization of ultraviolet-induced excision repair in the nucleus and the distribution of repair events in higher order chromatin loops in mammalian cells. *J. Cell Sci.* **Suppl. 6**, 243–262.

Z. Nackerdien, J. Michie, L. Bohm (1989) Chromatin decondensed by acetylation shows an elevated radiation response. *Radiat. Res.* **117**, 234–244.

G. L. Newton, J. A. Aguilera, J. F. Ward, R. C. Fahey (1996) Polyamine-induced compaction and aggregation of DNA—A major factor in radioprotection of chromatin under physiological conditions. *Radiat. Res.* **145**, 776–780.

M. I. Parker, J. D. Haan, W. Gevers (1986) DNA-hypermethylation in sodium-butyrate treated WI-38 fibroblasts. *J. Biol. Chem.* **261**, 2786–2790.

M. Perry, R. Chalkley (1982) Histone acetylation increases the solubility of chromatin and occurs sequentially over most of the chromatin. *J. Biol. Chem.* **257**, 7336–7347.

K. J. Pienta, D. S. Coffey (1984) A structural analysis of the role of the nuclear matrix and DNA loops in the organization of the nucleus and chromosome. *J. Cell Sci.* **Suppl. 1**, 123–135.

K. N. Prasad, P. K. Sinha (1976) Effect of sodium butyrate on mammalian cells in culture: a review. *In Vitro* **12**, 125–132.

P. R. Rezek, D. Weisman, P. E. Huvos, G. Fasman (1982) Sodium butyrate induced structure changes in HeLa cell chromatin. *Biochemistry* **21**, 993–1002.

M. R. Riehm, R. E. Harrington (1987) Protein dependent conformational behavior of DNA in chromatin. *Biochemistry* **26**, 2878–2886.

J. L. Roti-Roti, W. D. Wright, Y. C. Taylor (1993) DNA loop structure and radiation response. *Adv. Radiat. Biol.* **17**, 227–259.

M. J. Smerdon, S. Y. Lan, R. E. Calza, R. Reeves (1982) Sodium butyrate stimulates DNA repair in UV-irradiated normal and xeroderma pigmentosum human fibroblasts. *J. Biol. Chem.* **257**, 13441–13447.

P. J. Smith (1986) n-Butyrate alters chromatin accessibility to DNA repair enzymes. *Carcinogenesis* **7**, 423–429.

M. Spotheim-Maurizot, S. Ruis, R. Sabattier, M. Charlier (1995) Radioprotection of DNA by polyamines. *Int. J. Radiat. Biol.* **68**, 571–577.

G. Vidali, L. C. Boffa, E. M. Bradbury, V. G. Allfrey (1978) Butyrate suppression of histone deacetylation leads to accumulation of multiacetylated forms of histones H3 and H4 and increased DNase I sensitivity of the associated DNA sequences. *Proc. Natl. Acad. Sci.* **75**, 2239–2243.

R. L. Warters, B. W. Lyons (1992) Variation in radiation-induced formation of DNA double-strand breaks as a function of chromatin structure. *Radiat. Res.* **130**, 309–318.

E. C. Woudstra, J. M. Roesink, M. Rosemann, J. F. Brunsting, C. Driessen, T. Orta, A.W. T. Konings, J. H. Peacock, H. H. Kampinga (1996) Chromatin structure and cellular radiosensitivity: A comparison of two human tumor cell lines. *Int. J. Radiat. Biol.* **70**, 693–703.

L.-y. Xue, L. R. Friedman, N. L. Oleinick, S.-m. Chiu (1994) Induction of DNA damage in γ-irradiated nuclei stripped of nuclear protein classes: Differential modulation of double-strand breaks and DNA-protein crosslink formation. *Int. J. Radiat. Biol.* **66**, 11–21.

M. Yoshida, S. Horinouchi, T. Beppu (1995) Trichostatin A and trapoxin: Novel chemical probes for the role of histone acetylation in chromatin structure and function. *Bioessays* **17**, 423–430.

29

ESTIMATION OF FREE RADICAL INDUCED DNA BASE DAMAGES IN CANCEROUS- AND HIV INFECTED PATIENTS AND IN HEALTHY SUBJECTS

Ryszard Olinski, Pawel Jaruga, and Tomasz Zastawny

Department of Clinical Biochemistry
The Ludwik Rydygier Medical University in Bydgoszcz
Karlowicza 24, 85-092 Bydgoszcz, Poland

1. ABSTRACT

It was shown that free radical attack upon DNA generates a whole series of DNA damage among them modified bases. Hydroxyl radical attack on DNA leads to a large number of pyrimidine- and purine-derived base damages. Some of these modified DNA bases have considerable potential to damage the integrity of the genome; therefore they may be responsible for carcinogenesis.

We have investigated endogenous levels of typical free radical induced DNA base modification in chromatin of various human cancerous tissues and their cancer-free surrounding tissues. In all cases elevated amounts over control levels of modified DNA bases were found in cancerous and in precancerous tissues. Our data may indicate an important role of oxidative base damage in cancer development. Alternatively the increased level of base product may contribute to genetic instability and metastatic potential of tumor cells.

Similar increases in the base products were found in DNA isolated from lymphocytes of cancer patients undergoing chemo- and radiotherapy. These increases may be responsible for the development of secondary cancers such as leukemias, which are unavoidably associated with radiotherapy and chemotherapy.

In another set of experiments we have shown statistically significant increase of oxidatively modified DNA bases in AIDS patients when compared with control group. In the case of one base product namely 8-oxoguanine (8-oxoGua), which has high mutational potential, statistically significant changes were also found between control group and asymptomatic, HIV positive patients. We postulate that the observed increase in base product level may be one of the reasons responsible for apoptotic death of lymphocytes and cancers related to HIV infection.

Advances in DNA Damage and Repair, edited by Dizdaroglu and Karakaya. Kluwer Academic / Plenum Publishers, New York, 1999.

Our recent results demonstrated that only small fluctuation exist in the background level of the modified bases in the case of healthy individual over a longer period of time (several months).

2. INTRODUCTION

In view of the importance of DNA damage in carcinogenesis it is conceivable that any agent capable of reacting with DNA and chemically modifying it could be carcinogenic. It is very likely that reactive oxygen species (ROS) belong to this group of agents.

Reactive oxygen species are the products of partial reduction of oxygen. These species which include superoxide anion, hydrogen peroxide and hydroxyl radical are continuously produced in living cells as by-products of normal metabolism.

A considerable frequency and intensity of oxidative DNA damage in the cells of normal untreated animals or humans is now an accepted fact. (Ames and Gold, 1990), for example, estimated that the genome of adult rat liver cells contains about 1mln. Such oxidative DNA damage per cell and that 100 000 new or replacement lesions are added daily to this tremendous pool.

It was shown that free radical attack upon DNA generates a whole series of DNA damage among them modified bases. Hydroxyl radical (•OH) attack on DNA leads to a large number of pyrimidine- and purine-derived base damage. Some of these modified DNA bases have considerable potential to damage the integrity of the genome (Floyd et al., 1986, Shinegaga et al., 1989; Floyd, 1990; Dizdaroglu, 1991).

3. MUTAGENIC AND CARCINOGENIC PROPERTIES OF DNA BASE DERIVATIVE

8-oxoguanine is one of the most critical lesions. The presence of 8-oxoGua residues in DNA leads to GC to TA transversion unless repaired prior to DNA replication (Cheng et al., 1992). Therefore, the presence of 8oxoGua may lead to mutagenesis. Furthermore many observation indicate a direct correlation between 8-oxoGua formation and carcinogenesis in vivo (Floyd, 1990; Feig et al., 1994). Thus, it has been found that oxyradicals induced mutagenesis of hotspot codons of the human p53, and Ha-Ras genes (Hussain et al., 1994, Le Page et al., 1995). In agreement with this finding it was demonstrated that GC to TA transversions have been frequently detected in p53 gene and in ras protooncogene in the case of lung carcinomas and primary liver cancer (Cheng et al., 1992, Hussain et al., 1994, Le Page et al., 1995). In this context it is noteworthy that we have demonstrated elevated levels of typical free radical-induced DNA base modification, including 8-oxoguanine, in human cancerous lung tissues when compared with cancer free surrounding tissues (Olinski et al., 1992, Jaruga et al., 1994).

Mutagenic and carcinogenic potential of any modified DNA base is reflected in its miscoding properties. Recently, it has been demonstrated that several other bases have misscoding potential. Thus, presence of 2-hydroxyadenine (2-OH-Ade) in DNA may induce A to C and A to T transversions and A to G transition (Kamiya and Ueda et al., 1995). It was also shown that 2-hydroxy-deoxyadenosine-triphosphate (2-OH-dATP) is a substrate for DNA polymerase and may be incorrectly incorporated by DNA polymerase (Kamiya and Kasai, 1995). 8-hydroxyadenine (8-OH-Ade) also has miscoding properties and induces mutation in mammalian cells (Kamiya and Miura et al., 1995). 5-hydroxycytosine (5-OH-Cyt) has been shown to be potentially premutagenic lesion leading to GC to AT transition and GC to CG transversion. 5-OH-Cyt appears to be more mutagenic than any other product of oxidative DNA damage (Purmal et al., 1994; Feig et al., 1994). It is also possible that other than 8-oxo-G derivatives of guanine have miscoding properties (Ono et al., 1995). On the other hand biological consequences of other base modifications, like 4,6-diamino-5-formamido-pirymidine (FapyAde), 5,6-dihydroxyuracil (5,6-diOH-Ura) and 5-hydroxy-5-methylhydantoin (5-OH-5-MeHyd) have not been investigated. It is conceivable that these lesions may be premutagenic as well.

4. EXPERIMENTAL EVIDENCES SUGGESTING INVOLVEMENT OF MODIFIED BASES IN CARCINOGENESIS

A role of free radical modified DNA bases in development of cancer in human is supported by the abundant presence of oxidative base modification in cancer tissue (Malins and Haimanot, 1991; Olinski et al., 1992; Jaruga et al., 1994; Olinski et al., 1996).

We have investigated endogenous levels of typical free radical induced DNA base modification in chromatin of various human cancerous tissues and their cancer free surrounding tissues. In all cases elevated amounts over control levels of modified DNA bases were found in cancerous tissues (Olinski et al., 1992; Jaruga et al., 1994). The amounts of modified bases depended on the tissue type (Fig. 1.).

Lung tissues removed from smokers had the highest increases of modified bases over the control levels. Stomach, ovary, and brain cancerous tissues also had significantly higher amounts of lesions in their chromatin then their respective cancer free surrounding tissues. Colon cancerous tissues had significant increase of 8-oxoGua. Lung cancer patients whose tissues were examined were smokers. Oxygen derived species are known to be involved in causing DNA damage by cigarette smoke and tar, most likely by producing $^{\bullet}OH$ radical in metal ion catalyzed reaction.

Elevated levels of modified DNA bases in cancerous tissues may be due in part to the presence of large amounts of leukocytes in human tumors (Eccles and Alexander 1974). Activated leukocytes are a source of H_2O_2, which can cross-cellular and nuclear membranes and reach the nucleus to cause site specific DNA damage by producing $^{\bullet}OH$ radical in reaction with DNA bound metal ions. Direct proof of this suggestion comes from the work of Dizdaroglu et al. (1993). They demonstrated that exposure to activated leukocytes caused DNA base modifications in human cells typical of those induced by hydroxyl radical attack. Moreover, elevated levels of modified bases in cancerous tissues may be due to the production of the large amounts of hydrogen peroxide, which has been found to be characteristic of human tumor cells (Szatrowski and Nathan, 1991).

Furthermore, evidence exists that tumor cells have abnormal level and activities of antioxidant enzymes when compared with their respective normal cells, Sun, 1990. Low levels of antioxidant enzymes, such as superoxide dismutase or catalase in tumor cells may cause accumulation of superoxide anion and H_2O_2 with subsequent $^{\bullet}OH$ induced damage to DNA, resulting in greater amounts of modified bases in tumor cells then in normal cells

To check whether a relationship exist between oxidative DNA base modification and antioxidant enzyme activities in cancerous tissues and in cancer free tissues we have studied the base modification and activities of superoxide dismutase (SOD), catalase (CAT), and glutation peroxidase (GPx) in normal and cancerous human lung tissues removed from five lung cancer patients. In agreement with previous study higher levels of DNA lesions were observed in cancerous tissues than in cancer-free surrounding tissues (Fig. 2.).

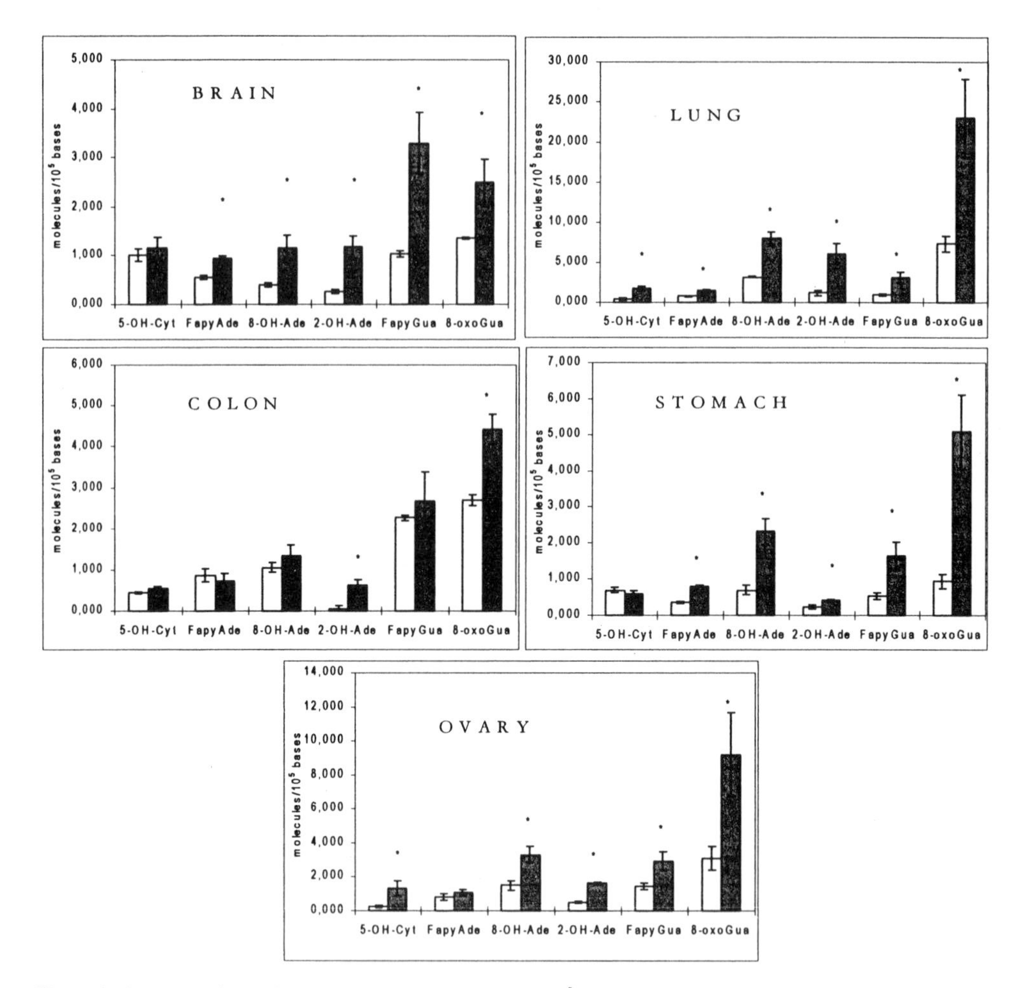

Figure 1. Amounts of modified DNA bases (molecules per 10^5 DNA bases) in human tissues. Each value represents the mean ± S.E.M. from measurements of chromatin samples isolated from five separate tissue samples. □ - normal; ■- cancerous. *Significantly different from controls ($P \leq 0,05$ by Student's t-test).

Most significant changes were observed in the case of 8-OH-Ade and 8-oxoGua in all five patients. Antioxidant enzymes levels were lower in cancerous tissues with respect to their surrounding cancer free tissues (Fig. 3.).

The highest decrease was observed in the activity of GPx, which amounted to 2 or even 3 fold. The activity of SOD in cancerous tissues was also significantly lower than in normal tissues. Only a modest decrease in the CAT activity was observed. The results indicated, for the first time, an association between decreased activities of antioxidant enzymes and increased levels of DNA lesions in cancerous tissues (Jaruga et al., 1994).

It is not known whether lower levels of antioxidant enzymes in cancerous lung tissues play a causative role in carcinogenesis or are merely the result of the desease. The same is true of the pyrimidine- and purine derived DNA lesions that were identified in cancerous tissues at higher levels then in control tissues. However, treatment of laboratory animals with carcinogenic agents causes similar pattern of oxidative base modification in

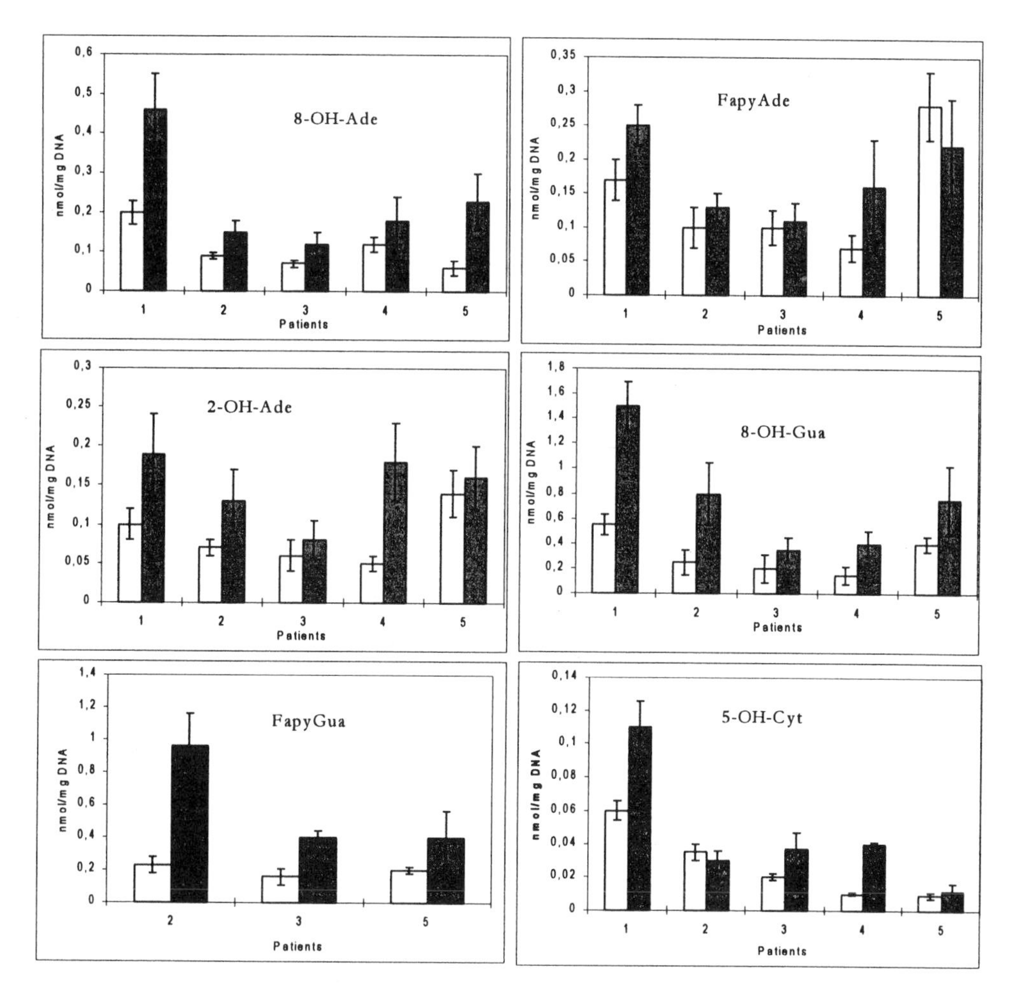

Figure 2. Amounts of modified DNA bases in human cancerous lung tissues and their surrounding cancer-free tissues. Each data point represents the mean ± S.D. from measurement of chromatin samples isolated from three separate tissue samples (1 nmol of a modified base/mg DNA ≈ 32 molecules of a modified base/10^5 DNA bases). □ - normal; ■- cancerous.

their target organs before tumor formation occurs (Kasprzak et al., 1997). Furthermore, as was mentioned before, several lesions identified in our work have been found to possess mutagenic properties (Wallace, 1987, Loeb et al., 1991, Grollman and Moriya, 1993). All these data may indicate an important role for these lesions in carcinogenesis. On the other hand, in fully developed cancer, increased levels of modified DNA bases may contribute to the genetic instability and metastatic potential of tumor cells. Our suggestion has been proved just recently. (Malins et al., 1996) showed that potential of metastasis of breast cancer tissues increased together with increases of oxidatively modified bases.

In our next step we wished to see whether precancerous tissues, had properties similar to cancerous tissues in terms of the levels of DNA base modifications and the activities of antioxidant enzymes. Recent data on nuclear matrix protein patterns in both benign prostatic hyperplasia (BPH) and prostate cancer have indicated that similar phenotypic events occurs in progression of cells from normal to BPH as from normal to cancer (Partin et al.,

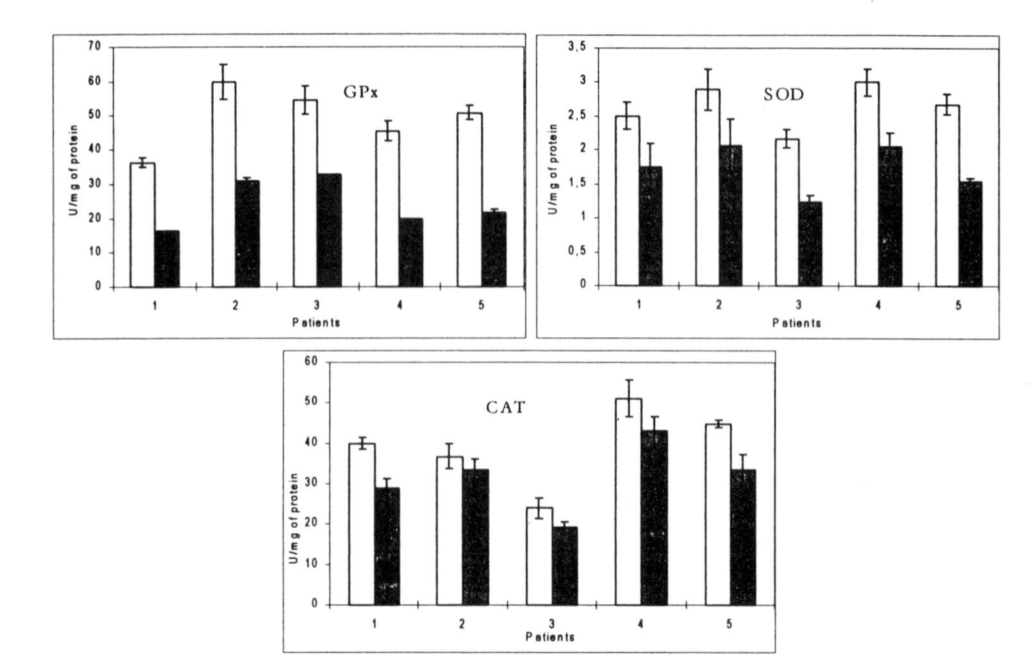

Figure 3. Activities of GPx, SOD and CAT in human cancerous lung tissues and their surrounding cancer-free tissues. Each data point represents the mean ± S.D. from three independent measurements. □ - normal; ■ - cancerous.

1993). Thus, it is possible that BPH represents premalignant condition which may predispose to prostate cancer. Therefore we decided to use prostate glands from BPH patients in our study. 15 patients were divided into three groups with respect to differences between the enzymes activities in BPH tissues and their surrounding disease free tissues. Patients no 1 - 5 had significantly lower activities of both SOD and catalase (Fig. 4.A.)

In this group of patients significant increases in the amounts of modified bases in BPH tissues were observed then in disease free tissues in most cases. In group II the activity of SOD was significantly lower in BPH tissues where there were no change in the activity of CAT. In this group, increases observed in BPH tissues were much less prominent then those in group I (Fig. 4.B.).

In group III the activities of SOD and CAT were similar in BPH and control tissues. Likewise, no increases in the amounts of modified bases in BPH tissues were observed. It is likely that patients with both decreased activity of antioxidant enzymes and increased levels of the modified bases are at greater risk of developing prostate cancer.

5. ANTICANCER THERAPY AS A SOURCE OF OXIDATIVELY MODIFIED DNA BASES. POSSIBLE INVOLVEMENT OF THE BASE DAMAGES IN DEVELOPMENT OF SECONDARY CANCERS

Most of the agents used in anticancer therapy are paradoxically responsible for induction of secondary malignancies and some of them may generate free radicals. Since free radical induced DNA damage may possess premutagenic properties and may play

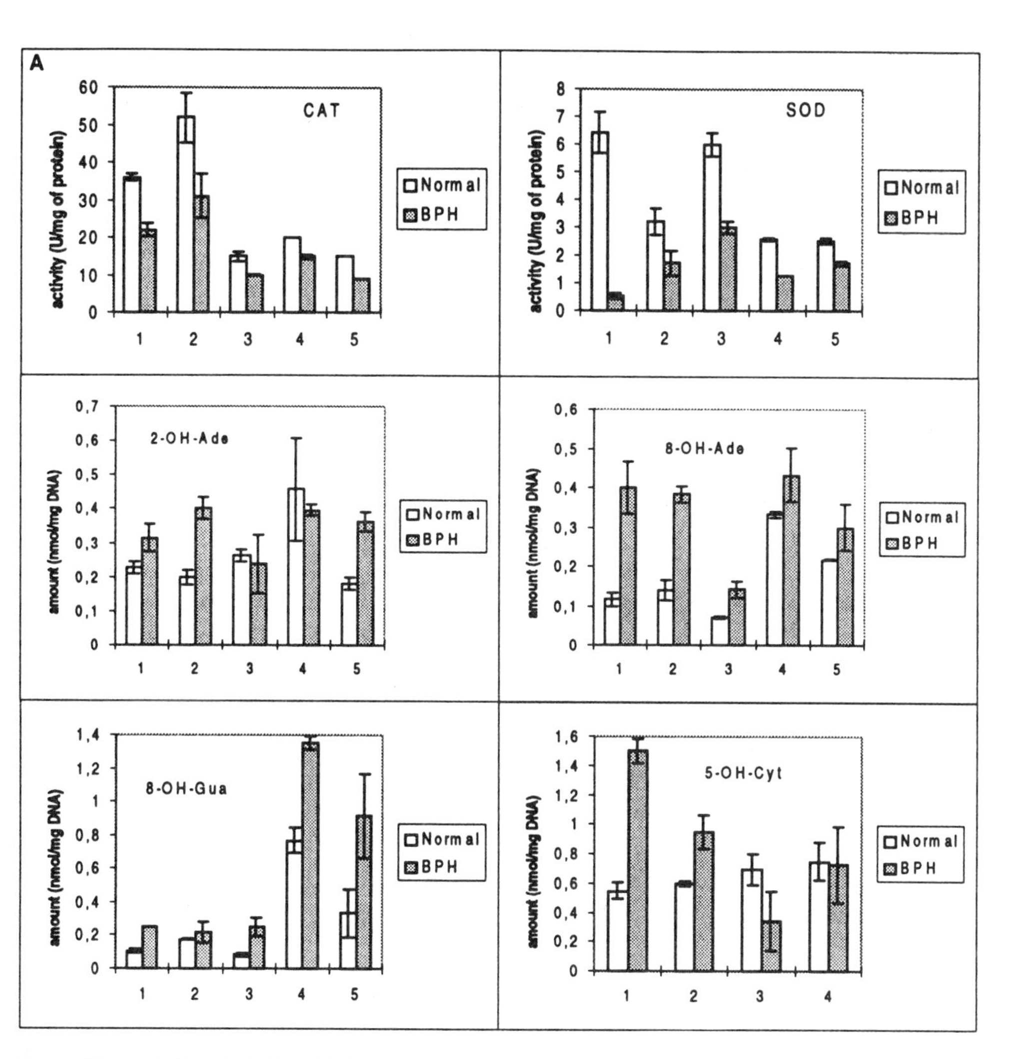

Figure 4A. Activities of CAT and SOD and the amounts of modified DNA bases in BPH and surrounding disease-free tissues of Group I of BPH patients. Each value represents the mean ± standard error from three independent measurements (1 nmol of modified base/ mg DNA ≈ 32 molecules of modified bases/10^5 bases).

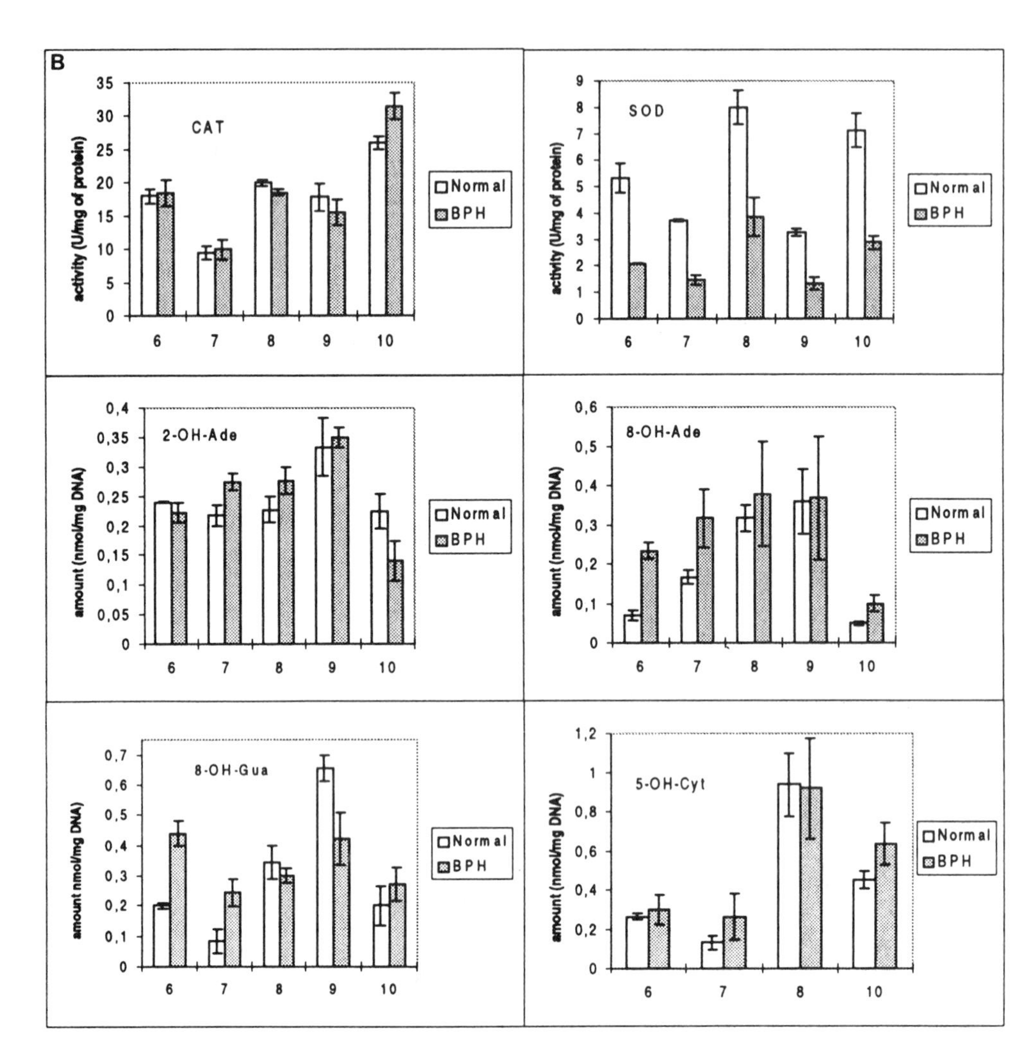

Figure 4B. Activities of CAT and SOD and the amounts of modified DNA bases in BPH and surrounding disease-free tissues of Group II of BPH patients. Each value represents the mean ± standard error from three independent measurements (1 nmol of modified base/ mg DNA ≈ 32 molecules of modified bases/10^5 bases).

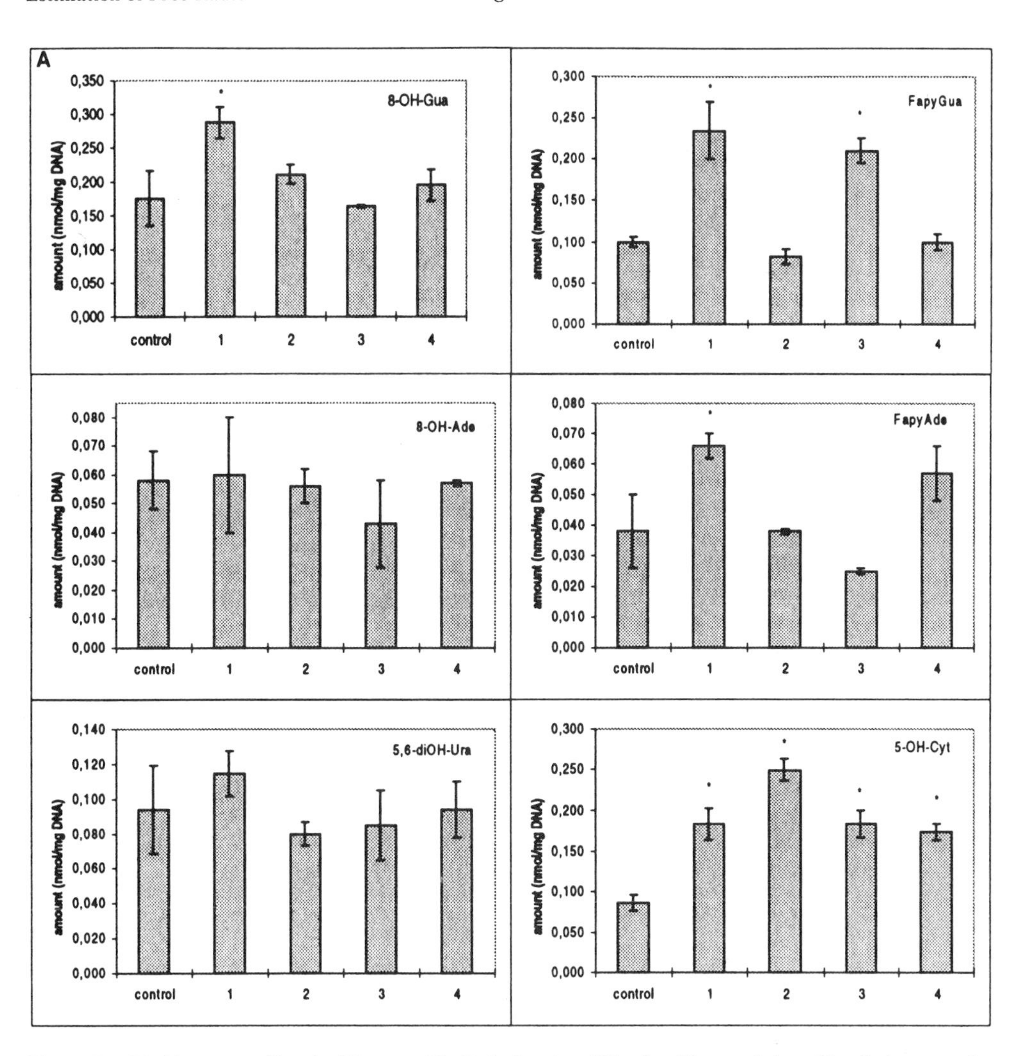

Figure 5A. Bladder cancer (female, 55 years old). Daily fraction, 180 cGy; (1) second day of irradiation, cumulative dose 360 cGy (180 + 180); (2) 5 consecutive days of irradiation totaling 900 cGy, cumulative dose 1260 cGy (360 + 900); (3) 5 consecutive days of irradiation totaling 900 cGy, cumulative dose 2160 cGy (360 + 900 + 900); (4) 10 consecutive days of irradiation totaling 1800 cGy, cumulative dose 3960 cGy (360 + 900 + 900 + 1800). Each data point represents the mean ± standard error from measurements of three chromatin samples, which were independently isolated from the same blood sample (1nmol of modified base/ mg DNA ≈ 32 molecules of modified bases/10^5 bases). *Significantly different from controls ($P \leq 0,05$ by Student's t-test).

some role in carcinogenesis we wised to see whether the modalities used in anticancer therapy are responsible for production of typical free radical induced base modifications in nuclear DNA of lymphocytes of cancer patients who are undergoing anticancer therapy.

Ionizing radiation is one of the most commonly used therapeutic agents of cancer. In general, approximately half of cancer patients receive radiation therapy for their disease management (Castro and Curtis, 1992). The result of our experiments provided evidence that exposure of cancer patients to therapeutic doses of ionizing radiation causes base modification in gnomic DNA of their lymphocytes (Fig. 5.A.and 5.B)

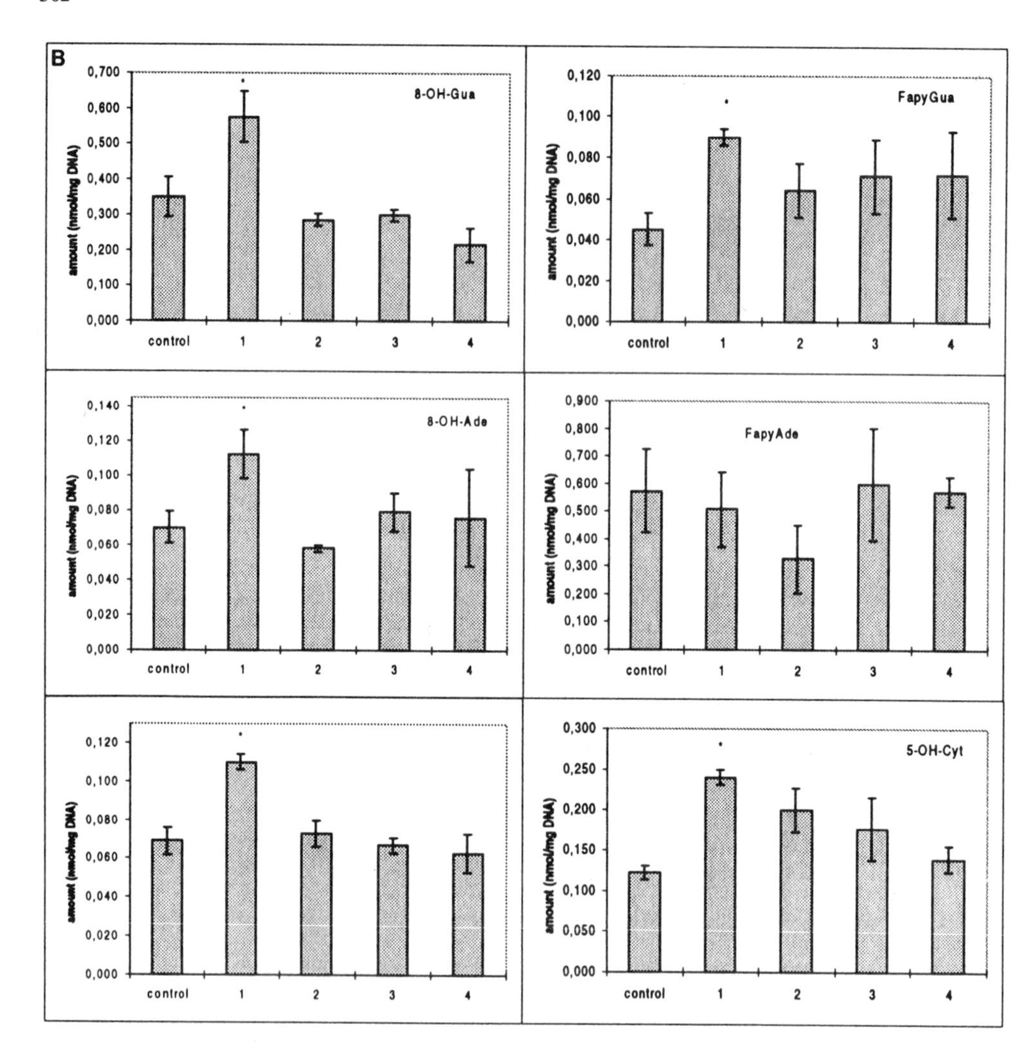

Figure 5B. Head and neck cancer (male, 38 years old). Daily fraction, 400 cGy; (1) second day of irradiation, cumulative dose 800 cGy (400 + 400); (2) days break; (3) 5 consecutive days of irradiation totaling 2000 cGy, cumulative dose 2800 cGy (800 + 2000); (4) 1 consecutive day of irradiation 400 cGy, cumulative dose 3200 cGy (800 + 2000 + 400). Other details are as in Figure 5A.

Anthracycline derivatives have been widely used in the treatment of several types of human malignancies. Cytotoxicity of these drugs has been attributed to inhibition of topoisomerase II as well as intracellular production of free radicals.

Recently using GC-MS method (Akman et. al, 1992) have shown that reactive oxygen reduction by the redox cycling of the doxorubicin quinone moiety is responsible for DNA base modification in isolated human chromatin. In our recent work using epirubicin (the analog of doxorubicin presenting a different configuration of the OH group in the C-4 position of the amino-sugar moiety) we observed similar base modification in chromatin isolated from lymphocytes of cancer patients undergoing chemotherapy (Fig. 6.).

The pattern of these modifications also suggests the involvement of •OH radical in their formation.

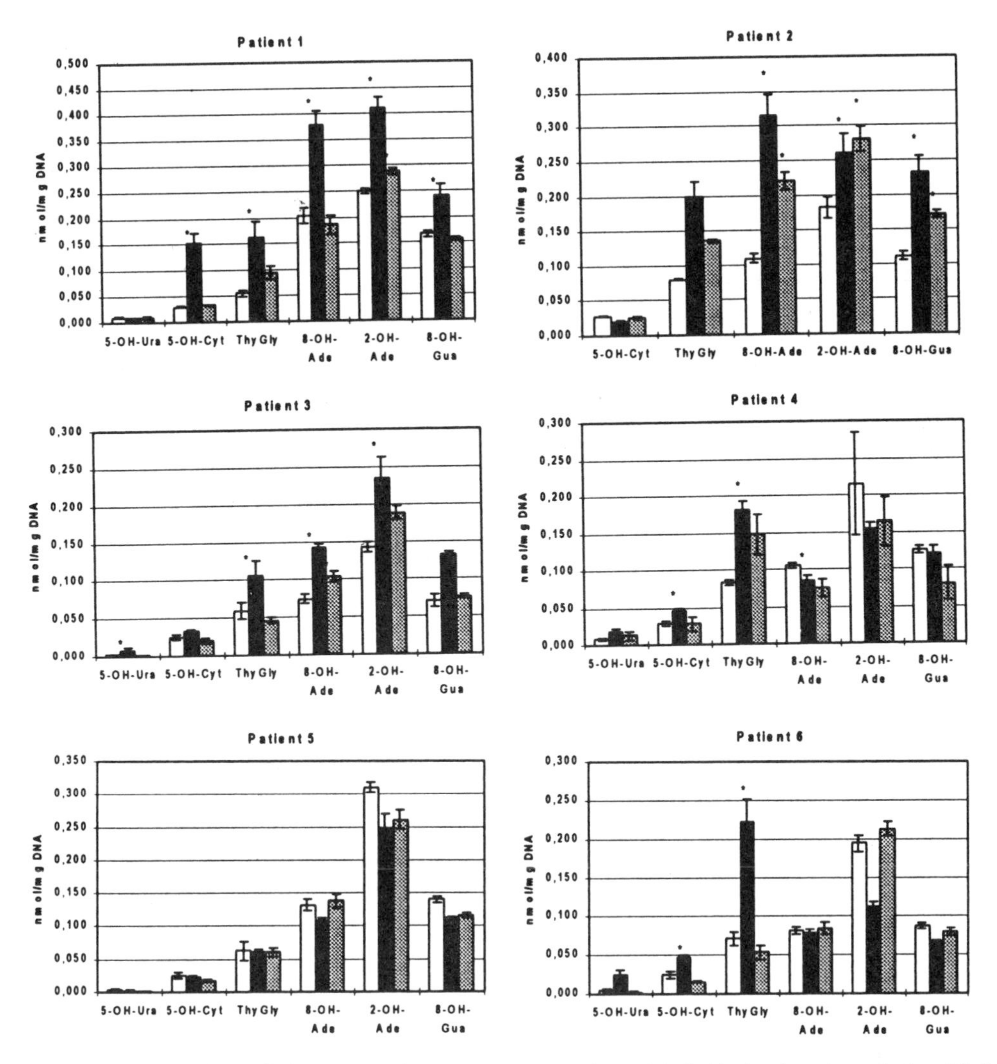

Figure 6. Amounts of modified DNA bases in lymphocyte DNA, for each individual patient. 1 nmol of modified DNA base/mg of DNA ≈ 32 modified bases/10^5 DNA bases. Each data point on the graph represents the mean ± standard error from the measurements of 3 chromatin samples isolated from the blood. Stars above bars indicate that the result is significantly different ($p \leq 0.05$ by Student's t-test) from control value. ■ - control, □- after 1 h, (shaded)- after 24 h.

Anticancer therapy caused significant increases in the amount of all four DNA base modifications over control levels for most of the patients (Fig. 5.A, 5.B and 6.). However, the extent of these modifications differs among patients. This interindividual variability may reflect individual differences in exposure, metabolism and repair capacity and may have, at least in part, genetic background (Setlow, 1993).

In the case of chemotherapy, for majority of patients, 24h after the infusion of the drug the base product returned to control value (Fig. 6). Likewise, in lymphocytes DNA of

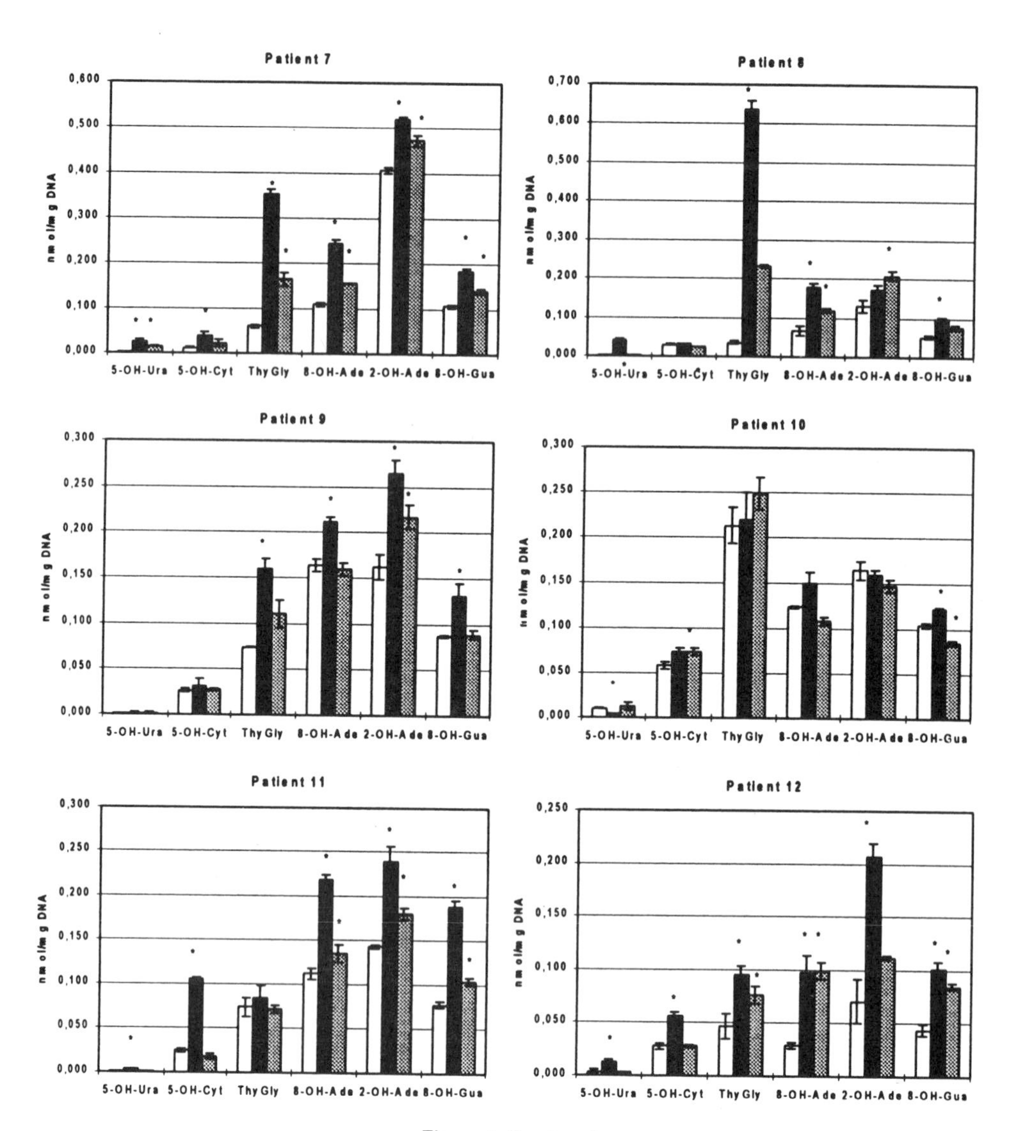

Figure 6. *Continued.*

the patients who were undergoing radiotherapy, in many cases, product levels decreased after they increased upon irradiation even reaching the control level (Fig. 5.A., 5.B.). Thus, the decrease in base products may well be indication of removal of these lesions by cellular repair processes. In the case of some patients and some modifications the level of damages stay high despite considerable period of time after their introduction.

There is a risk associated with developing secondary cancers after chemotherapy and radiotherapy (Inskip et al., 1993; Anderson and Berger, 1994). Long-lived B and T lymphocytes may serve as a target cells for carcinogens including some anticancer drugs and ionizing irradiation (Lajtha, 1981). Some of base modifications which escaped repair in lymphocytes DNA could lead to mutagenesis in critical genes and ultimately to secondary cancers such as leukemias.

6. ELEVATED LEVEL OF OXIDATIVELY MODIFIED DNA BASES IN HIV INFECTED PATIENTS AS A POSSIBLE CAUSE OF APOPTOTIC DEATH OF LYMPHOCYTES

It is believed that patients infected with human immunodeficiency virus (HIV) are under chronic oxidative stress. Recent evidence demonstrated that oxidative stress mediated by generation of ROS is one of the direct causes of programmed cell death (apoptosis) (Hockenbery et al., 1993).

Direct cause of apoptotic death of lymphocytes in HIV infected patients is still unclear. It is likely that oxidative stress in HIV infected cells could be responsible for elevated level of base modifications, which in turn could cause apoptotic death of lymphocytes. Therefore in our study we wished to see whether the level of oxidative DNA base modifications is elevated when compared with control group of patients.

One of the groups most endangered by HIV infection is injected drug users (IDUs). In Poland 70% of all AIDS patients are recruited from among this group (the most commonly used drug is heroin). In our study we compared the level of base product in lymphocytes of HIV-infected IDUs with that of seronegative IDUs (control group).

The results of this study showed higher background levels of typical free radical induced products of all four DNA bases in lymphocytes of HIV infected patients (Fig.7.).

Since there are no data that AIDS is associated with reduced ability to repair DNA damage , the observed increase should be a result of higher concentration of ROS in lymphocytes of HIV infected patients. Indeed wide variety of evidence support the theory that oxidative stress is involved in progression of AIDS (Ameisen and Capron 1991; Pace and Leaf, 1995).

Several investigators have recently proposed that apoptosis initiated by oxidative stress is the direct cause of lymphocytes loss in patients infected with HIV (Pace and Leaf,

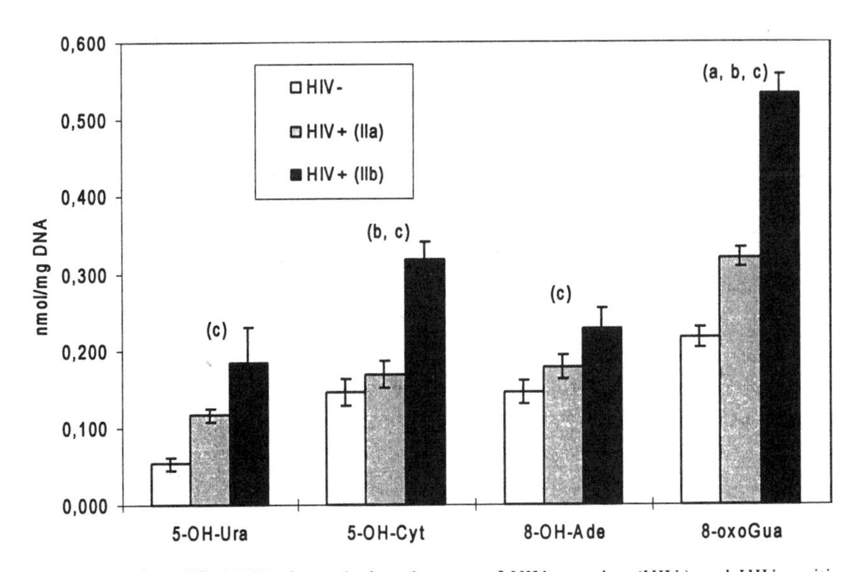

Figure 7. Amounts of modified DNA bases in lymphocytes of HIV negative (HIV-) and HIV positive IDU's: HIV+ (IIa) and HIV+ (IIb). Error bars indicated standard deviation of the mean. The data were analyzed by the Student *t*-test, $p<0.05$. a-significant difference between HIV-/HIV+ (IIa); b-significant difference between HIV+ (IIa)/HIV+ (IIb); c-significant difference between HIV-/HIV+ (IIb).

1995). However the mechanism(s) which directly triggered the apoptosis is (are) still unclear. One of the reasons, which may induce apoptotic death of cells, is unrepaired DNA damage. There are indications, that oxidative DNA damage induces apoptosis in murine T-cell hybridoma (Warters, 1992) in lymphocytes (Slatter et al., 1995). One kind of such lesions might be oxidatively modified DNA bases, which were found to persist in higher amounts in lymphocytes of HIV infected patients. Some of these modified base products are mutagenic and their increased concentration in lymphocytes of HIV infected patients may be related to malignancies associated with AIDS (it has been observed that HIV infection predisposes an individual to several neoplasies, especially non-Hodkin's lymphoma of B-cell origin (Beral et al., 1991; Schulz et al., 1996; Williams et al., 1994).

7. THE BACKGROUND LEVEL OF OXIDATIVELY MODIFIED DNA BASES IN HEALTHY SUBJECTS

The background level of these modified bases can be detected in normal tissues (Olinski et al., 1992; Jaruga et al., 1994; Olinski et al., 1996). This level may simply represent steady-state concentration reflecting the balance between their formation and repair. Our previous data proved that interindividual differences exist in the background levels of different modified bases in the case of cancer patients (Olinski et al., 1992; Jaruga et al., 1994; Olinski et al., 1996). However, nothing was known about individual fluctuation of this value in the case of single healthy individual over a longer period of time (several months). Since the background level may be meaningful parameter from the point of view of the etiology of cancer it is important to know the range of interindividual fluctuation and variation within the subject over a longer period of time of the base products. To ex-

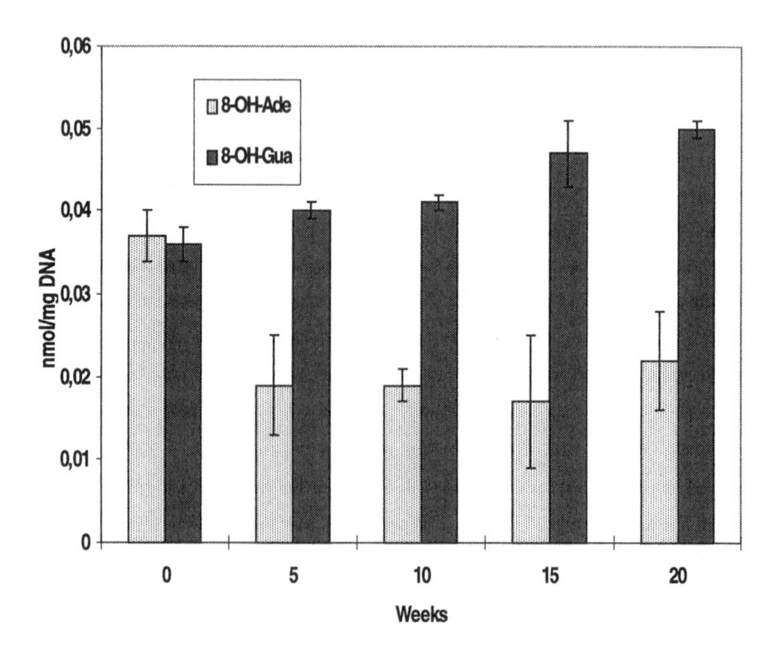

Figure 8. 8-OH-Ade and 8-oxoGua content in DNA of lymphocytes from healthy subjects (n=10), period over 20 weeks.

plain this, the blood samples was taken from 10 healthy volunteers (mostly recruited from among the stuff of our Department) and DNA was isolated from lymphocytes to determine the level of DNA base damage.

We have found that the level of 8-oxoGua and 8-OH-Ade, which have mutational potential is stable over 6 months, with mean value for 8-oxoGua - 0.043 nmol/mg DNA , with variation ranging from 0.036 - 0.050 while mean value of 8-OH-Ade was 0.030, ranging from 0.017 to 0.037) (Fig. 8.).

8. CONCLUSIONS

All these observations pointed out on meaningful role of oxidatively modified DNA bases in many of the biological processes responsible for cancer and AIDS development. Moreover clarifying the role of these damages in pathological processes could lead to novel therapeutic strategies and give insight into molecular events directly responsible for these pathologies.

ACKNOWLEDGMENTS

These studies were supported in part by grants from the Polish State Committee for Scientific Research (KBN) 4PO5A.121.08 and from the USA-Polish Maria Sklodowska-Curie Joint Fund II (MZ/NIST-97–298).

We appreciate the permission to use materials from publications in: FEBS Letters: R. Olinski et al. (1992) FEBS Lett. 309, 193–198 and P. Jaruga et al. (1994) FEBS Lett., 341, 59–64, in: Free Radicals in Biology and Medicine: R. Olinski et al. (1995) Free Rad. Biol. Med. 18, 807–813 and in: Cancer Letters: R. Olinski et al. (1996) Cancer Lett. 106, 207–215.

Authors are very grateful to Dr. Miral Dizdaroglu from Biotechnology Division, NIST, Gaithersburg, Maryland, USA, for opportunity to learn GC/MS technique, for the helpful discussion and for kind gift of the labeled internal standards.

REFERENCES

S.A. Akman, J.H. Doroshow, T.G. Burke and M. Dizdaroglu (1992) DNA base modifications induced in isolated human chromatin by NADH dehydrogenase-catalyzed reduction of doxorubicin. *Biochemistry* **31**, 3500–3506.

J.C. Ameisen and A. Capron (1991)Cell dysfunction and depletion in AIDS: The programmed cell death hypothesis. *Immunol Today* **12**, 102–105.

B.N. Ames and L.S. Gold (1990) Too many rodent carcinogens: mitogenesis increases mutagenesis, *Science* **249**, 970–971.

R.D. Anderson and N.A. Berger. (1994) Mutagenicity and carcinogenicity of topoisomerase- interactive agents. *Mutat. Res.* **309**, 109–142.

V. Beral, T. Peterman, R. Berkelman and H. Jaffe (1991) AIDS-associated non-Hodgkin lymphoma. *Lancet* **337**, 805–809.

J.R. Castro and S.B. Curtis (1992) The application of particle beams to radiation therapy. In: *Radiation Biology* (eds. D.J. Pizzarello and L.D. Colombetti), CRC Press, Boca Raton, pp. 265–285.

K.C. Cheng, D.S. Cahill, H. Kasai, S. Nishimura and L. Loeb (1992) 8-Hydroxyguanine, an abundant form of oxidative DNA damage, causes G-T and A-C substitutions. *J. Biol. Chem.* **267**, 166–172.

M. Dizdaroglu (1991) Chemical determination of free radical induced damage to DNA. *Free Rad. Biol. Med.* **10**, 225–242

M. Dizdaroglu (1992) DNA base modifications in chromatin of human cancerous tissues, *FEBS Lett.* **309**, 193–198.

M. Dizdaroglu, R. Olinski, J.H. Doroshow, S.A. Akman (1993) Modification of DNA bases in chromatin of intact target human cells by activated human polymorphonuclear leukocytes. *Cancer Res.* **53,** 1269–1272.

S.A. Eccles and P. Alexander (1974) Sequestration of macrophages in growing tumours and its effect on the immunological capacity of the host. *Br. J. Cancer* **30**, 42–49.

D.I. Feig, T.M. Reid and L.A. Loeb (1994) Reactive oxygen species in tumorigenesis. *Cancer Res.* (suppl.) **54**, 1890–1894

R.A. Floyd (1990) The role of 8-hydroxyguanine in carcinogenesis. *Carcinogenesis* **11**, 1447–1450.

R.A. Floyd, J.J. Watson, P.K. Wong, D.H. Altmiller and R.C. Rickard (1986) Hydroxyl free radical adduct of deoxyguanosine: sensitive detection and mechanizm of formation. *Free Rad. Res. Commun.* **1**, 163–172.

A.P. Grollman, M. Moriya (1993) Mutagenesis by 8-oxoguanine: an enemy within. *Trends Genet.* **9**, 246–249.

D.M. Hockenbery, Z.N. Oltval, X. Yin, C.L. Milliman and S.J. Kosmeyer (1993) Bcl-2 functions in an antioxidant pathway to prevent apoptosis. *Cell* **75**, 241–251.

D.J. Hu, T.J. Dondero, Rayfield MA, et al. (1996) The emerging genetic diversity of HIV. *JAMA* **275**, 210–216.

S.P. Hussain, F. Aguilar, P. Amstad and P. Cerutti (1994) Oxy-radical induced mutagenesis of hotspot codons 248 and 249 of the human p35 gene. *Oncogene* **9**, 2277–2281.

P.D. Inskip, R.A. Kleinerman, M. Stovall, D.L. Cookfair, O. Hadjimichael, W.C. Moloney, R.R. Monson, W.D. Thompson, J. Wactawski-Wende, J.K. Wagoner et al. (1993) Leukemia, lymphoma, and multiple myeloma after pelvic radiotherapy for benign disease. *Radiat. Res.* **135**, 108–24

P. Jaruga, T.H. Zastawny, J. Skokowski, M. Dizdaroglu and R. Olinski (1994) Oxidative DNA Base Damage and Antioxidant Enzyme Activities in Human Lung Cancer, *FEBS Lett.* **341**, 59–64.

H. Kamiya, H. Kasai (1995) Formation of 2-hydroxydeoxyadenosine triphosphate, an oxidatively damaged nucleotide, and its incorporation by DNA polymerases. Steady-state kinetics of the incorporation. *J. Biol. Chem.* **270,** 19446–19450.

H. Kamiya, H. Miura, N. Murata-Kamiya, H. Ishikawa, T. Sakaguchi, H. Inoue, T. Sasaki, C. Masutani, F. Hanaoka, S. Nishimura et al. 8-Hydroxyadenine (7,8-dihydro-8-oxoadenine) induces misincorporation in vitro DNA synthesis and mutations in NIH 3T3 cells. *Nucl. Acids Res.* **23,** 2893–2899

H. Kamiya, T. Ueda, T. Ohgi, A. Matsukage and H. Kasai (1995) Misincorporation of dAMP opposite 2-hydroxyadenine, an oxidative form of adenine. *Nucl. Acids. Res.* **23,** 761–766

K.S. Kasprzak, P. Jaruga, T.H. Zastawny, S.L. North, C. Riggs, R. Olinski and M. Dizdaroglu (1997) Oxidative DNA base damage and repair in kidneys and livers of nickel(II)-treated male F344 rats. *Carcinogenesis* **18,** 271–277.

L.G. Lajtha (1981) Which are the leukemic cells? *Blood Cells* **7**, 45–62.

F. Le Page, A. Marot, A.P. Grollman, A. Sarasin and A. Gentil (1995) Mutagenicity of a unique 8-oxoguanine in a human Ha-ras sequence in mammalian cells. *Carcinogenesis* **16**, 2779–2784.

D.C. Malins and R. Haimanot (1991) Major alterations in the nucleotide structure of DNA in cancer of the female breast. *Canser Res.* **51,** 5430–5432.

D.C. Malins, N.L. Polissar and S.J. Gunselman (1996) Progression of human breast cancers to the metastatic state is linked to hydroxyl radical-induced DNA damage. *Proc. Natl. Acad. Sci.* USA. **93**, 2557–2563.

R. Olinski, T.H. Zastawny, J. Budzbon, J. Skokowski, W. Zegarski and M. Dizdaroglu (1992) DNA base modifications in chromatin of human cancerous tissues. *FEBS Lett.* **309**, 193–198.

R. Olinski, T.H. Zastawny, M. Foksinski, A. Barecki and M. Dizdaroglu (1995) DNA base modifications and antioxidant enzyme activities in human benign prostatic hyperplasia. *Free Rad. Biol. Med.* **18,** 807–813.

R. Olinski, T.H. Zastawny, M. Foksinski, W. Windorbska, P. Jaruga, M. Dizdaroglu (1996) DNA base damage in lymphocytes of cancer patients undergoing radiation therapy. *Cancer Lett.* **106**, 207–215.

T. Ono, K. Negishi, and H. Hayatsu (1995) Spectra of superoxide-induced mutations in the lacI′gene of a wild-type and a mutM strain of Escherichia coli K-12. *Mutat. Res.* **326,** 175–183.

G.W. Pace and C.D. Leaf (1995) The role of oxidative stress in HIV disease. *Free Rad. Biol. Med.* **19**, 523–528.

A.W. Partin, R.H. Getzenberg, M.J. Carmichael, D. Vindivich, J. Yoo, J.I. Epstein and D.S. Coffey (1993) Nuclear matrix protein patterns in human benign prostatic hyperplasia and prostate cancer. *Cancer Res.* **534**, 744–746.

R.B. Setlow (1993) *Variations in DNA repair among humans. In: Human Carcinogenesis* Academic Press, New York.

T.F. Schulz, C.H. Boshoff and R.A. Weiss (1996) HIV infection and neoplasia. *Lancet*, **348**, 587–591.

A.F. Slatter, C.S.I. Nobel and S. Orrenius (1995) The role of intracellular oxidants in apoptosis. *Biochim. Biophys. Acta* **1271**, 59–62.

Y. Sun (1990) Free radicals, antioxidant enzymes, and carcinogenesis. *Free Rad. Biol. ed.* **8**, 583–599.

T.P. Szatrowski and C.F. Nathan (1991) Production of large amounts of hydrogen peroxide by human tumor cells. *Cancer Res.* **51,** 794–798.
S.S. Wallace (1987) The biological consequences of oxidized DNA bases. Br. *J. Cancer (Suppl.)* **8**, 118–28.
R.L. Warters (1992) Radiation-induced apoptosis in a murineT-cell hybridoma. *Cancer Res.* **52**, 883–890.
C.K.O. Williams et al. (1994) *AIDS-Associated Cancers. In: AIDS in Africa.* (eds. Max Essex et al.), Raven Press, Ltd., New York, pp. 325–371.

30

p53 TUMOR SUPPRESSOR GENE

Its Role in DNA Damage Response and Cancer

Mehmet Öztürk and Kezban Unsal

Bilkent University, Department of Molecular Biology and Genetics
06533 Bilkent, Ankara, Turkey

1. ABSTRACT

p53 is one of the tumor suppressor genes found to be mutated in great majority of human cancers. The protein product of p53 gene is a nuclear phosphoprotein with characteristic features of a transcription factor acting on different target genes in order to modulate their expression either positively or negatively. Normal cellular functions of p53 are closely related to DNA damage response inducible by a variety of agents. Upon DNA damage, p53 protein is 'activated' by an unknown mechanism. p53 activation is accompanied by an increased half life and nuclear accumulation. This is followed by a change of expression of different genes some of which are directly related to cell cycle arrest and apoptosis. The main cell cycle regulatory target of p53 is p21, a member of cyclin-dependent kinase inhibitory proteins. p53-mediated induction of p21 expression leads to a cell cycle arrest at the G1 phase of the cycle. Another target of p53 is bax, a protein well known for its activator role of apoptosis. A new family of p53 target genes have been recently identified. Their protein products may be involved in the oxydative state of cells. Some of the apoptotic activities of p53 may be mediated by these newly identified genes. Finally, a new gene encoding a close homolog of p53 protein has been identified. This relative of p53, named p73 shares strong homology with p53 at the DNA binding and transactivation domains. In addition, p73 has extra carboxyterminal sequences with no homology to known proteins. p73 appears to be insensitive to DNA damage, but it probably acts on known p53 target genes with equal efficiency. It remains to be determined whether p73 is also a tumor suppressor gene mutated in cancer cells.

2. p53 GENE, STRUCTURE AND EXPRESSION

p53 was discovered initially as a cellular protein forming complexes with the large T antigen of SV40 virus (Lane and Crawford, 1979; Linzer and Levine, 1979). The first

Advances in DNA Damage and Repair, edited by Dizdaroglu and
Karakaya. Kluwer Academic / Plenum Publishers, New York, 1999.

cloned murine p53 gene was found to immortalise cells in vitro (Jenkins *et al.,* 1984) and transform primary fibroblasts in cooperation with ras (Parada *et al.*, 1984). These finding, which were later shown to be conducted with a mutant form of p53, allowed to classify p53 as an oncogene. Indeed, wild-type p53 acts as a tumor suppressor gene rather than oncogene. The transforming ability of mutant p53 proteins is due to the functional inactivation of wild-type p53 which acts as a tetrameric protein.

The human p53 gene is located on the short arm of chromosome 17 (17p13.1) and spans 16–20 kb DNA. The gene has 11 exons coding for an mRNA of 2.2–2.5 kb and a protein of approximately 53 kDa of 393 amino acids. Both exon-intron organisation of the gene and amino acid sequence of the protein is conserved between species (Soussi *et al.*,1990).

p53 is a DNA binding protein with transcription regulatory activities, and can be divided into three domains, encompassing the amino-terminal domain containing the activation domain, the central core containing its sequence-specific DNA-binding domain, and multifunctional carboxy-terminal domain. The acidic activation domain lies within residues 1–43 (Unger *et al.*, 1992) and acts as a transcriptional activating domain. The central region between amino acids 100–300 is a DNA binding domain that recognises a motif containing two contiguous or close monomers of 5'-PuPuPu-C(A/T)(T/A)GPyPy-3' (El-Deiry *et al.*, 1992). The C-terminal basic domain of p53 lies within residues 300–393 and harbors several important motifs involved in tetramer formation, nuclear localization, non-specific DNA binding and regulation of specific DNA binding by the central region.

In vitro, p53 interacts with many transcription factors through its activation domain, such as the TATA box-binding protein (TBP) and several TBP-associated factors (TAF). The binding of p53 to TBP appears to be involved in p53-mediated inhibition of gene transcription. P53 also recognizes the eukaryotic single-stranded DNA-binding protein RP-A. RPA is required for unwinding DNA origins and its binding to single stranded DNA may be the initial step in DNA replication. It is presently unclear whether the binding of p53 to RPA has any significant function. The cooperation of p53 with TBP in binding to DNA and the interaction of TBP with both carboxyl and activation domain of p53 and the association of p53 with the product of mdm-2 gene are all unique properties of p53. These interactions may be very important for transcriptional regulation.

The central region of p53 contains the sequence-specific DNA-binding domain and the vast majority of p53 missense mutations in tumors are clustered within this central portion. These mutations affect properties of p53 whose alteration leads to increased tumorigenic potential of cells such as its ability to bind DNA. The purified core DNA-binding domain can bind cooperatively to DNA in a manner that may be mediated by DNA, core domains can form strong interactions with each other, which may facilitate DNA bending (Balagurumoorthy *et al.*, 1995) as well as looping (Stenger *et al.*, 1994). These properties may be relevant in promoters that contain two p53-binding sites spaced at a distance from each other, for example, in the p53 target genes *p21* and *cyclin G* (see Zauberman *et al.*, 1995, see below).

The carboxyl terminus of p53 is capable of nonspecifically binding to different forms of DNA and can be subdivided further into three regions, a flexible linker (residues 300–320) that connects the DNA-binding domain to the tetramerization domain, the tetramerization domain itself (residues 320–360), and, at the extreme carboxyl terminus, a stretch of 30 amino acids that is rich in basic residues (residues 363–393).

Mechanisms other than mutation can serve to inactivate the p53 gene product at the protein level in several cases where the gene sequence is wild-type. For example, a number of viral oncoproteins bind, and inactivate the p53 protein: the SV40 T antigen

(Kao *et al.,* 1993), the human papillomavirus E6 protein (Band *et al.*, 1993 and Kessis *et al.,* 1993), the adenovirus E1B (Levine and Momand, 1990), also the hepatitis B virus X protein (Feitelson *et al.,* 1993 and Wang *et al.,* 1994), and the Epstein-Barr virus BZLF-1 protein (Zhang *et al.,* 1994). In addition to viral mechanisms, cellular pathways also exist for the inactivation of p53 in the absence of mutation : for example gene amplification and concomitant overexpression of the MDM2 gene in soft tissue sarcomas (Leach *et al.*, 1993), which encodes a product that functions in a manner antagonistic to that of p53 function and normally operates in a "feedback loop" to limit p53-mediated growth arrest (Chen *et al.*, 1994).

3. p53 AND CANCER

Since the first documentation of human p53 mutations in colorectal cancers in 1990, extensive studies by many different laboratories on various human cancers showed that this gene is mutated or inactivated in a great majority of human cancers, independent of tissue origin and etiology. Presently, it is estimated that about 40% of human cancers display mutations on p53 gene. Almost all of the major cancers (cancers of the skin, lung, liver, breast, stomach, bladder etc.) display mutations scattered at the DNA binding domain. There are five hotspots (codons 175, 248, 249, 273 and 282) which were found to be more frequently affected by mutations. These residues are involved either directly or indirectly in specific binding of p53 to its target DNA sequences. The frequency of p53 mutations is low in some tumors. One of such tumors, cervical cancers are etiologically linked to infection with human papillomaviruses 16 and 18. These viruses encode a protein (E6) which is able to inactivate wild-type p53 by inducing a rapid degradation.

4. CELLULAR FUNCTIONS OF p53 PROTEIN

p53 protein was found to be present at very low levels in normal cells. At such low levels, the protein appears to have no major function. Under certain stress, cells are able to upregulate their p53 levels by a post-transcriptional mechanism not yet clearly definied. The major factors that induce p53 have in common the ability to cause DNA damage. According to the present hypothesis, the major role of p53 in normal cells is to protect them against deleterious mutations due to DNA damage. Although, p53 appears not to be involved directly in the repair of damages, it helps cells to minimise long term effect of DNA damage. p53 appears to act on two main aspects of DNA damage response; sensing of the DNA damage and the initiation of a series of responses.

There are two major p53-dependent responses:

a. Cell cycle arrest: DNA damage can be particularly dangerous for cells under proliferation. In order to avoid the replication of damaged DNA, cells have to arrest in their progression into the cell cycle. p53 has clear role in G1/S arrest following DNA damage. It may also have a role in the G2/M phase of the cycle. p53-mediated cell cycle arrest at the G1/S boundry involves the inhibition of G1 phase cyclin-dependent kinases (in particular cyclin D-dependent Cdk4 and cyclin E-dependent cdk2). These kinases can be inactivated by a family of cyclin-dependent kinase inhibitors. The p21 protein which is induced by p53 is one of the most potent kinase inhibitors. The induction of p21 protein by p53 is based on specific transcriptional activation of p21 gene. This gene displays p53-binding motifs in its 5'-untranslated sequences (El-Deiry *et al.*, 1992). The major pathway of p53-mediated cell cycle involves retinoblastoma protein and E2F transcription factor family. The initiation of S phase (initiation of DNA synthesis) requires the activation of a number of genes by E2F transcription factors. E2F transcription factors are found in two different states in cells. When bound to underphosphorylated retinoblastoma protein (and its relatives), E2Fs are inactive. Upon phosphorylation of retinoblastoma protein by G1 phase cyclin-dependent kinases, E2Fs are free and active. The inhibion of cyclin-dependent kinases by p21 result in inadequate phosphorylation of retinoblastoma protein which then remains associated with E2F proteins. In addition to this well defined pathway, p21 appears to be involved in other cell cycle regulatory functions(Michieli *et al.*, 1994; Macleod *et al.*, 1995; Parker *et al.*, 1995).

b. Programmed cell death or apoptosis: In some cell types (for example some cells of the lymphoblastoid origin), the induction of p53 by DNA damage lead to a p53-dependent apoptosis rather than cell cycle arrest. It is not clear how cells decide to undergo apoptosis versus cell cycle arrest after p53 activation. The choice of apoptotic cell death could be determined by the extent of DNA damage. If the damage is well beyond the repair capacity, p53 could lead cells towards an apoptotic death in order to eliminate the risks of replication of DNA damage. If not, a simple cell cycle arrest would give enough time to cells for efficient DNA repair. Although the mechanisms of decision making are not clear, the apoptotic effectors of p53 are known; namely bcl-2 and bax. These two proteins belong to a family involved in the regulation of apoptosis, bax and bcl-2 acting as positive and negative effectors. p53 was shown to be able to upregulate bax expression. It also appears to downregulate bcl-2 expression. The net outcome of these p53-induced changes is the induction of apoptosis in certain cells. In addition, p53 appears to be able to direclty induce apoptotic cell death.

Based on the above described functions of wild-type p53, it is expected that cancer cells which have lost wild-type p53 will display abnormal cell cycle regulation and resistance to apoptosis. Indeed, such cancer cells display defective G1 checkpoint control following DNA damage and some of them are resistant to p53-induced apoptosis. Cancer cells not expressing wild-type p53 are in general abnormal in their ploidy. Such changes which reflect abnormal DNA replication may be result from duplication of damaged DNA.

5. LATEST DEVELOPMENT IN THE p53 FIELD

One important development is the cloning of a new gene called p73. Since the discovery of p53 in 1979, most laboratories were after p53 homologous genes. After such a long gene hunting, the surprize came from a cloning artifact. During a screening program with the aim of finding new genes involved in insulin pathway, Kaghad *et al.* (1997) identified p73 as a p53 relative. The newly discovered gene encodes a protein of about 73 kDa. Most of the N-terminal portion of this protein shares strong homology with p53 protein. The highest homology was found in a region of about 200 amino acid scanning DNA binding portion of p53. All p53 residues previously identified as important for DNA binding are conserved in p73 protein. The five mutational hotspots of p53 protein are also identical in the new protein. p73 is also a nuclear protein capable of transactivating some of the known p53 target genes. There are three main aspects of p73 which differ from p53: a large C-terminal portion of p73 which is missing in p53, the apparent lack of induction of p73 following DNA damage, and finally p73 gene appears to be imprinted in contrast to p53. It is also presently unknown whether and at what degree p73 is mutated in human cancers. This new development brings a lot of excitement and new hopes in understanding the role of p53 family proteins in the development of cancer. It appears that both proteins trigger the same type of cellular response in terms of cell cycle arrest and apoptosis (Kaghad *et al.*, 1997). What differs so far is the upstram events leading to the 'activation' of these proteins. It is possible that p53 and p73 respond to different kinds of cellular signals, but the outcome of the response is similar if not identical. As in most cellular events, p53 and p73 may work as common conveyors of a network of signals which generate the same type of cellular response. It will be interesting to know what triggers p73 response and whether such triggering mechanism is also related to cellular stress.

ACKNOWLEDGMENTS

The authors work is supported by grants from TTGV, TUBA and Bilkent University. K.U. is supported by a Ph.D. fellowship from TUBITAK.

REFERENCES

V. Band, S. Dalal, L. Delmolino and E.J. Androphy (1993). Enhanced degradation of p53 protein in HPV-6 and BPV-1 E6-immortalized human mammary epithelial cells. *EMBO J* **12**, 1847–1852.

C.Y. Chen, J.D. Oliner, Q. Zhan, A.J., Fornace Jr., B. Vogelstein and M.B. Kastan (1994). Interactions between p53 and MDM2 in a mammalian cell cycle checkpoint pathway. *Proc. Natl. Acad. Sci. USA* **91**, 2684–2688.

W.S. El-Deiry, S.E. Kern, J.A. Pietenpol *et al.* (1992). Definition of a consensus binding site for p53. *Nat. Genet.* **1**, 45

M.A. Feitelson, M. Zhu, L.X. Duan and W.T. London (1993). Hepatitis B x antigen and p53 are associated in vitro and in liver tissues from patients with primary hepatocellular carcinoma. *Oncogene* **8**, 1109–1117.

J.R. Jenkins, K. Rudge, S. Redmond and A. Wade-Evans (1984). Cloning and expression analysis of full length mouse cDNA encoding the transformation associated protein p53. *Nucl. Acids Res.* **12**, 5609–5626.

M. Kaghad, H. Bonnet, A. Yang, L. Creancier, J. Biscan, A. Valent, A. Minty, P. Chalon, J. Lelias, X. Dumont, P. Ferrara, F. McKeon and D. Caput (1997). Monoallelically expressed gene related to p53 at 1p36, a region frequently deleted in neuroblastoma and other human cancers. *Cell* **90**, 809–819.

C. Kao, J. Huang, S.Q. Wu, P. Hauser and C.A. Reznikoff (1993). Role of SV40 T antigen binding to pRB and p53 in multistep transformation in vitro of human uroepithelial cells. *Carcinogenesis* **14**, 2297–2302.

T.D. Kessis, R.J. Slebos, R.J. Nelson, M.B. Kastan, B.S. Plunkett, S.M. Han, A.T. Lorincz, L. Hedrick and K.R. Cho (1993). Human papillomavirus 16 E6 expression disrupts the p53-mediated cellular response to DNA damage. *Proc. Natl. Acad. Sci. USA* **90**, 3988–3992.

D.P. Lane and L.V. Crawford (1979). T antigen is bound to a host protein in SV40-transformed cells. *Nature* **278**, 261

F.S. Leach, T. Tokino, P. Meltzer, M. Burrell, J.D. Oliner, S. Smith, D.E. Hill, D. Sidransky, K.W. Kinzler and B. Vogelstein (1993). P53 mutation and MDM2 amplification in human soft tissue sarcomas. *Cancer Res.* **53**, 2231–2234.

D.I. Linzer and A.J. Levine (1979). Characterization of a 54K dalton cellular SV40 tumor antigen present in SV40-transformed cells and uninfected embryonal carcinoma cells. *Cell* **17**, 43.

K.F.N. Macleod, G. Sherry, D. Hannon, T. Beach, K. Tokino, B. Kinzler, T. Vogelstein and T. Jacks (1995). P53-dependent and independent expression of p21 during cell growth, differentiation, and DNA damage. *Genes and Dev.* **9**, 935–944.

P. Michieli, M. Chedid, D. Lin, J.H. Peirce, W.E. Mercer and D. Givol (1994). Induction of WAF1/CIP1 by a p53-independent pathway. *Cancer Res.* **54**, 3391–3395.

L.F. Parada, H. Land, W.A. Weinberg, D. Wolf and V. Rotter (1984). Cooperation between gene encoding p53 tumor antigen and ras in cellular transformation. *Nature* **312**, 649–651.

S.B. Parker, G. Eichele, P. Zhang, A. Rawls, A.T. Sands, A. Bradley, E.N. Olson, J.W. Harper and S.J. Elledge (1995). P53-independent expression of p21 in muscle and other terminally differentiating cells. *Science* **267**, 1024–1027.

T. Soussi, C. Caron de Fomentel and P. May (1990). Structural aspects of the p53 protein in relation to gene evolution. *Oncogene* **5**, 945–952.

J.E. Stenger, P. Tegtmeyer, G.A. Mayr, M. Reed, Y. Wang, *et al.* (1994). P53 oligomerization and DNA looping are linked with transcriptional activation. *EMBO J.* **13**, 6011–6020.

T. Unger, M.M. Nau and S. Segal *et al.* (1992). P53: a transdominant regulator of transcription whose function is ablatedby mutations occuring in human cancer. *EMBO J.* **11**, 1383.

X.W. Wang, K. Forrester, H. Yeh, M. Feitelson, J.R. Gu and C.C. Harris (1994). Hepatitis B virus X protein inhibits p53 sequence-specific DNA binding, transcriptional activity, and association with transcription factor ERCC3. *Proc. Natl. Acad. Sci. USA* **91**, 2230–2234.

A.A. Zauberman, A. Lubp and M. Oren (1995). Identification of p53 target genes through immune selection of genomic DNA: The cyclin G gene contains two distinct p53 binding sites. *Oncogene* **10**, 2361–2366.

Q. Zhang, D. Gutsch and S. Kenney (1994). Functional and physical interaction between p53 and B2LF1; implications for Epstein-Barr Virus latency. *Mol. Cell Biol.* **14**, 1929–1938.

31

MOLECULAR MECHANISM OF NUCLEOTIDE EXCISION REPAIR IN MAMMALIAN CELLS

Joyce T. Reardon and Aziz Sancar

Department of Biochemistry and Biophysics
University of North Carolina School of Medicine
422 Mary Ellen Jones Building
Chapel Hill, North Carolina 27599-7260

1. ABSTRACT

Nucleotide excision repair is the major pathway for removal of damaged or abnormal nucleotides generated by both physical and chemical agents. Xeroderma pigmentosum (XP), an hereditary disease characterized by extreme sensitivity to sunlight and a predisposition to skin cancer due to defective nucleotide excision repair, can be caused by mutations in any one of seven genes, *XPA* through *XPG*. The basal excision repair reaction, which is functionally conserved from bacteria to man, can be conceptualized as occurring in five steps: (1) damage recognition, (2) dual incisions of the damaged strand, (3) release of the damage-containing oligomer to generate a gapped duplex molecule, (4) resynthesis of DNA (repair patch) using the undamaged strand as template, and (5) ligation of the newly synthesized DNA to regenerate an intact molecule. Excision and resynthesis are accomplished by six proteins in E. coli, but 25 polypeptides are required in human cells. In man the initial damage recognition is accomplished by the cooperative action of XPA and RPA. Transcription factor TFIIH, a multisubunit complex which contains both the XPB and XPD helicases, and XPC are recruited to the damage site and form a stable preincision complex with XPA and RPA. This complex first recruits the XPG nuclease which incises DNA 6±3 phosphodiester bonds 3' to the damage and then the XPF•ERCC1 complex which incises DNA 20±5 bonds 5' to the damage. As a result of these dual incisions, the damage is released in 24–32 nucleotide long oligomers. Resynthesis is performed by polymerases δ and ε, and ligation completes the basal excision reaction.

2. INTRODUCTION

2.1. Biological Relevance of Nucleotide Excision Repair

Carcinogenesis is a multistage event with DNA damage implicated in the initiation and promotion stages when DNA damage causes mutation of critical, target genes. The re-

Advances in DNA Damage and Repair, edited by Dizdaroglu and Karakaya. Kluwer Academic / Plenum Publishers, New York, 1999.

sulting inappropriate expression of these proteins, or the expression of abnormal polypeptides, may impair DNA metabolism and the resulting disruption of normal cellular events leads to cancer. DNA repair plays a key role in protection against cancers caused by either endogenous or exogenous DNA damaging agents.

2.1.1. Xeroderma Pigmentosum. Xeroderma pigmentosum (XP) is a rare, inherited disease caused by mutations in any of seven genes named XPA through XPG. Clinically, XP patients are extremely sensitive to sunlight and have an increased susceptibility for developing both skin and certain internal cancers at an early age (Kraemer et al., 1984). In addition, many XP patients exhibit neurological abnormalities and it has been suggested that unrepaired oxidative damage in DNA leads to the development of the neurological defects observed in XP patients (Robbins et al., 1983).

Cultured cells from XP patients are hypersensitive to both killing and mutation induction following UV exposure (reviewed in Cleaver and Kraemer, 1989). Biochemically, this hypersensitivity has been correlated with defects in nucleotide excision repair. Therefore, studies aimed at elucidating the molecular mechanisms of DNA repair include detailed analyses of the proteins encoded by XP genes, many of which were cloned as ERCC (excision repair cross complementing) genes on the basis of complementing the UV sensitive phenotype in rodent cell lines (reviewed in Thompson, 1998).

3. DNA REPAIR MECHANISMS

In both prokaryotic and eukaryotic systems there are three basic processes for correction of damaged bases, distinguished by both mechanistic details and substrate spectra. (1) In direct repair, the damage is rectified by proteins that enzymatically reverse rather than excise the damage. This type of repair includes photoreversal of UV-induced lesions by photolyase (Sancar, 1996a) and removal of alkyl groups by O^6-methylguanine DNA methyl transferase (Samson, 1992; Sekiguchi et al., 1996). (2) In base excision repair (Demple and Harrison, 1994), lesion-specific glycosylases initiate repair by releasing damaged bases and generating abasic sites which are further processed by apurinic/apyrimidinic (AP) endonuclease, polymerase and ligase; the repair patch may be as small as a single nucleotide. (3) In nucleotide excision repair (or simply excision repair) phosphodiester bonds are hydrolyzed on both sides of the lesion, the damaged base is released in short damage-containing oligonucleotides and polymerases fill-in the resulting gap (Sancar, 1995a, 1995b; Wood, 1996). Additionally, there is a repair system for the correction of mismatched bases (Modrich and Lahue, 1996).

3.1. Nucleotide Excision Repair Overview

3.1.1. The Basal Reaction. In both prokaryotes and eukaryotes, nucleotide excision repair constitutes a major pathway for the removal of damaged or abnormal nucleotides from DNA (Sancar and Sancar, 1988; Sancar, 1994). This versatile pathway recognizes and repairs a wide spectrum of DNA lesions ranging from bulky adducts to non-helix distorting lesions (Sancar, 1996b). The basic mechanism of excision repair may be divided into five steps: (i) damage recognition, (ii) dual incisions of the damaged strand, (iii) release of the damage-containing oligomer, (iv) resynthesis to fill-in the resulting gap, and (v) ligation to regenerate an intact DNA molecule. The mechanistically equivalent processes of excision and resynthesis are achieved by just six proteins in E. coli, but 25 polypeptides are required in human cells.

3.1.2. Preferential Repair. Preferential repair is the faster rate of repair observed for certain (usually transcribed) regions of the genome, or one of the two strands of the duplex, compared to the rest of the genome (genome-overall repair). Although the phenomenon of preferential repair of actively transcribing genes was first observed with human cells, the molecular details of the process are poorly understood for eukaryotic systems (Hanawalt, 1994; Friedberg, 1996). However, the molecular details of transcription-coupled repair have been well characterized in E. coli: a transcription repair coupling factor recruits the excision repair enzyme to RNA polymerase stalled at a lesion and, thus, lesions in the coding strand of actively transcribing genes are preferentially repaired (Selby and Sancar, 1993). Eukaryotic DNA is organized into chromatin which is generally inaccessible to most enzymatic activities, and there most likely are accessory proteins required specifically for access of repair proteins to these highly condensed regions of the genome. Within the context of this article, nucleotide excision repair refers to the basal enzymatic reaction which is transcription-independent and studied in vitro in the absence of nucleosomes.

3.2. Summary of Excision Repair in Escherichia coli

UvrA, UvrB and UvrC constitute the E. coli excision nuclease, (A)BC excinuclease (reviewed in Sancar and Sancar, 1988; Sancar, 1996b). UvrA binds specifically to both damaged DNA and UvrB, and by virtue of these interactions, delivers UvrB to DNA. UvrA, a molecular matchmaker (Sancar and Hearst, 1993), then dissociates from the UvrB-DNA complex. UvrC interacts with UvrB bound to DNA, induces UvrB to make the 3' incision and then UvrC makes the 5' incision; these dual incisions are at the 8th phosphodiester bond 5' to the lesion and at the 4th-5th bond 3' to the lesion. UvrD is a helicase that releases both UvrC and the 12–13 nt-long damage containing oligomer. Repair is completed by DNA polymerase I, which synthesizes the repair patch and displaces UvrB, and by DNA ligase.

3.3. Overview of Human Excision Repair

In humans, six of the XP proteins, as well as ERCC1, a trimeric replication protein (RPA, HSSB), and a multisubunit transcription factor (TFIIH), are required for the dual incision reaction (Table 1). We call this activity "excision nuclease" (from the Latin *excidere*, to cut out). In mammalian cells the excision nuclease incises the damaged strand at 20±5 phosphodiester bonds 5' and 6±3 bonds 3' to the lesion, releasing 24–32 mers containing the damaged nucleotide. Resynthesis is accomplished by DNA polymerase δ and ε.

3.3.1. Assays for Excision Repair. Various assays have been developed for the in vitro evaluation of nucleotide excision repair. The repair synthesis assay utilizing randomly damaged DNA and cell-free extracts (Wood et al., 1988; Sibghat-Ullah et al., 1989), which has been widely used, has high sensitivity but low specificity. Observations of damage-dependent DNA synthesis, that has no relation to excision repair and that occurs with equal efficiency in extracts from repair proficient and repair deficient cell lines (Figure 1), have led to numerous misleading reports in the literature. In contrast, the excision assay (Huang et al., 1992) has high sensitivity and absolute specificity and thus it is the assay of

Table 1. Minimal requirements for the excision step of human nucleotide excision repair

Repair factor	Proteins (M_r)	Activity	Role in repair
I. XPA	XPA (p31)	DNA binding	Damage recognition
II. RPA	p70	(a) DNA binding	Damage recognition
	p34	(b) Replication factor	
	p11		
III. TFIIH	XPB/ERCC3 (p89)	(a) Transcription factor	Preincision complex
	XPD/ERCC2 (p80)	(b) DNA-dependent ATPase	
	p62	(c) Helicase	
	p44		
	p34		
IV. XPC	XPC (p125)	DNA binding	Preincision complex
V. XPF	XPF/ERCC4 (p112)	Nuclease	5' Incision
	ERCC1(p33)		
VI. XPG	XPG/ERCC5 (p135)	Nuclease	3' Incision

choice for studying excision repair in humans (Figure 1). In this assay, radiolabeled circular plasmid DNA containing site-specific adducts (Huang et al., 1992) or linear DNA duplexes with centrally located lesions (Huang and Sancar, 1994) are used with either cell-free extracts or highly purified repair factors. Depending on the reaction conditions and the location of the radiolabel relative to the damage, we can use the incision assay to detect either the 5' or 3' nicking activity (Matsunaga et al., 1995), the excision assay to measure the release of excised fragments resulting from dual incisions (Huang et al., 1992; Svoboda et al., 1993), or the phosphorothioate repair patch assay to visualize the newly synthesized DNA (Sibghat-Ullah et al., 1990; Reardon et al., 1997b).

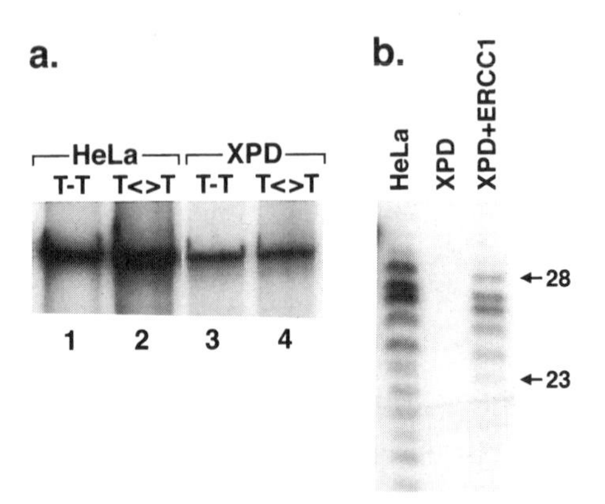

Figure 1. Comparison of repair synthesis and excision assays. Autoradiograph of agarose gel (panel a) shows results of a repair synthesis assay performed with extracts prepared from repair-proficient HeLa and repair-deficient XPD cells. The substrates, 4.3 kb plasmid pUNC1991–4 prepared and purified as described (Huang et al., 1992), were control plasmid, lanes 1 and 3, prepared with primers containing T-T and modified DNA, lanes 2 and 4, which contained four T<>T at unique positions. When normalized for the amount of DNA recovered, the repair synthesis signal observed with XPD extracts does not show a defect in excision repair. Panel b is an autoradiograph of a sequencing gel showing the results of excision assays with radiolabeled pUNC1991–4(T<>T) and extracts from HeLa, XPD and ERCC1 cell lines. Only the gel area including the excised fragments is shown. XPD, which has false "normal" repair synthesis activity (panel a), is totally defective in excision of T<>T photoproducts unless complemented with another repair deficient extract; ERCC1 extract alone has no excision activity (data not shown, but see Reardon et al., 1993).

Table 2. Substrates for human nucleotide excision repair[a] and repair system overlap

Lesion	Relative excision activity	Other repair systems
Acetylaminofluorene	++++	(mismatch)
Cisplatin-1,3-d(GpTpG)	++++	
6-4 Photoproduct	+++	(mismatch)
Cisplatin-1,2-d(GpG)	++	(mismatch)
8-oxoguanine	++	base excision
T<>T Photoproduct	++	(mismatch)
Psoralen monoadduct	+	
ABPD-Cholesterol	++++	
Thymine glycol	+	base excision
O^6meG, N^6meA	±	alkyltransferase, base excision
Mismatched bases	±	mismatch
Abasic site (synthetic)	++	base excision

[a] Levels of excision in repair assays utilizing cell-free extracts: ++++, > 10%; +++, 5-10%; ++, 3-5%; +, 1-3%; ±, < 1%. (mismatch) indicates recognition, but not repair, by proteins of the mismatch correction system.

3.4. DNA Lesions Recognized by Mammalian Systems

3.4.1. Substrates for Human Nucleotide Excision Repair. We have found that diverse lesions are excised in vitro by the mammalian excision nuclease, albeit with different efficiencies and kinetics (Table 2). These include the well-characterized UV-induced T<>T dimer (Huang et al., 1992) and (6–4) photoproduct (Kazantsev et al., 1996; Mu et al., 1997a) as well as bulky DNA lesions generated by chemical modification of bases including psoralen modified thymine (Svoboda et al., 1993; Huang and Sancar, 1994), AAF modified guanine (Mu et al., 1994) and GG, AG, and GTG intrastrand crosslinks generated by the chemotherapeutic agent cisplatin (Huang et al., 1994b; Zamble et al., 1996). When incorporated opposite mispaired bases, both UV-induced photoproducts and cisplatin intrastrand crosslinks are repaired more efficiently (Mu et al., 1997a). We have observed excision of non-bulky lesions generally repaired by other repair systems: e.g., mismatched bases and O^6-methylguanine (Huang et al., 1994a). We have also used synthetic AP sites or bulky adducts and have found that some of these "DNA lesions" are among the best substrates for the in vitro excision assay: the AP site analog, 2-aminobutyl-1,3-propanediol (ABPD), or ABPD with an attached cholesterol moiety (Huang et al., 1994a; Reardon et al., 1997b). Although they do not naturally occur in DNA, we routinely use these synthetic lesions for the sensitive assays required for rapid screening of chromatographic fractions or assaying the repair proficiency of cell-free extracts. Most recently we have found that oxidatively damaged nucleotides, specifically 8-oxoguanine, thymine glycol and urea, are repaired by the human excision nuclease. Strikingly, 8-oxoG is removed at a faster rate than thymine dimer by the mammalian excision nuclease (Figure 2; Reardon et al., 1997a).

3.4.2. Damaged Bases Recognized by Multiple Systems. It was once thought that each of the basic repair mechanisms had rather specific and exclusive substrate spectra, but recent reports indicate that there is considerable overlap among the various repair pathways (Table 2). For example, the human excision nuclease system recognizes and repairs non-bulky lesions once thought to be repaired only by direct repair or base excision (O^6meG), base excision (8-oxoguanine, thymine glycol and urea), or mismatch correction systems (G:A, G:G). It

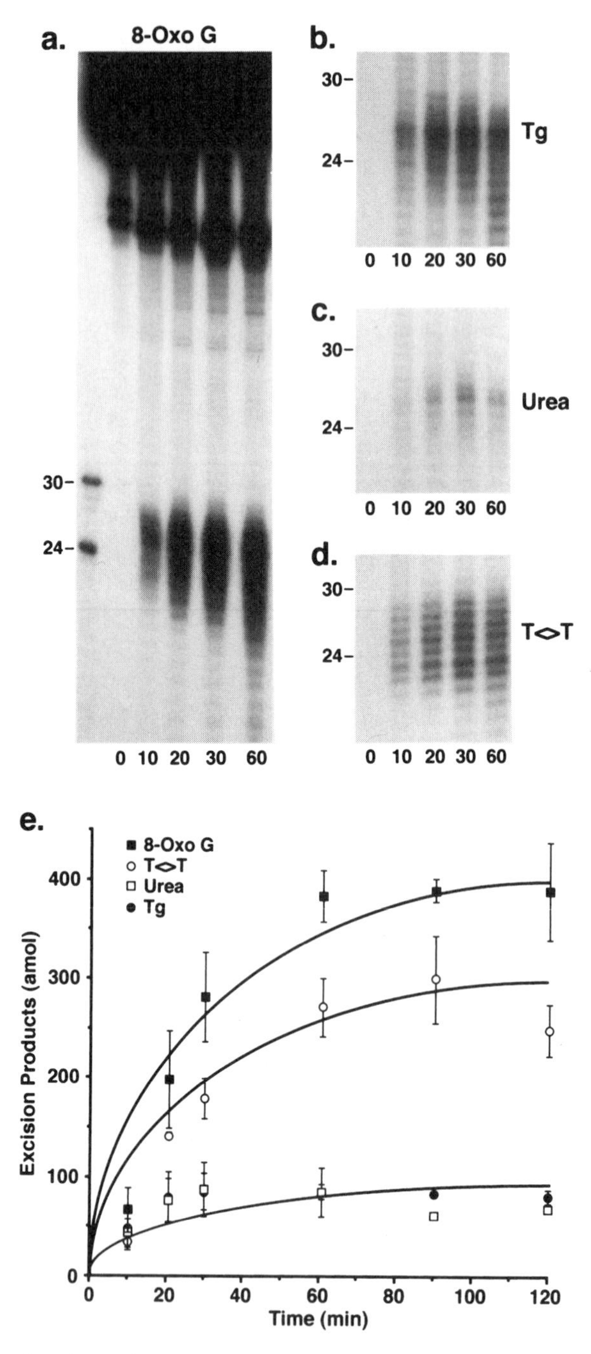

Figure 2. Time course of excision of oxidative base damage by mammalian excision repair factors present in CHO AA8 cell-free extracts. Sequencing gels (panels a-d) show the results of kinetic experiments with 8-oxoguanine (8-Oxo G), thymine glycol (Tg), urea and cis, syn-cyclobutane thymine dimer (T<>T). Quantitative analyses of multiple experiments conducted under identical conditions are shown in panel e; 200 amol of excision products represents about 1.4% excision of input DNA (Reardon et al., 1997a). Reprinted with permisission, Copyright (1997) National Academy of Sciences, U.S.A.

must be noted that, with the exception of 8-oxoguanine, these lesions are not efficiently repaired by the mammalian excision nuclease and, thus, it seems that nucleotide excision repair serves as a backup system for the pathways with primary responsibility for repair of these damaged or inappropriate bases. Likewise, there are recent reports that hMutSα, the DNA binding complex of the mismatch repair system, recognizes bulky lesions once thought to be in the exclusive domains of other repair pathways: AF and AAF (Li et al., 1996), thymine dimer and (6–4) photoproduct (Mu et al., 1997a), cisplatin-DNA adduct (Duckett et al., 1996; Yamada et al., 1997) and methylated bases (Duckett et al., 1996). Unlike with excision repair which removes, albeit inefficiently, non-bulky lesions, the mismatch repair system does not eliminate these base damages from DNA.

4. HUMAN EXCISION REPAIR FACTORS

In vivo characterization of excision repair with established XP cell lines defined the seven XP complementation groups and suggested a requirement for each of the XP gene products at an early stage of excision repair. Additionally, there are ten rodent UV-sensitive complementation groups (Collins, 1993) which we now know overlap to a large degree with human repair-deficient cell lines (Table 1). Interestingly, there is not a known human cell line defective in ERCC1 nor are there identified rodent homologs of XP-A or XP-C mutant cell lines.

Work by several laboratories in the past few years has resulted in the cloning of human DNA repair genes and an elucidation of the basic mechanism of the complex mammalian nucleotide excision repair system. With the exception of XPE, the genes for each of the XP (ERCC) proteins, as well as RPA and the non-XP components of TFIIH, have now been cloned and expressed as recombinant proteins in either bacterial or insect cell-based systems. The availability of cloned excision repair genes permits overexpression of proteins in heterologous systems, mutational analyses and a detailed molecular analysis of the repair process.

4.1. Characterization of the Subunits of the Human Excision Nuclease

The in vitro requirements for nucleotide excision repair proteins have been studied with the excision assay utilizing cell-free extracts prepared from repair proficient HeLa or repair deficient XP cells (Reardon et al., 1993), as well as the functionally equivalent rodent CHO AA8 and its derivative UV-sensitive cell lines (Reardon et al., 1993). The polypeptide requirements for basal excision repair have now been well characterized with highly purified or recombinant proteins (Mu et al., 1995; Mu et al., 1996) and the minimal requirements are summarized in Table 1.

4.1.1. XPA and RPA. XPA is a DNA binding protein with a modest preference for damaged DNA (Jones and Wood, 1993; Asahina et al., 1994). RPA is a trimeric DNA binding protein (Henrickson et al., 1994) with a preference for damaged as well as single-stranded DNA (He et al., 1995; Burns et al., 1996). XPA and RPA cooperate to form a damage recognition complex with a greater affinity for damaged DNA than that exhibited by either XPA or RPA alone (He et al., 1995; Li et al., 1995).

4.1.2. TFIIH. TFIIH is a multimeric factor with helicase activity that is required for basal transcription by RNA polymerase II and for excision repair; both XPB and XPD are components of TFIIH (Schaeffer et al., 1993; Drapkin et al., 1994), and the helicase activity is attributed to both XPB and XPD (Sung et al., 1993; Drapkin et al., 1994). Various forms of human TFIIH have been purified and it has been reported to contain 3–7 subunits in addition to XPB and XPD (Roy et al., 1994; Mu et al., 1996; Marinoni et al., 1997). These other subunits include p62, p52, p44, p34, and the trimeric CAK activity comprised of cdk7 (p41), cycH (p38), and Mat1 (p32); the CAK activity is found in three distinct complexes: CAK, CAK•XPD and TFIIH (Reardon et al., 1996a; Drapkin et al., 1996). Although there is biochemical and genetic evidence from both human and yeast studies suggesting a role for each of these additional subunits in excision repair, we have isolated a 5-subunit form of TFIIH, lacking detectable levels of both p52 and the trimeric CAK complex, which can fully reconstitute in vitro the basal damage recognition and excision steps of nucleotide excision repair (Mu et al., 1996). TFIIH functions to both unwind the DNA in the vicinity of DNA lesions and to recruit additional repair factors to the damaged DNA.

4.1.3. XPC•HHR23B. The excision repair function of XPC is not well defined, but it is a DNA binding protein which copurifies with HHR23B, the human homolog of the yeast RAD23 protein. Although there is a report that HHR23B enhances XPC-dependent excision repair (Sugasawa et al., 1996) and the XPC•HHR23B complex is used for most reconstitution assays, we have found that the basal excision reaction occurs efficiently even in the absence of HHR23B (Reardon et al., 1996b). This latter finding is consistent with the moderate UV sensitivity of S. cerevisiae Rad23 mutants (reviewed in Prakash et al., 1993) and the finding that XPC is not required for excision of certain lesions (Mu et al., 1996) or for removal of other lesions when present in partially denatured DNA (Mu and Sancar, 1997). It is quite possible that, under conditions of localized denaturation, i.e. "transcription bubbles", the need for XPC is abrogated by the DNA structure or by other proteins which substitute for XPC. Our current interpretation of available data is that XPC•HHR23B stabilizes unwound DNA and perhaps aids in unwinding DNA in preincision complexes (Mu et al., 1997b).

4.1.4. XPG and XPF•ERCC1. XPG is a member of the FEN-1 family of structure-specific endonucleases (Harrington and Lieber, 1994), incises at the single-stranded to double-stranded junction found at the 3' side of flap and bubble structures (O'Donovan et al., 1994) and makes the 3' incision in nucleotide excision repair (Matsunaga et al., 1995). The XPF•ERCC1 heterodimer has endonuclease activity with a preference for single-stranded DNA and the single-stranded region of duplex DNA within a 30-nucleotide bubble (Park et al., 1995a); this endonuclease activity is specific for the 5' junction of the bubble region (Bessho et al., 1997). The nuclease activity of both XPG and XPF•ERCC1 is stimulated by RPA (Park et al., 1995a; Matsunaga et al., 1996; Bessho et al., 1997).

4.2. Protein-Protein Interactions

An early suggestion of higher order complexes formed by the individual repair factors defined by the XP genes was the observed lack of in vitro complementation with certain mutant cell-free extracts (Reardon et al., 1993) which were interpreted as indications of strong interactions between ERCC1 and XPF (ERCC4) and weaker interactions between XPB (ERCC3) and XPD (ERCC2). Subsequently, these conclusions were confirmed by the discovery that both XPB and XPD are subunits of TFIIH and by isolation of XPF•ERCC1 as a tight complex. In addition, numerous other protein-protein interactions, which are of relevance to the functioning of human excision nuclease, have been reported.

4.2.1. XPA-RPA. The RPA-XPA interaction is crucial for damage recognition, since neither XPA nor RPA alone exhibits a strong affinity for damaged DNA. Affinity chromatography and immunoprecipitation were used to demonstrate an in vitro interaction of XPA with the p34 subunit (Matsuda et al., 1995) as well as both the p34 and p70 subunits of RPA (He et al., 1995; Li et al., 1995; Saijo et al., 1996). The report by Li et al. (1995) suggested that only the p70 interaction is essential for excision repair, but the interaction domains of both p34 and p70 were mapped to specific XPA residues.

4.2.2. XPA-RPA-XPG. Co-immunoprecipitation studies were used to demonstrate an interaction between XPG and RPA, and these protein contacts were reported to mediate formation of an XPA-RPA-XPG complex (He et al., 1995). The RPA interactive domain was mapped to a specific XPG domain, but the report did not indicate which RPA subunit(s) or XPA residues are involved in these interactions.

4.2.3. The XPF•ERCC1 Heterodimer and Interactions with XPA and RPA. Following the initial suggestion of an XPF•ERCC1 heterodimer (Reardon et al., 1993), analogous to the S. cerevisiae RAD1•RAD10 complex (reviewed in Prakash et al., 1993), support for the existence of this complex came from reports by Park and Sancar (1994) who used XPA affinity chromatography to detect in vitro formation of a ternary complex by the XPA, XPF(ERCC4) and ERCC1 proteins and by Li et al. (1994) who used the two-hybrid system and deletion mutagenesis combined with the pull-down assay to detect and map the interactive domains of XPA and ERCC1. Saijo et al. (1996) also detected a ternary complex of RPA, XPA and ERCC1, and Bessho et al. (1997) used recombinant proteins and affinity chromatography to demonstrate the XPF•ERCC1, XPF-XPA and XPF-RPA interactions.

4.2.4. TFIIH (XPC and XPG). We now know that the weak complementation observed in vitro (Reardon et al., 1993) which suggested an XPB-XPD interaction is explained by the presence of both polypeptides in the multisubunit repair and transcription factor TFIIH (Schaeffer et al., 1993; Drapkin et al., 1994). Additionally, during purification of repair factors from HeLa cell-free extracts, both XPC (Drapkin et al., 1994) and XPG (Mu et al., 1995) were observed to copurify with TFIIH through several fractionation steps. Deletion mutagenesis and immunoprecipitation analyses were used to map interactions within the TFIIH complex and the XPG interactions with multiple subunits, including XPB, XPD, p62 and p44 (Iyer et al., 1996). XPA interacts with at least one TFIIH subunit and, thus, recruits this complex to the damage site (Park et al., 1995b); the interactive domain was mapped to the carboxyl-terminal half of XPA but the identity of TFIIH interacting protein(s) is unknown.

4.2.5. XPC•HHR23B (XPA and TFIIH). XPC forms a heterodimer with HHR23B (Masutani et al., 1994) which copurifies with TFIIH, and the interactive domains of XPC and HHR23B have been mapped (Li et al., 1997). In the functionally analogous S. cerevisiae repair system RAD23 promotes formation of a RAD14 (XPA) complex with TFIIH (Guzder et al., 1995a) but, in the human system, XPA appears to interact with TFIIH even in the absence of XPC•HHR23B (Park et al., 1995b).

5. MOLECULAR MECHANISM OF DNA EXCISION REPAIR

As summarized above, DNA excision repair may be divided into five basic steps which are illustrated in Figure 3: (i) damage recognition, (ii) dual incisions of the dam-

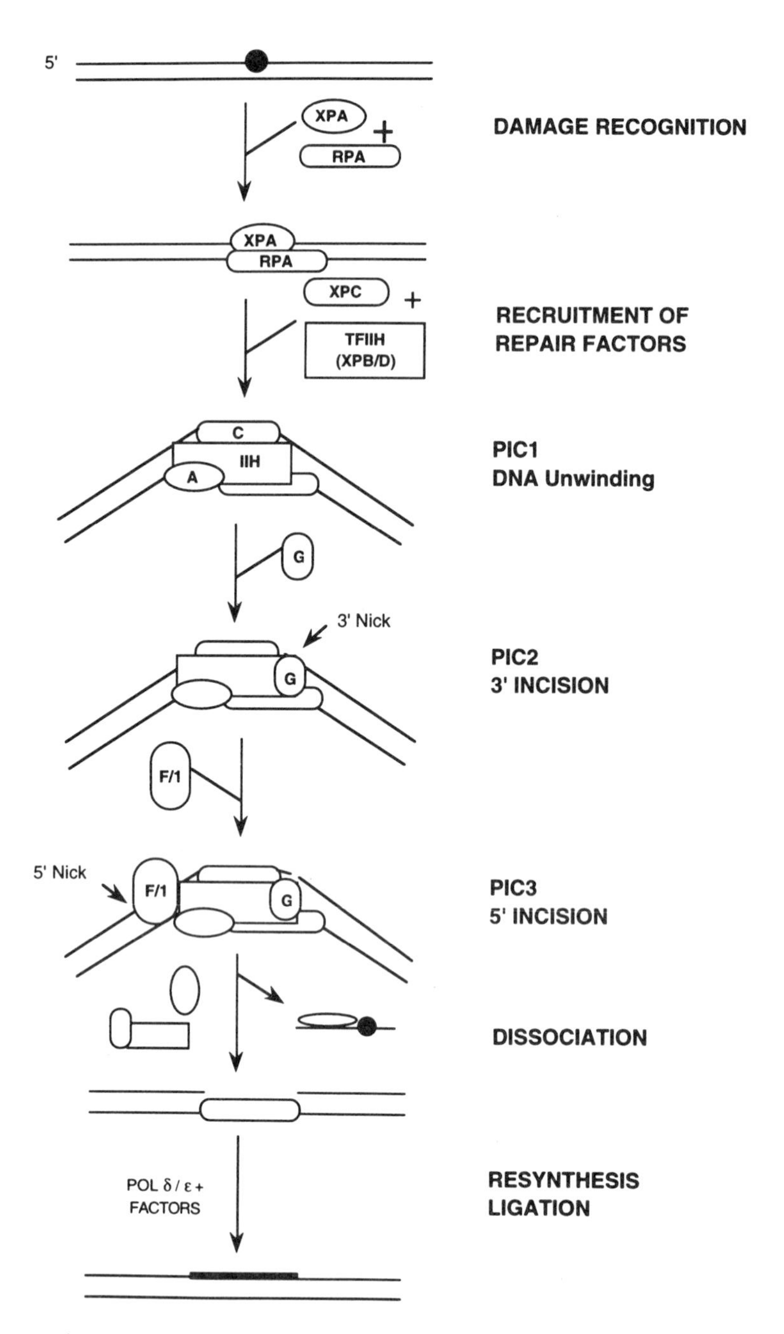

Figure 3. Model for transcription-independent excision repair of DNA. XPA and RPA cooperate to form the damage recognition factor, a short-lived complex which recruits TFIIH and XPC to the damage site, thus forming PIC1 (preincision complex). PIC2 is a stable intermediate formed by the association of XPG with PIC1; the XPF•ERCC1 heterodimer associates with other factors to form PIC3. The first of the dual incisions is made 3' to the lesion by XPG and then the 5' incision is made by XPF•ERCC1. The excised fragment and some repair factors dissociate prior to resynthesis by polymerase δ and ε.

aged strand, (iii) release of the damage-containing oligomer, (iv) resynthesis to fill-in the resulting gap, and (v) ligation to regenerate an intact DNA molecule.

5.1. Damage Recognition

This initial, and presumably rate-limiting, step of nucleotide excision repair is accomplished through a cooperative effort of XPA and RPA (He et al., 1995). The in vitro damage-dependent resynthesis assay was used to determine an absolute requirement for RPA in excision repair (Coverley et al., 1991) and reconstitution of basal excision repair with purified repair proteins established that RPA functions at the incision step in both the human (Mu et al., 1995) and yeast (Guzder et al., 1995b) systems. The biochemical basis for recognition of chemically and structurally diverse lesions by the XPA-RPA complex is poorly understood, but a plausible scenario is that XPA-RPA senses a localized helical distortion in the vicinity of damage, pauses at this destabilized region, and then probes the DNA backbone to discriminate between the damaged base and the normal base on the complementary strand (Gunz et al., 1996; Hess et al., 1997).

5.2. Preincision Complexes

Once the damage is recognized by XPA-RPA, other repair factors are recruited to and assemble at the damage site to form pre-incision complex 1. PIC1 (Mu et al., 1997b). PIC1, composed of XPA, RPA, TFIIH and XPC, assembles on the DNA in the vicinity of the damage. XPA and RPA interact to form a transient damage recognition complex and, by virtue of specific interactions, also recruit these other proteins to the damage site or direct them to their appropriate positions on the DNA molecule. The finding that this subassembly of repair proteins forms PIC1 is consistent with the XPA interaction with both RPA and TFIIH, as well as the TFIIH-XPC interaction.

The localized helical deformation due to the DNA lesion is enlarged in both the 5' and 3' directions by the helicase activity of TFIIH to generate a preincision bubble. Using cell extracts and platinated DNA, Evans et al. (1997) reported a bubble of 25 nucleotides that is formed even in the absence of XPA protein. In contrast, Mu et al. (1997b), using purified proteins and DNA containing a (6–4) photoproduct, observed XPA-dependent formation of an approximately 20 nucleotide bubble. XPC binds to these single-stranded regions to help maintain the open complex formation (Mu and Sancar, 1997).

5.3. Dual Incisions

XPG is recruited to PIC1 by virtue of its interactions with RPA and/or TFIIH and forms a stable complex, PIC2 (Mu et al., 1997b). The 3' incision may be made at this point and then XPF•ERCC1, which makes the 5' incision, assembles on the DNA to form PIC3. The 3' nick, made by XPG at 6±3 bonds 3' to the lesion, is the first of the dual incisions made in the vicinity of the base damage (Matsunaga et al., 1995). XPG also has a non-catalytic role in excision repair as its physical presence, but not its nuclease activity, is required for the 5' incision event (Wakasugi et al., 1997) and XPG stabilizes PIC2 independent of XPG nuclease activity (Mu et al., 1997b). Following the 3' incision by XPG, XPF•ERCC1 catalyzes the 5' incision at 20±5 phosphodiester bonds 5' to the lesion and the excised oligomer is released even in the absence of repair synthesis proteins (Mu et al., 1996).

5.4. Resynthesis and Ligation

The resulting excision gap is filled by DNA polymerases and ligated to restore the integrity of the DNA molecule. The results presented here (Figure 4) are based on experiments performed with whole cell extracts prepared from a rodent cell line, but were confirmed by qualitatively similar results obtained with HeLa extracts (Reardon et al., 1997b). The repair patch illustrated in this figure demonstrates a patch size consistent with the size of the excised fragment thus suggesting neither gap enlargement 5' to the lesion nor nick translation at the 3' end.

Repair synthesis assays using damaged plasmid DNA demonstrated a requirement for PCNA in the repair synthesis step (Nichols and Sancar, 1993; Shivji et al., 1993). This suggests an involvement of either polymerase δ or ε in resynthesis since, depending on the reaction conditions, both of these polymerases are stimulated by PCNA; indeed, in vivo studies with S. cerevisiae (Budd and Campbell, 1995) indicate that both polymerase δ and ε participate in resynthesis. An excision repair requirement for RFC, a multisubunit complex required for the loading of PCNA during replication, remains to be determined.

5.5. Comparison of E. coli Uvr(A)BC with the Human Excinuclease

Two striking differences between the prokaryotic and eukaryotic systems are immediately apparent when one examines both the size of the excised oligomer and the number of proteins involved in basal nucleotide excision repair. First, the (A)BC excinuclease makes dual incisions which result in the release of damage-containing 12 mers while the dual incisions mediated by the human excision nuclease remove damage in larger oligomers, generally 24–32 nucleotides in length. Second, these mechanistically similar dual incision events require three proteins in E. coli while mammalian cells utilize thirteen polypeptides assembled into six repair factors (Table 1). Additionally, there are differences in the mechanistic details: for example, the (A)BC excinuclease makes precise incisions on both sides of the damage, while the mammalian excinuclease shows variability with both the 3' and the 5' incisions; in E. coli, all three components of the excinuclease are required for dual incisions while human cells can assemble a partially functional excision nuclease even in the absence of XPF•ERCC1. Further, the human excinuclease releases the excised oligomer in the absence of polymerases while the E. coli pathway requires resynthesis for release of the damage-containing oligomer.

6. CONCLUDING COMMENTS

6.1. Damage Recognition and System Overlap: Cooperation and Interference

While many of the molecular details of human excision repair have been elucidated in recent years, we still do not have a clear picture of how DNA damage is recognized: what are the underlying physical chemical mechanisms for recognition of chemically and structurally diverse lesions and how are such adducts located (and excision repair initiated) in highly condensed regions of the genome? What is the molecular basis for preferential repair in human cells?

Recent reports indicate considerable overlap in substrate recognition by nucleotide excision and mismatch repair. Is this a physiologically relevant phenomenon or is it a pathological response where these two general repair pathways designed for base lesions and base mismatches, respectively, get in one another's way? In contrast, repair of nonbulky lesions by base excision and nucleotide excision systems represent functional redundancy which is beneficial to the cell.

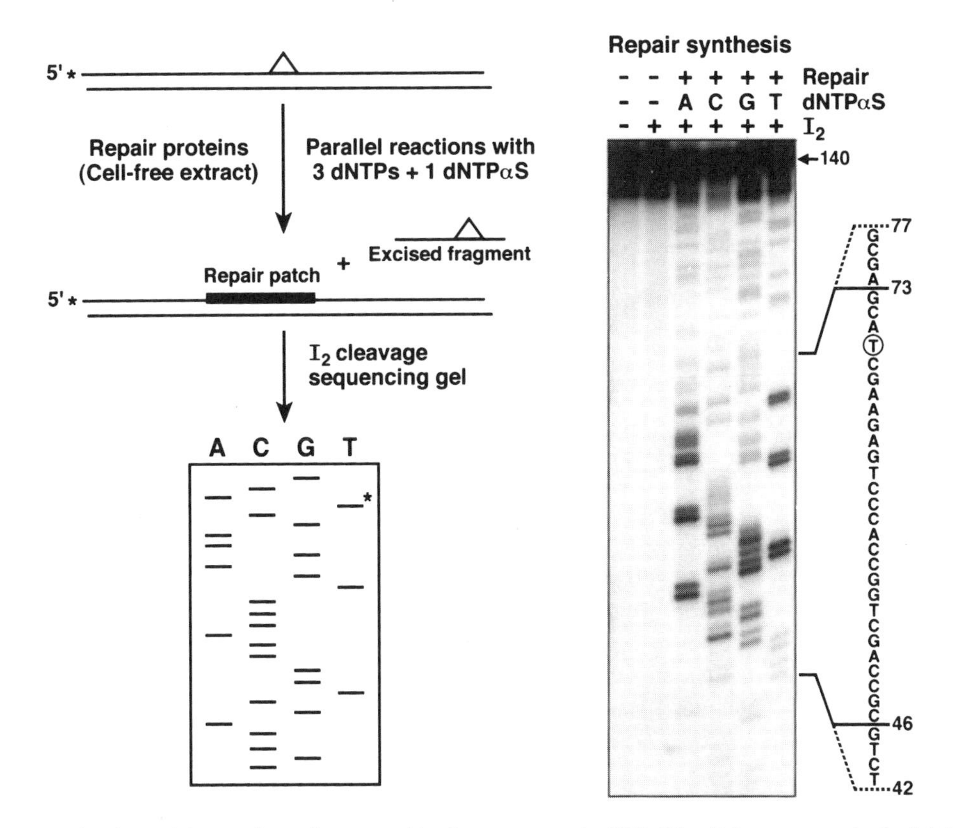

Figure 4. The excision repair patch generated by factors present in CHO K1 cell-free extracts. At the left is a schematic illustration of the phosphorothioate chemistry-based assay used to identify the repair patch (Reardon et al., 1997b). At the right is an autoradiograph of a sequencing gel showing the mammalian repair patch which coincides with the 5' and 3' incision sites and, in this example, extends for 28 nucleotides from position 46 to 73. The DNA damage is ABPD-cholesterol incorporated at nucleotide 70 opposite adenine (indicated by the circled T in the patch sequence). Reprinted by permission of Oxford University Press.

6.2. The Role of Nucleotide Excision Repair in Neurodegeneration

Based on our recent findings that repair of oxidative DNA damage is dependent on the activity of XP proteins (Reardon et al., 1997b), we suggested a reconsideration of the models proposed for the neurodegeneration associated with XP and Cockayne's syndrome to include XP-dependent nucleotide excision repair as a third repair pathway for removal of oxidative DNA lesions. Cooper et al. (1997) suggested that there are two XPG-dependent pathways for the removal of oxidative DNA damage: (1) transcription-coupled base excision repair and (2) XPG-dependent genome overall base excision repair; both pathways are distinct from nucleotide excision repair and require XPG but not its nuclease activity. We believe that the first two pathways play an important role during embryonic development, with a defect causing the neurodegeneration associated with Cockayne's syndrome, while the third pathway plays a more significant role in maintaining the long term genetic integrity of fully differentiated neurons, with defects leading to XP neurological disease caused by the abundant non-bulky oxidative base damages such as thymine glycol and 8-oxoguanine. Additional research on the relative activities of base and nucleotide excision repair in different cells and tissues at various stages of development are needed to further explore the relative involvement of these repair systems in neurodegenerative diseases as well as in the tissue damage caused by stroke or trauma.

6.3. Implications for Chemotherapy

Cisplatin, psoralen, melphalan and mitomycin C are among the DNA damaging agents commonly used as chemotherapeutic agents, but human cells have multiple systems to eliminate these agents from the cell and for the recognition and removal of DNA lesions generated by these agents. How do we circumvent this paradox, specifically deliver chemotherapeutic agents to tumor cells, and reduce or eliminate the capacity of cancerous cells to repair the very damage that is intended to kill the tumor?

NOTE ADDED IN PROOF

Very recent work has refined our model for nucleotide excision by demonstrating that the damage recognition subunit of the human excinuclease cannot be assigned to a single repair factor but rather to a four factor complex (PIC1) with the XPC function defined as that of a moleculat matchmaker (M. Wakasugi and A. Sancar [1998] Assembly, subunit composition, and footprint of human DNA repair excision nuclease. *Proc. Natl. Acad. Sci. USA* **95**, 6669–6674.)

REFERENCES

H. Asahina, I. Kuraoka, M. Shirakawa, E.H. Morita, N. Miura, I. Miyamoto, E. Ohtsuka, Y. Okada and K. Tanaka (1994) The XPA protein is a zinc metalloprotein with an ability to recognize various kinds of DNA damage. *Mutat. Res.* **315**, 229–237.

T. Bessho, A. Sancar, L.H. Thompson and M.P. Thelen (1997) Reconstitution of human excision nuclease with recombinant XPF-ERCC1 complex. *J. Biol. Chem.* **272**, 3833–3837.

M.E. Budd and J.L. Campbell (1995) DNA polymerases required for repair of UV-induced damage in Saccharomyces cerevisiae. *Mol. Cell. Biol.* **15**, 2173–2179.

J.L. Burns, S.N. Guzder, P. Sung, S. Prakash and L. Prakash (1996) An affinity of human replication protein A for ultraviolet-damaged DNA: implications for damage recognition in nucleotide excision repair. *J. Biol. Chem.* **271**, 11607–11610.

J.E. Cleaver and K.H. Kraemer (1989) Xeroderma pigmentosum. In The Metabolic Basis of Inherited Disease (eds. C.R. Scriver, A.L. Beaudet, W.S. Sly and D. Valle), *McGraw-Hill*, New York, Vol. 2, pp. 2949–2971.

A.R. Collins (1993) Mutant rodent cell lines sensitive to ultraviolet light, ionizing radiation and cross-linking agents: a comprehensive survey of genetic and biochemical characteristics. *Mutat. Res.* **293**, 99–118.

P.K. Cooper, T. Nouspikel, S.G. Clarkson and S.A. Leadon (1997) Defective transcription-coupled repair of oxidative base damage in Cockayne syndrome patients from XP group G. *Science* **275**, 990–993.

D. Coverley, M.K. Kenny, M. Munn, W.D. Rupp, D.P. Lane and R.D. Wood (1991) Requirement for the replication protein SSB in human DNA excision repair. *Nature* **349**, 538–541.

B. Demple and L. Harrison (1994) Repair of oxidative damage to DNA: enzymology and biology. *Annu. Rev. Biochem.* **63**, 915–948.

R. Drapkin, J.T. Reardon, A. Ansari, J.C. Huang, L. Zawel, K.J. Ahn, A. Sancar and D. Reinberg (1994) Dual role of TFIIH in DNA excision repair and in transcription by RNA polymerase II. *Nature* **368**, 769–772.

R. Drapkin, G. Le Roy, H. Cho, S. Akoulitchev and D. Reinberg (1996) Human cyclin-dependent kinase-activating kinase exists in three distinct complexes. *Proc. Natl. Acad. Sci. USA* **93**, 6488–6493.

D.R. Duckett, J.T. Drummond, A.I.H. Murchie, J.T. Reardon, A. Sancar, D.M.J. Lilley and P. Modrich (1996) Human MutSα recognizes damaged DNA base pairs containing O^6-methylguanine, O^4-methylthymine, or the cisplatin-d(GpG) adduct. *Proc. Natl. Acad. Sci. USA* **93**, 6443–6447.

E. Evans, J. Fellows, A. Coffer and R.D. Wood (1997) Open complex formation around a lesion during nucleotide excision repair provides a structure for cleavage by human XPG protein. *EMBO J.* **16**, 625–638.

E.C. Friedberg, G.C. Walker and W. Siede (1995) DNA Repair and Mutagenesis. *ASM Press*, Washington.

E.C. Friedberg (1996) Relationships between DNA repair and transcription. *Annu. Rev. Biochem.* **65**, 15–42.

D. Gunz, M.T. Hess and H. Naegeli (1996) Recognition of DNA adducts by human nucleotide excision repair: evidence for a thermodynamic probing mechanism. *J. Biol. Chem.* **271**, 25089–25098.

S.N. Guzder, V. Bailly, P. Sung, L. Prakash and S. Prakash (1995a) Yeast DNA repair protein RAD23 promotes complex formation between transcription factor TFIIH and DNA damage recognition factor RAD14. *J. Biol. Chem.* **270**, 8385–8388.

S.N. Guzder, Y. Habraken, P. Sung, L. Prakash and S. Prakash (1995b) Reconstitution of yeast nucleotide excision repair with purified Rad proteins, replication protein A, and transcription factor TFIIH. *J. Biol. Chem.* **270**, 12973–12976.

P.C. Hanawalt (1994) Transcription-coupled repair and human disease. *Science* **266**, 1957–1958.

J.J. Harrington and M.R. Lieber (1994) Functional domains within FEN-1 and RAD2 define a family of structure-specific endonucleases: implications for nucleotide excision repair. *Genes Dev.* **8**, 1344–1355.

Z. He, L.A. Henrickson, M.S. Wold and C.J. Ingles (1995) RPA involvement in the damage-recognition and incision steps of nucleotide excision repair. *Nature* **374**, 566–569.

L.A. Henrickson, C.B. Umbricht and M.S. Wold (1994) Recombinant replication protein A: expression, complex formation, and functional characterization. *J. Biol. Chem.* **269**, 11121–11132.

M.T. Hess, U. Schwitter, M. Petretta, B. Giese and H. Naegeli (1997) Bipartite substrate discrimination by human nucleotide excision repair. *Proc. Natl. Acad. Sci. USA* **94**, 6664–6669.

J.C. Huang, D.L.Svoboda, J.T. Reardon and A. Sancar (1992) Human nucleotide excision nuclease removes thymine dimers from DNA by incising the 22nd phosphodiester bond 5' and the 6th phosphodiester bond 3' to the photodimer. *Proc. Natl. Acad. Sci. USA* **89**, 3664–3668.

J.C. Huang and A. Sancar (1994) Determination of minimum substrate size for human excinuclease. *J. Biol. Chem.* **269**, 19034–19040.

J.C. Huang, D.S. Hsu, A. Kazantsev and A. Sancar (1994a) Substrate spectrum of human excinuclease: repair of abasic sites, methylated bases, mismatches, and bulky adducts. *Proc. Natl. Acad. Sci. USA* **91**, 12213–12217.

J.C. Huang, D.B. Zamble, J.T. Reardon, S.J. Lippard and A. Sancar (1994b) HMG-domain proteins specifically inhibit the repair of the major DNA adduct of the anticancer drug cisplatin by human excision nuclease. *Proc. Natl. Acad. Sci. USA* **91**, 10394–10398.

N. Iyer, M.S. Reagan, K.-J. Wu, B. Canagarajah and E.C. Friedberg (1996) Interactions involving the human RNA polymerase II transcription/nucleotide excision repair complex TFIIH, the nucleotide excision repair protein XPG, and Cockayne syndrome group B (CSB) protein. *Biochemistry* **35**, 2157–2167.

C.J. Jones and R.D. Wood (1993) Preferential binding of the xeroderma pigmentosum group A complementing protein to damaged DNA. *Biochemistry* **32**, 12096–12104.

A.Kazantsev, D. Mu, A.F. Nichols, X. Zhao, S. Linn and A. Sancar (1996) Functional complementation of xeroderma pigmentosum group E by replication protein A in an in vitro system. *Proc. Natl. Acad. Sci. USA* **93**, 5014–5018.

K.H. Kraemer, M.M. Lee and J. Scotto (1984) DNA repair protects against cutaneous and internal neoplasia: evidence from xeroderma pigmentosum. *Carcinogenesis* **5**, 511–514.

G.-M. Li, H. Wang and L.J. Romano (1996) Human MutSα specifically binds to DNA containing aminofluorene and acetylaminofluorene adducts. *J. Biol. Chem.* **271**, 24084–24088.

L. Li, S.J. Elledge, C.A. Peterson, E.S. Bales and R.J. Legerski (1994) Specific association between the human DNA repair proteins XPA and ERCC1. *Proc. Natl. Acad. Sci. USA* **91**, 5012–5016.

L. Li, X. Lu, C.A. Peterson and R.J. Legerski (1995) An interaction between the DNA repair factor XPA and replication protein A appears essential for nucleotide excision repair. *Mol. Cell. Biol.* **15**, 5396–5402.

L. Li, X. Lu, C. Peterson and R. Legerski (1997) XPC interacts with both HHR23B snd HHR23A in vivo. *Mutat. Res.* **383**, 197–203.

J.-C. Marinoni, R. Roy, W. Vermeulen, P. Miniou, Y. Lutz, G. Weeda, T. Seroz, D.M. Gomez, J.H.J. Hoeijmakers and J.-M. Egly (1997) Cloning and characterization of p52, the fifth subunit of the core of the transcription/DNA repair factor TFIIH. *EMBO J.* **16**, 1093–1102.

C. Masutani, K. Sugasawa, J. Yanagisawa, T. Sonoyama, M. Ui, T. Enomoto, K. Takio, K. Tanaka, P.J. van der Spek, D. Bootsma, J.H.J. Hoeijmakers and F. Hanaoka (1994) Purification and cloning of a nucleotide excision repair complex involving the xeroderma pigmentosum group C protein and a human homologue of yeast RAD23. *EMBO J.* **13**, 1831–1843.

T. Matsuda, M. Saijo, I. Kuraoka, T. Kobayashi, Y. Nakatsu, A. Nagai, T. Enjoji, C. Masutani, K. Sugasawa, F. Hanaoka, A. Yasui and K. Tanaka (1995) DNA repair protein XPA binds replication protein A (RPA). *J. Biol. Chem.* **270**, 4152–4157.

T. Matsunaga, D. Mu, C.-H. Park, J.T. Reardon and A. Sancar (1995) Human DNA repair excision nuclease: analysis of the roles of the subunits involved in dual incisions by using anti-XPG and anti-ERCC1 antibodies. *J. Biol. Chem.* **270**, 20862–20869.

T. Matsunaga, C.-H. Park, T. Bessho and A. Sancar (1996) Replication protein A confers structure-specific endonuclease activities to the XPF-ERCC1 and XPG subunits of human DNA repair excision nuclease. *J. Biol. Chem.* **271**, 11047–11050.

P. Modrich and R. Lahue (1996) Mismatch repair in replication fidelity, genetic recombination, and cancer biology. *Annu. Rev. Biochem.* **65**, 101–133.

D.Mu, E. Bertrand-Burggraf, J.C. Huang, R.P.P. Fuchs and A. Sancar (1994) Human and E. coli excinucleases are affected differently by the sequence context of acetylaminofluorene-guanine adduct. *Nucleic Acids Res.* **23**, 4869–4871.

D. Mu, C.-H. Park, T. Matsunaga, D.S. Hsu, J.T. Reardon and A. Sancar (1995) Reconstitution of human DNA excision nuclease in a highly defined system. *J. Biol. Chem.* **270**, 2415–2418.

D. Mu, D.S. Hsu and A. Sancar (1996) Reaction mechanism of human DNA repair excision nuclease. *J. Biol. Chem.* **271**, 8285–8294.

D. Mu and A. Sancar (1997) Model for XPC-independent transcription-coupled repair of pyrimidine dimers in humans. *J. Biol. Chem.* **272**, 7570–7573.

D.Mu, M. Tursun, D.R. Duckett, J.T. Drummond, P. Modrich and A. Sancar (1997a) Recognition and repair of compound DNA lesions (base damage and mismatch) by human mismatch repair and excision repair systems. *Mol. Cell. Biol.* **17**, 760–769.

D.Mu, M. Wakasugi, D.S. Hsu and A. Sancar (1997b) Characterization of reaction intermediates of human excision repair nuclease. *J. Biol. Chem.* **272**, 28971–28979.

A.F. Nichols and A. Sancar (1992) Purification of PCNA as a nucleotide excision repair protein. *Nucleic Acids Res.* **20**, 2441–2446.

A.O'Donovan, A.A. Davies, J.G. Moggs, S.C. West and R.D. Wood (1994) XPG endonuclease makes the 3' incision in human DNA nucleotide excision repair. *Nature* **371**, 432–435.

C.-H. Park and A. Sancar (1994) Formation of a ternary complex by human XPA, ERCC1, and ERCC4(XPF) excision repair proteins. *Proc. Natl. Acad. Sci. USA* **91**, 5017–5021.

C.-H. Park, T. Bessho, T. Matsunaga and A. Sancar (1995a) Purification and characterization of the XPF-ERCC1 complex of human DNA repair excision nuclease. *J. Biol. Chem.* **270**, 22657–22660.

C.-H. Park, D. Mu, J.T. Reardon and A. Sancar (1995b) The general transcription-repair factor TFIIH is recruited to the excision repair complex by the XPA protein independent of the TFIIE transcription factor. *J. Biol. Chem.* **270**, 4896–4902.

S. Prakash, P. Sung and L. Prakash (1993) DNA repair genes and proteins of Saccharomyces cerevisiae. *Annu. Rev. Genet.* **27**, 33–70.

J.T. Reardon, L.H. Thompson and A. Sancar (1993) Excision repair in man and the molecular basis of xeroderma pigmentosum syndrome. *Cold Spring Harbor Symp. Quant. Biol.* **58**, 605–617.

J.T. Reardon, H. Ge, E. Gibbs, A. Sancar, J. Hurwitz and Z.-Q. Pan (1996a) Isolation and characterization of two human transcription factor IIH (TFIIH)-related complexes: ERCC2/CAK and TFIIH*. *Proc. Natl. Acad. Sci. USA* **93**, 6482–6487.

J.T. Reardon, D. Mu and A. Sancar (1996b) Overproduction, purification, and characterization of the XPC subunit of the human DNA repair excision nuclease. *J. Biol. Chem.* **271**, 19451–19456.

J.T. Reardon, T. Bessho, H.S. Kung, P.H. Bolton and A. Sancar (1997a) In vitro repair of oxidative DNA damage by human nucleotide excision repair system: possible explanation for neurodegeneration in xeroderma pigmentosum patients. *Proc. Natl. Acad. Sci. USA* **94**, 9463–9468.

J.T. Reardon, L.H. Thompson and A. Sancar (1997b) Rodent UV-sensitive mutant cell lines in complementation groups 6–10 have normal general excision repair activity. *Nucleic Acids Res.* **25**, 1015–1021.

J.H. Robbins, R.J. Polinsky and A.N. Moshell (1983) Evidence that lack of deoxyribonucleic acid repair causes death of neurons in xeroderma pigmentosum. *Ann. Neurol.* **13**, 682–684.

R. Roy, J.P. Adamczewski, T. Seroz, W. Vermeulen, J.-P. Tassan, L. Schaeffer, E.A. Nigg, J.H.J. Hoeijmakers and J.-M. Egly (1994) The MO15 cell cycle kinase is associated with the TFIIH transcription-DNA repair factor. *Cell* **79**, 1093–1101.

M. Saijo, I. Kuraoka, C. Masutani, F. Hanaoka and K. Tanaka (1996) Sequential binding of DNA repair proteins RPA and ERCC1 to XPA in vitro. *Nucleic Acids Res.* **24**, 4719–4724.

L. Samson (1992) The suicidal DNA repair methyltransferases of microbes. *Mol. Microbiol.* **6**, 825–831.

A. Sancar (1994) Mechanisms of DNA excision repair. *Science* **266**, 1954–1956.

A. Sancar (1995a) DNA repair in humans. *Annu. Rev. Genet.* **29**, 69–105.

A. Sancar (1995b) Excision repair in mammalian cells. *J. Biol. Chem.* **270**, 15915–15918.

A. Sancar (1996a) No "end of history" for photolyases. *Science* **272**, 48–49.

A. Sancar (1996b) DNA excision repair. *Annu. Rev. Biochem.* **65**, 43–81.

A. Sancar and J.E. Hearst (1993) Molecular matchmakers. *Science* **259**, 1415–1420.

A. Sancar and G.B. Sancar (1988) DNA repair enzymes. *Annu. Rev. Biochem.* **57**, 29–67.

L. Schaeffer, R. Roy, S. Humbert, V. Moncollin, W. Vermeulen, J.H.J. Hoeijmakers, P. Chambon and J.-M. Egly (1993) DNA repair helicase: a component of BTF2 (TFIIH) basic transcription factor. *Science* **260**, 58–63.

M. Sekiguchi, Y. Nakabeppu, K. Sakumi and T. Tsuzuki (1996) DNA-repair methyltransferase as a molecular device for preventing mutation and cancer. *J. Cancer Res. Clin. Oncol.* **122**, 199–206.

C.P. Selby and A. Sancar (1993) Molecular mechanism of transcription-repair coupling. *Science* **260**, 53–58.

M.K.K. Shivji, M.K. Kenny and R.D. Wood (1993) Proliferating cell nuclear antigen is required for DNA excision repair. *Cell* **69**, 367–374.

Sibghat-Ullah, I. Husain, W. Carlton and A. Sancar (1989) Human nucleotide excision repair in vitro: repair of pyrimidine dimers, psoralen and cisplatin adducts by HeLa cell-free extract. *Nucl. Acids Res.* **17**, 4471–4484.

Sibghat-Ullah, A. Sancar and J.E. Hearst (1990) The repair patch of E. coli (A)BC excinuclease. *Nucl. Acids Res.* **18**, 5051–5053.

K. Sugasawa, C. Masutani, A. Uchida, T. Maekawa, P.J. van der Spek, D. Bootsma, J.H.J. Hoeijmakers and F. Hanaoka (1996) HHR23B, a human Rad23 homolog, stimulates XPC protein in nucleotide excision repair in vitro. *Mol. Cell. Biol.* **16**, 4852–4861.

P. Sung, V. Bailly, C. Weber, L.H. Thompson, L. Prakash and S. Prakash (1993) Human xeroderma pigmentosum group D gene encodes a DNA helicase. *Nature* **365**, 852–855.

D.L. Svoboda, J.-S. Taylor, J.E. Hearst and A. Sancar (1993) DNA repair by eukaryotic nucleotide excision nuclease: removal of thymine dimer and psoralen monoadduct by HeLa cell-free extract and of thymine dimer by Xenopus laevis oocytes. *J. Biol. Chem.* **268**, 1931–1936.

L.H. Thompson (1998) Nucleotide excision repair: its relation to human disease. In DNA Damage and Repair -- Biochemistry, Genetics and Cell Biology (eds. J.A. Nickoloff and M.F. Hoekstra), *Humana Press*, Totowa, Vol. 2, pp 335–394.

M. Wakasugi, J.T. Reardon and A. Sancar (1997) The non-catalytic function of XPG protein during dual incision in human nucleotide excision repair. *J. Biol. Chem.* **272**, 16030–16034.

R.D.Wood (1996) DNA repair in eukaryotes. *Annu. Rev. Biochem.* **65**, 135–167.

R.D. Wood, P. Robins and T. Lindahl (1988) Complementation of the xeroderma pigmentosum DNA repair defect in cell-free extracts. *Cell* **53**, 97–106.

M. Yamada, E. O'Regan, R. Brown and P. Karran (1997) Selective recognition of a cisplatin-DNA adduct by human mismatch repair proteins. *Nucleic Acids Res.* **25**, 491–495.

D.B. Zamble, D. Mu, J.T. Reardon, A. Sancar and S.J. Lippard (1996) Repair of cisplatin-DNA adducts by the mammalian excision nuclease. *Biochemistry* **35**, 10004–10013.

32

DETECTION OF 8-HYDROXY-2'-DEOXYGUANOSINE BY A MONOCLONAL ANTIBODY N45.1 AND ITS APPLICATION

Shinya Toyokuni

Department of Pathology and Biology of Diseases
Graduate School of Medicine, Kyoto University
Sakyo-ku, Kyoto, Japan

1. ABSTRACT

8-Hydroxy-2'-deoxyguanosine (8-OHdG), an oxidatively modified DNA product, is one of the most commonly used markers for the evaluation of oxidative stress. A monoclonal antibody (N45.1) specific for 8-OHdG was characterized and applied for quantitative immunohistochemistry and enzyme-linked immunosorbent assay (ELISA).

8-OHdG index of quantitative immunohistochemistry analyzed by NIH image freeware revealed a reasonable correlation with the 8-OHdG amount determined by high-performance liquid chromatography with electrochemical detector. Determination of urinary 8-OHdG by ELISA is applicable to clinical situations such as cancer chemotherapy.

2. INTRODUCTION

2.1. Oxidative Stress

Reactive oxygen species (ROS) are involved in a variety of biological phenomena such as mutation, carcinogenesis, aging, atherosclerosis, radiation or ultraviolet exposure, inflammation, ischemia-reperfusion injury and neurodegenerative diseases (Halliwell & Gutteridge, 1989). Furthermore, ROS is persistently produced in the cell as long as it utilizes oxygen as a major energy source. ROS, whether it is produced inside or outside of the cell, induces oxidative stress in the cell unless the antioxidant mechanisms overwhelm the ROS load.

Advances in DNA Damage and Repair, edited by Dizdaroglu and Karakaya. Kluwer Academic / Plenum Publishers, New York, 1999.

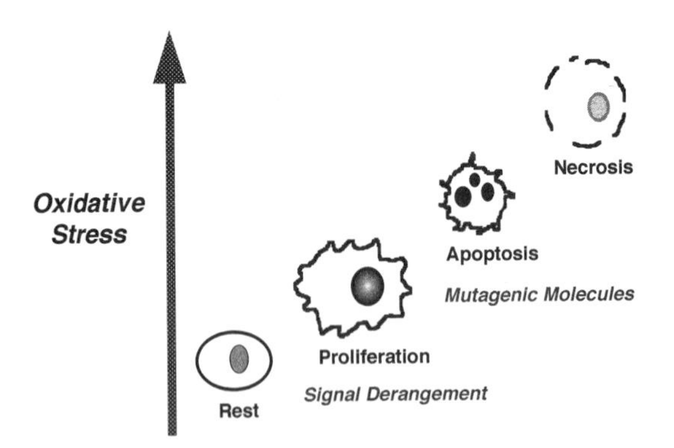

Figure 1. Significance of oxidative stress in cellular biology.

Recently, it is getting clear that relatively low load of oxidative stress promotes cellular proliferation rather than causing cell death (Fig. 1). For example, a large amount of hydrogen peroxide is produced without any stimulation in cultured cancer cells (Szatrowski & Nathan, 1991). Thus, cancer cells are usually exposed to more oxidative stress than normal cells (Burdon, 1995), but it appears that the stress is not strong enough to cause apoptosis or necrosis (Toyokuni et al., 1995). It was also shown in insulin-secreting RINm5F cells that different prooxidant levels by redox cycling quinone, 2,3-dimethoxy-1,4-naphthoquinone, either stimulate growth, trigger apoptosis, or produce necrosis (Dypbukt et al., 1994). Therefore, it is becoming more and more important to evaluate the cellular status of oxidative stress.

2.2. 8-Hydroxy-2'-Deoxyguanosine

Hydroxylation of C-8 position of 2'-deoxyguanosine (8-hydroxy-2'-deoxyguanosine, 8-OHdG) by ascorbic acid in the presence of oxygen was first reported by Kasai and Nishimura in 1984 when they were isolating mutagens and carcinogens in heated glucose (Kasai & Nishimura, 1984). It has been established that either hydroxyl radical (Kasai & Nishimura, 1984), singlet oxygen (Floyd et al., 1989; Devasagayam et al., 1991), or direct photodynamic action (Kasai et al., 1992) is responsible for the formation of 8-OHdG. Recently, it was shown that peroxynitrite also induces 8-OHdG through an active intermediate of which reactivity is similar to hydroxyl radical (Inoue & Kawanishi, 1995). Currently, 8-OHdG is one of the most popular markers for the evaluation of oxidative stress. So far, a large number of carcinogens including γ-irradiation, 4-nitroquinoline 1-oxide, ferric nitrilotriacetate (Fe-NTA) and asbestos fibers are reported to induce 8-OHdG formation in their target organs after treatment (Floyd, 1990). The number of scientific reports dealing with 8-OHdG is increasing every year, and there were approximately 100 reports in 1996 by Medline search (Fig. 2).

There are two reasons for the popularity of this DNA base-modified product; (1) sensitive and technically easy detection by the use of high performance liquid chromatography (HPLC) and electrochemical detector (ECD) is possible (Floyd et al., 1986), and (2) 8-OHdG itself has a biological significance of inducing G:C to T:A transversions upon DNA replication (Kuchino et al., 1987; Shibutani et al., 1991; Cheng et al., 1992; Wood et al., 1990). In spite of its popularity, there was some drawbacks regarding 8-OHdG. There has been a long discussion about the accuracy of the 8-OHdG amount, and the existence itself in vivo was even doubted; namely, a possibility of artifactual production of 8-OHdG

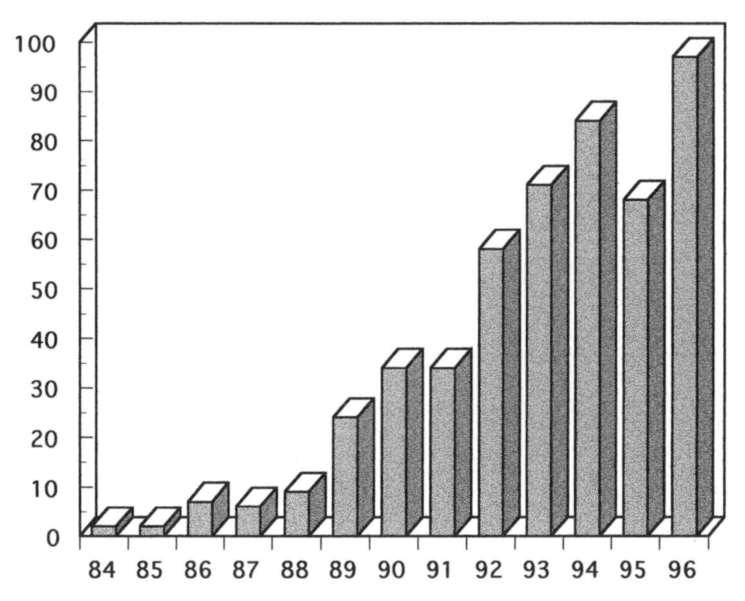

Figure 2. Reported number of papers on 8-hydroxy-2'-deoxyguanosine by Medline search.

during the sample preparation processes was pointed out (Lindahl, 1993). For example, phenol-chloroform extraction of DNA from tissues may contribute to the increased levels of 8-OHdG (Claycamp, 1992). The difference in the amount of 8-OHdG by methods (HPLC/ECD or gas chromatography/mass spectrometry [GC/MS]) has also been discussed (Halliwell, 1993; Cadet et al., 1997).

However, the situation has been dramatically changed recently by the cloning of human repair enzymes for 8-OHdG (human MutT, MutY and MutM homologues) (Sakumi et al., 1993; Slupska et al., 1996; Aburatani et al., 1997; Arai et al., 1997; Rosenquist et al., 1997; Radicella et al., 1997; Roldan-Arjona et al., 1997) and the production of specific monoclonal antibodies for 8-OHdG. These two advances have firmly established the significance of 8-OHdG in mammalian biology. Other chapters are dealing with the repair enzymes for 8-OHdG in detail.

3. RESULTS AND DISCUSSION

3.1. Production of Monoclonal Antibody for 8-Hydroxy-2'-Deoxyguanosine

There are four methods established for the detection of 8-OHdG: 1) HPLC/ECD, 2) GC/MS (Dizdaroglu, 1991), 3) ^{32}P postlabeling (Lutgerink et al., 1992), and 4) antibody-based methods. Merits and demerits of HPLC/ECD, GC/MS and antibody-based methods are summarized in Table 1. It is of note that the former three methods do not localize 8-OHdG in cells or tissues. Therefore, it is useful to develop a specific monoclonal antibody (mAb)-based method for the quantitation and localization of 8-OHdG. Thus far, five lines of research by the use of either polyclonal or monoclonal antibodies have been reported for the quantitation of 8-OHdG by immunoanalyses (Degan et al., 1991; Musarrat & Wani,

Table 1. Characteristics of 8-OHdG determination methods

	Merit	Demerit
HPLC/ECD	Absolute value is obtained	Needs DNA extraction and hydrolysis
	Completed in single day	Needs relatively expensive equipment
GC/MS	Absolute value is obtained	Needs chromatin extraction, hydrolysis and derivatization
	A variety of modified DNA bases quantified	Needs expensive equipment
Quantitative immunohistochemistry	No expensive equipment except ELISA reader is necessary	Only relative value is obtained
		Difficult to completely rule out cross-reaction
	No DNA extraction necessary	
	Localization can be evaluated	
	Applicable to old paraffin blocks	

HPLC/ECD, high performance liquid chromatography with electrochemical detector; GC/MS, gas chromatography with mass spectrometry

1994; Park et al., 1992; Yin et al., 1996; Yarborough et al., 1996; Bruskov et al., 1996; Toyokuni et al., 1997) (Table 2). However, antibodies except ours show a relatively high cross-reactivity with 8-hydroxyguanosine (8-OHG, RNA form). This is probably because 8-OHG was used as an antigen for raising specific antibodies for 8-OHdG. A mAb N45.1 specific for 8-OHdG was established by our research team using 8-OHdG-keyhole limpet hemocyanin as an antigen (Osawa et al., 1995). Characteristically, N45.1 1) recognizes both hydroxy function (keto function) of 8-hydroxyguanine and 2' portion of deoxyribose, 2) is scarcely reactive with other DNA adducts we have examined (21 molecules), thus highly specific for 8-OHdG, and 3) is applicable not only for immunoenzymatic assay (enzyme-linked immunosorbent assay) but also for immunohistochemistry in paraffin sections that is routinely used in hospitals for pathologic diagnosis (Toyokuni et al., 1997).

3.2. Quantitative Immunohistochemistry

We have constructed a quantitative immunohistochemical system for 8-OHdG determination in paraffin-embedded sections by the use of N45.1. Tissues were fixed with Bouin's solution (saturated picric acid:formaldehyde:acetic acid:H_2O = 15:5:1:10) (Luna, 1968) overnight, immersed sequentially for 24 hours in 50 % and 70 % ethanol to remove picric acid, dehydrated, embedded in paraffin, sectioned at 3.5 μm, and mounted on glass

Table 2. Published antibodies for 8-hydroxy-2'-deoxyguanosine

				Specificity			
Year	Author	Name	Antigen	8-OHdG/dG	8-OHdG/ 8-OHG	8-OHdG/ 8-OHGua	Applications
1991	Degan et al.	-, Po	8-OHG-BSA	68000	5.3	7600	IAP, IEA
1992	Park et al.	15A3, Mo	8-OHG-BSA	7100	4.4	19.6	IAP
1996	Yin et al.	1F7, Mo	8-OHG-KLH	100000	1	-	IEA, IH (F)
1996	Bruskov et al.	-, Mo	-	-	1.3	-	IEA
1997	Toyokuni et al.	N45.1, Mo	8-OHdG-KLH	≥100000	100	≥100000	IEA, IH (F,P)

8-OHdG, 8-hydroxy-2'-deoxyguanosine; 8-OHG, 8-hydroxyguanosine; 8-OHGua, 8-hydroxyguanine; Po, polyclonal antibody; Mo, monoclonal antibody; BSA, bovine serum albumin; KLH, keyhole limpet hemocyanin; IAP, immunoaffinity purification; IEA, immunoenzymatic assay; IH, immunohistochemistry; F, frozen section; P, paraffin section.

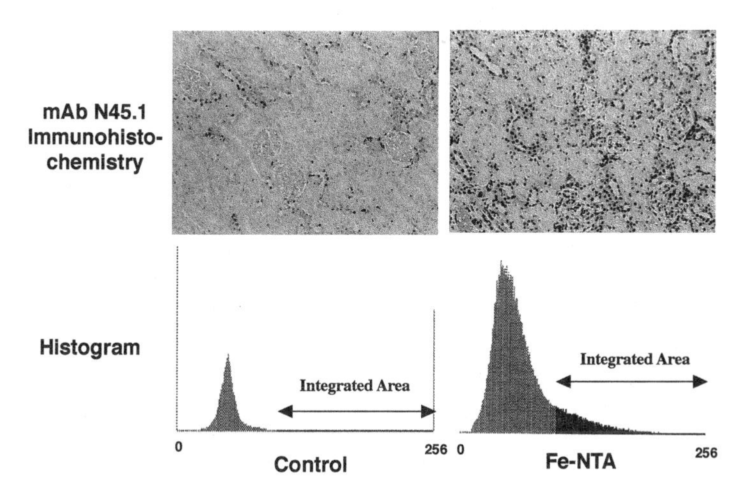

Figure 3. Quantitative immunohistochemistry by the use of monoclonal antibody N45.1 and NIH image freeware. Left, untreated control kidney of a male Wistar rat; right, kidney of a male Wistar rat 3 hours after ip administration of 15 mg Fe/kg of Fe-NTA.

slides coated with silane. The avidin-biotin complex method was used (Hsu et al., 1981). After deparaffinization with xylene and ethanol, normal rabbit serum (diluted to 1:75) for the inhibition of nonspecific binding of secondary antibody, mAb N45.1 (1–10 μg/ml), biotin-labeled rabbit anti-mouse IgG serum (diluted to 1:300), and avidin-biotin-alkaline phosphatase complex (diluted to 1:100) were sequentially applied. For quantitative analyses, black substrate is appropriate, and it is recommended that the final color presentation reaction be performed simultaneously. Further, some antigen retrieval method such as microwave or autoclave may be necessary in our experience for sections fixed with solely formaldehyde or of relatively hard tissues such as skin.

Quantitation of immunohistological data (8-OHdG Index) was calculated as shown below:

$$\text{8-OHdG Index} = \Sigma_{X>\text{threshold}} [(X - \text{threshold}) \times \text{area (pixels)}] / \text{total cell number}$$

where X is the staining density which is indicated by a number between 0 and 256 in gray scale. Hematoxylin and eosin staining specimens were used for the total cell count, and 8-OHdG immunohistochemistry specimens were used for densitometric analyses. Thirty five mm color slides of three appropriate locations for each specimen were prepared in a magnification of 20 × 5. This was approximately 670 × 440 μm^2 area. Color images were obtained as PICT files by a slide scanner in connection with a Macintosh computer. Brightness and contrast of each image file were uniformly enhanced by Adobe Photoshop software followed by analyses using NIH Image freeware which is available from the Internet by file transfer protocol from zippy.nimh.nih.gov. PICT image files were opened in gray scale mode by NIH image. Cell number was counted by "Analyze Particles" command after setting a proper threshold. For integration of nuclear density in immunohistochemistry, density slice of 100–150 to 256 was selected for "Measure" command (Fig. 3). Means of the integrated density obtained from three independently obtained files was used as a representative value.

Table 3. Selected reports on urinary 8-Hydroxy-2'-deoxyguanosine

Year	Author	Method	Subject
1989	Shigenaga et al.	HPLC/ECD	mouse>rat>human (per kg/day)
1990	Fraga et al.	HPLC/ECD	decrease with age in rats (per kg/day)
1992	Park et al.	mAb and HPLC/ECD	not affected by diet in rats (per kg/day)
1992	Loft et al.	HPLC/ECD	increased in smokers (per kg/day)
1993	Tagesson et al.	HPLC/ECD	increased in asbestos, rubber and azo-dye workers (per urinary creatinine)
1993	Loft et al.	HPLC/ECD	increased in smokers and lean and/or male (per kg/day)
1994	Lunec et al.	HPLC/ECD,GC/MS	decreased in systemic lupus erythematosus patients (per urinary creatinine)
1994	Lagorio et al.	HPLC/ECD	increased with exposure to benzene (per urinary creatinine)
1995	Tagesson et al.	HPLC/ECD	increased in cancer patients (per urinary creatitnine; breast, lung, colon, malignant lymphoma,malignant teratoma); increased after chemoterapy (not specified); increased during and after whole body irradiation
1996	Yamamoto et al.	HPLC/ECD	increased in patients with gynecologic cancer (per kg/day)
1997	Okamura et al.	HPLC/ECD	increased after repeated exercise (per kg/day)
1997	Erhola et al.	ELISA	increased in small-cell lung cancer patients; decreased in patients with remission after chemotherapy (per urinary creatinine)
1997	Leinonen et al.	ELISA	increased in patients with non-insulin-dependent diabetic patients (per urinary creatinine); association with the control status (HbA_{1c}) of diabetes mellitus patients

HPLC/ECD, high performance liquid chromatography with electrochemical detector; mAb, monoclonal antibody; ELISA, enzyme-linked immunosorbent assay.

This quantification system was applied to a prototype free radical-induced renal tubular damage and carcinogenesis model mediated by ferric nitrilotriacetate (Fe-NTA) (Toyokuni et al., 1990; Toyokuni et al., 1994b; Toyokuni, 1996). A GC/MS study revealed that 8-oxoguanine is one of the major oxidatively modified bases, and shows the maximal increase in this model (Toyokuni et al., 1994a). Major signals were observed in the nuclei of renal proximal tubules after Fe-NTA administration, and we confirmed a good proportional association between 8-OHdG index and 8-OHdG levels determined by HPLC/ECD method (Toyokuni et al., 1997). So far, immunohistochemistry using N45.1 was successfully applied to UV irradiation (Hattori et al., 1996), ischemia-reperfusion injury, and cancer tissues (unpublished data).

3.3. Enzyme-Linked Immunosorbent Assay (ELISA) System

3.3.1. Urinary 8-Hydroxy-2'-Deoxyguanosine. Though 8-OHdG or 8-oxoGTP is continuously generated in the cell, efficient repair systems remove 8-OHdG (MutM and MutY homologues) or hydrolyze to 8-oxoGMP (MutT homologue) as demonstrated by cloning of the repair enzymes (section 1.2 and other chapters). It is believed that the resulting deoxyribonucleoside, namely 8-OHdG, is excreted in the urine (Shigenaga et al., 1989; Park et al., 1992). Although many investigators have demonstrated the presence of 8-OHdG in the urine of humans and animals, the origin of 8-OHdG in the urine has not been fully elucidated. The important finding was that urinary 8-OHdG content is not affected by diet while urinary 8-OHG and 8-oxoguanine are greatly affected (Park et al., 1992). It is established that enzymatic activity of MutM is glycosylase/AP-lyase (apurinic, apyrimidic lyase) (Aburatani et al., 1997), so its product is not 8-OHdG. The possibilities

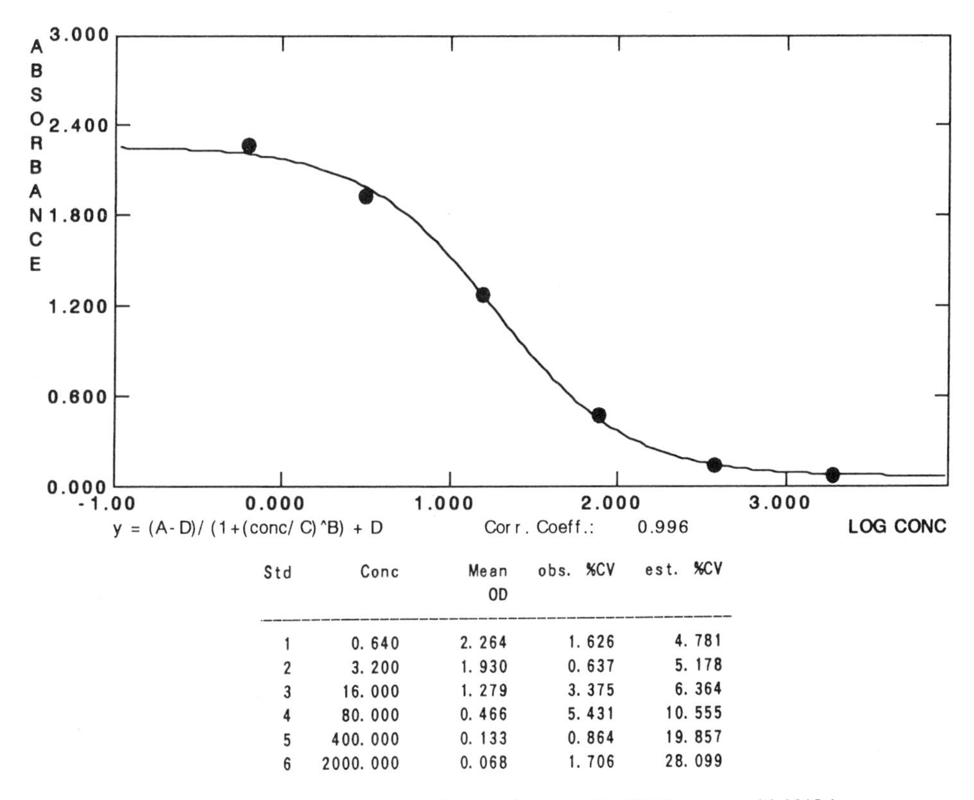

Std	Conc	Mean OD	obs. %CV	est. %CV
1	0.640	2.264	1.626	4.781
2	3.200	1.930	0.637	5.178
3	16.000	1.279	3.375	6.364
4	80.000	0.466	5.431	10.555
5	400.000	0.133	0.864	19.857
6	2000.000	0.068	1.706	28.099

Figure 4. A representative standard curve of competitive ELISA assay with N45.1.

are that 1) urinary 8-OHdG is derived from 8-oxoGMP, or 2) other repair enzyme(s) for 8-OHdG in DNA exist. How 8-OHdG comes out of the cell is not at all clear at present. We believe that 8-OHdG is not the result of passive flow into the blood of necrotic cells based on the finding that necrotized cells are not immunostained with N45.1 (Toyokuni et al., 1997), but rather that 8-OHdG is somehow actively excreted through membrane of the cell. This issue awaits further investigation.

3.3.2. ELISA System by N45.1. In spite of the above-mentioned problem, urinary 8-OHdG has been established as a marker of oxidative stress. Table 3 summarized the reports on the usefulness of urinary 8-OHdG. A competitive ELISA system was constructed by the use of N45.1 (Osawa et al., 1995). In this system, 8-OHdG conjugated to bovine serum albumin (BSA) is coated on 96-well plate, and the sample fluid competes with the coated 8-OHdG-BSA for N45.1. The detection range is 0.64–2000 ng/ml. Fig. 4 shows an example of standard curve. A proportional relationship between the levels of 8-OHdG determined by N45.1 ELISA and HPLC/ECD method is established (unpublished data). There is, however, a problem to be solved in the urinary 8-OHdG determination as discussed in the meeting. This is the difference in the absolute value of the two methods within one order of range. The possibilities are 1) detection of oligo-DNA by antibody, 2) cross-reaction of the antibody to unidentified urinary molecules, or 3) loss of 8-OHdG in the sample preparation of HPLC/ECD method. In spite of these situations, ELISA is getting popular since determination of urinary 8-OHdG by HPLC/ECD requires complicated procedures (Shigenaga et al., 1989; Degan et al., 1991; Park et al., 1992) or special HPLC configurations (Loft et al., 1993; Tagesson et al., 1995).

This ELISA system was applied to human urine samples of lung cancer patients receiving radiotherapy or chemotherapy, and non-insulin-dependent diabetic patients. Patients with small-cell carcinoma of lung showed higher levels of urinary 8-OHdG/creatinine than the controls. Furthermore, small-cell carcinoma patients with complete or partial response to the chemotherapy showed a significant decrease in urinary 8-OHdG/creatinine while patients with no change or progressive disease showed an increase (Erhola et al., 1997). In the other experiment, urinary excretion of 8-OHdG was markedly higher in non-insulin-dependent diabetic patients than in control subjects. High glycosylated hemoglobin which means a poor control of the disease was associated with a high level of urinary 8-OHdG (Leinonen et al., 1997).

4. CONCLUSION

Development of 8-OHdG-specific antibodies, especially monoclonal antibody N45.1, opened new insights in the field of oxidative DNA damage. Immunohistochemistry compensates for deficiencies of the other methods. ELISA is another useful method for the determination of urinary 8-OHdG.

ACKNOWLEDGMENTS

This work was supported in part by a grant-in-aid for scientific research from the Ministry of Education, Science, Sports, and Culture, Japan, and Program for Promotion of Basic Research Activities for Innovative Biosciences. I thank Japan Institute for the Control of Aging (Fukuroi, Shizuoka, Japan) for providing N45.1 monoclonal antibody and "8-OHdG Check" ELISA kits.

REFERENCES

H. Aburatani, Y. Hippo, T. Ishida, R. Takashima, C. Matsuba, T. Kodama, M. Takao, A. Yasui, K. Kamamoto and M..Asano. (1997). Cloning and characterization of mammalian 8-hydroxyguanine-specific DNA glycosylase/apurinic, apyrimidinic lyase, a functional mutM homologue. *Cancer Res.* **57**, 2151–2156.

K. Arai, K. Morishita, K. Shinmura, T. Kohno, S.R. Kim, T. Nohmi, M. Taniwaki, S. Ohwada and J. Yokota. (1997). Cloning of a human homolog of the yeast OGG1 gene that is involved in the repair of oxidative DNA damage. *Oncogene* **14**, 2857–2861.

V.I. Bruskov, A.I. Gaziev, L.V. Malakhova, IuA. Mantsygin and O.S. Morenkov, (1996). [Monoclonal antibodies to 8-oxo-2'-deoxyguanosine (8-hydroxyguanosine). Characteristics and use for determining DNA damage by active forms of oxygen]. *Biokhimiia* **61**, 737–744.

R.H. Burdon (1995). Superoxide and hydrogen peroxide in relation to mammalian cell proliferation. *Free Radic. Biol Med.* **18**, 775–794.

J. Cadet, J.L. Ravanat, and T. Douki (1997). Artifacts associated with the measurement of oxidized DNAbases. *Environmental and Health Perspective* **105**, 1034–1039.

K.C. Cheng, D.S. Cahill, H. Kasai, S. Nishimura and L.A. Loeb(1992). 8-Hydroxyguanine, an abundant form of oxidative DNA damage, causes G to T and A to C substitutions. *J Biol Chem.* **267**, 166–172.

H.G. Claycamp (1992). Phenol sensitization of DNA to subsequent oxidative damage in 8-hydroxyguanine assays. *Carcinogenesis* **13**, 1289–1292.

P. Degan, M.K. Shigenaga, E-M. Park, P.E. Alperin and B.N. Ames (1991). Immunoaffinity isolation of urinary 8-hydroxy-2'-deoxyguanosine and 8-hydroxyguanine and quantitation of 8-hydroxy-2'-deoxyguanosine in DNA by polyclonal antibodies . *Carcinogenesis* **12**, 865–871.

T.P.A. Devasagayam, S. Steenken, M.S.W. Obendorf, W.A. Schultz and H. Sies (1991). Formation of 8-hydroxy(deoxy)guanosine and generation of strand breaks at guanine residues in DNA by singlet oxygen. *Biochemistry* **30**, 6283–6289.

M., Dizdaroglu (1991). Chemical determination of free radical-induced damage to DNA. *Free Radic. Biol. Med.* **10**, 225–242.

J.M. Dypbukt, M. Ankarcrona, M. Burkitt, A. Sjoholm, K. Strom, S. Orrenius, and P. Nicotera (1994). Different prooxidant levels stimulate growth, trigger apoptosis, or produce necrosis of insulin-secreting RINm5F cells. The role of intracellular polyamines. *J Biol. Chem.* **269**, 30553–30560.

M. Erhola, S. Toyokuni, K. Okada, T. Tanaka, H. Hiai, H. Ochi, K. Uchida, T. Osawa, M.M. Nieminen, H. Alho and P. Kellokumpu-Lehtinen (1997a). Biomarker evidence of DNA oxidation in lung cancer patients: association of urinary 8-hydroxy-2'-deoxyguanosine excretion with radiotherapy, chemotherapy, and response to treatment. *FEBS Letters* **409**, 287–291.

R.A. Floyd, J.J. Watson, P.K. Wong, D.H. Altmiller and R.C. Rickard (1986). Hydroxyl free radical adducts of deoxyguanosine: sensitive detection and mechanisms of formation. *Free Radic. Res. Commun.* **1**, 163–172.

R.A. Floyd, M.S. West, K.L. Eneff and J.E. Schneider (1989). Methylene blue plus light mediates 8-hydroxyguanine formation in DNA. *Arch. Biochem. Biophys.* **273**, 106–111.

R.A. Floyd (1990). The role of 8-hydroxyguanine in carcinogenesis. *Carcinogenesis* **11**, 1447–1450.

C.G. Fraga, M.K. Shigenaga, J.W. Park, P. Degan and B.N. Ames (1990). Oxidative damage to DNA during aging: 8-hydroxy-2'-deoxyguanosine in rat organ DNA and urine. *Proc. Natl. Acad. Sci.* U S A, **87**, 4533–4537.

B. Halliwell (1993) Oxidative DNA damage: meaning and measurement. *In DNA and free radicals* (eds. B. Halliwell and O.I. Aruoma) Ellis Horwood Limited, Chichester, West Sussex, pp. 67–79.

B. Halliwell and J.M.C. Gutteridge (1989). *Free radicals in biology and medicine.* Oxford: Clarendon Press.

Y. Hattori, C. Nishigori, T. Tanaka, K. Uchida, O. Nikaido, T. Osawa, H. Hiai, S. Imamura and S. Toyokuni (1996). 8-Hydroxy-2'-deoxyguanosine is increased in epidermal cells of hairless mice after chronic UVB exposure. *J. Invest. Dermatol.* **107**, 733–737.

S-M. Hsu, N. Raine and H. Fanger (1981). A comparative study of the peroxidase-antiperoxidase method and avidin-biotin complex method for studying polypeptide hormones with radioimmunoassay antibodies. *Am. J. Clin. Pathol.* **75**, 734–738.

S. Inoue and S. Kawanishi (1995). Oxidative DNA damage induced by simultaneous generation of nitric oxide and superoxide. *FEBS Letters.* **371**, 86–88.

H. Kasai and S. Nishimura (1984). Hydroxylation of deoxyguanosine at the C-8 position by ascorbic acid and other reducing agents. *Nucleic Acids Res.* **12**, 2137–2145.

H. Kasai, Z. Yamaizumi, M. Berger and J. Cadet (1992). Photosensitized formation of 7,8-dihydro-8-oxo-2'-deoxyguanosine (8-hydroxy-2'-deoxyguanosine) in DNA by riboflavin: a non singlet oxygen mediated reaction. *J. Am. Chem. Soc.* **114**, 9692–9694.

Y. Kuchino, F. Mori, H. Kasai, H. Inoue, S. Iwai, K. Miura, E. Ohtsuka and S. Nishimura (1987). DNA templates containing 8-hydroxydeoxyguanosine are misread both at the modified base and at adjacent residues. *Nature* **327**, 77–79.

S. Lagorio, C. Tagesson, F. Forastiere, I. Iavarone, O. Axelson and A. Carere (1994). Exposure to benzene and urinary concentrations of 8- hydroxydeoxyguanosine, a biological marker of oxidative damage to DNA. *Occup. Environ. Med.* **51**, 739–743.

J. Leinonen, T. Lehtimagi, S. Toyokuni, K. Okada, T. Tanaka, H. Hiai, H. Ochi, P. Laippala, V. Rantalaiho, O. Wirta, A. Pasternack and H. Alho, H. (1997). New biomarker evidence of oxidative DNA damage in patients with non-insulin-dependent diabetes mellitus. *FEBS Letters* **417**, 150–152.

T. Lindahl (1993). Instability and decay of the primary structure of DNA . *Nature* **362**, 709–715.

S. Loft, K. Vistisen, M. Ewertz, A. Tjonneland, K. Overvad and H.E. Poulsen (1992). Oxidative DNA damage estimated by 8-hydroxydeoxyguanosine excretion in humans: influence of smoking, gender and body mass index. *Carcinogenesis* **13**, 2241–2247.

S. Loft, A. Fischer-Nielsen, I.B. Jeding, K. Vistisen and H.E. Poulsen (1993). 8-Hydroxydeoxyguanosine as a urinary biomarker of oxidative DNA damage. *J Toxicol Environ Health* **40**, 391–404.

L.G. Luna (1968). *Manual of histological staining methods of the Armed Forces Institute of Pathology.* New York: McGraw-Hill.

J. Lunec, K. Herbert, S. Blount, H.R. Griffiths and P. Emery (1994). 8-Hydroxydeoxyguanosine. A marker of oxidative DNA damage in systemic lupus erthematosus. *FEBS Letters* **348**, 131–133.

J.T. Lutgerink, E. Graaf, J.B. Hoebee, H.F.C. Stavenuitez, J.G. Westra and E. Kriek (1992). Detection of 8-hydroxyguanine in small amount of DNA by ^{32}P postlabeling . *Analytical Biochemistry* **201**, 127–133.

J. Musarrat and A.A. Wani (1994). Quantitative immunoanalysis of promutagenic 8-hydroxy-2'-deoxyguanosine in oxidized DNA. *Carcinogenesis* **15**, 2037–2043.

K. Okamura, T. Doi, K. Hamada, M. Sakurai, Y. Yoshioka, R. Mitsuzono, T. Migita, S. Sumida and Y. Sugawa-Katayama (1997). Effect of repeated exercise on urinary 8-hydroxy-deoxyguanosine excretion in humans. *Free Radical Research* **26**, 507–514.

T. Osawa, A. Yoshida, S. Kawakishi, K. Yamashita and H. Ochi (1995) Protective role of dietary antioxidants in oxidative stress. *In Oxidative stress and aging* (eds. R.G. Cutler, L. Packer, J. Bertram and A. Mori) Berkhauser Verlag, Basel, pp. 367–377.

E.M. Park, M.K. Shigenaga, P. Degan, T.S. Korn, J.W. Kitzler J.W., Wehr, C.M. Kolachana and B.N. Ames (1992). Assay of excised oxidative DNA lesions: isolation of 8-oxoguanine and its nucleoside derivatives from biological fluids with a monoclonal antibody column. *Proc. Natl. Acad. Sci.* U S A, **89**, 3375–3379.

J.P. Radicella, C. Dherin, C. Desmaze, M.S. Fox and S. Boiteux (1997). Cloning and characterization of hOGG1, a human homolog of the OGG1 gene of Saccharomyces cerevisiae. *Proc. Natl. Acad. Sci.* U S A, **94**, 8010–8015.

T. Roldan-Arjona, Y.F. Wei, K.C. Carter, A. Klungland, C. Anselmino, R.P. Wang, M. Augustus and T. Lindahl (1997). Molecular cloning and functional expression of a human cDNA encoding the antimutator enzyme 8-hydroxyguanine-DNA glycosylase. *Proc. Natl. Acad. Sci.* U S A, **94**, 8016–8020.

T.A. Rosenquist, D.O. Zharkov and A.P. Grollman (1997). Cloning and characterization of a mammalian 8-oxoguanine DNA glycosylase. *Proc. Natl. Acad. Sci.* U S A, **94**, 7429–7434.

K. Sakumi, M. Furuichi, T. Tsuzuki, T. Kakuma, S. Kawabata, H. Maki and M. Sekiguchi (1993). Cloning and expression of cDNA for a human enzyme that hydrolyzes 8-oxo- dGTP, a mutagenic substrate for DNA synthesis. *J. Biol. Chem.* **268**, 23524–23530.

S. Shibutani, M. Takeshita and A.P. Grollman (1991). Insertion of specific bases during DNA synthesis past the oxidation-damaged base 8-oxodG. *Nature* **349**, 431–434.

M.K. Shigenaga, C.J. Gimeno and B.N. Ames (1989). Urinary 8-hydroxy-2'- deoxyguanosine as a biological marker of in vivo oxidative DNA damage. *Proc. Natl. Acad. Sci.* U S A, **86**, 9697–9701.

M.M. Slupska, C. Baikalov, W.M. Luther, J.H. Chiang, Y.F. Wei and J.H. Miller (1996). Cloning and sequencing a human homolog (hMYH) of the Escherichia coli mutY gene whose function is required for the repair of oxidative DNA damage. *J. Bacteriol.* **178**, 3885–3892.

T.P. Szatrowski and C.F. Nathan (1991). Production of large amounts of hydrogen peroxide by human tumor cells. *Cancer Res.* **51**, 794–798.

C. Tagesson, D. Chabiuk, O. Axelson, B. Baranski, J. Palus and K. Wyszynska (1993). Increased urinary excretion of the oxidative DNA adduct, 8- hydroxydeoxyguanosine, as a possible early indicator of occupational cancer hazards in the asbestos, rubber, and azo-dye industries. *Pol. J. Occup. Med. Environ. Health* **6**, 357–368.

C. Tagesson, M. Kallberg, C. Klintenberg and H. Starkhammar (1995). Determination of urinary 8-hydroxydeoxyguanosine by automated coupled- column high performance liquid chromatography: a powerful technique for assaying in vivo oxidative DNA damage in cancer patients. *Eur. J. Cancer* **31**A, 934–940.

S. Toyokuni, S. Okada, S. Hamazaki, Y. Minamiyama, Y. Yamada, P. Liang, Y. Fukunaga and O. Midorikawa (1990). Combined histochemical and biochemical analysis of sex hormone dependence of ferric nitrilotriacetate-induced renal lipid peroxidation in ddY mice. *Cancer Res.* **50**, 5574–5580.

S. Toyokuni, T. Mori and M. Dizdaroglu (1994a). DNA base modifications in renal chromatin of Wistar rats treated with a renal carcinogen, ferric nitrilotriacetate. *Int. J. Cancer* **57**, 123–128.

S. Toyokuni, K. Uchida, K. Okamoto, Y. Hattori-Nakakuki, H. Hiai and E.R. Stadtman (1994b). Formation of 4-hydroxy-2-nonenal-modified proteins in the renal proximal tubules of rats treated with a renal carcinogen, ferric nitrilotriacetate. *Proc. Natl. Acad.Sci. USA*, **91**, 2616–2620.

S. Toyokuni, K. Okamoto, J. Yodoi and H. Hiai (1995). Persistent oxidative stress in cancer. *FEBS Letters.* **358**, 1–3.

S. Toyokuni (1996). Iron-induced carcinogenesis: the role of redox regulation. *Free Radic. Biol. Med.* **20**, 553–566.

S. Toyokuni, T. Tanaka, Y. Hattori, Y. Nishiyama, H. Ochi, H. Hiai, K. Uchida and T. Osawa (1997). Quantitative immunohistochemical determination of 8-hydroxy-2'-deoxyguanosine by a monoclonal antibody N45.1: its application to ferric nitrilotriacetate-induced renal carcinogenesis model. *Lab Invest.* **76**, 365–374.

M.L. Wood, M. Dizdaroglu, E. Gajewski and J.M. Essigmann (1990). Mechanistic studies of ionizing radiation and oxidative mutagenesis: genetic effect of a single 8-hydroxyguanine (7-hydro-8-oxoguanine) residue inserted at a unique site in a viral genome. *Biochemistry* **29**, 7024–7032.

T. Yamamoto, K. Hosokawa, T. Tamura, H. Kanno, M. Urabe and H. Honjo (1996). Urinary 8-hydroxy-2'-deoxyguanosine (8-OHdG) levels in women with or without gynecologic cancer. *J. Obstet. Gynaecol. Res.* **22**, 359–363.

A. Yarborough, Y.J. Zhang, T.M. Hsu and R.M. Santella (1996). Immunoperoxidase detection of 8-hydroxydeoxyguanosine in aflatoxin B1-treated rat liver and human oral mucosal cells. *Cancer Res.* **56**, 683–688.

B. Yin, R.M. Whyatt, F.P. Perera, M.C. Randall, T.B. Cooper and R.M. Santella (1996). Determination of 8-hydroxydeoxyguanosine by an immunoaffinity chromatography-monoclonal antibody-based ELISA. *Free Radic. Biol. Med.* **18**, 1023–1032.

33

MECHANISTIC STUDIES OF RADIATION-INDUCED DNA DAMAGE

Clemens von Sonntag

Max-Planck-Institut für Strahlenchemie
Stiftstr. 34-36, D-46468, Mülheim/Ruhr, Germany

1. ABSTRACT

In cells, the damage caused by ionizing radiation is mainly due to free-radical-induced alterations of their DNA. The reactions involved have been studied on a model level: DNA and DNA components in aqueous solution. OH radicals and solvated electrons yield short-lived intermediates such as nucleobase-OH-adduct radicals, radicals at the sugar moiety (and consequently strand breaks) and base radical anions. Their properties and reaction kinetics have been studied by pulse radiolysis and product analysis. Some aspects of the effect of O_2 in potentially enhancing the damage and the protective effects of thiols and bisbenzimidazole derivatives are reported.

2. INTRODUCTION

There are considerable differences between the absorption of a near-UV photon and ionizing radiation by, for example, a living cell. The component that absorbs most strongly the near-UV radiation is the cell's DNA. In contrast, the energy of ionizing radiation (e.g. high-energy electrons, α-particles, γ-quanta) is absorbed approximately proportionally to the weight of a given component, i.e. in a cell the membrane, the cytoplasm and the nucleus absorb the ionizing radiation in proportion to their relative weight. Nevertheless, the damage of the membrane and the cytoplasm are of little biological consequence as compared to DNA damage (*cf.* von Sonntag, 1987). There is an additional marked difference between UV and ionizing radiation: while UV-radiation is absorbed at random and in proportion to the absorption coefficient, the ionizing radiation is deposited in packages of typically 100eV in a sequence of ionization events which may be quite distant from one another but also can locally overlap leading to clustered lesions.

As a consequence of this, DNA damage can be caused by energy absorption by the DNA itself (direct effect) or via its reaction with the radicals formed by the decomposition of

Advances in DNA Damage and Repair, edited by Dizdaroglu and
Karakaya. Kluwer Academic / Plenum Publishers, New York, 1999.

energy in the water which surrounds the DNA (indirect effect) (*cf.* von Sonntag, 1987). The chemical consequences of the direct effect are not yet well understood in detail, but it is reasonable to assume that the ionizing radiation gives rise to DNA radical cations and electrons, whereby the radical cations may not only be formed in their ground states, but also in an excited state. At this point, the term DNA radical cation is rather unspecified, since the electron can be ejected from a base, the sugar moiety or a phosphate group. In addition, electron transfer from a component with a low ionization potential (e.g. guanine) to the radical cation of another moiety may occur. On the other hand, the radiolysis of water is well understood. The water radical cation loses rapidly a proton and hence forms an $^{\bullet}OH$ radical. After thermalization, the electron becomes solvated. In addition, excited water molecules decompose into $^{\bullet}OH$ radicals and $H^{\bullet}$ atoms. As a consequence, $^{\bullet}OH$ radicals, solvated electrons and $H^{\bullet}$ atoms are the primary free-radical species [in the case of sparsely ionizing radiation such as high-energy electrons or γ-quanta $G(^{\bullet}OH) \approx G(e_{aq}^-) \approx 2.8 \times 10^{-7}$ mol J^{-1}; $G(H^{\bullet}) \approx 0.6 \times 10^{-7}$ mol J^{-1}]. Some primary radicals recombine on the spot where the energy is deposited, thereby causing the formation of H_2O_2 and H_2, *cf.* reaction 1.

$$H_2O \xrightarrow[\text{radiation}]{\text{ionizing}} e_{aq}^-, {}^{\cdot}OH, H^{\cdot}, H^+, H_2O_2, H_2 \tag{1}$$

$$e_{aq}^- + N_2O + H_2O \rightarrow {}^{\cdot}OH + OH^- + N_2 \tag{2}$$

$$e_{aq}^- + S_2O_8^{2-} \rightarrow SO_4^{\bullet-} + SO_4^{2-} \tag{3}$$

The indirect effect can be mimicked by the radiolysis of DNA in dilute aqueous solution. For the investigation of particular aspects, e. g. the reaction of $^{\bullet}OH$ radicals with the thymine moiety, the DNA macromolecule itself is too complex, and it is useful to study smaller subunits, e.g. thymine, thymidine or thymidylic acid Such mechanistic studies will be reported in the present contribution. They are based on product studies and their kinetics have been followed by pulse radiolysis.

3. THE PULSE RADIOLYSIS TECHNIQUE

In a pulse radiolysis experiment a sample contained in a cuvette is submitted to high-energy electrons from an electron accelerator within ≤ 1 μs. The formation and decay of intermediates can be followed by various techniques. For the present systems detection by UV/Vis spectroscopy as well as by conductometry are the methods of choice (*cf.* von Sonntag and Schuchmann, 1994).

For the investigation of the reactions of the three water radicals, $^{\bullet}OH$, e_{aq}^- and $H^{\bullet}$, it would be useful to isolate each of these three reactive intermediates. Under certain conditions this can be done. Solvated electrons may be converted by N_2O into further $^{\bullet}OH$ radicals (reaction 2). The system then contains largely $^{\bullet}OH$ radicals (90%) and only a few $H^{\bullet}$ atoms (10%). When *t*-butanol is added in excess, the $^{\bullet}OH$ radicals are scavenged by *t*-butanol and the $H^{\bullet}$ atoms remain to react with the given substrate. For the study of the reactions of solvated electrons $^{\bullet}OH$ radicals and $H^{\bullet}$ atoms can be scavenged by 2-propanol (*cf.* von Sonntag, 1987).

$$e_{aq}^- + N_2O + H_2O \rightarrow {}^{\cdot}OH + OH^- + N_2 \tag{2}$$

In the reaction of the solvated electron with peroxodisulfate an $SO_4^{\bullet-}$ radical is formed (reaction 3). This radical has strongly oxidizing properties and can be used to generate nucleobase radical cations. It hence allows to mimic to some extent reactions that are also initiated by the direct effect (*cf.* von Sonntag, 1987).

$$e_{aq}^{-} + S_2O_8^{2-} \rightarrow SO_4^{\bullet-} + SO_4^{2-} \quad (3)$$

4. REACTIONS OF $^{\bullet}$OH AND $SO_4^{\bullet-}$ RADICALS WITH PYRIMIDINES AND PURINES

The rate of the addition of OH radicals to double bonds is close to diffusion-controlled, but due to their pronounced electrophilicity they show a marked regioselectivity. When they react with the pyrimidines, two kinds of radicals are formed. Upon addition to the C^5-position (reaction 4) a radical with reducing properties is formed which can be readily detected by its rapid reaction with tetranitromethane (reaction 5; the nitroform anion is characterized by a strong absorption at 350 nm). The C^6-OH-radical-adduct formed in reaction 6 has oxidizing properties (note that one of its mesomeric forms is an *O*-centered radical) and its yield can be assessed by its oxidation of *N,N,N',N'*-tetramethyl-*p*-phenylenediamine to its corresponding radical cation (reaction 7) (Fujita and Steenken, 1981; Hazra and Steenken, 1983).

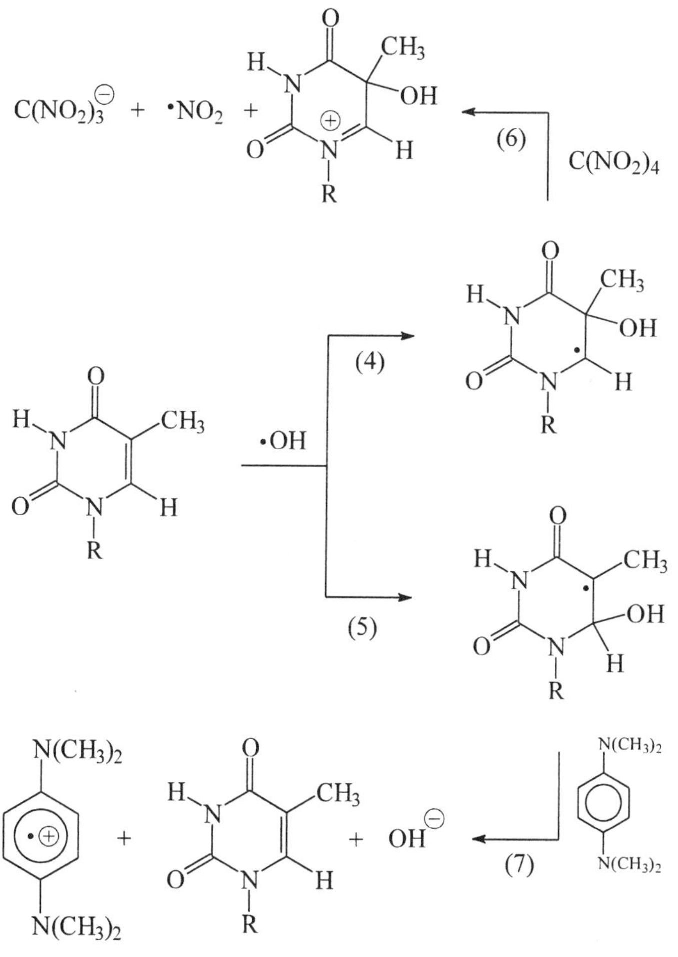

Using this approach, Fujita and Steenken, (1981) and Hazra and Steenken, (1983) have determined the yields of $^{\bullet}$OH-radical attack at the C^5- and C^6-positions for some substituted pyrimidines, and the noticeable regioselectivity of this radical can be seen from these data. It is noted that H-abstraction from methyl groups (where applicable) remains a minor process.

The $SO_4^{\bullet -}$ radical is generally believed in its reactions with pyrimidines and purines to produce the corresponding radical cations (*cf.* reaction 8), although an adduct cannot be excluded as a short-lived intermediate. In the case of N^1-substituted thymine-derivatives its fate has been elucidated by pulse radiolysis combining optical and conductometric detection techniques. As expected, it is a much stronger acid than thymidine itself (pK_a = 9.8) and deprotonates (reversibly) at N^3 already at low pH (pK_a = 3.6, equilibrium 9) (Deeble et al., 1990). In competition it deprotonates (irreversibly) at the methyl group (reaction 10) or reacts with water (reaction 11). As a result two radicals are formed: the C^5-$CH_2^{\bullet}$ radical and the C^5-OH-adduct radical. In the presence of O_2, the former is finally converted into 5-hydroxymethyl- and 5-formyluracil. In γ-irradiated cells these products have been observed as DNA lesions (Teebor et al., 1988), and it is conceivable that they have a thymine-derived radical cation as a precursor, i.e. they may be formed as result of the direct effect.

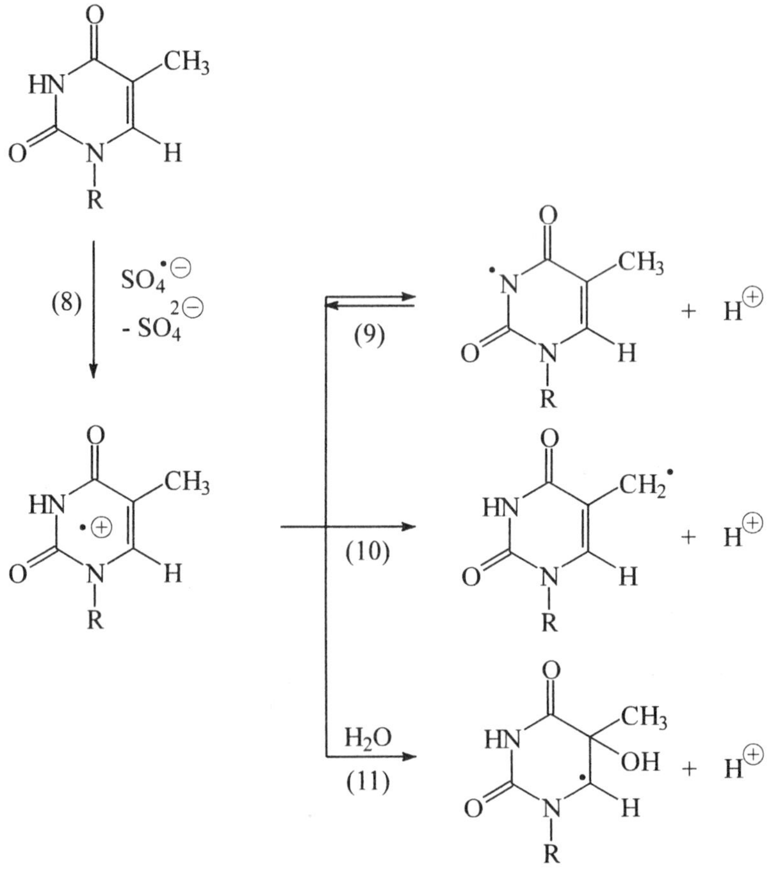

(8) $SO_4^{\bullet\ominus}$, $-SO_4^{2\ominus}$

(9) + $H^{\oplus}$

(10) + $H^{\oplus}$

H_2O (11) + $H^{\oplus}$

In the presence of O_2, most radicals are converted into the peroxyl radicals at close to diffusion-controlled rates (for a review see von Sonntag and Schuchmann, 1997). A case in point is the uracil (Schuchmann and von Sonntag, 1983) or thymine C^5-OH-adduct radical (*cf.* reaction 12). In analogy to similar radicals derived from cyclic anhydrides of amino acids (Mieden et al., 1993) these peroxyl radicals have relatively low pK_a values (equilibrium 13; for the strong electron-withdrawing effect of the peroxyl radical function see Schuchmann et al., 1989). The deprotonated peroxyl radical eliminates a superoxide radical (reaction 14) thereby forming an unstable isopyrimidine (Al-Sheikhly et al., 1984). The subsequent reactions of these interesting intermediates has also been elucidated by pulse radiolysis (Schuchmann et al., 1984). It is noted that such intermediates cannot be formed, when the pyrimidine is bound to the sugar moiety of DNA. This is quite an important aspect, because it clearly shows the limitations of model systems, or to put it somewhat more positively, a detailed kinetic analysis of the mechanism is required in order to allow to extrapolate a given finding to the next-higher level of complexity situation. For the present system it can be concluded that when the deprotonation is inhibited in acid solution, superoxide elimination no longer can occur and the peroxyl radicals have to decay in the same way as they would if they were bound to DNA, and indeed under such conditions ring fragmentation products are found (e.g. *N*-formylhydantoin from uracil, *cf.* Schuchmann and von Sonntag, 1983) which are also typical for more complex systems.

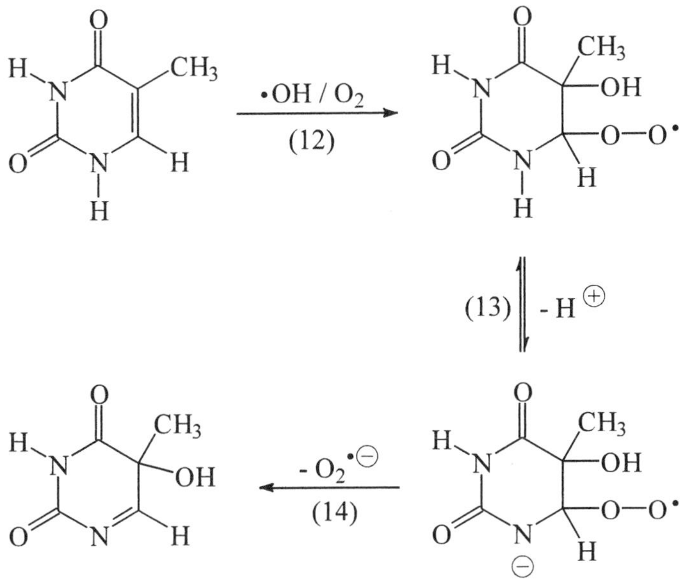

The $^{\bullet}OH$-adducts of purines in their redox properties do not show differences as marked as those of the pyrimidines. In addition, we still lack the wealth of information on the nature of the products and their yields that is available in the case of e.g. thymine; i.e. a material balance is still missing. This is partially due to damage amplification reactions, e.g. in the case of 2'-deoxyadenosine, which are not yet adequately understood at present (von Sonntag, 1994). The purine radical cations which may be produced with the help of the $SO_4^{\bullet-}$ radical are in equilibrium with their heteroatom-deprotonated forms (Steenken,

1989). These intermediates do not react with O_2, but do readily react with the superoxide radical (von Sonntag, 1994).

5. REACTIONS OF SOLVATED ELECTRONS WITH PYRIMIDINES AND PURINES

The solvated electron reacts with the pyrimidines and purines at close to diffusion-controlled rates (*cf.* reaction 15). In the case of thymine this electron adduct is more readily protonated at the heteroatom (equilibrium 16; $pK_a = 6.9$) than at C^6 (reaction 17). While the former reaction is reversible, the latter is not (within a reasonable pH range). When O_2 reacts with the radical anion or its heteroatom-protonated form, the original base is reformed (reactions 18 and 19) (Deeble and von Sonntag, 1987). However, O_2-addition to the C^6-centered radical leads to an irreversible fixation of the oxygen (reaction 20) and subsequent destruction of the thymine moiety (Deeble and von Sonntag, 1987).

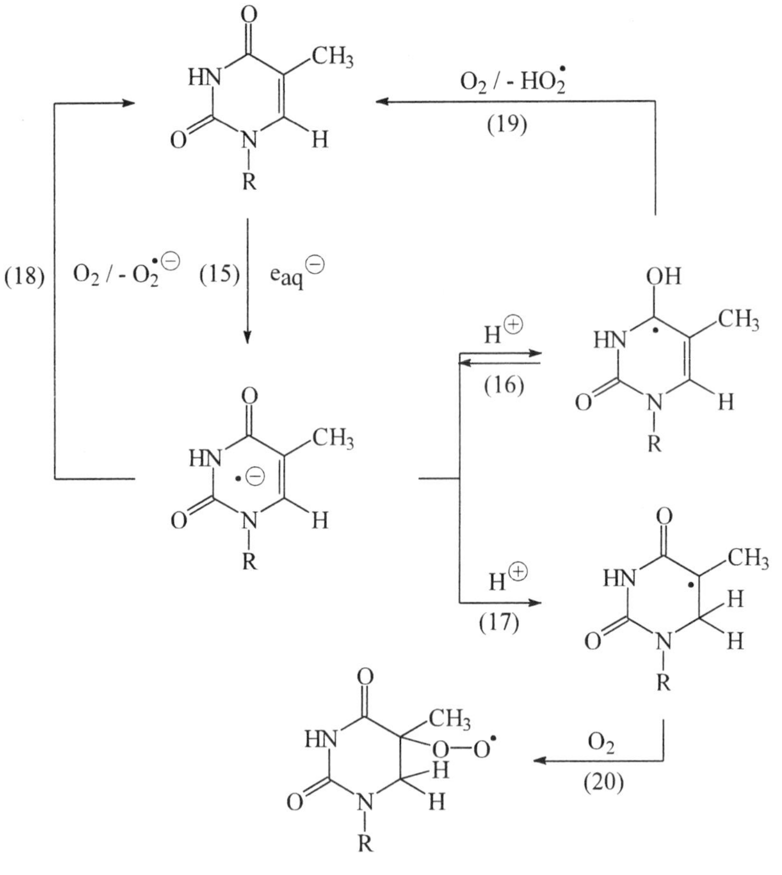

The electron adducts of cytosine and the purines are much more readily protonated at a carbon atom than thymine and (some of) these early intermediates undergo subsequent transformation reactions (Steenken, 1989; von Sonntag, 1991; Aravindakumar et al., 1994).

6. OH-RADICAL-INDUCED DNA STRAND BREAKAGE

A small percentage (≈ 15–20 %) of the $^{\bullet}OH$ radicals react with the sugar moiety by H-abstraction. Considerable information is available from the studies of model compounds about the subsequent reactions β-phosphatoalkyl radicals (Behrens et al., 1978; Behrens et al., 1982; Schuchmann et al., 1995). When adequately substituted (e.g. by an alkoxy group) to accommodate the positive charge they can eliminate the phosphate group heterolytically thereby forming an alkyl radical cation. In DNA the 4'-radical has such properties (*cf.* reaction 21) (Dizdaroglu et al., 1975; Beesk et al., 1979).

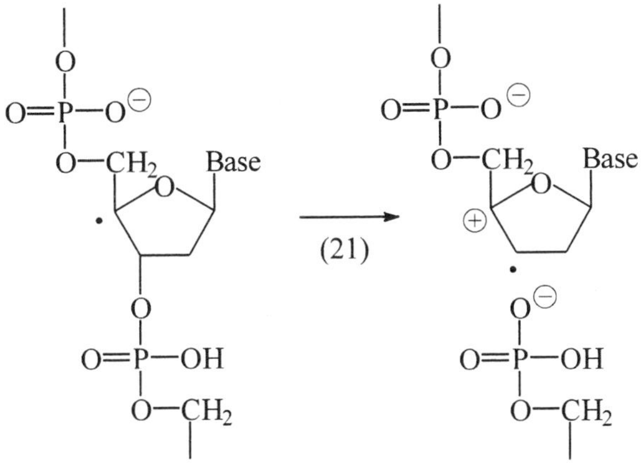

(21)

Here, this heterolytic phosphate elimination leads to a DNA strand break. In single-stranded polynucleotides such as poly(U) (Bothe and Schulte-Frohlinde, 1982) or single-stranded DNA (Bothe et al., 1983) the rate of strand breakage can be followed by pulse conductometry (*cf.* reactions 22–24). In large charged polymers some of the counterions are condensed at the surface of the polymer and thus do not contribute to the overall conductivity. When a scission has occurred and the broken parts diffuse apart, the overall electric field is reduced, and counterions are released (*cf.* reaction 24). In single-stranded DNA as many as 16 K^+ ions are released per strand break (Bothe et al., 1983).

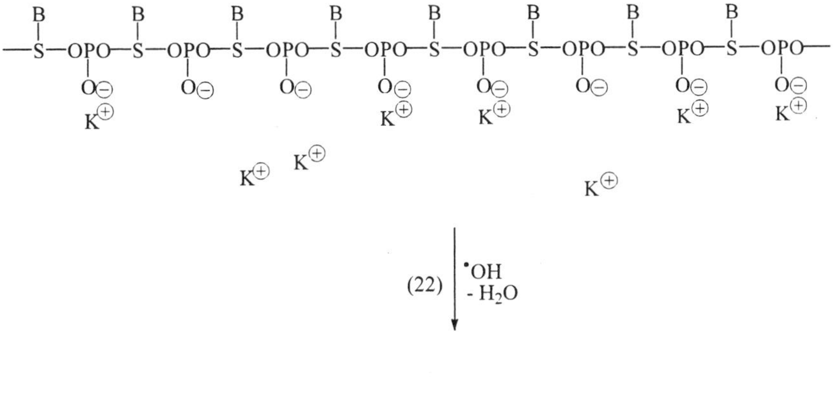

(22) $^{\bullet}OH$, $-H_2O$

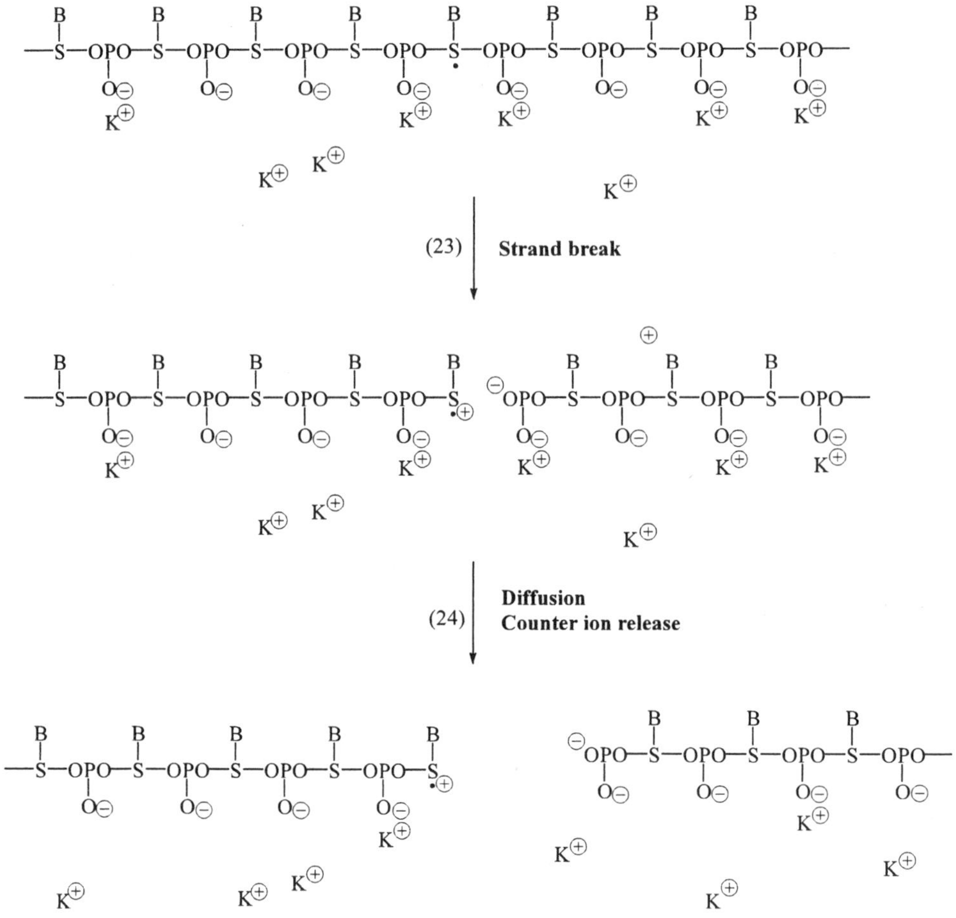

(23) **Strand break**

(24) **Diffusion**
Counter ion release

The measured rate of counter ion release varies between $\approx$ 300 s^{-1} (at pH 3) and $\approx$ 0.9 s^{-1} (at pH 8) (Schulte-Frohlinde et al., 1985). The observed kinetics cannot be linked to the reaction of the OH radical with the poly(U) molecule since its rate constant is 1.5×10^9 dm^3 mol^{-1} s^{-1} (Bothe and Schulte-Frohlinde, 1982), i.e. reaction 21 has occurred on the early μs time-scale in these experiments. This strong pH variation (which is typical for a process of the type shown in reaction 21, *cf.* Behrens et al., 1978; Behrens et al., 1982) supports the view that the observed rate of counterion release has to be associated with the rate of strand breakage (reaction 23) and not with the diffusion of the broken segments (reaction 24).

7. DAMAGE AMPLIFICATION REACTIONS

Most radiation-induced DNA damage is repaired by cellular repair enzymes. There is increasing evidence that clustered lesions ("locally multiply damaged sites; LMDS", Ward et al., 1990) contribute to a very large extent to the persisting damage. These clustered lesions can be caused by the enhanced energy deposition within a short DNA segment and its aqueous environment. This effect can be enhanced by damage amplification reactions, where via a radical transfer reaction one radical can cause the formation of more than one damaged site. The first kind of such a reaction was the formation of a chain break and concomitant base release by a base radical in poly(U). (Lemaire et al., 1984; Deeble and von Sonntag, 1984; Deeble et al., 1986) More recently in oligonucleotides two damaged neighboring bases (caused by the attack of one $^\bullet$OH radical) have been detected (Box et al. 1993). In the $^\bullet$OH-radical-induced reactions of 2'-deoxyadenosine damage-amplification reactions also occur (von Sonntag, 1994). Here it has been clearly shown (unpublished results) that the lifetime of the radicals is of a major importance. The lifetime of radicals sited on charged polymers is prolonged with respect to their bimolecular decay (*cf.* Ulanski et al., 1997), thus setting conditions that are favorable for damage amplification reactions. Peroxyl radicals are known to undergo chain reactions (von Sonntag and Schuchmann, 1997) and it may well be possible that in DNA the damage amplification reactions are more prominent in the presence of O_2 (see below).

8. CHEMICAL REPAIR BY THIOLS

The radiation-induced damage to cells is enhanced in the presence of O_2 ("*Oxygen effect*"; *cf.* von Sonntag, 1987). This is due to a competition between the "*repair*" of the radiation-induced DNA damage by glutathione (GSH). This thiol can reach cellular concentrations close to 10^{-2} mol dm^{-3}. In these "*repair*" reactions the DNA-derived radicals are reduced by H-donation (e.g. reaction 25).

GSH / - GS$^\bullet$ (25)

Thus in most cases this H-donation will not fully restore the original DNA. That this kind of damage is nevertheless less deleterious for the cell may be due to the fact that it can be restored by the cellular repair enzymes more effectively than the damage caused by the peroxyl radical reactions. On the other hand, it may also be speculated that this H-donation prevents damage amplification reactions and thus reduces the severity of clustered lesions.

The rate at which a given radical is reduced by a thiol depends strongly on its nature. This H-donation reaction is most likely a very complex sequence of reactions. Surprisingly, electron-rich ("reducing") radicals are more readily reduced than electron-deficient ("oxidizing") radicals (von Sonntag, 1987; Akhlaq et al., 1987). The latter are, however, rapidly reduced by the thiolate ions via electron donation (Simic and Hunter, 1986; Akhlaq et al., 1987). This is in line with the observation that peroxyl radicals also react very slowly (ca. 10^2 dm^3 mol^{-1} s^{-1}) with thiols (Lal et al., 1997; Hildenbrand and Schulte-Frohlinde, 1997).

9. RADIATION PROTECTION BY BISBENZIMIDAZOLE DERIVATIVES, E.G. HOECHST 33258

In contrast to thiols many bisbenzimidazole derivatives, e.g. Hoechst 33258, bind strongly to the minor groove of DNA. They have been shown to be effective radiation protectors (Smith and Anderson, 1984; Martin and Denison, 1992).

Hoechst 33258

The bisbenzimidazole derivatives are good electron donors, and in contrast to the thiols they act by electron donation. Their preferred targets are hence oxidizing radicals such as the heteroatom-centered purine radicals and peroxyl radicals (Adhikary et al., 1997). When Hoechst 33228 is one-electron-oxidized an *N*-centered radical is formed after deprotonation. Like the same type of radical formed in the case of tryptophan it does not react with O_2. When bound to double-stranded DNA Hoechst 33258 may protect the DNA by reacting with $^{\bullet}$OH radicals, but additional protection is provided by subsequent (slower) electron donation to DNA radicals. This reaction can be followed by pulse radiolysis either by the build-up of the Hoechst-33258-radical absorption or the concomitant bleaching of the Hoechst 33258 absorption itself. In the presence of O_2, this intramolecular electron transfer occurs with a (first) half-life of 1–2 ms (Adhikary et al., 1997). As the reaction proceeds its rate slows down, since the remaining damaged sites become more difficult to be reached by the (tightly bound) reductant.

REFERENCES

A. Adhikary, E. Bothe, C. von Sonntag, and V. Jain (1997) DNA protection by bisbenzimadozol derivative Hoechst 33258: model studies on the nucleotide level. *Radiation Res.*, **148**, 493–494.

A. Adhikary, E. Bothe, V. Jain, and C. von Sonntag (1997) Inhibition of radiation-induced DNA strand breaks by Hoechst 33258: OH-radical scavenging and DNA radical quenching. *Radioprotection* **32**, C1–89-C1–90.

M. S. Akhlaq, S. Al-Baghdadi, and C. von Sonntag (1987) On the attack of hydroxyl radicals on polyhydric alcohols and sugars and the reduction of the so formed radicals by 1,4-dithiothreitol. *Carbohydr. Res.* **164,** 71–83.

M. I. Al-Sheikhly, A. Hissung, H.-P. Schuchmann, M. N. Schuchmann, C. von Sonntag, A. Garner, and G. Scholes (1984) Radiolysis of dihydrouracil and dihydrothymine in aqueous solutions containing oxygen; first- and second-order reactions of the organic peroxyl radicals; the role of isopyrimidines as intermediates. *J. Chem. Soc. Perkin Trans. II*, 601–608.

C. T. Aravindakumar, H. Mohan, M. Mudaliar, B. S. M. Rao, J. P. Mittal, M. N. Schuchmann, and C. von Sonntag (1994) Addition of e_{aq}^- and H atoms to hypoxanthine and inosine and the reactions of α-hydroxyalkyl radicals with purines. A pulse radiolysis and product study. *Int. J. Radiat. Biol.* **66,** 351–365.

F. Beesk, M. Dizdaroglu, D. Schulte-Frohlinde, and C. von Sonntag (1979) Radiation-induced DNA strand breaks in deoxygenated aqueous solution. The formation of altered sugars as end groups. *Int. J. Radiat. Biol.* **36,** 565–576.

G. Behrens, G. Koltzenburg, A. Ritter, and D. Schulte-Frohlinde (1978) The influence of protonation or alkylation of the phosphate group on the e.s.r. spectra and on the rate of phosphate elimination from 2-methoxyethyl phosphate 2-yl radicals. *Int. J. Radiat. Biol.* **33,** 163–171.

G. Behrens, G. Koltzenburg, and D. Schulte-Frohlinde (1982) Model reactions for the degradation of DNA-4'radicals in aqueous solution. Fast hydrolysis of α-alkoxyalkyl radicals with a leaving group in β-position followed by radical rearrangement and elimination reactions. *Z. Naturforsch.* **37c,** 1205–1227.

E. Bothe, G.A. Qureshi and D. Schulte-Frohlinde (1983) Rate of OH radical induced strand break formation in single stranded DNA under anoxic conditions. An investigation in aqueous solutions using conductivity methods. *Z. Naturforsch.* **38c,** 1030–1042.

E. Bothe and D. Schulte-Frohlinde (1982) Release of K^+ and H^+ from poly U in aqueous solution upon γ and electron irradiation. Rate of strand break formation in poly U. Z. *Naturforsch.* **37c,** 1191–1204.

H.C. Box, E.E. Budzinski, H.G. Freund, M.S. Evans, H.B. Patrzyc, J.C. Wallace, and A.E. Maccubbin (1993) Vicinal lesions in X-irradiated DNA? *Int. J. Radiat. Biol.* **64,** 261–263.

D.J. Deeble, D. Schulz, and C. von Sonntag (1986) Reactions of OH radicals with poly(U) in deoxygenated solutions: sites of OH radical attack and the kinetics of base release. *Int. J. Radiat. Biol.* **49,** 915–926.

D.J. Deeble, M.N. Schuchmann, S. Steenken, and C. von Sonntag (1990) Direct evidence for the formation of thymine radical cations from the reaction of $SO_4^{\bullet-}$ with thymine derivatives: a pulse radiolysis study with optical and conductance detection. *J. Phys. Chem.* **94,** 8186–8192.

D.J . Deeble and C. von Sonntag (1984) γ-Radiolysis of poly(U) in aqueous solution. The role of primary sugar and base radicals in the release of undamaged uracil. *Int. J. Radiat. Biol.* **46,** 247–260.

D.J. Deeble and C. von Sonntag (1987) Radioprotection of pyrimidines by oxygen and sensitization by phosphate: a feature of their electron adducts. *Int. J. Radiat. Biol.* **51,** 791–796.

M. Dizdaroglu, C. von Sonntag, and D. Schulte-Frohlinde (1975) Strand breaks and sugar release by γ-irradiation of DNA in aqueous solution. *J. Am. Chem. Soc.* **97,** 2277–2278.

S. Fujita and S. Steenken (1981) Pattern of OH radical addition to uracil and methyl- and carboxyl-substituted uracils. Electron transfer of OH adducts with N,N,N',N'-tetramethyl-p-phenylenediamine and tetranitromethane. *J. Am. Chem. Soc.* **103,** 2540–2545.

D. K. Hazra and S. Steenken (1983) Pattern of OH radical addition to cytosine and 1-, 3-, 5-, and 6-substituted cytosines. Electron transfer and dehydration reactions of the OH adducts. *J. Am. Chem. Soc.* **105,** 4380–4386.

K. Hildenbrand and D. Schulte-Frohlinde (1997) Time-resolved EPR studies on the reaction rates of peroxyl radicals of poly(acrylic acid) and of calf thymus DNA with glutathione. Reexamination of a rate constant for DNA. *Int. J. Radiat. Biol.*, **71** 377–385.

M. Lal, R. Rao, X. Fang, H.-P. Schuchmann, and C. von Sonntag (1997) Radical-induced oxidation of dithiothreitol in acidic oxygenated aqueous solution: a chain reaction. *J. Am. Chem. Soc.* **119,** 5735–5739.

D. G. E. Lemaire, E. Bothe, and D. Schulte-Frohlinde (1984) Yields of radiation-induced main chain scission of poly U in aqueous solution: strand break formation via base radicals. *Int. J. Radiat. Biol.* **45,** 351–358.

R. F. Martin and L. Denison (1992) DNA lignads as radiomodifiers: studies with minor-groove binding bisbenzimidazoles. *Int. J. Radiat. Oncology Biol. Phys.* **23,** 579–584.

O. J. Mieden, M. N. Schuchmann, and C. von Sonntag (1993) Peptide peroxyl radicals: base-induced $O_2^{\bullet-}$ elimination versus bimolecular decay. A pulse radiolysis and product study. *J. Phys. Chem.* **97,** 3783- 790.

M. N. Schuchmann, M. Al-Sheikhly, C. von Sonntag, A. Garner, and G. Scholes (1984) The kinetics of the rearrangement of some isopyrimidines to pyrimidines studied by pulse radiolysis. *J. Chem. Soc. Perkin Trans. II*, 1777–1780.

M. N. Schuchmann, H.-P. Schuchmann, and C. von Sonntag (1989) The pK_a value of the $^{\bullet}O_2CH_2CO_2H$ radical: the Taft s* constant of the $-CH_2O_2^{\bullet}$ group. *J. Phys. Chem.* **93,** 5320–5323.

M. N. Schuchmann, M. L. Scholes, H. Zegota, and C. von Sonntag (1995) Reaction of hydroxyl radicals with alkyl phosphates and the oxidation of phosphatoalkyl radicals by nitro compounds. *Int. J. Radiat. Biol.* **68,** 121–131.

M. N. Schuchmann and C. von Sonntag (1983) The radiolysis of uracil in oxygenated aqueous solutions. A study by product analysis and pulse radiolysis. *J. Chem. Soc. Perkin Trans. II*, 1525–1531.

D. Schulte-Frohlinde, J. Opitz, H. Görner, and E. Bothe (1985) Model studies for the direct effect of high-energy irradiation on DNA. Mechanism of strand break formation induced by photoionization of poly U in aqueous solution. *Int. J. Radiat. Biol.* **48,** 397–408.

M. G. Simic and E. P. L. Hunter (1986) Reaction mechanisms of peroxyl and C-centered radicals with sulphydryls. *J. Free Radicals Biol. Med.* **2,** 227–230.

J. P. Smith and C. O. Anderson (1984) Modulation of the radiation sensitivity of human tumor cells by a bisbenzimidazole derivative. *Int. J. Radiat. Biol.* **46,** 331–344.

S. Steenken (1989) Purine bases, nucleosides and nucleotides: Aqueous solution redox chemistry and transformation reactions of their radical cations e^- and OH adducts. *Chem. Rev.* **89,** 503–520.

G. W. Teebor, R. J. Boorstein, and J. Cadet (1988) The repairability of oxidative free radical mediated damage to DNA: A review. *Int. J. Radiat. Biol.* **54,** 131–150.

P. Ulanski, E. Bothe, K. Hildenbrand, C. von Sonntag, and J. M. Rosiak (1997) The influence of repulsive electrostatic forces on the lifetimes of poly(acrylic acid) radicals in aqueous solution. *Nucleonika*, **42**, 425–436.

C. von Sonntag (1987) *The Chemical Basis of Radiation Biology.* London, Taylor and Francis.

C. von Sonntag (1991) *The chemistry of free-radical-mediated DNA damage.* In : Physical and Chemical Mechanisms in Molecular Radiation. (ed. W. A. Glass and M. N. Varma), pp. 287–321.

C. von Sonntag (1994) Topics in free-radical-mediated DNA damage: purines and damage amplification - superoxide reactions - bleomycin, the incomplete radiomimetic. *Int. J. Radiat. Biol.* **66,** 485–490.

C. von Sonntag, E. Bothe, P. Ulanski, and D. J. Deeble (1995) Pulse radiolysis in model studies toward radiation processing. *Radiat. Phys. Chem.* **46,** 572–532.

C. von Sonntag and H.-P. Schuchmann (1994) Pulse radiolysis. *Methods Enzymology* **233,** 3–20.

C. von Sonntag and H.-P. Schuchmann (1997) Peroxyl radicals in aqueous solution. *In Peroxyl Radicals.* (ed. Z. B. Alfassi), Wiley, London, pp. 173–234.

J. F. Ward, C. F. Webb, C. L. Limoli, and J. R. Milligan (1990) *DNA lesions produced by ionizing radiation: Locally multiply damaged sites. In Ionizing Radiation Damage to DNA: Molecular Aspects.* (ed. S. S. Wallace and R. B. Painter), pp. 43–50.

34

PROCESSING AND CONSEQUENCES OF OXIDATIVE DNA BASE LESIONS

Susan S. Wallace, Lynn Harrison,* Dongyan Jiang,† Jeffrey O. Blaisdell, Andrei A. Purmal,‡ and Zafer Hatahet

Department of Microbiology and Molecular Genetics
The Markey Center for Molecular Genetics
University of Vermont
Burlington, Vermont 05405

1. ABSTRACT

DNA base lesions produced by free radicals are common products of normal oxidative metabolism. These lesions are removed by base excision repair processing, the first step of which is recognition of the base lesion by a DNA-glycosylase. In general, the oxidative DNA base lesions are recognized by either a pyrimidine-specific or a purine-specific DNA-glycosylase. In this chapter, we describe the biochemical and biological properties of the newest of these activities the pyrimidine-specific oxidative DNA glycosylase, endonuclease VIII (*nei*) of *Escherichia coli*. We also describe an *in vitro* reconstitution of the base excision repair pathway using model DNAs that contain closely opposed lesions similar to multiply damaged sites produced by ionizing radiation. Here we show that when closely opposed lesions are more than three nucleotides apart, processing by base excision repair can lead to a potentially lethal double strand break. Finally, we describe the interaction between two ring saturation products of pyrimidines, one derived from cytosine, uracil glycol, the other from thymine, thymine glycol, with a model DNA polymerase, DNA polymerase I of *E. coli*. Both incorporation of the modified nucleoside triphosphate and translesion synthesis past the lesion in the DNA template, are considered

---

* Department of Physiology, LSU Medical Center, 1501 Kings Highway, Shreveport, Louisiana 71130

† Department of Molecular, and Cell Toxicology, Harvard University of Public Health, 665 Huntington Avenue, Boston, Massachusetts 02115-6021

‡ Pentose Pharmaceuticals, Inc., 45 Moulton St. , Cambridge Massachusetts, 02138

Advances in DNA Damage and Repair, edited by Dizdaroglu and
Karakaya. Kluwer Academic / Plenum Publishers, New York, 1999.

2. INTRODUCTION

Oxidative DNA base lesions are formed by a variety of agents that produce free radicals such as ionizing radiation and hydrogen peroxide, and as well, by normal cellular metabolism (for reviews see Breen and Murphy, 1995; Halliwell and Aruoma, 1991; and Cadenas, 1989). Over one hundred different products are produced by free radical interaction with DNA (for reviews see Kuwabara, 1991; Von Sonntag, 1987) and many of these are stable and thus have potential biological consequences. Most of the stable lesions have been identified by radiation chemists studying the radiolysis of DNA in solution (for reviews see Teoule, 1987; Von Sonntag, 1987; Dizdaroglu and Bergtold, 1986). The most well-studied include the pyrimidine ring saturation products, thymine glycol (Tg), uracil glycol (Ug), dihydrothymine (DHT) and dihydrouracil (DHU), the pyrimidine ring oxidation products, 5-hydroxycytosine (5-OHC) and 5 hydroxyuracil (5-OHU) and the purine oxidation products, 8-oxoguanine (8-oxoG), and 8-oxoadenine (8-oxoA). Urea and sites of base loss can also result from free radical attack on DNA bases. The structural characterization of many of these modified purines and pyrimidines has been accomplished and has been useful for interpreting the interactions between biologically important processing proteins and the lesion in DNA (for a review see Hatahet and Wallace, 1998). Because the spectrum of oxidative lesions are produced in DNA all at the same time by free radical-inducing agents such as ionizing radiation and hydrogen peroxide, it is difficult to assess the processing and consequences of individual lesions; that is, it is impossible to determine which lesions are responsible for which endpoint. Because of this, we have taken the approach of chemically synthesizing these modified bases as deoxynucleoside triphosphates and incorporating them into DNA or oligonucleotides to determine if they can be recognized by cellular enzymes, and as well, to investigate their interaction with DNA and RNA polymerases. Here we describe the enzymatic activities of the newest of the oxidative DNA glycosylases, *Escherichia coli*, endonuclease (endo) VIII and its role in the cell, the processing of multiply damaged sites by base excision repair, and the interactions of uracil glycol and thymine glycol with DNA polymerases.

3. RESULTS AND DISCUSSION

3.1. The Role of Endonuclease VIII in Base Excision Repair

In the bacterium *Escherichia coli,* oxidized DNA bases are recognized and removed by three DNA glycosylases, endonucleases III and VIII, which primarily recognize pyrimidine lesions, and formamidopyrimidine DNA-glycosylase (Fpg), which primarily recognizes purine lesions (for reviews see Wallace, 1997; Krokan et al., 1997; Cunningham, 1997; Demple and Harrison, 1994). These enzymes have broad substrate specificity, at least as measured *in vitro.* The oxidative DNA glycosylases cleave the damaged base at the N-glycosyl bond releasing the free base and then cleave the DNA backbone leaving either an α/β unsaturated aldehyde or a phosphate group attached to the 3' side of the resulting nick. The subsequent repair steps include removal of the block from the 3' terminus by the phosphodiesterase or phosphatase activities of the 5' apurinic (AP) endonucleases. This is followed by DNA polymerization and ligation. 8-oxoG opposite A is also recognized by the *E.coli* Mut Y protein, a DNA-glycosylase that removes the A opposite 8-oxoG preventing mutation fixation (for a review see Michaels and Miller, 1992).

Interestingly, there are significant sequence homologies between endo III and Mut Y (Michaels et al., 1990) and endoVIII and Fpg protein (Jiang et al., 1997a). Functional and structural homologs of endo III/Mut Y exist across phyla and members of this Nth superfamily have been identified in humans (for a review see Cunningham, 1997; Nash et al., 1996). Endo VIII and Fpg protein exhibit significant sequence homologies, especially at the N-terminal and C-terminal ends of the protein (Jiang et al., 1997a). In Fpg, the N-terminal proline has been implicated in the Schiff base formation required for the lyase activity of the protein (Tchou and Grollman, 1995) while the C-terminal end contains a putative zinc finger that has been shown to be involved in DNA binding (Castaing et al., 1993; Tchou et al., 1993). Both proteins share a number of enzymatic activities including βδ lyase activity (Jiang et al., 1997b; Melamede et al., 1994; Bailly et al., 1989; O'Connor and Laval, 1989;) and a dRPase activity (Jiang et al., 1997b; Graves et al., 1992) that removes a deoxyribose attached to the 5' side of the nick produced by a hydrolytic 5' AP endonuclease. Both proteins bind tightly to DNA containing a reduced AP (rAP) site, which is not a substrate, with a K_d of 4 nM for endo VIII (Jiang et al., 1997b) and 8 nM for Fpg protein (Castaing et al., 1992; Tchou et al., 1991). Although the primary *in vivo* substrate for Fpg protein appears to be 8-oxoG and the primary *in vivo* substrates for endo VIII appear to be oxidized pyrimidines (Jiang et al., 1997a), both proteins are capable of recognizing purines and pyrimidines *in vitro*. Several of the oxidized pyrimidines are reasonably good *in vitro* substrates for Fpg protein (Purmal et al., 1998b; Hatahet et al., 1994) while DNA containing 8-oxoA can be cleaved by endo VIII but it is a poor substrate for the enzyme (Purmal, Wallace and Dizdaroglu, unpublished observations). In addition to having different substrate specificities, the two proteins exhibit different binding patterns on a DNA substrate containing a rAP site. Fpg protein binds symmetrically around the lesion site with contacts primarily in the lesion-containing strand (Jiang et al., 1997b; Tchou et al., 1993). In contrast, endo VIII protein binds asymmetrically to the DNA about eight nucleotides 3' to the lesion again with contacts primarily on the lesion-containing strand (Jiang et al., 1997b). These data suggest that the active site positioning of the two proteins with respect to the lesions is different, as might be expected from their observed differences in substrate specificities.

Functional homologs of Fpg protein have been found across phyla but the eukaryotic homologs share no sequence relationship to Fpg protein (Nash et al., 1996). This differs from the eukaryotic functional homologs of endo III which share significant amino acid sequence identity with their bacterial counterparts (for a review see Cunningham, 1997). Interestingly, the eukaryotic versions of Fpg protein, called OGG, appear to have structural similarities to the Nth family of proteins (Cunningham, 1997; Nash et al.,1996). Since endo VIII shares substrate specificity with endo III, and eukaryotic cells have been shown to possess more than one endo III-like activity, it is possible that endo VIII homologs may be present in eukaryotes.

It has been known for some time from work reported from several laboratories that cells lacking Fpg and Mut Y exhibit a strong spontaneous mutator phenotype because of their reduced ability to respond to the potent premutagenic lesion, 8-oxoG (Michaels and Miller, 1992; Michaels et al., 1991, 1992; Cabrera et al., 1988; Radicella et al., 1988). However, cells lacking endo III, *nth* ,show only a minor mutator phenotype (Jiang et al. 1997a; Weiss et al., 1988). It was not until the isolation and cloning of the gene for endo VIII, *nei*, that the potential deleterious effects of the oxidized pyrimidines in genomic DNA were realized, that is, double mutants lacking both endos III and VIII, *nth nei,* show a strong mutator phenotype (Jiang et al., 1997a) significantly greater than that of *fpg* mutants at least as measured by forward mutation to rifampicin resistance (Figure 1). Further-

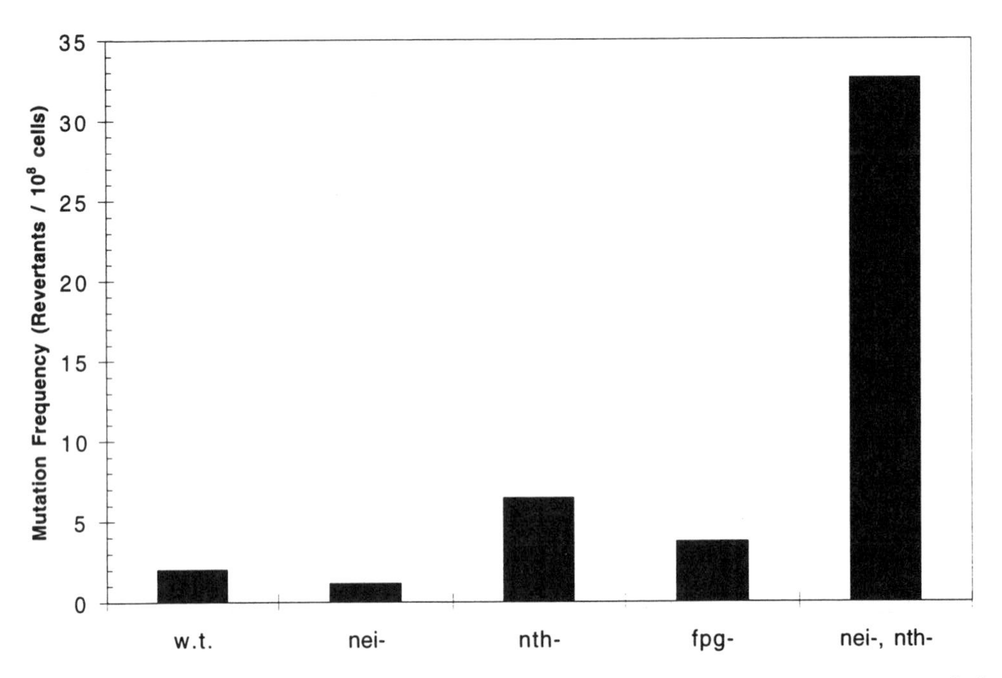

Figure 1. Spontaneous forward mutation frequency to rifampicin resistance of wild type *E. coli* and mutants lacking the oxidative DNA-glycosylsases.

more, the *nth nei* double mutants are also sensitive to the cytotoxic effects of ionizing radiation (Jiang et al.,1997a) and hydrogen peroxide (Figure 2). This is in contrast to *fpg* mutants which are not hypersensitive to hydrogen peroxide (Figure 2). Thus the lesions recognized by endos III and VIII are not only potentially mutagenic but potentially lethal.

3.2. Base Excision Repair Processing of Multiply Damaged Sites

When endogenous free radicals or hydrogen peroxide damage DNA, the products produced are generally random and affect a single strand. However, when DNA is traversed by ionizing radiation, a burst of free radicals can produce closely opposed lesions on both strands called multiply damaged sites (MDS) (Goodhead, 1994; Brenner and Ward, 1992; Ward, 1988; Ward et al., 1987). Since equi-damage doses of ionizing radiation are significantly more cytotoxic than, for example, hydrogen peroxide, the question has arisen as to whether the attempted repair of MDS converts otherwise non-lethal lesions into lethal double-strand breaks (Ward et al., 1987). We have addressed this question by site specifically placing modified bases or a site of base loss (AP site) opposite a single base gap or another modified base and using these MDS-containing DNAs as substrates for the *E. coli* oxidative DNA-glycosylases (Harrison et al., 1998). When either an oxidized base or an AP site is present one nucleotide away from a single base gap on the opposite strand, it is a poor substrate for the oxidative DNA glycosylases. In contrast, when the opposing gap is three or six nucleotides away from the target lesion, the lesion is readily recognized and the DNA is incised leading to a double strand break. When an oxidized pyrimidine (Tg or DHT) is placed opposite an oxidized purine (8-oxoG) and spaced three or more nucleotides apart, the pyrimidine is readily recognized and the DNA incised. When two bases are

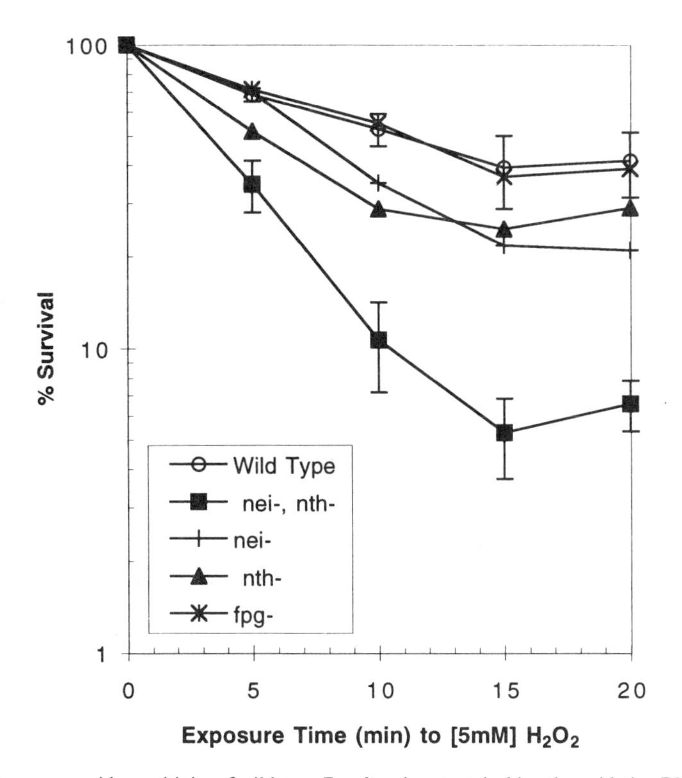

Figure 2. Hydrogen peroxide sensitivity of wild type *E. coli* and mutants lacking the oxidative DNA-glycosylases.

closely opposed one nucleotide apart, the first base is readily recognized but once the strand break is produced by the action of the DNA- glycosylase, the second base is not removed. From these (Harrison et al., 1998) and previously reported data (Chaudry and Weinfeld, 1997, 1995), it can be concluded that when two base damage sites or sites of base loss are opposed to each other on two DNA strands and are separated by one nucleotide, the lesion in one strand is readily cleaved but the lesion in the second strand is not. When the two lesions are separated by three or more nucleotides, a double strand break is formed because the lesion in the second strand can be cleaved.

The question arises as to whether, in the presence of the complete repair system, the damages in the MDS are sequentially repaired giving rise to an intact molecule. To address this question, we reconstituted the base excision repair system *in vitro* using a target 8-oxoG closely opposed to a single base gap (Harrison et al., submitted). We looked at sequential processing by Fpg protein, endonuclease IV, DNA polymerase I (pol I), and DNA ligase. As was stated above, when the 8-oxoG was one nucleotide from the opposed gap, little cleavage was observed but cleavage of the 8-oxoG was efficient if the opposed gap was positioned three or six nucleotides away from the target 8-oxoG. If the opposed gap was three or six nucleotides apart, the phosphate group left on the 3' side of the Fpg-induced nick was readily converted to a 3' hydroxyl by endonuclease IV, and polymerization by pol I in the presence of limiting deoxynucleoside triphosphates was observed; however, no ligation products were found. When repair of the opposing strand containing a single base gap was examined, repair of the strand break by endo IV, pol I and ligase was efficient and complete when the gap was one nucleotide removed from the opposing 8-oxoG.

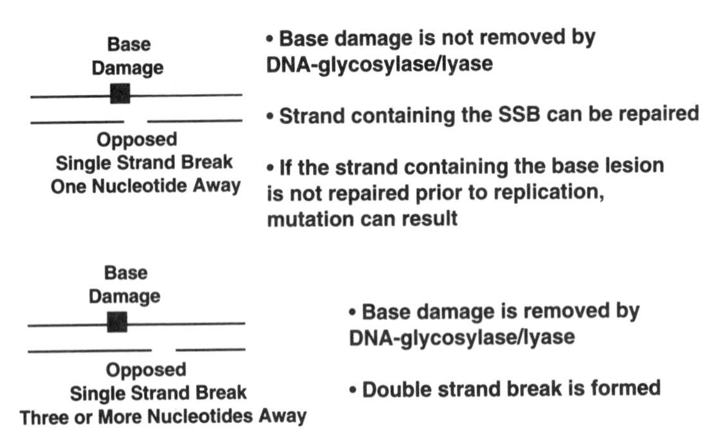

Figure 3. Possible consequences of multiply damaged sites in DNA.

This was true whether or not Fpg protein was present since 8-oxoG one nucleotide opposed to the strand break is not efficiently incised. In contrast, in the presence of Fpg protein and when the opposed gap was three or six nucleotides away, no ligation products were observed from repair of the gap and double strand breaks were formed. Repair of the gap-containing strand, however, could take place if Fpg protein was omitted from the reaction.

Taken together, these data show (see Figure 3) that in the presence of all the proteins required for repair of 8-oxoG and the single base gap, and when the 8-oxoG and the gap are separated from each other by three or more nucleotides, a double strand break is formed. In contrast, when the lesions are separated by only one nucleotide, repair of the gap-containing strand can occur even in the presence of Fpg protein. If these *in vitro* results hold true in the cell, the processing of an MDS by the complete base excision repair system could result in a double strand break or if the lesions are separated by a single base pair and if repair of the gap occurs, the strand containing 8-oxoG would be available to be repaired by base excision repair. If not repaired prior to replication translesion bypass of 8-oxoG could lead to mutation. Thus, processing of damages produced by ionizing radiation may convert an otherwise non-lethal or readily repaired lesion into a lethal double strand break or a mutation.

3.3. Interaction between DNA Lesions and DNA Polymerase

The biological consequences of oxidative DNA lesions have been predicted by studying the interactions between unique lesions and DNA polymerases *in vitro* (for a review see Hatahet and Wallace, 1998). If the lesion blocks the DNA polymerase *in vitro*, the lesion is potentially lethal. This can be tested *in vivo* by introducing the lesion into single stranded transfecting DNA, which cannot be repaired by base excision repair, and then measuring the survival of the virus, that is, its ability to be replicated by the cellular replicative polymerases. Of the oxidative lesions mentioned above, sites of base loss, strand breaks, ring fragmentation products such as urea, and very interestingly, thymine glycol, are lethal lesions. Most of the remaining oxidative lesions that have been well studied and retain their intact ring structure do not block DNA polymerases and are therefore not lethal (for reviews see Evans et al., 1993; Hatahet and Wallace, 1998).

Many of the oxidative DNA lesions are capable of mispairing *in vitro* and are thus potentially mutagenic. Mispairing *in vitro* is readily measured using two methods. In the first, a primer is used one base short of the lesion in the template strand and then each deoxynucleoside triphosphate is added one at a time to measure insertion of the correct versus the incorrect nucleotide (see for example Purmal et al., 1994a). Then the rate of elongation of the potentially mispaired lesion base can be determined. Alternatively, a specially designed template/primer system can be used where each product migrates slightly differently on a polyacrylamide gel depending on which particular nucleotide was inserted opposite the lesion (see for example Shibutani et al., 1996). The *in vitro* studies can be verified in the cell by any number of site-specific mutagenesis assays. Oxidized nucleotides can also be mutagenic if they are misincorporated from the nucleotide pools. Whether a particular oxidized nucleoside triphosphate can be incorporated into DNA can be easily determined by a replacement reaction (see for example Purmal et al., 1994b). Here it is determined whether or not the oxidized nucleotide in question can substitute for a normal nucleotide in a DNA polymerase reaction. To illustrate the principles that we and others have used to study the consequences of oxidative DNA lesions, uracil glycol will be compared to thymine glycol.

3.4. Deoxyuridine Glycol Triphosphate and Deoxythymidine Glycol Triphosphate as Substrates for DNA Polymerase

Deoxyuridine glycol triphosphate (dUgTP) was chemically synthesized and used as a substrate for DNA polymerase I Klenow fragment (Kf) (Purmal et al., 1998a). Unlike deoxythymidine glycol triphosphate (dTgTP), dUgTP, is a good substrate for Kf. Since uracil glycol is a major stable oxidation product of DNA cytosine and can be formed directly in DNA and presumably in the cellular nucleotide pools, it was of interest to determine the pairing characteristics of dUgTP during incorporation (Purmal et al., 1998a). dUgTP was incorporated always in place of dTTP and therefore, when incorporated from the nucleotide pools would not be expected to be mutagenic unless mispairing occurred during subsequent replications. Since dTgTP is such a poor substrate for DNA polymerase, it is unlikely to be incorporated from the nucleotide pools. However, when dTgTP was incorporated, it was always in place of dTTP, and thus would not be mutagenic.

We had previously examined the ability of 5-hydroxydeoxyuridine triphosphate (5-OHdUTP), 5-hydroxydeoxycytidine triphosphate (5-OHdCTP), 8-oxodeoxyadenosine triphosphate (8-oxodATP), and 8-oxodeoxyguanosine triphosphate (8-oxodGTP) as substrates for Kf (Purmal et al., 1994b). All were good substrates with 5-OHdUTP and 8-oxodATP being incorporated in place of dTTP and dATP, respectively. Thus 5-OHdUTP and 8-oxodATP would not be expected to be mutagenic when incorporated from the nucleotide pools. In contrast, 5-OHdCTP is incorporated in place of dTTP as well as dCTP, and 8-oxodGTP, as has also been shown by others (Akiyama et al., 1989), is incorporated in place of dGTP and dTTP. Thus both 5-OHdCTP and 8-oxodGTP have the ability to be mutagenic when misincorporated from the nucleotide pools. In fact, an enzyme, Mut T, has been identified across phyla which hydrolyzes 8-oxodGTP to 8-oxodGMP thus removing it from the nucleotide pools (Maki and Sekiguchi, 1992; Mo et al., 1992). Mutant *E. coli* cells unable to perform this task exhibit a high spontaneous mutation rate (Akiyama et al., 1987; Schaaper and Dunn, 1987). To summarize, of the oxidized lesions mentioned above, only 8-oxodGTP and 5-OHdCTP have the potential to be mutagenic when incorporated from the nucleotide pools. Several other oxidative lesions studied by others, including 2-oxodeoxyadenosine triphosphate (Kamiya and Kasai, 1995) and 5-formyl deoxyuridine triphosphate (Yoshida et al., 1997), also have the ability to mispair during incorporation and are thus potentially mutagenic.

3.5. The Fate of Uracil Glycol and Thymine Glycol in Template DNA

The potential of uracil glycol to be a potent premutagenic lesion would occur if Ug was present in the template since it is an oxidative product of DNA cytosine. We have shown that Ug in the template is not a strong blocking lesion to Kf (Purmal et al., 1998b) and therefore, presumably not a cytotoxic lesion. However, Ug always pairs with A and since it is derived from C, it suggests that it is a strong premutagenic lesion (Purmal et al., 1998b). In contrast, in most sequence contexts, Tg is a strong blocking lesion to Kf (for reviews see Hatahet and Wallace, 1998; Evans et al., 1993) and to other DNA polymerases as well, and therefore, potentially cytotoxic. This was confirmed by introducing Tg into single stranded phage tranfecting DNA and showing that it took one lesion to inactive the virus (Hayes et al., 1988; Achey and Wright, 1983). Tg always pairs with A *in vitro* and is a poor premutagenic lesion (Hayes et al., 1988), although in a bypass sequence context *in vivo*, some pairing with G was observed (Basu et al., 1989).

Molecular mechanics and dynamics calculations of cis-thymine glycol suggest that the most stable conformer has the two *cis* hydroxyl groups in the pseudoequatorial conformation (Clark et al., 1997; Miller et al., 1994). Since thymine glycol also has a methyl group attached to the C5 carbon, the methyl group must be pseudoaxial. Thus when Tg is present in the template strand, steric interactions cause significant distortion to the base pair 5' to the lesion which would significantly interfere with elongation of the Tg-A pair. The biochemical data, including steady state kinetic analysis, support the interpretations derived from the computational studies (Purmal et al., 1998b). Because uracil glycol has no methyl group attached to the C5 position on the pyrimidine ring, and even though the saturated ring assumes a half chair conformation, there is little predicted distortion either on the 5' or the 3' side of the lesion. In keeping with this prediction, full length product is observed with Kf when Ug is present in the template (Purmal et al., 1998b). Full length products have also been found with Kf when 8-oxoG, 8-oxoA, 5-OHC and 5-OHU are present in the template (Purmal et al. 1994a; Shibutani et al., 1991, 1993; Guschelbauer et al., 1991). For a summary of the *in vitro* interactions between free radical-modified purines and pyrimidines and pol I Kf, see Table 1.

Table 1. Interactions between free radical-induced DNA lesions and *Escherichia coli* DNA polymerase I Klenow fragment

Lesion	Incorporated as dNTP	Pairs with	Template block	Pairs with
Abasic Sites	no	—	yes	A>G>py
Urea	no	—	yes	A>G
UBA	no	—	yes	A>G>T
Tg	poor	A	yes	A
DHT	good	A	no	A
Ug	good	A	no	A
DHU	good	A	no	A
5-OHC	good	G>A	no	G>A>>C
5-OHU	good	A	no	A
8-oxoG	good	C>A	no	C>A
8-oxoA	good	T	no	T

4. PERSPECTIVES

It has become increasingly clear in recent years that oxidative DNA damage plays an important role in a number of disease processes. Not only has oxidative DNA damage been implicated in cancer (for a review see Loft and Paulson, 1996), but also in several neurodegenerative diseases such as Amytrophic Lateral Sclerosis (for a review see Brown, 1997) and Alzheimer's disease (Smith et al., 1997). Also, the accumulation of oxidative damage especially in non-dividing cells, has been postulated to be responsible for the degenerate effects of aging (for a review see Beckman and Ames, 1997). Accordingly, it is exceedingly important to be able to understand at the very fundamental level, how oxidative DNA damages are recognized and removed by DNA repair enzymes and as well, how any unrepaired lesions influence biological outcomes such as cytotoxicity and mutagenesis by their interaction with DNA and RNA polymerases. Because the responses to oxidative DNA damage evolved early in evolution and were important for the maintenance of genomic integrity, they have been amazingly conserved from bacteria to humans. This makes the study of simple systems such as bacteria and yeast, which are subject to elegant genetic analysis, important models for understanding how these basic processes work in higher eukaryotes. Interestingly, oxidative stress responses have been employed by higher eukaryotes in beneficial ways. For example, phagocytic cells respond to invading bacteria by releasing toxic amounts of the agents that produce free radicals and kill the bacteria (Babior, 1987). Oxidative stress responses have been also implicated in apoptosis (for a review see Fuchs et al., 1997). Thus understanding the interrelationship between oxidative DNA damage and repair and the mechanisms underpinning the cellular responses to oxidative stress are important areas for future investigation.

ACKNOWLEDGMENTS

This work was supported by National Institutes of Health grants R37 CA33657 and R01 CA52040 awarded by the National Cancer Institute and grant DE-FG02-88ER60742 from the U.S. Department of Energy.

REFERENCES

P. M. Achey and C. F. Wright (1983) Inducible repair of thymine ring saturation damage in ϕX174 DNA. *Radiat. Res.* **93**, 609–612.

M. Akiyama, H. Maki, M. Sekiguchi and T. Horiuchi (1989) A specific role of MutT protein: To prevent dG·dA mispairing in DNA reprelication. *Proc. Natl. Acad. Sci. USA* **86**, 3949–3952.

M. Akiyama, T. Horiuchi and M. Sekiguchi (1987) Molecular cloning and nucleotide sequence of the *mutT* mutator of *Escherichia coli* that causes A:T to C:G transversion. *Mol. Gen. Genet.* **206**, 9–16.

B. M. Babior (1987) The respiratory burst oxidase. *Trends Biochem. Sci.* **12**, 241–243.

V. Bailly, W. G. Verly, T. O'Connor and J. Laval (1989) Mechanism of DNA strand nicking at apurinic/apyrimidinic sites by *Escherichia coli* [formamidopyrimidine] DNA glycosylase. *Biochem. J.* **262**, 581–589.

A. K. Basu, E. L. Loechler, S. A.Leadon and J. M. Essigmann (1989) Genetic effects of thymine glycol: site-specific mutagenesis and molecular modeling studies. *Proc. Natl. Acad. Sci. USA* **86**, 7677–7681.

K. B. Beckman and B. N. Ames (1997) Oxidants, antioxidants, and aging. In *Oxidative Stress and the Molecular Biology of Antioxidant Defenses.* (ed. J. G. Scandalios) Cold Spring Harbor, NY, Cold Spring Harbor Press, pp. 201–246.

A. P. Breen and J. A. Murphy (1995) Reactions of oxyl radicals with DNA. *Free Rad. Biol. Med.* **18**, 1033–1077.

D. J. Brenner and J. F. Ward (1992) Constraints on energy deposition and target size of multiply damaged sites associated with DNA double-strand breaks. *Int. J. Radiat. Biol.* **61**, 737–748.

R. H. Brown, Jr. (1997) Superoxide dismutase and oxidative stress in amytrophic lateral sclerosis. In *Oxidative Stress and the Molecular Biology of Antioxidant Defenses* (ed. J. G. Scandalios) Cold Spring Harbor, NY, Cold Spring Harbor Press pp. 569–586.

M. Cabrera, Y. Nghiem and J. H. Miller (1988) mutM, a second mutator locus in *Escherichia coli* that generates G•C → T•A transversions. *J. Bacteriol.* **170**, 5405–5407.

E. Cadenas (1989) Biochemistry of oxygen toxicity. *Ann. Rev. Biochem.* **58**, 79–110.

B. Castaing, A. Geiger, H. Seliger, P. Nehls, J. Laval, C. Zelwer and S. Boiteux (1993) Cleavage and binding of a DNA fragment containing a single 8-oxoguanine by wild type and mutant FPG proteins. *Nucl. Acids Res.* **21**, 2899–2905.

B. Castaing, S. Boiteux and C. Zelwer (1992) DNA containing a chemically reduced apurinic site is a high affinity ligand for the *E. coli* formamidopyrimidine-DNA glycosylase. *Nucl. Acids Res.* **20**, 389–394.

M. A. Chaudhry and M. Weinfeld (1995) The action of *Escherichia coli* endonuclease III on multiply damaged sites in DNA. *J. Mol. Biol.* **249**, 914–922.

M. A. Chaudhry and M. Weinfeld (1997) Reactivity of human apurinic/apyrimidinic endonuclease and *Escherichia coli* exonuclease III with bistranded abasic sites in DNA. J. Biol. Chem. **272**, 15650–15655.

J. M. Clark, N. Pattabiraman, W. Jarvis and G. P. Beardsley (1987) Modeling and molecular mechanical studies of the cis-thymine glycol radiation damage lesion in DNA. *Biochemistry* **26**, 5404–5409.

R. P. Cunningham (1997) Caretakers of the genome? *Curr. Biol.* **7**, R576–R579.

B. Demple and L. Harrison (1994) Repair of oxidative damage to DNA: enzymology and biology. *Ann. Rev. Biochem.* **63**, 915–948.

M. Dizdaroglu and D. S. Bergtold (1986) Characterization of free radical-induced base damage in DNA at biologically relevant levels. *Anal. Biochem.* **156**, 182–188.

J. Evans, M. Maccabee, Z. Hatahet, J. Courcelle, R. Bockrath, H. Ide and S. Wallace (1993) Thymine ring saturation and fragmentation products: lesion bypass, misinsertion and implications for mutagenesis. *Mutat. Res.* **299**, 147–156.

D. Fuchs, G. Baier-Bitterlich, I. Wede and H. Wachter (1997) Reactive oxygen and apoptosis. In *Oxidative Stress and the Molecular Biology of Antioxidant Defenses* (ed. J. G. Scandalios) Cold Spring Harbor, NY, Cold Spring Harbor Press pp. 139–167.

D. T. Goodhead (1994) Initial events in the cellular effects of ionizing radiations: clustered damage in DNA. *Int. J. Radiat. Biol.* **61**, 7–17.

R. J. Graves, I. Felzenszwalb, J. Laval and T. R. O'Connor (1992) Excision of 5'-terminal deoxyribose phosphate from damaged DNA is catalyzed by the Fpg protein of *Escherichia coli*. *J. Biol. Chem.* **267**, 14429–14435.

W. Guschlbauer, A. M. Duplaa, A. Guy, R. Teoule and G. V. Fazakerley (1991) Structure and *in vitro* replication of DNA templates containing 7,8-dihydro-8-oxoadenine. *Nucl. Acids Res.* **19**, 1753–1758.

B. Halliwell and O. I. Aruoma (1991) DNA damage by oxygen-derived species. *FEBS Lett.* **281**, 9–19.

L. Harrison, Z. Hatahet, A. A. Purmal and S. S. Wallace (1998) Multiply damaged sites in DNA: Interactions with *Escherichia coli* endonuclease III and VIII. *Nucl. Acids Res.* **26**, 932–941.

L. Harrison, Z. Hatahet and S. S. Wallace (1998) In vitro repair of synthetic ionizing radiation-induced multiply damaged DNA sites. Submitted.

Z. Hatahet, Y. W. Kow, A. A. Purmal, R. P. Cunningham and S. S. Wallace (1994) New substrates for old enzymes. 5-Hydroxy-2'-deoxycytidine and 5-hydroxy-2'-deoxyuridine are substrates for *Escherichia coli* endonuclease III and formamidopyrimidine DNA N-glycosylase, while 5-hydroxy-2'-deoxyuridine is a substrate for uracil DNA N-glycosylase. *J. Biol. Chem.* **269**, 18814–18820.

Z. Hatahet and S. S. Wallace (1998) Translesion DNA synthesis. In *DNA Damage and Repair, Vol 1: DNA Repair in Prokaryotes and Lower Eukaryotes* (eds. J. A. Nickoloff and M. F. Hoekstra), Totowa, NJ, Humana Press, Inc., pp. 229–262.

R. C. Hayes, L. A. Petrullo, H. M. Huang, S. S. Wallace and J. E. LeClerc (1988) Oxidative damage in DNA. Lack of mutagenicity by thymine glycol lesions. *J. Mol. Biol.* **201**, 239–246.

D. Jiang, Z. Hatahet, J. O. Blaisdell, R. J. Melamede and S. S. Wallace (1997a) *Escherichia coli* endonuclease VIII: Cloning, sequencing and overexpression of the *nei* structural gene and characterization of *nei* and *nei nth* mutants. *J. Bacteriol.* **179**, 3773–3782.

D. Jiang, Z. Hatahet, R. J. Melamede, Y. W. Kow and S. S. Wallace (1997b) Characterization of *Escherichia coli* endonuclease VIII. *J. Biol. Chem.* **272**, 32230–32239.

H. Kamiya and H. Kasai (1995) Formation of 2-hydroxydeoxyadenosine triphosphate, an oxidatively damaged nucleotide, and its incorporation by DNA polymerases. Steady-state kinetics of the incorporation. *J. Biol. Chem.* **270**, 19446–19450.

H. E. Krokan, R. Standal and G. Slupphaug (1997) DNA glycosylases in the base excision repair of DNA. *Biochem. J.* **325**, 1–16.

M. Kuwabara (1991) Chemical processes induced by OH attack on nucleic acids. *Radiat. Phys. Chem.* **37**, 691–704.

S. Loft and H. E. Poulsen (1996) Cancer risk and oxidative DNA damage in man. *J. Mol. Med.* **74**, 297–312.

H. Maki and M. Sekiguchi (1992) MutT protein specifically hydrolyses a potent mutagenic substrate for DNA synthesis. *Nature* **355**, 273–275.

R. J. Melamede, Z. Hatahet, Y. W. Kow, H. Ide and S. S. Wallace (1994) Isolation and characterization of endonuclease VIII from *Escherichia coli*. *Biochemistry* **33**, 1255–1264.

M. L. Michaels, C. Cruz, A. P. Grollman and J. H. Miller (1992) Evidence that MutY and MutM combine to prevent mutations by an oxidatively damaged form of guanine in DNA. *Proc. Natl. Acad. Sci. USA* **89**, 7022–7025.

M. L. Michaels and J. H. Miller (1992) The GO system protects organisms from the mutagenic effect of the spontaneous lesion 8-hydroxyguanine (7,8-dihydro-8-oxoguanine). *J. Bacteriol.* **174**, 6321–6325.

M. L. Michaels, L. Pham, C. Cruz and J. H. Miller (1991) MutM, a protein that prevents G•C → T•A transversions, is formamidopyrimidine-DNA glycosylase. *Nucl. Acids Res.* **19**, 3629–3632.

M. L. Michaels, J. Tchou, A. P. Grollman and J. H. Miller (1992) A repair system for 8-oxo-7,8-dihydrodeoxyguanine. *Biochemistry* **31**, 10964–10968.

M. L. Michaels, L. Pham, Y. Nghiem, C. Cruz and J. H. Miller (1990) MutY, an adenine glycosylase active on G-A mispairs, has momology to endonuclease III. *Nucl. Acids Res.* **18**, 3841–3845.

J. Miller, K. Miaskiewicz and R. Osman (1994) Structure-function studies of DNA damage using ab initio quantum mechanics and molecular dynamics simulation. *Ann. N Y Acad. Sci.* **726**, 71–91.

J.-Y. Mo, H. Maki and M. Sekiguchi (1992) Hydrolytic elimination of a mutagenic nucleotide, 8-oxodGTP, by human 18-kilodalton protein: sanitization of nucleotide pool. *Proc. Natl. Acad. Sci. USA* **89**, 11021–11025.

H. M. Nash, S. D. Bruner, O. D. Scharer, T. Kawate, T. A. Addona, E. Spooner, W. S. Lane and G. L. Verdine (1996) Cloning of a yeast 8-oxoguanine DNA glycosylase reveals the existence of a base-excision DNA-repair protein superfamily. *Curr. Biol.* **6**, 968–980.

T. R. O'Connor and J. Laval (1989) Physical association of the 2,6-diamino-4-hydroxy-5N-formamidopyrimidine-DNA glycosylase of *Escherichia coli* and an activity nicking DNA at apurinic/apyrimidinic sites. *Proc. Natl. Acad. Sci. USA* **86**, 5222–5226.

A. A. Purmal, Y. W. Kow and S. S. Wallace (1994a) Major oxidative products of cytosine, 5-hydroxycytosine and 5-hydroxyuracil, exhibit sequence context-dependent mispairing *in vitro*. *Nucl. Acids Res.* **22**, 72–78.

A. A. Purmal, Y. W. Kow and S. S. Wallace (1994b) 5-Hydroxypyrimidine deoxynucleoside triphosphates are more efficiently incorporated into DNA by exonuclease-free Klenow fragment than 8-oxopurine deoxynucleoside triphosphates. *Nucl. Acids Res.* **22**, 3930–3935.

A. A. Purmal, J. P. Bond, B. A. Lyons, Y. W. Kow and S. S. Wallace (1998a) Uracil glycol deoxynucleoside triphosphate is a better substrate for DNA polymerase I klenow fragment than thymine glycol deoxynucleoside triphosphate. *Biochemistry* **37**, 330–338.

A. A. Purmal, G. W. Lampman, J. P. Bond, Z. Hatahet and S. S. Wallace (1998b) Enzymatic processing of uracil glycol, a major oxidative product of DNA cytosine. *J. Biol. Chem.*, In press.

J. P. Radicella, E. A. Clark and M. S. Fox (1988) Some mismatch repair activities in *Escherichia coli*. *Proc. Natl. Acad. Sci.* **85**, 9674–9678.

R. M. Schaaper and R. L. Dunn (1987) Spectra of spontaneous mutations in *Escherichia coli* strains defective in mismatch correction: the nature of *in vivo* DNA replication errors. *Proc. Natl. Acad. Sci. USA* **84**, 6220–6224.

S. Shibutani, N. Sizuki, Y. Matsumoto and A. P. Grollman (1996) Miscoding properties of 3,N4-etheno-2'-deoxycytidine in reactions catalyzed by mammalian DNA polymerases. *Biochemistry* **35**, 14992–14998.

S. Shibutani, V. Bodepudi, F. Johnson and A. P. Grollman (1993) Translesional synthesis on DNA templates containing 8-oxo-7,8-dihydrodeoxyadenosine. *Biochemistry* **32**, 4615–4621.

S. Shibutani, M. Takeshita and A. P. Grollman (1991) Insertion of specific bases during DNA synthesis past the oxidation-damaged base 8-oxodG. Nature **349**, 431–434.

M. A. Smith, P. L. R. Harris, L. M. Sayre and P. George (1997) Iron accumulation in Alzheimer disease is a source of redox-generated free radicals. *Proc. Natl. Acad. Sci. USA* **94**, 9866–9869.

J. Tchou and A. P. Grollman (1995) The catalytic mechanism of Fpg protein. Evidence for a Schiff base intermediate and amino terminus localization of the catalytic site. *J. Biol. Chem.* **270**, 11671–11677.

J. Tchou, H. Kasai, S. Shibutani, M. H. Chung, J. Laval, A. P. Grollman and S. Nishimura (1991) 8-Oxoguanine (8-hydroxyguanine) DNA glycosylase and its substrate specificity. *Proc. Natl. Acad. Sci. USA* **88**, 4690–4694.

J. Tchou, M. L. Michaels, J. H. Miller and A. P. Grollman (1993) Function of the zinc finger in *Escherichia coli* Fpg protein. *J. Biol. Chem.* **268**, 26738–26744.

R. Teoule (1987) Radiation-induced DNA damage and its repair. *Int. J. Radiat. Biol. Relat. Stud. Phys. Chem. Med.* **51**, 573–589.

C. von Sonntag (1987) *The Chemical Basis of Radiation Biology*. London, Taylor and Francis.

S. S. Wallace (1997) Oxidative damage to DNA and its repair. In *Oxidative Stress and the Molecular Biology of Antioxident Defenses* (ed. J. Scandalios), New York, Cold Spring Harbor Press pp. 49–90.

J. F. Ward (1988) DNA damage produced by ionizing radiation in mammalian cells: identities, mechanisms of formation, and reparability. *Prog. Nucl. Acid Res. Mol. Biol.* **35**, 95–125.

J. F. Ward, J. W. Evans, C. L. Limoli and P. M. Calabro-Jones (1987) Radiation and hydrogen peroxide-induced free radical damage to DNA. *Br. J. Cancer Suppl.* **8**, 105–112.

B. Weiss, R. P. Cunningham, E. Chan and I. R. Tsaneva (1988) AP endonucleases of *Escherichia coli*. In *Mechanisms of Consequences of DNA Damage Processing* (eds. E. C. Friedgerg and P. Hanawalt), New York, A. R. Liss pp. 133–142.

M. Yoshida, K. Makino, H. Morita, H. Terato, Y. Ohyama and H. Ide (1997) Substrate and mispairing properties of 5-formyl-2'-deoxyuridine 5'- triphosphate assessed by *in vitro* DNA polymerase reactions. *Nucl. Acids Res.* **25**, 1570–1577.

IONIZING RADIATION DAMAGE TO DNA

A Challenge to Repair Systems

John F. Ward

Department of Radiology, 0610
University of California, San Diego
La Jolla, California 92093

1. ABSTRACT

At doses at which mammalian cells are killed, ionizing radiation produces 3×10^3 - 3×10^4 altered moieties in the genomic DNA of the cell. Mechanisms whereby these alterations are produced are well described, i.e., from ˙OH radicals, direct ionization of DNA (von Sonntag, 1987) and perhaps from peroxyl radicals arising from ˙OH reactions with other cellular molecules (Ward et al., 1997). Enzymatic repair of altered bases is well documented, and since these altered bases are produced during endogenous oxidation it is not surprising that enzymatic processes have evolved for their removal (Lindahl,1990). Ionizing radiation also produces strand breaks, but it is not known if these are produced by endogenous oxidation. The repair of singly damaged sites is well known. However, ionizing radiation also produces multiply damaged sites (MDS): It deposits energy non-uniformly; increments of energy (average amount 60eV) in small (nanometer) volumes. Thus a resulting radical reacting with DNA frequently does so in the presence of other radicals which can also react in close proximity. MDS have three variables: 1. Types of lesions involved (base damage, strand break); 2. Total number of lesions involved (for low LET radiation the most frequent is two); 3. Size of the site over which the damages are distributed. MDS represent a challenge to the cellular repair systems, they represent sites in the DNA at which base identity has been destroyed on both strands; if the damaged sites are closely opposed there is no complimentary strand to guide repair.

2. INTRODUCTION

Radiation damage to intracellular DNA is initiated by the deposition of energy in the macromolecule itself or with neighboring molecules such as water. These two sources of

Advances in DNA Damage and Repair, edited by Dizdaroglu and
Karakaya. Kluwer Academic / Plenum Publishers, New York, 1999.

damage have been termed "direct" and "indirect" effects respectively. Since it is generally accepted that radiation induced excitation is not a source of biologically significant damage we need concern ourselves only with ionizations of molecules. Ionization of water by radiation to produce reactive species is generally written:

$$H_2O \rightarrow {}^{\bullet}OH, H^{\bullet}, e_{aq}, H_2, H_2O_2 \quad (1)$$

The major reactive species is the hydroxyl radical, OH, since in oxygenated systems the reducing species are rapidly scavenged and converted to the relatively unreactive superoxide anion radical $O_2^{\bullet-}$.

The yields of these reactive species are quoted in mM per Joule absorbed, which is equivalent to concentration in mMolar per Gray and in the case of $^{\bullet}$OH radicals is 0.24 (Spinks and Woods, 1976). This is the yield of $^{\bullet}$OH radicals at times of about 1 ms after the radiation has been delivered, if the yield is measured at times corresponding to the life time of an OH radical in a mammalian cells, 1 ns (Roots and Okada, 1972) the yield is about twice as high (Jonah, 1977).

To discuss the types of damage produced in cellular DNA we need to consider both origins, "direct" and "indirect". It is possible that the damage produced in DNA from both origins are the same (Ward, 1975). Indeed those measurements of assays of base damage produced by irradiation of DNA intracellularly (Mori and Dizdaroglu, 1994 Mori et al. 1993, Nackerdien et al. 1992) has not uncovered any base damage structures which are not produced by irradiation in dilute aqueous solution (Fuciarelli et al.1990), i.e. where $^{\bullet}$OH radicals are the predominant originators of the damage.

3. YIELDS OF RADIATION-INDUCED BASE DAMAGE

The yields of damage produced by irradiation of DNA in dilute oxygenated aqueous solution (i.e. where the only reactive species initiating the damage is the OH radical) taken from the literature are shown in Table 1. Admittedly the measurements of base damage and of strand breaks were made in different laboratories, but in both situations care was taken to ensure that the water was free of competing OH radical scavengers so that the yields of damage should be comparable. It is clear that the major type of damage produced is to the base moieties, and the yield of their damage is 5–10 times that of single strand

Table 1. Yields of damage in DNA irradiated in aqueous oxygenated solution

Molecular change	% of $^{\bullet}$OHs reacting	Reference
Strand break		Henner et al. (1973) Milligan et al. (1993)
5' Phosphate +3'Phosphate	8	
5'P. +3'Phosphoglycolate	4	
Base damage		Fuciarelli et al. (1990)
8-hydroxyguanine	24	
8-hydroxyadenine	4	
FaPy-guanine	3.4	
FaPy-adenine	1.4	
Thymine glycol	15	
Cytosine glycol	8	

Table 2. Measured radiation yields of DNA base damage in mammalian cells relative to that calculated from the measured yield of single strand breaks

Calculated yield of base damage	Measured amounts of base damage		
	Human cells *in vitro*	Mouse liver *in vivo*	Mouse L5178Y *in vitro*
1	40	21	9
Reference	Mori and Dizdaroglu (1994)	Mori et al. (1993)	Nackerdien et al (1992)

breaks. This ratio of yields was indicated earlier before such detailed measurements were possible (Scholes et al. 1960, Ward, 1985). Thus these yields fit those expected from considerations of mechanism - •OH radicals react 4–5 times faster with the unsaturated base moieties than with the deoxyribose (Buxton et al. 1988), the latter reaction being that which leads to DNA strand breaks.

There is controversy over the yields of damaged base produced upon irradiation of DNA intracellularly (Dizdaroglu and Halliwell, 1992, Douki et al. 1996). Different yields are found by different measurement techniques. The major criticism of the technique of gas chromatography mass spectroscopy with selected ion monitoring is that the procedure of derivatization of the bases prior to chromatography produces altered bases. While it is clear that such artifacts can be formed under specific conditions, their presence has been recognized (Dizdaroglu, 1993, Swarts et al. 1996) and either avoided by modifying the hydrolysis conditions or adjusted for in the treatment of the data.

It should be noted that the GCMS/SIM technique was used to analyze the DNA irradiated in dilute aqueous solution (as described above). In those experiments the measured yields agree with those projected from mechanistic considerations and from earlier more gross measurements of base damage. Thus it would seem that the GCMS/SIM technique does give the correct yields for ionizing radiation induced damage. For the purposes of the present discussion we will utilize data using this measurement technique which gave the correct results for base damage yield in the conditions in which the mechanism is more clearly known. The yield of DNA single strand breaks (SSB) produced in mammalian cells is known to be 1,000 per cell (6 x 10^{-12}gm DNA) per Gray (Elkind and Redpath, 1977). From this yield the expected yield of damaged bases in mammalian cells can be calculated (Ward 1985) making reasonable estimates of the relative amounts of sugar damage and strand break damage from "direct" ionization of DNA and from OH radical reactions. This calculated yield (2,700 per cell per Gray) can be compared to the actual yields measured in Dizdaroglu's laboratory for DNA irradiated within mammalian cells (Mori and Dizdaroglu, 1994, Mori, et al., 1993, Nackerdien, et al., 1992) and the comparison is shown in Table 2.

One caveat regarding these relative yields is that they do not represent the total yields of damaged sugars: damage (probably in the sugar moiety) can lead to strand breaks under alkaline conditions (e.g. in which conditions the cellular yields of SSB are measured and from which the expected base damage yield is calculated). Alkali treatment (equivalent to that used in the measurement of SSBs in cells) increases the yield of strand breaks by a factor of 1.75 (Lafleur et al. 1979). If this factor is taken into account the excess yields of base damage in cells (shown in Table 2) should be increased by the same factor. It is clear that the measured yields of radiation induced base damage are much greater than those predicted from our knowledge of the chemical reactions which give rise to such damage. Thus we must search for an alternate mechanism whereby the 'extra' base damage is produced. We have suggested (Ward and Milligan, 1997) that peroxyl radicals could be responsible. These are formed by reactions (2) and (3):

$$RH + {}^{\bullet}OH \rightarrow R^{\bullet} + H_2O \quad (2)$$

$$R^{\bullet} + O_2 \rightarrow RO_2^{\bullet} \quad (3)$$

Where RH represent a cellular molecule. Thus the hypothesis is that radicals produced elsewhere in the cell, being less reactive than an $^{\bullet}OH$ radical can migrate to the DNA and initiate reactions that cause the base damage. The frequency of this reaction will depend on several factors e.g. the presence of other molecules capable of scavenging peroxyl radicals. We have shown that such a radical, the methyl peroxyl produced as a consequence of the reaction of an $^{\bullet}OH$ radical with dimethylsulfoxide does cause damaged bases in DNA (Milligan et al. 1996). An important corollary is that the same methyl peroxyl radical does not react to cause single strand breaks in DNA.

Oxidized bases are found in unirradiated cellular DNA and are excreted in the urine of unirradiated animals (see e.g. Ames, et al. 1995). The mechanism whereby endogenous oxidation of DNA bases occurs is not defined. If peroxyl radicals caused by radiation can cause oxidized bases, then it is possible that such radicals, formed by autoxidation may be the precursors of the oxidized bases produced endogenously. Again, such radicals being less reactive than $^{\bullet}OH$ radicals can be produced at some distance from the macromolecule and still migrate and react with it. There is an additional corollary to this hypothesis, if it is correct then endogenous oxidative processes (caused by peroxyl radicals) would not cause strand breaks, whereas radiation would since it is clear that $^{\bullet}OH$ radicals are the strand break precursors (Roots and Okada 1972). This could be the explanation for the effectiveness of low doses of radiation in inducing a variety of genes. The question has been: if radiation is to cause induction of genes then it must form a product in significant yield at low doses which is not present from oxidative processes.

4. BIOLOGICALLY SIGNIFICANT RADIATION DAMAGE

Elsewhere we have argued that singly damaged sites induced by ionizing radiation are unimportant as far as cellular end-points are concerned (Ward, 1995). This conclusion was based on three lines of evidence:

I. Endogenous damage to DNA bases would required 10s of Gray to produce therefore why should doses of less than 0.5 Gray cause a doubling of the mutation frequency.
II. Oxidized bases are also produced by treatment of mammalian cells with hydrogen peroxide yet the presence of base damage induced by this means, in amounts which would require radiation doses of hundreds of Gray, cause no increase in mutation frequency.
III. High LET radiation is more effective than low LET in causing cell death, mutations, etc., yet there is no reason to suppose that it is more effective in causing singly damaged sites in cellular DNA.

We concluded that it is the multiply damaged sites (MDS) produced by ionizing radiation which are the source of the biological consequences. These MDS, sometimes called clusters of damage, are a consequence of the non-uniform deposition of ionizing radiation energy. From low linear energy transfer (LET) radiation (e.g. γ-rays or X-rays) energy is deposited in amounts averaging 60 eV, i.e., enough energy to produce several radical pairs in a volume of nanometer dimensions. These 'clusters' of radicals react locally producing several damaged moieties in proximity on the DNA double strand (For review see Ward 1988). We have measured the yields of the MDS which include base damages (Milligan et al. 1996). After irradiation of supercoiled plasmid DNA, the yields of single and double strand breaks) were measured after separation of the open circle and linear forms of the molecule by agarose gel electrophoresis. Aliquots of the irradiated DNA were treated with the base damage specific enzymes endonuclease III and formamido-pyrimidine glyclosylase, the former measuring the damaged pyrimidines and the latter damaged purines. We found that enzyme treatment increased the yields of both single and double strand breaks (J.R. Milligan et al., in preparation).

The increase in SSB by the enzymes individually is approximately that expected from the yields which are quoted in Table 2, although Laval (These Proceedings) has indicated that the base damage specific endonucleases do not cut DNA at the site of every base damage. Treatment with both enzymes produces an increase which is approximately the sum of the yields for treatment with the enzymes individually. This indicates that there is little overlap in the activities of the enzymes. Therefore the yield of abasic sites, which are sensitive to both enzymes, must be minimal (as previously indicated by LaFleur et al. 1979). The increase in yield of DSB is more than that for SSB and is approximately the square of that value. This is the increase in yield which is expected considering the various double damaged sites which can contribute to this yield (Milligan et al, in preparation).

A corollary of the large increase in DSB produced by treatment with both enzymes is that the yield of DSBs represents only a small fraction of the total yield of MDS (less than 10%). Therefore consideration should be given to the biological significance of these MDS as well as to DSB. Previous calculations of the yield of point mutations in the exons of HPRT genes with respect to the yields of DSB indicated an almost 1:1 agreement (Ward et al. 1994). Since it is known that a majority (~65%) of DSBs are rejoined accurately (Löbrich et al. 1995) it could be inferred that lesions other than DSBs are also responsible for causing mutations; MDS involving base damage could be such lesions.

5. OTHER VARIABLES OF MULTIPLY DAMAGED SITES

As well as the composition in terms of base damages and strand breaks, MDS can vary in two more aspects:

I. The multiplicity of damage can range upward from two constituent individual lesions. This aspect will be more apparent as the LET of the radiation increases (Ward, 1981). A corollary of this aspect is that these more complex lesions would be even more difficult to repair- this may be the explanation for the lower amount of DSB repair occurring after high LET radiation (Heilman et al. 1996.)

II. The spacing of the lesions can vary even for low LET radiation. The size of the initial energy deposition event is of the order of 3nm diameter, this coupled with the limited diffusional range of the consequent OH radicals before it reacts (around 3nm) indicates that damages could be as much as 25 base pairs apart. It should be remembered that the standard laboratory methods for the quantitation of DSB are not sensitive to the distance of separation of the two SSBs. Thus the observation that a large proportion of the DSBs which are rapidly repaired (Ward et al. 1991) may reflect those DSBs in which the constituent lesions are far from each other and are repaired as individual lesions. It could be suggested that only DSB, and indeed MDS, which have their constituent lesions closely opposed would be the forerunners of biological damage.

6. EXTRAPOLATION FROM SOME RADIATION RESULTS TO OXIDATION DAMAGE

The conclusion that singly damaged sites produced by radiation are biologically unimportant indicates that the idea stated in terms such as "Oxidation by products of normal metabolism cause extensive damage to DNA, proteins and lipids. We argue that this damage, the same as that produced by radiation is a major contribution to degenerative processes of aging" (Ames et al. 1995) must be interpreted carefully. It is clear that radiation does produce the same single lesions as oxidation (although SSB have not been shown to be produced by oxidation in mammalian cells). However the yield of these single lesions from biologically significant doses of radiation are inconsequential as far as biological damage is concerned. Thus equating oxidation damage to radiation damage can lead to misinterpretation: One mistaken conclusion, which has been voiced, is "since radiation produces so little damage compared to endogenous oxidation the safe levels of radiation exposure can be increased." Clearly the other lesions- those which are biologically effective at low radiation doses (MDS) - and which are not produced by oxidation must be taken into account.

A second contribution of radiation studies to oxidation damage is in the identification of oxidative lesions. The types of oxidized bases produced by radiation have been known for many years and molecular mechanisms whereby they are produced have been determined (von Sonntag, 1987). However, some of the newly discovered radiation lesions are not considered in the oxidation field. Budzinski et al. (1995) have studied the radiation chemistry of model oligomers and have found by examination of the products by NMR spectroscopy that in addition to the lesions produced by attack of a single •OH radical (discussed above) formamido products of the pyrimidines are produced as well as tandem base lesions. Tandem lesions are made up an 8-hydroxyguanine next to a formamido residue. The experiments indicate that these tandem lesions are not formed as a consequence of the attack of two radicals (since they are formed linearly with radiation dose). The production of such a double lesion is suggested to occur by reaction of the base radical formed by reaction of the •OH radical with the neighboring base. If oxidative damage occurs through the intermediary of the same initial base radical then tandem lesions may also be produced in autoxidative damage. Budzinski et al. (1995) also found the formation of single formamido lesions from both pyrimidines, each of these tandem lesions is formed in yield equal to about the yield of 8-OHG moieties in the absence of a neighboring damaged base. These lesions have not been detected by the GC/MS assay.

Another double lesion which has been hypothesized to be formed by the attack of a single •OH radical is a DNA DSB (Siddiqi and Both, 1987). These investigators showed the production of DSBs by radiation even under conditions where each damaged site induced could only be due to the attack of a single •OH radical. One might expect therefore that if oxidizing damage is initiated by the same oxidizing species as that operating in radiation damage that DSBs should be induced. However, as mentioned above, the DNA damage produced by oxidation (altered bases) may result from reactions of peroxyl radicals, not •OH radicals. If this is correct then no DSBs would result since peroxyl radicals would not be expected to react with the deoxyribose moieties and hence would not produce strand breaks. These DSB produced by single •OH radicals are formed in yield too low to account for the measured intracellular DSB yields hence the MDS mechanism is invoked.

7. REPAIR OF RADIATION DAMAGE

The majority of the singly damaged sites produced by radiation are expected to be readily repaired since they are the types of damage produced by endogenous oxidation (Lindahl, 1990). However, radiation induced MDS, which are not expected to be produced by oxidation, may present problems for the cells repair systems. Chaudhry and Weinfeld (1995, 1997) and Wallace and Harrison (these proceedings) have investigated the abilities of repair enzymes to process a lesion on one strand in the presence of a lesion on the other strand. They have found that the enzymes have difficulty reacting with target damage if another damaged site is within a few base pairs.

Another consequence of the induction of a DSB in DNA is the possibility that the two ends of the break are not rejoined with each other, misrejoining occurs with an end of another DSB. The standard measures of DNA DSB rejoining can not distinguish accurate rejoining from misrejoining. Recently Löbrich et al (1995) devised a method of measuring accurate rejoining and were able to show that for GM38 human fibroblasts ~65% of the DSB were accurately rejoined in 2h but that misrejoining continued for a much longer time period. Thus the crucial time period for the rejoining of DSB is 2 h which corresponds with the time during which the repair of sub-lethal damage occurs (Hall 1995). Of course this repair (amount and kinetics) may be dependent on many factors (cell line, stage in cell cycle etc.).

The severity of MDS (e.g. number of lesions within a single site) has been suggested as the cause of the greater radiobiological effectiveness (RBE) of radiation depositing its energy at a greater density (high linear energy transfer - LET) (Ward, 1981). It was hypothesized that the MDS become more complex as a consequence of the greater amount of energy deposited in a local volume by high LET radiation (e.g. producing a greater concentration of radicals).

Of course this is not the case for DNA irradiated in dilute solution with high LET e.g. α-particles. We showed (Milligan et al. 1996) that the SSB yield in these circumstances followed faithfully the yield of •OH radicals - as with low LET radiation.

However, within a cell radicals do not move far from their place of origin and consequently the high local concentration of radicals from high LET radiation causes complex MDS. Because of the complexity of this damage it was hypothesized that the cell would have greater difficulty repairing this damage (Ward, 1981). Evidence of this effect has now appeared (Heilman et al. 1996). In an ingenious experiment these authors embedded CHO-K1cells in an agarose strip prior to irradiation. The strip was then aligned so that the radiation beam entered one end and deposited its energy along the length of the strip. The energy of the particles was such that the Bragg peak (corresponding to the region of highest LET) occurred within the strip. After irradiation the DNA was electrophoresed from the strip transversely into an agarose gel. If the electrophoresis was carried out immediately after irradiation the yield of DSB was almost independent of the cells' original positions in the strip. Hence the yield of initial damage does not change with LET. However, if the cells are allowed time to repair prior to electrophoresis, then the DNA of those irradiated with particles before the high LET region, i.e., prior to the Bragg peak, were capable of repairing their DSB, whereas those irradiated within the Bragg peak were not. In fact the yield of unrepaired DSB as a function of depth in the agarose strip tracks the RBE of the irradiating particle. This experiment is an excellent demonstration of the difficulty which cells have in repairing DSB produced by high LET radiation.

ACKNOWLEDGMENTS

Work carried out in the author's laboratory is supported by grant CA46195 from the National Institutes of Health.

REFERENCES

B.N. Ames, L.S. Gold and W.C. Willett (1995) The causes and prevention of cancer. *Proc. Nat. Acad. Sci.* (U.S.A.) **92**, 5258–65.

E.E. Budzinski, J.D. Dawidzik, J.C. Wallace, H.G. Freund and H.C. Box (1995) The radiation chemistry of d(CpGpTpA) in the presence of oxygen. *Radiat. Res.* **142**, 107–09.

G.V. Buxton, C.L. Greenstock, W.P. Helman and A.B. Ross (1988) Critical review of rate constants for reactions of hydrated electrons, hydrogen atoms and hydroxyl radicals (OH/O^-) in aqueous solution. *J. Phys. Chem. Ref. Data* **17**, 513–886.

M.A. Chaudhry and M. Weinfeld (1995) The action of the Escherichia Coli endonuclease III on multiply damaged sites in DNA. *J. Molec. Biol.* **23**, 914–922.

M.A. Chaudhry and M. Weinfeld (1997) Reactivity of human apurinic/apyrimidinic endonuclease and Escherichia Coli exonuclease III with bistranded abasic sites in DNA. *J. Biol. Chem.* **272**, 15650–55.

M. Dizdaroglu (1993) Quantitative determination of oxidative base damage in DNA by stable isotope-dilution mass spectrometry *FEBS Lett.* **315,** 1–6.

T. Douki, T. Delatour, F. Bianchini and J. Cadet (1996) Observation and prevention of an artefactual formation of oxidized DNA bases and nucleosides in the GC-EIMS method. *Carcinogenesis* **17**, 347–353.

M.M. Elkind and J.L. Redpath (1977) Molecular and cell biology of radiation lethality. In Cancer, A Comprehensive Treatise, Becker, F.F. editor *Plenum Press* **6**, 51–99.

A.F. Fuciarelli, B.J. Wegher, W.F. Blakely and M. Dizdaroglu (1990) Yields of radiation-induced base products in DNA: effects of DNA conformation and gassing conditions. *Int. J. Radiat. Biol.* **58**, 397–415.

E.J. Hall (1994) Radiobiology for the Radiologist. J.P. Lippincott, Philadelphia, PA.

B. Halliwell and M. Dizdaroglu (1992) The measurement of oxidative damage to DNA by HPLC and GC/MS techniques. *Free Rad. Res. Commun.* **16**, 15–28.

J. Heilman, G. Taucher-Scholz, T. Haberer, M. Scholz and G. Kraft (1996) Measurement of intracellular DNA double-strand break induction and rejoining along the track of carbon and neon particle beams in water. *Int J. Radiat. Oncol. Biol. Phys.* **34**, 599–608.

W.D. Henner, L.O. Rodriguez, S.M. Hecht and W.A. Hazeltine (1973) γ-Ray induced deoxyribonucleic acid strand breaks. *J. Biol. Chem.* **258**, 713–716.

C.D. Jonah and J.R. Miller (1977) Yield and decay of the OH radical from 200 ps to 3ns. *J. Phys. Chem.* **81**, 1974–1976.

M.V.M. Lafleur, J. Woldhuis and H. Loman (1979) Alkali-labile sites and post-irradiation effects in gamma-irradiated biologically active double-stranded DNA in aqueous solution. *Int. J. Radiat Biol.* **36**, 241–247.

T. Lindahl, (1990) Repair of intrinsic DNA lesions. *Mutat. Res.* **238**, 305–311.

M. Löbrich, B. Rydberg, and P. K. Cooper (1995) Repair of x-ray-induced DNA double strand breaks in specific *Not* I restriction fragments in human fibroblasts: Joining of correct and incorrect ends. *Proc. Nat. Acad. Sci.* (U.S.A.) **92**, 12050–54.

J.R. Milligan, J.A. Aguilera and J.F. Ward (1993) Variation of single strand break yield with scavenger concentration in plasmid DNA irradiated in aqueous solution. *Radiat. Res.* **133**, 151–157.

J.R. Milligan, J.A. Aguilera, C.C.L. Wu, J.Y-Y. Ng and J.F. Ward (1996) The difference that LET makes to the precursors of DNA strand breaks. *Radiat. Res.* **145**, 442–448.

J.R. Milligan, J.Y. Ng., C.C.L. Wu, J.A. Aguilera, J.F. Ward, Y.W. Kow, S.S. Wallace, S.S. and R.P. Cunningham (1996). Methyperoxyl radicals as intermediates in DNA damage by ionizing radiation. *Radiat. Res.* **146**, 436–444.

T. Mori and M. Dizdaroglu (1994) Ionizing radiation causes greater DNA base damage in radiation-sensitive mutant M10 cells than in parent mouse lymphoma L5178Y cells. *Radiat. Res.* **140**, 85–90.

T. Mori, Y. Hori and M. Dizdaroglu (1993) DNA base damage generated *in vivo* in hepatic chromatin of mice upon whole body γ-irradiation. *Int. J. Radiat. Biol.* **64**, 645–650.

Z. Nackerdien, R.Olinski and M. Dizdaroglu (1992) DNA base damage in chromatin of irradiated cultured human cells. *Free Radical Res. Commun.* **16**, 259–273

R. Roots and S. Okada (1972) Protection of DNA molecules of cultured mammalian cells from radiation-induced single strand scissions by various alcohols and SH compounds. *Int. J. Radiat. Biol.* **21**, 329–342.

G. Scholes, J.F. Ward and J. Weiss (1960) Mechanisms of the radiation induced degradation of nucleic acids. *J. Molec. Biol.* **2**, 379–381.

J.W.T. Spinks and R.J. Woods (1976) *An Introduction to Radiation Chemistry* (2nd Edition) John Wiley and Sons, New York.

S.G. Swarts, G.S. Smith, L. Miao and K.T. Wheeler (1996) Effects of formic acid hydrolysis on the quantitative analysis of radiation-induced DNA base damage products assayed by gas chromatography/mass spectrometry. *Radiat. And Environ. Biophys.* **35**, 41–54.

C. von Sonntag (1987) *The Chemical Basis of Radiation Biology*, (Taylor and Francis, London)

J.F. Ward (1975) Molecular mechanisms of radiation induced damage to nucleic acids. *Adv. Radiat. Biol.* **5**, 182–240.

J.F. Ward (1981) Some biochemical consequences of the spatial distribution of ionizing radiation produced free radicals. *Radiat. Res.* **86**, 185–95.

J.F. Ward (1985) Biochemistry of DNA lesions. *Radiat. Res.* **104**, S103-S111.

J.F. Ward (1988) DNA damage produced by ionizing radiation in mammalian cells: identities, mechanisms of formation and repairability. *Progress in Nucleic Acids and Molecular Biology*, **35**, 95–125.

J.F. Ward, C.L. Limoli and P.M. Calabro-Jones (1991) An examination of the repair saturation hypothesis for describing shouldered survival curves. *Radiat. Res.* **127**, 90–96.

J.F. Ward, J.R. Milligan, and G.D.D. Jones (1994) Biological consequences of non-homogeneous energy depositions by ionizing radiation. *Radiat. Protec. Dosim.* **52**, 271–276.

J.F. Ward, (1995) Radiation Mutagenesis: The initial DNA lesions responsible. *Radiat. Res.* **142**, 362–368.

J.F. Ward and J.R. Milligan (1997).. Four mechanisms for the production of complex damage. *Radiat. Res.* **148**, 481–83.

ABSTRACTS

PRODUCTION AND CHARACTERIZATION OF MONOCLONAL ANTIBODIES AGAINST THE PROTEIN ANPG; HUMAN 3-METHYLADENINE DNA GLYCOSYLASE

N. Aarsæther, V. Andresen, H. Eknes, O. Sidorkina* and J. Laval*

Department of Molecular Biology, University of Bergen, Norway and *Institute Gustave Roussy, 94805 Villjuif Cedex, France

Chemical alkylation of DNA is the consequence of exposure to environmental agents or cellular metabolism. In mammalian cells repair of cytotoxic lesions associated with DNA methylation like 3-methyladenine, 7-methylguanine and methylguanine as well as ethylated bases is performed by 3-methyladenine DNA glycosylase (3MeAde).

We have produced monoclonal antibodies (Mabs) against a 26kDa truncated human 3-MeAde (1,2). By performing Western blot using extracts from human cells each of 4 Mabs was immunological reactive with a protein of 34 kDa which is identical to the predicted size of the wild type 3-MeAde.

We are in progress to characterize these Mabs by epitope mapping and to elucidate the intracellular localization of human 3-MeAde and its importance by performing cytoimmunofluorescence and immunoprecipitation using these Mabs.

References

1. O´Connor T.R. and Laval J. Human cDNA expressing functional DNA glycosylase excising 3-methyladenine and 7-methylguanine. (1991) BBRC 176, 1170–1177.
2. O´Connor T.R. Purification and characterization of human 3-metyhyladenine DNA glycosylase. (1993) Nucleic Acids Res. 21, 5561–5569.

DNA OXIDATION AND REPAIR FOLLOWING TREATMENT OF PRIMARY RAT HEPATOCYTE CULTURE WITH THE FLAVONOID MYRICETIN

Valerie Abalea, Josiane Cillard, Odile Sergent, Pierre Cillard and Isabelle Morel

Laboratoire de Biologie Cellulaire, INSERM U456, Groupe Détoxication Réparation Tissulaire UFR des Sciences Pharmaceutiques, 2 av. du Pr. L. Bernard, Rennes, France

Oxidative DNA damage and repair in primary rat hepatocyte cultures was investigated following 4 hours incubation with the flavonoid myricetin (100 and 300 μM), in the presence or absence of iron (Fe-NTA 100 μM). Seven DNA oxidation products were identified and quantified in cellular DNA by gas chromatography-mass spectrometry (GC-MS) in selected ion monitoring mode. Concomitantly, DNA repair capacity of hepatocytes was estimated by the elimination of oxidized-base products in the culture media, using the same GC-MS method. In hepatocyte cultures supplemented with higher concentration of myricetin (300 μM) or with iron alone, intense DNA oxidation was noted, whereas no DNA repair could be detected. However, supplementation of hepatocyte cultures with the lower concentration of myricetin (100 μM) alone or with both myricetin (100 or 300 μM) and iron did not result in accumulation of oxidation products in cellular DNA but rather led to their elimination in culture media showing an induction of DNA repair mechanisms. A high level of DNA oxidation in hepatocytes could then be related to a low activity of their DNA repair mechanisms. This revealed that myricetin presented a dual effect of being either pro- or antioxidant towards DNA bases depending on the concentration used and on the presence of added iron. Moreover, mechanisms of adverse effects of myricetin towards DNA might be explained by modifications in DNA repair capacity. With regard to lipid peroxidation and enzyme leakage as indices of cytotoxicity in hepatocyte cultures, myricetin (100 and 300 μM) presented antilipoperoxidant and cytoprotective activities in the presence or absence of iron. This showed that induction of DNA oxidation by myricetin was independent of lipid peroxidation and did not create immediate cytotoxicity. The oxidative genotoxic potential of myricetin at high concentration, modifying DNA repair mechanisms, should be keep in mind in further understanding mutagenesis and carcinogenesis processes involving dietary phenolic antioxidants.

SPONTANEOUS MUTATION PATTERNS IN EXON 3 OF HUMAN *HPRT* EXAMINED IN A *SACCHAROMYCES CEREVISIAE* SYSTEM

Elisabeth A. Bailey (1), William Thilly (2), and Bruce Demple (1)

(1) Harvard School of Public Health, 665 Huntington Ave, Boston MA 02115, USA; (2) Massachusetts Institute of Technology, 77 Massachusetts Ave, Cambridge MA 02139, USA

Increased levels of spontaneous mutations have been associated with many human genetic diseases. The cellular processes that influence spontaneous damage and the spontaneous mutation rate, however, are largely unknown. Metabolic by-products are a significant factor contributing to spontaneous mutagenesis, and perhaps the most predominant lesions generated by endogenously-produced agents are AP sites. AP endonucleases constitute a set of enzymes that initiate repair of potentially lethal and/or mutagenic 3'-blocking lesions and AP sites in DNA. The ultimate goal of the work presented here involves examination of the role of the major yeast *S. cerevisiae* AP endonuclease (Apn1) in the control of spontaneous mutations in exon 3 of the human hprt gene.

First, a region containing exon 3 of *hprt* and 1kb of flanking DNA on either side was integrated into the yeast chromosome downstream from the *ARS309* origin of replication. The mutational target is positioned downstream from a *GAL1* promoter. The orientation of the target sequence relative to the promoter and to the origin of replication has been varied. This allows us to determine whether endogenously induced AP sites are repaired in a replication- or transcription-coupled fashion. Next, to identify and isolate mutations occurring in the target sequence of wild-type *S. cerevisiae*, at a rate estimated to be as low as 10^{-8} per cell per generation, cells were grown for 200 generations in order to enrich the total mutant population. The mutation types and frequencies of exon 3 of *hprt* were then examined by using CDCE/hifi PCR (Constant Denaturing Capillary Electrophoresis / hi-fidelity PCR). The work presented here describes the analysis of mutations in exon 3 of hprt as determined for the wild-type *S. cerevisiae* strain for eventual comparison to the mutator strain apn1-.

MECHANISM OF KINETIN FORMATION IN DNA

J. Barciszewski[1)], S.I.S. Rattan[2)], G.E. Siboska[2)], B.O. Pedersen[2)] and B.F.C. Clark[2)]

[1)]Institute of Bioorganic Chemistry, Polish Academy of Sciences, Poznan, Poland; [2)]Aarhus University, Aarhus, Denmark

For about 40 years, kinetin (furfuryladenine) has been isolated from autoclaved herring sperm DNA and was shown to act as a cytokinin plant growth hormone. It has been assumed to be an artificial rearrangement product of the autoclaving of DNA and not to occur naturally. Since its discovery, this compound has been widely used as a cytokinin in various aspects of plant science, including its application in plant biotechnology and cell biology. Recently, kinetin has been shown to delay the onset of many age-related characteristics that appear in normal human skin fibroblasts undergoing ageing in vitro and prolongs the lifespan of the fruit fly. Because of intriguing effects of kinetin, we decided to reinvestigate the question concerning its origin and its biological properties using electrochemical (EC) analysis and mass spectrometric detection with very high sensitivity.

A mechanism of kinetin synthesis involves a reaction of furfural with an adenine residue at a DNA level. It has been shown that furfural is formed during the oxidative damage of DNA by hydroxylation of the C5' deoxyribose. These findings clearly explain why kinetin has been found only in stored samples of DNA after autoclaving but not in freshly prepared DNA. The proposed mechanism for the formation of kinetin links, for the first time, this modified base with oxidative damage processes occuring in the cell.

THE MODULATING EFFECTS OF QUERCETIN ON FOOD MUTAGENS IN HUMAN BLOOD AND SPERM SAMPLES IN THE COMET ASSAY

Nursen Basaran[1], A. Ahmet Basaran[2], and Diana Anderson[3]

Hacettepe University, Faculty of Pharmacy, [1]Department of Pharmaceutical, Toxicology, [2]Department of Pharmacognosy, Ankara, Turkey, [3]BIBRA, International, Woodmanterne Road, Carshalton, Surrey, UK

The chemical and biological activities of many flavonoids such as quercetin have been the subject of extensive studies for many years. Considering the widespread occurrence of quercetin in foods relatively little work has been done on the toxicity of the compound and identified as as either mutagen, antimutagen or both. The present study was carried out the investigate the modulating effects of quercetin in human lymphocytes and human sperms on two known food mutagens Trp (3-amino-1-metyl-5H pyrido (4,3b) indole acetate) and IQ (2-amino-3-metyl-3H imidazol(4,5f) quinoline) in the new Comet technique which provides a relatively simple assay for measuring DNA strand breakage in single cells. A computerized image analysis system (Comet 3.0 Kinetic imaging Ltd.) was used to measure Comet parameters. The parameters chosen for presentations are the tail moment and the percentage head of DNA for the lymphocytes and sperms respectively. In human lymphocytes, positive increases in tail moments hence DNA damage were both seen with quercetin with quercetin with and without metabolic activation. The food mutagens, Trp and IO also induced positive responses alone. At highest dose of 500 μM quercetin in combination with Trp and IQ there was a protective, antigenotoxic effect. At its lowest dose of 100 μM there was an exacerbating effect with IQ. In human sperm over a similar dose range there was no DNA damage and there was a protective effect in combination with the food mutagens. In conclusion, quercetin can modulate the effects of food mutagens and exhibit antigenotoxic effects since DNA damage was reduced in the Comet say in human lymphocytes and sperms.

DNA DIVERSITY IN THE SEPARATED HIGHLAND ISOLATE

Rita Balamirzoeva

Department of Ethnic Population Genetics, Institute of History, Archeology and Ethnography, Daghestanian Center of Russian Academy of Sciences, Russia

The problems of gene pool preservation under ongoing environmental pollutions are urgent both for small and large ethnic populations. An obvious and effective solution of this problem is to examine the population rates of genetic changes in the gene pool of isolated and inbred populations of the ethnic minorities, because changes here are more transparent during environmental changes than in the more heterogeneous populations of large outbred ethnic groups [Chakraborty, Chakravarti, 1977; Bittles, Makov, 1992; Bulayeva, 1991; Bulayeva, Pavlova,1993]. Among the most interesting ethnic isolates for the study of human genetic adaptation to the changing environments are the populations located in the Daghestan (Northern Caucasus,Russia). This region comprises at least 30 different ethnic groups and exhibits a considerable diversity of languages and environmental habitats. In the long-term Daghestanian population-genetic study [Bulayeva et al, 1974–1997] were studied genetic adaptation of highland isolates to the new for them lowland area. The paper describes the results of study of the DNA polymorphism among the members of the separated in 1944 highland isolate. DNA was analyzed with two single-locus multiallelic probes. One of this is Miotonin Protein Kinase (MPK) and other is Angiotenzinogene (ANG). The most polymorphic loci known in the human genome are tandemly repeated microsatellites (MCS) [Ryskov et al., 1996], which were used in our study. The MPK and ANG were characterized by polymerase chain reaction (PCR). The results obtained showed that among highland and migrant parts of the separated isolate were found a statistical significant differences by frequencies of the repeats of ANG and MPK. It was shown the significant relationships between inbreeding and polymorphism of the microsatellites under study.

GREEN FLUORESCENT PROTEIN EXPRESSION AFTER UV-IRRADIATION

C. Baumstark-Khan, M. Palm, and G. Horneck

DLR, Institute of Aerospace Medicine, 51147 Köln, Germany

Green fluorescent protein (GFP) is a reporter system for monitoring in-vivo gene expression superior to other assays as GFP fluorescence does not require specific substrates or cofactors. CHO cells were co-transfected with the pCX-GFP (β-actin/β-globin gene promoter) and a neomycin resistence transmitting plasmid. Resulting stable cell clones were selected and maintained using G418. CX-GFP codes for a red-shifted GFP-variant (488 nm excitation, 509 nm emission) enabeling visualization of GFP expression by spectroscopy, fluorescence microscopy or FACS-analysis (488 nm excitation, FITC filters).

GFP expressing cells were subjected to different fixation/staining protocols. Formaldehyde-fixed (FA) cells display GFP fluorescence comparable to living cells. Co-staining of DNA is possible in FA-fixed cells using Hoechst 33258 or DAPI. Propidium iodide (PI) staining of DNA is difficult, due to the fact that GFP-fluorescence in ethanol-fixed samples is not stable and in FA-fixed samples there is some overlapping of emission spectra from GFP and PI.

DNA repair proficient and deficient CHO strains were used for survival tests after UVC-irradiation. Cells carrying the GFP-construct show no difference in cellular survival compared to cells not carrying the construct. After certain post-irradiation growth periods cells were harvested and the numbers of GFP expressing cells and fluorescence were determined by FACS analysis. Generally, GFP fluorescence in irradiated cells is not seen when cell membranes are damaged (leak-out of the soluble GFP). Irradiated cells without membrane damage express GFP continuously (increasing GFP contents).

IS FREE RADICAL SCAVENGING ACTION OF ACE INHIBITORS RELATED TO THEIR PROTECTIVE EFFECT ON REPERFUSION ARRHYTHMIAS IN RATS?

Mustafa Birincioglu, Tijen Aksoy, Ercument Ölmez, and Ahmet Acet

Department of Pharmacology, Faculty of Medicine, Inönü University, Malatya, Turkey

The antiarrhythmic effects of captopril, a sulphydryl-containing angiotensin converting enzyme (ACE) inhibitor, were compared with those of the non-sulphydryl-containing ACE inhibitor lisinopril and the sulphydryl-containing agent glutathione in an *in vivo* rat model of coronary artery ligation. To produce arrhythmia, the left main coronary artery was occluded for 7 min, followed by 7 min of reperfusion. Captopril (3 mg kg^{-1}) and lisinopril (0.1, 0.3 or 1 mg kg^{-1}) caused marked decreases in mean arterial blood pressure (BP) and heart rate, whereas glutathione (5 mg kg^{-1}) had no effect on them. The incidence of ventricular tachycardia (VT) on ischemia and reperfusion was significantly reduced by captopril and lisinopril. Captopril and 1 mg kg^{-1} lisinopril also significantly decreased the number of ventricular ectopic beats (VEB) during occlusion and the duration of VT on reperfusion, respectively. These drugs also attenuated the incidence of reversible ventricular fibrillation (VF) and the number of VEB during reperfusion. However, glutathione only reduced the incidence of VT on reperfusion, significantly. These results suggest that, in this experimental model, ACE inhibitors limit the arrhythmias following ischemia-reperfusion and free radical scavenging action of these drugs does not have a major contributory role in their protective effect.

HETEROLOGOUS EXPRESSION, PURIFICATION AND CHARACTERIZATION OF THE PREFORMS OF MITOCHONDRIAL AND NUCLEAR HUMAN URACIL-DNA GLYCOSYLASES

Sangeeta Bharati, Geir Slupphaug and Hans E. Krokan

UNIGEN Center for Molecular Biology, Norwegian University of Science and Technology, N-7005 Trondheim, Norway

The mitochondrial (UNG1) and the nuclear (UNG2) forms of human UDG are generated from the human *UNG* gene by alternative splicing and different transcriptional start points. These enzymes are synthesized as preproteins differing in their N-terminal sequences, which contain the signal sequences for their respective intracellular transport. Although a shorter and highly active form of UDG, lacking 84 aminoacids from the N-terminal in UNG1 (UNGΔ84), has been cloned, purified and characterized extensively, not much is known about the significance of the two presequences. The present work aims at the cloning, expression and purification of the full-length forms of human UDG using two high expression systems: Baculovirus and *Pichia pastoris*. UNG1 was cloned in the baculovirus vector, pBacPAK8 and expressed and purified to homogeneity from the baculovirus-infected insect cells. However, amino-acid sequencing showed that the purified protein lacked 28 amino acids from the NH_2-terminal (UNGΔ28). This preform of UNG1 has been characterized biochemically and has been found to have similar properties as the UNGΔ84 form. UNGΔ28 is cleaved by HeLa mitochondrial extracts to aprotein of similar size as the UDG isolated from mitochondria. No such cleavage is observed with nuclear extracts suggesting that UNGΔ28 is a partially processed form of the mitochondrial protein, UNG1. The second alternate expression system that was used is the methylotropic yeast, *Pichia pastoris*, in an attempt to express and isolate the full-length forms of both UNG1 and UNG2. We have successfully cloned the two forms in *Pichia* and are presently in the process of identifying the optimal conditions for the expression of the two proteins. Subsequently, UNG1 and UNG2 will be purified and characterized both structurally and functionally with the objective of understanding the significance of the two presequences and also the differential intracellular localization of the two proteins.

5-FORMYLDEOXYURIDINE-INDUCED MUTAGENESIS IN BACTERIA AND MAMMALIAN CELLS

Svein Bjelland[1], Arne Klungland[2], Yoshihito Ueno[3], Akira Matsuda[3] and Erling Seeberg[2]

[1]School of Science and Technology, Stavanger College, Ullandhaug, PO Box 2557, N-4004 Stavanger, [2]Division for Molecular Biology, Institute of Medical Microbiology, National Hospital, University of Oslo, N-0027 Oslo, Norway, [3]Faculty of Pharmaceutical Sciences, Hokkaido University, Kita 12, Nishi-6, Kita-ku, Sapporo 060, Japan

Oxidation of the methyl group of thymine yields 5-formyluracil (5-foU) as a major product. When free 5-foU and 5-formyldeoxyuridine (5-fodUrd) are supplied to the growth medium of Chinese hamster fibroblast cells, an increased mutation frequency is observed at the hypoxanthine-guanine phosphoribosyltransferase locus, 5-fodUrd being more potent than 5-foU. These results indicate that 5-foU and 5-fodUrd serve as nucleotide precursors and are incorporated in the DNA during replication, exhibiting other base pairing characteristics than thymine itself. Addition of 5-fodUrd to the Miller mutagenicity tester strains of *Escherichia coli* indicates that all possible base substitutions are formed - further suggesting that 5-foU is capable of forming base pairs with all the normal base residues in DNA. By measuring mutations to rifampicin resistance in various repair defective *E. coli* mutants, it was found that 5-fodUrd induced significantly more mutations in wild-type bacteria than in the excision repair defective mutants *alkA*, *uvrA*, *alkA uvrA* or *xth nfo* implying that mutations are fixed by excision repair. Work is in progress to establish the mutational spectra for 5-fodUrd-induced mutagenesis in the various repair defective mutants.

This work was supported by the Norwegian Cancer Society.

THE EFFECT OF INHIBITING ADENOSINE DEAMINASE ACTIVITY DURING RENAL ISCHEMIA, UPON THE POSTISCHEMIC RESTORATION OF ATP AND FORMATION OF LIPID PEROXIDATION IN RATS

Mustafa Vakur Bor*, Cemal Çevik**, Osman Durmus§, and Nurten Türközkan**

*Department of Clinical Biochemistry, Aarhus University Hospital, Norrebrogade 44, DK-8000 Aarhus C, Denmark, Department of Clinical Chemistry** and Surgery§, Gazi University School of Medicine

Tissue damage as a consequence of ischemia is a major medical problem in an industrialized society. Whereas the conventional view has attributed this injury process to ischemia itself, recent studies have found that a variable, but often substantial proportion of the injury is caused by toxic oxygen metabolites at the time of reperfusion. Renal ischemia results in a rapid decrease in tissue ATP and rise in the ATP degradation products adenosine, inosine, and hypoxanthine. When xanthine oxidase converts hypoxanthine to xanthine in the presence of molecular oxygen superoxide radical is generated. In the present study the effect of blocking the degradation of ATP during ischemia, upon the postischemic restoration of ATP and formation of superoxide mediated lipid peroxidation was investigated. The effect of adenosine deaminase inhibitor, deoxycoformisin, on renal malondialdehyde (MDA) concentrations, as an index of lipid peroxidation and on ATP levels were studied after 45 minute renal ischemia and 15 minute post ischemic renal perfusion in rats. ATP content of ischemic kidney dropped immediately after renal ischemia. The content of MDA in the kidney was increased 15 minute after reperfusion following renal ischemia. Both in ischemia and reperfusion period, deoxycoformisin treatment (2mg/kg) increased the ATP levels and decreased the MDA levels significantly compared to untreated animals. We concluded that, the application of adenosine deaminase inhibitor is thought to exert a two phased protective action in renal ischemia-reperfusion induced damage. The reduction of adenosine catabolism elevates tissue levels of adenosine and facilitating purine salvage for ATP resynthesis. By decreasing hypoxanthine formation, it deprives xanthine of its substrate and thus reduces free radical formation.

CHROMOSOME DAMAGE IN LYMPHOCYTES AND IN EXFOLIATED BUCCAL AND NASAL MUCOSA CELLS OF WELDERS

S. Burgaz[1], B. Tecimer[1], O. Erdem[1], M. Yilmazer[1], Y. Kemaloglu[2] and A.E. Karakaya[1]

Gazi University, [1]Fac. Pharmacy, Dept.Toxicol., [2]Fac. Medicine, Dept. Otolaryngol., 06330, Ankara, Turkey

It is known that welders have been exposed to fumes and gases such as ozone, NOx, and metal oxides generated from the welding processes, and to electric fields, magnetic fields and UV radiation.Welding fumes and gases are responsible for a high percentage of cancers of the lung, nasal, larynx, pancreas, myleoid leukaemia and non-hodgkin lymphoma. In this study micronucleus (MN) test was carried out in peripheral lymphocytes and exfoliated cells from buccal and nasal samples of welders (n=32) for the detection of probable genotoxic effects of welding fumes involved in welding processes. Twenty-five controls matched for age and smoking habits were used in the study. Urinary concentrations of nickel and chromium of welders were not significantly different from those of controls. The mean (±S.D.) MN frequencies (‰) in peripheral lymphocytes were 1.38±0.10 among the welders and 0.91±0.15 among the controls, a statistically significant difference ($p<0.05$). Micronucleated cells (‰) in the nasal cavity 1.28±0.20 in welders and 0.99±0.12 in control group, the difference being statistically significant ($p<0.001$). The mean frequency (‰) of MN in buccal cells in exposed group (0.85±0.20) was similar to that founded in control group (1.01±0.19). No correlation was found between the frequencies of MN in peripheral lymphocytes and two types of exfoliated cells in workers and between MN in buccal and nasal cells in all the exposed subjects. Our data may suggest that exposure to chemicals during welding processes induces genotoxic effects, however, it is difficult to determine with certainty which compund(s) were responsible for this observed genotoxic damage.

DNA OXIDATION BY ACTIVATED BLEOMYCIN

Richard M. Burger

Laboratory of Chromosome Biology, Public Health Research Institute, 455 First Avenue, New York, NY 10016, USA

Bleomycin, a *Streptomyces* product with clinically useful antitumor activity, initiates an oxidative attack on DNA deoxyribose which appears responsible for the drug's cytotoxicity. The bleomycins are water-soluble, basic glycopeptides averaging 1550 D. Domains common to all bleomycins bind to transition metal ions and to DNA. Iron-bleomycin can be activated to DNA-cleaving competence by pathways analogous to those of cytochrome P-450 activation. Bleomycin, Fe(II) and O_2 react to form the kinetically competent species, called activated bleomycin. It can also form from Fe(III)-bleomycin and peroxides without requiring a reductant or O_2. A ferric peroxide complex, it initates one- and two-electron oxidations. The proximate active drug species which mediates the actual DNA attack may be a shorter-lived product; peroxide cleavage is the rate-limiting reaction of activated bleomycin. The DNA products of activated bleomycin attack are associated with two kinds of DNA lesions, only one of which requires O_2 for formation, and this breaks DNA directly. *In vivo* experiments indicate that the induction of lysogenic bacteriophage by bleomycin requires O_2, unlike induction by mitomycin. *In vitro* each DNA cleavage is accompanied by the release of one equivalent of a malondialdehyde-like compound (base propenal) consisting of nucleic base and deoxyribose carbons 1–3. The O_2-independent reaction releases free base only, leaving the DNA polymer continuous but alkali-labile. With O_2 present, both products form in similar amounts, but the anaerobic yield of free base is no greater than that produced aerobically. A scheme accounting for this product partition is proposed.

FREE RADICAL CHEMISTRY ASSOCIATED WITH C-1' POSITION OF NUCLEOSIDES

Chryssostomos Chatgilialoglu

I.Co.C.E.A., Consiglio Nazionale delle Ricerche, Via P. Gobetti 101, 40129 Bologna, Italy

As research progresses in the area of the mechanism of attack of DNA and RNA oxidative cleavers, it has become evident that hydrogen abstraction from the C-1' position is involved in most cases. The fate of this originally formed C-1' radical species is not yet known. Furthermore, it is envisaged from the recent literature that C-1' radicals may constitute useful intermediates which can generate rich chemistry, currently unexplored but potentially important in medicinal chemistry. These considerations prompted us to undertake a systematic investigation of the radical chemistry associated with the C-1' position by utilizing modified nucleosides as models.

The production of C-1' radicals in model ribo- and 2'-deoxyribonucleosides has been achieved through either a direct modification of the C-1' position which can function as a precursor of a C-1' radical species, or through remote functionalization and utilization of known types of radical migrations to the C-1' position. Product studies coupled with time resolved spectroscopies (laser flash photolysis and pulse radiolysis) have been used to study the fate of C-1' radical species under anoxic or aerobic conditions. Thus, kinetic data from various steps involving these radicals have been obtained for the first time. The influence of the nature of the base residues on their reactivity has been studied and a mechanistic picture of the 2'-deoxyribonolactone formation under aerobic conditions has been proposed. These data can be further applied to the DNA oxidative cleavage mechanism which involves the hydrogen abstraction from the C-1' position.

A PILOT STUDY OF THE EFFECTS OF CIGARETTE SMOKING AND CYP2D6 GENOTYPE ON DNA STRAND BREAKAGE

Suzanne Cholerton, Carol Boustead, Michelle Wilson, and Iain Willits

Department of Pharmacological Sciences, University of Newcastle upon Tyne, NE2 4HH, UK and Jeffrey R Idle, Institute for Cancer Research and Molecular Biology, Norwegian University of Science and Technology, 7005 Trondheim, Norway

Inter-individual variation in expression of enzymes which mediate the activation and detoxification of chemical carcinogens is likely to be an important determinant of cancer susceptibility. The lack of CYP2D6 activity seen in individuals homozygous for mutant CYP2D6 alleles (poor metabolizers; PM) may contribute to their reduced risk of cigarette smoking-associated lung cancer. In this study we measured DNA damage in lymphocytes from 11 cigarette smokers and 7 non-smokers of known CYP2D6 genotypes using the single cell gel electrophoresis (comet) assay. The analysis of 395 comets from the cigarette smokers and 231 comets from the non-smokers revealed significantly ($P = 0.018$) more DNA damage in the smokers. However, within the cigarette smokers there was no significant difference in DNA damage between individuals homozygous for the CYP2D6 wild type allele (wt/wt), heterozygous (wt/v) and homozygous for mutant CYP2D6 alleles (v/v) (wt/wt: wt/v P = 0.513; wt/wt: v/v P = 0.297; wt/v:v/v P = 0.881) and no significant (P = 0.584) difference between DNA damage in those individuals with CYP2D6 genotypes consistent with the extensive metabolizer phenotype (wt/wt and wt/v) and those with the CYP2D6 genotype consistent with the PM phenotype (v/v). Similarly, within the non-smokers there was no significant (P = 0.699) difference in DNA damage between wt/wt and wt/v individuals. These preliminary results suggest that CYP2D6 genotype is not a major factor which determines the DNA damage associated with cigarette smoking.

GENE-SPECIFIC DAMAGE AND REPAIR DURING REPLICATIVE CELL SENESCENCE

Mette Christiansen[1], Suresh I.S. Rattan[1], Vilhelm A. Bohr[2] and Brian F.C. Clark[1]

Center for Molecular Gerontology, [1]Laboratory of Cellular Aging, Institute of Molecular and Structural Biology, University of Aarhus, Denmark, [2]Laboratory of Molecular Genetics, National Institute on Aging, Baltimore, USA

DNA damage accumulate during aging of cells *in vitro* and *in vivo*, and may lead to mutations and altered gene expression. Since accumulation of DNA damage is necessarily a signal of inefficient DNA repair, it is important to determine what happens during cellular aging to various repair mechanisms involved in maintaining the stability of the genome. Using Cockayne´s syndrome as a model of aging, decreased gene-specific repair has been shown in these cells as compared with normal human cells, substantiating the theory that there is a relation between decreased DNA repair and aging.

In our laboratory the cell culture of human osteoblasts has been established as an experimental system for cellular aging and senescence of bone cells. This system will be used for studying gene-specific repair of UV-induced damage during aging of fibroblasts and osteoblasts *in vitro*. Using the method developed by Bohr et al. (Cell 40, 359–369, 1985) for measuring the repair of UV-induced damage, we will be analyzing a selected set of genes. UV-damage is chosen as the first approach as it is a well-characterized system, but as oxidative damage is physiologically more important, repair of 8-hydroxyguanine will be measured in the same model systems. Furthermore, a comparison of DNA repair of damage in aging fibroblasts with that in aging osteoblasts will be undertaken, to determine the extent of cell-type-specific DNA repair.

STUDIES ON CHROMATIN STRUCTURE IRRADIATED WITH FAST NEUTRONS

Bogdan Constantinescu, Liliana Radu*, Roxana Bugoi and Doina Gostian*

Institute of atomic Physics, POB MG-6, Bucharest, Romania, *Molecular Genetics Department, Victor Babes Institute, Spl. Independentei, 99–101, Bucharest, 76201, Romania

Growing interest in neutrontherapy and radioprotection requires complex studies on the mechanism of neutron action on biological systems, especially on chromatin. Study of fast neutrons irradiation effects on Walker tumor chromatin structure are presented. The

Bucharest U-120 classical variable energy cyclotron was employed as an intense source of fast neutrons using 13.5 MeV d + Be thick target reaction. Thermal transition, intrinsic fluorescence of chromatin-ethidium bromide complexe behavior versus irradiation dose were determined. The treatment with thiotepa accelerates the damage process, but combined cure thiotepa-thyroxine or thiotepa thiotepa plus (+) D3 (D index 3) vitamin, partially recuperates the negative effect. the anticancer drug thiotepa determine structural and functional changes of the basic and acidic chromatin proteins, but a recovering effect of thyroxine and thyroxine + D3 (D index 3) vitamin (as protectors) are evident. The results suggest a combination of fast neutron irradiation with anticancer drug thiotepa for tumor destruction enhancement and the use of hormonal compound thyroxine and D3 (D index 3) vitamin for normal cells better protection during neutrontherapy. such results could constitute an indication for associated chemotherapy-radiotherapy schedule in clinical applications.

ROLE OF THE HAMSTER RAD51 PROTEIN IN DNA DAMAGE RESPONSE AND HOMOLOGOUS RECOMBINATION

Martine Defais, Stéphane Vispé, Christophe Cazaux, and Claire Lesca

Institut de Pharmacologie et de Biologie Structurale, UPR 9062 CNRS, Toulouse, France

Among the various DNA repair processes developped by mammalian cells, homologous recombination has still an hypothetical role. In an attempt to understand the participation of this mechanism to repair we cloned the cDNA of the hamster Rad51 protein (CgRad51), an homolog of E. coli RecA protein. The protein is highly identical to its human and mouse couterparts. No induction of the protein could be detected after various DNA damaging treatments. This reveals a difference between the regulation of bacterial RecA and S. cerevisiae ScRad51 on one hand, and higher eukaryote Rad51 on the other hand. We are currently establishing CgRad51 overexpressing and underexpressing CHO cell lines. The overexpression of the protein leads to an increased level in homologous recombination as measured by the restoration of a functional *lacZ* gene from two genes carrying non overlapping deletions. The implication of CgRad51 in DNA repair and homologous recombination after various genotoxic treatments of these cell lines will be dicussed.

EVOLUTION OF DNA REPAIR PROTEINS AND PROCESSES: ANALYSIS OF COMPLETE GENOMES

Jonathan A. Eisen and Philip C. Hanawalt

Stanford University, Department of Biological Sciences, Stanford, CA, U.S.A. 94305–5020

A new source of information about the potential DNA repair capabilities of different species is comparative genomic analysis. This genomic information also provides a wealth of data relating to the evolution of DNA repair proteins and processes. Complete genomes are now available for six bacterial species, two Archaea, and one eukaryote. We have made a database of all proteins with well established roles in DNA repair in any species. The sequences of these proteins were compared to all the complete genomes as well as to all sequences in Genbank using a variety of database searching algorithms. Proteins and open reading frames with significant similarity to any known repair protein were compiled. By studying the presence and absence of homologs of known repair genes across the main domains of life (with particular emphasis on those species for which complete genomes are available), and by using molecular phylogenetic analysis of the proteins, we have attempted to reconstruct the evolutionary history of DNA repair proteins and DNA repair processes. Specific repair genes likely to have been present in the ancestor of all living organisms are identified. In addition, the presence and absence of specific repair genes from certain species (e.g., the mycoplasmas have lost many repair genes) is used to predict the likely repair capabilities of these species. The database of repair proteins, alignments, links to sequence databases, and other information will be available at http://www-leland.stanford.edu/~jeisen/Repair/Repair.html.

EUKARYOTIC DNA REPAIR ENZYMES WITH DEOXRIBOPHOSPHODIESTERASE (dRpase) ACTIVITIES

William A. Franklin, Margarita Sandigursky, Walter A. Deutsch*, Adly Yacoub* and Mark R. Kelley§

Departments of Radiology and Radiation Oncology, Albert Einstein College of Medicine, Bronx, New York 10461, USA; *Pennington Biomedical Research Center, Louisiana State University, Baton Rouge, LA 70808, USA; §Department of Pediatrics, Section of Pediatric Endocrinology, Wells Center for Pediatric Research and Department of Biochemistry and Molecular Biology, Indiana University School of Medicine, Indianapolis, IN 46202, USA

The DNA deoxyribophosphodiesterase (dRpase) activity was first described in *E. coli* as an activity that removes sugar-phosphate groups at incised apurinic/apyrimidinic (AP) sites. All enzymes with dRpase activity can remove the terminal 2-deoxyribose-5-phosphate at a 5' incised AP site either via hydrolysis or via a β-elimination mechanism. Some enzymes are also able to remove the product *trans*-4-hydroxy-2-pentenal-5-phosphate at a 3' terminus produced by treatment of AP site-containing DNA with an AP lyase. In *E. coli*, the enzymes exonuclease I, RecJ, and the Fpg protein all have associated dRpase activities. In eukaryotes, it has been demonstrated that DNA polymerase β contains a dRpase activity that removes a 5' terminal deoxyribose-phosphate via β-elimination. We have recently characterized dRpase activities associated with the Drosophila S3 protein and the yeast Ogg1 protein. Both of these enzymes contain DNA glycosylase activities that remove the modified DNA base 8-oxoguanine. A glutathione S-transferase fusion protein of Ogg1 (GST-Ogg1) was purified and subsequently found to efficiently remove sugar-phosphate residues at incised 5' AP sites. A dRpase activity was also demonstrated for the removal of *trans*-4-hydroxy-2-pentenal-5-phosphate at 3' incised AP sites and from intact AP sites which was dependent on the presence of Mg^{2+}. Previous studies have shown that DNA repair proteins that possess AP lyase activity leave an inefficient DNA terminus for subsequent DNA synthesis steps associated with base excision repair. However, our findings suggest that in the presence of $MgCl_2$, Ogg1 can efficiently process 8-oxoguanine so as to leave a one nucleotide gap that can be readily filled-in by a DNA polymerase, and importantly, does not therefore require additional enzymes to process *trans*-4-hydroxy-2-pentenal-5-phosphate left at a 3' terminus created by a β-elimination catalyst. Similar dRpase actives were found associated with the *Drosphila* S3 protein.

GENE SPECIFIC FORMATION AND REPAIR OF 8-HYDROXYGUANINE IN MAMMALIAN CELLS IN VIVO

Tanja Thybo Frederiksen[1], Tinna Stevnsner [1] and Vilhelm A. Bohr[2]

[1]Danish Centre for Molecular Gerontology, Department of Molecular and Structural Biology, Aarhus University, C.F.Mollers Alle, Bldg. 130, 8000 Aarhus C, Denmark.
[2]Laboratory of Molecular Genetics, National Institute on Aging, NIH, 4940 Eastern Ave., Baltimore, MD 21224, USA

In the past years, there has been extensive investigation of the repair of 8-hydroxyguanine (8-OH-Gua). This lesion has been shown to cause GC to TA transversions and is believed to cause cancer, but the principal mechanism and the activity of the repair are still unclear. Using a Gene Specific Repair Assay we are measuring the repair of 8-OH-Gua in mammalian cells in vivo. The Escherichia coli enzyme Fpg-glycosylase specifically recognizes and removes 8-OH-Gua from DNA. By combining treatment of oxidative damaged DNA with this enzyme and Southern transfer followed by hybridization with specific probes, we will measure the repair of 8-OH-Gua in specific regions of the genome. The major problem to be resolved is the induction of a detectable level of lesions, and we are therefore currently testing the properties of different chemicals and their influence on the cell viability as well as the extent of gene specific oxidative damage. We are planning to investigate whether the accumulation of oxidative damage seen with age is due to a decrease in repair activity.

HYPERBARIC OXYGEN TREATMENT EFFECTS ON OXIDATIVE DAMAGE ASSOCIATED WITH ISCHEMIA/REPERFUSION INJURY

Kevin T. Geiss[1], Kenneth L. Hensley[2], and Richard A. Henderson[3]

Department of Zoology, Miami University, Oxford, OH and GEO-CENTERS, INC., Toxicology Division, AL/OET, Wright-Patterson Air Force Base, OH, USA[1], Oklahoma Medical Research Foundation, University of Oklahoma, Norman, OK, USA[2], 74th MED GRP/SGPH, Wright-Patterson Air Force Base, OH, USA[3]

Interruption of blood supply (ischemia) results in tissue injury, however, most damage occurs as blood rapidly returns (reperfusion). Reperfusion generally results in marked increases in oxygen radicals, causing damage to DNA and other cellular components. Along with superoxide, increased nitric oxide (NO) levels ensue. NO, a messenger molecule normally present in low concentrations, is produced enzymatically and at higher concentrations may cause cellular damage. Particularly, NO production in the presence of superoxide generates peroxynitrite, which has been associated with cell damage and death. Recently, hyperbaric oxygen (HBO) treatment was shown to suppress NO production.

Thus, HBO's decrease of available NO may inhibit the production of peroxynitrite. We used Mongolian gerbils as our cerebral ischemia model. Animals underwent transient global ischemia via bilateral carotid ligation with or without HBO (3 atm) and allowed to recover for 1,2 or 24 hrs. Brain tissue was analyzed for specific mRNAs, which respond to ischemic events, for heat shock protein 70 (HSP70), ornithine decarboxylase (ODC), and inducible NO synthase (iNOS). Ischemia produced up to a 50% decrease in both iNOS and ODC mRNA levels in animals treated with HBO before ischemia. We detected time-dependent increase in HSP70 mRNA for all treatment groups, except when HBO was provided before ischemia (10% decrease). HBO appears to impact the mRNA levels of ischemia-related genes and likely the production of NO and radical species. These results are relevant to cases of ischemic injury, as well as potentially in operational exposures of personnel to radical producing chemicals in non-isobaric conditions.

INFLUENCE OF α- AND γ-RADIATION INDUCED NICKS ON THE PHYSICAL PROPERTIES OF CALF THYMUS DNA

A. G. Georgakilas[1], L. C. Margaritis[2], T. Katshorhis[2], L. Sakelliou[3] and E. G. Sideris[1]

[1]Institute of Biology, NCSR 'DEMOKRITOS', Athens 153 10, Greece, [2]Department of Biology, University of Athens, Athens 157 73, Greece, [3]Nuclear Physics and Elementary Particles Section, Athens 106 80, Greece

Exposing DNA solutions to ionizing radiations, strand breaks and modifications of nucleotides (base and/or sugar) are induced in the DNA double helix. The radiation induced damage has a considerable impact on the physical and structural properties of the DNA molecule, leading (after relatively high radiation doses) to severe conformational changes and destabilization through the disruption of base stacking and hydrogen bonding. At relatively low doses and low dose rates, however, where nicked DNA molecules are mainly produced (maximum 1–2 Single Strand Breaks-SSBs-/molecule and/or alkali revealed sites), different studies suggest that the structural integrity of the molecule is maintained and very slight conformational changes are detected relative to the native (non-irradiated) B-form while some of their physical properties (e.g. thermal stability, mobility) change significantly. Many of the above studies have shown that nicked DNA molecules have an increased ability to form stacking interactions with their neighboring molecules and it seems that nicked sites act as the precursors for these interactions. One of the intriguing questions associated with this type of damages and alterations is its consequence on DNA conformation. Up to what extent the double helix retains its structural integrity at the nicked site? How does the repair enzyme recognize the structural anomaly? We try to make an approach to these questions by studying some of the physical properties (thermal stability, size and mobility and dielectric behavior) of macromolecular calf thymus DNA irradiated in buffered solutions at relatively low doses of γ-irradiation (0–4 Gy) and α-particles (0–12 Gy) respectively.

By applying the methods of UV-Thermal Transition Spectrophotometry (TTS), Pulsed Field Gel Electrophoresis (PFGE) and ac-Dielectric Relaxation Spectroscopy (DRS) we studied the changes in the physical properties of nicked DNA molecules and also of the ones exposed to high doses. Our results suggest that increased internal (and external) base stacking interactions in (and between) nicked DNA molecules lead to more (thermally) stable and of reduced mobility molecules which overcome the loss of connectivity associated with the disruption of the covalent backbone of DNA.

INDUCTION OF OXIDANT BY INSECTICIDES AND PROTECTION BY VITAMIN E

Belma Giray and Filiz Hincal

Hacettepe University, Faculty of Pharmacy, Department of Pharmaceutical Toxicology, Ankara, TR-06100, Turkey

Oxidative stress is associated with a disturbance in the pro-oxidant-antioxidant balance in favor of the pro-oxidant. Alteration in the balance by xenobiotics is a key feature of many physiological and pathophysiological phenomena and processes as diverse as inflammation, ageing, carcinogenesis, drug action and drug toxicity. Insecticides are the most common environmental pollutants. We have therefore, interested in the possible oxidative stress inducing effects of widely used three differemt types insecticides. We have examined the effect of this insecticides on the induction of lipid peroxidation in rats by single or repeated oral doses. Endosulphan, a chlorinated hydrocarbon insecticide of cyclodiene group, induced the production of thiobarbituric acid reactive substance (TBARS) in hepatic and cerebral tissues of rat significantly ($p<0.001$), when given orally with a single dose of 30 mg/kg and repeated doses of 10 and 15 mg/kg/day. Monocrotophos, an organophosphorus insecticide, significantly ($p<001$)

Increased TBARS in both tissues with a single dose of 1.7 mg/kg. Same results were observed with the single (170 mg/kg) and the repeated doses (75 mg/kg/day) of cypermethrin, a pyrethroid. The free radical scavenger and chain-breaking antioxidant, (α-tocopherol significantly ($p<0.001$) inhibited the lipid peroxidation in both tissues when given 100 mg/kg/day for three days, prior the insecticide treatments. Our results suggest that free radicals hence the induction of oxidative stress might be involved in the toxicity of these insecticides.

USE OF INFRARED SPECTRAL MODELS IN CANCER RESEARCH AND THEIR POTENTIAL CLINICAL APPLICATIONS

Sandra J. Gunselman*, Nayak L. Polissar† and Donald C. Malins‡

*Geo-Centers, Inc., CETT/CSU Foothills Campus, Fort Collins, CO 80523, USA, †The Mountain-Whisper-Light Statistical Consulting and Dept. of Biostatistics, University of Washington, Seattle, WA 98195, USA, ‡Pacific Northwest Research Foundation, 720 Broadway, Seattle, WA 98122, USA

The application of statistical analysis to Fourier transform infrared (FT-IR) spectra was only recently developed and has been shown to be a sensitive tool to assess DNA structural modifications thought to be involved in the development of carcinogenesis (1–4). DNA was extracted from normal and cancerous tissues and analyzed by FT-IR. Subtle modifications in DNA structure were readily identified by principal components analysis (PCA) of FT-IR spectra that were not discernible using traditional spectroscopic techniques. After baselining and normalizing each spectrum, the samples underwent statistical analysis including unequal variance *t*-tests, PCA and discriminant analysis. The FT-IR/PCA technology provided a perspective of the changes in the entire DNA structure -- nucleotide bases, phosphodiester groups and the deoxyribose moiety. Results indicated distinct structural differences in DNA of normal and cancerous tissues. Comparisons of normal prostate and prostate cancer by P-values showed statistically significant ($P \leq 0.05$) differences in the areas 1590–1510 cm^{-1}, which is assigned to C-O stretching and NH_2 bending vibrations, and 1060–1010 cm^{-1}, which is generally assigned to deoxyribose vibrations. PCA of the prostate samples revealed clustering, an indication of structural similarity between tissue types and discriminant analysis was able to classify the samples with high sensitivity and specificity. The FT-IR/PCA technology was also applied to normal and cancerous breast tissue with similar results, evidence that the discriminatory ability of the technology is not limited to a single tissue type.

Acknowledgments

This study was supported by US Army Medical Research and Material Command Contract No. DAMD17–95–1–5062. The views, opinions and/or findings presented here are those of the authors and should not be construed as an official Dept. of the Army position, policy or decision.

References

1. Malins, D.C., Polissar, N.L., & Gunselman, S.J. (1996) *Proc. Natl. Acad. Sci., USA* **93**(6), 2557–25563.
2. Malins, D.C., Polissar, N.L., Nishikida, K., Holmes, E.H., Gardner, H.S. & Gunselman, S.J. (1995) *Cancer* **75**(2), 503–517.
3. Malins, D.C., Polissar, N.L. & Gunselman, S.J. (1997) *Proc. Natl. Acad. Sci., USA* **94**(1), 259–264.
4. Malins, D.C., Polissar, N.L., Su, J., Gardner, H.S. & Gunselman, S.J. (1997) *Nature Medicine* **3**(8), 927–930.

RESONANT DISSOCIATION OF DNA BASES BY LOW ENERGY ELECTRONS

Ina Hahndorf*, Michael A. Huels**, Eugen Illenberger* and Leon Sanche**

*Institut für Physikalische und Theoretische Chemie, Freie Universität Berlin, Takustraße 3, D- 14195 Berlin, Deutschland, **Dep. de Medecine Nucleaire et de Radiobiologie, Faculte de Medecine, Universite de Sherbrooke, Sherbrooke, Quebec, Canada J1H 5N4

Many of the harmful consequences of radiation exposure on living matter can be traced to modifications of the cellular DNA. It has been shown that one of the main steps that lead to DNA strand lesions, is the localization of low energy electrons on pyrimidine bases. On our poster, we present measurements which demonstrate the sensitivity of specific DNA bases to resonant molecular dissociation at electron energies below ionization. The experiments are peformed in a crossed beam apparatus. An effusive molecular beam emanates from an oven and intersects a monochromatic electron beam. Anions which are formed in the reaction region are focused by a weak electric field into a high resolution mass spectrometer , mass selected and detected.

We have measured the electron energy dependence (0 to 10 eV) for the production of a vast variety of anion (and neutral) fragments, induced by resonant electron attachment to canonical thymine and cytosine.The most abundant stable negative ions produced are monomer thymine and cytosine that are observed at electron energies below 1.7 eV. The measured anion yield functions display a strong dependence on incident electron energy. This suggests that the electron-molecule reactions are resonant in nature and lead to dissociation into chemically different fragments. We propose that models of radiation damage to living matter must not neglegt processes such as dissociative electron attachment, which can be induced by the numerous sub-ionization electrons produced along the radiation tracks.

EFFECTS OF TRIS AND PHENOL IN γ-IRRADIATED DNA SAMPLES

K. S. Haveles[1], Th. Katsorchis[2], L. H. Margaritis[2], V. Sophianopoulou[1], E. G. Sideris[1]

[1] Institute of Biology, N.C.S.R. “DEMOKRITOS”, Aghia Paraskevi 153 10 Athens, Greece. [2] Division of Cell Biology and Biophysics, Department of Biology, University of Athens, 157 01 Athens, Greece

Single strand breaks (ssb’s) and double strand breaks (dsb’s), their interrelation and the degree of their repair are of extreme importance in understanding the early reactions involved in the radiation-induced DNA damage. To study these phenomena, many laboratories have used Tris [tris(hydroxymethyl)aminomethane] and Phenol as free radical scavengers in various *in vitro* and/or *in vivo* biological systems. These studies have attributed the observed decrease of induced dsb’s solely to the scavenging of the free radicals and the decrease on the frequency of their reactions with the DNA molecule. On the basis of the obtained results the theories of the Free Radical Transfer and the Locally Multiply Damaged Sites, concerning the introduction of dsb’s on the DNA molecule, were formulated.

We have used Thermal Transition Spectrophotometry (TTS) combined with Pulsed Field Gel Electrophoresis (PFGE) to study the effects of Tris and Phenol in γ-irradiated DNA samples. Both scavengers affect the stability of the DNA molecules *per se* as it is measured by the helix to single coil mean thermal transition temperature (T_m) during TTS and the overall migration of the DNA molecules during the PFGE analysis. When DNA samples exposed to gamma rays in the absence of these radical scavengers, the expected decrease at the T_m temperature and at the apparent mean molecular weight (MW) was observed. In the presence of these scavengers, both their T_m temperature and their apparent mean MW were almost restored to the respective values of the non-irradiated samples.

These observations suggest that the effect of Tris and Phenol on the DNA, might be attributed not only to the scavenging features of free radicals, but also to possible alterations of the dynamic properties of the DNA samples.

OXIDATIVE DNA BASE DAMAGE IN LYMPHOCYTES OF HIV INFECTED DRUG USERS

Pawel Jaruga[1] , Barbara Jaruga[2] ,Waldemar Halota[2] and Ryszard Olinski[1]

[1]Department of Clinical Biochemistry, University School of Medical Sciences, Karlowicza 24, 85–092 Bydgoszcz, Poland, [2]Provincial Hospital of Infectious Diseases, Floriana 12, 85–061 Bydgoszcz, Poland

It is believed that patients infected with a human immunodeficiency virus (HIV) are under chronic oxidative stress. Oxidative stress is defining as a condition characterized by increased production of reactive oxygen species (ROS), e.g., hydroxyl radical and hydrogen peroxide. ROS may be an important factor in mutagenesis and carcinogenesis, because can cause extensive DNA modification, including modified bases.Therefore, if not repaired on time, they can contribute to carcinogenesis. Alternatively they can trigger apoptotic death of the cell. An inappropriate induction of apoptosis may play a central role in pathogenesis of AIDS.

Direct cause of apoptotic death of lymphocytes in HIV infected patients is still unclear. Therefore in present study we wish to see whether the level of oxidative DNA base modifications, which can contribute to apoptotic death of lymphocytes in HIV infected patients, is elevated when compared with a control group of patients. One of the group most endangered by HIV infection are injection drug users (IDUs). In our study we compared the level of a base product in lymphocytes of HIV-infected IDUs with that of seronegative IDUs (control group). The results show statistically significant increases of oxidatively modified DNA bases in AIDS patients when compared with control, seronegative patients.

HYPOCHLOROUS ACID-INDUCED BASE DAMAGE IN ISOLATED CALF THYMUS DNA

Andrew Jenner, Matthew Whiteman and Barry Halliwell

International Antioxidant Research Centre, Pharmacology Group, University of London, Kings College, Manresa Road, London, SW3 6LX, UK

The enzyme myeloperoxidase is released by phagocytic cells at sites of inflammation and catalyzes the formation of hypochlorous acid (HOCl) from hydrogen peroxide and chloride ions. HOCl is capable of oxidising many important biological molecules and has been shown to attack nucleotides and individual DNA bases *in vitro* and to inactivate various DNA repair enzymes. In addition, chlorinated bases have been shown to be mutagenic in the Ames test. These effects could lead to mutation and contribute to the increased risk of carcinogenesis that is associated with chronic inflammation. Exposure of calf thymus DNA to HOCl lead to extensive DNA base modification, measured by GC-MS. Large concentration-dependent increases in pyrimidine oxidation products (thymine glycol [*cis/trans*], 5-hydroxycytosine, 5-hydroxyuracil, 5-hydroxyhydantoin) but not purine oxidation products (8-hydroxyguanine, 2-hydroxyadenine and 8-hydroxyadenine, FAPy guanine or FAPy adenine) were observed at pH 7.4. In addition, large increases in 5-chlorouracil (formed from 5-chlorocytosine during sample preparation), a novel chlorinated base, were observed. Time course studies suggested that the formation of purine oxidation products in isolated DNA by hypochlorous acid was not a major oxidation pathway. We also observed that HOCl causes a concentration and pH dependent loss of certain base damaged species from oxidatively damaged DNA, including all the oxidised purines: 8-hydroxyguanine, 2-hydroxyadenine and 8-hydroxyadenine, FAPy guanine and FAPy adenine. In particular the most common biomarker of *in vivo* DNA base damage, 8-hydroxyguanine was rapidly and extensively lost. Consequently, measurements of 8-hydroxyguanine levels in inflammatory conditions where local HOCl production would be large such as rheumatoid arthritis, could be underestimates of the true extent of oxidative DNA damage. The pH dependence of the formation and degradation of DNA base lesions suggest that both HOCl and OCl^- (hypochlorite) are involved in base modification. The pattern of HOCl induced DNA damage is completely different from that produced by other reactive species so far investigated. Measurement of this unique pattern of pyrimidine-derived DNA base damage, particularly the specific lesion 5-chlorocytosine (measured after acid hydrolysis as 5-chlorouracil) might act as a specific biomarker to examine possible HOCl induced DNA damage *in vivo*.

GENOTOXICITY TESTS: APPLICATION TO OCCUPATIONAL EXPOSURE AS BIOMARKERS

Ali E. Karakaya, Semra Sardas, and Sema Burgaz

Gazi University, Faculty of Pharmacy, Department of Toxicology, Hipodrom 06330 Ankara, Turkey

Epidemiologists have tentatively attributed about 4 per cent of cancer deaths in industrialized countries such as the United States, to occupational causes. Certain occupational groups are exposed at much higher concentrations than is the general population to potentially hazardous genotoxins. The possibility exists that these exposure may significantly increase the risk of cancer for some of the workers. In industrialized countries, with strict governmental regulations of actual and potential industrial health hazards during the last two decades, it is likely that this figure will decrease even further in the future. But on the other hand while cancer deaths due to occupational exposure may be only a small portion of all cancer deaths in the developing countries, the risk may increase owing to rapid industrialization, uncontrolled transfer of technology and lack of regulations controlling hazardous substances. Screening workers who have had known or suspected contact with genotoxic chemicals can be useful in quantifying exposures and assessing risks. Recently we have used some biomarkers to evaluate and assess some occupational genotoxic chemical exposure in various workplaces in Turkey. In our research the genotoxic risk of the following occupational groups were evaluated: Styrene exposed furniture workers, nurses handling antineoplastic drugs, operating room personnel, hospital sterilising staff exposed to ethylene oxide, engine repair workers exposed to polycyclic aromatic hydrocarbons, road-paving workers exposed to bitumen, car painting workers and hair colorists exposed to oxidation hair dyes. Results of some markers of exposure and cytogenetic biomarkers (sister chromatid exchange, micronucleus assay and comet assay) in the above mentioned ocupational groups will be presented.

IDENTIFICATION OF 2 NOVEL MUTATIONS IN TURKISH CYSTIC FIBROSIS PATIENTS

O. Kilinç,(1), F. Köprübasi (2), E. Dagli (3), M. Demirkol (4) and A. Tolun (1)

(1) Boðaziçi University, Department of Molecular Biology and Genetics, 80815 Bebek, Istanbul, Turkey; (2) Institute of Child Health, Ege University Medical School, Izmir, Turkey;(3) Department of Pediatrics, Marmara University Medical School, Istanbul, Turkey (4) Institute of Child Health, Ýstanbul University Medical School, Istanbul, Turkey

Cystic fibrosis (CF) is a relatively severe and frequent autosomal recessive genetic disorder in the populations studied. 713 different mutations of the CFTR gene have been reported to Cystic Fibrosis Genetic Analysis Consortium so far. We analyzed all of the CFTR gene except for the first and the last exons in 160 CF chromosomes using denaturing gradient gel electrophoresis (DGGE) and direct sequencing techniques. Fifteen of the exons were analyzed in five triplex assays. In addition, some exons with different Tm values were amplified individually and coelectrophoresed in pairs on a single gel. Two novel mutations are detected. K68E is a substitution mutation in a patient who is heterozygous for another yet unidentified variation in exon 4. A $\rightarrow$ G transversion at amino acid 68 in exon 3 causes the substitution of Glu for Lys. The homozygous presence of 3849+5 G $\rightarrow$ A, the other mutation, was detected in a CF patient who has a chloride sweat test score of 100mEq/l, gastrointestinal symptoms and lung diseases. We believe that this is a splice site mutation in intron 19. Mutations in 30 percent of the CF alleles are characterized, with frequencies of 12 per cent for DF508, 5 per cent for 1677delTA, 4 per cent for 2789+5 G $\rightarrow$ A, 2 per cent for R1070Q, 1 per cent each for G542X, W1282X, F1052V, 1525–1 G $\rightarrow$ A and 0.5 per cent each for S466X, 2184Ains N1303K, 306delTA and 3129del4. The recent data will be discussed with respect to mutations, DNA polymorphisms, and contribution of consanguinous marriages to the incidence of the disorder in our population. Mutation data will be compared to those reported for the populations in some neighboring countries.

RADIATION-INDUCED DNA STRAND BREAKS IN HUMAN HEMATOPOIETIC CELLS MEASURED USING THE COMET ASSAY

Maarit Lankinen, Leena Vilpo and Juhani Vilpo

Tampere University Medical School and Tampere University Hospital, Finland

Human blood lymphocytes, granulocytes and bone marrow mononuclear cells (BMMNCs) were separated by density gradient centrifugation. A subpopulation of progenitor cells was further isolated from the BMMNCs using anti-CD34-coated magnetic beads. The cells were irradiated with UV- or γ-rays. The extent of DNA damage, i.e. single-strand breaks (SSBs) and alkali-labile lesions as well as double-strand breaks (DSBs) of individual cells, was investigated using the single-cell gel electrophoresis technique also known as comet assay. In this method damaged or intact cells are embedded in agarose on a microscope slide and lysed by detergents and salts. Then electrophoresis is performed under alkaline or neutral conditions (Singh et al., Exp. Cell. Res. 175, 1988). Increased extension of the DNA from the nucleus towards the anode is observed in cells with increased damage. DNA with strand breaks forms a comet like image where the nucleus is the head of the comet and the damaged DNA is the tail of the comet. With γ-irradiation we did not find any major differences between the cell types, so this indicates an approximately similar extent of formation and repair of γ-irradiation-induced DNA strand breaks in immature and mature human hematopoietic cells. Also with UV-irradiation, no significant differences between the cell types were observed. Furthermore, we have studied malignant cells from patients who have chronic lymphocytic leukemia. Interestingly, the comet formation in some cases was different from normal lymphocytes.

THE USE OF EXCISION ENZYMES TO PROBE FOR THE MIGRATION OF OXIDATIVE DAMAGE IN DNA

Tracy Melvin,[1] Peter O'Neill,[1] AntonyW. Parker,[2] and Teresa Roldan-Arjona,[3]

MRC Radiation and Genome Stability Unit, Chilton, Oxon., U.K. [1] Lasers for Science, Rutherford Appleton Laboratory, Chilton, Oxon., U.K. [2]Departamento de Genetica, Universidad de Cordoba, 14071-Cordoba, Spain [3]

The contribution of direct ionisation of cellular DNA by radiation to the damage profile, as compared to damage resulting from water radiolysis products, is dependent on the radiation quality. Direct ionisation yields both an electron and a radical cation within the DNA matrix. Migration of the radical cation towards the guanine residue might be predicted upon the basis of its ionisation potential. Water is essentially transparent at 193nm; photolysis of aqueous solutions of DNA results in monophotonic ionisation of the nucleic acid bases and prompt strand breakage in low yields 3' to the guanine residue.[1] Some of our previously reported data,[2] from transient optical absorption laser flash photolysis experiments, on a series of single stranded oligonucleotides with ionising light (193 nm) indicate that although migration occurs, the migration distance is limited.

To probe for electron hole migration in double stranded DNA a different approach to that used for the single-stranded oligonucleotide samples was needed since selective photooxidation of a single nucleic acid base within a DNA helix using 193 nm light is not possible. Various samples of DNA were irradiated with 193 nm light, then using a series of excision enzymes, modifications at different residues within various sequences were then probed; our results indicate that guanine is a target for direct ionisation damage as a result of electron hole migration from neighbouring bases.

References

1. Melvin, T., Botchway, S.W., O'Neill, P., and Parker, A.W. (1996) *J.Am.Chem.Soc.*, 118, 10031–10036; Melvin, T., Plumb, M.A., Botchway, S.W., O'Neill, P., and Parker, A.W. (1995) *Photochem.Photobiol.*, 61, 584–591.
2. Melvin, T., Botchway, S.W., Parker, A.W., and O'Neill, P. (1995) *J.Chem.Soc., Chem.Commun.*, 653–654.

STUDY OF REPAIR PROCESSES INTENSITY AND CYTOGENETICAL DAMAGES YIELD AT LOW DOSES OF GAMMA-RADIATION

A. A. Oudalova, D. V. Vasiliev, V. G. Dikarev, N. S. Dikareva, Z. A. Simonova, and S. A. Geraskin

Russian Institute of Agricultural Radiology & Agroecology, 249020 Obninsk, Russia

An absence of clear understanding of mechanisms underlying biological effect of low dose radiation does not allow a reliable quantitative estimation to be made about hazard of low-level radiation. Linear non-threshold concept which is often used for a decision of this problem has not any serious scientific justification, is of extrapolation nature and contradicts to available experimental data. To reveal a genuine shape of dose curve and mechanisms of its forming, a series of experiments on the study of cytogenetic disturbances yield at the range of low doses was carried out. It was shown that the relationship between the yield of cytogenetic damages and dose is non-linear with a site at low doses within which the frequency of aberrant cells increases over the control level significantly and practically independent on dose value. The limits of this site are dependent on experimental object. So, the yield of aberrant cells did not dependent on dose over the range from 1 to 10 cGy in root meristem of seedlings of barley irradiated seeds, and over the range from 5 to 30 cGy in leaf meristem of irradiated barley seedlings. The most likely explanation of this "plateau" existence in dose dependence is connected with a hypothesis about a mutagenic repair activity induced by low level radiation. This repair proves an enhanced genetic disturbances in comparison to control. Study of an unscheduled DNA synthesis revealed that a repair intensity in barley seedlings non-linearly depends on dose in the dose range from 1 to 100 cGy. Two ranges were separated (about 5 cGy and 50–70 cGy), where an intensity of DNA synthesis increased over spontaneous one significantly. Comparison of the form of the empirical curve, derived during these investigations, with results obtained on other objects allows the conclusion that the relationship between the yield of genetic disturbances and dose at low doses is non-linear and universal in character varying for different objects only in dose values at which changes in the nature of the relationship occur.

GENOMIC GENE ENRICHMENT FOR LMPCR ANALYSIS

Henry Rodriguez[1] and Steven A. Akman[2]

[1]Department of Biology, Beckman Research Institute & [2]Department of Cancer Biology, Wake Forest Cancer Center

Ligation-Mediated PCR (LMPCR) is a powerful fragment amplification technique that is several thousand fold more sensitive than other genome mapping methods. Its unique aspect is a blunt-end ligation of an asymmetric double-stranded linker, permitting exponential PCR amplification. An important factor limiting the sensitivity of LMPCR is the representation of target gene DNA relative to non-targeted genes; therefor, we developed a method to eliminate excess non-targeted genomic DNA. Restriction enzyme-digested genomic DNA is fractionated by Continuous Elution Electrophoresis (CEE), capturing the target sequence of interest. The amount of target DNA in the starting material for LMPCR is enriched, resulting in a stronger amplification signal. CEE provided a 24-fold increase in the signal strength attributable to strand breaks plus modified bases caused by reactive oxygen species, detected by LMPCR, in the human *p53* and *PGK1* genes. We are currently taking advantage of the enhanced sensitivity of target gene-enriched LMPCR to map DNA damage induced in human breast epithelial cells by exposure to non-cytotoxic concentrations of H_2O_2. CEE provides a rapid means of genomic gene enrichment for LMPCR analysis, and is easily applicable to other techniques requiring enrichment of a target gene.

REPAIR OF OXIDATIVE DAMAGE IN TRANSCRIBED AND NON-TRANSCRIBED DNA STRANDS IN HUMAN CELLS

Graciela Spivak and Philip C. Hanawalt

Department of Biological Sciences, Stanford University, Stanford, CA 94305–5020, USA

Certain DNA lesions are removed preferentially from the transcribed strands of expressed genes in bacteria, yeast and mammalian cells. Initially it was thought that only lesions repaired by nucleotide excision repair were subject to transcription-coupled repair, but recent studies have shown that some lesions caused by reactive oxygen species, which are recognized by glycosylases and are subject to base excision repair, are preferentially repaired in the transcribed strands of active genes. Those results were obtained using antibodies to repair patches containing BrdUrd or to thymine glycol, one of the major oxidative DNA lesions. Our aim is to study the incidence and repair of oxidative lesions in transcribed and non-transcribed strands of specific DNA fragments, utilizing glycosylases, endonucleases or treatments that recognize and nick DNA at the sites of different lesions, such as purine and pyrimidine modifications, AP sites and other alkali-sensitive sites and single strand breaks. We have adapted the technology originally developed for gene level analysis of cyclobutane pyrimidine dimer repair. Once the method is established for normal cells, cells from patients with repair-deficient syndromes will be tested for their ability to repair particular oxidative lesions in transcribed and non-transcribed DNA strands, as well as in the genome overall. These studies will provide information on pathways and proteins required for the repair of each lesion or lesion group, and should further the understanding of relationships between DNA repair deficiencies, carcinogenesis and developmental abnormalities in human genetic disease.

FREE RADICAL REACTIONS IN MECHANISMS OF DNA DAMAGE IN BLOOD-FORMING ORGANS

V. L. Sharygin and M. K. Pulatova

Semenov Institute of Chemical Physics of Russian Academy of Sciences, Kosygin's str. 4, GSP-1, Moscow, 117977 Russia

This paper will cover the following problems on: 1.The nature and the yields of primary radiation-induced free radicals of DNA, lipids, protein and cell water in blood-forming tissues and blood of animals have been established using the electron paramagnetic resonance technique. 2. To evaluate the role of radiation-induced DNA lesions to organismic injury the two effective radioprotectors have been used to modify in vivo the yields of radicals of DNA, lipids, protein and cell water. The protectors administrated prior to the irradiation of animals with their LD100/45 have provided with 97–99% animal survival. It has been shown that the radioprotectors have not practically decreased the yields of radicals in animal organs. The data demonstrate the importance of biochemically induced DNA lesions in cell killing. 3. To assess the injury and biochemical adaptation of blood system to radiation and chemical effects the changes in contents of cell-free DNA in blood plasma, the changes of antiproteolytic activities of a 2-macroglobulin and a 1-antitrypsin, of antioxidant activities of superoxide dismutase and glutathione peroxidase in blood and blood plasma of animals have been studied. The averaged values of these indexes have been obtained also for children of different ages and from inhabitancy regions with different radioactive pollution suffered as the result of Chernobyl atomic power station accident. Cases of abnormally high or low DNA concentration in child blood plasma have been given great attention, and this biochemical indicator has been used to reveal children who may be attributed to risk groups.

DRUGS AND DNA SYNTHESIS SYSTEM: THE BIOCHEMICAL MECHANISMS OF DNA DAMAGE, REPAIR AND PROTECTION

M. K. Pulatova and V. L. Sharygin

Semenov Institute of Chemical Physics of Russian Academy of Sciences, Kosygin's str., 4, GSP-1, Moscow, 117977 Russia

The accuracy of DNA copying during replication and repair depends not only on the activity of the enzymes involved in these processes, but also on the concentrations and ratio of their substrates: deoxyribonucleotides (dNTP). The paper will cover some problems on indirectly forming of DNA lesions due to the suppress of activity of radical enzyme ribonucleotide reductase (NDPR) catalyzing the biosynthesis of dNTP in cells. The imbalance of dNTP is responsible of defective replication and incomplete DNA repair due to errors in base incorporation, mispair. A close correlation between NDPR activity and DNA synthesis rate in animal organs has been shown. In vivo regulatory factors of NDPR activity have been studied to elucidate the mechanisms of radioprotective, radiosensitizing and therapeutic actions of anticancer drugs (nitrofuran derivatives, nitroimidazoles, alkylnitrosourea) and radioprotectors (indralin, indometaphen). They are the protein synthesis, redox status of SH groups in catalytic site, content of effects (ATP), molecular oxygen, iron content. The drug and radiation effects caused: 1. The early SOS-activation of NDPR in organs 10–30 min later after drug injections or irradiation at low dose. 2.The inhibition of NDPR in organs and tumour cells 3–12 h later after radiation and chemical effects was responsible of the DNA synthesis suppression. The destruction of iron-containing submit of NDPR and imbalance of DNA precursors have been shown. 3. The radioprotectors (provided with 97% survival of animals with their LD100/45) stimulated activation of NDPR within the time, when irradiation of unprotected animals caused essential and prolonged inactivation of NDPR. A high level of NDPR activity was keeping for long time providing with DNA precursors needed for the complete repair.

GERMLINE MUTATION ANALYSIS OF DNA REPAIR RELATED BRCA1 AND BRCA2 TUMOR SUPRESSOR GENES

Hilal Özdag, Emre Öktem, Marie Ricciardone, Mehmet Öztürk, and Tayfun Özcelik

Bilkent University, Department of Molecular Biology and Genetics, 06533, Bilkent, Ankara, Turkey

Hereditary breast cancer (HBC) is one of the most common inherited malignancies that affect the female population. It is estimated that 1/200 individuals in the general population, and 1/20 women with breast cancer carry germline mutations in HBC genes. The two recently identified genes, BRCA1 and BRCA2, are shown to be mutated in almost 90% of HBC patients in different populations. The BRCA1 gene is located on chromosome 17q21 with 24 exons encoding a protein of 1863 amino acids. The BRCA2 gene is located on chromosome 13q12 with 27 exons and encoding a protein of 3418 amino acids. Although the function of these tumor suppressor genes is still unclear, recent evidence suggests that they have a role in transcriptional activation and RAD52-mediated DNA repair. In addition, BRCA1 is thought to be a component of RNA polymerase II holoenzyme. We have started a nationwide collaborative project to determine the molecular bases of HBC in the Turkish population.with the following objectives: to 1) determine what portion of high risk families and individuals with early onset disease have mutations in BRCA1 and BRCA2 genes, 2) identify whether there is a common mutation and/or mutation hot spots in Turkish patients, 3) correlate the age of onset of the disease with the mutation type, 4) and obtain population genetics data to provide information on ancient mutations and migration patterns. Heteroduplex and single strand conformation analyses techniques followed by automated DNA sequencing using dye-terminator chemistry have been established as our two step mutation detection strategy. Preliminary results of mutation analysis will be presented.

GENE AND PROMOTER STRUCTURE OF THE MURINE URACIL-DNA GLYCOSYLASE

Kristin Solum, Hilde Nilsen and Hans Krokan

UNIGEN Center for Molecular Biology, The Medical Faculty, Norwegian University of Science and Technology, N-7005 Trondheim Norway

Uracil-DNA glycosylase (UDG) is the first enzyme in base excision repair (BER) for removal of uracil from DNA (Lindahl 1974). The gene for human uracil-DNA glycosylase *(UNG)* contains seven exons and encodes a nuclear (UNG2) and mitochondrial (UNG1) form of UDG (Haug *et al.*, 1996, Nilsen *et al.* 1997). The two forms of UDG differ in their N-terminal amino acid sequences which are involved in the direction of the proteins to the nucleus and mitochondria. We have also found the murine cDNAs that corresponds to human UNG1 and UNG2. UNG1 and UNG2 are generated by using alternative transcription starts from two different promoters (P_A and P_B) and alternative splicing (Nilsen *et al.*). We have sequenced the genomic murine UDG gene (mUNG) and found that it has a similar organisation as *UNG*. On the nucleotide level the mUNG and *UNG* are very conserved in the exons and exon-intron boundaries, while they show little homology in the introns except intron 3. The length of introns varies significantly for intron 2 and 5, 0,98 kb longer and 5,55 kb shorter respectively. In P_A and P_B we find several conserved cis-acting element as Ap2, Sp1, Myc and E2F, for binding of various transcription factors. Haug *et al.* have demonstrated a differential regulation of P_A and P_B in *UNG*, with SP1 and MYC as strong positive regulators and E2F as a negative regulator. Results shows that E2F in P_B can modulate the activity of both promoters. A striking difference between *UNG* and mUNG is a duplication in the murine P_B that contains E2F and Sp1 elements.

References

Haug, T *et al.* (1996). *Genomics*, **36:** 408–416
Lindahl, T. (1974). *Proc. Natl. Acad. Sci.* **71:** 3649–3654
Nilsen, H. *et al.* (1997). *Nucleic Acids Research* **25:** 750–755

CLONING AND CHARACTERIZATION OF MOUSE 8-OXOGUANINE DNA GLYCOSYLASE/AP LYASE (MOGG1)

Dmitry O. Zharkov, Thomas A. Rosenquist, and Arthur P. Grollman

Department of Pharmacological Sciences, State University of New York at Stony Brook, Stony Brook, NY, 11794–8651, USA

Putative mouse and human cDNAs for a DNA repair enzyme, 8-oxoguanine DNA glycosylase, were identified by searching the GenBank EST database for homology to *S.cerevisiae* Ogg1. Both human and mouse genes conserve sequence homologies to members of the endonuclease III family. cDNA for the mouse Ogg1 gene was expressed in *E.coli*. Introduction of the mouse Ogg1 cDNA in *mutM⁻mutY⁻ E.coli*, a strain deficient in oxidative damage repair, suppresses G:C→T:A transversions in a rifampicin resistance assay. *mutM⁻mutY⁻ E.coli* bearing a plasmid expressing mouse Ogg1 protein possess an activity that nicks oligonucleotide duplexes containing a single 8-oxoguanine lesion. The enzyme shows a strong preference for the 8-oxodG:dC pair with little or no activity on other mismatches containing 8-oxoguanine, guanine or adenine. Unlike MutM protein of *E.coli*, mouse Oggl protein shows no significant activity on duplex substrates containing formamidopyrimidine. The mechanism of reaction involves the formation of a Schiff base between the enzyme and C1¢ of the deoxyribose moiety, indicated by KCN inhibition and borohydride crosslinking. The reaction proceeds by b, but not d elimination, removing the damaged base and nicking DNA 3¢ to the site of the lesion.

Acknowledgments

This work was supported by NIH grants CA17395 and ES04086.

NUCLEAR LOCALISATION OF THE XPD PROTEIN: A SEARCH FOR THE TARGETING DOMAIN

Antonia M. Pedrini, Fabio Santagati, Carlo Rodolfo, Elena Botta, Tiziana Nardo, and Miria Stefanini

Istituto di Genetica Biochimica ed Evoluzionistica-CNR, Via Abbiategrasso, 207 - 27100 Pavia, Italy

The XPD gene product, a DNA helicase component of the basal transcription factor TFIIH, is involved also in nucleotide excision repair (NER). Mutations in this protein give rise to genetic diseases characterised by defective NER. Aminoacid sequence analysis has revealed the presence of a highly basic fourteen aminoacid region (aa 682–695) indicated as a putative nuclear location signal (NLS) in spite of the fact that it does not conform to a classic NLS.

The aim of this study was to verify whether the *XPD* gene product contains NLS and whether this signal corresponds to the putative NLS. To identify the determinant of XPD subcellular localisation, we have used a mammalian expression vector containing a small reporter protein (green florescent protein-GFP) that was fused in frame with the XPD protein. The location of the green fluorescent signal of the GFP component was directly detected after transient expression of the fusion product in transfected cells; the repair function was analysed by UDS and UV survival. The results indicated that only a fraction of the overexpressed XPD-GFP fusion protein is localised in the nucleus while most of the protein remains in the cytoplasm; the amount of protein traslocated in the nucleus was sufficient to fully restore its functional activity. However, we were unable to identify any sequence specific for nuclear localisation. Our data tend to suggest that the XPD protein does not have any specific NLS making it likely to be translocated into the nucleus by binding to other components of the TFIIH complex.

STRUCTURAL INFORMATION ON SMALL OLIGONUCLEOTIDES OBTAINED BY ELECTROSPRAY IONIZATION TANDEM MASS SPECTROMETRY

Allan Weimann[1,2], Paula Iannitti[1] and Margaret M. Sheil[1]

[1]University of Wollongong, Northfields Ave., Wollongong, NSW, Australia, [2]Department of Clinical Pharmacology, Q7642, Rigshospitalet, Tagensvej 20, DK-2200 Copenhagen N, Denmark

Most papers on mass spectrometry of oligonucleotides have focused on sequencing the oligonucleotides. In this work the main focus has been to try and get as much detailed information as possible from the oligonucleotides. The goal has been to locate as precisely as possible where a change in the oligonucleotides have occurred, with the main focus on adduct formation. Because many different product ions can be formed by electrospray ionization mass spectrometry-mass spectrometry fragmentation of oligonucleotide anions and cations, a lot of valuable information can be gained. This can make it possible to elucidate where a given change in DNA have occurred (with which base, sugar or phosphate the adduct has been formed), but first the product ions need to be identified. The product ions have here been identified in detail by using high resolution and higher order mass spectrometry. By using in-source collision activation it is furthermore possible to select and fragment the bases (or a stable base-adducts). From this fragmentation it may be possible to locate where on a given base the adduct formation has occurred.

THE INFLUENCE OF CADMIUM ON DIFFERENT PATHWAYS OF DNA DAMAGE AND REPAIR INDUCED BY UV- AND γ-IRRADIATION

C. V. Privezentzev, V. G. Bezlepkin, N. P. Sirota, and A. I. Gaziev

Institute of Biology of the Southern Seas, Sevastopol, Ukraine. Institute of Theoretical and Experimental Biophysics, Puschino, Russia

It has been found that when mice are treated by Cadmium (Cd), which itself induces rather low increase in the frequency of micronucleated bone marrow polychromatic erythrocytes (PCEs), and then exposed to γ-irradiation, fewer micronuclei are induced than by the irradiation alone. Perhaps, effect observed may be a case of induced repair like an adaptive response to very low doses of radiation. To verify this suggestion influence of Cd on DNA damage and repair induced acute γ-irradiation was evaluated in mouse lymphoid cells. Single injection of Cd chloride given before the irradiation is shown to decrease initial level of DNA lesions in thymocytes and to activate DNA repair in lymphocytes and splenocytes. This effect seems to be dependent of time interval between Cd pretreatment and the irradiation. Further the effect of Cd on adaptive response induced in vivo by low doses of radiation was assessed using micronucleus test. Cd is found to inhibit the adaptive response. Nevertheless, Cd is not shown to suppress SOS-response induced by UV-radiation. The effect does not depend on Cd concentration in incubation medium.

Conclusion has been made that Cd itself seems to induce a DNA repair network, but the exact mechanisms of this process are still unclear.

SUBSTRATE SPECIFICITY OF EUKARYOTIC Nth HOMOLOGUES

Teresa Roldán-Arjona[1], Bensu Karahalil[2,3] and Miral Dizdaroglu[3]

[1]Departamento de Genética, Facultad de Ciencias, Universidad de Córdoba, Spain.
[2]Department of Toxicology, Faculty of Pharmacy, Gazi University, Ankara, Turkey.
[3]National Institute of Standards and Technology, Gaithersburg, MD20899, USA

E. coli endonuclease III protein (Nth-Eco) removes damaged pyrimidine residues from DNA by base excision-repair. It is an iron-sulfur enzyme possessing both DNA glycosylase and apurinic/apyrimidinic lyase activities. *Schizosaccharomyces pombe* and human genes encoding Nth homologues (Nth-Spo and hNTH1) have been recently identified. The corresponding cDNAs have been subcloned in appropriated expression vectors and the proteins purified to apparent homogeneity. The substrate specificity of both eukaryotic enzymes for modified bases in oxidatively damaged DNA has been investigated, using the gas chromatography/isotope-dilution mass spectrometry (GC/IDMS) technique. DNA substrates were prepared by γ-irradiation or by treatment with H_2O_2 in the presence of Fe(III)-EDTA or Cu(II). Up to 17 modified bases were detected in these substrates. The DNA substrates were incubated with Nth-Spo and hNTH1 proteins, followed by precipitation. The pellets and supernatant fractions were analyzed by GC/IDMS. The results revealed an efficient excision of a number of pyrimidine-derived lesions by Nth-Spo and hNTH1. These were 5-hydroxycytosine, thymine glycol, 5-hydroxy-6-hydrothymine, 5,6-dihydroxycytosine, and 5-hydroxyuracil. None of the other pyrimidine lesions or purine lesions were excised. Excision was measured as a function of time, enzyme concentration, substrate concentration, and temperature. Kinetic constants were determined. Although some DNA base lesions removed by Nth-Spo and hNTH1 proteins were similar to those previously described substrates of *E. coli* Nth-Eco protein, differences between substrate specificities of these enzymes were noted.

REPAIR OF CYCLOBUTANE PYRIMIDINE DIMERS IN THE O^6-METHYLGUANINE-DNA METHYLTRANSFERASE (MGMT) GENE OF MGMT PROFICIENT AND DEFICIENT HUMAN CELL LINES AND COMPARISON WITH THE REPAIR OF OTHER GENES AND A REPRESSED X-CHROMOSOMAL LOCUS

Frank Skorpen, Camilla Skjelbred, Bente Alm, Per Arne Aas, Svanhild A. Schønberg, and Hans E. Krokan

UNIGEN Center for Molecular Biology, Norwegian University of Science and Technology, N-7005 Trondheim, Norway

We examined the removal of cyclobutane pyrimidine dimers (CPDs) from the O^6-methylguanine-DNA methyltransferase (MGMT) gene in a human MGMT-proficient cell line (HaCaT) and three deficient cell lines (A-172, A-253 and WI-38 VA13) in which the *MGMT*-gene was not transcribed. Repair rates in the *MGMT* gene were compared with those of the active uracil-DNA glycosylase (*UNG*) and *c-MYC* genes as well as in the X-chromosomal 754 locus and ribosomal gene cluster. In HaCaT cells, CPDs were removed faster and to a greater extent from the transcribed strand (TS) than from the non-transcribed strand (NTS) of the active *MGMT* gene. In contrast, A-172 and A-253 cells showed no strand bias in repair of CPDs in the *MGMT* gene, while WI-38 VA13 cells displayed a more general deficiency in repair. Although repair of the NTS of the *MGMT* gene varied considerably between cell lines, the rates approached the poor repair rates of the ribosomal gene cluster and the repressed 754 locus in the MGMT deficient cell lines. Both strands of the *c-MYC* gene were relatively efficiently repaired in all cell lines, probably due to an observed transcription of both strands in the *c-MYC* region. In HaCaT and WI-38 VA13 cells, the NTSs of the *MGMT* and/or *UNG* genes were both repaired somewhat faster than the 754 locus. Our results indicate that the relative rates of repair of CPDs in transcriptionally inactive *MGMT* genes are similar to the rates in repressed loci and lower than repair rates in the NTSs of active genes.

EFFECT OF BILE ACIDS ON LIPID PEROXIDATION: THE ROLE OF IRON

Nair Sreejayan and Christoph von Ritter

Department of Medicine II, Klinikum Grosshadern, Ludwig-Maximilians-University, D-81377 Munich, Germany

The toxic effect of hydrophobic bile acids is claimed to be mediated through lipid peroxidation. Conversely, antioxidant properties of tauroursodeoxycholic acid (TUDC), a hydrophilic bile acid, have been suggested as a possible mechanism by which TUDC confers its beneficial effect in a variety of diseases. We have investigated the effect of taurodeoxycholic acid (TDCA), a hydrophobic bile acid and TUDCA on lipid peroxidation employing a pure lipid system both in the presence and absence of iron ions. Neither TDC nor TUDC at a concentration range of 0–10 mM showed any effect on spontaneous lipid peroxidation of phosphatidyl choline liposomes or sodium arachidonate solution. This excludes the possibility of direct prooxidant/antioxidant properties for TDC and TUDC. Addition of ferrous ions (0.1 mM) to the lipid system brought about a linear increase in lipid peroxidation with time, reaching a plateau in 30 min. The presence of TDC (1 mM) caused an increase in the rate and extend of iron stimulated lipid peroxidation, while TUDC did not have any influence. The propensity of bile acids to increase iron induced lipid peroxidation was related to polarity, ie lithocholic acid > taurochenodeoxycholic acid > TDC > taurocholic acid > TUDC. The increase in the iron stimulated lipid peroxidation by TDC was concentration dependent and no further increase was observed beyond 2 mM which is below its criticle micellar concentration. TUDC (5 mM) completely abolished the effect of TDC on iron induced lipid peroxidation. This suggests that TUDC does not function as an antioxidant *per se* but may prevent lipid peroxidation caused by TDC by displacing the hydrophobic bile acid from its lipid sites. In conclusion, enhancement of iron induced lipid peroxidation may mediate the toxicity of hydrophobic bile acids. Furthermore, our data suggests that TUDC may exert its beneficial effect by inhibiting TDC induced lipid peroxidation.

EVIDENCE FOR OXIDATIVE DNA DAMAGE AS MOLECULAR MECHANISM OF CHROMIUM(VI) MUTAGENESIS IN SACCHAROMYCES YEAST, AND BIG BLUE TRANSGENIC MOUSE MODELS

David M. Sonntag[*†], Lei Cheng[**], and Kathleen Dixon[*]

[*]University of Cincinnati, Department of Environmental Health, Toxicology Division, [**]Emory University, Department of Internal Medicine, [†]United States Air Force, AFIT/CIMI, Wright-Patterson AFB, Ohio

Hexavalent chromium(Cr^{6+}) compounds are carcinogenic as determined by epidemiologic and long-term animal studies, and mutagenic in a wide variety of test systems. Mutagenesis resulting from the intracellular reduction of Cr(VI) is the focus of multiple studies in our laboratory. Other workers have identified various cellular enzymatic and non-enzymatic pathways which participate in the reduction of Cr(VI), including cytochrome P-450, ascorbate, glutathione(GSH), DT-diaphorase and aldehyde oxidase. Of these, glutathione and ascorbate appear to generate a variety of reactive intermediates, such as Cr(V) and oxygen radicals, during the reduction of Cr(VI), and which may be ultimately responsible for the mutagenic and carcinogenic effects of Cr^{6+}. We tested the hypothesis that *the reduction of Cr(VI) by glutathione is a major pathway through which reactive intermediates are generated, and that these reactive intermediates can give rise to oxidative DNA damage that is responsible, at least partially, for Cr mutagenesis and carcinogenesis.* We studied this hypothesis in both Saccharomyces cerevisiae yeast, as well as the transgenic Big Blue mouse. Mutational frequency and spectra were obtained in the *Sup-4o*, and *lacI* target genes for yeast and mice, respectively. Yeast and mice were exposed to Cr^{6+}, along with a variety of GSH-depleting and GSH-supplementing agents. In the yeast model, several novel Ty retrotransposon insertions were identified. In the mouse model, tissue-specific mutational frequency and spectra were obtained from genomic

DNA in lung, liver, and kidney. In addition, production of 8-OH-deoxyguanosine in genomic DNA was measured by an ELISA, as well as an HPLC-ECD method.

IRON OVERLOAD INDUCES OXIDATIVE DNA DAMAGE IN RAT SPERM CELLS IN VIVO AND IN VITRO

Anja Wellejus[1], Henrik E. Poulsen[2], and Steffen Loft[1]

Department of Pharmacology, Panum Institute, Health Science Faculty, University of Copenhagen, Denmark

Iron overload induce DNA oxidation via the Fenton reaction as earlier shown in rat testes in vivo. We have investigated the effect of iron overload on oxidative damage in terms of 8-oxodG in nuclear DNA in sperm and testicular cells in vitro and in vivo in rats. Sperm cells were isolated from the epididymis and testes cells were isolated after homogenization. The cells were incubated with increasing concentrations of iron(II)chloride and iron sulfate. DNA was isolated by ethanol precipitation after pronase incubation and 8-oxodG was measured by HPLC-EC. Sperm cells incubated with iron(II)chloride showed an increased 8-oxodG level from 0.75 to 16 per10^5 dG at 0 and 50 mM, respectively, whereas iron sulfate increased the level in a dose dependent manner from 3 to 45 8-oxodG per 10^5 dG at 0 and 300 mM, respectively. In testicular cells incubated with iron sulfate the level of 8-oxodG rose dose dependently from 0.25 to 7 at 0 and 300 mM, respectively. Three groups of 7–8 rats received 0, 50 and 100 mg iron/kg body weight as dextran ip. After 24 hours epididymal sperm and testes cells were collected for analysis. Sperm DNA showed a significant increase in 8-oxodG in the animals treated with iron, the mean values were 1.3, 4.0 and 3.5, respectively, whereas 8-oxodG values in the testes showed no significant change. In conclusion, iron overload induces oxidative DNA damage in epididymal sperm cells in vivo and in vitro, whereas it only produces damage in vitro in testicular cells.

WY-14,643, A POTENT PEROXISOME PROLIFERATOR, INDUCES EXPRESSION OF N-METHYLPURINE-DNA GLYCOSYLASE (MPG) IN RAT LIVER AND INCREASES URINARY EXCRETION OF 1,N^6-ETHENOADENINE (εA)

I. Roussyn[1], T. Y. Yen[2], V. A. Wong[3], R. C. Cattley[3], M. L. Cunningham[4] and J. A. Swenberg[1,2]

[1]Curriculum in Toxicology and [2]Department of Environmental Sciences and Engineering, University of North Carolina, Chapel Hill, NC, [3]Chemical Industry Institute of Toxicology, and [4]National Institute of Environmental Health Sciences, Research Triangle Park, NC, USA

Peroxisome proliferators (PPs) are known to cause a high incidence of hepatocellular carcinoma in rodents, with WY-14,643 being one of the most potent compounds. The "oxidative stress hypothesis" has been proposed as a mechanism of PPs hepatocarcinogenicity. However, a poor correlation between hepatic 8-hydroxy-2'-deoxyguanosine

(OH^8dG) levels and the tumorogenic potency of PPs was found. In this study, the ability of WY-14,643 to induce expression of MPG and urinary excretion of DNA repair products was examined. Urine and livers from control and WY-14,643 treated animals were used. MPG expression was assessed by Quantitative RT-PCR. Immunoaffinity purification and LC/MS analysis of the εA in urine was employed as an index of DNA repair by MPG. WY-14,643 induced expression of MPG 250–300% above controls 3 weeks after dietary feeding of the highest dose. Both time- and dose-dependent increases in expression of MPG were observed. An elevated εA excretion in the urine of treated animals was detected as early as 1 week of feeding, with a 2-fold increase above the levels of endogenous εA excretion. The elevation of urinary excretion of εA may be the result of induced expression of MPG acting on endogenous etheno adducts, increased *de novo* etheno adduct formation due to increased lipid peroxidation, or both. Since MPG is known to also repair OH^8dG, its induction of expression might provide further evidence supporting a role of oxidative DNA damage by PPs. These data suggest that DNA base excision repair may be an important factor in PP-induced carcinogenesis.

INDUCTION OF DNA DAMAGE AND CHROMOSOMAL ABERRATIONS BY NUCLEOLYTIC ENZYMES INTRODUCED INTO LIVING CELLS

O. M. Rozanova, S. I. Zaichkina and E. E. Ganassi

Laboratory of Cytogenetics, Institute of Theoretical and Experimental Biophysics, Pushchino, Moscow Region, Russia, 142292

Many models for production of chromosome aberrations have been proposed. They are based on the conception that the certaily types of DNA lesion initiates chromosomal aberrations. Ionizing radiation and chemical mutagens induce a wide variety of DNA lesions: base damage, single- and double-strand breakes. The majority suggest that double-strand break DNA is critical DNA lesions. As compared to ionizing radiation, nucleolitic enzymes (restriction endjnucleases and DNAase I) offer the advantage of inducing only one specific type of DNA lesion either double- or single-strand breakes. The aim of our investigation is to study the molecular basis of chromosome aberration induction by different nucleolitic enzymes. An asynchronous culture of Chinese hamster fibroblasts, cultured by a standard technique was used. 24 h after seeding, the cells were permeabilized with trypsin and treated with enzymes. Cells were fixed after 7,5 and 18 h to analyze metaphases and after 24 h to analyze cells with micronuclei. The action of restriction enzymes Sai I, Hind III, Eco RV, Hae III, and DNAase I was studed. The yield of DNA damage in individual cells was determined spectrophotometrically in situ by a specially elaborated cytochemical method. The results obtained demonstrate that the damage to specific DNA damage sequences leads to the formation of chromosome aberrations rather the type of DNA damage (single-strand and blund- or cohesive-end double-strand DNA breaks) and induction of chromosome aberrations by nucleolitic enzymes introduced into the living cells is a perspective method of studying the molecular target of chromosome damage initition.

ANTIOXIDATIVE ACTION OF IMIDAZOLOPYRAZINES: *IN VITRO* DETERMINATION OF RADICAL-SCAVENGING EFFICIENCIES

Bertrand de Wergifosse[#], Ingrid Devillers[°], Jacqueline Marchand-Brynaert[°], André Trouet[#], Fernand Baguet[#] and Jean-François Rees[#]

Université Catholique de Louvain, # Animal Biology Unit, °Organic Chemistry UnitPlace Croix du Sud, 4; B - 1348 Louvain-la-Neuve - Belgium

Reactive Oxygen Species (ROS) and other free radicals have long been recognised to play a key role in oxidative genetic damages. Due to their high content in Cytochrome P_{450}, hepatocytes are particularly sensitive to deleterious effects of free radicals, because of central role in the metabolism of xenobiotics and carcinogens, or as a result of various metabolic processes. Coelenterazine, a marine luciferin, and other chemically-derived analogs (CLZ's) have already emerged for their interesting antioxidative properties to protect oxidatively stressed fibroblasts and hepatocytes in culture. Following a chemiluminescent method (Gotoh and Niki, 1994), we report here the measures of CLZ's reactivity with superoxide anion radical ($O_2^{\cdot-}$), and structure-activity relationships. Competitive kinetic studies lead us to calculate high rate constants, ranged from 0.60 to 2.63 x 10^5 M^{-1} s^{-1} at pH 7.8. This *in vitro* study is the first step to understand how these lipophile radical scavengers could react with ROS and protect cells against radical and oxidative DNA damages, in order to prevent carcinogenic processes, especially in liver.

B. de Wergifosse is granted by the F.R.I.A.

J.-F. Rees is Senior Researcher of the National Funds for Scientific Research (FNRS).

DNA AND CHROMOSOMAL DAMAGE INDUCED BY X-RAYS AND I131 IN CELLS OF RATS IMMUNIZED WITH TULAREMIA VACCINE

Armen K. Nersesyan

Cancer Research Centre, Yerevan 375052, Armenia

Recently we have shown that in somatic cells of rats immunized with tularemia live vaccine (TLV) the chromosomal aberrations (CAs) levels induced by some chemical agents were significantly decreased. Immunization of rats with TLV also decresed the incidence of tumors induced by some chemical carci nogens, prolonged the latent period and reduced the mean weight of tumors. It would be of interest to know if immunization of rats with TLV can protect against CAs and DNA damage induced by X-rays and I131. These agents are known to generate genotoxicity through oxigen-radical mechanism. The experiments were performed with male Wistar rats immunized with TLV and non-immunized (controls). Rodents were irradiated with X-rays (0.5 and 1.0 Gy) or received I131 orally (3.7 kBq/g), and bone marrow cells were studied for CAs. DNA damage and repair in somatic cells of irradiated with X-rays (1.0 Gy) immunized and control rats were studied using comet assay immediatly, 15, 30 and 60 min after irradiation.

The results have shown that in bone marrow cells of immunized rats the CAs levels induced by the both agents were significantly decreased compared with corresponding controls. Most of the repair (approximately 50% in cells of controls and 70% of immunized rats) occured within first 15 min. DNA repair was essentially complete by the end of one h in the cells of immunized rats but not in the cells of controls. Thus, immunization of rats with TLV leads to protaction of DNA and chromosomes of somatic cells against some physical and chemical agents. Further investigations of the antigenotoxic/anticarcinogenic effects of TLV are certainly warranted, if only because the tularemia vaccination process in rats is quite similar to that in humans.

THE EFFECTS OF 3-MC ON THE ANTIOXIDANT ENZYME ACTIVITIES AND TOTAL GLUTATHIONE LEVEL DURING THE AGING PERIOD IN RAT LIVER

Ismet Yilmaz, and *Kayahan Fiskin

Inönü Üniversity, Faculty of Arts and Sciences, Department of Chemistry Laboratory of Biochemistry, Malatya, Turkey. *Akdeniz Üniversity, Faculty of Arts and Sciences, Department of Biology Laboratory of Molecular Biology, Antalya, Turkey

The role of free radicals as a source of incessant cellular damage during aging was a originally postulated by Harman. It has also been put forward that reactivity oxygen species (ROS) are involved in aging some degenerative disease processes and age related disease like cancer. The production of reactive oxygen specieses during lipid peroxidation can also possibly lead to damaging reactions due to indiscriminate attack on cellular substances such as proteins and nucleic acids.

The purpose of this study has been investigated the carcinogenetic relations with aging which has attracted great attention in the field of mammalian systems.The effects of 3-methylcholantren (3-MC) on the antioxidant enzyme (Superoxide Dismutase (SOD), Se-dependent and Se-independent Glutathione Peroxidase (GSH-Px), Catalase (CAT) activities and total glutathione level (GSH) were found in 2, 12, 17 month-old wistar rats during 2, 8 and 12 hours periods by intraperitoneally injection.

Among mentioned enzymes; CAT, Se-dependent and Se-independent GSH-Px enzyme activities were determined as significantly higher in old rats than the youngs. On the other hand , SOD activity and total glutathione level were decreased in 17 month-old rats after treatment with 3-MC.

HIGHER NUMBER OF MAST CELLS IN ATRIUM THAN IN VENTRICLE OF RAT HEART: A POSSIBLE ANTIAPOPTOTIC INVOLVEMENT OF NERVE GROWTH FACTOR

Anton B. Tonchev, Olawale A.R. Sulaiman, Kamen P. Valchanov, Peter I. Ghenev, and George N. Chaldakov

Laboratory of Electron Microscopy, Department of Anatomy and Histology, Medical University of Varna, BG-9002 Varna, Bulgaria

Nerve growth factor (NGF) is the paradigm of effector cell-derived neurotrophins responsible for neuronal suvival. Recently, it was shown that NGF exerts its neurotrophic potential *via* an antiapoptotic action (Arumae U, *Biomed Rev* 1995; 4: 15). Moreover, NGF is a stimulator of mast cell (MC) proliferation (Aloe L, Levi-Montalcini R, *Brain Res* 1997; 133: 358), and MC are both source of and target to NGF (Leon et al, *Proc Natl Acad Sci USA* 1994; 91: 3739). Therefore, we aimed at knowing whether MC presence is influenced by and/or contributes to a higher NGF concentration, corresponding to the higher sympathetic nerve density, found in the atrium than in the ventricle of rat heart (Korsching, Thoenen, *Proc. Natl. Acad. Sci. USA* 1983; 80: 3513). We numeriaclly analyzed toluidine blue-stained MC in both atria (5.01 ± 0.50 MC/mm^2) and ventricles (2.05 ± 0.10 MC/mm^2) in sections of formalin fixed/paraffin embedded hearts of adult male Wistar rats (n=7). Our present data of a significantly higher ($p<0.01$) cardiac MC number in the atrium than in the ventricle may thus represent a further evidence of involvement of immune cells in neural-effector interaction. In neuroimmunology, the term "trophic" may actually mean antiapoptotic, rather than nutritional. For testing such a hypothesis, some antiapoptotic markers, e.g. Bcl-2, are currently planned to be applied.

P53-MEDIATED GROWTH ARREST IN RB-NEGATIVE CELLS

Kezban Ünsal[1], Anne Pierre Morel[2], Frederique Ponchel[2], Denise Glaise[3], Christiane Guguen-Guillouzo[3], Brian Carr[4] and Mehmet Öztürk [1,2]

[1]Department of Molecular Biology and Genetics, Bilkent University, Ankara, Turkey. [2]INSERM U453, Centre Leon Berard, Lyon, France. [3]INSERM U49, Hopital Pontchaillou, Rennes, France. [4]Pittsburgh Transplantation Institute, University of Pittsburgh, Pittsburgh, PA, USA

We have explored the role of wild-type p53 in regulating the proliferation, apoptosis and differentiation of a hepatocellular carcinoma cell line. Hep 3B-TR cells lacking p53, retinoblastoma and TGF-β receptor type II proteins were used to generate stable clones expressing temperature-sensitive murine p53–135val mutant protein. The induction of transcriptionally active p53 resulted in an accumulation of p21 (also known as *waf1*, *cip1*, *sdi1*), *bax* and *mdm-2* proteins. Under these conditions, cells responded by a sustained growth arrest at G1 phase and a loss of colony formation ability. p53-dependent apoptosis and differentiation were not observed. These results demonstrate that the major outcome of wild-type p53 activation in these hepatocellular carcinoma cells is growth arrest. Because Hep 3B-TR cells lack both retinoblastoma protein and type II receptor to TGF-β, the p53 effect was independent of both retinoblastoma- and TGF-β mediated growth suppressive pathways. p53-induced growth arrest appears to be mediated by p21. Although p21 is able to inhibit retinoblastoma protein function, it may also interfere with the functions of other growth-regulatory proteins. We are generating cell lines that express inducible p21 protein in order to investigate whether p53-mediated growth arrest in Hep 3B-TR cells is induced by p21. We will also study the Rb-related p107 and p130 proteins as potential targets of p21 protein in these cells.

CHARACTERIZATION OF THE ACTIVE SITE OF ENDONUCLEASE III BY MASS SPECTROMETRY

Monica McTigue, Charles Iden, Robert Rieger, Richard P. Cunningham* and Arthur P. Grollman

Department of Pharmacological Sciences, State University of New York at Stony Brook, Stony Brook, NY 11794–8651, USA and Department of Biological Sciences, SUNY at Albany, Albany, NY 12222*, USA

E. coli Endonuclease III is a base excision repair enzyme with both N-glycoslyase and AP lyase activites (Cunningham, R. P., Mutation Research 383, 189–196, 1997). This enzyme releases a variety of modified pyrimidines from oxidatively-damaged DNA, then cleaves the phosphodiester bond adjacent to the abasic site through a beta-elimination reaction. Oligonucleotides were treated with osmium tetroxide in order to generate a DNA substrate containing thymine glycol. Two thymine glycol isomers were separated by HPLC and a marked difference in rates of cleavage by endonuclease III was observed. Reductive crosslinking with sodium borohydride was used to trap the Schiff base intermediate formed between C1' of the abasic site deoxyribose moiety and endonuclease III. Negative ion ESI-MS (Zharkov, D. O. et al., J. Biol. Chem. 272, 5335–5341, 1997) is being used to characterize the crosslinked complex. This approach should be generally applicable for characterization of other DNA glycosylases with AP lyase activity.

Research supported by NIH grants CA17395 and GM46312. (1)

POSSIBLE ROLE OF OXYGEN RADICALS ON CARCINOGENESIS AND LIPOSOMAL ENZYME THERAPY

A. Tezcaner*, M.Y. Özden**, and V. Hasirci*

*METU, Department of Biological Sciences, Biotechnology Research Unit, 06531 Ankara, Turkey, **METU, Department of Science Education, 06531 Ankara, Turkey

Clinical and epidemiological findings imply a possible role of oxidative stress on many cellular events (mutagenicity, cytotoxicity and change in gene expression). Oxidant carcinogens cause structural damage to DNA and may mutate cancer-related genes. Lipid peroxide mediated DNA damage may play a role in carcinogenesis and arthrogenesis. The possible relation between antioxidant enzyme activity and the levels of DNA base lesions, the effect of lipid peroxidation product-DNA adducts in relation to carcinogenesis are aimed to be studied with a liposomal system. In this study, optimization of parameters that affect the stability of the liposomal system, were carried out. The least permeable solute was lysine regardless of liposome composition. Anionic liposomes seemed to be the least stable. Liposomes composed DPPC released at a slower rate than DMPC. DSC confirmed that phospholipids with high T_c or high concentrations of cholesterol (ca. 50%) increased stability. Addition of charge to liposomes changed the Tc and dimensions. Addition of dicetylphosphate and stearylamine to cholesterol poor DPPC and DMPC multilamellar vesicles caused an increase in Tc.

In future, free radical-mediated DNA damage will be studied with plasmid DNA using DRV. Lipid peroxidation end products and DNA damage will be investigated via and PAGE, respectively. Demonstration of a balance between the antioxidant enzymes (SOD and CAT) rather than action of one provides protection against DNA is seeked.

PROCESSING OF OXIDATIVE DNA DAMAGE IN FAMILIAL ALZHEIMER DISEASE

Leonora Lipinski , Nelci Hoehr, Sharlyn Mazur, Tomasz H. Zastawny[1], Pawel Jaruga[1], Miral Dizdaroglu[1], and Vilhelm Bohr

Laboratory of Molecular Genetics, National Institute on Aging, National Institutes of Health, Baltimore, MD 21224, USA, [1]Chemical Science and Technology Laboratory, National Institute of Standards and Technology, Gaithersburg, MD 20899, USA

Recently, Parashad and coworkers (PNAS, 1996, 93, 5146–5150) demonstrated that cells from patients with familial Alzheimer disease (AD) had an altered response to fluorescent light (FL) exposure as compared to cells from unaffected individuals, suggesting that there may be some deficiency in the DNA repair processing of oxidative lesions induced by FL in AD. This project was initiated to develop a better understanding of the DNA repair capacity of AD for FL induced lesions. The formation and repair of oxidative base lesions in AD and normal fibroblasts was measured after exposure to FL using gas chromatography/mass spectrometry (GC/MS) analysis. Several oxidative DNA lesions were induced in AD and normal cells by exposure to FL. The oxidative lesions induced by FL exposure include 5-hydroxycytosine, formamidopyrimidines, and 8-hydroxyguanine. Additionally, the capacity of AD cells to repair oxidative DNA damage is being investigated using an *in vitro* DNA repair synthesis assay. *In vitro* DNA repair synthesis was carried out by whole cell extracts from AD and normal cells on plasmid DNA damaged by γ-irradiation. This substrate was repaired efficiently by whole cell extracts from both AD and normal cells. Similar *in vitro* assays using plasmid DNA damaged by various oxidative sources are currently in progress.

OPTIMISATION AND VALIDATION OF RESTRICTION SITE MUTATION ASSAY FOR ras AND p53 GENES IN THE RAT

Sinan Suzen* and James M. Parry**

* University of Ankara, Department of Toxicology, Faculty of Pharmacy, Ankara, Turkey, **University of Wales, Department of Molecular Biology, School of Biological Sciences, Swansea, UK

The Restriction Site Mutation (RSM) technique or Restriction Fragment Length Polymorphism/Polymerase Chain Reaction (RFLP/PCR) is a DNA-based method for detecting mutations in animals or cell cultures (1,2). Mutations are determined and identified as alterations (base changes and insertions or deletions) of the DNA sequence at a chosen restriction endonuclease recognition sequence. At the first step of the assay, genomic DNA that is exposed to physical or chemical mutagen is exhaustively digested with the restriction endonuclease without the selection of mutant phenotype. At the second step, resistance sequences containing the mutated target site are specifically amplified using the PCR. As the last step, the RSM assay products are subjected to further restriction endonuclease digestion in order to remove any amplified products containing sensitive restriction endonuclease recognition sequences. In contrast to the most of the traditional mutation analyses, the RSM assay does not rely upon the selection of a mutated phenotype and thus is not limited to mutational analysis in only a few genes. In this study, we have developed protocols to analyse mutations in restriction endonuclease recognition sequences of the ras and p53 genes (3,4) of the rat by the RSM assay. The optimization and validation of the assay have been performed using a variety of tissues from individual laboratory rats. Several suitable restriction endonucleases (HindIII, CfoI, HinfI, NlaIV, AluI, DdeI, NcoI, MspI, BslI) and primer pairs have been identified for detecting mutations in the ras proto-oncogene and p53 tumour suppressor gene of the rat.

References

1. Parry, J. M.; Shamsher, M.; Skibinski, D. O. F.; Mutagenesis; 5; 209–212; 1990.
2. Felley-Bosco, E.; Pourzand, C.; Ziljstra, J.; Amstad, P.; and Cerutti, P.; Nucleic Acid Research; 19; 2913–2919; 1991.
3. Bos, J. L.; Cancer Research; 46; 4682; 1989.
4. Hollestein, M.; Sidransky, D.; Vogelstein,B.; Harris,C.; Science; 253; 49–53; 1991.

DNA BASE DAMAGE IN HUMAN LYMPHOCYTES

K. J. Lenton and J. R. Wagner

Centre de Recherche – IUGS, L'Hôpital d'Youville, 1036 Bélvèdere Sud, Sherbrooke, Québec, Canada J1H 4C4

Intracellular antioxidants help provide a highly reductive environment within cells and therefore protect target molecules against reactive oxygen species. However, normal metabolic processes, as well as exogenous agents , are still able to induce oxidative damage in DNA. The inefficiency of repair for this damage can result in mutations, which could lead to cellular transformation or ageing processes.

The DNA lesions 5-hydroxydeoxyuridine, 5-hydroxydeoxycytidine and 7,8-dihydro-8-oxo-deoxyguanosine can be detected by HPLC methods sensitive enough to detect endogenous levels in human cells. The repair kinetics of these lesions can also be measured by determining residual levels in irradiated calf thymus DNA after incubation with whole cell extracts.

Human blood cell populations contain differing levels of the aqueous soluble antioxidants ascorbic acid and glutathione, with monocytes generally containing higher levels and granulocytes lower as compared with T and B lymphocyte populations. These differences in antioxidant defense may result in variations of endogenous DNA base damage in these cells.

Data for antioxidant levels in lymphocyte populations will be compared with background levels of DNA base damage in monocyte, granulocyte and B,T cell populations, and may provide a link between endogenous antioxidants and DNA damage.

THE OPTIMIZATION OF THE PCR-RESTRICTION ISOTYPING FOR APO-E GENOTYPE

Ibrahim Pirim

Genetic Department, Medical School, Ataturk University, 25240 Erzurum, Turkey

Apo-E, a 34kDa protein consisting of 299 residues, has the function of mediating the clearance of plasma lipoproteins. The length of the apo-E gene is approximately 3.7kb which include 4 exons and 3 introns. Exon 4 contains the codons for the amino acids at position 112 and 158 of the mature protein which determine the apo-E genotype. The three most most common types of apo-E protein designed E2, E3 and E4 have in fact been shown to be the products of three alleles at a single gene locus. The possible three homozygous genotypes (E2/2, E3/3 and E4/4) and the three heterozygous genotypes (E3/2, E3/4 and E2/4) result from the inheritance of any two of the above apo-E alleles. Apo-E contains a cys residue at position 112 and arg residue at position 158 (cys112, arg158). Apo-E2 isoform (cys112, cys158) has a cys for arg substitution at position 158, whilst the apo-E4 isoform (arg112, arg158) has an arg for cys substititition at position 112. DNA has been isolated from buccal cells of randomly choosen subjects. Amplification of apo-E target sequence was performed. The sequence of the 5' primer is 5'TCCAAG-GAGCTGCCAGGCGGGCA3' and the 3' primer is 5'ACAGAATTCGCCCCGGCCTGGTACACTGCCA3'. This primer pair amplifies a 227bp region of DNA that spans both apo-E polymorphic sites. About 15ul of PCR products were loaded onto 3% agarose gel to examine the efficiency of amplification. 15ul of amplification products was subjected to restriction by CfoI. The digested products were subjected to electrophoresis, stained with ethidium bromide and viewed under UV light. Digestion of the PCR products with the restriction enzyme CfoI results in a possible four fragment lengths. Thus providing six apo-E genotypes. Interpretation of these fragments is as follows: Apo-E genotype 2/2, 3/2, 4/2, 3/3, 4/3, 4/4 Size of band fragments 91/81, 91/81/48, 91/81/72/48, 91/48, 91/72/48, 72/48 The purpose of this work is to know the apo-E genotype of people who are living around Erzurum and to find out most common apo-E genotype in the region of people with a given diseases such as heart disease and Alzheimer disease.

ANTITUMOR AND GENOTOXIC EVALUATION OF NORGAMEM, REVERCAN AND TAT ON THE BIOLOGICAL SYSTEMS

Vitalija Simkeviciene and Juozapas Straukas

Institute of Biochemistry, Mokslininku 12, 2600 Vilnius, Lithuania

Norgamem (N) and revercan (R) induce a reverse transformation of tumor cells, restoring contact inhibition in cell cultures. N is not active in experimental tumor systems of mice, but it is an active inhibitor by chemically induced tumor development and effective in human cancer. The present report describes a studies of the efficacy of N, R and their new derivative - 2-amino-2-thiazoline L-thiaprolinate (TAT) in experimental tumor systems of mice (L 1210, Nk/Ly, S 180, La and EAC) and on the bacterial systems by using test microbes E.coli, B.subtilis, Salmonella typhimurium, bacteriophage ?11 and ? 105. The analysis of our data demonstrates no significant difference in TAT, N and R toxicity and antitumor activity. They don't cause a primary tumor inhibition. The survival wild type strains and inactivation of phages suggest that TAT, N and R shows a low toxicity and a similar biological effect, too. N and R don't show genotoxic effect in the reversion assay (Ames) as well as on the reparation of bacterian DNA using reparation deficient strains. N and R show an effect on the growth of culture B.subtilis: R - suppresses insignificantly and N - increases biosynthesis of intracellular neutral proteases. The data of our research suggest that N, R and TAT don't inhibit mice tumors, are not mutagenic in examined systems, and haven't antimutagenic properties. TAT shows a more expressed genotoxic effect.

STUDIES ON GENOTOXICITY OF OCCUPATIONAL EXPOSURES

Uleckiene Saule and Kanapieniene Ramune

Lithuanian Oncology Center, Vilnius, Lithuania

Significant increases in the incidences of sister chromatid exchange, micronucleus formation and chromosomal aberrations in peripheral lymphocytes were observed in study of workers engaged in the manufacture of sulphuric acid (SA). It is known, that occupational exposure to strong-inorganic-acid mists containing sulphuric acid is carcinogenic (IARC, 1992). However, there are almost no data on the genetic and related effects of exposure to sulphuric acid mists in experimental systems; only the effects of pH reduction have been investigated. Environmental polycyclic hydrocarbons are of major concern because of their potential carcinogenicity and mutagenicity (Kliesch U. et al., 1982). Benzo(a)pyrene (BP) has been used as the prototype for this class of compound. The aim of this study was to investigate some genotoxic effects of benzo(a)pyrene and sulphuric acid in vivo. Chromosomal aberrations in bone marrow and sperm-shape abnormalities were studied in DBF1 mice male. The animals (5 in each group) were treated with BP (single dose - 100 mg/kg, i.p.) and (or) SA (2 % or 0.2 % p.o.- single or multiple treatment) in acute or subacute experiment. Chromosomal aberrations analysis showed that BP as well as SA gave a positive response in this test. At 24 h after treatment with SA increase in percentage of chromosome aberrations was observed ($P<0.05$ in comparison with control group). The chromosomal aberrations results for combined treatment with BP and SA were greater than those for BP only ($P<0.05$). BP also induced the sperm-shape abnormalities, however SA did not show effect on this. The study is continuing using micronucleous test. The results on investigation of protective role of some original peptide-like compounds against BP induced genotoxicity and carcinogenicity also will be presented and discussed.

RAT LIVER MITOCHONDRIA CONTAIN A SPECIFIC INCISION ACTIVITY FOR THYMINE GLYCOL

Robertus H. Stierum, Deborah L. Croteau and Vilhelm A. Bohr

Laboratory of Molecular Genetics, National Institute on Aging, National Institutes of Health, 4940 Eastern Avenue, Baltimore, Maryland 21224, USA

Mitochondrial (mt) DNA may be vulnerable to radicals produced during oxidative phosphorylation or to chemical carcinogens. Structural alterations in mtDNA and associated mt dysfunction have been correlated with aging and diseases. Accumulation of mtDNA damage may be responsible for these mtDNA alterations. mtDNA repair could therefore be important in prevention of mt dysfunction.

Previously, DNA repair enzymes have been partially purified from mt. Tomkinson *et al.* [*N.A.R., **18**, 929–935, (1990)*] observed three mt UV-endonuclease activities, which could be thymine glycol DNA glycosylases/AP lyases. Further, it is controversial whether bulky lesions induced by for example *N*-acetoxy-*N*-acetylaminofluorene (3-AF) are repaired in mt.

As an initial approach, we tested DEAE-fractionated extracts from purified rat liver mt for the ability to nick plasmid DNA treated with several DNA damaging agents including osmium tetroxide (OsO_4) and 3-AF. Results show that DEAE-fractionated mt extracts show incision activity specific for OsO_4-damaged DNA, but not for 3-AF-damaged DNA. To further characterize this incision activity, we used a radio-labelled oligonucleotide containing one thymine glycol. Damage-specific incision at the modified thymine was observed. Using subsequent cation exchange and size exclusion chromatography, preliminary results indicate that the activity has a ≈75 mM KCl optimum. The activity is EDTA-resistant and has a MW of approximately 25 kDa. Further purification and characterization of this putative mitochondrial thymine glycol DNA glycosylase/AP lyase is ongoing.

OXIDATIVE DNA BASE DAMAGE AND ANTIOXIDANT ENZYME LEVELS IN CHILDHOOD ACUTE LYMPHOBLASTIC LEUKEMIA

Sema Sentürker[a], Bensu Karahalil[b,c], Mine Inal[a], Hülya Yilmaz[d], Hamza Müslümanoglu[e], Gündüz Gedikoglu[f] and Miral Dizdaroglu[b]

[a]Department of Biochemistry and [e]Department of Genetics, Faculty of Medicine, Osman Gazi University, Eskisehir, Turkey, [b]Chemical Science and Technology Laboratory, National Institute of Standards and Technology, Gaithersburg, MD 20899, USA, [c]Department of Toxicology, Faculty of Pharmacy, Gazi University, Ankara, Turkey, [d]Department of Molecular Medicine, Research Institute for Experimental Medicine, University of Istanbul, Istanbul, Turkey, [f]Foundation for Our Children with Leukemia, Faculty of Medicine, University of Istanbul, Istanbul, Turkey

We have investigated the levels of several antioxidant enzymes and the level of oxidative DNA base damage in lymphocytes of children with acute lymphoblastic leukemia (ALL) and in disease-free children. Children with ALL had just been diagnosed with the disease and had received no therapy prior to obtaining blood samples. A multitude of typical hydroxyl radical-induced base lesions in lymphocyte DNA of children were identified and quantified by gas chromatography-isotope dilution mass spectrometry. Greater levels of DNA base lesions were observed in patients with ALL than in children without the disease. The levels of antioxidant enzymes glutathione peroxidase, catalase and superoxide dismutase in lymphocytes of ALL patients were lower than in lymphocytes of controls. These findings are in agreement with earlier observations in various types of adulthood cancer. Some of the identified DNA base lesions are known to possess premutagenic properties and may play a role in carcinogenesis. The results may indicate a possible link between decreased activities of antioxidant enzymes and increased levels of DNA base lesions due to oxidative damage, and support the notion that free radical reactions may be increased in malignant cells.

FACTORS AFFECTING 5-METHYLCYTOSINE TO THYMINE TRANSITIONS IN THE P53 GENE OF COLORECTAL CANCERS

Stephen B. Waters and Steven A. Akman

Department of Cancer Biology, Comprehensive Cancer Center of Wake Forest University, Winston-Salem, NC 27157, USA

Mutations in the p53 gene of colorectal cancers exhibit a high frequency of cytosine-to-thymine(C->T) transitions. 65% of the mutations are transitions. 50% of the mutations occur at methylated CpG sites and actually represent 5-methylcytosine to thymine(^{5m}C->T) transitions. In contrast, mutagenesis model systems typically exhibit frameshift mutations and the p53 mutational spectra of other types of cancers have significantly lower percentages of these transitions. Possible explanations for this phenomenon are an increased frequency of lesions producing transitions in colorectal cells or a diminished capacity to repair the lesions. This research addresses the problem by assaying the repair of intermediates that lead to these mutations. GT mismatches represent the deamination-driven intermediate of the ^{5me}C:G->T:A transitions and extrahelical bases act as intermediates of insertions and deletions. Repair will be assayed by introducing these mispairs in a mutation reporting plasmid. After passing this plasmid through colorectal carcinoma cells, the efficiency of repair will be determined indirectly by mutagenesis and directly using the Single Nucleotide Primer Extension(SNuPE) assay.

The ^{5m}C->T transitions occur with greater frequency in 5 hotspots. There is a positive correlation between the number of mutations at a hotspot and the GC-richness of the surrounding sequence. Therefore, the repair of GT mispairs will be assayed in sequence with varying GC content. This seeks to establish a correlation between GC-richness and inefficient repair. Finally, the susceptibility of the G in GT mispairs to secondary damage is being determined. This work will provide information important to determining the mechanism involved in the ^{5m}C->T transition.

REPAIR OF ETHENOADDUCTS IN DNA. PURIFICATION AND CHARACTERISATION OF *E. COLI* 3,N^4-ETHENOCYTOSINE DNA GLYCOSYLASE

Murat Saparbaev and Jacques Laval

Groupe « Reparation des lesions Radio- et Chimio-Induites », URA 147 CNRS, Institut Gustave Roussy, 94805 Villejuif Cedex, France

Exocyclic DNA adducts are generated in cellular DNA by industrial pollutants such as the carcinogen vinyl chloride, by the widespread environmental compound ethyl carbamate and by endogenous sources like lipid peroxidation. The etheno (ε) derivatives of purine and pyrimidine bases 1, N6-ethenoadenine (εA), N^2, 3-ethenoguanine (N^2,3εG) and 3,N^4-ethenocytosine (εC) have been shown to cause mutations in vitro and in vivo. It has been shown that εA is excised by human and *E.coli* 3-methyladenine-DNA-glycosylases (ANPG and AlkA proteins) [Saparbaev et al, (1995) Nucleic Acids Res., 23, 3750]. However, the enzyme responsible for εC-repair, when present in DNA was not yet identified and until now there was a complete lack of information about the repair εC in prokaryotic organisms.

In the present study, using a 34-mer duplex oligonucleotide containing a single εC we have detected, to the best of our knowledge, for the first time, a DNA-glycosylase activity in *E.coli* cell extract. After incubation of εC-containing oligonucleotides labelled with ^{32}P at the 5'-end with crude extracts from *E. coli*, we observed on denaturing gel a product of incision migrating at position of 19-mer. When εC-containing oligonucleotide labelled with ^{32}P at 3'-end was used, a product migrating on denaturing gel at position 14-mer was observed. Therefore we propose that incision occurs at position of εC adduct by action of a DNA glycosylase and followed by cleavage at abasic site by an endonuclease. In order to identify the enzyme responsible for the εC-repair in DNA, we investigated whether this lesion was a substrate for a previously characterised DNA repair enzyme. We checked εC-repair activity in crude extracts of various *E.coli* strains deficient in DNA repair, nevertheless all tested strains contained εC-incision activity, leading to conclusion that this activity belongs to an uncharacterised enzyme.

The protein carrying the enzymatic activity excising εC from the duplex oligonucleotide has been partially purified from E.coli and its properties inveεstigated. The enzyme was specific for εC when present in double-stranded oligonucleotide and no excision of εC was observed when present in the single-stranded oligonucleotide. The base-pair specificity of the enzyme was investigated using oligonucleotides with a sequence derived from part of natural gene. We have measured the activity of the repair protein towards duplex oligonucleotide containing mismatches generated by each of the four different bases opposite εC. Excision of εC by the enzyme does not show any obvious preference, although εC/G is a best substrate.

In vitro demonstration that εC is recognised and excised by an *E.coli* DNA glycosylase, implies that this enzyme may be responsible for repair of this mutagenic lesion *in vivo*.

This work was supported by grants from CNRS, ARC, Fondation pour la Recherche Médicale and European Communities.

EXPRESSION OF FRAGILE SITES IN LYMPHOCYTES OF COLON CANCER PATIENTS AND THEIR RELATIVES

Berrin Tunca*, Ünal Egeli**, Abdullah Zorluoglu***, Tuncay Yilmazlar***, and Ayhan Kizil***

* Department of General Biology, Faculty of Science, Uludað University, Bursa, Turkey** Department of Medical Biology , Uludað University Medical School, Bursa, Turkey***Department of Surgery, Uludað University Medical School, Bursa, Turkey

Fragile site expression and chromosome aberrations rate were determined in blood lymphocytes of 45 individuals, including 15 colon cancer patients, 15 their clinically healthy family members, and corresponding age-and sex matched controls. Aphidicolin, BrdU and Caffine were used for fragile site induction. For determining of fragile sites, we chose the sites that appeared one or more times in 50 cells analyzed for each subject, and at least three of the all subject and at least three of the all subjects tested in both colon cancer patients and their relatives. In both colon cancer patients and their relatives were found 1p21–22, 1p31–33, 1q21–22, 1q25, 1q44, 2q31–33, 2q37, 3p14, 5q13–15, 5q21–22, 5q31–33 and 14q24 fragile sites. As a result of cytogenetic and statistic evaluation, both the chromosomal aberration rates and expression frequencies of fragile site observed in patients and their relatives were significantly higher than those in healthy control subjects.

The possible relation between enhanced expression of fragile site and the inheritance of a genetic predisposition to colon cancer requires further examination. We suppose that fragile site expression may be an appropriate marker that shows a genetic susceptibility in the formation of colon cancer.

EVALUATION OF DNA DAMAGE IN LYMPHOCYTES OF CANCER PATIENTS UNDER GAMMA-RADIATION THERAPY BY SINGLE CELL GEL ELECTROPHORESIS

Semra Sardas[1], Yelda Unal[1], Cemil Kusoglu[2], and Ali Esat Karakaya[1]

[1]Gazi University, Faculty of Pharmacy, Toxicology Department 06330, Hipodrom-Ankara/Turkey [2]Ankara Oncology Hospital, Ankara/Turkey

There has been increasing interest in the comet assay (single cell gel electrophoresis technique) in past two to three years. The preceding last months in particular has brought a rapid increase in the number of papers and reports published using the comet assay, as well as increased expansion into new fields. The single cell gel assay has also been used to examine DNA damage and repair under a variety of experimental conditions. Comets form as broken ends of the negatively charged DNA molecule become free to migrate in the electric field towards the anode and the number of broken ends which can still migrate a short distance from the comet head. Tail length initally increases with damage but reaches a maximum that is largely defined by the electrophoresis conditions, not the size of the fragments . At low damage levels, streching of attached strands of DNA, rather than migration of individual pieces, is likely to occur. With increasing number of breaks, DNA pieces migrate freely into the tail of the comet, and at the extreme (the apoptotic cell), the head and tail are well separeted. Stretching and migration of separated strands, are generally accepted to explain the DNA migration patterns observed in the comet assay. In this study the results of 12 Hodgkin's lymphoma or bone metastases cancer patients were analyzed before and after receiving 4000c Gy of irradiation in 20 fractions and 3000c Gy in 10 fractions respectively, were analyzed. Normal cell counts (spherical, undamaged lymphocytes) of pretreated and posttreated patients were extremely significant ($p<0.0001$). Also irrespective irradiation all patients had various grades of damages and it was more apparent that the grade of damage increases after irradiation therapy.

CLONING OF HUMAN ENDONUCLEASE III cDNA AND CHARACTERIZATION OF THE ENZYME

Shogo Ikeda[1], Tapan Biswas[1], Rabindra Roy[1,2], Tadahide Izumi[1], Istvan Boldogh[3], Alex Kurosky[2], Altaf H. Sarker[4], Shuji Seki[4], and Sankar Mitra[1,2]

[1]Sealy Center for Molecular Science, [2]Department of Human Biological Chemistry and Genetics, and [3]Department of Microbiology and Immunology, University of Texas Medical Branch, Galveston, TX 77555, USA and [4]Department of Molecular Biology, Institute of Cellular and Molecular Biology, Okayama University Medical School, Okayama 700, Japan

Endonuclease III (Nth) of E. coli is a DNA glycosylase/AP lyase specific for oxidatively damaged bases including thymine glycol (TG), dihydrouracil (DHU), and 5-hydroxycytosine. The cDNA of the human homolog of this enzyme (hNTH) was cloned by screening a human cDNA library with the mouse NTH cDNA that was first isolated by PCR cloning with primers based on conserved sequences in EST-tagged cDNA's.

The nucleotide sequence of the hNTH cDNA is identical to the published sequence of Hilbert et al. (JBC 272:6733, 1997) and is different from the sequence published by Aspinwall et al. (PNAS 94:109, 1997) at 2 amino acid residues as well as in the untranslated region. The full-length enzyme with 304 amino acid residues was expressed as a glutathione-S-transferase fusion polypeptide in E. coli, whose cleavage with thrombin resulted in production of the wild type protein with 2 additional amino acid residues and a truncated protein with deletion of 22 residues at the amino terminus. Kinetic analysis showed that the two polypeptides have identical enzymatic activity. Optical absorbance spectrum of pure yellow brown hNTH showed absorbance peak at 280 nm and a second peak at 410 nm due to the presence of [4Fe-4S] cluster. The enzyme cleaves TG-containing form I plasmid DNA and DHU-containing oligonucleotide duplex. The AP lyase activity of hNTH involves predominantly β-elimination. The pH optimum of the enzyme is about 8 with a 55-mer DHU-containing oligonucleotide. It has a similar reaction rate with either DHU•A or DHU•G base pairs in the oligonucleotide substrate and an optimum requirement for 75 mM NaCl. The enzyme has a Km of 47 nM and kcat of ~ 0.6/min.

As expected, the DNA glycosylase/AP lyase forms a transient Schiff's base intermediate with the deoxyribose residue that is generated after base removal, and can be stabilized by reduction with NaCNBH3. Quantitation of the covalent complex indicated that about 25% of the enzyme molecules in a typical preparation were active. DNase I footprinting studies with both stable and transient enzyme DHU-oligonucleotide complexes showed protection of 10–11 nucleotides centered around DHU in the damaged strand, and a span of 9 nucleotides with G opposite DHU at the 5′ boundary in the complementary strand. (Research supported by U.S. Public Health Service grants ES08457, CA53791 and AG10514.)

AN OXIDATIVE DAMAGE-SPECIFIC ENDONUCLEASE FROM RAT LIVER MITOCHONDRIA

Deborah L. Croteau, Colette M. J. ap Rhys, Edgar K. Hudson, Grigory L. Dianov, Richard G. Hansford, Vilhelm A. Bohr

Laboratory of Molecular Genetics, National Institute on Aging, NIH, Baltimore, MD 21224, USA

Reactive oxygen species have been shown to generate mutagenic lesions in DNA. One of the most abundant lesions in both nuclear and mitochondrial DNA is 7,8-dihydro-8-oxoguanine, (8-oxoG). We report here the partial purification and characterization of a mitochondrial oxidative damage endonuclease (mt ODE) from rat liver that recognizes and incises at 8-oxoG and abasic sites in duplex DNA. Rat liver mitochondria were purified by differential and Percoll gradient centrifugation and mt ODE was extracted from Triton X-100 solubilized mitochondria. Incision activity was measured using a radiolabeled double-stranded DNA oligonucleotide containing a unique 8-oxoG and reaction products were separated by polyacrylamide gel electrophoresis. Gel filtration chromatography predicts mt ODE's molecular weight to be between 25 – 30 kDa. Mt ODE has a monovalent cation optimum between 50 – 100 mM KCl and a pH optimum between 7.5 and 8. Mt ODE does not require any co-factors and is active in the presence of 5 mM EDTA. It is specific for 8-oxoG and preferentially incises at 8-oxoG:C base pairs. Mt ODE is a putative 8-oxoG glycosylase/lyase enzyme because it can be covalently linked to the 8-oxoG oligonucleotide by sodium borohydride reduction. In addition, MtODE's activity increases in older rats compared to younger rats.

A MODEL OF REPAIR FOR TUMOR INDUCTION BY HIGH AND LOW LET RADIATION

Z. Hiz and F. J. Burns

New York University Medical Center, New York, NY 10016, USA

Solution from the developed model for cancer induction in rat skin based on two stages involving mutational changes and biological repair of premutagenic damage followed by cell death and cell proliferation fit the cumulative cancer incidence very well. Particular solution ignoring cell death and proliferation for split dose electron exposures in general fit the data reasonably well, but exceptions occur for young irradiation ages. Discrepancy between the approximate formula and the incidence is more uniform for larger split times at old irradiation age. Split does exposure data fitted to the approximate formula gives the average repair rates; λ=0.116 hrl, λ=1.146 hrl, λ=1730 hrl for 28, 113 and 182 day-old rats respectively. This suggests that repair actually becomes more rapid in older ages, however cell death and proliferation may alter this interpretation. Splitting dose for high LET argon ion exposure causes more tumors in comparison to single doses. This suggests no repair, however, stimulation of cell proliferation may increase tumor yield. All these facts are critically important to understand how repair, cell death and mutation compete in the pathway to cancer induction. More detailed investigations of the mechanisms of each phenomenon both theoretically and experimentally are needed.

PARTICIPANTS

DIRECTOR

Dr. Miral Dizdaroglu, National Institute of Standards and Technology, 100 Bureau Dr., Stop 8311, Gaithersburg, MD 20899-8311, USA

CO-DIRECTORS

Dr. Jacques Laval, URA 147 CNRS, U 140 INSERM, Institut Gustav-Roussy, 94805 Villejuif Cedex, FRANCE
Dr. Hans Krokan, UNIGEN Center for Molecular Biology, University of Trondheim, N-7005 Trondheim, NORWAY
Dr. Ali E. Karakaya, Department of Toxicology, Faculty of Pharmacy, Gazi University, 06330 Hipodrom, Ankara, TURKEY
Dr. Barry Halliwell, Pharmacology Group, University of London King's College, Chelsea Campus, London SW3 6LX, UK

LECTURERS

Dr. Steffen Loft, Department of Pharmacology, Panum Institute, University of Copenhagen, Blegdamsvej 3, DK-2200 Copenhagen N, DENMARK
Dr. Serge Boiteux, UMR217 CNRS/CEA, Dpt. Radiobiologie et Radiopathologie BP6, Bat.05, 92265-Fontenay aux Roses , FRANCE
Dr. Jean Cadet, CEA/Department de Recherche Fondamentale, sur la Matiere Condensee, SCIB/LAN, F-38054 Grenoble Cedex 9, FRANCE
Dr. Hiroshi Ohshima, Unit of Endogenous Risk Factors, International Agency for Research on Cancer, 69372 Lyon Cedex 08, FRANCE
Dr. Clemens von Sonntag, Max-Planck-Institut für Strahlenchemie, Stiftstr. 34–36, PO Box 101365, D-45413 Mülheim a.d. Ruhr, GERMANY
Dr. Etsuo Niki, Research Center for Advanced Science and Technology, University of Tokyo, 4–6–1 Komaba, Meguro-Ku, Tokyo 153, JAPAN

Dr. Susumu Nishimura, Banyu Tsukuba Research Institute, Okuba-3, Tsukuba 300–33, JAPAN
Dr. Shinya Toyokuni, Department of Pathology, Graduate School of Medicine, Kyoto University, Sakyo-ku, Kyoto 606, JAPAN
Dr. Ryszard Olinski, Department of Clinical Biochemistry, University School of Medical Sciences, Karlowicza 24, 85–092 Bydgoszcz, POLAND
Dr. Mehmet Öztürk, Department of Molecular Biology and Genetics, Bilkent University, 08533 Bilkent, Ankara, TURKEY
Dr. Okezie Aruoma, Pharmacology Group, University of London King's College, Chelsea Campus, Manresa Road, London SW3 6LX, UK
Dr. Tomas Lindahl, Imperial Cancer Research Fund, Clare Hall Laboratories, South Mimms, Hertfordshire EN6 3LD, UK
Dr. Joseph Lunec, Division of Chemical Pathology, Centre for Mechanisms of Human Toxicity, Hodgkin Bldg., PO Box 138, University of Leicester, Leicester, LE1 9HN, UK
Dr. Steve Akman, Department of Cancer Biology, Wake Forest Comprehensive Cancer Center, Medical Center Boulevard, Winston-Salem, North Carolina 27157, USA
Dr. Vilhelm A. Bohr, Laboratory of Molecular Genetics, National Institute on Aging, NIH, Baltimore, MD 21224, USA
Dr. Bruce Demple, Department of Molecular and Cellular Toxicology, Harvard School of Public Health, Boston, MA 02115, USA
Dr. Walter A. Deutsch, Pennington Biomedical Research Center, 6400 Perkins Rd., Baton Rouge, LA 70808, USA
Dr. Paul Doetsch, Department of Biochemistry, Emory University School of Medicine, Atlanta, GA 30322, USA
Dr. Errol C. Friedberg, Department of Pathology, University of Texas Southwestern Medical Center, Dallas, TX 75235–9072, USA
Dr. Matthew B. Grisham, Department of Physiology and Biophysics, Louisiana State University Medical Center, Shreveport, Louisiana 71130–3932, USA
Dr. Arthur P. Grollman, School of Medicine, Department of Pharmacological Sciences State University of New York at Stony Brook, Stony Brook, NY 11794–8651, USA
Dr. Lawrence Grossman, Department of Biochemistry, Johns Hopkins University School of Hygiene and Public Health, Baltimore, MD 21205, USA
Dr. Philip C. Hanawalt, Department of Biological Sciences, Stanford University, Stanford, CA 94305–5020, USA
Dr. Kazimierz S. Kasprzak, Laboratory of Comparative Carcinogenesis, National Cancer Institute, FCRDC, Frederick, MD 21702, USA
Dr. Yoke W. Kow, Department of Radiation Oncology, Emory University School of Medicine, Atlanta, GA 30335, USA
Dr. Stuart M. Linn, Division of Biochemistry and Molecular Biology, University of California, Berkeley, CA 94720–3202, USA
Dr. Sankar Mitra, The University of Texas Medical Branch at Galveston, Sealy Center for Molecular Science, 6.136 Medical Res. Bldg., 301 University Blvd., Galveston, TX 77555–1079, USA
Dr. Nancy Oleinick, Division of Radiation Biology, School of Medicine (BRB), Case Western Reserve University, Cleveland, OH 44106–4942, USA
Dr. Joyce Reardon, Department of Biochemistry, University of North Carolina at Chapel Hill, Chapel Hill, NC 27599–7260, USA

Dr. Dennis Reeder, National Institute of Standards and Technology, 100 Bureau Dr., Stop 8311, Gaithersburg, MD 20899-8311, USA
Dr. Susan S. Wallace, Department of Microbiology and Molecular Genetics, University of Vermont, Burlington, VT 05405, USA
Dr. John F. Ward, Department of Radiology, University of California San Diego, La Jolla, CA 92093, USA

ASI STUDENTS

Armen K. Nersesyan, Cancer Research Centre, Yerevan State University, Yerevan, ARMENIA
Bertrand de Wergifosse, Cellular Biology Laboratory, Université Catholique de Louvain - UCL, Place Croix du Sud, 4 - Bâtiment Carnoy, B - 1348 Louvain-la-Neuve, BELGIUM
Anton Tontchev, Laboratory of Electron Microscopy, Department of Anatomy and Histology, Medical University of Varna, BG-9002 Varna, BULGARIA
Kevin Lenton, Centre de Recherche en gerontologie et gereatrie, Universite de Sherbrooke, 1036 Belvedere Sud, Sherbrooke, (Quebec), J1H 4C4, CANADA
M. Vakur Bor, Department of Clinical Biochemistry, Aarhus University Hospital, Norrebrogade 44, DK-8000 Aarhus C, DENMARK
Mette Christiansen, Danish Center for Gerontology, University of Aarhus, Langelandsgade 140, bygn. 510, 8000 Aarhus C, DENMARK
Tanja Thybo Frederiksen, Danish Center for Gerontology, University of Aarhus, C.F. Mollers Alle, Bldg. 130, 8000 Aarhus C, DENMARK
Allan Weiman, Department of Clinical Pharmacology, Q7642, Rigshospitalet, 20 Tagensvej, DK-2200 Copenhagen N, DENMARK
Anja Wellejus, Department of Pharmacology, University of Copenhagen, 18.5 Panum Institute, Blegdamsvej 3, DK-2200 Copenhagen N, DENMARK
Maarit Lankinen, Department of Clinical Biochemistry, Laboratory of Molecular Hematology, Tempere University Hospital, PO Box 200, 33521 Tempere, FINLAND
Valerie Abalea, Laboratoire de Biologie Cellulaire, INSERM U456, 2, Avenue du Pr. L. Bernard, 35043 Rennes Cedex, FRANCE
Martine Defais, CNRS, Institut de Pharmacologie et Biologie Structurale, 205 route de Narbonne, 31077 Toulouse Cedex, FRANCE
Francoise Laval, URA 147 CNRS, U 140 INSERM, Institut Gustav-Roussy, 94805 Villejuif Cedex, FRANCE
Murat Saparbaev, URA 147 CNRS, U 140 INSERM, Institut Gustav-Roussy, 94805 Villejuif Cedex, FRANCE
Christa Baumstark-Khan, DLR, Institute of Aerospace Medicine, Radiobiology Unit Linder Hoehe, D-51147 Köln, GERMANY
Nair Sreejayan, University of Munich, Klinikum Grosshadern, Department of Medicine D-81377 München, GERMANY
Alexander Georgakilas, Laboratory for Rad. Mol. Genetics, National Centre for Scientific Res. "Demokritos" Aghia Paraskevi, 153 10 Athens, GREECE
Kostas Haveles, Laboratory for Rad. Mol. Genetics, National Centre for Scientific Res. "Demokritos" Aghia Paraskevi, 153 10 Athens, GREECE
Lampros Papandreou, University of Ionnina, Medical School , Department of Obstetrics and Gynaecology, 2nd Obstetrics Clinic, GR-451 10 Ionnina, GREECE

Chryssostomos Chatgilialoglu, Consiglio Nazionale Delle Richerche,Via P. Gobetti, 101 40129 Bologna, ITALY
Antonia M. Pedrini, Instituto di genetica Biochimica, ed Evoluzionistica del CNR Via Abbiategrasso, 207, 27100 Pavia, ITALY
Vitalija Simkeviciene, Institute of Biochemistry, Department of Bioorganic Compounds Technology, Mokslininku 12, 2600 Vilnius, LITHUANIA
Saule Uleckiene, Laboratory of Environmental Carcinogens, Lithuanian Oncology Center, Polocko 4, 2007 Vilnius, LITHUANIA
Niels Aarsaether, Department of Molecular Biology, HIB-University of Bergen, 5020 Bergen, NORWAY
Svein Bjelland, Stavanger College, School of Science and Technology, Ullandhaug, PO Box 2557, N-4004 Stavangar, NORWAY
Kristi Kvaloy, UNIGEN Center for Molecular Biology, Norwegian University for Science and Technology, N-7005 Trondheim, NORWAY
Hilde Nilsen, UNIGEN Center for Molecular Biology, Norwegian University for Science and Technology, N-7005 Trondheim, NORWAY
Kristin Solum, UNIGEN Center for Molecular Biology, Norwegian University for Science and Technology, N-7005 Trondheim, NORWAY
Jan Barciszewski, Institute of Bioorganic Chemistry, Polish Academy of Sciences, Noskowskiego 12/14, 61–704 Poznan, POLAND
Pawel Jaruga, Department of Clinical Biochemistry, Medical Academy, 85–094, Bydgoszcz, POLAND
Rita Balamirzoeva, Institute of History, Arch. and Ethnogr. Daghestanian Center of Russian Academy of Sciences, 367030, Makhachkala, Daghestan, RUSSIA
Alla A. Oudalova, Russian Institute of Agricultural Radiology & Agroecology, 249020 Obninsk, Kaluga Region, RUSSIA
Bogdan Constantinescu, Institute of Atomic Physics, PO BOX MG-6, Bucharest, ROMANIA
Teresa Roldan-Arjona, Faculdad de Ciencias, Departamento de Genetica, Universidad de Cordoba, 14071-Cordoba, SPAIN
Rafael Rodriguez, Faculdad de Ciencias, Departamento de Genetica, Universidad de Cordoba, 14071-Cordoba, SPAIN
Elizabeth Bilsland, Gothenburg University, Lundsbergs Laboratory, Department of Molecular Biology, Medicinaregatan 9C, S-413 90 Gothenburg, SWEDEN
Bulbin Akbasak, Department of Biology and Genetics, Faculty of Medicine, Inonu University, Malatya, TURKEY
Mustafa Birincioglu, Department of Pharmacology, Faculty of Medicine, Inonu University, Malatya, TURKEY
Sema Burgaz, Department of Toxicology, Faculty of Pharmacy, Gazi University, Ankara, TURKEY
Belma Giray, Department of Toxicology, Faculty of Pharmacy, Hacettepe University, Ankara, TURKEY
Bensu Karahalil, Department of Toxicology, Faculty of Pharmacy, Gazi University, Ankara, TURKEY
Asuman Karakaya, Department of Toxicology, Faculty of Pharmacy, University of Ankara, Ankara, TURKEY
Okyay Kilinc, Department of Molecular Biology and Genetics, Bogazici University, Istanbul, TURKEY

Hilal Ozdag, Department of Molecular Biology and Genetics, Bilkent University, Ankara, TURKEY

Ibrahim Pirim, Department of Genetics, Faculty of Medicine, Ataturk University, Erzurum, TURKEY

Semra Sardas, Department of Toxicology, Faculty of Pharmacy, Gazi University, Ankara, TURKEY

Sema Senturker, Biochemistry Department, Faculty of Medicine, Osman Gazi University, Eskisehir, TURKEY

Sinan Suzen, Department of Toxicology, Faculty of Pharmacy, University of Ankara, Ankara, TURKEY

Aysen Tezcaner, Department of Biological Sciences, Middle East Technical University, Ankara, TURKEY

Berrin Tunca, Department of Molecular Biology, Faculty of Medicine, Uludag University, Bursa, TURKEY

Kezban Unsal, Department of Molecular Biology and Genetics, Bilkent University, Ankara, TURKEY

Ismet Yilmaz, Department of Biochemistry, Faculty of Medicine, Inonu University, Malatya, TURKEY

Berran Yucesoy, Department of Toxicology, Faculty of Pharmacy, University of Ankara, Ankara, TURKEY

Suzanne Cholerton, Department of Pharmacological Sciences, University of Newcastle Upon Tyne, Newcastle NE2 4HH, UK

Andrew Jenner, Pharmacology Department, King's College London University, London, SW3 6LX, UK

Tracy Melvin, MRC Radiation and Genomic Stability Unit, Chilton, Oxon, OX11 ORD, UK

Cyril Privezentzev, National Academy of Science of Ukraine, The A.O. Kovalevsky Inst. of Biology of the Southern Sea, Nakhimova Ave. 2, 335011 Sevastopol, UKRAINE

Elizabeth Bailey, Laboratory of Toxicology, Harvard School of Public Health Boston, MA 02115, USA

George J. Broder, Fox Cancer Center, Radiology, 1553 Cherry Lane, Rydal, PA 19046, USA

Richard M. Burger, Laboratory of Chromosome Biology, The Public Health Research Institute, New York, NY 10016, USA

Jonathan A. Eisen, Department of Biological Sciences, Stanford University, Stanford, CA 94305–5020, USA

Kevin Geiss, Department of Zoology, Miami University, Oxford, OH 45056, USA

Sandra Gunselman, Colorado State University, Department of Environmental Health Fort Collins, CO 80523, USA

Zekiye Hiz, New University Medical Center, Department of Environmental Epidemiology, New York, NY 10016, USA

Leonora Lipinski, Laboratory of Molecular Genetics, National Institute on Aging, NIH Baltimore, MD 21224, USA

Monica McTigue, Health Sciences Center, Department of Pharmacology, University of New York at Stony Brook, Stony Brook, NY 11794–8651, USA

Carol W. Moore, Department of Microbiology, Science Bldg., Room 919A, Convent Ave. at 138th Street, New York, NY 10031, USA

Henry Rodriguez, Department of Biology, Beckman Research Institute of the City of Hope, Duarte, CA 91010, USA

Ivan Roussyn, Curriculum in Toxicology, University of North Carolina at Chapel Hill Chapel Hill, NC 27599–7400, USA
David M. Sonntag, University of Cincinnati, Department of Environmental Health Cincinnati, OH 45220, USA
Robertus Stierum, Laboratory of Molecular Genetics, National Institute on Aging, NIH Baltimore, MD 21224, USA
Graciela Spivak, Department of Biological Sciences, Stanford University, Stanford, CA 94305–5020, USA
Julien L. Van Lancker, UCLA Department of Pathology, AL-209 CHS, Los Angeles, CA 90095, USA
Steve Waters, Department of Cancer Biology, Wake Forest Comprehensive Cancer Center, Winston-Salem, NC 27157, USA
Dmitry Zharkov, Department of Pharmacology, State University of New York at Stony Brook, Stony Brook, NY 11794–8651, USA

Participants Photo: First row from the bottom, from left: 1. Walter A. Deutsch, USA; 2. Paul W. Doetsch, USA; 3. John F. Ward, USA; 4. Clemens von Sonntag, Germany; 5. Graciela Spivak, USA; 6. Philip C. Hanawalt, USA; 7. Berran Yücesoy, Turkey; 8. Bensu Karahalil, Turkey; 9. Suzanne Cholerton, UK; 10. Asuman Karakaya, Turkey; 11. Nursen Basaran, Turkey; 12. Belma Giray, Turkey; 13. Ismet Yilmaz, Turkey; 14. Sema Sentürker, Turkey; 15. Susan S. Wallace, USA; 16. Miral Dizdaroglu, USA; 17. Jacques Laval, France. Second row, from left: 1. unknown; 2. Bruce Demple, USA; 3. Allan Weiman, Denmark; 4. Saule Uleckiene, Lithuania; 5. Vitalija Simkeviciene, Lithuania; 6. Vakur Bor, Denmark; 7. Pawel Jaruga, Poland; 8. Alla A. Oudalova, Russia; 9. Rita Balamirzoeva, Russia; 10. Zekiye Hiz, USA; 11. Chryssostomos Chatgilialoglu, Italy; 12. Hiroshi Ohshima, France; 13. Mustafa Birincioglu, Turkey; 14. Ryszard Olinksi, Poland; 15. Steffen Loft, Denmark; 16. Francoise Laval, France; 17. Carol Moore, USA; 18. Armen Nersesyan, Armenia; 19. Nair Sreejayan, Germany. Third row, from left: 1. Kristi Kvaloy, Norway; 2. Hans Krokan, Norway; 3. Kristin Solum, Norway; 4. Niels Aarsaether, Norway; 5. Dmitry Zharkov, USA; 6. Serge Boiteux, France; 7. Svein Bjelland, Norway; 8. Susumu Nishimura, Japan; 9. Cyril Privezentzev, Lithuania; 10. Bulbin Akbasak, Turkey; 11. Sandra Gunselman, USA; 12. Okezie I. Aruoma, UK; 13. Bertrand de Wergifosse, Belgium; 14. Errol C. Friedberg, USA; 15. Lawrence Grossman, USA; 16. Rafael Rodriguez, Spain; 17. Richard Burger, USA; 18. Kazimierz S. Kasprzak, USA; 19. Teresa Roldan-Arjona, Spain; 20. Sinan Süzen, Turkey. Fourth row, from left: 1. Hilde Nilsen, Norway; 2. Julien L. Van Lancker, USA; 3. Ibrahim Pirim, Turkey; 4. Christa Baumstark-Khan, Germany; 5. Maarit Lankinen, Finland; 6. Berrin Tunca, Turkey; 7. Aysen Tezcaner, Turkey; 8. Lampros Papandreou, Greece; 9. Kevin Geiss, USA; 10. Dennis Reeder, USA; 11. Antonia Pedrini, Italy; 12. David M. Sonntag, USA; 13. Ivan Roussyn, USA; 14. Elizabeth Bilsland, Sweden; 15. Anja Wellejus, Denmark; 16. Valerie Abalea, France; 17. Yoke W. Kow, USA; 18. Sankar Mitra, USA; 19. Arthur P. Grollman, USA; 20. Ali E. Karakaya, Turkey; 21. Nancy L. Oleinick, USA; 22. Steve A. Akman, USA. Fifth row, from left: 1. Alexander Georgakilas, Greece; 2. Kostas Haveles, Greece; 3. Jonathan Eisen, USA; 4. Elizabeth Bailey, USA; 5. Anton Tontchev, Bulgaria; 6. Tracy Melvin, UK; 7. Robertus Stierum, USA; 8. Leonora Lipinski, USA; 9. Mette Christiansen, Denmark; 10. Tanja Thybo Frederiksen, Denmark; 11. Henry Rodriguez, USA; 12. Steve Waters, USA; 13. Andrew Jenner, UK; 14. Murat Saparbaev, France; 15. Jan Barciszewski, Poland; 16. Monica McTigue, USA; 17. Bogdan Constantinescu, Romania; 18. Hilal Ozdag, Turkey; 19. Mehmet Öztürk, Turkey; 20. Kezban Ünsal, Turkey; 21. Kevin Lenton, Canada; 22. Vilhelm A. Bohr, USA; 23. Stuart M. Linn, USA; 24. Joyce Reardon, USA; 25. Spouse of Chryssostomos Chatgilialoglu, Italy.

NATO ADVANCED STUDY INSTITUTE
DNA DAMAGE AND REPAIR
October 14-24, 1997 - ANTALYA / TURKEY

AUTHOR INDEX

SUBJECT INDEX